Current crop models

State-of-the-art and future developments

It is widely recognised that agriculture is a significant contributor to global warming and climate change. Agriculture needs to reduce its environmental impact and adapt to current climate change whilst still feeding a growing population, i.e. become more 'climate-smart'. Burleigh Dodds Science Publishing is playing its part in achieving this by bringing together key research on making the production of the world's most important crops and livestock products more sustainable.

Based on extensive research, our publications specifically target the challenge of climate-smart agriculture. In this way we are using 'smart publishing' to help achieve climate-smart agriculture.

Burleigh Dodds Science Publishing is an independent and innovative publisher delivering high quality customer-focused agricultural science content in both print and online formats for the academic and research communities. Our aim is to build a foundation of knowledge on which researchers can build to meet the challenge of climate-smart agriculture.

For more information about Burleigh Dodds Science Publishing simply call us on +44 (0) 1223 839365, email info@bdspublishing.com or alternatively please visit our website at www.bdspublishing.com.

Related titles:

Advances in crop modelling for a sustainable agriculture
Print (ISBN 978-1-78676-240-5); Online (ISBN 978-1-78676-242-9, 978-1-78676-243-6)

Improving data management and decision support systems in agriculture
Print (ISBN 978-1-78676-340-2); Online (ISBN 978-1-78676-342-6, 978-1-78676-343-3)

Modelling climate change impacts on agricultural systems
Print (ISBN 978-1-80146-174-0); Online (ISBN 978-1-80146-176-4, 978-1-80146-177-1)

Precision agriculture for sustainability: Second edition
Print (ISBN 978-1-80146-881-7); Online (ISBN 978-1-80146-882-4, 978-1-80146-883-1)

Smart farms
Print (ISBN 978-1-80146-382-9); Online (ISBN 978-1-80146-383-6, 978-1-80146-384-3)

Chapters are available individually from our online bookshop:
https://www.bdspublishing.com/webshop/

climate•SMART•publishing
IN AGRICULTURAL SCIENCE

BURLEIGH DODDS SERIES IN AGRICULTURAL SCIENCE
NUMBER 170

Current crop models

State-of-the-art and future developments

Edited by Professor Gerrit Hoogenboom, University of Florida, USA

burleigh dodds
SCIENCE PUBLISHING

Published by Burleigh Dodds Science Publishing Limited
82 High Street, Sawston, Cambridge CB22 3HJ, UK
www.bdspublishing.com

Burleigh Dodds Science Publishing, 1518 Walnut Street, Suite 900, Philadelphia, PA 19102-3406, USA

First published 2025 by Burleigh Dodds Science Publishing Limited

Library of Congress Control Number: 2025946835

British Library Cataloguing in Publication Data
A catalogue record for this book is available from the British Library

ISBN 978-1-80146-972-2 (print)
ISBN 978-1-80146-974-6 (ePub)
ISBN 9 78-1-80146-973-9 (PDF)
ISSN 2059-6936 (print)
ISSN 2059-6944 (online)

DOI: 10.19103/AS.2025.0155

Typeset by Deanta Global Publishing Services, Dublin, Ireland

Contents

UMR1391 ISPA, France and CIRAD, University of Montpellier, UPR Recyclage et Risque, France; Loïc Strullu, Strullu Company, France; Remi Vezy, CIRAD, University of Montpellier, CNRS, INRAE, IRD, UMR AMAP, France; Tiphaine Vidal, Université Paris-Saclay, INRAE, UR BIOGER, France; Eric Justes, CIRAD, France; and Jean-Louis Durand, INRAE UR P3F, France

Myriam Adam, UMR AGAP Institut and Université de Montpellier, Institut Agro, France and University of Battambang, Cambodia; Gregory Beurier, Lauriane Rouan and Michael Dingkuhn, CIRAD, UMR AGAP Institut, Université de Montpellier, CIRAD, INRAE, Institut Agro, France; Florian Larue, Université de Montpellier, France; and Bertrand Muller, CIRAD, UMR AGAP Institut, Université de Montpellier, Institut Agro, France and CIRAD, Madagascar

Series list

Title	Series number
Achieving sustainable cultivation of maize - Vol 1 From improved varieties to local applications Edited by: Dr Dave Watson, CGIAR Maize Research Program Manager, CIMMYT, Mexico	001
Achieving sustainable cultivation of maize - Vol 2 Cultivation techniques, pest and disease control Edited by: Dr Dave Watson, CGIAR Maize Research Program Manager, CIMMYT, Mexico	002
Achieving sustainable cultivation of rice - Vol 1 Breeding for higher yield and quality Edited by: Prof. Takuji Sasaki, Tokyo University of Agriculture, Japan	003
Achieving sustainable cultivation of rice - Vol 2 Cultivation, pest and disease management Edited by: Prof. Takuji Sasaki, Tokyo University of Agriculture, Japan	004
Achieving sustainable cultivation of wheat - Vol 1 Breeding, quality traits, pests and diseases Edited by: Prof. Peter Langridge, The University of Adelaide, Australia	005
Achieving sustainable cultivation of wheat - Vol 2 Cultivation techniques Edited by: Prof. Peter Langridge, The University of Adelaide, Australia	006
Achieving sustainable cultivation of tomatoes Edited by: Dr Autar Mattoo, USDA-ARS, USA and Prof. Avtar Handa, Purdue University, USA	007
Achieving sustainable production of milk - Vol 1 Milk composition, genetics and breeding Edited by: Dr Nico van Belzen, International Dairy Federation (IDF), Belgium	008
Achieving sustainable production of milk - Vol 2 Safety, quality and sustainability Edited by: Dr Nico van Belzen, International Dairy Federation (IDF), Belgium	009
Achieving sustainable production of milk - Vol 3 Dairy herd management and welfare Edited by: Prof. John Webster, University of Bristol, UK	010
Ensuring safety and quality in the production of beef - Vol 1 Safety Edited by: Prof. Gary Acuff, Texas A&M University, USA and Prof. James Dickson, Iowa State University, USA	011
Ensuring safety and quality in the production of beef - Vol 2 Quality Edited by: Prof. Michael Dikeman, Kansas State University, USA	012
Achieving sustainable production of poultry meat - Vol 1 Safety, quality and sustainability Edited by: Prof. Steven C. Ricke, University of Arkansas, USA	013
Achieving sustainable production of poultry meat - Vol 2 Breeding and nutrition Edited by: Prof. Todd Applegate, University of Georgia, USA	014

Introduction

Whilst crop models have made great progress during the past three decades, all models face trade-offs in how they deal with the huge complexity of the agricultural system they seek to simulate. Individual models have developed in differing ways and for different objectives, with different strengths and weaknesses. All continue to face challenges in accurately predicting the complex effects of genotype, environmental and management (GxExM) interactions, particularly GxE interactions. This volume reviews the development and status of the current crop models and how they deal with a wide range of challenges such as accounting for and identifying options for adapting and mitigating the effects of climate change on crop production. Reviewing each model provides the opportunity to take stock of what each model does best and how it might help enhance other models. The book builds on the successful earlier volume published by Burleigh Dodds Science: *Advances in crop modelling for a more sustainable agriculture*, edited by Emeritus Prof. Ken Boote (published in 2019).

In making useful comparisons between models, it is not always clear how particular models handle variables such as genetics and environmental factors, as well as differences among varieties, cultivars, hybrids etc. for a given crop. With this in mind, authors of each chapter have been encouraged to provide an overview of the origin and development of their model; explain how crop and genetic differences are simulated; summarise recent model improvements; provide examples of current applications with emphasis on GxE interactions; and identify areas for further model enhancement and how these are being addressed, for example, the incorporation of genetics into the model.

Part 1 North America

The first group of chapters evaluates the different crop models that have been developed in North America.

Chapter 1 reviews the well-known Decision Support System for Agrotechnology Transfer (DSSAT) Cropping System Model. DSSAT is an integrated platform that combines models for many crops with software tools that facilitate the simulation of cropping conditions over a wide range of environments and management regimens. The chapter begins by discussing this model's origin and development before focussing on recent advances. These include addition of models for new grain, vegetable, forage and other crops. The chapter also explores how the model accounts for genetic differences and how it models water balance, nitrogen (N) balance and soil and plant phosphorus (P) processes. These are then used to evaluate options for genetic improvement and adaptation to climate change to maintain or improve crop yield. Finally, the chapter covers practical issues such as the DSSAT user interface, model visualization tools, application programs, software development, distribution and training options.

Chapter 2 explores the USDA-ARS Adaptive Cropping Systems Laboratory (ACSL) Crop, Land and Soil SIMulation (CLASSIM) group of models. Based on current understanding of crop and soil processes, CLASSIM simulators incorporate mechanistic representations of gas exchange, the energy balance, relationships between crop development and temperature, and responses to short- and long-term stresses. Unique sets of genetic coefficients account for phenotypic differences among crop varieties, including responses to temperature, photoperiod, fertility and drought, as well as inherent growth rates. These models are then integrated with a soils module that allows for representation of row topography and vertical / lateral movement of roots, water, solutes, gases, and heat through the soil profile. The chapter discusses how CLASSIM can be used in practice through a number of case studies covering on-farm applications, climate change impacts, crop quality characteristics and soil carbon sequestration. The chapter also discusses recent improvements to provide a broader, more accurate range of agricultural system assessments.

Chapter 3 discusses Cycles, a process-based model capturing interactions in the soil-plant-atmosphere continuum. The model integrates energy, water, carbon and nitrogen cycles, coupling plant photosynthesis with soil heterotrophic respiration. The chapter outlines the model structure, including a streamlined input/output system to improve usability and to enable high computing performance. The chapter highlights the use of Cycles in in-silico trait exploration and modeling of biogenic carbon and water cycles. Three case studies show Cycles in action: exploring maize root phenotypes under variable hydric regimes; comparing greenhouse gas (GHG) emissions in till and no-till systems; and evaluating maize–sorghum polycultures for stabilizing silage yield. These examples illustrate how advances in agroecosystem modeling can support both advances in scientific understanding and improved landscape management.

Chapter 4 describes the role and use of Simple Simulation Models (SSM). A fundamental principle in the development of SSM models is to strive for simplicity without resorting to excessive reductionism or oversimplification. To achieve this goal, conservative relationships and summary expressions are used to build more comprehensible and transparent models. Such an approach enables users to grasp the underlying principles of the model without being overwhelmed by excessive detail. The chapter reviews the development of SSM and explains how they account for environmental variables, genetic parameters, plant growth characteristics (such as phenology, leaf area development and senescence, crop mass production and partitioning as well as nitrogen balances) as well as soil nitrogen and water balances. It then shows their application in such areas as identifying desirable plant genetic traits, assessing the effects of management practices such as irrigation and rotations, as well as the impact of weather variables on crop performance.

Chapter 5 explores SALUS (Systems Approach to Land Use Sustainability), a process-based model designed to simulate interactions between soil, climate, genetics and agronomic management, focusing on their effects on crop growth and yield, soil organic matter and nitrogen dynamics. The SALUS model is structured into three main components: a set of crop growth/development and residues modules; a soil organic matter and nitrogen cycling module; and a soil water balance and temperature module. Operating on a daily time step, SALUS explores how different components of an agricultural field system interact and influence plant growth and soil nutrient dynamics. A particular strength is the ability to simulate multiple years of a cropping system. The chapter includes three case studies showing the use of SALUS: to evaluate optimal locations for bioenergy crop implementation while quantifying both productivity and environmental impacts; to determine optimal variable rate fertilizer applications; and to assess the long-term impact of differing tillage practices across the US Midwest.

Chapter 6 reviews the DeNitrification DeComposition (DNDC) model, a process-based, soil biogeochemical model designed to assess the impact of agricultural management practices on soil carbon (C) and nitrogen (N) dynamics. One of DNDC's greatest strengths is its ability to simulate the complex biogeochemical interactions that govern the nitrogen cycle and produce results that can be used to estimate greenhouse gas (GHG) emission/sequestration on a particular field, and how those emissions vary with different inputs. The chapter shows how DNDC accounts for C and N cycling by simulating the impacts of major ecological drivers (climate, soil, vegetation, and management) on soil climate conditions, plant growth, and decomposition. The chapter also shows how calibration techniques can be used to offset limitations in data availability in simulating differing scenarios. It also highlights weaknesses such as, given the focus on biogeochemical interactions, the fact that soil moisture and cropping growth sub-models are not as advanced as other models. This is an area where other models can provide options for improvement.

Chapter 7 describes the Agricultural Land Management Alternative with Numerical Assessment Criteria (ALMANAC) model. ALMANAC is a well-established process-based model which has proven particularly useful for applications such as simulating the potential future impacts of climate change on crop production. The chapter describes the basic processes simulated by the ALMANAC model, including integration of soil and weather data and the process of deriving parameters for various plant types. The chapter describes examples of successful applications of ALMANAC in: predicting cellulosic biofuel production with switchgrass over a wide range of environments and climatic conditions; assessing the potential performance of adapted maize hybrids across latitudes and in both high and low-yielding conditions; modelling the impact of climate warming on forage and grain sorghum yields; and investigating potential competition for resources facing coffee or cocoa plants grown in agroforestry systems. The chapter also identifies areas for future development.

Chapter 8 reviews the USDA Root Zone Water Quality Model (RZWQM), initially developed to simulate detailed mechanistic processes in the root zone. RZWQM is equipped with several interrelated modules, including water balance, soil carbon and nitrogen cycles, plant growth, surface energy balance, and solute transport. It also features a generic crop module which has primarily been applied to corn (maize), soybean and winter wheat. With appropriate calibration, all three crop modules in the current RZWQM2 can adequately simulate crop production under various soil, weather and management conditions. This has allowed the model to be used, for example, to simulate the effects of differing irrigation practices on soybean, nitrogen treatments on winter wheat and planting dates on corn.

Part 2 Europe

The next group of chapters discusses the crop models found and developed in Europe.

Chapter 9 explores the STICS (Simulateur mulTIdisciplinaire pour les Cultures Standard) model. STICS is a deterministic functional process-based soil-crop model that is able to account for energy, water, carbon and nitrogen dynamic balances across the crop cycle. The STICS model, thus, allows for the simulation of a wide variety of annual crops, intercropping of two species, and perennial crops at the cropping scale or over multiple cropping seasons. The chapter describes the key features of STICS, including how it accounts for genetic differences at species and cultivar levels, as well as crop species parameterized (including perennials such as grapevines a well as multi-species grasslands). It also describes applications such as: assessing the sensitivity of different alfalfa cultivars to varying environmental conditions; optimising nitrogen application in wheat; and simulating different sowing dates for adapted soybean genotypes to minimise water and heat stress. The chapter also summarise recent enhancements in both crop and soil processes as well as areas for further improvement.

Chapter 10 investigates the SAMARA (Simulator of Crop Trait Assembly, MAnagement Response and Adaptation) crop model. SAMARA was developed to study plant types (particularly rice and sorghum) as affected by different climatic and soil conditions. This model relies on integrating substantial morphological plant organ information as well as processes such as partitioning of biomass to organs on the basis of inter-organ competition.

SAMARA's particular strength lies in its ability to simulate plant plasticity in response to environmental conditions such as elevated CO2 levels or drought stress. The chapter evaluates the main characteristics of the model, recent developments, and key applications that illustrate the strength of the model to capture genotype by environment (GxE) interactions. Applications include assessing drought stress responses in different sorghum varieties in West Africa and the potential of different adaptive management practices for rice cropping systems in Madagascar

Chapter 11 examines the Genotype by Environment interaction on Crop Growth Simulator (GECROS). This model is a generic model that addresses multiple crop responses to interactions such as carbon-nitrogen, root-shoot, and source sink interactions. The chapter discusses the development of the GECROS model and its range of potential applications to make agriculture more resilient and sustainable. This chapter also evaluates weaknesses and how these might be addressed.

Chapter 12 provides a detailed review of the WOFOST (WOrld FOod STudies) cropping system model. The chapter explores the development and different applications of this model in such areas as yield forecasting, climate impact assessments, yield gap analysis and agronomic decision support. The chapter also summarises recent enhancements that allow for more realistic representation of genotype x environment x management (GxExM) interactions. The chapter discusses those enhancements and their impact in improving WOFOST functionality.

Chapter 13 investigates the Light INTerception and UTilisation (LINTUL) family of crop models. All LINTUL models are based on the relationship between leaf area index and light interception, and the concept of light or radiation use efficiency to convert intercepted radiation into dry matter. This chapter provides an overview of key relationships in the LINTUL models that describe how crops might respond to potential water and nutrient limited production situations. This chapter also discusses recent developments and improvements to account for previous limitations in the LiNTUL model family.

Chapter 14 discusses the Soil Water Atmosphere Plant (SWAP) model and its combination with the WOFOST Model (discussed in Chapter 12). SWAP was developed to simulate water, solute and temperature distribution and movement through soil while, as noted, WOFOST simulates crop growth particularly in response to environmental and management variables. The combination of the two into SWAP-WOFOST has resulted in an integrated, more robust ecohydrological and crop growth simulation model. The chapter discusses key SWAP-WOFOST functions such as: plant phenological development and CO2 assimilation; respiration/partioning and growth/senescence; carbon balance and crop nitrogen; transpiration and stress factors; and soil nitrogen. The chapter summarises applications of SWAP-WOFOST such as modelling the interplay between groundwater levels, climate and land use in the Pampas region of Argentina, and the effects of soil and water salinity on rice production in south-central Bangladesh.

Chapter 15 reviews the Scientific Impact assessment and Modelling Platform for Advanced Crop and Ecosystem management (SIMPLACE) framework. SIMPLACE consists of discrete and interchangeable software modules. Model complexity differs according to the user needs and input data availability. One application may, for example, need a detailed description of soil water, nitrogen and phosphorus dynamics, while another may need to be more general with respect to the description of soil processes but must have a detailed description of carbon assimilation, respiration and biomass partitioning. As a flexible modelling framework, SIMPLACE is designed to meet these and other needs. Based in part on the LINTUL family of crop models (discussed in Chapter 13), SIMPLACE is able to simulate crop phenology and development, shoot growth, biomass accumulation, yield formation, crop nitrogen demand, crop water demand, root growth, soil and canopy temperature, root water uptake, soil water, carbon and nitrogen dynamics. This comprehensive but flexible modelling framework has made it possible to use SIMPLACE to simulate optimal management practices, such as irrigation, fertilisation, sowing, crop rotations, in conjunction with genotype and environment conditions, i.e. a range of GxExM interactions.

Chapter 16 explores the MOdel for Nitrogen and Carbon in Agroecosystems (MONICA) agroecosystem model and its development from the original HERMES model (named after the Greek God). This chapter describes the MONICA model's main features and its various applications. This chapter then evaluates recent advances in model development. The chapter also highlights the model's strengths and weaknesses, future developments and strategies for successful application of this model.

Chapter 17 discusses the AquaCrop Model, a water-focussed crop simulation model. The strength of the model lies in its ability to take account of the differential sensitivity of key physiological processes to water stress. As an example, AquaCrop uses three different levels of soil water content to track the onset of the reduction of canopy expansion and stomatal conductance, and the triggering of early senescence in response to water deficit. AquaCrop is able to accurately simulate green canopy cover, crop transpiration, above-ground biomass and crop yield on each day of the growing cycle. The structure of AquaCrop, thus, makes it possible to assess the combined effect of increased temperature and evapotranspiration, altered rainfall patterns, and elevated atmospheric CO2 concentrations on crop growth and yield.

Chapter 18 evaluates the Daisy crop model. Daisy simulates complex crop traits in different environments by taking account of variables such as water, solute or nitrogen transport, crop growth characteristics, different agroecosystems and management practices. The chapter discusses how the model manages genotype x environment x management interactions for particular crops. Daisy has been extensively used to study transport phenomena for soil water and nitrate leaching out of the root zone as well as to study organic matter turnover and soil carbon sequestration. Applications include: modelling the potential use of "catch crops" to prevent nitrate leaching; the effects of intercropping on soil nitrogen dynamics; the effects of agroforestry practices such as short rotation coppice and alley cropping on soil carbon and nitrogen pools, soil carbon sequestration and nitrate leaching; the effects of continuous multi-year crop rotations; and nitrate leaching risk assessment

Part 3 Asia and the Pacific

The last Part of the book explores two important crop models developed by Asian researchers.

Chapter 19 reviews the InfoCrop model. The InfoCrop model is a process-based simulation model developed by Indian scientists to simulate the growth, development, and yield of major tropical crops such as wheat, rice, maize, sorghum, pearl millet, soybean, chickpea, pigeon pea, groundnut, mustard, potato and cotton. The chapter examines the development, key components, and applications of the InfoCrop model. The chapter also includes a case study and explores the future trends in further enhancing model functionality.

As the last chapter in the book, Chapter 20 explores the CropGrow series model, comprising WheatGrow and RiceGrow, developed by researchers at Nanjing Agricultural University in China. The model framework consists of multiple sub-models that comprehensively address wheat or rice growth and development processes, including phasic development and phenology, organ development and population establishment, photosynthetic production and biomass accumulation, assimilate partitioning, yield and quality formation, nutrient dynamics, and water balance. This holistic approach provides a detailed, dynamic understanding of wheat and rice response to different variables, supporting precision agriculture applications and optimized crop management. The chapter explores applications such as wheat response to low and high temperature stresses as well the ways the model can support decision support systems for farmers.

It is important to note that, whilst the Editor and Publisher aimed for as comprehensive coverage as possible, we were not able to include a number of other models as originally planned. This was, for example, due to the inability of some authors to participate at the time we approached them given other commitments. Omission from this book does not reflect in any way on the importance and value of a particular model. We look forward to the opportunity to include other models in the future.

Part 1

North America

Chapter 1

The Decision Support System for Agrotechnology Transfer (DSSAT) crop modeling ecosystem: evolution and advances

Kenneth J. Boote[1], Gerrit Hoogenboom, Cheryl H. Porter, Willingthon Pavan and Vakhtang Shelia, University of Florida, USA; Upendra Singh, International Fertilizer Development Center, USA; Jeffrey W. White, USDA-ARS, USA; Senthold Asseng, Technische Universität München, Germany; Jon I. Lizaso, Universidad Politécnica de Madrid, Spain; L. Patricia Moreno, CIAT, Kenya; Santiago Cuadra, EMBRAPA, Brazil; Bruce Kimball, USDA-ARS, USA; Alwin Hopf, Fabio A. A. Oliveira, Thiago B. Ferreira and Henrique Boriolo Dias, University of Florida, USA; Phillip Alderman, Oklahoma State University, USA; William D. Batchelor, Auburn University, USA; Diego Pequeno, CIMMYT, Mexico; Kelly R. Thorp, USDA-ARS, USA; James W. Jones, University of Florida, USA; Richard Ogoshi, University of Hawaii, USA; L. Anthony Hunt, University of Guelph, Canada; and Gordon Y. Tsuji, University of Hawaii, USA

1 kjboote@ufl.edu

http://dx.doi.org/10.19103/AS.2025.0155.01

1 Introduction

The Decision Support System for Agrotechnology Transfer (DSSAT) ecosystem is a combination suite of software created to facilitate the development and use of integrated crop, soil, weather, and management components and the datasets required for this development and applications. The DSSAT Cropping System Model (CSM) is the standalone executable component of DSSAT that simulates the dynamic performance of different crops and makes use of DSSAT's tools for input, output, analysis, and applications for simulating crop and soil performance for specific climates and management conditions.

In this chapter, we discuss the evolution of the DSSAT-CSM as it has been improved to simulate different crop and soil processes, examples of adding capabilities to simulate additional crops, model adaptation for new crop species, and examples of model applications for genetic improvement in the face of climate change. This chapter reviews the crop C, water, N, and P balance processes in the DSSAT-CSM and the tools and utilities in DSSAT.

2 The origins of the Decision Support System for Agrotechnology Transfer (DSSAT)

DSSAT originated out of the International Benchmark Sites Network for Agrotechnology Transfer (IBSNAT) project funded by the United States Agency for International Development (USAID). In the early 1980s, the USAID made a bold step to support a project based on systems analysis of agricultural production to address food security in developing countries. The IBSNAT project for improving agricultural production was led by Professor Goro Uehara, a soil physicist at the University of Hawaii, and co-led by Fred Beinroth, a soil scientist from the University of Puerto Rico. The IBSNAT project was a follow-up to the International Benchmark Soils Project also funded by USAID. Professor Goro Uehara, the leader of the initial soil project, concluded that predicting yields only based on soil type (similarity) was insufficient and that there was a need to consider the weather, plant genetics, crop and soil management aspects in addition to soil information. This led to the combined soil-crop-systems modeling approach in the IBSNAT project with an emphasis on field experimentation and minimum data required to be used in this integrated systems approach. The IBSNAT Project was funded from 1982 through 1992. However, in later years, USAID and other federal agencies were less supportive of funding for basic model development and improvement, although there was some funding for model applications.

At the start of the IBSNAT Project, crop modeling teams from the University of Florida (SOYGRO and PNUTGRO models) and from USDA-ARS in Temple, Texas (CERES-Maize and CERES-Wheat models) were invited to collaborate with scientists from the University of Puerto Rico, the University of Edinburgh (Scotland), the University of Guelph (Canada), and the International Fertilizer Development Center in Muscle Shoals, Alabama (Wilkerson et al., 1983; Ritchie et al., 1985; Boote et al., 1986; Jones and Kiniry, 1986; Jones et al., 1987). The SOYGRO model was developed by Wilkerson et al. (1981, 1983) and the PNUTGRO model by Boote et al. (1986). The CERES-Maize model had been developed by Allen Jones and Jim Kiniry (Jones and Kiniry, 1986) with guidance by Dr. J. T. Ritchie. The CERES-Wheat model had been developed by Ritchie et al. (1985) on a project funded by the US Central Intelligence Agency. Their goals were to develop relatively simple models that could simulate wheat and maize production in the Cold War regions based on satellite mapping of crop area presence and presumably satellite sensing of temperature, solar radiation, and likely rainfall. The CERES-Maize model is simpler in structure than SOYGRO or CROPGRO. It uses a simple radiation use efficiency approach that goes directly to dry matter, whereas CROPGRO considers leaf photosynthesis and growth respiration costs (Penning de Vries et al. 1974). CERES models use approaches for leaf area progression and partitioning to leaf and stem, based on hard-wired algorithms in the source code. The CROPGRO models are more mechanistic for partitioning and leaf area growth and are more generic, with parameters and relationships for processes being placed in external "read-in" files.

Early during the IBSNAT project, the investigators realized that the CERES-Maize, CERES-Wheat, and SOYGRO models had different input file types (structure, variable names, and placements), which required different subroutines for inputting weather, soils, management, and cultivar information. There was a lack of common file structures and a lack of common input "read" methods across the different crop models. Dr. Jim Jones solved these issues during his 1983 sabbatical at the University of Hawaii, working with Professor Uehara and his team to set up common file formats for weather, soil, and management information. This required standardizing the exact structures of these three types of files (*.WTH, *.SOL, and various management input files), along with creating common read subroutines for both models. This effort was successful and led to the first official release of the DSSAT software Version 2.1 by 1989 (IBSNAT, 1989). DSSAT Version 3.0 (IBSNAT, 1993) was released at the end of the IBSNAT Project. As other models were developed and added to DSSAT, such as CERES-RICE and SUBSTOR-Potato, those models were required to use the same input file structure.

A strong emphasis of the IBSNAT project and current DSSAT modelers is the collection and preservation of data on soil water, soil N, final yield, and crop growth analyses collected during experiments. It became clear that such data were needed to develop, improve, and document model performance. Jim Jones participated in a 1984 meeting at the ICRISAT Center in India, where the modelers recognized the need for a common Minimum Data Set (MDS) for all inputs to crop models (ICRISAT, 1984). This led to the creation of the MDS system in IBSNAT and DSSAT that standardized the inputs required for the crop models as well as the file formats used. In addition, strong emphasis was placed on developing protocols for collecting the minimum data set (MDS). This standardization of protocols and standardized inputs led to the development of data utility programs for processing weather, soil, management input data, and experimental observation files, as well as tools for the application and display of output data for the models, forming the basis for the Decision Support System for Agrotechnology Transfer (DSSAT) software. Manuals were created to document the best procedures for data collection, what type of data, suggested frequencies of data collection, and documentation of models and tools in DSSAT (Hunt et al., 1994; Tsuji et al., 1994; Hunt et al., 2001; Hoogenboom et al., 2010; White et al., 2013). These manuals and documents remain available in the latest DSSAT shell: DSSAT Version 3, Volumes 1, 2, 3, and 4 (Tsuji et al., 1994) and DSSAT Version 4.5, Volumes 1, 2, 3, and 4 (Hoogenboom et al., 2010). For further details of IBSNAT activities and outcomes, see IBSNAT (1993), IBSNAT (1994),Uehara and Tsuji (1998), and Jones et al. (2016).

During the IBSNAT project (1980s), we explored available software for entering and storing such data. In this case, dBase2 software was initially used, which turned out to be an unfortunate choice because the company went out of business and no longer provided any support. After learning this lesson, we concluded that the best way to save long-term data was as ASCII files (the default for the past 40 years). While it may be fashionable and convenient to use various other methods of data entry and storage, those methods must be able to convert to, and from, ASCII files which are what the CSM model reads and outputs.

3 Advances in model development in the 1990s

Beginning about 1990, we encountered the drawbacks of having three different crop legume models (different source codes for soybean, peanut, and dry bean) that were somewhat different for the crop C and N balance aspects but barely different for soil water balance or soil N balance. We already had a crop parameter file for each of the three species, where that crop parameter file had a similar structure but different values for parameters for given crop species. But when we needed to repair or improve code in the soil water balance, for example, we found it necessary to make the same improvements in the three sets of code. Therefore, we decided to create one common source code for the three crops to avoid those issues. After reviewing the three models (SOYGRO, PNUTGRO, and BEANGRO) (Hoogenboom et al., 1990a, 1990b, 1992, 1994), we developed source code to create a single generic model that would simulate all three crops (Hoogenboom et al., 1991, 1992). This required us to remove any remaining crop-specific hard-coded parameters and relationships from the crop C and N balance subroutines and to put those parameters into an external species file (formerly called the "crop parameter file"). Gerrit Hoogenboom accomplished this complex task. The first trial version was named LeGro in 1993 (Hoogenboom et al. 1999), which later became CROPGRO. Also, during this period of code improvement, we added new code to the legume module in CROPGRO to simulate explicit nodule growth and N-fixation (Hoogenboom et al., 1990a, 1991). Boote et al. (2009) provide details on simulating nodule growth and N-fixation.

3.1 Introduction of leaf-level hedgerow photosynthesis and rubisco kinetics response to CO_2 into CROPGRO

During his career at the University of Florida, Ken Boote researched leaf and canopy photosynthesis and the modeling of those processes (Boote et al., 1983a, Boote et al., 1984; Boote and Loomis, 1991; Boote and Pickering, 1994). We developed a mechanistic model of leaf-to-canopy assimilation coupled with hedgerow light interception, where the leaf-level photosynthesis simulates on an hourly basis for sunlit and shaded leaves (Boote and Pickering, 1994; Pickering et al., 1995; Boote et al., 1998). This model captures the behavior of rubisco kinetics of Farquhar and von Caemmerer (1982) and mechanistically simulates photosynthesis response to CO_2, O_2, temperature, solar radiation, leaf area index, specific leaf area, and leaf N concentration. The simulated response to CO_2 is thus an emergent outcome of the rubisco-kinetics, rather than resulting from an externally prescribed CO_2 response curve as often used in other models. The simulated leaf-level and canopy-level photosynthesis were tested against observed data and shown to be accurate by Alagarswamy et al. (2006). Since then, this hourly leaf-level photosynthesis method has been the default for all the CROPGRO model crops, in place of the older, but still available, daily canopy photosynthesis option. The CERES-based models in DSSAT are based on radiation use efficiency (RUE), and they use an externally prescribed CO_2 response modifier based on observed CO_2 response data.

The need to account for the CO_2 responsiveness of the models was recognized in the 1990s (Rosenzweig et al., 1995) and led to model evaluations against available CO_2-response data along with CO_2 impact assessment (Curry et al., 1995). In the late 1990s, we created DSSAT code and parameters that set CO_2 level for the actual years of the observed crop experiments based on the Mauna Loa Observatory CO_2 record, first established by Keeling et al. (1976). The need to do this became obvious as the CO_2 increased over decades from 315 ppm in 1959 to 427 ppm in 2024, which has increased simulated soybean yield by 19% (CROPGRO-Soybean simulations for 2 years in Iowa and 2 seasons in Ohio). Boote (2011) subsequently illustrated how elevated CO_2 is assisting soybean breeders to increase yield because there are beneficial G x E responses. In retrospect, a good portion of the soybean yield increase over this period can be attributed to rising CO_2 in addition to genetic improvement and technology. Boote et al. (2010) examined the CO_2 response curves for C3 and C4 crops simulated by the CERES-based models in DSSAT and evaluated CERES and CROPGRO modeled crops against metadata on observed CO_2 response. In addition, simple ozone impact subroutines that interact with CO_2 levels were introduced into the NWheat model (Guarin et al., 2019) and the CROPGRO-Soybean model (Guarin et al., 2024).

Further collaboration with Dr. Nigel Pickering (Pickering et al., 1990, 1993, 1995) led to the development of a stand-alone program for leaf-to-canopy assimilation that also simulated hourly canopy energy balance and evapotranspiration (ET) based on first principles and thus do not require the FAO-56 engineered ET equations of Allen et al. (1998). This code was placed into the SOYGRO/CROPGRO model to enhance the ability of the model to simulate response to CO_2 (Pickering et al., 1995). The hourly energy balance code was unused for many years but was later revived, improved, and evaluated by Cuadra et al. (2021). In addition to simulating ET, this option simulates soil and canopy temperature, uses crop temperature to drive crop processes, and uses CO_2- and vapor pressure deficit-sensitive control of stomatal conductance. This is included as an option in the latest DSSAT release.

3.2 Introduction of pest damage capabilities into DSSAT

Most crop models do not simulate pest damage from insects, diseases, or nematodes, primarily because of the difficulty of linking two models (the crop and the pest) together, thus requiring communication and feedback. We pursued a different approach whereby pest damage was described as the type of action on rates or organs (state variables) as described by Boote et al. (1981, 1983b). This was implemented in DSSAT by Bill Batchelor (Batchelor et al.,1993). For example, pest damage such as observed percent defoliation or percent necrosis of existing leaves on given dates was entered into the observed data file (File T) and read by a coupling routine in the model with interpolation between dates, and then applying that given damage to the crop to reduce a state variable (leaf area, leaf mass, leaf N, etc.) or rate (leaf photosynthesis). This approach works for all the CROPGRO-style models and some of the CERES-style models, such as CERES-Rice, CERES-Maize, CERES-Millet, CERES-Sorghum, CERES-Sugarbeet, and NWheat (Memic et al., 2020; Ferreira et al., 2021). The pest damage effect has been used frequently to reproduce leafspot damage effects on peanut growth and yield (Naab et al., 2004, 2005; Singh, M. et al., 2013). But this approach only works to explain past performance where observed damage is recorded (e.g. via scouting). There is one example of direct linkage of the PNUTGRO model to a leafspot model (Bourgeois, 1989), but unfortunately, that code did not persist. New developments in model coupling have directly linked a wheat fusarium model to the wheat simulation model for the purpose of predicting pest damage effects on yield (Fernandes et al., 2017; Pequeno et al., 2024). In this and other ongoing efforts, we are currently evaluating various methods that allow for the coupling of two or more models in order to provide opportunities for running multiple instances of two models in parallel.

3.3 Transitions from 1998 onward

DSSAT Version 3.5 was the first stable and widely-used version of CSM and the associated tools for data entry and analysis of simulations. The individual models were well-calibrated and documented in the published DSSAT book (Tsuji et al., 1998). Four volumes of reference manuals (Tsuji et al. 1994) were released with DSSAT V3.5. In 2010, four additional volumes of reference manuals were placed inside the DSSAT software shell (Hoogenboom et al., 2010). Models for individual crops and processes of water balance and N balance were well documented by chapters in the Tsuji et al. 1998 book: e.g. CROPGRO model for grain legumes (Boote et al., 1998); CERES models for cereals (Ritchie, 1998), water balance (Ritchie, 1998), soil-crop N balance (Godwin and Singh, 1998), root-tuber crops (Singh et al., 1998), DSSAT structure (Jones et al., 1998), and pests (Teng et al., 1998).

Between 1998 and into the next century, DSSAT software managed the Y2K transition beyond the 2-digit year code plus day of year to 4-digit year code plus day of year. There was a transition from DOS-based software to use Windows-style of point-and-click, even though some of the individual tools remained unchanged. Also, during this time, the code was modularized to allow the combination of multiple models into a single code base and to improve the process of adding or replacing self-contained scientific capabilities or modules. The modularization organized the DSSAT source code into the initiation of state variables, process rate calculations, integration of state variables, and output of module predictions. Modularity applied well for many subroutines (modules) such as the photosynthesis method (leaf-level versus older RUE-style), soil organic matter (CENTURY or older Godwin-style), soil water balance, soil N balance, and evapotranspiration (several methods to be discussed later). However, we had to maintain separate growth-process modules for many crops because processes for leaf area expansion, grain addition, grain growth, and tissue N mobilization differed considerably in many cases. The CROPGRO model is an exception, using a single common plant growth module for all its crop species.

4 The creation of the Cropping Systems Model (CSM)

One of the advantages of modularization is that it now allows all the crop models, including CERES-style models, to be placed into the same executable file. The Cropping System Model (CSM) was created as a single executable program, with an integrated set of computer code representing more than 40 crops. Previously, there was a different executable code for CROPGRO models separate from executables for each of the CERES models (CERES-Maize different from CERES-Rice, for example). Once all crop modules were in the same executable code, then subroutines such as soil organic matter, soil water balance, soil inorganic N balance, ET method, etc., were shared and accessible across all DSSAT-plant modules. Figure 1 shows the modular structure of the Cropping System Model, its important linkages, and how plant, soil, management, soil-plant-atmosphere, and management modules connect to a site-specific land unit module. This structure was particularly useful for incorporating the CENTURY-style of soil organic matter simulation, which was incorporated into DSSAT by Gijsman et al. (2002). With the addition of the CENTURY-style soil organic carbon (SOC) module and use of a single executable for the entire system, the capability of crop rotation/sequence was enhanced, allowing multiple crops to be simulated on the same land unit in sequence to evaluate no-tillage practices, soil C sequestration, and greenhouse

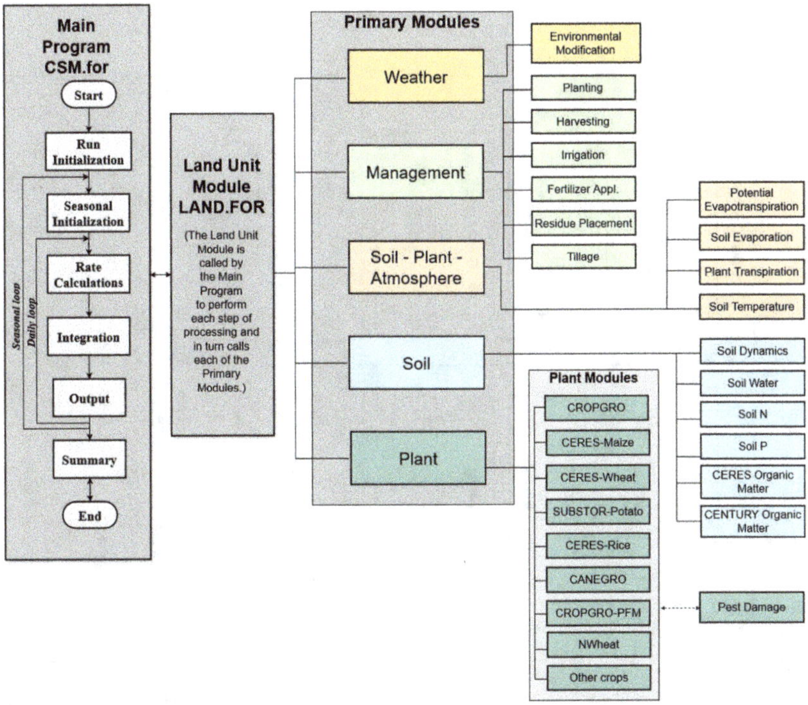

Figure 1 The modular structure of the Cropping System Model, showing important linkages, and how plant, soil, soil-plant-atmosphere, management, and weather modules connect to a site-specific land unit module.

gas (GHG) emissions. In 2004, with the release of DSSAT v4.0 (Jones et al., 2003), the new executable model was named the CSM. For example, the new naming of crop species is described by both model and species, such as CSM-CROPGRO-Soybean model or CSM-CERES-Maize. Figure 2 lists the many crops that are currently simulated by the Cropping System Model, with CROPGRO annual crops, CERES-style annual crops, CROPGRO-Perennial-Forage-Model crops, and other crop modules. Modeled crops shown in broken line boxes are under development.

Another advantage of modularization is that the source code was organized in such a way that each module represented a discrete scientific process or function. For example, previously in the non-modularized code, soil water content was initialized in one location of the code, rates were calculated in various places, the soil water content was updated in numerous locations in the code, and outputs were done in a separate location. With modularization, all functions associated with soil water content were done in a single module. Isolation of these processes improved transparency and allowed new

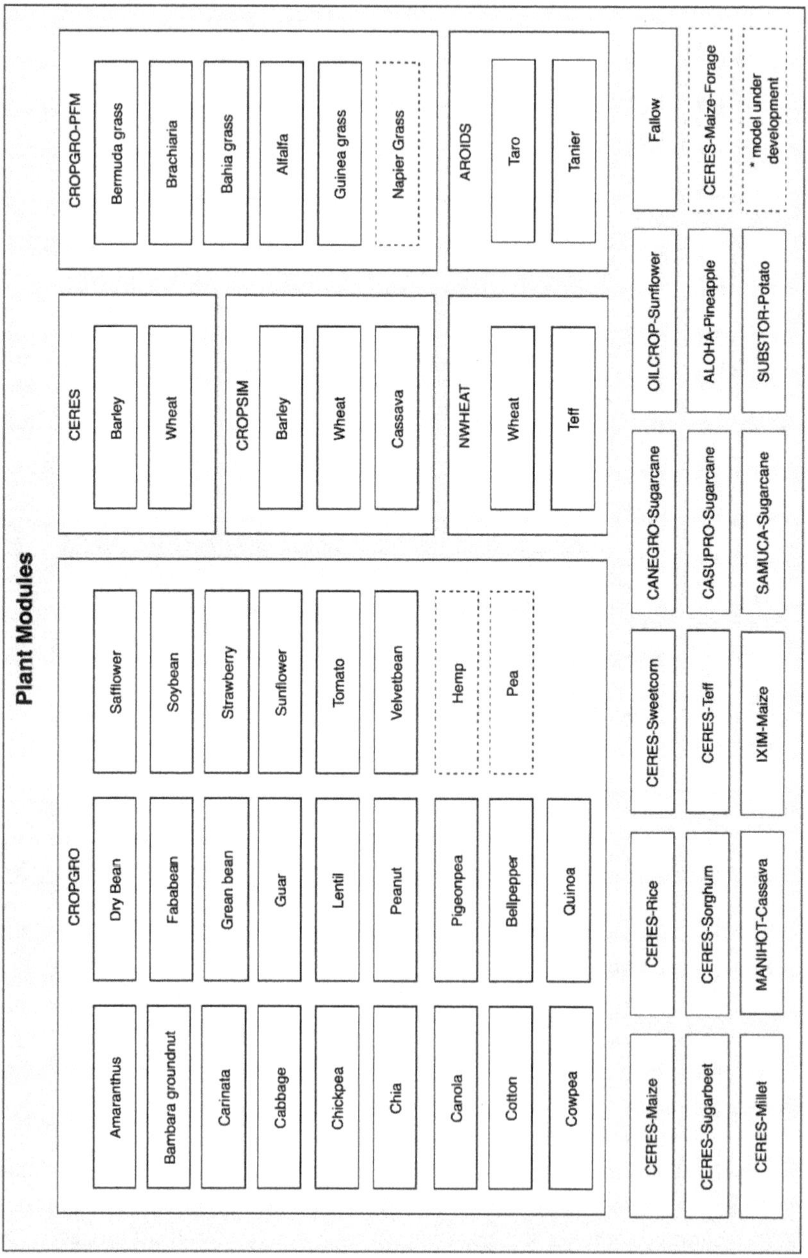

Figure 2 A listing of all the crop (plant) modules in the CSM.

Published by Burleigh Dodds Science Publishing Limited, 2025.

programmers to more easily navigate, understand, and make advancements to the code.

5 Adapting the CROPGRO crop growth model to simulate additional new grain crops

The generic structure of the CROPGRO model allows the addition of new crop species based only on the modification of the species, ecotype, and cultivar parameter files which are "read-in" input files. There are two informative examples of CROPGRO adaptation for grain legume crops: faba bean (Boote et al., 2002) and pigeonpea (Alderman et al., 2015). These two publications carefully describe the approach and methods to follow for adapting CROPGRO to a new species: (1) modifying tissue composition, cardinal temperature parameterization, leaf photosynthesis, etc. based on available literature information, (2) calibrating time-series growth and final yield outputs against observed time-series growth analyses (LAI, total crop mass, reproductive mass) using inverse model optimization of parameters (in the species file). Ideally, the observed growth and yield data are collected in multiple environments under conditions without water and N limitations. This approach has been quite successful. One of the concepts learned, particularly relative to the faba bean model, was the need to reparameterize many temperature responses because faba bean has much lower cardinal and optimum temperatures for processes as compared to soybean, which provided the initial template for species, ecotype, and cultivar files. It also showed that the CROPGRO model style of phenology can function for long-day as well as short-day species. The CROPGRO-chickpea model, also adapted from CROPGRO (Singh and Virmani, 1996; Hoogenboom et al., 1997; Singh et al., 2014a), is a long-day cool-season crop like faba bean. Singh et al. (2014a) updated and tested the CROPGRO-Chickpea model, particularly for temperature parameterization.

Adapting CROPGRO for a new crop requires both extracting crop-specific information from the scientific literature, as well as inverse optimization of model parameters to best simulate time-series growth and development of the crop. Figure 3 shows simulated and observed LAI, canopy height, canopy width, leaf, stem, pod, and total aboveground mass of faba bean cultivar Alameda grown under irrigation and relying on N-fixation during 1986–1987 season in Cordoba, Spain (Boote et al., 2002). The GBuild graphical program of DSSAT not only plots the simulated and observed data, but also computes goodness-of-fit statistics (RMSE and D-statistic) as shown in Fig. 3. The model calibration process, after setting compositions, cardinal temperatures and approximate life cycle, involved iteratively modifying species, ecotype, and cultivar parameters to maximize the D-statistic averaged over all crop output variables over six treatments in this case (including N-fertilized and second season treatments).

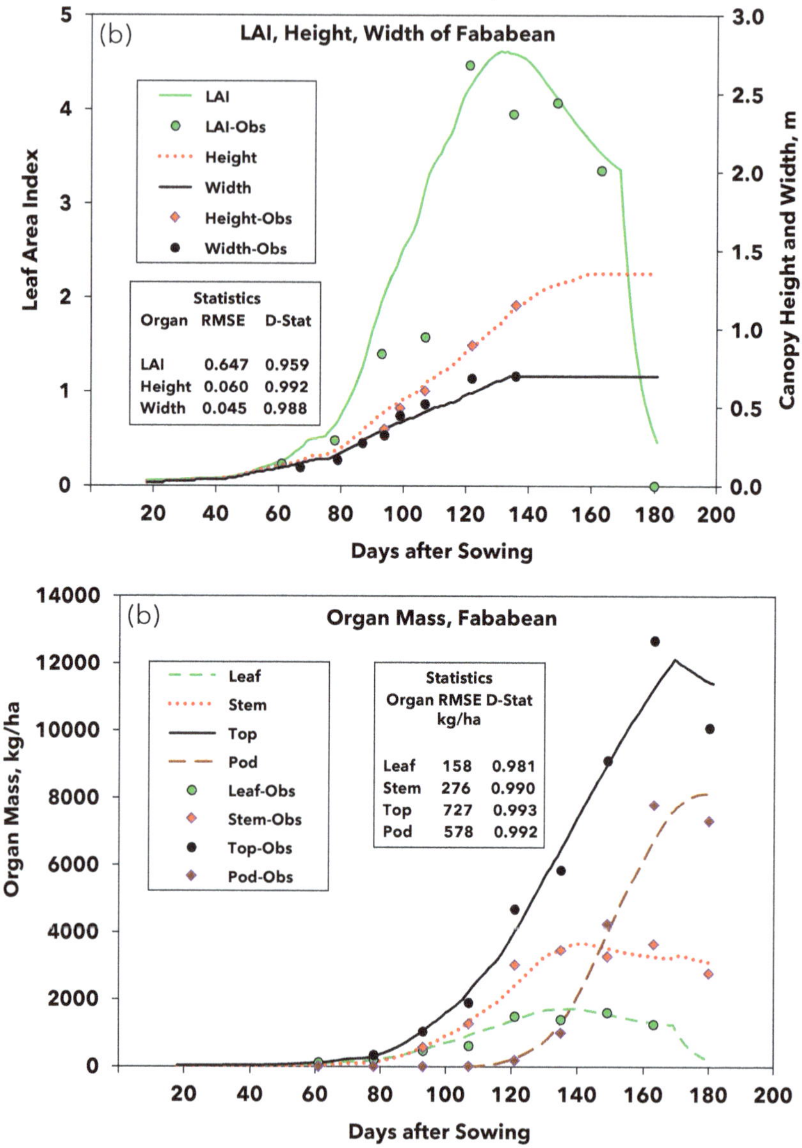

Figure 3 Comparison of simulated (lines) and observed (symbols): (A) leaf area index, canopy height, canopy width, (B) leaf, stem, total aboveground mass, and pod mass over time for faba bean cultivar Alameda sown December 10, 1986 and irrigated at Cordoba, Spain as simulated by CROPGRO-Fababean model. Statistics (RMSE and D-statistics) are shown.

Averaging is acceptable for the D-statistic because it is a type of normalized statistic. The RMSE was evaluated per output variable averaged over all treatments because of different magnitudes (LAI different from leaf mass, for example). The goal was to minimize RMSE for each output variable as well as maximize the D-statistic and to approach the observed time-series means. The regression coefficient R^2 is not used and should never be used for time-series evaluation as it suffers from "part-whole" effect going from small to large masses over time. Final yield and its components are included with the time-series data and are only sparsely and separately evaluated. Obviously, modifying given species and genetic parameters affects different aspects of the time-series output. Even cardinal temperatures and day-length sensitivity on life cycle and cardinal temperatures on processes can be modified away from sparse literature values based on simulations of sowing date effects (see Confalone et al., 2011).

Crops added to DSSAT since v3.5 that use the CROPGRO-template and adaptation approach include chickpea (Hoogenboom et al., 1997), tomato (Scholberg et al., 1997; Boote et al., 2012), cowpea (Boote unpublished, 1998), mucuna or velvetbean (Hartkamp et al., 2002), faba bean (Boote et al., 2002), cotton (Pathak et al., 2007; Boote unpublished, 2010; Thorp et al. 2014; Boote unpublished, 2024), canola (Deligios et al., 2013), pigeonpea (Alderman et al., 2015), safflower (Singh et al., 2015; Steberl et al. 2019), quinoa (Präger et al., 2019), chia (Mack et al., 2020), carinata (Boote et al., 2021), strawberry (Hopf et al., 2022), amaranthus (Nkebiwe et al., 2022), guar (Boote et al., 2023), lentil (Qi et al., 2024), bambara groundnut (Karunaratne et al., 2024) sunflower (Boote, unpublished, 2014), green bean (Boote unpublished, 2009), cabbage, and pepper. See the list of crops in Fig. 2. All the CROPGRO-type crops share the same source code, with differences in physiology and genetics being specified by different species, ecotype, and cultivar files.

Adapting CROPGRO to the various new crops basically followed the same approaches as described above. Minor issues with new data and new crops have suggested over time where minor code modifications were needed. For example, chia is a very small-seeded crop (Mack et al., 2020), which required a code change in the model to allow very small seed size. That change in itself was relatively easy, but the modified code resulted in a conflict with a hard-coded limit on how small early LAI was allowed to be, and when that coded limit was removed, then seed size (and plant population) resulted in having a greater impact on initial leaf area index. Consequently, one of the older parameters (SIZELF) no longer has any important effects on early LAI.

Quinoa provides another example of a recently adapted crop following the CROPGRO template. Figure 4A, 4B, and 4C illustrate LAI, total aboveground crop mass, and grain harvest index (HI) as affected by four sowing dates in 2016 at Stuttgart, Germany. For this region, early sowing was considerably delayed

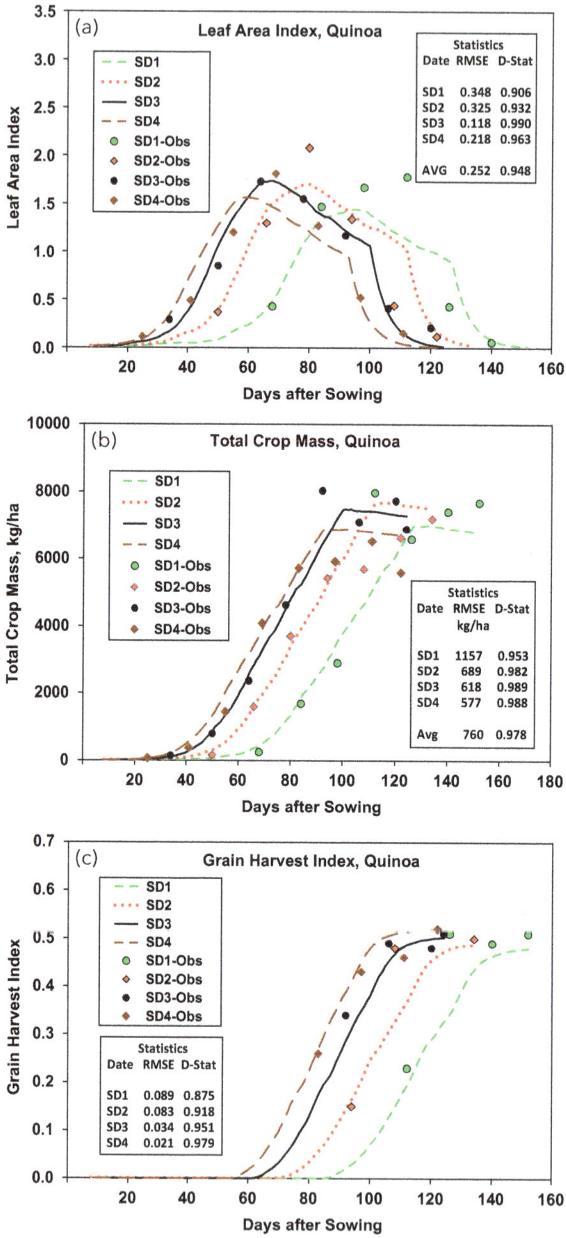

Figure 4 Comparison of simulated (lines) and observed (symbols): (A) leaf area index, (B) total aboveground crop mass, and (C) grain harvest index for quinoa cultivar Jesse sown on four dates (SD) (March 17, April 4, April 20, and April 27) in 2016 and grown under irrigation at Stuttgart, Germany as simulated by CROPGRO-Quinoa model. Statistics (RMSE and D-statistics) are shown. Simulations based on model adaptation and data of Präger et al. (2019).

by cool temperatures, while the crop progressed and grew more rapidly with later sowing dates. This exercise demonstrated the need and value of setting cardinal temperature values for crop developmental phases (germination, time to anthesis and time to maturity) as well as leaf expansion, photosynthesis, and grain growth processes. See Boote et al. (2018) for discussion of cardinal temperatures for different species.

6 Adaptation of CROPGRO for vegetables, multi-harvest, and fresh weight production

One of the early attempts was to adapt CROPGRO to simulate tomato (Scholberg et al., 1997; Boote and Scholberg, 2006; Boote et al., 2012). This was unique in three ways: (1) tomato is a non-legume, (2) it produces fruits that are fleshy with very small seeds, and (3) it is marketed on the basis of fruit fresh weight. The latter required adding code to CROPGRO to predict dry matter concentration (water concentration is the complement) of the fruit, which allowed simulated fruit dry matter yield to be converted to fresh weight (Boote and Scholberg, 2006).

The adaptation of CROPGRO for strawberry (Hopf et al., 2022) required the creation of a multi-harvest feature to allow repeated harvest of fruits concurrent with continued flowering and fruit set. CROPGRO already had the capability for continued flowering and fruit-set (it simulates daily addition of cohorts of flowers and fruits until the crop reaches a full fruit load or carrying capacity), but it did not allow simulation of periodic harvesting. New source code was added to allow harvesting of fruits as they reached harvestable fruit age on a 3 or 4-day harvesting interval typical of strawberry (or any chosen interval). An important aspect is how this uses the already existing XFRUIT cultivar parameter, which is defined as the limit on the fraction of daily assimilate allocated to reproductive growth. XFRUIT was coded to differentiate indeterminate cultivars of peanut and cotton that continue leaf growth through maturity. An XFRUIT less than 1.0 allows the CROPGRO-Strawberry model to continue to add new reproductive sinks for 6 to 7 months because a full-fruit load is not attained. Continued harvesting is accounted for so that the model never stops adding flowers and fruits. The multi-harvest feature can be used for periodic harvests for CROPGRO-Green Bean, or for CROPGRO-Tomato for indeterminate 18-month tomato production in greenhouses, although, as of this publication, it has not been tested for long-cycle greenhouse tomato production.

Four DSSAT models for vegetable crops, CROPGRO-Tomato, CROPGRO-Strawberry, CROPGRO-Green Bean, and CERES-Sweet-Corn feature the ability to simulate water concentration of fruits or sweet corn ears and fresh weight yield prediction. The three CROPGRO-style models simulate water concentration of tomatoes or fruits as a function of fruit age and output total fresh weight production and dry matter concentration as well as fresh weight per fruit, which is a quality aspect. The CERES-Sweet Corn model simulates a fixed dry matter concentration of ears and outputs total fresh market yield at the time of harvest. Other vegetable models in DSSAT include cabbage and green pepper, although fresh weight output has not yet been developed for these models.

7 The CROPGRO Perennial Forage Model (CROPGRO-PFM)

The CSM was originally designed to only simulate annual crops, where the only carryover variables were for soils (e.g. soil water and nitrogen) if simulated in crop sequence/rotation. The CSM remained that way until code was added by Stuart Rymph, Ph.D. graduate student at the University of Florida, to simulate perennial forage growth (Rymph, 2004). He modified code of the CROPGRO model by adding state variables to represent reserve storage tissue (rhizome, taproot, stem base) and reserve concentrations, with rules for partitioning to storage tissue and rules for mobilizing reserves (total non-structural carbohydrates (TNC) and Nitrogen) to drive re-growth as well as rules for re-filling reserves. The mobilization and re-fill functions depend on LAI, current photosynthesis rate, current reserve TNC concentration, and day-length-linked dormancy. The model is designed for multi-harvest (few to many harvests over decades), which requires an MOW file that specifies the harvest dates, the amount of living stubble remaining, the percentage leaf of remaining stubble, and a "re-staged" node number on tillers. The CROPGRO-PFM model has been adapted for brachiaria (Pedreira et al., 2011; Pequeno et al., 2014, 2018; Santos M. G. et al., 2019, Santos M. L. et al., 2022), bermudagrass (Pequeno et al., 2018), guineagrass (Lara et al., 2012; Brunetti et al. 2021), and bahiagrass (Rymph, 2004; Smith et al., 2023). These are all tropical grasses with C-4 photosynthesis, which required minor adaptation to leaf photosynthesis equations. For C-4 photosynthesis, the leaf internal CO_2 to ambient CO_2 (C_i/C_a) ratio was reduced from 0.7 (C-3) to 0.4 (C-4) and the specificity for CO_2 versus O_2 (*tau* in literature) was increased by a factor of 10 (C-4) which greatly reduced the response to rising CO_2. The model was also adapted for alfalfa a C-3 legume (Malik et al., 2018, Jing et al. 2020).

The CROPGRO-PFM operated as a standalone executable for more than 10 years and was brought into DSSAT V4.7 in 2017. Introducing CROPGRO-PFM into the existing DSSAT code required code adaptation to bring in the unique subroutines needed for perennial forages and their harvesting. There is no final "end-of-season" output for the CROPGRO-PFM model because outputs occur periodically for few to many harvests, which list harvest dates, herbage mass, percent leaf of herbage, and crude protein of herbage, along with residual stubble and LAI. Those outputs go to the standard PLANTGRO.OUT, but they are also summarized in a new FORAGE.OUT file. The model does not mature or die but rather must be terminated in the harvest section of the management File X, although there are species-file rules for freezing temperatures that kill above-ground leaf and stem tissue. The CROPGRO-PFM is recommended to be simulated only with the CENTURY-style of soil organic matter mineralization because dying organs (aboveground organs and roots) contribute to fresh organic matter whether on the surface or in the soil. Santos et al. (2022) used

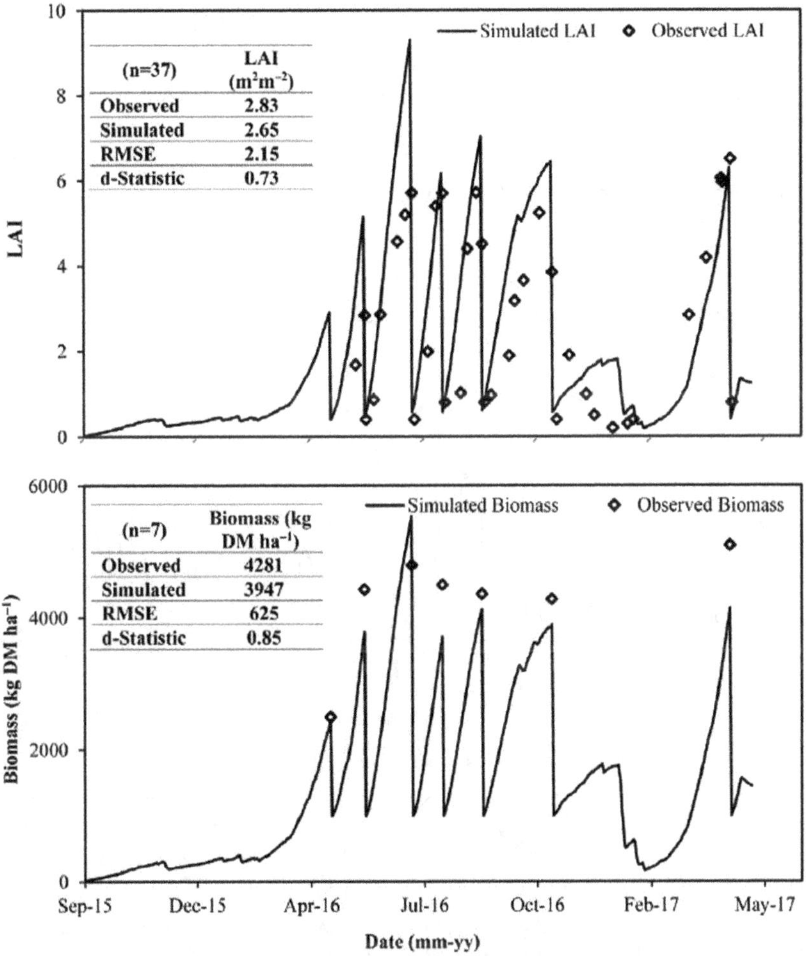

Figure 5 Comparison of simulated (lines) and observed (symbols) leaf area index (LAI) and crop biomass simulated over time at Almudevar, Spain, with the CROPGRO-PFM-Alfalfa model (with permission, Malik et al., 2018).

Published by Burleigh Dodds Science Publishing Limited, 2025.

the model to evaluate decades-long simulations of brachiaria pasture in Brazil, testing various scenarios of fertilization and animal manure return to the soil.

Figure 5 illustrates the dynamics of LAI and aboveground biomass of alfalfa over multiple cutting harvest cycles during two years in Spain (Malik et al., 2018). In this example, alfalfa was sown from seed in the fall and harvested six times in the following season. Alfalfa has considerable taproot mass (up to 5000 kg ha^{-1}) that contains up to 20% total non-structural carbohydrates and relies considerably on rapid TNC mobilization to drive re-growth because the residual LAI is very small after cutting. Freeze kill occurred about February 2017, thus partially reducing above-ground biomass.

For some perennial tropical grasses such as Tifton-85 bermudagrass (Pequeno et al., 2018), establishment is by transplanting. Tifton-85 bermudagrass is a type of hybrid that has both above and below-ground stolons/rhizomes that store considerable TNC reserves and is harvested on approximately 4-week intervals during warm-season to provide high quality forage. As shown in Fig. 6, rapid re-growth is dependent on both residual LAI as well as mobilization of TNC reserves as reported by Alderman et al. (2011a, 2011b) and simulated by Pequeno et al. (2018).

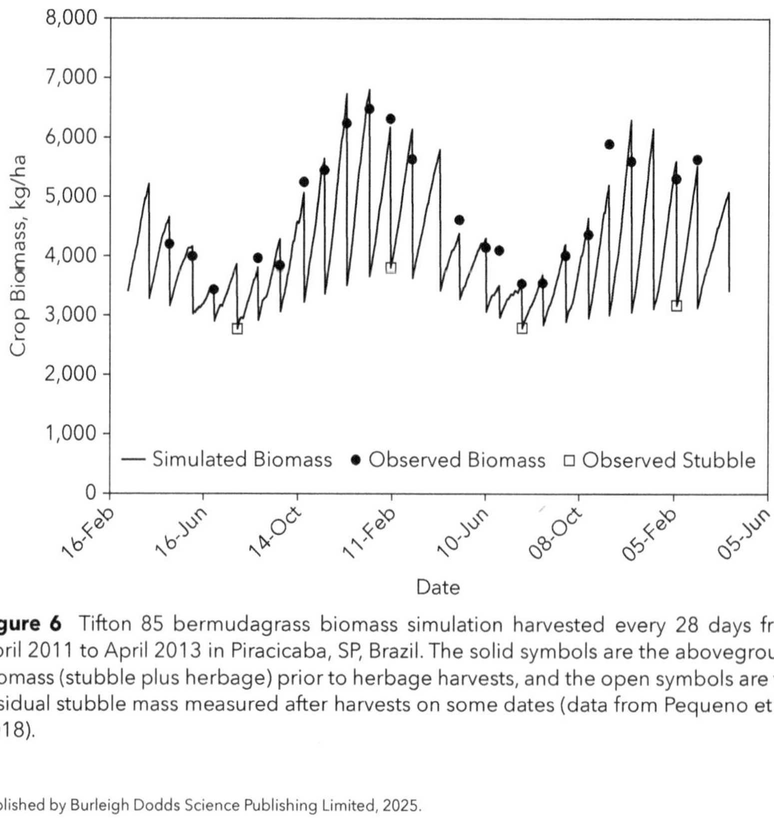

Figure 6 Tifton 85 bermudagrass biomass simulation harvested every 28 days from April 2011 to April 2013 in Piracicaba, SP, Brazil. The solid symbols are the aboveground biomass (stubble plus herbage) prior to herbage harvests, and the open symbols are the residual stubble mass measured after harvests on some dates (data from Pequeno et al., 2018).

The CROPGRO-PFM template approach should be applicable for new forage crops such as ryegrass and clover. However, each forage species requires its own unique species file, which additionally contains the relationships and parameters for growing storage organs, rate of mobilization of reserves, rate of re-fill of reserves, and degree of daylength-induced shift in partitioning to storage. The cultivar and ecotype files have less importance than for the typical DSSAT annual crops.

8 Adding CERES-style and non-CROPGRO models for new crops

Another approach for adding new crops to the CSM code is to add completely new crop modules. This approach is more appropriate when the growth or phenological characteristics of a new crop differ greatly from those described in the CROPGRO template. In this case, both model coding and calibration of parameters are required. The CERES-Sweet Corn model (Lizaso et al., 2007) and the CERES-Sugarbeet model (Anar et al., 2019) are examples of crop modules added to CSM following the style of CERES models. Other models in CSM were adapted from an existing model to use the modular format of CSM (Jones et al., 2001), such as ALOHA pineapple (Zhang and Bartholomew, 1993; Zhang et al., 1997; Vásquez-Jiménez et al., 2024), CANEGRO sugarcane (Singels et al., 2008; Jones and Singels, 2018), and NWheat (Asseng et al., 2000; Kassie et al., 2016). NWheat was used as a template for a teff model (Paff et al. 2019), and there is also a CERES-Teff model (U. Singh, unpublished, 2024). The CROPSIM model (Hunt and Pararajasingham, 1995) was added to DSSAT-CSM as a template model for wheat, barley, and cassava. The CROPSIM template was used to develop a new crop model specific to cassava called MANIHOT (Moreno-Cadena et al., 2020; Moreno-Cadena, 2021).

9 Modeling genetics in DSSAT-CSM

CERES-style crops externalize genetic parameters in species, ecotype, and cultivar files; however, some species' genetic attributes remain in the source code, such as allometric relationships of partitioning to growth stage. The genetic attributes of the CROPGRO crops are contained only in the species, ecotype, and cultivar files. For CROPGRO-style crops with its single generic source code, the species file contains parameters and relationships for sensitivity of processes (leaf appearance rate, rate of reproductive progress, photosynthesis, respiration, leaf area expansion, protein mobilization, pod addition, and seed growth rate) to temperature, along with compositions, N effects on photosynthesis, and many other parameters. The species file and the ecotype file are reserved for modification by the model developers, and model users need only modify the cultivar file to mimic different cultivars within a crop species.

The cultivar file, for example, contains critical photoperiod parameters, photothermal durations (or degree-day units) required to reach given growth stages, along with other traits such as photosynthesis, determinacy, leaf appearance rate, seed size, seed fill duration, and seed composition. The number of cultivar coefficients varies, depending on the crop module that is being used. For instance, the CERES-Maize model includes 6 cultivar coefficients, while the CROPGRO model includes 18 cultivar coefficients. For the CERES models, phase durations (photoperiod sensitivity and the duration of various phases in the life cycle) affecting time to flowering and maturity are given as degree-day units. For CROPGRO models, phase durations affecting reproductive stages and maturity are computed as photothermal days (ptd). The photothermal days accumulated in one calendar day are typically less than one, modified when the hourly temperature is less than optimum and the day length is greater than the critical short day length (CSDL) in the case of short-day species. Genotype-specific parameters (GSPs) in crop models are artificial constructs to mimic genetic traits; at present, they are not well linked to real genes or QTL, although some preliminary efforts are underway as described later in the chapter.

Typical GSPs are shown in Table 1 for three life cycle type cultivars of CERES-Maize and IXIM-Maize. A longer life cycle with later flowering and later maturity appear to give somewhat greater grain yield and biomass for this location in Iowa. The values of P1, P2, and P5 determine the life cycle duration of maize, although there are minimal effects of daylength (P2) for the Iowa location. The G2 is the maximum possible grain number per plant, and G3 defines the kernel growth rate during linear grain filling. PHINT is phyllochron interval per leaf, and smaller values cause faster leaf appearance and greater leaf area index (LAI). IXIM-Maize is an alternate maize model in DSSAT developed by Lizaso et al. (2011, 2022) which requires two additional GSPs for its more mechanistic leaf area growth. IXIM also has a leaf-to-canopy hedgerow light interception and photosynthesis and someday may replace CERES-Maize as the DSSAT default maize model.

The GSPs for four different CROPGRO species (Table 2) play a major role in setting the life cycle and productivity of these legumes, as shown in Fig. 7A and Fig. 7B, but crop differences are also a function of parameters and relationships in the species file. Cowpea had the shortest life cycle with rapid onset of pods and grain, with shorter time from first seed to physiological maturity (SDPM) as well as very short individual seed growth duration (SFDUR), and for those reasons, cowpea had the lowest grain yield. Three of the crops are short-day types, while peanut is not sensitive to daylength (PPSEN=0.00). Pigeonpea had the latest onset of grain growth because of its strong daylength sensitivity

Table 1 Definitions of genotype-specific parameters (GSP) for CERES-Maize and IXIM-Maize models with example values and simulated life cycle and yield for Short (2500 growing degree-days (GDD)), Medium (2600 GDD), and Long (2700 GDD) season cultivars (when sown in Ames, Iowa).

GSP Name	Cultivar coefficient definition	Short 2500 GDD	Medium 2600 GDD	Long 2700 GDD
P1	Thermal time from emergence to end of juvenile phase during which plant is not sensitive to photoperiod (GDD†)	160	190	215
P2	Days delay in development per hour increase in photoperiod above the critical short photoperiod (d/h)	0.400	0.500	0.500
P5	Thermal time from silking to physiological maturity (GDD)	800	840	880
G2	Maximum possible number of grains (kernels) per plant	730	800	840
G3	Grain (kernel) filling rate during linear grain filling under optimum conditions (mg $kernel^{-1}$ day^{-1})	8.6	8.5	8.4
PHINT	Phyllochron interval, thermal time between successive leaf tip appearance (GDD)	39.0	39.0	39.0
AX††	Leaf area of largest leaf (cm^2 $leaf^{-1}$)	680	720	800
LX††	Leaf longevity of most long-lived leaf, from 50% expansion to 50% senescence (GDD)	760	780	800
Simulated Output for CERES-Maize†††				
Time to anthesis (days after sowing)		70	73	78
Time to maturity (days after sowing)		122	129	140
Maximum leaf area index		4.01	4.60	5.13
Grain yield (kg ha^{-1})		11504	12957	13601
Final total biomass (kg ha^{-1})		21470	23315	24645

† Degree days above Tb = 8C. Computed with T_{mean} = (Tmax + T_{min})/2.
†† Additional GSPs for IXIM-Maize.
††† Simulated anthesis, maturity, maximum leaf area index, grain yield, and final biomass, for three hypothetical cultivars, sown April 27, 1999 at 8 plants m^{-2}, 224 kg N ha^{-1} on a Clarion soil at Ames, Iowa.

(PPSEN=0.71). Despite a longer life cycle than soybean, pigeonpea yield was similar to that of soybean. By contrast, peanut had the highest grain yield attributed not only to its long-life cycle, but also to its extended duration from first seed to physiological maturity (long SDPM) along with long indeterminate pod addition (PODUR) and long single grain growth duration (SFDUR). These traits, along with its higher leaf photosynthesis led to higher grain yield for

Table 2 Definitions of GSP for the CROPGRO model with values for soybean (cv. Bragg), pigeonpea (cv. 76W), peanut (cv. Florunner), and cowpea (cv. CAL # 5).

GSP name	Genotype-specific parameter definition	Soybean MG 7 Bragg	Pigeon pea 76W	Peanut Georgia Green	Cowpea CAL # 5
CSDL	Critical short daylength below which reproductive development progresses rapidly with no daylength effect (h)	12.33	12.94	11.84	12.80
PPSEN	Slope of the relative response of development to photoperiod with time (1/h)	0.320	0.71	0.000	0.264
EM-FL	Time from emergence to first flower appearance (ptd †)	19.5	32.7	21.2	23.5
FL-SH	Time from first flower to first pod (ptd †)	10.0	17.4	9.2	3.8
FL-SD	Time from first flower to first seed (ptd †)	15.2	24.1	18.8	6.0
SD-PM	Time from first seed to physiological maturity (ptd †)	37.6	37.7	74.3	24.8
FL-VS	Time from first flower to last leaf on main axis – from Ecotype file (ptd †)	9.0	11.3	68.0	8.0
FL-LF	Time from first flower to end of leaf expansion (ptd †)	19.0	28.9	88.0	7.0
TRIFOL	Rate of node appearance at opt. temp (node td^{-1}) – in Ecotype file	0.32	0.80	0.35	0.45
LFMAX	Maximum leaf photosynthetic rate at 30 °C, 350 ppm CO_2, and high light (mg CO_2 m^2 s^{-1})	1.00	1.10	1.40	1.00
SLAVR	Specific leaf area of cultivar under standard growth conditions (cm^2 g^{-1})	355.	320.	260.	250.
SIZLF	Maximum size of full leaf (compound leaf) (cm^2)	170.	171.	18.	260.
XFRT	Maximum fraction of daily growth partitioned to seed + shell	1.00	0.95	0.92	1.00
WTPSD	Genetic potential weight per seed (g)	0.17	0.18	0.69	0.14
SFDUR	Seed filling duration for pod cohort (ptd †)	24.0	24.6	40.0	9.0

(Continued)

Table 2 (*Continued*)

GSP name	Genotype-specific parameter definition	Soybean MG 7 Bragg	Pigeon pea 76W	Peanut Georgia Green	Cowpea CAL # 5
SDPDV	Seeds per pod, standard growth conditions (# pod^{-1})	2.00	3.89	1.65	6.4
PODUR	Duration of pod addition (ptd †)	10.0	17.3	24.0	16.0
THRSH	Threshing percentage, maximum % of seed to seed+shell	78.0	74.0	80.0	83.0
SDPRO	Potential seed protein (fraction)	0.400	0.224	0.270	0.300
SDLIP	Potential seed lipid (fraction)	0.200	0.015	0.510	0.065

† ptd, photo-thermal days.

peanut compared to soybean and pigeonpea. The peanut and pigeonpea are non-senescing types of legumes that retain green LAI to maturity, whereas soybean has rapid grand senescence of its LAI at the end of its life cycle (Fig. 7A). Notice also that cultivar types within species are allowed to have different seed protein and oil concentrations (Table 2).

10 Major processes modeled in DSSAT-CSM: water balance processes

This and the following sections on processes modeled in the CSM are adapted from Hoogenboom et al. (2019a) with minor updates.

All the crop models in DSSAT-CSM share the same soil water balance subroutines. On a daily basis, the soil water balance is computed by adding irrigation and rainfall and subtracting surface runoff, drainage, plant transpiration, and soil evaporation. Within a soil column, soil water is redistributed by vertical drainage, capillary rise, and tillage. Rainfall is supplied as a user input in weather files. Irrigation is specified in the experimental details input file, which supplies information about the type of irrigation, the efficiency of water supply, and the amount of irrigation applied. Partitioning of rainfall to infiltration and surface runoff is based on the Natural Resource Conservation Service (NRCS) curve number approach (Ritchie, 1998).

Drainage of soil water follows the tipping bucket approach for layered soils with only one-dimensional flow (Ritchie, 1998). Successive soil layers are defined by the lower limit (wilting point), drained upper limit (field capacity), and saturated volumetric soil water content. Downward water movement within the soil depends on a soil drainage factor (fraction per day), limited by

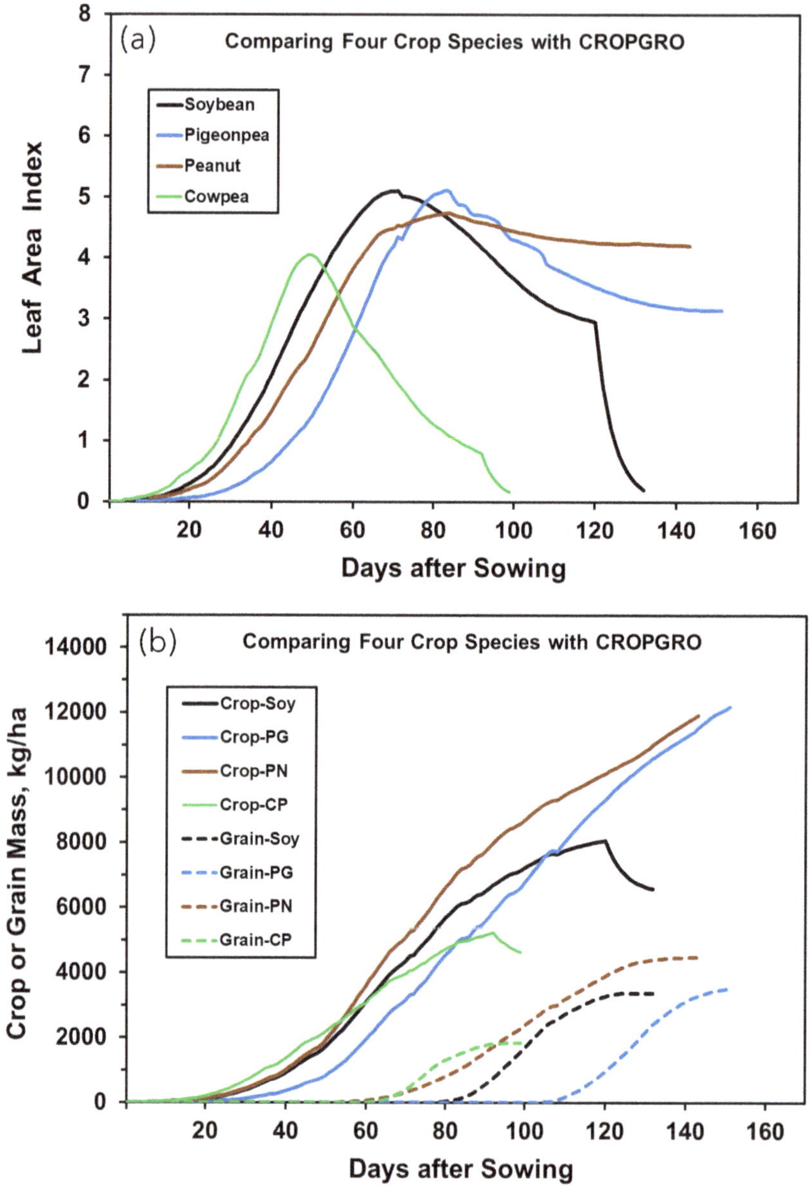

Figure 7 (A) Leaf area index and (B) total crop mass and grain mass, as simulated by CROPGRO for four grain legumes (soybean, pigeonpea, peanut, and cowpea) in 1984 at Gainesville, FL (files and observed data are in DSSAT release, data courtesy of C. E. Maliro (1986)).

the saturated hydraulic conductivity of soil layers. There is a present effort to develop a two-dimensional soil water model in DSSAT for use particularly with plastic mulch bed horticultural crops.

Evapotranspiration (ET) depends on total potential evapotranspiration (ETp) demand using any one of several options: Priestley-Taylor (Priestley and Taylor, 1972), FAO-56 (Sau et al., 2004 modification of Allen et al., 1998), ASAE-standard FAO-56 for short grass or tall grass (DeJonge and Thorp, 2017) and a new hourly energy balance option (Cuadra et al., 2021). The Priestley-Taylor option requires only standard weather data input, whereas the FAO-56 and hourly energy balance options additionally require wind speed and dewpoint temperature as input. After ETp is calculated, it is partitioned to the potential transpiration of the crop canopy (Tp) or potential evaporation of the soil (Ep) as a function of the LAI and an energy extinction coefficient (Kep). Kep is different from, and should be lower than, the extinction coefficient of LAI for photosynthetically active radiation. The parameter Kep can differ for each crop in CROPGRO, but it is more complex for the CERES crops where a "mixed" function is derived from the extinction of photosynthetically active radiation. The actual soil water evaporation depends on the potential Ep and the soil water content of the top layer, using the Ritchie (1972) method with constant and square root of time falling rate stages. The former soil water evaporation method (Suleiman and Ritchie (2003); Ritchie et al. (2009)) has been discontinued as the DSSAT default after experience testing against flux site ET experiments showed simulated soil Es that was too high (Kimball et al., 2023). Soil Es is also influenced (decreased) by the degree of prior crop residue cover, which must be set correctly in the crop management file. Moreover, decomposition of surface residue occurs when the CENTURY-style of SOC is used, but decomposition (disappearance) does not occur with the Godwin and Singh SOC method. Actual simulated transpiration of the crop, Ts, is the minimum of the potential Tp or the root water uptake. Potential root water uptake from successive layers follows the approach described by Ritchie (1998), and it is dependent on root length density and the fraction of available soil water content in each layer. Total root water uptake is then integrated over all layers, and transpiration is reduced if potential root water uptake is less than potential Tp. The daily photo-assimilation is reduced as a function of the ratio of actual transpiration (Ts or root uptake) to potential Tp, using a 0-1 drought stress factor called SWFAC. Expansive processes (leaf area expansion, height increase, canopy width increase) are reduced somewhat sooner by a similar factor called TURFAC. The SWFAC and TURFAC factors also influence partitioning to roots, nodule growth, N-fixation, and phenology (accelerate progress after seed-set). See Boote et al. (2008) for a review of water balance, evapotranspiration, and simulation of water stress effects in the CROPGRO model.

11 Major processes modeled in DSSAT-CSM: nitrogen (N) balance processes

Soil nitrogen dynamics in DSSAT's CSM (Godwin and Singh, 1998; Singh and Porter, 2019) are handled in the soil inorganic N module and in two soil organic matter modules. In the inorganic N module, a mass balance accounts for all additions of inorganic N to the soil, all processes transforming N from one type to another, and all removals of inorganic N from the soil column. Additions of inorganic N are from fertilizer applications and from mineralized N resulting from the decomposition of organic matter. Fertilizer applications are defined in the experimental details file and include the date applied, fertilizer type, amount of N applied, application method, and the depth and percentage of incorporation into the soil. Slow-release fertilizers, nitrification inhibitors, and urease inhibitors can also be specified.

Daily transformations of nitrate, ammonium, and urea are computed based on process rates of nitrification, denitrification, ammonia volatilization, and urea hydrolysis. Removals of inorganic N from the system are based on plant uptake, immobilization related to organic matter processes, leaching, and N gas losses due to ammonia volatilization, denitrification, and nitrification. Gaseous emissions of N_2O, NO, and CO_2 are computed based on organic matter decomposition, nitrification, and denitrification processes. The partitioning of denitrification and nitrification into N_2O, NO, and N_2 comes from DayCent (Del Grosso et al., 2001), as adapted by Peter Grace (Cheryl Porter, personal communication, 2024). For flooded rice systems, in addition to the processes listed above, the model simulates the chemical and biological processes occurring in the floodwater.

Two options are available in DSSAT for the computation of soil organic matter dynamics: the original CERES-based module (Godwin and Singh, 1998) and the CENTURY-based module (Gijsman et al., 2002). See also the review by Singh and Porter (2019). The main difference is the inclusion of surface fresh organic matter and three pools of soil organic matter in the CENTURY model. The more complex CENTURY model allows more control over initialization of stable C pools and, therefore, overall decomposition dynamics, but it also requires additional input data, which are difficult to obtain. These routines interact with the inorganic N and P modules by computing the transformation of organic N and P into inorganic forms as a product of mineralization. Conversely, immobilization can remove inorganic N and P from the soil and reduce plant-available nutrients.

11.1 Plant N processes in DSSAT-CSM

The details of modeling plant N uptake in DSSAT-CSM vary between the individual crop modules, but for all crop modules, N uptake is computed as the minimum of N demand and N supply (Singh and Porter, 2019). The potential N supply from the soil profile is a function of root length density, nitrate and ammonium concentrations, and soil water in each soil layer (Godwin and Singh, 1998). Soil N supply is influenced by environmental factors such as soil temperature, soil moisture, soil pH, mineralization of organic matter, and management of N fertilizers and organic amendments. Root morphology, root architecture, and root length density may limit the ability of the crop to access the N supply.

Crop N demand differs with the growth stage of a crop, with higher critical tissue N concentrations early, and therefore, N demand per unit of growth is higher during early crop growth and development. N demand is driven by plant growth rate, growth stage, and tissue N status as a function of the growth stage (Godwin and Singh, 1998). Total crop N demand is the summation of all deficiency demands from various plant organs plus the demand by new growth. For legumes simulated with the CROPGRO model, when the supply of N is less than the demand, carbohydrates are directed to nodule growth and N-fixation to meet the crop N demand via N-fixation (Boote et al., 1998; Boote et al., 2009). The N-fixation rate is influenced by soil temperature, plant water status (TURFAC), and soil aeration (Boote et al., 2009). When the N supply is less than N demand, vegetative tissues are grown at lower N concentrations. If this condition persists, N deficiency symptoms arise, resulting in a reduction in LAI, reduced photosynthesis (growth and yield reduction), and accelerated senescence.

12 Major processes modeled in DSSAT-CSM: inorganic soil phosphorus (P) processes

The CSM inorganic soil P module maintains state variables for labile, active, and stable forms of phosphorus (Singh and Porter, 2019). Transformation between the pools assumes first-order kinetics with rate constants computed based on soil chemical and physical properties. Additions to the system are from fertilizer application and mineralization due to decomposition of organic matter. Removal of P from the system is from plant uptake and immobilization of P by microbes to meet decomposition demand. Tillage events will redistribute soil P in the layers affected.

The computation of plant available P assumes that soil P is relatively immobile and that only soluble P in close proximity to roots is available for uptake. The soil column is partitioned into root and non-root volume zones using a species-dependent "effective" root radius parameter and the dynamically-varying root length density to define the root zone volume for P uptake. Pools of labile, active, and stable P are maintained separately for root and non-root soil zones. As roots proliferate, mass of soil and nutrients are transferred from non-root to root soil zones, making more P available to the plant with higher root density. Soluble P is a proportion of labile P calculated daily and is dependent on soil water content, labile P, and soil texture in each layer. This soluble P in the root zone is the daily P supply available for potential root uptake. Soil P initialization is critical to a successful simulation of P processes, but data are often difficult or expensive to obtain. Labile P is computed from measured extractable P soil tests using an expert system that depends on the laboratory extraction method used and soil characteristics.

12.1 Plant P processes in DSSAT-CSM

Modeling P demand is similar to that of crop N demand (Singh and Porter, 2019). Each day demand for each plant part is calculated as the amount of P required to bring tissue concentration up to a stage-dependent optimal concentration, plus the demand for new growth. This demand can be met through the soil's P supply and by mobilization of P to grain from vegetative, pod, or root tissue. P uptake is defined as the minimum of supply and demand. The amount of P taken up by roots may be further limited by a species-dependent minimum vegetative N:P ratio, thus limiting P uptake when vegetative N concentrations are low. When supply falls short of demand, P stresses occur affecting rates of photosynthesis, vegetative and reproductive growth, and senescence.

13 Using DSSAT-CSM to evaluate genetic improvement and adaptation to climate change

The GSP, and even ecotype and species parameters can be varied to evaluate hypothetical genetic improvement, often relative to future climate changes, such as drought stresses, rising temperatures, and/or CO_2 increases. Examples of simulating the effects of different cultivar traits have been described for soybean (Boote and Tollenaar, 1994; Boote et al., 1998, 2001, 2002; Boote, 2011), maize (Boote et al., 2001; Tesfaye et al., 2018), peanut (Putto et al., 2013; Singh, P. et al., 2013; Narh et al., 2015; Boote et al., 2021), chickpea (Boote et al., 2013; Singh et al., 2014a), sorghum (Singh et al., 2014b), and pearl millet (Singh et al., 2017).

Several important principles must be followed for evaluating genetic improvement: (1) First, can the model simulate a given trait and is variation in a given GSP within a reasonable range relative to literature information or plant breeder advice (Boote et al., 2002; Boote, 2011)? For example, RUE is often considered a genetic trait (Martre et al., 2024), but many scientists consider RUE to be constant for a given species. So, increasing RUE often shows large yield increases, but is that reasonable? (2) Comparison should be made against a standard cultivar that is already calibrated for a given location (Tesfaye et al., 2018; Martre et al., 2024). (3) Do reasonable traits exist, and are they appropriate for the desired environmental factors and has the model been previously tested against those factors such as elevated CO_2 or temperature?

13.1 Genotype by environment Interactions (sensitivity to climate change, temperature, and CO_2)

Yield improvement with single and multiple GSP trait combinations were simulated with the CROPGRO-Peanut model using 15 years of rainfed weather in Ghana under a hypothesized increase in CO_2 and +3 °C increase in temperature (Table 3). The short-season cultivar Chinese had been well calibrated for multiple seasons in Ghana, and the soil water-holding characteristics were also well established by comparison to observed data (Naab et al., 2004; Naab et al. 2005). Because Chinese is a short-season cultivar with a relatively low LAI, yield increased 4.2% with delayed flowering (+10% EM-FL) and increased 9.0% with longer grain-filling duration (+10% SD-PM), and 12.4% with both traits combined. Likewise, with a relatively low LAI, an increase in photosynthesis (+10% LFMAX) caused a 7.3% increase in yield. Moreover, combinations of four traits, such as delayed flowering, longer grain fill, greater partitioning, and higher photosynthesis, were additive and increased yield by 24.7%. In fact, the response to these combined GSPs is modest, as genetic improvement of newly-released cultivars in Ghana indicates 76–80% yield improvement over this Chinese cultivar (Narh et al., 2015). A strong G x E interaction is present for temperature. Three of the traits (delayed flowering, longer grain-filling, and greater partitioning) increased yield more under the +3 °C temperature scenario than under ambient conditions. Under future climate change, a temperature increase of this magnitude is probable and would cause a shorter life cycle, which the increased EM-FL and increased SD-PM help to offset. The peanut model simulates that the rate of pod addition and partitioning are reduced under elevated temperature (Prasad et al. 2003; Boote et al., 2018), and the genetic increase in partitioning helps to recover pod addition under +3 °C scenario. The peanut yields of this cultivar in this region are low for multiple reasons: un-improved short-cycle cultivar, water stress, low soil fertility (SLPF <1.0), as well as leafspot occurrence.

Strong G x E interactions of GSP traits with water supply (Table 4) are simulated with the CSM-CROPGRO-Chickpea model for crops grown under full irrigation or under water-limited terminal drought (Boote et al., 2013). Chickpea in India is typically sown at the end of the monsoon and depends on stored residual water on high-clay soils (Singh et al., 1999a, 1999b). In this case, the target environment, irrigated versus rainfed, is critical because the responses to GSPs were often opposite and large for contrasting soil water availability. For example, the effect of increased specific leaf weight (SLW) was to decrease yield 11% under irrigation but to increase yield 18% under water limitation. Increased

Table 3 Simulated peanut seed yield response to GSP trait combinations under two CO_2 levels, +3 °C temperature increase, and combination of 500 ppm CO_2, +3 °C and 20% less rainfall (RF), using CSM-CROPGRO-Peanut model. The crop was grown rainfed with 15 years of weather at Wa, Ghana. With permission from Boote et al. (2021).

GSP trait modified†	Ambient CO_2 380 ppm	CO_2 500 ppm	+3 °C temp	+3 °C 500 ppm −20%RF
Chinese (90-d std) (kg ha^{-1})	1841	2248	1553	1770
		% change in yield		
+10% LFMAX (leaf Ps)	7.3	6.4	6.9	5.4
+10% Specific Leaf Area	0.5	0.5	0.5	0.0
+10% EMFL	4.2	3.6	4.3	3.9
+10% SDPM	9.0	8.3	**12.6**	**12.4**
+10% EMFL and +10% SDPM	12.4	11.1	**16.1**	**14.6**
+10% WTPSD (wt/sd)	0.7	0.8	0.7	0.5
-10% Pod Adding Duration	0.1	0.7	3.9	3.6
+10% XFRT (partitioning)	3.4	3.8	5.3	5.7
+10% SIZLF (veg. vigor)	1.5	1.4	0.9	0.8
EMFL, SDPM, XFRT	16.5	15.5	**21.3**	**19.7**
EMFL, SDPM, LFMAX	20.4	18.0	22.7	20.3
EMFL, SDPM, XFRT, LFMAX	24.7	22.6	**28.6**	**25.8**

Bold font signifies genotype by environment interaction for some GSPs.

Table 4 Grain yield response to varying GSPs, simulated for 22 years for Annigeri chickpea grown under either rainfed or irrigated conditions at Patancheru, India. Sown on September 29 (day of year 302) on a very fine montmorillonitic clay soil, starting at field capacity. Simulated with the CSM-CROPGRO chickpea model as developed by Singh and Virmani (1996) and modified by Singh et al. (2014a). Table used with permission from Boote et al. (2013).

Cultivar GSP modified	Rainfed		Irrigated	
	Mean yield	Percent change	Mean yield	Percent change
	kg/ha	%	kg/ha	%
Standard simulation (Annigeri)	773		2614	
Rooting/SLW/LFMAX traits				
+10%, rate of root depth progression	791	2.3	2614	0.0
+10% leaf photosynthesis (LFMAX)	783	1.3	2961	13.3
+10% specific leaf weight (SLW)	916	**18.5**	2328	**−10.9**
Life cycle traits				
10% longer from emergence to anthesis	669	**−13.5**	3016	**15.4**
10% longer seed-fill (first seed to maturity)	787	1.8	2893	10.7
Seed size/partitioning				
10% larger potential seed size	746	**−3.5**	2709	**3.6**
10% faster pod addition	774	0.2	2717	3.9

SLW had a negative effect on yield under irrigation because it reduced LAI and light interception. But the same higher SLW trait was beneficial under rainfed conditions because it reduced LAI, light interception, and transpiration, thus conserving water for grain growth later in the life cycle. Later flowering acted to increase LAI and gave 15.4% yield increase under irrigation but reduced yield by 13.5 % under the terminal drought. The common factor was the amount of LAI produced and the amount of soil water conserved and left at the time seed growth began. Likewise, higher photosynthesis had a much larger benefit under irrigation; while under terminal drought, it had little benefit because it increased early LAI and water extraction too much.

In a project funded by the CGIAR Research Program on Policies, Institutions and Markets and the Global Futures for Agriculture Project funded by the Bill and Melinda Gates Foundation, Tesfaye et al. (2018) used the CERES-Maize model to evaluate the yield benefits from traits for drought tolerance and heat tolerance for six regions in Eastern and Southern Africa and South Asia. The CERES-Maize model was first calibrated for local varieties at each of these sites. The drought tolerance trait was defined as greater root length density at depth along with a slight reduction in the lower limit of soil water content for the soil horizons, the latter assuming slightly more complete soil water extraction. The heat tolerance trait was accomplished by pushing up the Topt2 (highest temperature for optimal growth) and Tfailure (temperature above which no growth occurs) values on rate of grain fill (RGFILL in species file) by 2 °C. The benefits of drought tolerance, heat tolerance, and combined drought/heat tolerance were simulated over long-term weather for the six sites (Table 5). The benefits of drought tolerance varied considerably by location, being greatest in Hyderabad. Benefits of heat tolerance were generally small except for the Kiboko site, although the benefit of heat tolerance was greater under future

Table 5 Simulated maize yields and relative yield changes due to drought, heat, and combined drought and heat tolerance compared to benchmark maize varieties under baseline climate in Eastern and Southern Africa and South Asia (used with permission, Tesfaye et al., 2018).

Site	Benchmark	Drought tolerant		Heat tolerant		Drought and heat tolerant	
	Yield kg/ha	Yield kg/ha	% change	Yield kg/ha	% change	Yield kg/ha	% change
Chiredzi	1327	1558	17.4	1383	4.2	1626	22.5
Hyderabad	3908	6177	58.1	3930	0.6	6179	58.1
Kiboko	2395	2456	2.5	3976	66.0	4330	80.8
Matopos	3124	3156	1.0	3147	0.7	3179	1.7
Meiso	4589	4874	6.2	4615	0.6	4863	6.3
Melkasa	6026	6358	5.5	6030	0.1	6370	5.7

warmer climate. Benefits of combined traits depended mostly on the value of the dominant trait.

13.2 Gene-based modeling

One of the most challenging aspects of applying the DSSAT crop models is the estimation of the GSPs, which, for all models, is required for local cultivars and hybrids prior to any real-world application. There have been efforts to bridge the gap between biotechnology, genetics, plant breeding and crop modeling using either genes or quantitative trait loci (QTLs). The first, simple, gene-based model GENEGRO, was developed by White and Hoogenboom (1996) more than 25 years ago, linking several genes to the GSPs of the dry bean model BEANGRO. Predictions of phenology were as accurate as the original model, while final yield and biomass predictions were more challenging. A similar gene-based approach was applied by Messina et al. (2006) for the E-genes of soybean, which determine the time to maturity. Like dry bean, this was successful for life cycle and somewhat less successful for grain yield. Recent developments have been based on QTLs that are directly or indirectly linked to GSPs or plant traits (Wallach et al., 2018). As the cost of mapping QTLs/genes becomes cheaper, it is expected that rapid advances can be made in this area. A future model might link QTLs/genes to traits affecting one or more growth and development processes via modules that would allow for the input of genetic data, typically from molecular markers, directly into crop models (Hoogenboom and White, 2003; White and Hoogenboom, 2003). If successful, this improvement should reduce the requirements for calibration of a new cultivar for local conditions, assuming that marker data are public and that proper phenotyping of marker effects has been done (Hoogenboom et al., 2003).

An example of common bean GSPs from genes defined in the CSM-GeneGro model (Hoogenboom et al., 2004) is illustrated here. They worked with seven genes, referred to herein as Ppd (long daylength delay), Hr (enhances effect of Ppd, requires Ppd to be present), Fin (indeterminate), Fd (early flowering and maturity), Ssz1 (seed size), Ssz2 (seed size), and Ssz3 (seed size). The following listed GSPs were computed (described by Hoogenboom et al., 2004) as a function of these genes using 0 to indicate recessive and 1 to indicate dominance nature of a given loci (the effect is implemented by entry of 0 or 1 in the allele positions in the following equations).

$$\text{PPSEN} = 0.004 + 0.0154\ \text{Ppd} + 0.036\ \text{Hr} - 0.0104\ \text{Ppd Hr}$$
$$\text{EM-FL} = 26.77 + 4.886\ \text{Fin} - 5.88\ \text{Fd}$$
$$\text{FL-SH} = 4.63 + 0.972\ \text{Ssz1} - 0.98\ \text{Ssz2} - 1.8\ \text{Ssz3}$$
$$\text{FL-SD} = 10.61 + 2.028\ \text{Ssz2} - 2.1\ \text{Ssz3}$$
$$\text{SD-PM} = 21.027 - 0.11\ \text{Ssz1} + 4.13\ \text{Hr}$$

FL-VS = 7.00 + 4.76 Fin – 2.75 Ssz2 – 1.02 Fin Ssz2
FL-LF = 18.0 + 3.8 Fd – 6.9 Ssz2
SLAVR = 322 + 41 Ssz1 – 38 Ssz2 – 25.3 Ssz3
WTPSD = 0.22 + 0.21 Ssz1 + 0.07 Ssz2
SDPDVR = 5.14 - 0.2 Fin - 1.9 Ssz1 + 0.24 Ssz3

Eight less important GSPs not shown are also a function of the genes. Cultivar Porrillo Sintetico (mostly Meso-American) is Ppd, hr, Fin, fd, ssz1, ssz2, and Ssz3, while cultivar Manitou is Ppd, hr, fin, Ssz1, Ssz2, and Ssz3. Cultivar Bayo Madero is Ppd, Hr, Fin, Ssz1, Ssz2, and Ssz3. [As used here, a capital letter indicates dominant, and lower-case indicates recessive.] By changing 'Fin' to 'fin', Boote et al. (2016) created a Calima-like cultivar with Ppd, hr, fin, Ssz1, Ssz2, and Ssz3. With these genes for different cultivars, we were able to demonstrate different modeled growth dynamics with the CROPGRO-Dry Bean model based on the presence or absence of the dominant allele for those genes (where first-letter capitalization indicates the dominant allele). Fig. 8, Fig. 9, and Fig. 10 illustrate the emergent behavior of the genes when compared to the observed growth of Porrillo Sintetico and Jatu Rong (Calima-like) cultivars grown at Palmira in 1990 by Sexton et al. (1994). Fig. 8A illustrates how genes for time to flower and determinacy affect main stem node number. Calima is the only cultivar with recessive fin, whereas the other three gene-based cultivars have dominant allele Fin, along with different photo-thermal times to flower. The simulated Calima-like gene-based cultivar closely matches the GSP-based simulations and data of main stem node number for the Jatu Rong cultivar while the simulated gene-based Porrillo Sintetico fits well the observed data and GSP-based simulations for Porrillo Sintetico (Fig. 8B). The gene-based cultivars differed in their pattern of LAI (Fig. 9A), reaching a peak sooner for the Calima-like (fin) with its early flowering (Fd), followed by the gene-based cultivars with Fin but relatively mid-cycle flowering, and later peak with highest LAI for gene-based Porrillo Sintetico. Comparison to GSP-based simulations and real data for Jatu Rong and Porrillo Sintetico at Palmira (Fig. 9B) illustrates very similar predictions of LAI patterns for the gene-based Calima-like and Porrillo Sintetico. The four gene-based cultivars had very different total seed growth patterns (Fig. 10A), with earliest onset and lowest yield associated with the Calima-like cultivar, followed by later onset and higher seed yield for the other gene-based cultivars. For Jatu Rong and Porrillo Sintetico at Palmira, there were quite comparable simulated patterns of seed growth for the gene-based Calima-like and Porrillo-Sintetico cultivars compared to the GSP-based cultivars (Fig. 10B). These comparisons well illustrate how the resulting widely varying growth patterns are attributed to genes that control life cycle progression and other aspects of dry bean growth.

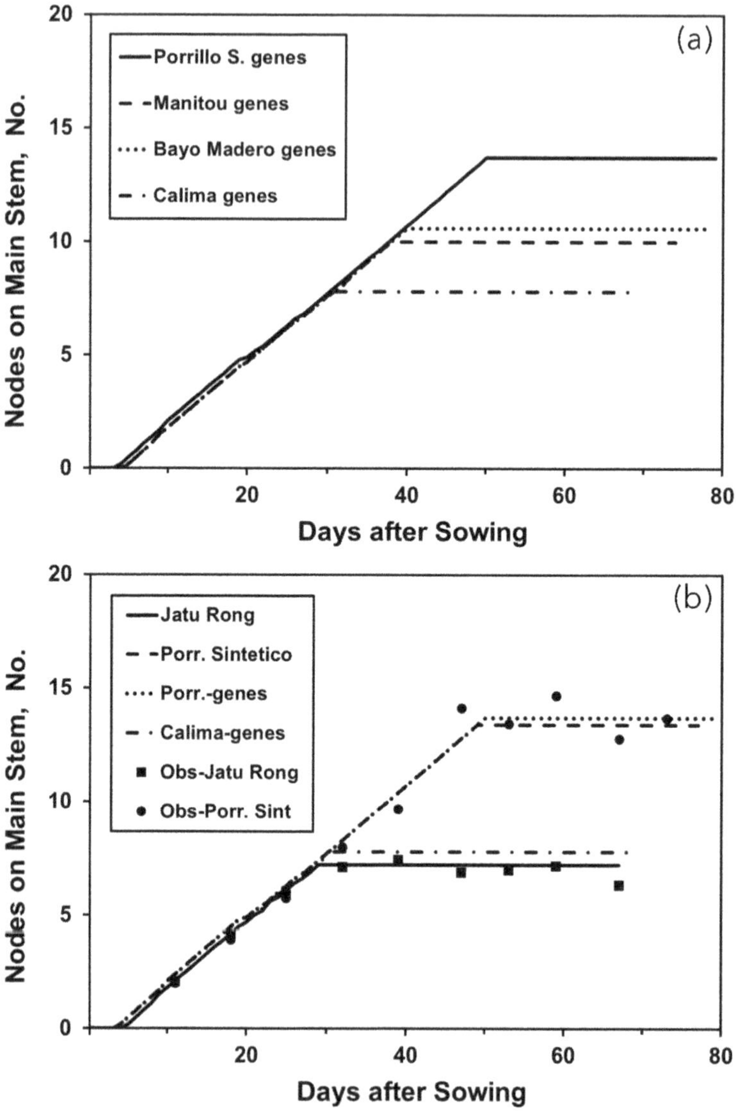

Figure 8 (A) Node number on main stem as affected by genes in four gene-based dry bean cultivars at Palmira, (B) Node number on main stem as affected by genes in two gene-based dry bean cultivars compared to data for Jatu Rong (Calima-like) and Porrillo Sintetico at Palmira (used with permission from Boote et al., 2016).

There is a recent example of quantitative trait loci (QTLs) linked to the simulation of flowering date in dry bean (Oliveira et al., 2021). The authors investigated the effects of QTLs, environment (E), and QTL × E interactions on the rate of progress to first flower of 187 recombinant inbred lines

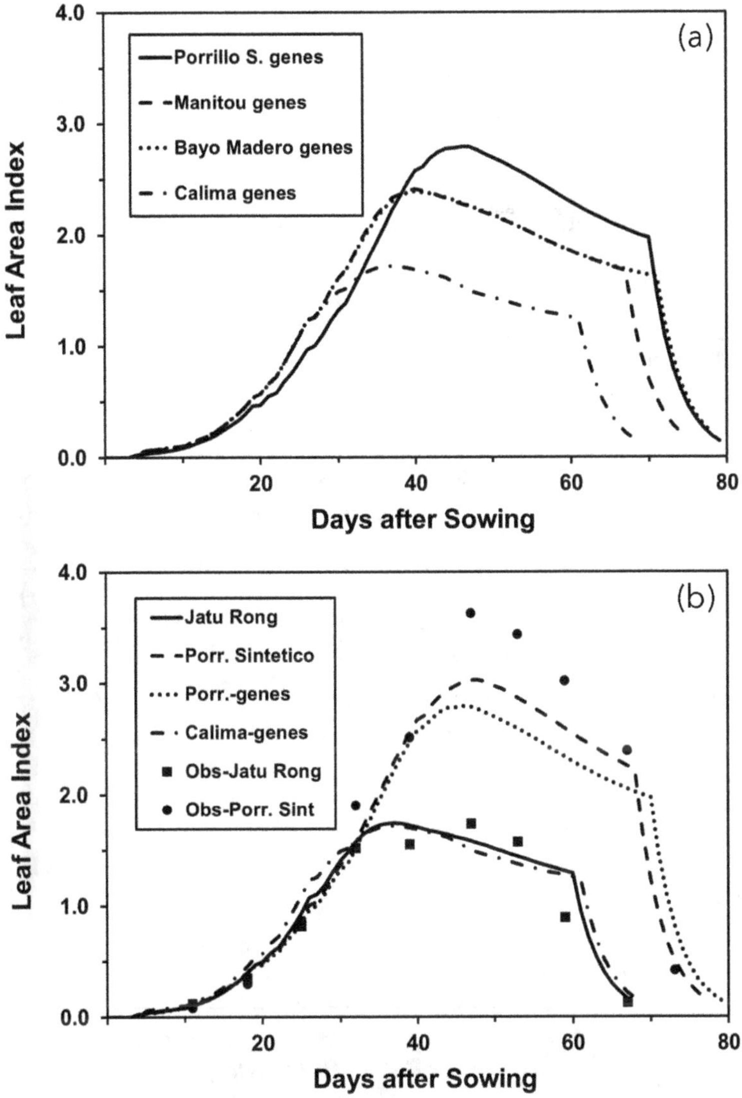

Figure 9 (A) Leaf area index as affected by genes in four gene-based dry bean cultivars at Palmira, (B) Leaf area index as affected by genes in two gene-based dry bean cultivars compared to data for Jatu Rong (Calima-like) and Porrillo Sintetico at Palmira (used with permission from Boote et al., 2016).

from a bi-parental bean family grown in two seasons across five sites highly different for daylength and temperature. They identified 12 QTLs responding to daylength, maximum temperature, minimum temperature, and solar radiation that influenced rate of progress to first flower in dry

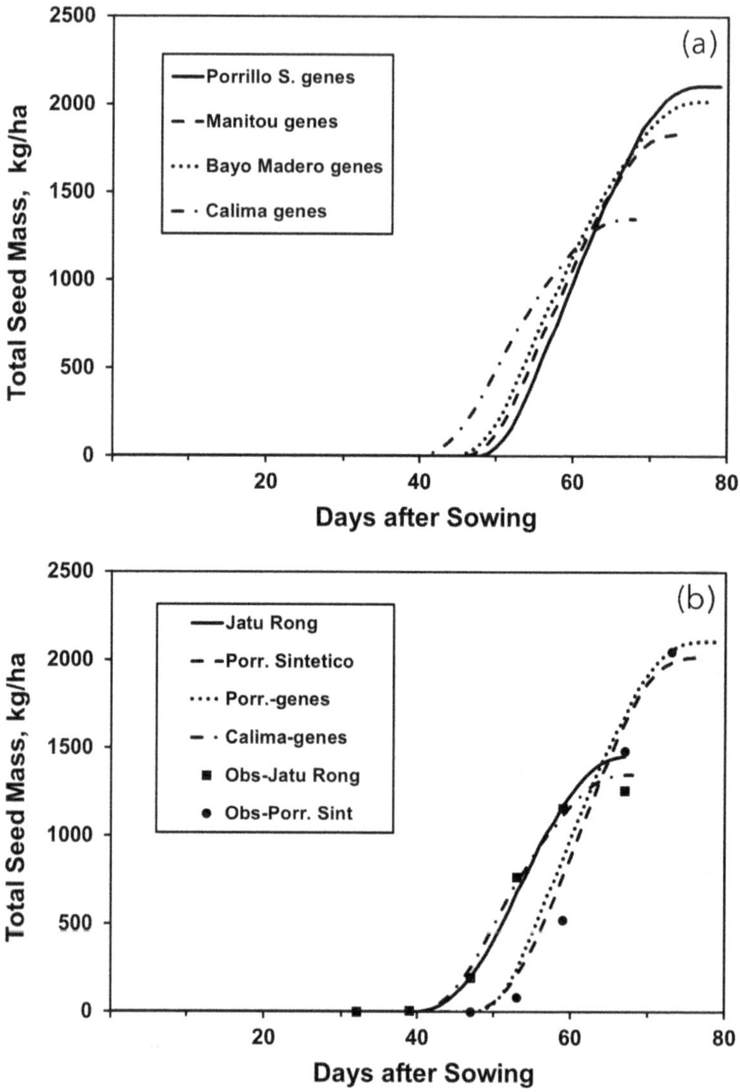

Figure 10 (A) Grain mass over time as affected by genes in four gene-based dry bean cultivars at Palmira, (B) Grain growth pattern as affected by genes in two gene-based dry bean cultivars compared to data for Jatu Rong (Calima-like) and Porrillo Sintetico at Palmira (used with permission from Boote et al., 2016).

bean. The allelic presence or absence of each QTL, along with regression magnitudes of each QTL and their interactions with E were set up similar to the Hoogenboom et al. (2004) example above. The integrated gene-based module in the CROPGRO-Dry bean model simulated days to first flower that agreed closely with observed values (root mean square error of 2.73 days and model efficiency of 0.90) across the five sites and 187 genotypes. The gene-based module was developed to operate in a hybrid mode in the DSSAT-CSM where the user can decide to use the original GSPs or the QTL data to simulate flowering date.

14 The DSSAT user interface

This section is adapted from Hoogenboom et al. (2019a) with updates.

The combination of different models, tools, utilities, and applications requires the development of a unique interface that provides easy access for users who may not be familiar with crop models in general, especially with the challenges of formatting input and output files. The CSM model is at the core of the DSSAT shell (Fig. 11). The shell provides the user interface to connect the CSM to the various databases (weather, soils, genetics, experiments-data, economics, pest information, and prior outputs), utility tools, and application tools. The support software allow users to create the experiments (XBuild and XB2), to create soil files (SBuild), to enter weather data (WeatherMan), to enter field observations (ATCreate), to plot model outputs (GBuild), and to estimate genetic coefficients from data (GLUE, GENCALC, and TSE). In addition, there are application programs for Sensitivity Analysis, Seasonal Analysis, Crop Rotation Analysis, Yield Forecasting, and Spatial Analysis. Yield Forecast and Climate Change applications require careful user initiative to link to their own spatial tools, climate change scenarios, and long-term weather records. DSSAT provides tools to assist users in preparing the different input files that are needed for running a model, defining the experiments and treatments or scenarios a user wants to simulate, and conducting an analysis of crop model outputs from the simulations, including comparisons with observed data for model evaluation and strategic analyses for model scenarios (Fig. 11).

In order to facilitate the interaction between the crop models, the data tools, the utilities, and the application programs, a very strict protocol is required for file naming conventions, specific file formats, and system settings that define the location and names of the model input and output files. The strict protocol has allowed programmers to develop new tools and utilities that can be easily incorporated into DSSAT based on standard input and output formats. This approach was first presented to DSSAT users in DSSAT Version 2.1 (IBSNAT, 1989) and DSSAT Version 3.0 (IBSNAT, 1993). The original design and concept are still viable in the most current version of DSSAT Version 4.8.2 (Hoogenboom et al., 2023).

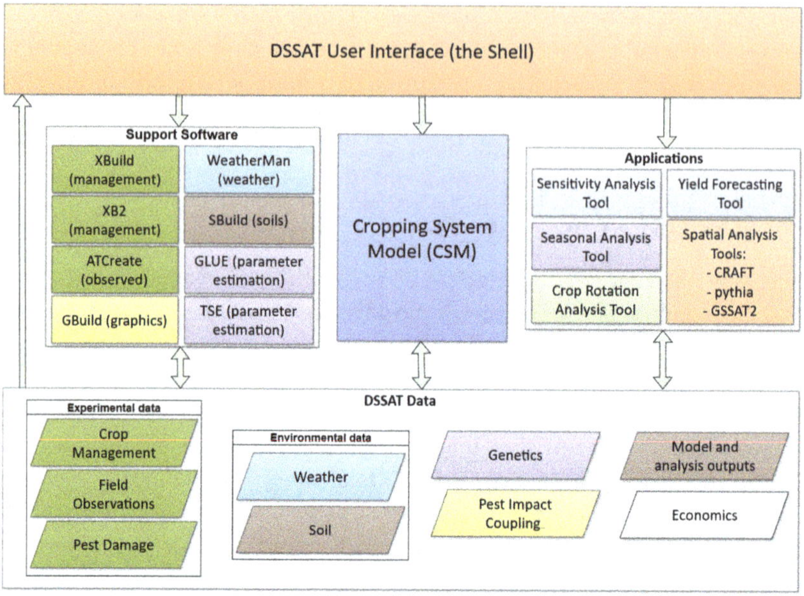

Figure 11 The DSSAT Crop Modeling Ecosystem, linking the CSM to databases, support software, and applications.

14.1 Minimum data set (MDS) for crop modeling

In order to run a crop simulation model, a minimum set of input data is required. This minimum data set includes daily weather data, soil surface and soil profile information, crop management, and initial conditions at the start of the simulation. The minimum weather data include the metadata for the weather station, especially latitude, longitude, elevation, sensor height, and daily maximum and minimum temperature, rainfall, and solar radiation.

The minimum soil data include the metadata for the location where the soil conditions were measured: soil surface color, slope, drainage, and permeability, as well as soil texture, bulk density, and soil organic carbon for each individual soil horizon. Boundary or initial conditions at the start of the simulation are also very important, especially for the soil environment, requiring initial soil moisture, nitrate and ammonia for each horizon or soil layer, as well as the soil organic C, surface biomass residue, and roots of the previous crop.

The crop management data includes the crop and cultivar selection, planting date, plant density, row spacing, sowing depth, irrigation, and fertilizer inputs. For irrigated treatments and scenarios, the dates, amounts, and the type of irrigation system must be defined; for fertilized treatments and scenarios, the dates, amounts, and types of inorganic fertilizer must be defined, as well as the depth of incorporation. For organic fertilizer using plant or animal material, the type and composition of the organic fertilizer must be defined. If a crop, such as rice, tomatoes, other vegetables, and vegetatively-propagated forage, is started using transplants, the initial weight of the transplant material, age, and the temperature of the nursery must be defined. For potatoes, the weight of the seed potato is an input, for cassava, the weight and length of the stick and orientation of planting are defined; and for sugarcane, the initial cane mass is defined.

The previously listed input data for weather, soil, crop management, and initial conditions are the MDS required for running the model. But for model calibration, evaluation, and improvement, measurements of the crop and soil are required so that comparisons can be made between simulated and observed data. Depending on the research goals and objectives, measurements can include yield and yield components, detailed crop phenology, time-series crop growth analyses, and soil profile measurements such as soil moisture, nitrate and ammonia, organic carbon, and other information (Hoogenboom et al., 2012). See Boote et al. (2015) and Kersebaum et al. (2015) for discussion of quality-classification of experimental data sets for crop modeling.

14.2 Input data tools

Most researchers have their own individual standard methodology for recording experimental data in field books, spreadsheets, and other electronic media. These individual differences make it somewhat challenging to convert the measured data into a format that can be directly applied in a crop modeling system. DSSAT, therefore, provides specific tools for entering weather, soil, crop management, and observational data.

14.2.1 XBuild

XBuild is the tool for entering crop management data that are stored in a crop management file (Fig. 11). The tool is designed so that the user first enters information that defines the field, specifically the weather station and soil profile that are associated with that experiment, followed by the crop and cultivar selection, and planting information. Initial conditions are defined in the Environmental Section of XBuild. The user can enter different levels for each management scenario, such as multiple cultivars or hybrids, different planting dates, and different input levels and application dates for irrigation and fertilizer.

Published by Burleigh Dodds Science Publishing Limited, 2025.

14.2.2 WeatherMan

The tool called WeatherMan allows for the entry and formatting of weather data into DSSAT weather files (Fig. 11). A user can import weather data preferably from spreadsheets, but WeatherMan can also import other formats, including CSV and ASCII text files. Once the data have been imported into WeatherMan, quality control procedures can be applied to identify extreme values or extreme changes in values for two consecutive days and any missing values. A new internal database is created in WeatherMan with what is referred to as the "corrected" data. The final procedure is to export the data back into DSSAT format weather files.

14.2.3 SBuild

The soil water balance simulation in DSSAT is based on the tipping bucket approach with three key soil moisture variables, including Saturated Water Content (SAT), Drained Upper Limit (DUL), and the Lower Limit of plant extractable water (LL). There are procedures for measuring these, but such measurements are not commonly available for all experiments. The SBuild program of DSSAT allows a user to enter soil surface information, including soil color, slope, permeability, and drainage characteristics, and soil texture (percent clay and percent silt), bulk density, and organic carbon for each soil horizon (Fig. 11). SBuild then uses internal pedotransfer functions to calculate SAT, DUL, and LL for each soil horizon or layer, and it saves the information for that particular soil profile in the soil input file.

14.2.4 ATCreate

Measurement data for model evaluation can be differentiated into two types in DSSAT. The first type is referred to as the summary data and includes the key phenological stages, yield and yield components at final harvest, and other measurements that can be obtained at critical stages, such as maximum leaf area index (LAI) or grain nitrogen concentration. The summary data are stored in File A as a single line per treatment. The second type of measurement data is referred to as time series data for growth analysis, tissue N concentrations, soil moisture content, soil nitrogen measurements, and other relevant data that can be used for model evaluation. The time series data are stored in File T and are organized by treatment and then observation date. There is also the ability to store the observations for the individual replicates. The ATCreate program allows users to enter observations either manually or by importing a spreadsheet or text file, thereby creating File A and File T for each experiment. It is important to select the appropriate header for each column of data so that the other programs within DSSAT are able to recognize the observed variables.

Power-users can use spreadsheets to directly create the File A and File T, by saving as *.PRN files but correctly setting required column widths and headers for data columns, and then correctly naming the file and its internal lead line. The file called DATA.CDE holds the names of these variables (short name, long name, and units) so that header names are shared and are also readable by the graphics program and other programs in DSSAT.

15 Methods for estimating genotype-specific parameters

The DSSAT modeling system defines genotype-specific parameter (GSPs) inputs that allow a user to define characteristics of cultivars, hybrids, clones, and transplant material. As model developers, the DSSAT group is unable to provide local cultivar-specific parameters beyond those with specific experiments included in DSSAT, which means that a model must be calibrated first for local genetics, requiring some of the critical crop observations associated with the MDS described previously. Once the crop management and observational data have been entered, the specific cultivar is then calibrated, either manually (looking at the means and other statistical outputs of GBuild) or using optimization tools. The ultimate goal is to minimize the error between simulated and observed phenological dates, yield, and yield components.

Within the DSSAT Ecosystem, there are several tools that can be used for crop cultivar calibration, the *GLUE* tool, *GENCALC* tool, and TSE tool. In addition, the Sensitivity Analysis utility can also be used to improve the value of one or more GSPs by setting the range and increment for a particular GSP and by comparing simulated with observed data. A comparison of the performance of the first two tools for rice was conducted by Buddhaboon et al. (2018).

15.1 GLUE

The General Likelihood Uncertainty Estimation (GLUE) is a statistical approach that results in multiple sets of parameter values that are equally as likely as a final solution, *but it only gives the most probable set of GSPs*. The initial evaluation of GLUE for DSSAT was made for the CSM-CERES-Sweetcorn model by He et al. (2009, 2010). To estimate the most likely values for the GSPs for a new cultivar, a user first must provide the required input files associated with weather, soil, and crop management and basic observations, especially for phenology, yield, and yield components. For more robust outcomes, at least two non-stressed treatments from different environments representing different locations, planting dates, or years are recommended. Once the model runs properly, the GSPs can be estimated with GLUE, first for phenological GSPs and then for the yield and yield component GSPs. GLUE stochastically generates each parameter set used in the optimization, therefore, each new GLUE optimization on the same data will give a slightly different set of GSPs. Running a high number of simulations gives greater confidence in the results, and it is recommended to simulate at least 10,000 runs. Caution: GLUE uses only final end-of-season data for the optimization algorithm. Therefore, if in-season growth analyses data are available, users should also look at time-series simulations and statistics (RMSE and d-statistic) available with the GBuild tool. A new version of GLUE has recently been released that allows for the use of multiple cores for parallel processing and significantly speeds up the calibration process (Ferreira et al., 2024).

15.2 GENCALC

The Genetic Coefficient Calculator or GENCALC uses a rule-based approach to determine the value for one or more GSPs (Hunt et al., 1993). In the input file for GENCALC, one or more GSPs are associated with one or more particular plant output variables. During the calibration process, these GSPs are varied to minimize the error between the simulated and observed variables. GENCALC normally optimizes the phenological GSPs first, followed by growth, and then yield components and yield for final optimization. Most crop models do not have a specific GSP that controls only yield; rather, GSPs can affect multiple yield components. GENCALC is for more advanced DSSAT users who are familiar with the GSPs of a particular model and are comfortable editing the GENCALC rules file (Anothai et al., 2008).

15.3 TSE

A third tool, TSE (Time-Series-Estimator), developed by Memic et al. (2021), makes use of time-series growth analysis data from File T as well as File A data and uses normalized root mean square error methods. The TSE optimizes user-selected GSPs against user-selected targeted crop growth variables. For an advanced user and where time-series growth data are available, the TSE Tool may offer greater capability for optimizing GSPs. The GLUE and GENCALC tools can only use end-of-season File A data, which can result in good final yield simulations but potentially with poor outcomes for time-series simulations.

16 Model visualization tools: GBuild

For performance evaluation of the model with experimental data, visualization tools that not only provide a visual comparison between simulated and observed data but also statistical analysis are critical (Yang J. M. et al., 2014a; Yang J. Y. et al., 2014b). The main tool in DSSAT for visualization and comparison of model simulations with observed data is GBuild.

Published by Burleigh Dodds Science Publishing Limited, 2025.

GBuild is an analysis utility for the visualization of simulated and experimental data (Uryasev et al., 2004). It gives a user the ability to easily plot graphs that are routinely used during crop model development and evaluation. The basic design of GBuild is based on a set of codes that use headers for each column of data that represents different variables. The file selection in GBuild allows a user to select one or more output files for plotting any combination of the variables and runs/treatments and then proceed to display the graph. The graphic type selection options provide different views of the simulated results and include time series, e.g. displaying the simulated data as a function of date or days after planting (DAP), and simulated data vs. experimental data. To better compare the simulated results with observations from experiments, GBuild includes statistics for time series data with emphasis on the d-statistics (Willmott et al., 1985), Root Mean Square Error (RMSE) and means for time-series data, phenological and end-of-season data, such as flowering and maturity dates, yield and yield components. The graphic output of simulated and observed data can be visualized, printed, and exported to an Excel spreadsheet along with the statistics.

17 Application programs

Crop simulation models embedded in decision support systems are very powerful tools for scenario analyses. So far, this chapter has provided an overview of the science of the crop simulation models that are included in DSSAT, as well as the tools and utilities for weather, soil, experimental inputs, observational data entry, crop model calibration, and evaluation. Once a crop model has been calibrated, the most important and useful aspects are associated with the applications. DSSAT, therefore, includes several application programs, as shown in Fig. 11.

17.1 Sensitivity analysis tool

In addition to evaluating the model with real-world data, it is also important to understand the response of the model to variation in a given specific input, such as weather data, cultivars, soil data, and values for individual GSPs. This approach, in which all inputs are kept constant except for one input or GSP, is called sensitivity analysis. This DSSAT tool is called *Sensitivity Analysis* and enables the user to evaluate the model sensitivity to changes of cultivars, single GSPs, soil profiles, weather inputs for different locations or years, plant and row spacing, and various other options. Variables that have numeric values, such as planting date, can be varied using a starting value, an increment value, and the number of iterations. The program automatically creates a new experimental file ready to run with the selected sensitivity input variation. Following the simulations, the linked GBuild graphics program allows for visual analysis of simulation results and associated statistics.

17.2 Seasonal analysis tool

The *Seasonal Analysis* application allows a user to explore the effects of weather variability and to evaluate the uncertainty and risk factors associated with various management and genetic inputs (Thornton and Hoogenboom, 1994). The *Seasonal Analysis* application enables a user to simulate many hypothetical scenarios quickly and efficiently using long-term historical weather records or stochastically generated weather data. The model simulates a distribution of yields and other outputs, converting uncertainty in weather into uncertainty in yield for the specified management scenarios (Fig. 11). The seasonal analysis application works on a field scale and emphasizes weather uncertainty. Economic risks can also be estimated using the cost of inputs and prices of products, including the variability associated with these costs and prices. The application can be used to select optimal crop and variety, planting options, irrigation options, application of fertilizer, marketing options, insurance risks, policy advisement, and investments in equipment, technology, and diversification of land use. For each scenario, many weather years are simulated, with re-initialization of soil variables done at the beginning of each simulation so that the results reflect the variation in model outputs due to interannual weather variability. A graphical interface allows a user to explore distributions of outcomes for variables, including crop yields, farm profits, and environmental factors such as nitrogen leaching and irrigation requirements. The application includes graphical options such as box plots, cumulative function plots, and mean-variance plots for biophysical and economic variables. This feature allows the user to optimize management practices that will benefit the farmer and to select best management practices relative to maximum profit, minimum risk of low profit or yield, minimum degradation of the environment, or other criteria.

17.3 Crop rotation analysis tool (sequence analysis)

The *Crop Rotation Analysis Tool* (Fig. 11) is used for the analysis of crop rotation, and, in addition to weather and economic uncertainties, takes into account the effects of long-term cropping systems including changes in the soil system with respect to carry-over of soil water, carbon, nitrogen, and other nutrient components. Crop rotation analysis (or Sequence Analysis) allows a user to produce long-term simulations of a given cropping system for

Published by Burleigh Dodds Science Publishing Limited, 2025.

predictions of farming system sustainability such as soil carbon loss, soil fertility degradation, decreasing yields, and increased greenhouse gas emissions (Thornton et al., 1995). Users can explore the sustainability of various options over a long period of time and optimize options for managing the land to sustain productivity, soil health, and natural resources. Crop rotation analysis is used to explore cropping system options based on a pattern of crops planted in sequence. Sequence Analysis/Crop Rotation differs from Seasonal Analysis in several ways. In Seasonal Analysis, soil variables are re-initiated every season, but in Sequence Analysis, soil variables are initialized only once at the beginning of a long-term, continuous simulation. In Sequence Analysis, the soil C, soil N, and soil water carry over into the next fallow or crop, thus running continuously without re-initialization. The crop rotation sequencing or patterns are flexible to create and can mimic a real observed experiment of 10-year duration, for example, with real weather. But the most common option is to use generated weather and "repeat" multiple times (30 or more) over generated weather.

The dynamics of soil organic carbon are of prime importance in these long-term simulations. In sequence simulations, soil organic carbon is "carried over" from one season to the next, and these long-term soil fertility dynamics can be analyzed. It is best to use the CENTURY-style SOC method for sequence rotations because it handles the decomposition of surface residue and senesced plant tissue, which facilitate no-tillage simulations. The initialization of soil organic carbon state variables and pools for CENTURY can be difficult, but it is very important to the predictive accuracy of the model. Methods have been developed to estimate the amount of stable, intermediate, and microbial soil carbon present in the system (Basso et al., 2011; Porter et al., 2010). The best way to estimate (optimize) the stable pool for the CENTURY-style SOC is to have a zero N fertilized treatment with knowledge of initial soil nitrate and ammonium at sowing.

17.4 Spatial analysis

The crop models in DSSAT are point-based models in that the inputs are based on site-specific information such as the weather data from a local weather station, the soil data from a local soil profile at the experimental site, and crop management for a plot or field. For many applications, there is significant interest in understanding the variability across space for crop growth and yield. The strength of the CSM is that it can simulate at a spatial scale as small as 1 m or less for precision agriculture to ~100 km (1 arc-degree) for global simulations. The current DSSAT software does not include a specific tool for the preparation of input files and visualization of output files. However, the underlying crop model and associated input and output files with GPS coordinates can be easily integrated into other systems if the DSSAT file naming convention and file system structure are maintained. Several examples of spatial applications that use DSSAT in the background include MINK for global gridded simulations developed by the International Food Policy Research Institute (Robertson, 2017), pDSSAT for global gridded climate change applications in agriculture (Elliott et al., 2014), and CCAFS Regional Agricultural Forecasting Toolbox (CRAFT) for yield and production forecasting, climate change impact studies and agricultural risk assessment (Shelia et al., 2019).

17.5 Climate change applications

Climate change applications typically use the Seasonal Analysis Tool to evaluate 30+ years of weather (baseline weather (1990–2020), near-term future weather (2020–2050), and long-term future weather (2050–2080)). The future climate weather can be generated by multiple Global Climate Models, which vary the temperature, atmospheric CO_2, and rainfall based on hypothesized rates of GHG increase and socio-economic pathways. This exercise has been repeated many times over the past decades for many regions in the world. For a simple evaluation, a user can also use the Environmental Modifications section of the File X to vary input temperature, atmospheric CO_2, rainfall, and solar radiation relative to a baseline weather file. Climate change applications can also be done using Sequence Analysis if one wishes to evaluate long-term consequences on yield, soil degradation, and soil C sequestration (or lack thereof).

17.6 Yield forecast application

The Yield Forecast tool within DSSAT involves creating a unique type of File X designed for forecasting. The Tool allows users to estimate in-season yield forecasts using observed weather data combined with multiple realizations of future weather data to predict a range of possible outcomes for the remainder of the season for yield and other model outputs (Shelia et al., 2019; Tesfaye et al., 2023). Simulations use the observed weather data for the forecast year up through the last observed date or a date specified by the user. After that date, the model uses either generated daily weather data or all available historical weather data to simulate an ensemble of predictions for the remainder of the season. The Yield Forecast Application synthesizes the simulations to present the range of outcomes predicted and the most probable.

17.7 Example applications

See the chapter by Hoogenboom et al. (2019) which provides several illustrative case studies of application approaches: (1) Interaction of nitrogen and water management on performance of maize (1982 maize experiment in DSSAT), (2) Impact of irrigation management on performance of soybean (1978 soybean experiment in DSSAT), (3) Seasonal strategy analysis of irrigation management on soybean over 30 seasons, (4) Sequence analysis that considers the carry-over of the prior crop's extraction of soil water on next crop or the degradation or sequestration of soil organic carbon and effect on long-term soil fertility.

18 DSSAT-CSM software development, distribution, and training

One of the unique aspects of the original IBSNAT Project was its participatory approach that encouraged scientists from different disciplines and different organizations to collaborate in support of the development of one unique DSSAT ecosystem. This network of model developers has expanded since the end of the IBSNAT Project in 1993 to include many scientists from Brazil, China, South Korea, France, Nigeria, India, South Africa, Kenya, and many other countries across the globe. The mutual support and collaboration are facilitated using an Open Source approach, in which the source code of the crop simulation models along with the tools and utility programs are freely available. Currently, the source code for the Cropping System Model can be obtained from GitHub (https://github.com/dssat/dssat-csm-os). The CSM source code is open source using the 3-Clause BSD License (BSD-3-Clause). Since 2014, the DSSAT Development Team has also facilitated biannual DSSAT Development Sprints, which are hackathons for crop modelers. During these one-week, hands-on "working" workshops, the emphasis is on crop model improvement and the advancement of application programs, tools, and utilities. These Development Sprints also provide a forum for communications and information exchange among developers, users, and others interested in modeling and decision support systems. They allow new DSSAT developers to learn the protocols and methods used.

In the early years, the DSSAT software was sold to cover some of the administrative and development costs. In 2011, the distribution of DSSAT was converted to a free download system from the DSSAT portal (www.DSSAT.net), and since then, the interest in DSSAT has exponentially increased. There were 6,630 downloads of DSSAT Version 4.5, 6,500 downloads of DSSAT Version 4.6, and over 18,500 downloads of DSSAT Version 4.7 since November 2017. DSSAT Version 4.8 was released in April 2021 with 11,100 downloads until present (September, 2024). For the dissemination of information to the DSSAT User Network, a DSSAT Listserv is used, and currently, there are over 29,000 unique e-mail addresses of registered users. User support and documentation are also provided through the DSSAT portal (www.DSSAT.net).

Because DSSAT is such a comprehensive software program, it requires training of agricultural scientists who traditionally specialize in single disciplines and who may not be very familiar with the systems approach that encompasses multiple disciplines. Annual crop modeling training workshops have been held in the US since 1985, initially as two-week workshops and then condensed to intensive, six-day training workshops. Since 2002, these workshops have been hosted at the University of Georgia's Griffin campus, with an average of 55 international participants from universities, governmental organizations, and private industries. The rapid expansion of the DSSAT user network has also resulted in requests for international training programs across the globe, and recent workshops were held in Argentina, Australia, Germany, Indonesia, Kazakhstan, Nigeria, Philippines, Singapore, South Africa, Tanzania, Thailand, Tunisia, Türkiye, and Vietnam. In addition, some of our expert users are now DSSAT trainers, facilitating workshops in Pakistan, Indonesia, Türkiye, Brazil, China, and other countries. Ideal capacity building requires multiple workshops, starting with the basics of crop modeling and data requirements, followed by data collection for model evaluation, and finishing with model applications, the most critical part of crop modeling and decision support (Kihara et al., 2012).

19 Future trends and conclusion

19.1 Collaboration

One of the challenges of maintaining and developing scientific software platforms in agriculture is the limited availability of resources and data for software development and improvement. Many agricultural scientists are not good programmers, especially in computer languages such as Fortran, Delphi, and Visual Basic that are currently used in DSSAT. Early in the development of DSSAT and the crop models, the emphasis had been on scientists developing the crop models rather than relying on professional programmers who develop the code based on input provided by the scientists. This has shifted in recent years with the addition of more programmers, along with an increased number of model users who want to apply models but who are lacking the hard-core scientific skills to improve model relationships. There is a general lack of emphasis on scientific model improvement in the current overall crop modeling community. Sustainable model development will require the collaboration of scientists with state-of-the-art crop-soil-systems expertise with programmers who can write scientifically sound code. DSSAT source code is Open Source and is shared on GITHUB which has allowed the DSSAT Development group to expand to a community of those interested in advancing and improving the DSSAT, including the models as well as the tools and utility programs.

19.2 Opportunities

The current agricultural production system is challenged by the increase of climate extremes and economic risks associated with climate change. At the same time, there is pressure to grow more and healthy food using sustainable practices. Among these practices, the reduction of greenhouse gas emissions and increase of soil carbon stocks are some of the major challenges that can be supported by biogeophysical models. Measuring, Monitoring, Reporting, and Verification (MMRV) efforts are demanding the scaling up of carbon stocks measured from point to plot, farm, and country. Crop models can be used to translate historical and actual crop management while assimilating remote sensing and local soil observations. Artificial Intelligence (AI) can play a valuable role in those tasks, and AI can also be trained by leveraging crop models' outputs. At the same time, technology is rapidly improving with new sensor technologies, the Internet of Things, remote sensing, genetic markers, and artificial intelligence. AI can be potentially useful to search for information and data on which to parameterize crop models. The amount of data that is being collected for agricultural production systems is exponentially expanding, providing opportunities for data analytics and the use of AI for strategic and actionable decisions. Unfortunately, much of that data is relatively superficial, e.g. remotely sensed biomass, and is lacking in depth and quality of crop physiology needed for model improvement. The promise of genetic technologies and QTL markers needs to be better linked to the genetic attributes used for cultivars in the crop models, and if done well, may reduce the need for detailed experiments for every new cultivar released. When AI is used in collaboration with these massive data sets, the DSSAT-CSM can play a major role in helping to understand the physiological relationships and interactions between Genotype, Environment, and Management (G * E * M). This can provide alternative management options that increase crop yield and quality, optimize resource use, and minimize environmental impact for long-term sustainable agricultural production. Boote (2019) reviewed the future of crop modeling relative to potential uses and applications and the needs for model improvement. There are great opportunities as well as challenges. A new concern since that review is whether Machine Learning and AI because of their newness and popularity will short-circuit true learning and understanding of mechanistic processes in the models.

19.3 Conclusion

The DSSAT crop modeling ecosystem is one of the oldest and most widely used crop modeling platforms across the world. The success of DSSAT and the CSM is based on the inclusiveness and participatory approach employed since the original development of the CERES and CROPGRO models that led to the much-expanded Cropping System Model. The modularity of the CSM with its linkage of multiple crop models and process components in a single executable code offers much promise for future enhancement of the CSM and DSSAT. We continue to add new crops to CSM, including perennial forages and vegetables, and have added new capabilities to the DSSAT software, such as yield forecasting. The emphasis on sharing data and model code in an open-source environment has also facilitated continued model improvement. The CSM and DSSAT offer multiple application possibilities, including crop management, minimizing environmental pollution, optimizing genetic yield improvement, and evaluating the impacts of future climate change.

DSSAT with the CSM is not just a software program but an ecosystem of:

- Crop model users
- Crop model trainers
- Crop model developers/scientists
- Models for important food, feed, fiber, pasture, and fuel crops
- Tools and utilities for data preparation
- Minimum data for model calibration and evaluation
- Application programs for assessing real-world problems

20 Acknowledgments

The authors appreciate the interest of the DSSAT Users' Community in the use of the DSSAT Cropping System Model for a very wide range of applications as shown by the large number of download requests for DSSAT and modeling publications based on DSSAT. Your encouragement motivates us to improve DSSAT continuously!

21 Where to look for further information

- K. J. Boote (ed.). Advances in crop modelling for a sustainable agriculture. Burleigh Dodds Science Publishing, United Kingdom.
- Boote, K. J. 2019. The future of crop modelling for sustainable agriculture. In: K. J. Boote (ed.). *Advances in crop modelling for a sustainable agriculture*. Burleigh Dodds Science Publishing, United Kingdom.

- Hoogenboom, G., C. H. Porter, K. J. Boote, V. Shelia, P. W. Wilkens, U. Singh, J. W. White, S. Asseng, J. I. Lizaso, L. P. Moreno, W. Pavan, R. Ogoshi, L. A. Hunt, G. Y. Tsuji, and J. W. Jones. 2019. The DSSAT crop modeling ecosystem. In: K. J. Boote (ed.). *Advances in crop modelling for a sustainable agriculture*. Burleigh Dodds Science Publishing, United Kingdom.
- Go to www:DSSAT.NET for free download of latest release. Hoogenboom, G., C. H. Porter, V. Shelia, K. J. Boote, U. Singh, W. Pavan, F. A. A. Oliveira, L. P. Moreno-Cadena, T. B. Ferreira, J. W. White, J. I. Lizaso, D. N. L. Pequeno, B. A. Kimball, P. D. Alderman, K. R. Thorp, S. V. Cuadra, M. S. Vianna, F. J. Villalobos, W. D. Batchelor, S. Asseng, M. R. Jones, A. Hopf, H. B. Dias, A. Jintrawet, R. Jaikla, E. Memic, L. A. Hunt, and J. W. Jones. 2024. Decision Support System for Agrotechnology Transfer (DSSAT) Version 4.8.5 (www .DSSAT.net). DSSAT Foundation, Gainesville, Florida, USA.
- Jones, J. W., G. Hoogenboom, C. H. Porter, K. J. Boote, W. D. Batchelor, L. A. Hunt, P. W. Wilkens, U. Singh, A. J. Gijsman, and J. T. Ritchie. 2003. The DSSAT cropping system model. *European Journal of Agronomy* 18:235-265 (https://doi.org/10.1016/S1161-0301(02)00107-7).

22 References

Alagarswamy, G., et al. 2006. Evaluating the CROPGRO-Soybean model ability to simulate photosynthesis response to carbon dioxide levels. *Agronomy J.* 98:34–42.

Alderman, P. D., et al. 2015. Adapting the CSM-CROPGRO model for pigeonpea using sequential parameter estimation. *Field Crops Res.* 181:1–15.

Alderman, P. D., et al. 2011a. Regrowth dynamics of 'Tifton 85' bermudagrass as affected by nitrogen fertilization. *Crop Sci.* 51:1716–1726.

Alderman, P. D., et al. 2011b. Carbohydrate and nitrogen reserves relative to regrowth dynamics of 'Tifton 85' bermudagrass as affected by nitrogen fertilization. *Crop Sci.* 51:1727–1738.

Allen, R. G., et al. 1998. *Crop Evapotranspiration. Guidelines for Computing Crop Water Requirements.* FAO Irrigation and drainage paper no 56. FAO, Rome, Italy.

Anar, M., et al. 2019. Modeling growth, development and yield of sugarbeet using DSSAT. *Agric Syst.* 169:58–70.

Anothai, J., et al. 2008. A sequential approach for determining the cultivar coefficients of peanut lines using end-of-season data of crop performance trials. *Field Crops Res.* 108:169–178.

Asseng, S., et al. 2000. Performance and application of the APSIM Nwheat model in the Netherlands. *Eur. J. Agron.* 12: 37–45.

Basso, B., et al. 2011. Procedures for Initializing Soil Organic Carbon Pools in the DSSAT-CENTURY model for agricultural systems. *Soil Sci.* Soc. Am. J. 75:69–78.

Batchelor, W. D., et al. 1993. Extending the use of crop models to study pest damage. Trans. *ASAE* 36:551–558.

Boote, K. J. 1981. Concepts for modeling crop response to pest damage. Paper No. 81-4007. ASAE, St. Joseph, MI.

Boote, K. J. 2011. Improving soybean cultivars for adaptation to climate change and climate variability. Chapter 17. In: S. S. Yadav, et al. (eds.), *Crop Adaptation to Climate Change*. Wiley-Blackwell, West Sussex, UK, pp. 370–395.

Boote, K. J. 2019. The future of crop modelling for sustainable agriculture. In: K. J. Boote (ed.). *Advances in Crop Modelling for a Sustainable Agriculture*. Burleigh Dodds Science Publishing, United Kingdom.

Boote, K. J., et al. 2010. Testing effects of climate change in crop models. In: D. Hillel and C. Rosenzweig (eds.). *Handbook of Climate Change and Agroecosystems*. Imperial College Press, London UK.

Boote, K. J., et al. 1983a. Canopy photosynthesis: Apparent, net, or gross? In: Advances in Photosynthesis Research, Proceedings, 6th Int'l Congress on Photosynthesis, August 1–6, 1983, Brussels, Belgium. Dr. W. Junk Publishers, The Hague. 4 pp.

Boote, K. J., et al. 2023. Adapting the CROPGRO model to simulate growth and yield of guar, *Cyamopsis tetragonoloba* L, an industrial legume crop. *Industrial Crops & Products* 197 (2023) 116596. https://doi.org/10.1016/j.indcrop.2023.116596.

Boote, K. J., et al. 2009. Modeling N-Fixation and Its Relationship to N Uptake in the CROPGRO Model. In: L. Ma, L. Ahuja, and T. Bruulsema (eds.) *Quantifying and Understanding Plant Nitrogen Uptake for Systems Modeling*. Taylor & Francis Group LLC, Boca Raton, Florida.

Boote, K. J., et al. 1984. Factors influencing crop canopy CO_2 assimilation of soybean. In Proceedings, World Soybean Research Conference III, Westview Press, Boulder, CO, pp. 780–788.

Boote, K. J., et al. 2021. Incorporating realistic trait physiology into crop growth models to support genetic improvement. In silico Plants https://doi.org/10.1093/insilicoplants/diab002

Boote, K. J., et al. 1987. PNUTGRO V1.0. Peanut Crop Growth Simulation Model. User's Guide. Department of Agronomy and Department of Agricultural Engineering, University of Florida, Gainesville, Florida. 48 pp.

Boote, K. J., et al. 1983b. Coupling pests to crop growth simulators to predict yield reductions. *Phytopathology* 73:1581–1587

Boote, K. J., et al. 1986. Modeling growth and yield of groundnut. In: Agrometeorology of Groundnut. Proc. Int. Symp. Aug 21–26, 1985, ICRISAT Sahelian Center, Niamey, Niger. Patancheru, India: ICRISAT, pp. 243–254.

Boote, K. J., et al. 1998. The CROPGRO Model for Grain Legumes. pp. 99–128. In: G. Y. Tsuji, et al. (eds.). *Understanding Options for Agricultural Production*. Kluwer Academic Publishers, Dordrecht.

Boote, K. J., et al. 2013. Putting mechanisms into crop production models. *Plant Cell & Environ*. 36:1658–1672.

Boote, K. J., et al. 2001. Physiology and modelling of traits in crop plants: implications for genetic improvement. *Agric Syst.* 70:395–420.

Boote, K. J. and Loomis, R. S. 1991. The prediction of canopy assimilation. Chapter 7. IN K. J. Boote and R. S. Loomis (eds.) *Modeling Crop Photosynthesis – From Biochemistry to Canopy*. CSSA Special Publication No. 19. American Society of Agronomy, Madison, WI, pp. 109–140.

Boote, K. J., et al. 2002. Adapting the CROPGRO legume model to simulate growth of faba bean. *Agron J.* 94:743–756.

Boote, K. J. and Pickering, N. B. 1994. Modeling photosynthesis of row crop canopies. *HortScience* 29:1423–1434.

Boote, K. J., et al. 2015. Sentinel Site Data for Crop Model Improvement–Definition and Characterization. In: *Improving Crop modeling Tools to Assess Climate Change Effects on Crop Response*. Advances in Agricultural Systems Modeling 07. American Society of Agronomy, Crop Science Society of America, and Soil Science Society of America, Madison, Wisconsin.

Boote, K. J., et al. 2012. Improving the CROPGRO-Tomato model for predicting growth and yield response to temperature. *HortSci* 47:1038–1049.

Boote, K. J., et al. 2008. Experience with Water Balance, Evapotranspiration, and Prediction of Water Stress Effects in the CROPGRO Model. In: L. R. Ahuja, et al. (eds.) *Response of Crops to Limited Water: Modeling Water Stress Effects on Plant Growth Processes, Volume 1 of Advances in Agricultural Systems Modeling*. ASA-CSSA-SSSA, Madison, Wisconsin.

Boote, K. J. and Scholberg, J. M. S. 2006. Developing, parameterizing, and testing of dynamic crop growth models for horticultural crops. *Acta Hort*. 718:23–34.

Boote, K. J., et al. 2021. Adapting the CROPGRO model to simulate growth and production of *Brassica carinata*, a bio-fuel crop. *GCB-Bioenergy* 13:1134–1148. https://doi.org/10.1111/gcbb.12838.

Boote, K. J. and Tollenaar, M. 1994. Modeling genetic yield potential. In: K. J. Boote, et al. (eds.). *Physiology and Determination of Crop Yield*. ASA-CSSA-SSSA, Madison, WI, pp. 533–565.

Boote, K. J., et al. 2016. Crop modeling approaches for predicting phenotype of grain legumes with linkage to genetic information. Chapter 8. Pp. 163–192. In: X. Yin & P.C. Struik (eds.). *Crop Systems Biology*, Springer International Publishing, Switzerland.

Boote, K. J., et al. 2018. Modeling sensitivity of grain yield to elevated temperature in the DSSAT crop models for peanut, soybean, dry bean, chickpea, sorghum, and millet. *European J Agron*. 100: 99–109.

Bourgeois, G. 1989. Interrelationships between late leafspot disease and Florunner peanut: a modeling approach. Ph.D. dissertation, Univ. of Florida.

Brunetti, H., et al. 2021. Improving the CROPGRO Perennial Forage Model for simulating growth and biomass partitioning of guineagrass. *Agron J*. 13:3299–3314. https://doi.org/10.1002/agj2.20766.

Buddhaboon, C., et al. 2018. Methodology to estimate rice genetic coefficients for the CSM-CERES-Rice model using GENCALC and GLUE genetic coefficient estimators. *J Agric Sci*. 156:482–492.

Confalone, A., et al. 2011. Temperature and photoperiod effects on *Vicia faba* phenology simulated by CROPGRO-Fababean. *Agron J*. 103:1036–1050.

Cuadra, S. V., et al. 2021. Energy balance in the DSSAT-CSM-CROPGRO model. *Agric For Meteorol*. 297:Article 108241. https://doi.org/10.1016/j.agrformet.2020.108241.

Curry, R. B., et al. 1995. Response of soybean to predicted climate change in the USA. In: C. Rosenzweig, et al. (eds.). *Climate Change and Agriculture: Analysis of Potential International Impacts*, ASA Spec. Pub. No. 59, ASA-CSSA-SSSA, Madison, WI, pp. 163–182.

DeJonge, K. C. and Thorp, K. R. 2017. Implementing standardized reference evapotranspiration and dual crop coefficient approach in the DSSAT cropping system model. *Trans ASABE* 60:1965–1981.

Deligios, P. A., et al. 2013. Predicting growth and yield of winter rapeseed in a Mediterranean environment: Model adaptation at a field scale. *Field Crops Res.* 144:100–112.

Del Grosso, S. J., et al. 2001. Simulated interaction of carbon dynamics and nitrogen trace gas fluxes using the DAYCENT model. In: M. Schaffer, et al. (eds.). *Modeling Carbon and Nitrogen Dynamics for Soil Management*. CRC Press, Boca Raton, Florida, pp. 303–332.

Elliott, J., et al. 2014. The parallel system for integrating impact models and sectors (pSIMS). *Environ Model Softw.* 62:509–516.

Farquhar, G. D. and von Caemmerer, S. 1982. Modelling of photosynthetic response to environmental conditions. In: O. L. Lange, et al. (eds.). *Encyclopedia of Plant Physiology, NS vol. 12B: Physiological Plant Ecology II*, Springer-Verlag, Berlin, Germany, pp. 549–587.

Fernandes, J. M. C., et al. 2017. A weather-based model for predicting early season inoculum build-up and spike infection by the wheat blast pathogen. *Tropical Plant Pathol* 42:230–237. https://doi.org/10.1007/s40858-017-0164-2

Ferreira, T. B., et al. 2021. Coupling a pest and disease damage module with CSM-NWheat - A wheat crop simulation model. *Trans ASABE* 64:2061–2071.

Ferreira, T. B., et al. 2024. Enhancing crop model parameter estimation across computing environments: utilizing the GLUE method and parallel computing for determining genetic coefficients. Computers and Electronics in Agriculture. (Under Review).

Guarin, J. R., et al. 2024. Modeling the effects of tropospheric ozone on the growth and yield of global staple crops with DSSAT v4.8.0. *Geosci Model Dev* 17:2547–2567.

Gijsman, A. J., et al. 2002. Modifying DSSAT crop models for low-input agricultural systems using a soil organic matter-residue module from CENTURY. *Agron J.* 94:462–474.

Godwin, D. C. and Singh, U. 1998. Nitrogen balance and crop response to nitrogen in upland and lowland cropping systems. In G. Y. Tsuji, et al. (eds.). *Understanding Options for Agricultural Production*. Kluwer Academic Publishers, Dordrecht, the Netherlands, pp. 55–77.

Guarin, J. R., et al. 2019. Modeling the effects of tropospheric ozone on wheat growth and yield. *Eur J Agron*. 105:13–23.

Hartkamp, A. D., et al. 2002. Adaptation of the CROPGRO growth model to velvet bean (*Mucuna pruriens*): I. Model development. *Field Crops Res* 78:9–25.

He, J., et al. 2009. Applying GLUE for estimating CERES-Maize genetic and soil parameters for sweet corn production. *Trans ASABE* 52:1907–1921.

He, J., et al. 2010. Influence of likelihood function choice for estimating crop model parameters using the generalized likelihood uncertainty estimation method. *Agric Syst*. 103:256–264.

Hoogenboom, G. and White, J. W. 2003. Improving physiological assumptions of simulation models by using gene-based approaches. *Agron J*. 95:82–89.

Hoogenboom, G., et al. 1990a. Nitrogen fixation, uptake, and remobilization in legumes: a modeling approach. In: *Proceedings of IBSNAT Symposium: Decision Support System for Agrotechnology Transfer, Las Vegas, 18 October, 1989. Part II: Posters. Department of Agronomy and Soil Science, College of Tropical Agriculture and Human Resources*, University of Hawaii, Honolulu, Hawaii, pp. 138–186.

Hoogenboom, G., et al. 1990b. *BEANGRO V1.00. Dry Bean Crop Growth Simulation Model. User's Guide*. Florida Agricultural Experiment Station Journal No. N-00379. University of Florida, Gainesville, Florida. 120 pp.

Hoogenboom, G., et al. 1991. *Predicting Growth and Development of Grain Legumes with a Generic Model*. ASAE Paper 91–4501, American Society of Agricultural Engineers, St. Joseph, Michigan.

Hoogenboom, G., et al. 1992. Modeling growth, development and yield of grain legumes using SOYGRO, PNUTGRO, and BEANGRO: A Review. *Trans ASAE* 35:2043–2056.

Hoogenboom, G., et al. 1994. BEANGRO, a process oriented dry bean model with a versatile user interface. *Agron J.* 86:182–190.

Hoogenboom, G., et al. 1997. *Prediction of Yield and Water Use for Chickpea-Based Cropping Systems*. Resource Management Program, International Crops Research Institute for the Semi-Arid Tropics (ICRISAT), Patancheru, Andhra Pradesh, India. 17 pp.

Hoogenboom, G., et al. 1999. Decision Support System for Agrotechnology Transfer v3.5. p. 1–36. In: G. Hoogenboom, et al. (eds.). *DSSAT* University of Hawaii, Honolulu, Hawaii.

Hoogenboom, G., et al. 2004. From genome to crop: integration through simulation modeling. *Field Crops Res.* 90:145–163.

Hoogenboom, G., et al. (eds.). 2010. *Decision Support System for Agrotechnology Transfer Version 4.5*. University of Hawaii, Honolulu, HI.

Hoogenboom, G., et al. 2012. Experiments and data for model evaluation and application. In: J. Kihara, et al. (eds.). *Improving Soil Fertility Recommendations in Africa using the Decision Support Systems for Agrotechnology Transfers (DSSAT)*. Springer, Dordrecht, the Netherlands, pp. 9–18.

Hoogenboom, G., et al. 2019a. The DSSAT crop modeling ecosystem. In: K. J. Boote (ed.). *Advances in Crop Modelling for a Sustainable Agriculture*. Burleigh Dodds Science Publishing, United Kingdom.

Hoogenboom, G., et al. 2019b. Decision Support System for Agrotechnology Transfer (DSSAT) Version 4.7 (www.DSSAT.net). DSSAT Foundation, Gainesville, Florida.

Hoogenboom, G., et al. 2023. Decision Support System for Agrotechnology Transfer (DSSAT) Version 4.8.2 (www.DSSAT.net). DSSAT Foundation, Gainesville, Florida, USA.

Hopf, A., et al. 2022. Development and improvement of the CROPGRO-Strawberry model. *Scientia Horticulturae* 291:110538. https://doi.org/10.1016/j.scienta.2021.110538.

Hunt, L. A. and Pararajasingham, S. 1995. CROPSIM – WHEAT: a model describing the growth and development of wheat. *Can J Plant Sci.* 75:619–632.

Hunt, L. A., et al. 2001. Agronomic data: advances in documentation and protocols for exchange and use. *Agric Syst*. 70:477–492.

Hunt, L. A., et al. 1993. GENCALC - Software to facilitate the use of crop models for analyzing field experiments. *Agron J.* 85:1090–1094.

Hunt, L. A., et al. 1994. General input and output files structures for crop simulation models. In: P. F. Uhlir and G. C. Carter (eds.). *Crop Modeling and Related Environmental Data. A Focus on Applications for Arid and Semiarid Regions in Developing Countries*. CODATA, Paris, France (ISBN 1-884893-00-7), pp. 35–73.

International Crops Research Institute for the Semi-Arid Tropics (ICRISAT). 1984. *Proceedings of the International Symposium on Minimum Data Sets for Agrotechnology Transfer*. ICRISAT Center, Patancheru, India.

International Benchmark Sites Network for Agrotechnology Transfer (IBSNAT). 1989. *The Decision Support System for Agrotechnology Transfer Version 2.1 (DSSAT v2.1) User's*

Guide. Department of Agronomy and Soil Science, College of Tropical Agriculture and Human Resources, University of Hawaii, Honolulu, Hawaii.

International Benchmark Sites Network for Agrotechnology Transfer (IBSNAT). 1993. *The IBSNAT Decade*, p. 178. Department of Agronomy and Soil Science, College of Tropical Agriculture and Human Resources, University of Hawaii, Honolulu, Hawaii.

International Benchmark Sites Network for Agrotechnology Transfer (IBSNAT). 1994. *The Decision Support System for Agrotechnology Transfer Version 3.0 (DSSAT v3) User's Guide*. Department of Agronomy and Soil Science, College of Tropical Agriculture and Human Resources, University of Hawaii, Honolulu, Hawaii.

Jing, Q., et al. 2020. Simulating alfalfa regrowth and biomass in eastern Canada using the CSM-CROPGRO-Perennial forage model. *Eur J Agron*. 113:125971. https://doi.org/10.1016/j.eja.2019.125971

Jones, C. A. and Kiniry, J. R. (eds.). 1986. *CERES-Maize: A Simulation Model of Maize Growth and Development*. Texas A & M. University Press, College Station, Texas.

Jones, J. W., et al. 1987. *SOYGRO V5.4. Soybean Crop Growth Simulation Model. User's Guide*. Department of Agricultural Engineering and Department of Agronomy, University of Florida, Gainesville, Florida. 50 pp.

Jones, J. W., et al. 2001. Approaches to modular model development. *Agric Syst*. 70:421–443.

Jones, J. W., et al. 1998. Decision Support System for Agrotechnology Transfer: DSSAT v3. In: G. Y. Tsuji, et al. (eds.). *Understanding Options for Agricultural Production. Systems Approaches for Sustainable Agricultural Development*, vol 7. Springer, Dordrecht, the Netherlands.

Jones, J. W., et al. 2003. The DSSAT cropping system model. *Eur J Agron*. 18:235–265.

Jones, J. W., et al. 2016. Brief history of agricultural systems modeling. *Agric Syst*. 155: 240–254. https://doi.org/10.1016/j.agsy.2016.05.014

Jones, M. R. and Singels, A. 2018. Refining the Canegro model for improved simulation of climate change impacts on sugarcane. *Eur J Agron*. 100: 76–86.

Kassie, B. T., et al. 2016. Performance of DSSAT-Nwheat across a wide range of current and future growing conditions. *Eur J Agron*. 81:27–36.

Karunaratne, A. S., et al. 2024. Adapting the CROPGRO model to simulate growth, development, and yield of Bambara groundnut (*Vigna subterranea* L. Verdc), an underutilized crop. *Eur J Agron*. 159. https://doi.org/10.1016/j.eja.2024.127279.

Keeling, C. D., et al. 1976. Atmospheric carbon dioxide variations at Mauna Loa Observatory, Hawaii. *Tellus* 28: 538–551.

Kersebaum, K. C., et al. 2015. Analysis and classification of data sets for calibration and validation of agro-ecosystem models. *Environ Model Softw*. 72:402–417.

Kihara, J., et al. (eds.). 2012. *Improving Soil Fertility Recommendations in Africa using the Decision Support for Agrotechnology Transfer (DSSAT)*. Springer, Dordrecht, the Netherlands. ISBN 978-94-007-2959-9. 195 pp.

Kimball, B. A., et al. 2023. Simulation of evapotranspiration and yield of maize: an inter-comparison among 41 maize models. *Agric For Meteorol*. 333:109396. https://doi.org/10.1016/j.agrformet.2023.109396

Lara, M. A. S., et al. 2012. Predicting growth of *Panicum maximum*: an adaptation of the CROPGRO-perennial forage model. *Agron J*. 104:600–611.

Lizaso, J. L., et al. 2007. Developing a sweet corn simulation model to predict fresh market yield and quality of ears. *J Amer Soc Hort Sci*. 132:415–422.

Lizaso, J. I., et alt. 2011. CSM-IXIM: a new maize simulation model for DSSAT version 4.5. *Agron J.* 103:766–779. https://doi.org/10.2134/agronj2010.0423

Lizaso, J. I., et al. 2022. Chapter 9. Simulating Maize Crop Processes in DSSAT: CSM-CERES-Maize vs. CSM-IXIM-Maize. In: L. R. Ahuja, et al (eds.). *Modeling Processes and Their Interactions in Cropping Systems: Challenges for the 21st Century.* Book Series:Advances in Agricultural Systems Modeling. https://doi.org/10.1002/9780891183860.ch9.

Mack, L., et al. 2020. Adapting the CROPGRO model to simulate chia growth and yield. *Agron J.* 2020:1–19. https://doi.org/10.1002/agj2.20305

Malik, W., et al. 2018. Adapting the CROPGRO model to simulate alfalfa growth and yield. *Agron J.* 110:1777–1790.

Maliro, C. E. 1986. *Physiological Aspects of Yield in Four Grain Legume Crops Under Two Water Regimes.* M.S. Thesis, University of Florida.

Martre, P., et al. 2024. Global needs for nitrogen fertilizer to improve wheat yield under climate change. *Nature Plants* 10:1081.

Memic, E., et al. 2021. Cultivar coefficient estimator for the cropping system model based on time-series data: a case study for soybean. *Trans ASABE* 64:1391–1402.

Memic, E., et al. 2020. Extending the CSM-CERES-beet model to simulate impact of observed leaf disease damage on sugar beet yield. *Agronomy* 2020-12-08. https://doi.org/10.3390/agronomy10121930

Messina, C. D., et al. 2006. A gene-based model to simulate soybean development and yield responses to environment. *Crop Sci.* 46:456–466.

Moreno-Cadena, P. 2021. *Modeling the Dynamics of Starch Content in Cassava.* Ph.D. dissertation, University of Florida, Gainesville, Florida.

Moreno-Cadena, L. P., et al. 2020. Importance of genetic parameters and uncertainty of MANIHOT, a new mechanistic cassava simulation model. *Eur J Agron.* 115 (2020) 126031. https://doi.org/10.1016/j.eja.2020.126031.

Naab, J. B., et al. 2004. Using the CROPGRO-peanut model to quantify yield gaps of peanut in the Guinean savanna zone of Ghana. *Agron J.* 96:1231–1242.

Naab, J. B., et al. 2005. Effects of sowing date and fungicide application on yield of early and late maturing peanut cultivars grown under rainfed conditions in Ghana. *Crop Prot* 24:325–332.

Narh, S., et al. 2015. Genetic improvement of peanut cultivars for West Africa evaluated with the CSM-CROPGRO-Peanut model. *Agron J.* 106:2213–2229.

Nkebiwe, P. M., et al. 2022. Adapting the CROPGRO-faba bean model to simulate the growth and development of Amaranthus species. *Agron J.* https://doi.org/10.1002/agj2.21090

Oliveira, A. A., et al. 2021. Incorporating a dynamic gene-based process module into a crop simulation model. *In Silico Plants* 3(1):1–18. https://doi.org/10.1093/insilicoplants/diab011

Paff, K. and Asseng, S. 2019. Comparing the effects of growing conditions on simulated Ethiopian tef and wheat yields. *Agric For Meteorol.* 266:208–220.

Pathak, T. B., et al. 2007. Use of global sensitivity analysis for CROPGRO cotton model development. *Trans Am Soc Agric Eng.* 50:2295–2302.

Penning De Vries, F. W. T., et al. 1974. Products requirements and efficiency of biosynthesis a quantitative approach. *J Theor Biol.* 45:339–377.

Pedreira, B. C., et al. 2011. Adapting the CROPGRO perennial forage model to predict growth of *Brachiaria brizantha. Field Crops Res.* 120:370–379.

Pequeno, D. N. L., et al. 2024. Production vulnerability to wheat blast disease under climate change, Nature Climate Change. https://doi.org/10.1038/s41558-023-01902-2

Pequeno, D. N. L., et al. 2014. Simulating forage production of Marandu palisade grass (*Brachiaria brizantha*) with the CROPGRO-Perennial Forage model. *Crop Pasture Sci.* 65:1335–1348.

Pequeno, D. N. L., et al. 2018. Species-genotypic parameters of the CROPGRO perennial forage model: implications for comparison of three tropical pasture grasses. *Grass Forage Sci.* 73:440–455. https://doi.org/10.1111/gfs.12329

Pickering, N. B., et al. 1990. *A Moisture- and CO2-Sensitive Model of Evapotranspiration and Photosynthesis.* Paper No. 90-2519. American Society of Agricultural and Biological Engineers, St. Joseph, MI.

Pickering, N. B., et al. 1993. *Modeling Soybean Growth Under Climate Change Conditions.* Paper No. 93-4510. American Society of Agricultural and Biological Engineers, St. Joseph, MI.

Pickering, N. B., et al. 1995. Adapting SOYGRO V5.42 for prediction under climate change conditions. In: C. Rosenzweig, et al. (eds.). *Climate Change and Agriculture: Analysis of Potential International Impacts*, ASA Spec. Pub. No. 59, ASA-CSSA-SSSA, Madison, WI, pp. 77–98.

Porter, C. H., et al. 2010. Modeling organic carbon and carbon-mediated soil processes in DSSAT v4.5. *Oper Res Int J.* 10(3):247–278. https://doi.org/10.1007/s12351-009-0059-1

Präger, A., et al. 2019. Simulating growth and development processes of quinoa (*Chenopodium quinoa* Willd.): adaptation and evaluation of the CSM-CROPGRO model. *Agronomy* (published: 2 December 2019).

Prasad, P. V. V., et al. 2003. Supra-optimal temperatures are detrimental to peanut (*Arachis hypogaea* L) reproductive processes and yield at ambient and elevated carbon dioxide. *Global Change Biol.* 9:1775–1787.

Priestley, C. H. B. and Taylor, R. J. 1972. On the assessment of surface heat and evaporation using large scale parameters. *Mon Weather Rev.* 100:81–92.

Putto, C., et al. 2013. Determination of plant traits that affect genotype x location (G x L) interaction in peanut using the CSM-CROPGRO-Peanut model. *Intl J Plant Prod.* 7:537–568.

Qi, J., et al. 2024. Simulating the development and growth of lentil using the CSM-CROPGRO model. *Agron J.* 116:2391–2410.

Ritchie, J. T. 1972. Model for predicting evaporation from a row crop with incomplete cover. *Water Resour Res.* 8:1204–1213.

Ritchie, J. T. 1998. Soil water balance and plant water stress. In G. Y. Tsuji, et al. (eds.). *Understanding Options for Agricultural Production.* Kluwer Academic Publishers, Dordrecht, the Netherlands, pp. 41–54.

Ritchie, J. T., et al. 1985. *CERES-Wheat: A User-Oriented Wheat Yield Model.* Preliminary Documentation. AGRISTARS Publication No. YM-U3-04442-JSC-18892. Michigan State University, East Lansing, Michigan.

Ritchie, J. T., et al. 2009. Extension of an existing model for soil water evaporation and redistribution under high water content conditions. *Soil Sci Soc Am J.* 73:792–801.

Robertson, R. D. 2017. *Mink: Details of a Global Gridded Crop Modeling System.* Washington, DC: International Food Policy Research Institute (IFPRI). http://ebrary.ifpri.org/cdm/ref/collection/p15738coll2/id/131406

Rosenzweig, C., et al. (eds.). 1995. *Climate Change and Agriculture: Analysis of Potential International Impacts*. ASA Spec. Pub. No. 59, ASA-CSSA-SSSA, Madison, WI.

Rymph, S. J. 2004. *Modeling Growth and Composition of Perennial Tropical Forage Grasses*. Ph.D. dissertation, University of Florida, Gainesville, Florida.

Santos, M. G., et al. 2019. Simulation of productivity and soil moisture under Marandu palisadegrass using the CSM-CROPGRO-Perennial Forage model. *Crop Pasture Sci.* 70:159–168.

Santos, M. L., et al. 2022. Applying the CROPGRO perennial forage model for long-term estimates of Marandu palisadegrass production in livestock management scenarios in Brazil. *Field Crops Res.* 286. https://doi.org/10.1016/j.fcr.2022.10862

Sau, F., et al. 2004. Testing and improving evapotranspiration and soil water balance of the DSSAT crop models. *Agron J.* 96:1243–1257.

Scholberg, J. M. S., et al. 1997. Adaptation of the CROPGRO model to simulate the growth of field-grown tomato. pp. 133–151. In M. J. Kropff, et al. (eds.). *Systems Approaches for Sustainable Agricultural Development: Applications of Systems Approaches at the Field Level*. Kluwer Academic Publishers, Dordrecht, the Netherlands.

Sexton, P. J., et al. 1994. Yield-determining processes in relation to cultivar seed size of common bean. *Crop Sci.* 34:84–91.

Shelia, V., et al. 2019. A multi-scale and multi-model gridded framework for forecasting crop production, risk analysis, and climate change impact studies. *Environ Model Softw.* 115:144–154.

Singels, A., et al. 2008. *DSSAT v4.5 Canegro Sugarcane Plant Module: Scientific Documentation*. SASRI, Mount Edgecombe, South Africa. pp 34. (https://sasri.sasa.org.za/agronomy/icsm/documents/DSSAT%20Canegro%20SCIENTIFIC%20documentation_20081215.pdf)

Singh, M. P., et al. 2013. Using the CSM-CROPGRO-Peanut model to simulate late leaf spot effects on peanut cultivars of differing resistance. *Agron J.* 105:1307–1316.

Singh, P., et al. 1999a. Soybean-chickpea rotation on Vertic Inceptisols: 2. Long-term simulation of water balance and crop yields. *Field Crops Res.* 63:225–236.

Singh, P., et al. 1999b. Soybean-chickpea rotation on Vertic Inceptisols: 1. Effect of soil depth and landform on light interception, water balance and crop yields. *Field Crops Res.* 63:211–224.

Singh, S., et al. 2015. Adapting the CROPGRO model to simulate growth and yield of spring safflower in semiarid conditions. *Agron J.* 108:64–72.

Singh, P., et al. 2017. An assessment of yield gains under climate change due to genetic modification of pearl millet. *Sci Total Environ.* 601–602:1226–1237.

Singh, P., et al. 2013. Potential benefits of drought and heat tolerance in groundnut for adaptation to climate change in India and West Africa. *Mitig Adapt Strateg Glob Change* 19:509–529. https://doi.org/10.1007/s1027-012-9446-7

Singh, P., et al. 2014a. Climate change impacts and potential benefits of drought and heat tolerance in chickpea in South Asia and East Africa. *Europ J Agron.* 52:123–137.

Singh, P., et al. 2014b. Quantifying potential benefits of drought and heat tolerance in rainy season sorghum for adapting to climate change. *Agr Forest Met.* 185:37–48.

Singh, P. and SVirmani, S. M. 1996. Modelling growth and yield of chickpea (*Cicer arietinum* L). *Field Crop Res.* 46:41–59.

Singh, U., et al. 1998. Modeling growth and development of root and tuber crops. pp. 129–156. In G. Y. Tsuji, et al. (eds.). *Understanding Options for Agricultural Production*. Kluwer Academic Publishers, Dordrecht, the Netherlands.

Singh, U. and Porter, C. H. 2019. Improving modelling of nutrient cycles (e.g. nitrogen cycles) in crop cultivation. In: K. J. Boote (ed.). *Advances in Crop Modelling for a Sustainable Agriculture*. Burleigh Dodds Science Publishing, United Kingdom.

Smith, H. D., et al. 2023. Adapting the CROPGRO Perennial Forage model to predict growth and development of Pensacola bahiagrass. *Field Crops Res.* 302:109095. https://doi.org/10.1016/j.fcr.2023.109095.

Steberl, K., et al. 2019. Modifying the CROPGRO Safflower model to simulate growth, seed and floret yield under field conditions in Southwestern Germany. Agronomy (published 19 December 2019).

Suleiman, A. A. and Ritchie, J. T. 2003. Modeling soil water redistribution during second stage evaporation. *Soil Sci Soc Am J.* 67:377–386.

Teng, P. S., et al. 1998. Simulation of pest effects in crops using coupled pest-crop models: the potential for decision support. pp. 221–266. In G. Y Tsuji, G. Hoogenboom, and P. K. Thornton (eds.) *Understanding Options for Agricultural Production*. Kluwer Academic Publishers, Dordrecht, the Netherlands.

Tesfaye, K., et al. 2018. Potential benefits of drought and heat tolerance for adapting maize to climate change in tropical environments. *Clim Risk Manag.* 19:106–119. https://doi.org/10.1016/j.crm.2017.10.001.

Tesfaye, K., et al. 2023. High spatial resolution seasonal crop yield forecasting for heterogeneous maize environments in Oromia, Ethiopia. *Clim Serv.* 32:100425. https://doi.org/10.1016/j.cliser.2023.100425

Thornton, P. K. and Hoogenboom, G. 1994. A computer program to analyze single-season crop model outputs. *Agron J.* 86:860–868.

Thornton, P. K., et al. 1995. A computer program to analyze multiple-season crop model outputs. *Agron J.* 87(1):131–136.

Thorp, K. R., et al. 2014. Evaluation of CSM-CROPGRO-COTTON for simulating effects of management and climate change on cotton growth and evapotranspiration in an arid environment. *Trans ASABE* 57:1627–1642. https://doi.org/10.13031/trans.57.10612

Tsuji, G. Y., et al. (eds.). 1998. *Understanding Options for Agricultural Production*. Kluwer Academic Publishers, Dordrecht, the Netherlands.

Tsuji, G. Y., et al. (eds.). 1994. *Decision Support System for Agrotechnology Transfer Version v3*. University of Hawaii, Honolulu, Hawaii.

Uehara, G. and Tsuji, G. Y. 1998. Overview of IBSNAT. In: Tsuji G.Y., Hoogenboom G., Thornton P.K. (eds.). *Understanding Options for Agricultural Production. Systems Approaches for Sustainable Agricultural Development*, vol 7. Springer, Dordrecht, the Netherlands.

Uryasev, O., et al. 2004. Graphical Display Program (GBuild). Chapter 6 in DSSAT v.4. Vol 2. Data Management and Analysis Tools. In: P. W. Wilkens, et al. (eds.). *International Consortium for Agricultural Systems Applications*, University of Hawaii, Honolulu, Hawaii. pp. 155–174.

Vásquez-Jiménez, J., et al. 2024. Predicting growth and development of the pineapple cultivar "MD-2" with the DSSAT Aloha Pineapple Model. *Fruits* 79(1):1–12. https://doi.org/10.17660/th2024/002.

Wallach, D., et al. 2018. A dynamic model with QTL covariables for predicting flowering time of common bean (*Phaseolus vulgaris*) genotypes. *Eur J Agron*. 101:200–209.

White, J. W. and Hoogenboom, G. 1996. Simulating effects of genes for physiological traits in a process-oriented crop model. *Agron J*. 88:416–422.

White, J. W. and Hoogenboom, G. 2003. Gene-based approaches to crop simulation: past experiences and future opportunities. *Agron J*. 95:52–64.

White, J. W., et al. 2013. Integrated Description of Agricultural Field Experiments and Production: the ICASA Version 2.0 Data Standards. *Comput Elect Agric*. 96:1–12.

Wilkerson, G. G., et al. *1981. Modeling Soybean Growth for Crop Management*. Paper No. 81-4014. ASAE, St. Joseph, MI.

Wilkerson, G. G., et al. 1983. Modeling soybean growth for crop management. *Trans. ASAE*. 26:63–73.

Willmott, C. J., et al. 1985. Statistics for the evaluation and comparison of models. *J Geophys Res*. 90:8995–9005. https://doi.org/10.1029/JC090iC05p08995

Yang, J. M., et al. 2014a. An evaluation of the statistical methods for testing the performance of a crop simulation model with observed data. *Agric Syst*. 127:81–89.

Yang, J. Y, et al. 2014b. EasyGrapher: software for data visualization and statistical evaluation of DSSAT cropping system model and the CANB model. *Inter J Com Theory Eng*. 6:210–214.

Zhang, J. and Bartholomew, D. P. 1993. Simulation of pineapple growth, development, and yield. *Acta Horticulturae* 334:205–219. https://doi.org/https://doi.org/10.17660/ActaHortic.1993.334.21

Zhang, J., et al. 1997. ALOHA-Pineapple v.2.1: a computer model to predict the growth, development and yield of pineapple. *Acta Horticulturae* 425:287–296. https://doi.org/10.17660/ActaHortic.1997.425.32

Chapter 2

The Crop, Land, and Soil Simulation (CLASSIM) group of models[1]

D.H. Fleisher, D.J. Timlin, S. Li, J. Barnaby, S. Yesilkoy, E. Han and V.R. Reddy, USDA-ARS, USA; Z. Wang, University of Maryland, USA; S. Beegum, A. Mitra, University of Nebraska-Lincoln, USA; and W. Sun, Colorado State University, USA

1 Introduction

Expected increases in global population by 2050 will require an increase in food production between 30% and 62% over today's output to combat rising food insecurity (Mirzabaev et al., 2023; Searchinger et al., 2018; van Dijk et al., 2021). Pressure among industrial, commercial, and public sectors regarding land availability and natural resources places constraints on production capacity

1 **Note:** The following abbreviations are used in this chapter: A: atmosphere; ACSL: Adaptive Cropping Systems Laboratory; C: atmospheric carbon dioxide concentration; E: environment; EWE: extreme weather event; G: genetics; GxExM: genetic x environment x management; M: management; N: nitrogen; P: plant; S: soil; SOC: soil organic carbon; SPA: soil-plant-atmosphere; T: surface air temperature; W: water.

http://dx.doi.org/10.19103/AS.2025.0155.02
Published by Burleigh Dodds Science Publishing Limited, 2025.

and is likely to continue into the future (Farooq et al., 2022; Folberth et al., 2020; Foley et al., 2011). Climate projections indicate that the combination of changes in atmospheric carbon dioxide concentration (C), air temperature (T), and water availability (W) are likely to negatively impact current/future yields of many agriculturally important crops without adaptation (Benitez-Alfonso et al., 2023; Neupane et al., 2022; Rezael et al., 2023). The occurrence of extreme weather events during crop growing seasons, which have the potential to reduce yields, is also occurring with increased frequency (Anapalli et al., 2019; Kimball, 2016; Walker and Van Loon, 2023). Annual increases in yields per unit land area for many crops have plateaued or decreased in the U.S. and most developed nations (Arata et al., 2020; Cassman et al., 2010). These factors underly the crucial need to identify methods to boost production efficiency and yield per unit area while protecting ecosystem functions and natural resources.

Mechanistic models that integrate an understanding of soil-plant-atmospheric (SPA) components with specific representations for crop genetics, environment, and management (G x E x M) provide decision support to tackle these issues. These tools can assist and, in some cases, supplement the need to conduct site-specific field trials (Di Paola et al., 2016) as they allow exploration of a wide-breadth of M options as well as G, or phenotypic, variations in response to cropping systems needs and challenges (Fleisher et al., 2021; Resop et al., 2014; Sun et al., 2023a). Crop and soil models have traditionally been used to assist farmers with on-site questions associated with pre- and in-season M questions (Haskett et al., 1995; Reddy et al., 1995, 2002; Timlin et al., 2002b). They are more recently used to better understand system behaviors and identify impacts and responses to climate changes at on-farm, local, regional, national, or even broader spatial scales (Jones et al., 2017; Timlin et al., 2024). Models are also being used to assess cropping rotations, cover crop utilization, and soil health metric studies (e.g. Doltra et al., 2019; Wang et al., 2025; Zhang et al., 2020). Most of these tools mimic daily or hourly SPA interactions over the course of a crop growing season and are comprehensive in the sense that critical aspects of E and M on crop growth and development are incorporated. Crop genetics are typically accounted for in terms of phenotype using 'genetic coefficients,' which are adjusted during the calibration process (Fleisher et al., 2020). Soil components are included to address W and nutrient surface and sub-surface dynamics, nitrogen (N) transformations, root growth, and crop residue during and after the growing season (Beegum et al., 2023b; Dathe et al., 2014).

The characterization of models as mechanistic or process level indicates a focus on using mathematics to describe SPA components at a fundamental physiological, biochemical, and/or physics-based level. In contrast to correlative or statistical approaches, mechanistic approaches focus on understanding causality among relationships (Reynolds and Acock, 1985).

As such, users have confidence when using the model to extrapolate outside of the range of data used to build, test, and/or validate the model, including what-if type assessments such as those associated with projected climate impacts, management strategies, and phenotype questions (Jones et al., 2017). The United States Department of Agriculture (USDA) – Agricultural Research Service (ARS) Adaptive Cropping Systems Laboratory (ACSL) Crop, Land, and Soil Simulation group of models (henceforth referred to as 'ACSL models') focus at the organ or plant level of spatial scale. As early as the 1990s, the USDA recognized the need to identify potential responses of agricultural systems to environmental changes and better understand agricultural system behavior (Schram and Wallace, 1996). With this in mind, ACSL models now describe certain processes, such as plant gas exchange (photosynthesis and transpiration) at a mechanistic level using biochemical approaches for plant photosynthesis (Farquhar et al., 1980; von Caemmerer, 1981) at the leaf level. Other processes, such as allocation of photosynthate to different organs or leaf expansion rate, tend to remain empirical as there is less experimental data to validate these approaches. The relationship between plant developmental progress and accumulated T values and photoperiod is an example of this (Fig. 1). The integration of process with empirical modeling incorporated in the structure of crop model source code is frequently referred to as 'explanatory'. These models are also classified as time-varying and deterministic (as opposed to stochastic or probabilistic) simulators. An example of state variables

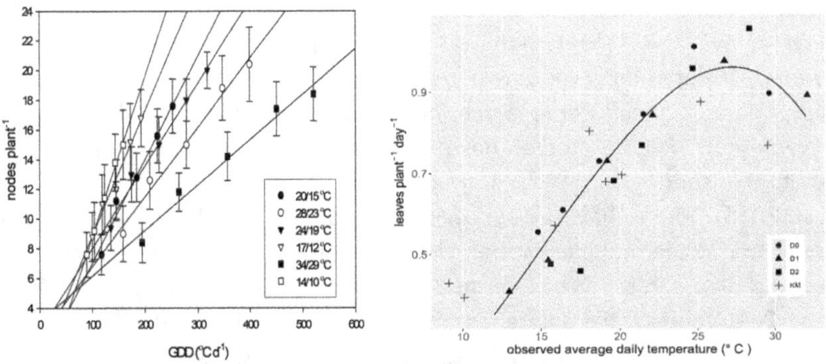

Figure 1 Linear (left) and nonlinear (right) T-based approaches for describing leaf appearance rates in potato. The relationship between leaf (node) number versus cumulative degree days in the left plot illustrates significantly different linear regression lines for six different growth T treatments (adapted with permission from Fleisher et al., 2006). This implies a simple one-fit growing degree day approach will lead to errors. A nonlinear curve fit based on the beta function (Yin et al., 2003), which is used in our SPUDSIM and MAIZSIM crop models more accurately simulated these T responses (right plot). Note the shape of the curve more appropriately captures the nonlinear trends after about 26°C. Symbols on this plot indicate data sources where D0, D1, and D2 were published by Fleisher et al. (2006) and KM by Kirk and Marshall (1992).

related to plant growth rate and the influence of weather and soil W status as implemented in ACSL models is shown in Fig. 2.

Developing crop and soil models for agricultural decision support has been a focus for decades (Fleisher, 2009). Initially, factors which limited crop yield were identified and relationships were established such as the correlation between light interception and whole plant growth rate, denoted as radiation use efficiency (Monteith, 1977) (Fig. 3). During the 1970s and 1980s, additional factors such as weather, soils, physiology, crop genetics, and, in some cases, pests and diseases were integrated into modeling frameworks with varying levels of sophistication (El-Sharkawy, 2011). Advances in scientific instrumentation resulted in an increase in experimental data quantity and quality (e.g. microclimate information and leaf/canopy gas exchange, for example), which were used to develop and/or improve submodules providing a more detailed, explanatory range of SPA components. Such models depicted bio-physico-chemical processes and nonlinear feedback mechanisms between crop growth and development, soil processes, and atmospheric/weather conditions. Today's explanatory crop models use hundreds of variables and thousands of lines of computer source code to simulate different aspects of agricultural systems. However, significant knowledge gaps remain (Boote et al., 2013; Silva and Giller, 2021), particularly with respect to extreme weather events, crop stress, genetics, and modeling of various soil health metrics as influenced by different M practices.

Model improvement efforts are usually based on using the expertise stored in existing models as a starting point. As new data and understanding become available, older methods are enhanced, revised, or replaced as needed. Multiple crop models currently exist for the same crop because of independent agricultural research initiatives across different universities and governments. Recent studies have identified 32 wheat (Liu et al., 2023), 9 potato (Fleisher et al., 2017), 41 maize (Kimball et al., 2023), and 13 rice (Li et al., 2015) models that are peer-reviewed and considered as explanatory-type simulators. While multiple models developed for the same crop share similar building blocks in terms of their mathematical and conceptual bases (Wallach et al., 2016), substantial differences remain in terms of how different responses are captured in source code. Subsequently, such differences result in significant uncertainty, or variability, among models when evaluated against common datasets (Wallach et al., 2021).

Considerable uncertainties in model predictions tend to be associated with effects of abiotic stresses such as those linked with heat, W, and nitrogen (N) components. These uncertainties indicate substantial knowledge gaps remain not only in the mechanisms by which crop development, growth, and quality components are influenced by stresses but also in terms of how such information is embedded into the structure of the model such that the correct

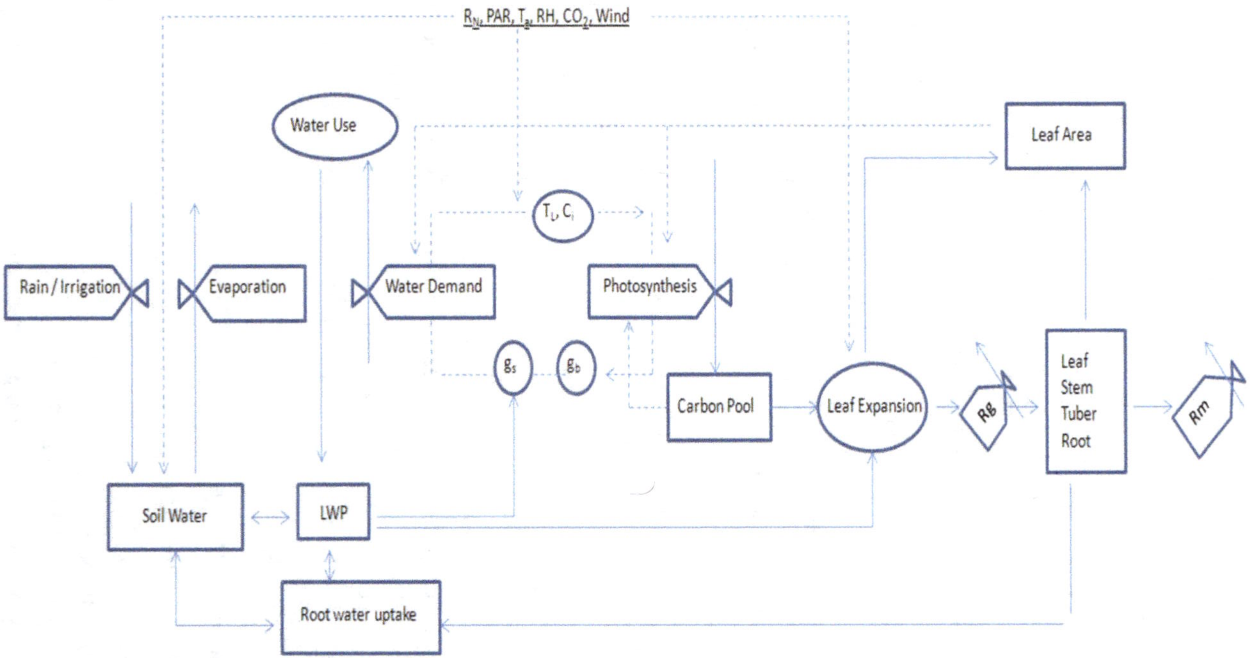

Figure 2 Variables for whole plant (boxes) and individual leaf (circles) time-varying states, and associated C-, T-, and W-driven rate processes used in most ACSL crop models. Each box represents hundreds of lines of computer source-code utilizing algorithmic and mathematical relationships to approximate process-level mechanisms. This includes, for example, a coupled biochemical leaf level and stomatal conductance model (inner boxes) within an energy balance that links plant soil and water status, weather, and phenotype with gas exchange processes. Variables include: T_L – leaf temperature; C_i – intercellular C concentration; g_s, g_b – stomatal conductance, and boundary layer conductance, to water vapor and C; R_g – growth respiration; R_m – maintenance respiration; LWP – leaf water potential.

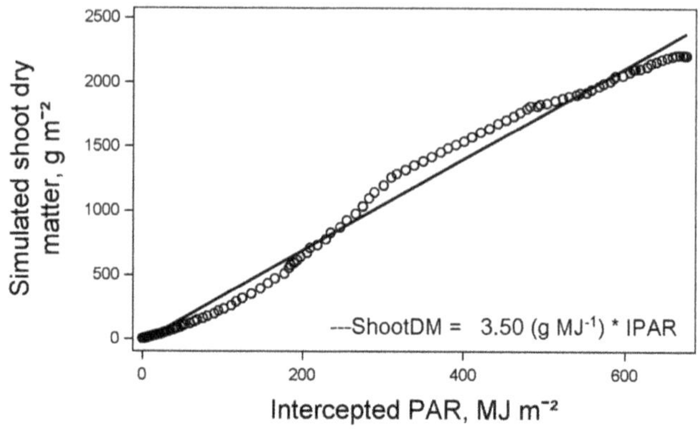

Figure 3 Simulated relationship between maize shoot dry mass and canopy intercepted photosynthetically active radiation (PAR) for a farm site located in Iowa, U.S. as simulated by the MAIZSIM model. The slope of this relationship, shown with the solid line, was 3.5 grams of dry matter per MJ of intercepted PAR and represents radiation use efficiency (RUE). This value is similar to the measured values. As with other ACSL models, more mechanistic methodologies are utilized to simulate growth processes in place of RUE. However, this example shows how emergent behavior can arise from more complex models of plant physiological processes.

result is not necessarily obtained for the wrong reason (Wallach et al., 2016). Many existing approaches also do not account, or have not been adequately tested, for the impacts of the interactions among high T, C, and/or W on crop response (Stockle and Kemanian, 2020). There is a need to also address quality of grain or product, particularly when such is influenced by growth conditions. Given such challenges, ACSL models have recently been developed and/or improved for crops of economic and environmental importance including cereal rye, cotton, maize, potato, rice, soybean with plans to revamp approaches for wheat and sorghum models. Over the next several sections, we discuss the uniqueness of ACSL crop and soil models and indicate how such models are well-suited for improved on-farm applications and studies involved in today's agricultural systems challenges.

2 Crop model structure

ACSL research has resulted in five primary crop models:

- cotton (GOSSYM) (Beegum et al., 2023c)
- maize (MAIZSIM) (Yang et al., 2009)
- potato (SPUDSIM) (Fleisher et al., 2010b)
- soybean (GLYCIM) (Sun et al., 2022; Sun et al., 2023b)
- cereal-rye (RYESIM) (Wang et al., 2025)

Each model is integrated with a two-dimensional soil model, 2DSOIL (Timlin and Pachepsky, 1997), as described later in this section. A sixth model for rice (RICESIM) (Li et al., 2021) has not yet been integrated with 2DSOIL but includes other new components, as described later. These models can be executed as individual, stand-alone programs but are also embedded in a desktop-based graphical user interface, Crop, Land, and Soil Simulation (CLASSIM), which simplifies use of the models and enhances functionality with regards to database manipulation, input data access and scenario evaluations including multi-year rotation studies and sensitivity analyses (Timlin et al., 2023).

Each model runs on an hourly time-step. Certain sub-components, such as plant developmental rate, may run on a 24-hour basis, depending on the model. Input data regarding site geographic location, soil properties, weather, phenotype, and M are required to drive the models (Table 1). Model simulations integrate this information with mathematical descriptions of crop, soil, and their interactions with the atmosphere, and generate time-series responses. The scope is at the single plant level, sometimes referred to as 'point' scale, although the spatial extent could be viewed as the field plot, with an assumption that all variables are homogenous. This is distinguished from geospatial or landscape output, although multiple independent model runs can be conducted on adjacent parcels of landscape and linked to characterize spatial responses as well (Resop et al., 2012). Model runs can be conducted across multiple seasons by linking the outputs from one simulation into the inputs of the subsequent run. This allows for the simulation of multi-year cropping rotations. Output data from each simulation covers the response of plant growth and development, yield, resource (W and N) use, plant stress, and soil components including W, heat, nutrient movement, N leaching, surface residue decomposition, surface gas fluxes, and related responses (Table 2).

Each crop module has functional components, usually represented as separate methods, functions or subroutines in the computer source code. Most of ACSL models follow the order indicated in Fig. 4, although specific processes can vary as indicated (Table 3). The phenology routine keeps track of plant physiological age, stress conditions, and a tabulation of prior exposure to climatic conditions (e.g. accumulation of heat units), and, along with current weather for the given time-step, estimate progression through developmental stages. For example, a pre-defined number of hourly or daily accumulation of thermal units may be needed for a particular potato variety to initiate a new leaf (Fig. 1) or progress through different vegetative or growth stages. Once this number is met, the next leaf appears or developmental stage occurs.

Potential organ growth rates, such as the aerial expansion rate of leaves in the plant canopy, are determined next, and these are frequently estimated by thermal conditions and plant phenotype. This also includes the initiation of new organs such as individual leaves, lateral branches, storage, or reproductive

Table 1 Primary input data information needed to execute ACSL crop models organized by type.

Variable	Description	Units
Weather*		
Tair	Hourly or daily maximum and minimum temperatures (T)	°C
SRAD	Solar radiation measured at 2-m above ground level	MJ m^{-2}
PREC	Hourly or daily rainfall	cm or mm
WS	Windspeed	km h^{-1}
CO_2	Atmospheric carbon dioxide concentration	ppm
RH	Relative humidity	%
Soils**		
Layer	Vertical depth of given soil layer	cm
Texture	Sand/Silt/Clay components	%
BD	Bulk density	g cm^{-3}
OM	Organic matter concentration	%
NO_3	Nitrate concentration	ppm
NH_4	Ammonium concentration	ppm
Tsoil	Average soil T in soil layer	°C
TH33	Soil volumetric W content at field capacity	cm^3 cm^{-3}
TH1500	Soil volumetric W content at wilting point	cm^3 cm^{-3}
Ks	Saturated hydraulic conductivity	cm d^{-1}
Other hydraulic parameters	Hydraulic values associated with van Genuchten equation (van Genuchten, 1980)	various
Management		
Variety	Crop and variety selection	–
Tillage	Tillage type and date	–
Surface residue	Amount, quality, and termination date of cover crop residue or other material	various
Date: planting	Calendar date for planting	date
Date: emergence	Calendar date for emergence (model dependent)	date
Date: harvest	Optional Calendar date for harvest	date
Plant density	Number of plants per unit area at emergence	plants m^{-2}
Row spacing	Spacing between rows	cm
Planting depth	Depth of seed or seed piece at planting	cm
Fertilizer	Type, amount, and date of, application	–
Irrigation	Type, amount, and date of application	–

*Weather: Minimum set of data includes solar radiation, T, and rainfall. Values for wind speed, humidity and CO_2 concentration can be supplied as constant by the user if time-series information is not available.

**Soil: Each set of data is described for each soil layer.

Table 2 Overview of representative ACSL crop model outputs organized by type.

Type	Description	Units
Production (time-series or end-of-season)*		
Yield	Mass of grain, pod, ear, tuber, boll or other yield bearing organ	g $plant^{-1}$
Organ mass	Mass of leaf, stem, root, and senesced weight	g $plant^{-1}$
Leaf area index	Ratio of leaf area to covered ground area	m^2 m^{-2}
Nitrogen	N uptake and N content in each plant organ	g $plant^{-1}$ or %
Resource use efficiencies	W or N use efficiency	g $plant^{-1}$ or mm $plant^{-1}$
Transpiration/Evapotranspiration/Evaporation	W use measures	mm
Phenology	Dates of important developmental stages	date
Physiological (time-series, hourly or daily)		
Canopy temperature	Average canopy leaf surface T	°C
Photosynthesis	Daily or hourly rate of net or gross photosynthesis	g C $plant^{-1}$ d^{-1}mol C m^{-2} h^{-1}
Respiration	Daily or hourly rate of total growth and maintenance respiration	g C $plant^{-1}$ d^{-1}mol C m^{-2} h^{-1}
Stress indices	Relative (0 to 1) index indicating severity of carbon, N, T, or W stress	–
Stomatal conductance	Maximum daily leaf stomatal conductance	mm m^{-2} s^{-1}
Leaf water potential	Bulk average leaf W potential	MPa
Soil (time-series, hourly or daily)		
Water content	W content by soil layer or full profile	mm or cm^3 cm^{-3}
Nitrogen content	Total NNO_3 as N content by profile or layer	kg ha^{-1}
2D resource plots	2D soil profile of soil W, T, or N content	various
2D root plots	2D soil profile of root mass or length density	g cm^{-2}
Soil surface fluxes	Evaporation, infiltration, runoff, CO_2, O_2	various
Soil C, N storage	Amount of N, C in humus, litter, manure or roots	kg ha^{-1}

* Crop values are frequently scaled to homogenous farm unit from g $plant^{-1}$ to kg ha^{-1}. All mass units are in terms of dry weight.

Published by Burleigh Dodds Science Publishing Limited, 2025.

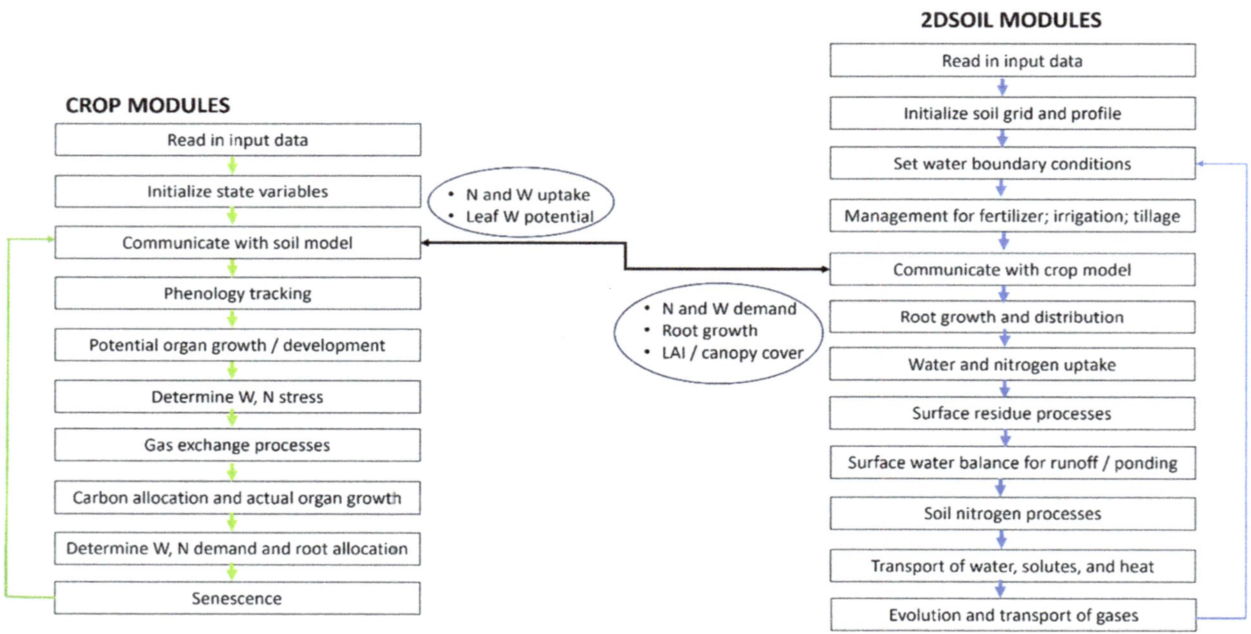

Figure 4 Flow chart of 2DSOIL and general crop modules describing the basic subroutines for each ACSL group model. Information is passed to and from crop and soil modules, particularly as pertaining to soil N and W availability, actual uptake, and carbon used for root growth within the soil. Simulation completes when the crop has reached maturity or at a pre-determined harvest date supplied as input to the model.

Table 3 Major components of ACSL crop models along with types and differences with respect to various soil, plant, atmosphere component representations. Primary reference is provided where appropriate.

Component	General approach	Primary reference*
Phenology[1]	Developmental stage progression is influenced by T, W, N, photoperiod and plant stress.	–
Leaf area[2]	Single or big-leaf approach influence by T, W, N, and nutrient supply as well as crop genetics	–
Light interception[3]	Sunlit/Shaded canopy partition approach	(Campbell and Norman, 2000)
Light utilization[3]	Coupled leaf photosynthesis x stomatal conductance nested within energy balance	(Kim and Lieth, 2003)
Organ growth formation	Potential growth determined from T, genetics and reduced by carbon availability, W, T, and N stress. Carbon partitioning coefficients which are influenced by age, plant stress, and soil W status determine priority among organs for photosynthate.	–
Yield formation	Pod, ear, or panicle simulated along with grain for soybean, maize, and rice; Individual bolls simulated for cotton; One tuber class for potato	–
Root Distribution	Algorithmic allocation based on soil element suitability factors determined by existing root mass, mechanical resistance, W and N content, oxygen, and other factors	(Dathe et al., 2014) (Timlin and Pachepsky, 1997; Wang et al., 2021b)
N Stresses	Direct influence on leaf expansion, gas exchange, carbon allocation	–
W Stress	Direct influence on leaf expansion, gas exchange, carbon allocation	** (Fleisher et al., 2015; Yang et al., 2009)
Heat Stress	Influence on leaf senescence, gas exchange	–
ET	Obtained from energy balance at leaf surface	(Kim and Lieth, 2003)
CO_2 effects	Direct influence on leaf gas exchange	(Kim and Lieth, 2003)

[1] Cereal rye model only considers vegetative growth.
[2] Single leaf approach refers simulation of individual leaf and/or branches. Big leaf approach indicates the entire canopy is depicted as one overall leaf and stem class. SPUDSIM can use either approach; RICESIM and GLYCIM use a big leaf approach; GOSSYM and MAIZSIM use single leaf.
[3] All models use a coupled gas exchange methodology nested within an energy balance at the leaf level. Canopy leaf area, and light interception, is divided into sunlit and shaded leaf classes at each hour.

organs. These rates are then reduced by plant stressors, including N or W deficiency, non-optimal T, or limited carbon status. Whole plant carbon assimilation rates are calculated in all ACSL models using leaf-level gas exchange approaches, which are scaled up to the whole canopy (de Pury and Farquhar, 1997) via determination of sunlit and shaded leaf class (Campbell and Norman, 2000). This determines hourly values of photosynthate available to support new growth and respiratory needs as well as W demand to support transpiration. After removing maintenance respiration costs, this photosynthate is then allocated to different organ classes based on their potential growth demand and empirically derived allocation coefficients.

When carbon is limited, each crop model uses different methods to determine a growth priority. In SPUDSIM, for example, potato tuber growth demand takes precedent over other organs in varieties that are more determinate (i.e. early maturing). In cereal crops, after flowering, nearly all assimilate will go towards grain fill. The next step involves communication from each crop module to 2DSOIL to indicate the plant W and N demand required for the current time step as well as the carbon available to support new root growth (Fleisher et al., 2015; Yang et al., 2009). The soil module uses that information to determine root growth distribution, including diffusive movement of root length through lateral and vertical depths and actual uptake rates for N and W. A final step is to predict leaf and stem senescence rates.

General components simulated in the soil program are outlined in Table 4. The 2DSOIL model allows simulations of:

- transport/movement of W, heat, N, oxygen (O_2), and C;
- surface/atmospheric and boundary layer processes including evaporation and O_2 and C fluxes;
- root processes, including growth and distribution within the soil profile, water uptake, ammonium and nitrate uptake, root respiration;
- M practices including subsoil irrigation, tillage, surface residue decomposition, and organic and inorganic fertilizer applications; and
- biotransformations including denitrification, N mineralization, and other processes.

The soil profile is described as a spatial grid of vertical and lateral (between row) grid cells, or elements, and is mathematically represented as a finite element system. In general, the soil model reads in input data and initializes the soil profile (which involves, among other activities, setting values for soil state variables including heat, N, W, and gas concentrations) before calling the crop module at each time-step.

The soil model also partitions carbon allocated to root growth from the crop module into different elements of the soil profile. This is done via a diffusive growth routine (Dathe et al., 2014) (Wang et al., 2021b), which also allows for preferential root growth and root length density based on favorability indicators for each element (e.g. mechanical resistance as determined by bulk density, O_2 concentration, W or N concentration, temperature, and existing root

Table 4 Major components of the 2DSOIL model. Specific details of each component can be investigated via the primary reference provided.

Component	General description	Additional references*
Bio-transformations	Includes mineralization and immobilization of organic matter in the soil considering nitrification and denitrification. Recent improvements include surface mulch layer decomposition.	(Wang et al., 2021a)
Gas transport	Molecular diffusion equation is used for gas transport of CO_2, H_2O, and O_2 in air-filled pore space.	(Beegum et al., 2023b) (Wang et al., 2022)
Grid and coordinates	Determines spatial discretization of the soil domain, including surface and bottom soil profile boundary conditions and soil layers.	–
Heat transport	Involves 2D convective-dispersive equation for transport of heat in variably saturated soil.	(Wang et al., 2022)
Time control/ output of modules	Coordinates time steps to synchronize soil modules. Time-steps can vary, particularly for transport processes and intra-soil interactions, due to differences in numerical integration.	–
Soil transport processes	Includes water and solute transport, both involving finite-element solutions. For water transport, Richard's equation for two-dimensional Darcian flow in variably saturated soil is concerned; For solute transport, 1D and 2D convective-dispersion equations in variably saturated soil are used.	–
Soil–atmosphere boundary	Includes hourly estimate of potential fluxes of water, solutes, and heat at soil surface driven by meteorological variables.	(Acock and Trent, 1991)
Water and chemical uptake by roots	Includes uptake of water and solute based on cell location of state variables, determination of root characteristics, and leaf and soil-water potentials.	–
Root growth processes	Includes determination of soil-based growth preference factors, diffusive movement of roots in vertical and horizontal directions including root length density and mass proportioned to mature and young root classes.	(Dathe et al., 2014) (Wang et al., 2021b)

*Timlin and Pachepsky (1997) is the underlying reference for all modules. Additional references are as indicated.

length density). This distribution has an indirect influence on the amount of W and N, which can be taken up by the plant at each time step. Root W uptake is a function of leaf and soil W potential and root surface and soil resistances. N transformations include mineralization and immobilization of organic matter, including nitrification and denitrification. Other components which have recently been incorporated in the soil module include methods for simulating the decomposition of surface residue (Wang et al., 2021a), such as that from a terminated cover crop and other organic sources, surface W movement, including runoff, infiltration, exfiltration, and W ponding (Wang et al., 2020), and transport and surface C fluxes from root and microbial respiration (Timlin et al., 2023).

ACSL models were developed in either FORTRAN or C++ and compiled for Windows-based systems. Several models are also being tested for Linux systems. The 2DSOIL model was written in FORTRAN. MAIZSIM, RYESIM, and SPUDSIM models were written in C++ while GLYCIM, GOSSYM, and RICESIM were developed in FORTRAN. Each crop model is individually compiled as dynamic link library which is fully integrated with versions of 2DSOIL specific to each model. All source code for GLYCIM, GOSSYM, MAIZSIM, SPUDSIM, and 2DSOIL is fully available to the general public at this underlined location. Code for RICESIM and RYESIM is undergoing rigorous testing but will be released to the same location. The CLASSIM interface was developed in Python version 3. The graphical user interface and relational databases can be downloaded and installed via the same link as above. Crop model manuals and visual tutorials for using the CLASSIM interface are included as part of the installation kit.

3 Input data

3.1 Weather and soil

All ACSL crop models function on an hourly time-step with respect to crop growth and plant W and N uptake. Soil processes can operate at even finer time-steps depending on numerical convergence needs, but outputs are represented at an hourly level. Hourly or daily meteorological data (Table 1) regarding air T, solar radiation, and rainfall are required, and data for windspeed, C, and relative humidity can be provided or estimated by the model (Timlin et al., 2002a). When daily values are supplied, algorithms based on celestial mechanics (Campbell and Norman, 2000) are used to estimate hourly values depending on day of year, time of day and geographic location.

Soil inputs describe each soil layer or horizon (Table 1). These include soil texture (sand, silt, clay composition) and bulk density, as well as initial organic matter content, nitrate and ammonium concentrations, soil moisture content, and soil temperature. Hydraulic information is also required, including water contents at field capacity and wilting points, along with information needed for running the Richard's equation for two-dimensional water flow in a variably saturated soil (Šimůnek et al., 1992). These values include saturated and unsaturated hydraulic conductivities and related properties, which can be measured or estimated through the use of pedo-transfer functions (Schaap et al., 2001).

Published by Burleigh Dodds Science Publishing Limited, 2025.

3.2 Management data

Management data include pre and in-season farm operations. These include irrigation and fertilizer type, amount, and date of application, cover crop residue amount and termination date, and tillage operations. Initial conditions such as crop type, variety, planting dates, depth of sowing, plant density and row spacing, emergence date, if available can be supplied. Users simulate tillage and different irrigation and fertilizer amounts and options. Automatic irrigation based on soil water availability can also be selected (Table 1).

4 Model calibration: genetic coefficients determining plant growth characteristics

The accuracy of model predictions is strongly influenced by the structure of the model's source code, which describes how aspects of the SPA components are mathematically described, quality of the input data, and the manner in which phenotypic traits of the given cultivar are represented (Fleisher et al., 2017; Wallach et al., 2014). Obtaining values for the variables used to describe this latter category, typically referred to as 'genetic' coefficients, is usually what is meant by model calibration. Genetic coefficients (Table 5) characterize a specific variety's developmental response to Tair, Tsoil, N or W stress, photoperiod, as well as maturity class. These parameters also include potential growth characteristics, such as the number of spikelets per rice panicle, the maximum daily growth rate of potato tubers, potential leaf extension rates, etc. Although most ACSL models share similar code structure, each model utilizes a different set of coefficient parameters. These are developed by the modeler and are frequently established from empirically observed differences among phenotypes for a given crop. Default values are provided for these parameters for each model. The number of genetic coefficients and their corresponding grouping in terms of stress, development, or growth characteristics are summarized in Table 5 for all ACSL models and discussed in more detail below.

Most genetic coefficients have a biological meaning. However, many cannot be obtained from direct measurements but are inferred, such as the determinacy coefficient in SPUDSIM, which influences the sink strength of tubers after initiation occurs and is thus related to maturity class. In practice, mathematical or statistical approaches are used to estimate values, minimizing differences between observed and simulated data (Fleisher et al., 2020; Wallach et al., 2016). While the crop modeling community uses the term 'genetic' in practice, such values are more representative of crop phenotype and do not necessarily have a direct, or simple, relationship with gene expression (Messina et al., 2022), although there are some exceptions.

Table 5 Primary genetic coefficients in each ACSL model. Values in parentheses indicate the number of coefficients most frequently modified during calibration. The number of coefficients directly associated with developmental rate, stress response, and growth potential are indicated as well as what external factors, weather (T - air or soil temperature; Pp - photoperiod), water (W), or nitrogen (N) status, influence the response. A primary reference for each model detailing the calibration parameters is provided.

Crop	Model	Num	Develop[1]	Stress[2]	Growth[3]	Response[4]	Reference
Soybean	GLYCIM	27 (23)	16	2	9	T; Pp	(Sun et al., 2022)
Cotton	GOSSYM	35 (3)	15	9	11	T; Pp; W	(Beegum et al., 2023c)
Maize	MAIZSIM	6 (2)	5	0	1	T; Pp	(Kim et al., 2012)
Cereal rye	RYESIM	6 (6)	6	0	0	T; N; W	(Wang et al., 2025)
Potato	SPUDSIM	9 (7)	4	2	3	T; Pp; N	(Fleisher et al., 2010b)
Rice	RICESIM[5]	18 (6)	9	0	9	T; Pp; N; W	(Li et al., 2017a)

[1]Development category includes values relating progression through crop developmental stages as related to thermal time, photoperiod or other environmental dependency.
[2]Stress category refers to values that directly integrate varietal differences to stress response, such as N or W deficiency or excess. All ACSL models include N, W, T, and Carbon stresses, but these responses are assumed conserved across all varieties unless noted above.
[3]Growth category includes values that directly influence canopy morphological characteristics and potential growth rates as constrained by genetics.
[4]Response indicates environment or plant/soil status variables that are directly encompassed in the varietal parameters.
[5]Adapted from the ORYZA crop model and utilizes a similar set of genetic coefficients except with respect to plant gas exchange processes and heat stress (Li et al., 2021, 2022a).

Values for these coefficients should remain consistent for the same variety regardless of site location or M factors to avoid confounding genetic factors with E and M components. Confounding can reduce confidence in model predictions, particularly when evaluating what-if type questions which involve making simulations outside of the range of data originally used for calibration/validation (Fleisher et al., 2020). However, due to limitations in model structure and perhaps the quality of available experimental data regarding field conditions, modelers can be challenged to keep these values consistent across different production environments (Heidmann et al., 2008). ACSL crop models focus on including process-level depictions where possible to improve stability of the calibration process (Fleisher et al., 2020) and reduce these confounding issues.

4.1 GLYCIM (soybean) genetic coefficients

GLYCIM has 27 genetic coefficients (Table 5), 16 of which describe differences in varietal developmental rates (Table 6) with respect to temperature and photoperiod. Node additions during vegetative growth are predicted as a function of hourly thermal time and modified by the P2 coefficient. Progress from vegetative to start of floral induction depends on the maximum number of mainstem leaves (P3) and photoperiod. Multiple coefficients regulate progression through reproductive (R) stages, each of which has a different sensitivity to thermal time and photoperiod. For example, progress from floral induction to full bloom stages (R0 to R2) is described by P5, P6, P7, and P8 and is primarily photoperiod-driven. The relationships change with respect to solstice date and W stress. Movement from full bloom to full pod (R2 to R4 stages) is related to photoperiod and day of year. Seed filling stage durations (up to R7) are influenced by the rate of seed-fill, thermal time, and plant stress as adjusted by P11 through P14. Physiological maturity depends on MG, P14, and P15, which account for T and stress responses during seed filling. To account for less determinate varieties, a P16 coefficient also allows for overlap between leaf addition rates and R stages.

Potential yield is influenced by total number of seeds per unit mass and the relationship of rate of seed-fill to T as indicated by S1 and S2. Other parameters influence leaf area expansion (P25) and relationships between dry weight and morphological characterization, such as plant height and stem elongation rates (P17, P24). The P23 parameter also accounts for the influence of plant density and row spacing on branching potential. For further details on stability of these coefficient values, refer to Reddy et al. (1995). A recent study by Sun et al. (2023b) simplified the calibration approach by first identifying the vegetative node production (V2) value and modifying reproductive progression rates (R1 to R3, and R7 to R12) if necessary. Changes in growth parameters can be adjusted if warranted and sufficient field data is available to check in-season responses. Changes in MG and solstice dependent progression rates (P6 ...P11) can be used to account for photoperiod differences among cultivars.

4.2 GOSSYM (cotton) genetic coefficients

GOSSYM has 50 coefficients, 35 of which (Table 5) address species-specific and other cultivar-dependent parameters (Boone et al., 1993; Reddy and Baker, 1988). The other parameters describe E and M responses. These parameters (Table 7) modify various functional relations including processes such as time to first square (R6), time from first square to first flower (R7), time from first flower to first open boll (R8), pre-fruiting mainstem node time interval (R3), and post-fruiting mainstem node time interval (R4) are the major cotton developmental

Table 6 Representative genetic coefficients and value ranges for GLYCIM (soybean) model from the total number indicated in Table 5. The assigned ID terms are for convenience and not necessarily in the official model documentation. Abbreviations: V - vegetative growth; R - reproductive growth; TT - thermal time.

ID	Description	Units	Range
Phenology			
MG	Maturity group	–	0 to 10
P2	V: Leaf appearance rate versus TT	TT^{-1}	0.0067 to 0.0132
P3	V: Maximum leaf number before flowering	#	11 to 24
P5	R: Progress rate to floral initiation before solstice	d^{-1}	0.0192 to 0.0525
P6	R: Progress rate to floral initiation at solstice	d^{-1}	0.0022 to 0.0501
P7	R: Progress rate to floral initiation after solstice	d^{-1}	-0.197 to -0.003
P8	R: Progress rate from floral initiation towards full bloom	d^{-1}	0.0882 to 0.3
P11	R: Progress rate from full bloom to full seed	TT^{-1}	0.0039 to 0.0053
P12	R: Duration from first to full seed	TT	13 to 213
P13	R: Potential duration from first to full seed without stress	TT	250 to 700
P14	R: Duration decline from first to full seed with stress	TT^{-1}	0.5 to 2
P15	R: Progress rate towards physiological maturity	TT^{-1}	0.0018 to 0.0075
P16	R: Reproductive stage where vegetative growth ceases	#	2 to 5
Growth			
S1	G: Number of seeds per pound weight	# lb^{-1}	2400 to 3800
S2	G: Seed fill rate at 24°C	mg d^{-1}	7.5 to 8.5
P17	G: Petiole potential elongation and dry weight ratio	–	0.0024 – 0.003
P18	G: Potential rate of current root weight increase	%	0 to 100
P19	G: Increase in pod weight with R stages	–	0.5 to 1.5
P20	G: Influence of T on potential seed weight and seed fill	–	0.5 to 1.5
P23	G: Potential branch production	# plt^{-1} m^{-2}	0.05 to 0.5
P24	G: Stem weight to stem elongation ratio	g cm^{-1}	1 to 2
P25	G: Ratio of leaf area growth versus leaf number increase	cm^2 $leaf^{-1}$	0.5 to 2

processes in the model (Beegum et al., 2024). These processes are described as T and/or physiological age dependencies which are then modified by the genetic coefficients. Water or N stresses can modify these rates as well. Multiple growth and stress coefficients are also used to distinguish varietal responses. These include maximum potentials for cotton boll size (G2), the square growth rate (G3), and leaf area expansion duration and rates (G10, G11, G12, and G14–G18), and sensitivities to leaf W potential (G4, G9) and W and T stresses

Published by Burleigh Dodds Science Publishing Limited, 2025.

(G23–G25). In practice, calibration usually focuses on identifying values for T based developmental rate coefficients (R3–R8) (Beegum et al., 2024).

4.3 MAIZSIM (maize) genetic coefficients

MAIZSIM has six genetic coefficients (Table 5) primarily focused on developmental responses to T and photoperiod. Values for P3 and P4 (Table 8) are usually not changed during the calibration process and influence the rate of leaf tip appearance and initiation rate, respectively. These rates are primarily driven by T using a non-linear relationship as per Kim et al. (2012). The P5 coefficient effectively delays the time of silking. For example, a value of three means this delay would be equivalent to the time it takes for three leaves to appear. The stay green value, P2, represents the persistency or duration of the canopy and is thus related to the length of the crop life cycle. Photoperiod sensitivity can be accounted for in these equations (P6). Juvenile leaf number, P1, represents the minimum number of leaves at tasseling and is related to the relative maturity of the variety. Values for P1 and P2 are often published by breeders but can be manually fit to observed data as needed.

4.4 RICESIM (rice) genetic coefficients

The RICESIM model primarily uses 18 coefficients (Table 5), nine of which are developmental and the other nine growth-related, to represent varietal differences. Additional coefficient values for organ respiration rates, leaf gas exchange characteristics, and W stress components are conserved among varieties and not generally adjusted during the calibration process. Similar to the ORYZA model (Li et al., 2017a), developmental rates are based on thermal time computations, which are driven by cardinal temperatures as represented by T1 to T5 (Table 9). However, in RICESIM, progress through developmental stages occurs on an hourly versus daily basis (Li et al., 2020) using values for D1 through D4. Growth coefficients include the maximum number of spikelets per area (G1), individual grain weight (G2), specific leaf area (G3), and age-dependent parameters which influence maximum leaf area expansion rates (G4), and changes in carbon allocation among plant organs (G5...G8) in response to physiological aging. A new coefficient was developed, G9, to account for varietal differences to spikelet heat stress tolerance as per Li et al. (2024a). Photoperiod sensitivity among varieties during vegetative growth is described by an optimum threshold (P1) above which developmental progress can be delayed as modified by P2. Coefficients most frequently calibrated for individual varieties account for thermal (D1, D2, D3, D4, G1 and G9), and photoperiod (P1, P2) sensitivity.

Table 7 Representative genetic coefficients and value ranges, or typical value, for GOSSYM (cotton) model from the total number indicated in Table 5. The assigned ID terms are for convenience and not necessarily in the official model documentation. Abbreviations: C - carbon; N - nitrogen; T - temperature; V - vegetative; W - water.

ID	Description	Units	Range
Phenology			
R1	Days after emergence until C partitioning to stem reduces	days	35
R2	Relative morphogenetic delay due to N-stress	–	1.5
R3	Relative time interval between pre-fruiting nodes	–	1.1–1.35
R4	Relative time interval between nodes in the main stem	–	0.9–1.0
R5	Relative time interval between nodes in V branches	–	1.0–1.25
R6	Relative time interval between emergence to first square	–	0.9–1.1
R7	Relative time interval between first square and bloom	–	0.9–1.15
R8	Relative time interval between emergence to open boll	–	0.9–1.1
R9	Relative increase in leaf age	–	0.75–1.0
R10	Relative influence of physiological days on leaf abscission	–	1.0–1.2
Growth and Stresses			
G1	Related to minimum boll age when loss can occur	–	4.0
G2	Maximum cotton boll size	g	6.0–7.0
G3	Relative increase in carbon allocation to stem	–	0.85–1.2
G4	Min leaf W potential associated with stem growth W stress	MPa	–0.9
G5	Relative square growth potential	–	0.9–1.0
G6	Relative square growth potential at T_{night} > 35°C	–	0.0
G7	Correction factor for root/shoot ratio	–	1.0
G8	Relative potential change in leaf growth due to W stress	–	1.0
G9	Minimum leaf W potential for leaf growth W stress	MPa	–0.85
G10	Relative change in pre-fruiting leaf area	–	1.0
G11	Relative change in mainstem leaf area	–	1.0–1.25
G12	Relative change in mainstem and pre-fruiting leaf weight	–	0.7–1.1
G13	Minimum LAI that affects boll temperature	$m^2\ m^{-2}$	3.0
G14	Relative influence of duration of pre-fruiting leaf expansion	–	1.0–2.5
G15	Relative rate of pre-fruiting leaf area at unfolding	–	1.0
G16	Relative rate of mainstem leaf area at unfolding	–	1.0
G17	Relative rate of fruiting branch leaf area at unfolding	–	0.9–1.0
G18	Relative rate of internode elongation duration for height	–	1.0
G19	Relative rate of initial internode length for height	–	0.8–1.1
G20	Relative initial internode length reduction < 14 stem nodes	–	0.9–1.15
G21	Relative initial internode length reduction > 13 stem nodes	–	0.95–1.15
G22	Relative current internode length	–	0.7–1.0
G23	Relative loss of bolls due to heat injury	–	0.5–1.0
G24	Relative potential fruit growth with T stress	–	1.1–1.35
G25	Relative potential fruit growth with W and T stress	–	0.8–1.85

Table 8 Representative genetic coefficients and value ranges for MAIZSIM (corn) model from the total number indicated in Table 5. The assigned ID names are for convenience and not necessarily in the official model documentation.

ID	Description	Units	Range
Phenology			
P1	Number of juvenile leaves	# $plant^{-1}$	13–20
P2	Stay green value controlling the onset of senescence	–	0–8
P3	Maximum leaf tip appearance rate	leaves d^{-1}	0.42–0.61
P4	Maximum leaf tip initiation rate	leaves d^{-1}	0.8–0.99
P5	Number of phyllochrons to tassel after appearance of last leaf	#	2–4
P6	Daylength sensitivity	–	Yes/No

4.5 RYESIM (cereal rye) genetic coefficients

The RYESIM model uses six genetic coefficients (Table 5) to account for varietal differences in developmental rates. These rates are primarily thermal time driven in which the thermal increments are based on cardinal temperature values P1, P2, and P3 (Table 10). Transitions among developmental stages are based on phyllochron units as indicated in P4, P5, and P6. Growth parameters for leaf-level photosynthesis are assumed to be conserved among varieties and are not included here. The model was developed to account for vegetative growth stages and thus does not include reproductive growth stages (Wang et al., 2025).

4.6 SPUDSIM (potato) genetic coefficients

SPUDSIM has nine genetic coefficients that account for differences in varietal responses to T, photoperiod, and plant N status (Table 5). Developmental parameters C1 through C3 account for the influence of daily air T (C1), daily T amplitude (C2), and photoperiod (C3) on tuber induction rate, while C4 accounts for cultivar determinacy and thus accounts for the extent to which vegetation continues to grow following tuber initiation (Table 9). Adjustments to C4 are used to account for differences in cultivar maturity class (e.g. early, mid, or late maturing). Growth parameters account for potential canopy leaf area expansion rate (G1), maximum tuber growth rate (G2), and specific leaf weight (G3). The model also accounts for varietal differences in response to the influence of excessive or deficient plant N status on tuber initiation rate (G4, G5) (Fleisher et al., 2020; Fleisher et al., 2010b). In general, values for photoperiod sensitivity (C3) and determinacy (C4) establish differences among maturity groups.

Table 9 Representative genetic coefficients and value ranges for RICESIM (rice) model from the total number indicated in Table 5. The assigned ID names are for convenience and not necessarily in the official model documentation. Abbreviations: C - carbon; T - temperature.

ID	Name and Description	Units	Range
Phenology			
D1	DVRJ; Developmental rate during juvenile stage*	°C h^{-1}	0.016–0.036
D2	DVRI; Developmental rate during photoperiod sensitive stage*	°C h^{-1}	0.016–0.036
D3	DVRP; Developmental rate during panicle development*	°C h^{-1}	0.009–0.013
D4	DVRR; Developmental rate during reproductive stages*	°C h^{-1}	0.025–0.026
T1	TBD; base T for development *	°C	9–12
T2	TBLV; base T for juvenile leaf growth*	°C	9–12
T3	TMD; maximum T for development*	°C	40–42
T4	TOD; optimum T for development*	°C	30–33
T5	TODN; optimum night-time T for development*	°C	30–33
P1	MOPP; optimum photoperiod*	h	10.0
P2	PPSE; photoperiod sensitivity*	h^{-1}	0.0
Growth and Stresses			
G1	SPGF; Spikelet growth factor*	# g^{-1}	28–50
G2	WGRMX; maximum individual grain weight	g $grain^{-1}$	0.000025
G3	SLA; initial specific leaf area*	cm^2 g^{-1}	222–565
G4	RGRLMX; max relative leaf area expansion rate*	°C h^{-1}	0.0085
G5	FSHTB; initial fraction of C allocated to shoot**	–	0.5
G6	FLVRB; initial fraction of shoot C allocated to leaves**	–	0.6
G7	FSTTB; initial fraction of shoot C allocated to stems**	–	0.4
G8	FSOTB; initial fraction of shoot C allocated to panicles**	–	0.6
G9	T_{crit}; Critical temperature for spikelet fertility	°C	33–34.2

*Parameter adapted from ORYZA model as discussed in Li et al. (2020, 2021, 2022a, 2024b)
**C allocation fractions change with physiological age. Values supplied are typical start values.

Table 10 Representative genetic coefficients and typical value for the RYESIM (cereal rye) model from the total number indicated in Table 5. The assigned ID names are for convenience and not necessarily in the official model documentation. Abbreviations: T - air temperature.

ID	Name and Description	Units	Typical Value
Phenology			
P1	T_{base}; base T for thermal time calculation	°C	0
P2	T_{opt}; optimum T for thermal time calculation	°C	20
P3	T_{max}; maximum T for thermal time calculation	°C	30
P4	P_{hy}; phyllochron for mainstem leaves	°Cd^{-1}	106
P5	Number of phyllochrons to reach single ridge stage	#	1.7
P6	Number of phyllochrons to reach stem elongation stage	#	3.7

5 Other data inputs and output data

In addition to those shown in the preceding sub-sections, each crop model also uses other parameters that are typically not adjusted as part of the calibration practice unless there is evidence to support changing them. These are frequently referred to as 'conserved' values and include leaf photosynthetic and stomatal conductance parameters, root properties [e.g. W and N uptake rates for young or mature roots, root length density, root growth preference (vertical versus lateral distribution)], carbon allocation coefficients, and similar characteristics (Table 11).

Additional input parameters related to the soil module are common among all crop models, which typically are not adjusted by the user. The geometry of the spatial two-dimensional grid can be specified to represent differences in topography, such as a raised bed or hill (Wang et al., 2020), and it is possible to represent different soil materials and surface management configurations between row and interrow positions. Different boundary conditions with respect to W, solute, heat, and gas movement can be set at the soil-atmosphere layer as well as the bottom soil profile. This can be useful, for example, for simulation of a ponded surface layer as well as bottom soil profile layers to account for impermeable surfaces, water tables, and other types of boundaries.

2DSOIL also includes a carbon and N transformation module based on the SOILN module (Bergström and Jarvis, 2007; Johnsson et al., 1987), as well as recent improvements for C production and transport from roots and microbial respiration (Beegum et al., 2023b). Three organic pools (i.e. fast cycling of plant

Table 11 Representative genetic coefficients and value ranges for SPUDSIM (potato) model from the total number indicated in Table 5. The assigned ID names are for convenience and not necessarily in the official model documentation. Abbreviations: N- Nitrogen; TI - tuber initiation.

ID	Description	Units	Range
Phenology			
C1	Max influence on TI rate when daily air temperature = 15°C	–	1.5–2.5
C2	Max influence on TI rate when temperature amplitude = 25°C	–	1.5–3.0
C3	Max influence on TI rate when photoperiod ≤ 12 h	–	1.5–3.0
C4	Degree of cultivar determinacy	–	0.1–1
Growth and Stress			
G1	Maximum canopy expansion rate	cm^2 $plant^{-1}$ d^{-1}	100–450
G2	Maximum tuber growth rate	g $plant^{-1}$ d^{-1}	5–35
G3	Specific leaf area	cm^2 g^{-1}	227–450
G3	Influence of high N status on TI	–	0.8–1.2
G4	Influence of low N status on TI	–	0.8–1.2

residues, slow cycling of humus, and organic fertilizer) are considered in the module. Inputs describing dehumidification, mineralization, nitrification, and denitrification rates based on soil moisture and T are also required but are typically unchanged unless more information becomes available.

Output data include time-series data regarding plant growth and development, resource utilization values, fluxes from the soil surface, and soil variables (Table 2). Production data are summarized as end-of-season variables (Fig. 5) as well as time-series (hourly or daily) values. Two-dimensional soil

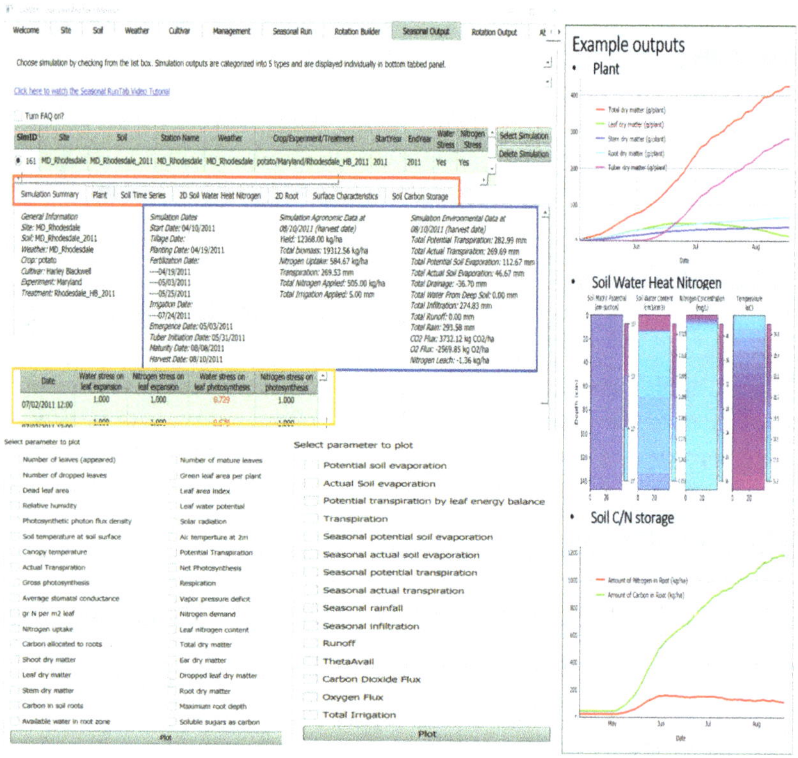

Figure 5 Example model outputs from a single season run from the ACSL GLYCIM crop model. These results are shown from the Crop, Land, And Soil SIMulation (CLASSIM) graphical user interface (Timlin et al., 2023), which facilitates end-user access to our crop models along with visualization of the outputs. TOP: Red box indicates the type of single-season model execution conducted; Blue box indicates a summary of end-of-season variables, including simulation dates such as emergence and flowering, biomass, nitrogen uptake and leaching, as well as water use parameters; Yellow box indicates hourly N and W stresses. Stress associated with water shown to influence leaf photosynthesis on July 2 is shown in this example; BOTTOM: Partial list of time-series plant and soil variables which can be selected by the user; RIGHT: Example of x-y plots for plant biomass over time, soil 2D values in mid-season for W content, N concentration, and T; example of soil carbon and N distribution associated with root mass over time.

profile information regarding W, N, T, and root distributions, as well as C and O_2 fluxes from the soil/atmosphere interface, are also available. Additional information not shown includes soil W content, T, and N per soil layer over time, as well as soil carbon storage, including changes in N and carbon content of different organic pools.

6 Distinctive features of ACSL crop models

6.1 Advantages of an hourly time-step

Accounting for diurnal patterns in atmospheric variables can be important to capture when simulating crop and soil responses, particularly to account for the influence of short-term extreme weather events. For example, exposure to high Tair during flowering in cereals, such as rice, is known to result in pollen sterility and reduce grain yield (Impa et al., 2021; Madan et al., 2012). This response can be difficult to capture accurately for models which utilize daily data, as flowering occurs during a few hours. Many models with daily time steps estimate hourly Tair using sinusoidal functions that incorporate predicted sunrise time and daylength (Bouman et al., 2001; McMaster et al., 2008; van Oort et al., 2015). Such approximations can be accurate but can still introduce errors compared to using directly measured data (Fig. 6a). Similarly, the sinusoidal methods used in this way can be location specific such that extending model predictions

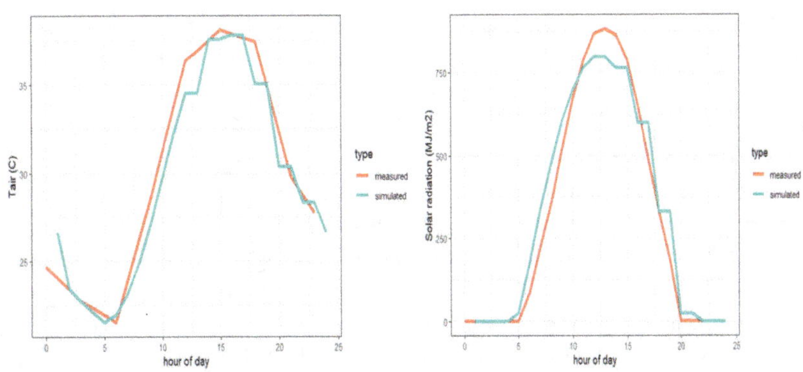

Figure 6 Differences in observed versus predicted hourly input data: (a) for T at a location in Sutherland, Nebraska, USA, on July 4, 2022, when daily values were supplied as input to the SPUDSIM model. Average T for all daytime hours was underestimated by about 0.5°C. During mid-day, however, estimates were underestimated by as much as 2°C; (b) for solar radiation the same location where values between observed and predicted hourly values were over-estimated by 5%. Due to the nonlinear sensitivity of photosynthesis to light, this resulted in a leaf-level photosynthesis over-prediction of about 6% to 12%. Over the course of the growing season, such errors can result in as much as a 10% difference in crop biomass as compared to using directly observed hourly values.

to other production regions further reduces the predictive accuracy (Li et al., 2022a).

Diurnal fluctuations in other weather variables such as vapor pressure deficit, solar radiation (due to varying cloud cover and rainfall), and wind speed can also be important to capture. As with T, methods to estimate diurnal values can be for models that use an hourly time-step. However, variations in measured versus estimated hourly solar radiation profiles can result in differences in simulated crop growth rate even though the daily light integrals are identical (Fig. 6b). Such differences can result in significant model predictive differences due to the nonlinear relationship between light intensity and leaf level photosynthesis (Xin et al., 2016; Gross, 1982). Simulated soil T can be underpredicted by up to 2°C when using estimated hourly T as well (Timlin et al., 2002a). Use of daily input data can result in up to a 10% difference in final crop biomass as compared to using directly measured hourly values (Fig. 6).

6.2 Accounting for nonlinear temperature responses

Air temperature is a primary climatic driver of plant growth and development. The relationship between T and phenology is most often simulated with linear thermal time relationships (de Wit et al., 2019; Holzworth et al., 2014; Hoogenboom et al., 2015). These thermal indices, such as growing degree days, can relate daily T summation to crop development rates, dry matter gain, and other responses (Zhou and Wang, 2018). Such relationships work well over an intermediate range of T. However, crop growth and developmental response to T is nonlinear at upper and lower ranges (Fleisher et al., 2006; Kumudini et al., 2014), and lack of compensating for this response can result in significant under- or over- estimations. Given recent and projected increases in T during the growing season of many crops (IPCC, 2023), it is particularly relevant to incorporate methods that account for this nonlinear response (Fleisher and Timlin, 2006). Several ACSL models have now integrated nonlinear functions (such as a beta function (Yin et al., 2003), Fig. 1) into crop development predictions, including leaf appearance and leaf or canopy area expansion rates (Fleisher et al., 2010b; Kim et al., 2012; Paff et al., 2023). Potential errors using linear thermal time estimates versus nonlinear approaches are illustrated in Fig. 1.

6.3 Leaf-level gas exchange processes

Leaf gas exchange refers to photosynthetic, respiratory, and transpiration processes involving the exchange of C, O_2, and W vapor between the leaf surface and atmosphere. ACSL models directly simulate these processes and fluxes via coupling a leaf-level biochemical and stomatal conductance model within an energy balance approach (Fleisher et al., 2010a; Kim et al., 2012; Kim and Lieth, 2003; Li et al., 2021; Sun et al., 2022; Yang et al., 2009). This is illustrated in Fig. 7. Net photosynthetic rate is described mathematically using biochemically derived relationships (Farquhar et al., 1980; von Caemmerer, 1981) which, in addition to response to intercellular C, include absorbed photosynthetically active radiation, and leaf and air T as direct inputs (Fig. 7a). Intercellular C in turn is influenced by leaf stomatal conductance which is also regulated by ambient C levels, humidity, leaf T (Fig. 7b) and net photosynthesis (Ball et al., 1987). Equations for these processes are nested within an energy balance at the leaf surface, which accounts for wind, leaf shape, and boundary layer conductances to derive leaf T and resulting transpiration rates (Kim and Lieth, 2003). The upshot of this process is a mechanistic determination of leaf photosynthetic and transpiration rates which accounts for direct influence of micro-climatic factors and biochemical parameters.

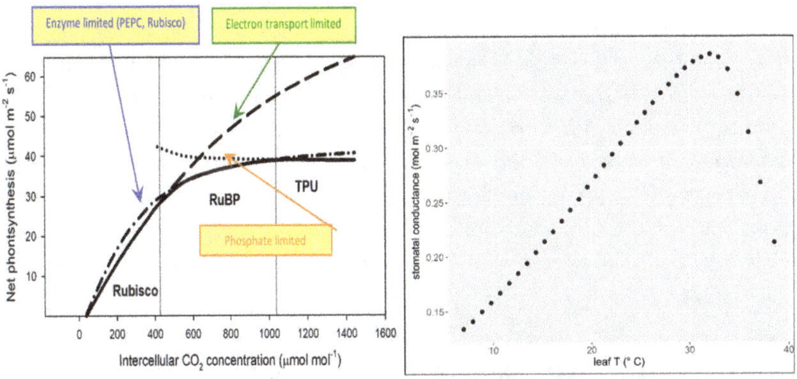

Figure 7 Example of the leaf gas exchange methodology used in ACSL crop models. (Left): Leaf-level photosynthesis is estimated using biochemical models to describe the nonlinear relationship between net photosynthetic rate versus intercellular C supply and photosynthetically active radiation (Farquhar et al., 1980; von Caemmerer, 1981). This representative curve is shown for a C3 crop response. The black line is the minimum of the three rate-limiting steps. Dashed or dotted lines represent each of these limitations, including enzyme (Rubisco for C3 or pep-carboxylase (PEPC) for C4 crops) mediated, and electron transport rate limitations (RuBP), on C assimilation rate, as well as a final restriction based on availability of phosphate (TPU). (Right): The supply of C into the leaf is regulated by stomatal conductance at the leaf surface which is a function of ambient C, humidity, T, and photosynthesis using simple mathematical relationships (Ball et al., 1987). This plot shows simulated conductance versus leaf T levels for a representative rice leaf. These two processes are nested within an energy balance equation (Campbell and Norman, 2000) which accounts for other micro-climatic conditions as well as leaf surface conductances. The end-result is a mechanistic, process level depiction of hourly photosynthesis and transpiration rates.

This approach allows simulation of direct and indirect influences of elevated C on rubisco activity, photosynthesis, conductance and evapotranspiration, and thus provide more mechanistic insight, and potentially higher accuracy (Boote et al., 2013), than traditional radiation use efficiency or other empirical approaches. For example, an advantage of this method is that the observed decrease in stomatal conductance to elevated C reported for most crops (Tausz-Posch et al., 2020) is directly simulated in ACSL models. This in turn interacts with local meteorological variables including T, solar radiation and other inputs to exert mechanistic response directly on tissue T, photosynthesis and transpiration. Methods which do not directly incorporate these responses use empirical functions to account for high T, CO_2 and other factors on carbon assimilation rate.

Despite the focus on more modeling complexity, emergent behavior arising from the more fundamental approach, such as radiation use efficiency, are produced (Fig. 2) and can be used as check on model reasonableness. Linking energy balance approaches to gas exchange may also be crucial for predicting interactive impacts of high Tair with the warming canopy effect due to elevated C on stomatal conductance (Boote et al., 2016). As with other mechanistic components, putting more processes into the model in theory provides more confidence in making model projections outside of the range of data from which it was usually tested. In comparison with models against common datasets, ACSL models and others that utilize similar leaf-level gas exchange methods provided more realistic response to water use and yield under higher C for some crops (e.g. Durand et al., 2018). Replacing empirical equations with this approach has also resulted in improved simulations in ACSL models for cotton (Beegum et al., 2023c), rice (Li et al., 2021), and soybean (Sun et al., 2022).

6.4 Two-dimensional soil profile

ACSL crop models use a two-dimensional (2D) soil profile to account for the vertical and lateral movement of materials (W and solutes), heat, roots, and gases. This approach accounts for non-heterogeneity in row topography and surface and soil characteristics (Timlin and Pachepsky, 1997). The 2D approach also allows for physics-based simulations of transport of these components as well as allowing for temporal changes in lateral and vertical distribution profiles of root growth in response to plant and soil state variables. A majority of crop models utilize 1D approaches to simulating only the vertical movement of the soil component. These typically assume multiple homogenous soil layers. Methods for simulating W transport in 1D models most commonly use what is referred to as a tipping-bucket approach (similar to a finite volume method). In this case, W fills up each successive layer until field capacity (Brisson et al.,

2003) before it starts to fill the next subsequent layer. These methods also use more empirical relationships to describe root depth and density profiles along with soil N distributions as opposed to convective, diffusive process of root growth (Wang et al., 2021b). Water and N uptake simulations in both 1D and 2D approaches tend to follow similar principles (e.g. root W and N absorption, including passive and active processes), but 2DSOIL also accounts for the differences in root age and can perform more sophisticated surface processes, such as surface W flow and row plant shading.

Recent effort is also focusing on incorporation of:

- tillage effects on soil structure, deformation, and porosity,
- flooding effects on soil O_2 content and its influence on root activity, organic matter decomposition, and above-ground plant responses,
- effects of irrigation methods, surface runoff, and drainage on soil structure and function,
- improved biogeochemistry to simulate greenhouse gas emissions

Other additions to the 2DSOIL module include a mechanistic soil C production and transport module to improve evaluation of soil respiration and carbon fluxes (Beegum et al., 2023b), residue decomposition of surface mulch including terminated cover crop biomass (Wang et al., 2021a), a vapor transport module to improve coupled soil W and heat simulations (Wang et al., 2022), a combined model of surface runoff, ponding, infiltration/exfiltration and sub-surface W movement to improve soil water balance with more complex topographies (Wang et al., 2020), and diffusive root growth routines to account for spatial variation in soil properties (Dathe et al., 2014; Wang et al., 2021b). Such changes allow for a better understanding of the influence of cropping-rotation and management considerations (e.g. conventional versus no-tillage production) on soil health metrics.

6.5 Relationship between soil water availability and bulk leaf water potential

Soil and plant W status have direct effects on leaf and canopy gas exchange, organ growth rates, and carbon allocation in all ACSL models (Fleisher et al., 2015). Bulk leaf W potential is calculated dynamically each hour based on plant W demand and soil W availability (Fig. 8). Potential plant W demand is obtained from the transpiration estimate of the hourly gas exchange process. Actual soil W uptake rate is then computed from the soil as follows. First, the bulk leaf W potential and soil/root resistances needed to satisfy this transpiration demand are determined by soil moisture level, root distribution,

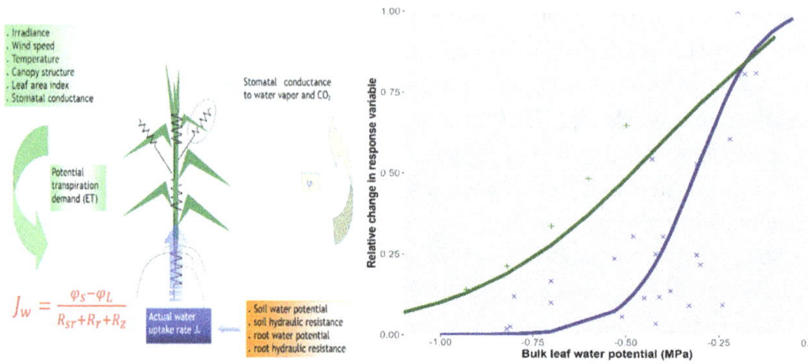

Figure 8 Estimation and implementation of W stress in ACSL crop and soil models. (Left): Plant W demand is derived as a function of the coupled gas exchange approach from Fig. 7. Stomatal conductance is linked with photosynthetic rate and embedded within an energy balance such that influence of solar irradiance, wind, T, and other components are used to provide hourly estimates for transpiration and evaporation. This amount is compared with the quantity of W, J_w, which can be extracted by roots from the soil profile. This is calculated as the ratio between the difference in soil W, Ψ_s, and bulk leaf W potential, Ψ_L, versus soil to root interface, R_{sr}, root, R_r, and xylem, R_z, resistances as shown in the equation. Stomatal conductance is adjusted each time-step based on this value as a hydraulic feedback signal. (Right): the relationship of stomatal conductance (blue) or leaf expansion (green) response versus bulk leaf W potential used for the SPUDSIM model. As leaf W potential decreases, the plant is increasingly W stressed. Stomatal conductance exhibits a sharper decline than leaf expansion, indicating that gas exchange processes will decrease as the initial response to declining soil W availability. A final response to more severe W stress is to allocate more carbon to the roots versus the shoot of the plant. Experimental data summarized from Gandar and Tanner (1976), Liu et al. (2009), and Vos and Oyarzun (1988).

soil W potential, soil to root hydraulic resistance, root hydraulic resistance, and xylem resistance (Acock and Trent, 1991) (Fig. 8). As leaf W potential decreases, more W can be extracted at a given soil W potential level. However, leaf W potential is nonlinearly correlated with stomatal conductance such that declining W potential will decrease conductance (Fig. 8). This decline reduces transpiration demand as well as photosynthesis rate, resulting in increased leaf T. Sustained reduction in soil W availability accompanied by low bulk leaf W potential will thus reduce both W uptake and growth rate. Decreased leaf W potential will also decrease leaf expansion rates like that shown for stomatal conductance. While these expansion rates are primarily driven by thermal time and crop phenotype, reduced leaf W potential decreases leaf cell expansion because of cell turgor pressure changes, which slow growth rate. A third W stress component is associated with shifts in carbon allocation between shoot and root growth. Allocation will be adjusted for preferential growth of roots, depending on growth stage and degree of W stress. For the SPUDSIM model, the sink strength, and thus priority, for carbon is also increased for storage (i.e.

tubers) organs as opposed to vegetative components. Among these factors, photosynthetic response is more sensitive, while changes in leaf expansion rate and re-allocation of carbon to roots occur at higher levels of W stress and over longer periods of time. Because the models are physiologically based, the timing and degree of a W-based stressed event will have different impacts on carbon allocation and yield depending on the growth cycle of the plant (Dathe et al., 2014; Fleisher et al., 2015).

7 Case studies: on-farm applications

In the 1980s and 1990s, GOSSYM and GLYCIM models were utilized in multiple on-farm studies involving farmers and extension agents. The primary goals were to assist in-season grower decisions for risk reducing M practices. We summarize examples of how the modeling tool was used:

- Scientists worked with farmers and extension agents to run the GOSSYM model in the states of Mississippi and South Carolina for use as an aid for in-season M decisions (Whisler et al., 1986). Using the model, growers were informed that their cotton crop would shortly run out of N fertilizer, resulting in declines in photosynthesis and eventually yield. Using the model, the farmers themselves identified that an addition of between 11 kg N ha^{-1} and 22 kg N ha^{-1} would result in a yield increase of 0.14–0.22 kg ha^{-1}, resulting in improved net farm income.
- GOSSYM was integrated with an expert system, COMAX, to provide optimized M recommendations (McKinion et al., 1989). GOSSYM/COMAX provided a single best M solution in one model run based on current and projected meteorological conditions. Most users ran the model once or twice each week to estimate their crop development and identify potential N or W stresses before visual symptoms were apparent. Farmers used the model to optimize their late season N application, identify the best planting dates and seeding rates, and even spray for insect prevalence. Users indicated the value of using this tool on-farm ranged between $100 to as much as $450 per hectare (U.S. 1980s dollars).
- Growers used GLYCIM for a variety of pre-season and in-season decision-making. These included optimizing input, particularly W use, resulting in a yield increase of up to 29% and a 4-fold improvement in W use (Remy, 1994). In locations in Mississippi, the soils form large cracks as they dry. Growers typically did not irrigate until these cracks appeared. Using the GLYCIM model, growers found that they were watering too late in the season and cutting off irrigation too soon (Timlin et al., 2002b). Based on GLYCIM results, farmers irrigated earlier in the season, reducing W stress and W loss via deep drainage (Manning, 1996).

From these experiences, growers reported the model's estimates of the relative differences in growth or yield associated with M strategies were more valuable than the absolute yield predictions (Whisler et al., 1986). Growers also found it useful to cross-check observed growth stages and compare with the model's phenology prediction which helped improve their recognition of the relationship between plant stress development and soil moisture levels (Ahuja et al., 2002).

8 Case studies: climate change impacts

Projected increases in extreme weather events and climate variability may destabilize production systems without sufficient adaptation responses (Li et al., 2024b; Mirzabaev et al., 2023; Walker and Van Loon, 2023). Increases in C may offset some of these negative climate impacts (Rezael et al., 2023), but some studies suggest elevated C can exacerbate heat stress due to its influence on leaf stomatal closure (Cai et al., 2016). More recent climate projections indicate substantial changes in regional and local spatial patterns (IPCC, 2023), and thus, the extent to which crop yield will respond to such change can vary significantly. This spatial variability has important implications for understanding climate impacts and developing adaptation strategies (Leng and Huang, 2017; Martinez-Feria and Basso, 2020), which can either benefit (Rezael et al., 2023) or compensate for these variations. In this section, we summarize a case study using the RICESIM model (Li et al., 2022a) to assess projected climate impacts, environmental constraints to production, and phenotypic adaptation strategies to compensate, or take advantage of, future climate projections.

Rice (*Oryza sativa* L.) is one of the most important global cereals, and the U.S. is the fourth largest rice exporter, contributing to about 10% of world trade (FAO, 2015). Production primarily occurs in the Sacramento Valley in California, the Mississippi Delta, and the Gulf Coast. The thermal environment varies among these three locations, and there is an increasing concern that T is already impacting grain yield and quality. Projected changes in rice yield obtained from the RICESIM model were summarized in Fig. 9A for two inbred rice lines in the 2040s and 2070s throughout the multi-state area. The Wells variety was grown in the Gulf Coast and Mississippi Delta regions, while CH201 was grown in California. Yields generally declined between 12% and 25% for the 2040s depending on which shared socioeconomic pathway (SSP2-4.5 or SSP5-8.5), representing different greenhouse gas emission scenarios (IPCC, 2023), was considered. These declines increased to 22–41% in the 2070s. The spatial variability of the response was readily apparent. For example, the Sacramento Valley showed uniform declines throughout its spatial extent while Gulf coast regions exhibit less of a yield decline compared to northern latitudes in the Mississippi Delta. These yield losses were partially offset when the response to

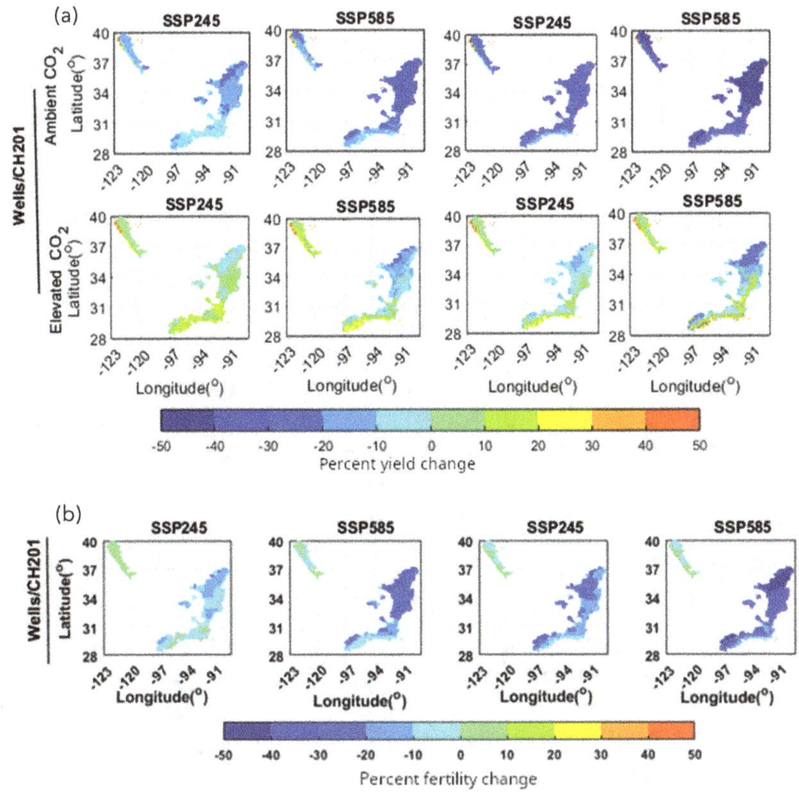

Figure 9 (A) Predicted percent change in rice yield for cultivars Wells and CH201. The Wells cultivar was grown between -97 and -91° longitudes, which includes the U.S. Gulf Coast region (parts of Texas and Louisiana) as well as the lower and upper Mississippi Delta (parts of Louisiana, Mississippi, and Arkansas). The CH201 cultivar was grown between -120 and -123° longitudes, which encompasses the Sacramento Valley in California. Percent yield changes were computed with respect to historical yields (2000 to 2020) in response to two different greenhouse gas emission scenarios, SSP245 and SSP585 in the 2040s and 2070s. (B) Predicted changes in spikelet fertility with respect to historical values using the same cultivars and regions as identified in (A). Spikelet fertility was unaffected by CO_2 concentration. All plots were modified with permission from Li et al. 2024b).

elevated C was incorporated. Yield projections along the Gulf Coast, southern Delta, and most of California were the same or higher than observed yields under current climate conditions, indicating that climate change impacts can be offset and beneficial for certain locations. Model results suggest these responses occurred primarily due to spatial differences associated with weather patterns during rice anthesis and grain-fill. Projected air T increase results in a reduction in grain yield because of increased spikelet sterility (Fig 9B). Despite increased sterility, incorporating the CO_2 enrichment offset some of these impacts, presumably due to gains in photosynthetic rate (Li et al., 2022a; Li et al., 2024b).

Such geographic differences suggest crop breeders may want to breed for different phenotypic traits based on different regions as an adaptation strategy. Three of the genetic coefficients were selected in RICESIM that were shown to have a genetic linkage (Cooper et al., 2009; Li et al., 2016). These traits represented (1) increased yield potential via increased spikelet number per panicle (Masuduzzaman et al., 2016), (2) enhanced leaf photosynthetic response to high T (Adachi et al., 2011; Adachi et al., 2017; Makino, 2021), and (3) increased spikelet heat tolerance during flowering (Huo et al., 2017). Clear spatial patterns emerged, reflecting the importance of spatially focused breeding strategies (Fig. 10). For example, improved thermal tolerance for photosynthesis showed the most benefit for the Gulf coast and eastern edges of the Delta region, while decreased heat sterility mechanisms were most impactful for the Sacramento Valley and northern Delta.

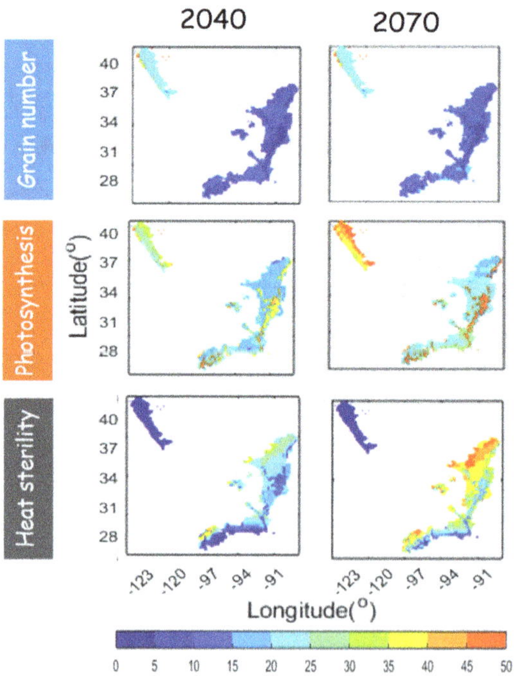

Figure 10 Projected percent changes in historical rice yield in comparison with non-adapted rice (color bar) in response to one of three improvements in rice phenotype include: grain number – higher spikelet number per panicle; photosynthesis – increased heat tolerance to C3 photosynthetic biochemistry; heat sterility – increased heat tolerance of spikelet pollen during anthesis. Figure adapted with permission by (Li et al., 2022b).

9 Case studies: crop quality

Despite its importance, there is little integration of crop quality characteristics into models, likely because factors that influence quality frequently occur during post-harvest operations (Burton, 1982). However, cotton fiber quality is strongly influenced by T, W, and N factors during the growing season (Conaty et al., 2015; Lokhande and Raja Reddy, 2014a, 2014b, 2015; Mauget et al., 2019). Thus, there is potential for including in-season weather and M factors on cotton quality as well as agronomic yield in crop models. Fiber quality factors, including length, strength, micronaire, uniformity, color grade, and trash grade are major indices that affect grower economic returns. Relationships between quality and growing season factors are thus useful to the industry but few efforts have been made in this area (Sequeira et al., 1994). This case study (Beegum et al., 2023a) focused on the integration and evaluation of a new cotton fiber quality simulation module with the GOSSYM crop model.

Developing a fiber quality model requires formulating functional relationships between quality indicators along with abiotic stresses and M conditions. These relationships were developed and parameterized with respect to variations in abiotic stress, N, W and M, and cultivar sensitivity. Potential fiber quality can be estimated as a function of daily T between plant flowering through open boll stages. Actual fiber quality is then derived using modifications for plant W and N status using empirical reduction factors as illustrated in Fig. 11. GOSSYM outputs an average weighted value among all bolls per plant for each quality metric. The integration of these relationships allows for the estimation of multiple growth factors on yield and quality factors, which can assist cotton farmers in terms of their farm management. Farmers can adjust variety, planting date, and timing and amount of irrigation and fertilizer to optimize these responses. In this case study, multiple simulations were conducted in Stoneville, Mississippi, USA, over a 10-year period (2011–2020). Scenarios consisted of (a) six different growing season T (from –2 to +4°C minimum and maximum daily T increases over baseline 2015 daily values), (b) six variations in N fertilization from 60 to 160 kg ha^{-1}, (c) six irrigation regimes from 0% to 100% W demand, and (d) early (–21, –14, or –7 days) or delayed (+7, +14, or +21 days) planting dates. Two C concentrations (420 and 860 ppm) were also considered.

The influence of these abiotic and M factors on quality components was linked to growth and development responses. These include the number and addition rates of fruiting branches as well as the timing of flowering and formation of cotton bolls. These are processes mostly driven by T. Leaf N concentration and W potential are also influenced by T as well as the M conditions. These combinations of factors give rise to differences in fiber characteristics, as shown in Fig. 12. Temperature had the most substantial influence with cotton fiber strength and length declining as heat increased. These factors increased with

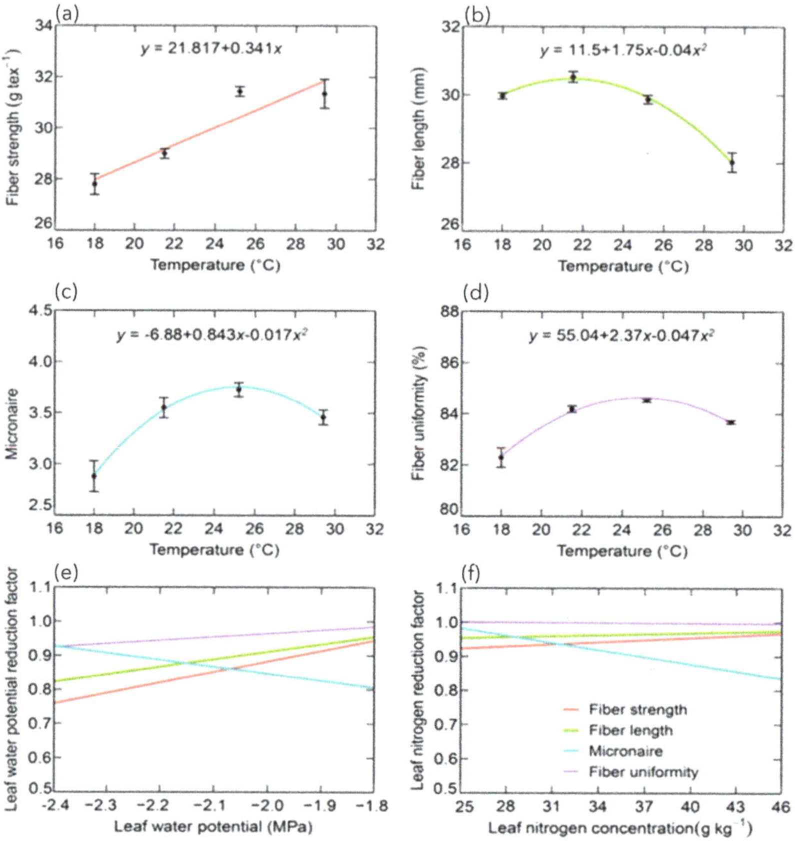

Figure 11 Relationship (a)–(d) between average T from flowering to open ball stage and four cotton fiber quality indices and (e)–(f) reduction of this potential response as a function of leaf W potential and N. The reduction values are used to reduce the level of each index as indicated on the graphs. Adapted, with permission from Beegum et al. (2023a).

higher N and irrigation amounts. Ambient C (420 ppm) had a more positive influence on fiber strength, but other fiber responses were not influenced by either C level. The GOSSYM model can now evaluate trade-offs among fiber quality with resource uses and yield. Further work will expand on the functional relationships indicated here to account for varietal dependent parameters to expand on the capability of the model to help with end-user decision support.

10 Case studies: soil carbon sequestration

Carbon dioxide released from agricultural soils is a major source of carbon in the global carbon cycle (Bahn et al., 2010). The amount of carbon in the soil is

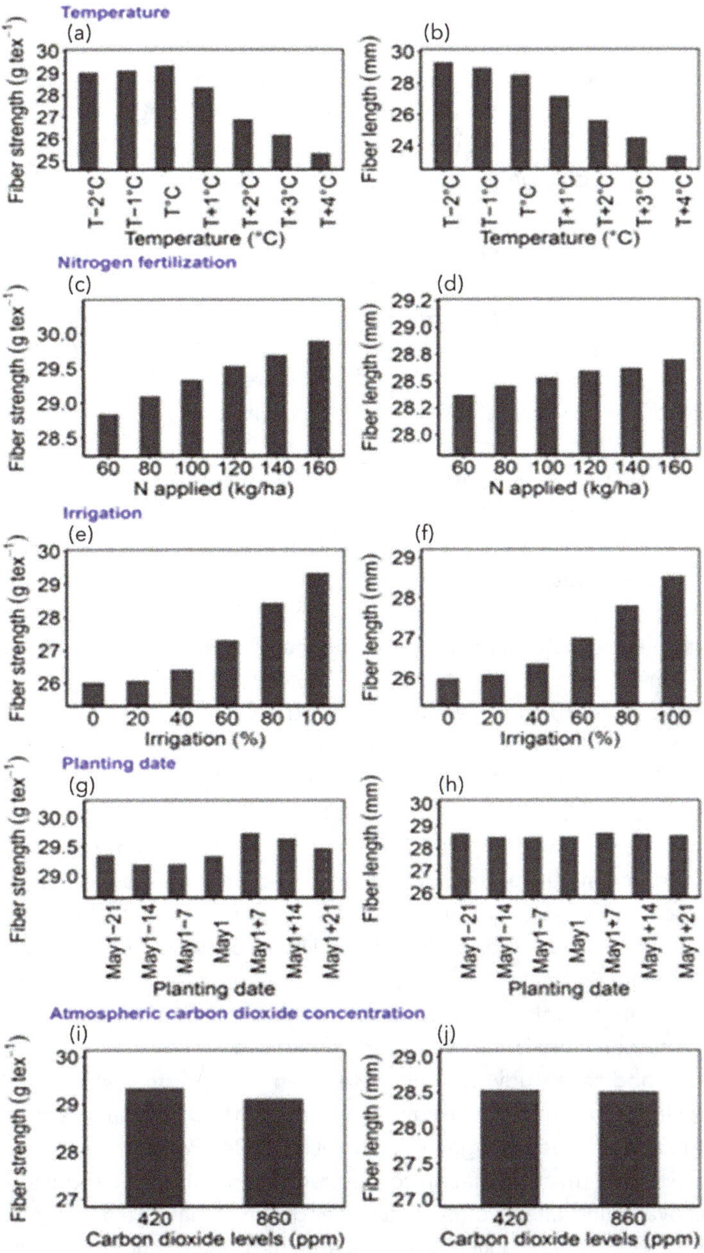

Figure 12 Variation in simulated fiber quality factors using the modified GOSSYM crop model for strength (a, c, e, g, i) and length (b, d, f, h, j) versus different abiotic treatment levels for temperature, N, irrigation, plant dates, and C levels. Adapted with permission from Beegum et al. (2023a).

roughly twice that in the atmosphere (Köchy et al., 2015; Scharlemann et al., 2014), of which soil respiration is a higher source than fossil fuel use (Ballantyne et al., 2015; Raich and Tufekcioglu, 2000). More accurate estimation of soil C storage and C respiration are thus needed for improved agricultural M options (Bastos and Fleischer, 2021). Soil respiration involves multiple processes, including C production and transport and the influence of soil T, W content, roots, and microbial activity (Davidson et al., 2004). Current biogeochemical models such as SOILCO2, DNDC, and DAYCENT (Li et al., 1992; Parton et al., 1998; Šimůnek and Nimmo, 2005; Šimůnek and Suarez, 1993) simulate these processes but utilize empirically defined crop growth modules along with other simplifications which can limit accuracy. The interaction of the plant with the atmosphere and soil is an important component as soil respiration is highly correlated with crop metabolic and photosynthetic rates. A C production and transport module was recently integrated into 2DSOIL (Beegum et al., 2023b), which addresses transient gas transport as influenced by porosity, W content, gas diffusion, and C production in the soil. This latter response includes microbial respiration, carbon and N ratio transformations, and root C release through respiratory processes. This approach takes advantage of root physiological age and distribution throughout the soil profile. The following case studies evaluated 2DSOIL performance as integrated with MAIZSIM and GLYCIM models.

MAIZSIM simulations were compared with surface C flux estimates (Beegum et al., 2023b) from two-year maize crop rotation studies in Ontario, Canada (Drury et al., 2012; He et al., 2018; Li et al., 2017b) and in California, U.S. (Lee et al., 2006; Rolston et al., 2006). Soil type, irrigation, weather, and soil carbon additions (as litter) varied among locations. A comparison between measured and simulated responses (Fig. 13) confirmed the model accurately reproduced changes in soil surface C flux, W content, and T. These responses were partly driven by the model's capability to simulate litter, microbial, and humus respiration rates. With these improvements, ACSL crop models can simulate more complex studies evaluating the interface of crop M with respect to yield, resource use optimization, and C storage.

A second case study was conducted using long-term soybean and maize data in Illinois, U.S. to evaluate soil organic carbon (SOC) dynamics in response to tillage and residue management (Sun et al., 2024). Part of this initiative was to evaluate long-term crop rotation consequences on soil carbon sequestration as well as evaluating effects of potential changes in climate. This information can inform conservation practice M in a C-enriched environment with respect to increasing soil C storage by increasing surface residue input (Prior et al., 2005) or practicing conservation tillage. In this case study, GLYCIM and MAIZSIM models were used at multiple locations in Illinois, USA during the 1900 to 2020 period. Simulations from GLYCIM and MAIZSIM models were evaluated

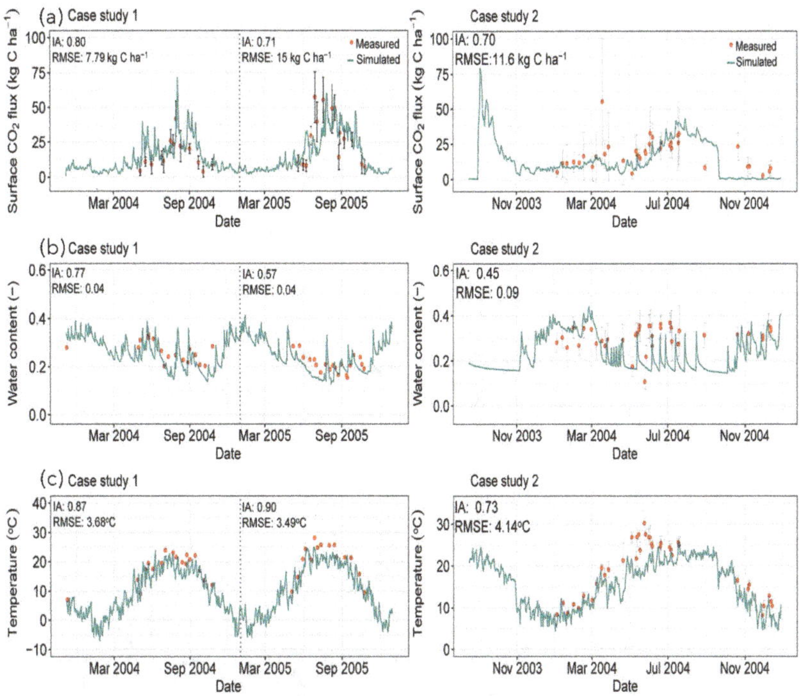

Figure 13 Measured and simulated (a) soil surface C flux, (b) soil W content, and (c) soil T in two years of study for maize crop grown in Woodslee, Ontario Canada (case study 1, left) (Drury et al., 2012; He et al., 2018; Li et al., 2017b) and California (case study 2, right) (Lee et al., 2006; Rolston et al., 2006). Abbreviations: IA = index of agreement and RMSE – root mean square error. Symbols – red dot – measured data; blue line – simulated data. Figures modified with permission from Beegum et al. (2023b).

against measured data from (Black et al., 2017) (Fig. 14) and showed both models accurately replicated observed soil C flux, T, and W content in response to multi-year air T and C treatments. Models were then used to evaluate the impacts of long-term crop rotations on SOC dynamics under current (1920–2020) and future projected climate scenarios from 2020 to 2100 (Fig. 15). Prior to 1960s, no clear trend in SOC or soil N were observed from the available experimental data but model simulations showed a decline in both values from 1900 to 1960. After 1960s, simulated data showed a slight increasing annual trend. Pre-growing season surface residues and fertilizer inputs were roughly 4x less on average during the 1900–1960 time period compared to post-1980s. This suggested the models may over-estimate the influence of crop residue input on SOC.

Projected changes in SOC were also simulated for different climate scenarios (Fig. 16). Rising T and elevated C may profoundly affect SOC storage

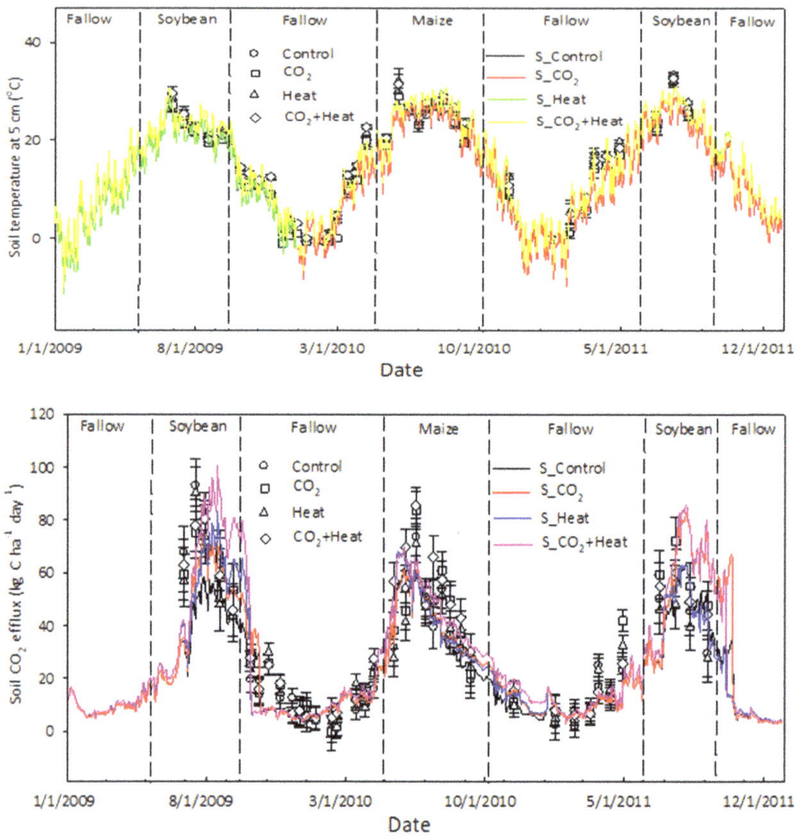

Figure 14 Predicted and simulated soil T and soil CO_2 surface flux during a three-year period at free-air CO_2 enrichment (FACE) experimental plots in Illinois, USA (Black et al., 2017). Symbols represent measurements from different FACE treatments including a control plot, control with elevated C, control +2°C, and control +CO_2 and T as indicated. Figures modified with permission from Sun et al. (2024).

and M approaches. Daily maximum air T increased by about 0.03–0.07 °C year^{-1} (compared with 0.005 °C year^{-1} from observed 1900–2020), and summer rainfall amount slightly decreased (not shown). Without the C response, SOC values plateaued by 2040 and subsequently declined for both greenhouse gas emission scenarios. This response was largely due to increased heterotrophic respiration rate, similar to Black et al. (2017), Huang et al. (2020), and Lu et al. (2013). However, SOC changes have been observed to vary considerably and depend on the size of initial soil carbon stocks (Crowther et al., 2016). Adding the C response to the crop models had a positive effect on SOC stock, with a 5–6% increase from 2020 levels or an average 0.05 (versus –0.09) Mg C ha^{-1}

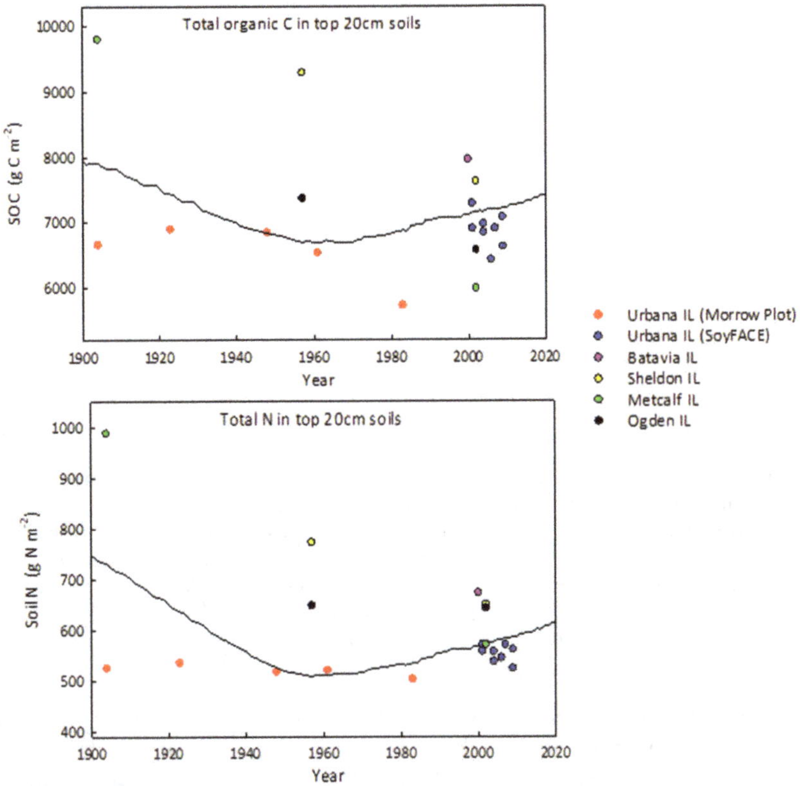

Figure 15 Measured and predicted (a) organic carbon (SOC) and (b) total N levels in soils under conversion from tall grass prairie to agricultural practice at different field plots in Illinois, USA. Lines show the average of GLYCIM and MAIZSIM modeled soil C or total N. Points are the observed values from each of six experimental sites (Black et al., 2017).

year^{-1} increase similar to Huang et al. (2020). This increase was likely due to higher above and below-ground crop and root biomass formation.

11 Conclusion and future trends

ACSL modeling research trends focus on improving and increasing the accuracy, scope, and utility of crop and soil models. One major theme is the need to continuously update models and tools to reflect the current state of empirical knowledge and incorporate the latest theories. The laboratory focuses on providing a suite of comprehensive crop and soil modeling tools for assessing the consequences of M options on the resiliency of the agricultural systems. A significant effort of ACSL includes the execution of in-house experimental research underlying many of the empirical databases needed for

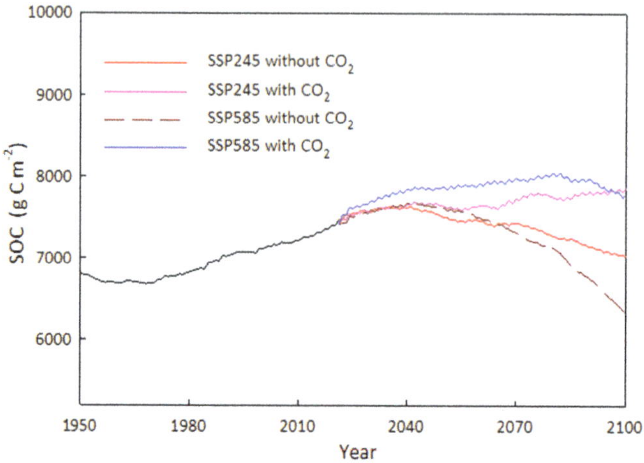

Figure 16 Mean annual values for total soil organic matter (g C $^{--2}$) at multiple field plots in Illinois as predicted by GLYCIM and MAIZSIM models in the top 20 cm under two IPCC6 shared socioeconomic pathways (SSPs) (SSP245 and SSP585) which reflect different future greenhouse gas emission scenarios. Historical weather data were used from 1950 through 2020 and an ensemble of five general circulation models were used for annual predictions from 2021 through 2099. Field M conditions were identical for each year of study. Figure modified with permission from Sun et al. (2024).

testing and improving many modeling components. Controlled environment facilities [including growth chambers, greenhouses, and soil-plant-atmosphere research (SPAR) chambers (Baker, 2004)] along with field-based enclosures or systems [such as open top chambers (e.g. Schapendonk et al., 2000) and free-air-CO_2 enrichment facilities (FACE) (e.g. Bunce, 2012)] are frequently used to evaluate the response of crops to elevated C, Tair, W and N availabilities and other variables. However, much of this data focused historically on single-factor experiments. As a result, many crop models have been tested with and/or incorporate relationships from studies where single climate factors are held constant over the life cycle (Cai et al., 2016; Fleisher et al., 2022; Kimball, 2016). ACSL future research plans thus include continued coupling of model development, improvement, and testing with the generation of new experimental data to evaluate multi-climate and nutritional (e.g. N and W availability) responses against selected phenotypes of economically important crops at different developmental stages.

Stress conditions during crucial growth stages, including anthesis and grainfill, also remain poorly characterized and represented in crop models (Fleisher et al., 2017; Silva and Giller, 2021). As such, impacts of episodic drought and high heat exposures are frequently underestimated (Heinicke et al., 2022). Such deficiencies limit the ability of modeling tools to provide accurate predictions of crop yield and resource use efficiency under multi-factor abiotic stresses as well as how stress is influenced by elevated C (Boote et al., 2017). Short-term heat stress effects in particular need to be improved for all models (Eitzinger et al., 2013; Van Der Velde et al., 2012), which indicates the need to utilize hourly time-steps to improve diurnal responses.

The frequency and impact of extreme weather events (EWE) are also of crucial importance to agricultural systems. EWEs are defined here as short-term weather events occurring over a few hours or even multiple days which exceed (a) optimum T thresholds for crop development or yield during a critical developmental stage, (b) conditions of optimum soil W availability due to high or low rainfall, and (c) conditions of soil O_2 depletion associated with flooding. EWE can also include hail or wind damage, but such factors are difficult to incorporate in crop models. EWEs are typically experienced at a local spatial scale, are increasing in frequency, and significantly impact farm yields and operations (Burke and Emerick, 2016; Hatfield et al., 2011). Potential long-term seasonal stresses associated with projected climate changes, such as prolonged drought, flooding, or heat waves, are also expected to continue to increase in local regions (Ray et al., 2019). Crops have different sensitivities to the timing and magnitude of an EWE (Anapalli et al., 2019; Fleisher et al., 2014; Korres et al., 2016), and efforts to test and improve models generally need to be done on an individual plant basis as a consequence. Model end-users frequently assume SPA relationships embedded in crop and soil models are state-of-the-art, but extensive testing has not been conducted (Silva and Giller, 2021). The interaction of climate factors and events, such as the assessment of high T during extended drought under varying different C levels, particularly at tasseling, anthesis, and grain fill developmental stages, are topics that need to be evaluated and improved in crop and soil models. Improvements associated with simulating photosynthetic acclimation under elevated C (Elliott et al., 2014) and evaluating impacts of higher night-time T (Xu et al., 2021) on grainfill periods will be addressed. ACSL will also be incorporating responses of short-term flooding on soils with respect to O_2 deprivation on root activity and associated influence with high T on soil N dynamics. The effects of high T, drought, and flooding on maize, rice, and soybean under elevated C will also be addressed in our modeling program.

Improved representation of soil biogeochemical processes is needed to better understand farm M assessments. Improving model capabilities to account for the influence of crop genetics and M on root and soil processes, including N and W movement and availability, N transformations, greenhouse gas emissions, and vapor and gas (O_2, C, H_2O) transport (Folberth et al., 2019) can better account for the consequences of different M strategies on soil health responses, surface mulch decomposition, and carbon sequestration (Dathe et al., 2014; Heinicke et al., 2022). ACSL research will include nitrous oxide, carbon, and O_2 dynamics and gas transport to improve soil model response to flooding and organic matter dynamics. Related components currently under way include improved representation of tillage simulation processes, including adjustment of soil bulk density and porosity, and the integration of soluble phosphorus dynamics and geochemical reactions to improve the capacity of the modeling system to handle additional crop nutrients.

A significant focus also involves model application work. ACSL has developed new collaborations with farmers and extension agents to utilize versions of our models to provide in-season decision support, particularly in terms of evaluating the potential impact of EWE and the likelihood of crop recovery. Other applications focus on the continued identification of crop phenotypes (Hsiao et al., 2024) as a genetic-based, regional adaptation strategy for breeders to account for warmer, potentially W-limited production. Our efforts also include the assessment of crop rotation strategies that will confer yield resilience and sustainable production efforts.

12 Where to look for further information

Due to the breadth of this subject matter, there are multiple resources available to provide further background. While there is not one single, seminal source of reference, there are various peer-review papers, texts, scientific conferences, and websites available of which we indicate several in this section. For recent peer-review studies of crop modeling and efforts to link with plant genetics, see the following:

- Muller, B., Martre, P. (2019) Plant and crop simulation models: powerful tools to link physiology, genetics, and phenomics. J Exp Bot 70:2339-2344. https://doi.org/10.1093/jxb/erz175
- Hammer, G., Messina, C., Wu A., Cooper M. (2019) Biological reality and parsimony in crop models–why we need both in crop improvement! in silico Plants 1. https://doi.org/10.1093/insilicoplants/diz010
- Stockle, C.O., Kemanian, A.R. (2020) Can Crop Models Identify Critical Gaps in Genetics, Environment, and Management Interactions? Front Plant Sci 11:737. https://doi.org/10.3389/fpls.2020.00737
- Pasley, H., Brown, H., Holzworth, D., Whish, J., Bell, L., Huth, N. (2022) How to build a crop model. A review. Agronomy for Sustainable Development 43. https://doi.org/10.1007/s13593-022-00854-9 --García-Tejera, O., López-Bernal, Á., Villalobos, F.J. (2024). Crop Models. In: Villalobos,F.J., Fereres, E. (eds) Principles of Agronomy for Sustainable Agriculture. Springer, Cham. https://doi.org/10.1007/978-3-031-69150-8_40:

Several peer-review highy relevant regarding such principles application:

- Peng, B., Guan, K., Tang, J., Ainsworth, E.A., Asseng, S., Bernacchi, C.J., Cooper, M., Delucia, E.H., Elliott, J.W., Ewert, F., Grant, R.F., Gustafson, D.I., Hammer, G.L., Jin, Z., Jones, J.W., Kimm, H., Lawrence, D.M., Li, Y., Lombardozzi, D.L., Marshall-Colon, A., Messina, C.D., Ort, D.R., Schnable, J.C., Vallejos, C.E., Wu, A., Yin, X., Zhou, W. (2020) Towards a multiscale crop modelling framework for climate change adaptation assessment. Nature Plants 6:338-348. https://doi.org/10.1038/s41477-020-0625-3
- Timlin, D., Paff, K., Han, E. (2024) The role of crop simulation modeling in assessing potential climate change impacts. Agrosystems, Geosciences & Environment 7. https://doi.org/10.1002/agg2.20453 No single textbook covers the breadth of this topic. Below is a recent survey of current model components followed by two classical texts that are still available online:-- Ahuja L.R., Kersebaum K.C., Wendroth O. (2022) Modeling Processes and Their Interactions in Cropping Systems: Challenges for the 21st Century American Society of Agronomy,Inc. / Crop Science Society of America, Inc. / Soil Science Society of America, Inc. and John Wiley & Sons, Inc., Hoboken, NJ USA. https://doi.org/10.1002/9780891183860
- The following two text books are somewhat older but represent some of the fundamental principles for quantifying soil, plant, and atmospheric components in crop and soil models that are frequently used as building blocks for all of our modern modeling tools:
- Campbell, G.S., Norman, J.M. (2000) An Introduction to Environmental Biophysics, 2nd edition Springer-Verlag, New York, Inc., New York, USA. (p. 283).
- Thornley, J.H., Johnson, I.R. (1990) Plant and crop modelling. (p. 669). Oxford: Clarendon. (Available through Amazon.com)
- Several professional scientific conferences also occur on an annual basis that include poster and oral presentations by academic and industry experts on model development and application. These conferences are frequently attended by many of the authors and are include:
- CANVAS, formerly the American Society of Agronomy (ASA), Crop Science Society of America (CSSA), Soil Science Society of America (SSSA) International Annual Meeting. Locations vary each year. https://www.acsmeetings.org/
- American Geophysical Union (AGU) Annual Meeting, locations very each year. https://www.agu.org/annual-meeting

While there are not dedicated websites to crop modeling per se, some international groups have useful websites that discuss ongoing activities.

- The Agricultural Model Intercomparison and Improvement Project is a highly interactive, international group of scientists with a focus on improving agricultural models and scientific and technological capabilities. The reader is referred to their website for more information which includes relevant modeling articles and current research projects: https://agmip.org

- All of our USDA-ARS ACSL crop and soil models, including sourcecode and the CLASSIM interface, are freely available at the following link: https://www.ars.usda.gov/research/software/?modeCode=80-42-05-20

13 References

Acock, B. and Trent, A. (1991) *The Soybean Crop Simulator GLYCIM: Documentation for the Modular Version 91*. University of Idaho, Department of Plant, Soil and Entomological Sciences, Moscow, Idaho.

Adachi, S., et al. (2011) Identification and characterization of genomic regions on chromosomes 4 and 8 that control the rate of photosynthesis in rice leaves. *J Exp Bot* 62:1927–1938. https://doi.org/10.1093/jxb/erq387.

Adachi, S., et al. (2017) Fine mapping of carbon assimilation rate 8, a quantitative trait locus for flag leaf nitrogen content, stomatal conductance and photosynthesis in rice. *Front Plant Sci* 8:60. https://doi.org/10.3389/fpls.2017.00060.

Ahuja, L.R., et al. (2002) Agricultural system models in field research and technology transfer. https://doi.org/10.1201/9781420032413.

Anapalli, S.S., et al. (2019) Quantifying water and CO2 fluxes and water use efficiencies across irrigated C3 and C4 crops in a humid climate. *Sci Total Environ* 663:338–350. https://doi.org/10.1016/j.scitotenv.2018.12.471.

Arata, L., et al. (2020) A worldwide analysis of trend in crop yields and yield variability: evidence from FAO data. *Econ Model* 90:190–208. https://doi.org/10.1016/j.econmod.2020.05.006.

Bahn, M., et al. (2010) Soil respiration at mean annual temperature predicts annual total across vegetation types and biomes. *Biogeosciences* 7:2147–2157. https://doi.org/10.5194/bg-7-2147-2010.

Baker, J. (2004) A method for estimating carbon dioxide leakage rates in controlled-environment chambers using nitrous oxide. *Environ Exp Bot* 51:103–110. https://doi.org/10.1016/j.envexpbot.2003.08.001.

Ball, J.T., et al. (1987) A model predicting stomatal conductance and its contribution to the control of photosynthesis under different environmental conditions, in: I. Biggins (Ed.), *Progress in Photosynthesis Research*, Martinus-Nijhoff Publishers, Dordrecht, The Netherlands. pp. 221–224.

Ballantyne, A.P., et al. (2015) Audit of the global carbon budget: estimate errors and their impact on uptake uncertainty. *Biogeosciences* 12:2565–2584. https://doi.org/10.5194/bg-12-2565-2015.

Bastos, A. and Fleischer, K. (2021) Fungi are key to CO2 response of soil. *Nature* 591:532–534. https://doi.org/10.1038/d41586-021-00737-1.

Beegum, S., et al. (2023a) Development of a cotton fiber quality simulation module and its incorporation into cotton crop growth and development model: GOSSYM. *Comput Electron Agric* 212. https://doi.org/10.1016/j.compag.2023.108080.

Beegum, S., et al. (2023b) Incorporation of carbon dioxide production and transport module into a soil-plant-atmosphere continuum model. *Geoderma* 437. https://doi.org/10.1016/j.geoderma.2023.116586.

Beegum, S., et al. (2023c) Improving the cotton simulation model, GOSSYM, for soil, photosynthesis, and transpiration processes. *Sci Rep* 13:7314. https://doi.org/10.1038/s41598-023-34378-3.

Beegum, S., et al. (2024) Algorithm for estimating cultivar-specific parameters in crop models for newer crop cultivars. *Eur J Agron* 160. https://doi.org/10.1016/j.eja.2024.127308.

Benitez-Alfonso, Y., et al. (2023) Enhancing climate change resilience in agricultural crops. *Curr Biol* 33:R1246–R1261. https://doi.org/10.1016/j.cub.2023.10.028.

Bergström, L. and Jarvis, N.J. (2007) Prediction of nitrate leaching losses from arable land under different fertilization intensities using the SOIL-SOILN models. *Soil Use Manage* 7:79–84. https://doi.org/10.1111/j.1475-2743.1991.tb00854.x.

Black, C.K., et al. (2017) Elevated CO(2) and temperature increase soil C losses from a soybean-maize ecosystem. *Glob Chang Biol* 23:435–445. https://doi.org/10.1111/gcb.13378.

Boone, M.Y.L., et al. (1993) Calibration of GOSSYM: theory and practice. *Comput Electron Agric* 9:193–203. https://doi.org/10.1016/0168-1699(93)90038-3.

Boote, K.J., et al. (2013) Putting mechanisms into crop production models. *Plant Cell Environ* 36:1658–1672. https://doi.org/10.1111/pce.12119.

Boote, K.J., et al. (2016) Testing approaches and components in physiologically based crop models for sensitivity to climatic factors, in: J. Hatfield and D. Fleisher (Eds.), *Improving Modeling Tools to Assess Climate Change Effects on Crop Response*, ASA,CSSA,SSSA, 5585 Guilford Rd., Madison, WI, USA.

Boote, K.J., et al. (2017) Modeling sensitivity of grain yield to elevated temperature in the DSSAT crop models for peanut, soybean, dry bean, chickpea, sorghum, and millet. *Eur J Agron*. https://doi.org/10.1016/j.eja.2017.09.002.

Bouman, B.A.M., et al. (2001) *ORZYA2000: Modeling Lowland Rice*. International Rice Research Institute and Wageningen University and Research Centre, Los Banos, Philippines.

Brisson, N., et al. (2003) An overview of the crop model STICS. *Eur J Agron* 18:309–332.

Bunce, J.A. (2012) Responses of cotton and wheat photosynthesis and growth to cyclic variation in carbon dioxide concentration. *Photosynthetica* 50:395–400. https://doi.org/10.1007/s11099-012-0041-7.

Burke, M. and Emerick, K. (2016) Adaptation to climate change: evidence from US agriculture. *AEJ Econo Policy* 8:106–140. https://doi.org/10.1257/pol.20130025.

Burton, W.G. (1982) *Post-Harvest Physiology of Food Crops*. Longman Group, LTD., Harlow, UK.

Cai, C., et al. (2016) Responses of wheat and rice to factorial combinations of ambient and elevated CO2 and temperature in FACE experiments. *Glob Chang Biol* 22:856–874. https://doi.org/10.1111/gcb.13065.

Campbell, G.S. and Norman, J.M. (2000) *An Introduction to Environmental Biophysics*, 2nd edition. Springer-Verlag, New York, NY, USA.

Cassman, K.G., et al. (2010) Crop yield potential, yield trends, and global food security in a changing climate, in: C. Rozensweig and D. Hillel (Eds.), *Handbook of Climate Change and Agroecosystems*, Imperial College Press, London. pp. 37–51.

Conaty, W.C., et al. (2015) The relationship between cotton canopy temperature and yield, fibre quality and water-use efficiency. *Field Crops Res* 183:329–341. https://doi.org/10.1016/j.fcr.2015.08.010.

Cooper, M., et al. (2009) Modeling QTL for complex traits: detection and context for plant breeding. *Curr Opin Plant Biol* 12:231–40. https://doi.org/10.1016/j.pbi.2009.01.006.

Crowther, T.W., et al. (2016) Quantifying global soil carbon losses in response to warming. *Nature* 540:104–108. https://doi.org/10.1038/nature20150.

Dathe, A., et al. (2014) Modeling potato root growth and water uptake under water stress conditions. *Agric For Meteorol* 194:37–49. https://doi.org/10.1016/j.agrformet.2014.03.011.

Davidson, E.A., et al. (2004) Effects of an experimental drought on soil emissions of carbon dioxide, methane, nitrous oxide, and nitric oxide in a moist tropical forest. *Glob Change Biol* 10:718–730. https://doi.org/10.1111/j.1365-2486.2004.00762.x.

de Pury, D.G.G. and Farquhar, G.D. (1997) Simple scaling of photosynthesis from leaves to canopies without the errors of big-leaf models. *Plant Cell Environ* 20:537–557.

de Wit, A., et al. (2019) 25 years of the WOFOST cropping systems model. *Agric Syst* 168:154–167. https://doi.org/10.1016/j.agsy.2018.06.018.

Di Paola, A., et al. (2016) An overview of available crop growth and yield models for studies and assessments in agriculture. *J Sci Food Agric* 96:709–714. https://doi.org/10.1002/jsfa.7359.

Doltra, J., et al. (2019) Simulating soil fertility management effects on crop yield and soil nitrogen dynamics in field trials under organic farming in Europe. *Field Crops Research* 1–11. https://doi.org/10.1016/j.fcr.2018.12.008.

Drury, C.F., et al. (2012) Nitrogen source, application time, and tillage effects on soil nitrous oxide emissions and corn grain yields. *Soil Sci Soc Am J* 76:1268–1279. https://doi.org/10.2136/sssaj2011.0249.

Durand, J.-L., et al. (2018) How accurately do maize crop models simulate the interactions of atmospheric CO2 concentration levels with limited water supply on water use and yield? *Eur J Agron* 100:67–75. https://doi.org/10.1016/j.eja.2017.01.002.

Eitzinger, J., et al. (2013) Sensitivities of crop models to extreme weather conditions during flowering period demonstrated for maize and winter wheat in Austria. *J Agric Sci* 151:813–835. https://doi.org/10.1017/s0021859612000779.

Elliott, J., et al. (2014) Constraints and potentials of future irrigation water availability on agricultural production under climate change. *Proc Natl Acad Sci* 111:3239–3244. https://doi.org/10.1073/pnas.1222474110.

El-Sharkawy, M.A. (2011) Overview: Early history of crop growth and photosynthesis modeling. *Biosystems* 103:205–211. https://doi.org/10.1016/j.biosystems.2010.08.004.

FAO. (2015) FAOSTAT. Production of commodity (World).

Farooq, M.S., et al. (2022) Uncovering the research gaps to alleviate the negative impacts of climate change on food security: A review. *Front Plant Sci* 13:927535. https://doi.org/10.3389/fpls.2022.927535.

Farquhar, G.D., et al. (1980) A biochemical model of photosynthetic co2 assimilation in leaves of c3 species. *Planta* 149:78–90.

Fleisher, D.H. (2009) Crop models - in open field, in: L. Rodriguez, et al. (Eds.), *Systems Analysis and Modeling in Food and Agriculture*, UNESCO, Eolss Publishers, Oxford, UK. p. 486.

Published by Burleigh Dodds Science Publishing Limited, 2025.

Fleisher, D.H. and Timlin, D. (2006) Modeling expansion of individual leaves in the potato canopy. *Agric For Meteorol* 139:84–93. https://doi.org/10.1016/j.agrformet.2006.06.002.

Fleisher, D.H., et al. (2006) Approaches to modeling potato leaf appearance rate. *Agron J* 98:522. https://doi.org/10.2134/agronj2005.0136.

Fleisher, D.H., et al. (2010a) Effects of CO2 and Temperature on crops: Lessons from SPAR growth chambers, in: D. Hillel and C. Rosenzweiig (Eds.), *Handbook of Climate Change and Agroecosystems: Impacts, Adaptation, and Mitigation*. World Scientific Publishing Co., Hackensack, NJ. pp. 55–86.

Fleisher, D.H., et al. (2010b) Simulation of potato gas exchange rates using SPUDSIM. *Agric For Meteorol* 150:432–442. https://doi.org/10.1016/j.agrformet.2010.01.005.

Fleisher, D.H., et al. (2014) Potato gas exchange response to drought cycles under elevated carbon dioxide. *Agron J* 106:2024. https://doi.org/10.2134/agronj14.0220.

Fleisher, D.H., et al. (2015) Improving potato drought simulations: Assessing water stress factors using a coupled model. *Agric For Meteorol* 200:144–155. https://doi.org/10.1016/j.agrformet.2014.09.018.

Fleisher, D.H., et al. (2017) A potato model intercomparison across varying climates and productivity levels. *Glob Chang Biol* 23:1258–1281. https://doi.org/10.1111/gcb.13411.

Fleisher, D.H., et al. (2020) Cultivar coefficient stability and effects on yield projections in the SPUDSIM model. *Agron J* 112:828–843. https://doi.org/10.1002/agj2.20070.

Fleisher, D.H., et al. (2021) Yield response of an ensemble of potato crop models to elevated CO2 in continental Europe. *Eur J Agron* 126. https://doi.org/10.1016/j.eja.2021.126265.

Fleisher, D.H., et al. (2022) Response of a U.S. rice hybrid variety to high heat at two CO2 concentrations during anthesis and grainfill. *Agric For Meteorol* 323. https://doi.org/10.1016/j.agrformet.2022.109058.

Folberth, C., et al. (2019) Parameterization-induced uncertainties and impacts of crop management harmonization in a global gridded crop model ensemble. *PLoS One* 14:e0221862. https://doi.org/10.1371/journal.pone.0221862.

Folberth, C., et al. (2020) The global cropland-sparing potential of high-yield farming. *Nat Sustain* 3:281–289. https://doi.org/10.1038/s41893-020-0505-x.

Foley, J.A., et al. (2011) Solutions for a cultivated planet. *Nature* 478:337–342. https://doi.org/10.1038/nature10452.

Gandar, P.W. and Tanner, C.B. (1976) Potato leaf and tuber water potential measurements with a pressure chamber. *Am Potato J* 53:1–13.

Gross, L.J. (1982) Photosynthetic dynamics in varying light environments: A model and its application to whole leaf carbon gain. *Ecology* 63:84–93.

Haskett, J.D., et al. (1995) Estimation of soybean yields at county and state levels using GLYCIM: A case study for Iowa. *Agron J* 87:926–931. https://doi.org/10.2134/agronj1995.00021962008700050025x.

Hatfield, J.L., et al. (2011) Climate impacts on agriculture: Implications for crop production. *Agron J* 103:351–370. https://doi.org/10.2134/agronj2010.0303.

He, W., et al. (2018) Estimating the impacts of climate change on crop yields and N2O emissions for conventional and no-tillage in Southwestern Ontario, Canada. *Agric Syst* 159:187–198. https://doi.org/10.1016/j.agsy.2017.01.025.

Heidmann, T., et al. (2008) Calibration procedure for a potato crop growth model using information from across Europe. *Ecol Model* 211:209–223. https://doi.org/10.1016/j.ecolmodel.2007.09.008.

Heinicke, S., et al. (2022) Global gridded crop models underestimate yield responses to droughts and heatwaves. *Environ Res Lett* 17. https://doi.org/10.1088/1748-9326/ac592e.

Holzworth, D.P., et al. (2014) APSIM – Evolution towards a new generation of agricultural systems simulation. *Environ Model Softw* 62:327–350. https://doi.org/10.1016/j.envsoft.2014.07.009.

Hoogenboom, G., et al. (2015) *Decision Support System for Agrotechnology Transfer (DSSAT) Version 4.6.* DSSAT Foundation, Prosser, Washington. www.dssat.net

Hsiao, J., et al. (2024) Model-aided climate adaptation for future maize in the U.S. *Environ Res Food Syst*. https://doi.org/10.1088/2976-601X/ad3085.

Huang, Y., et al. (2020) Assessing synergistic effects of no-tillage and cover crops on soil carbon dynamics in a long-term maize cropping system under climate change. *Agric For Meteorol* 291. https://doi.org/10.1016/j.agrformet.2020.108090.

Huo, X., et al. (2017) NOG1 increases grain production in rice. *Nat Commun* 8:1497. https://doi.org/10.1038/s41467-017-01501-8.

Impa, S.M., et al. (2021) High night temperature effects on wheat and rice: Current status and way forward. *Plant Cell Environ* 44:2049–2065. https://doi.org/10.1111/pce.14028.

IPCC. (2023) Climate Change 2023: Synthesis Report. Contribution of Working Groups I, II, and III to the Sixth Assessment Report of the Intergovernmental Panel on Climate Change, in: H. L. a. J. R. e. [Core Writing Team (Ed.), IPCC, Geneva, Switzerland. p. 184.

Johnsson, H., et al. (1987) Simulated nitrogen dynamics and losses in a layered agricultural soil. Agriculture, Ecosystems & Environment 18:333–356. https://doi.org/10.1016/0167-8809(87)90099-5.

Jones, J.W., et al. (2017) Brief history of agricultural systems modeling. *Agric Syst* 155:240–254. https://doi.org/10.1016/j.agsy.2016.05.014.

Kim, S.H. and Lieth, J.H. (2003) A coupled model of photosynthesis, stomatal conductance and transpiration for a rose leaf (Rosa hybrida L.). *Ann Bot* 91:771–781. https://doi.org/10.1093/aob/mcg080

Kim, S.H., et al. (2012) Modeling Temperature Responses of Leaf Growth, Development, and Biomass in Maize with MAIZSIM. *Agron J* 104:1523. https://doi.org/10.2134/agronj2011.0321.

Kimball, B.A. (2016) Crop responses to elevated CO2 and interactions with H2O, N, and temperature. *Curr Opin Plant Biol* 31:36–43. https://doi.org/10.1016/j.pbi.2016.03.006.

Kimball, B.A., et al. (2023) Simulation of evapotranspiration and yield of maize: An inter-comparison among 41 maize models. *Agric For Meteorol* 333. https://doi.org/10.1016/j.agrformet.2023.109396.

Kirk, W.W. and Marshall, B. (1992) The influence of temperature on leaf development and growth in potatoes in controlled environments. *Ann Appl Biol* 120:511–525.

Köchy, M., et al. (2015) Global distribution of soil organic carbon – Part 1: Masses and frequency distributions of SOC stocks for the tropics, permafrost regions, wetlands, and the world. *Soil* 1:351–365. https://doi.org/10.5194/soil-1-351-2015.

Korres, N.E., et al. (2016) Cultivars to face climate change effects on crops and weeds: a review. *Agron Sustain Dev* 36. https://doi.org/10.1007/s13593-016-0350-5.

Kumudini, S., et al. (2014) Predicting Maize Phenology: Intercomparison of Functions for Developmental Response to Temperature. *Agron J* 106:2087–2097. https://doi.org/10.2134/agronj14.0200.

Lee, J., et al. (2006) Tillage and field scale controls on greenhouse gas emissions. *J Environ Qual* 35:714–25. https://doi.org/10.2134/jeq2005.0337.

Leng, G. and Huang, M. (2017) Crop yield response to climate change varies with crop spatial distribution pattern. *Sci Rep* 7:1463. https://doi.org/10.1038/s41598-017-01599-2.

Li, C., et al. (1992) A model of nitrous oxide evolution from soil drive by rainfall events: 1. model structure and sensitivity. *J Geophys Res* 97:9759–9776. https://doi.org/10.1029/92JD00509.

Li, S., et al. (2020) Evaluation of different crop models for simulating rice development and yield in the U.S. Mississippi Delta. *Agronomy* 10. https://doi.org/10.3390/agronomy10121905.

Li, S., et al. (2021) Application of a coupled model of photosynthesis, stomatal conductance and transpiration for rice leaves and canopy. *Comput Electron Agric* 182. https://doi.org/10.1016/j.compag.2021.106047.

Li, S., et al. (2022a) Improving simulations of rice in response to temperature and CO2. *Agronomy* 12. https://doi.org/10.3390/agronomy12122927.

Li, S., et al. (2022b) Model-based directed breeding strategy for climate resilience in U.S. rice production locations, ASSA-CSA-SSSA 2022 Annual Meeting, Baltimore, Maryland.

Li, S., et al. (2024a) The impact of climate change and extreme climate on rice in the U.S. Agricultural and Forest Meteorology in review.

Li, S., et al. (2024b) Quantifying the impact of climate change and extreme heat on rice in the United States. *Agric For Meteorol* 355. https://doi.org/10.1016/j.agrformet.2024.110145.

Li, T., et al. (2015) Uncertainties in predicting rice yield by current crop models under a wide range of climatic conditions. *Glob Chang Biol* 21:1328–1341. https://doi.org/10.1111/gcb.12758.

Li, T., et al. (2016) Combining limited multiple environment trials data with crop modeling to identify widely adaptable rice varieties. *PLoS One* 11:e0164456. https://doi.org/10.1371/journal.pone.0164456.

Li, T., et al. (2017a) From ORYZA2000 to ORYZA (v3): an improved simulation model for rice in drought and nitrogen-deficient environments. *Agric For Meteorol* 237–238:246–256. https://doi.org/10.1016/j.agrformet.2017.02.025.

Li, Z., et al. (2017b) Evaluation of the DNDC model for simulating soil temperature, moisture and respiration from monoculture and rotational corn, soybean and winter wheat in Canada. *Ecol Model* 360:230–243. https://doi.org/10.1016/j.ecolmodel.2017.07.013.

Liu, B., et al. (2023) AgMIP-wheat multi-model simulations on climate change impact and adaptation for global wheat. *Open Data J Agric Res* 9:10–25. https://doi.org/10.5281/zenodo.4027033.

Liu, F., et al. (2009) Capability of the 'Ball-Berry' model for predicting stomatal conductance and water use efficiency of potato leaves under different irrigation regimes. *Scientia Horticulturae* 122:346–354. https://doi.org/10.1016/j.scienta.2009.05.026.

Lokhande, S. and Raja Reddy, K. (2015) Cotton reproductive and fiber quality responses to nitrogen nutrition. *Int J Plant Prod* 9:191–210.

Lokhande, S. and Reddy, K.R. (2014a) Quantifying temperature effects on cotton reproductive efficiency and fiber quality. *Agron J* 106:1275–1282.

Lokhande, S. and Reddy, K.R. (2014b) Reproductive and fiber quality responses of upland cotton to moisture deficiency. *Agron J* 106:1060–1069.

Lu, M., et al. (2013) Responses of ecosystem carbon cycle to experimental warming: a meta-analysis. *Ecology* 94:726–738. https://doi.org/10.1890/12-0279.1.

Madan, P., et al. (2012) Effect of elevated CO2 and high temperature on seed-set and grain quality of rice. *J Exp Bot* 63:3843–3852. https://doi.org/10.1093/jxb/ers077.

Makino, A. (2021) Photosynthesis improvement for enhancing productivity in rice. *Soil Sci Plant Nutr* 67:513–519. https://doi.org/10.1080/00380768.2021.1966290.

Manning, E. (1996) GLYCIM: crop model for soybeans. *Progress Farmer* 11:33.

Martinez-Feria, R.A. and Basso B. (2020) Unstable crop yields reveal opportunities for site-specific adaptations to climate variability. *Sci Rep* 10:2885. https://doi.org/10.1038/s41598-020-59494-2.

Masuduzzaman, A., et al. (2016) Evaluation of rice lines tolerant to heat during flowering stage. *Rice Res Open Access* 4. https://doi.org/10.4172/2375-4338.1000170.

Mauget, S., et al. (2019) Planting date effects on cotton lint yield and fiber quality in the U.S. southern high plains. *Agriculture* 9(4). https://doi.org/10.3390/agriculture9040082.

McKinion, J.M., et al. (1989) Application of the GOSSYM/COMAX system to cotton crop management. *Agric Syst* 31:55–65.

McMaster, G.S., et al. (2008) Simulating the influence of vernalization, photoperiod and optimum temperature on wheat developmental rates. *Ann Bot* 102:561–569. https://doi.org/10.1093/aob/mcn115.

Messina, C.D., et al. (2022) Crop Improvement for Circular Bioeconomy Systems. *J ASABE* 65:491–504. https://doi.org/10.13031/ja.14912.

Mirzabaev, A., et al. (2023) Severe climate change risks to food security and nutrition. *Clim Risk Manag* 39. https://doi.org/10.1016/j.crm.2022.100473.

Monteith, J. (1977) Climate and efficieny of crop production in Britain. *Philos Trans R Soc B Biol Sci* 281:277–294.

Neupane, D., et al. (2022) Does climate change affect the yield of the top three cereals and food security in the world? *Earth* 3:45–71. https://doi.org/10.3390/earth3010004.

Paff, K., et al. (2023) A comparison of wheat leaf-appearance rate submodules for DSSAT CROPSIM-CERES (CSCER). *Ecol Model* 482. https://doi.org/10.1016/j.ecolmodel.2023.110406.

Parton, W.J., et al. (1998) DAYCENT and its land surface submodel: description and testing. *Glob Planet Change* 19:35–48.

Prior, S.A., et al. (2005) Elevated atmospheric CO2 effects on biomass production and soil carbon in conventional and conservation cropping systems. *Glob Change Biol* 11:657–665. https://doi.org/10.1111/j.1365-2486.2005.00935.x.

Raich, J.W. and Tufekcioglu A. (2000) Vegetation and soil respiration. *Biogeochemistry* 48:71–90.

Ray, D.K., et al. (2019) Climate change has likely already affected global food production. *PLoS One* 14:e0217148. https://doi.org/10.1371/journal.pone.0217148.

Reddy, K.R., et al. (2002) Applications of a Cotton Simulation Model, GOSSYM, for Crop Management, Economic, and Policy Decisions. https://doi.org/10.1201/9781420032413.ch3

Reddy, V.R. and Baker, D.N. (1988) Estimation of parameters for the cotton simulation model GOSSYM: cultivar differences. *Agric Syst* 26:111–122. https://doi.org/10.1016/0168-1699(93)90038-3.

Reddy, V.R., et al. (1995) Crop management and input optimization with GLYCIM: differing cultivars. *Comput Electron Agric* 13:37–50. https://doi.org/10.1016/0168-1699(95)00013-T.

Remy, K. (1994) *GLYCIM Soybean Model Proves Its Worth, Research Highlights.* MAFES, Publ., Mississippi State University, Mississippi.

Resop, J.P., et al. (2012) Combining explanatory crop models with geospatial data for regional analyses of crop yield using field-scale modeling units. *Comput Electron Agric* 89:51–61. https://doi.org/10.1016/j.compag.2012.08.001.

Resop, J.P., et al. (2014) Biophysical constraints to potential production capacity of potato across the U.S. Eastern Seaboard Region. *Agron J* 106:43. https://doi.org/10.2134/agronj2013.0277.

Reynolds, J. and Acock, B. (1985) Predicting the response of plants to increasing carbon dioxide - a critique of plant-growth models. *Ecol Model* 29:107–129.

Rezael, E.E., et al. (2023) Climate change impacts on crop yields. *Nat Rev Earth Environ* 4:831–846. https://doi.org/10.1038/s43017-023-00491-0.

Rolston, D.E., et al. (2006) *Predicting Changes in Landscape-Scale Soil Organic C Following the Implementation of Minimum TTillage*. University of California, Davis.

Schaap, M.G., et al. (2001) ROSETTA: a computer program for estimating soil hydraulic parameters with hierarchical pedotransfer functions. *J Hydrol* 251:163–176.

Schapendonk, H.C.M., et al. (2000) Effects of elevated CO2 concentration on photosynthetic acclimation and productivity of two potato cultivars grown in open-top chambers. *Aust J Plant Physiol* 27:1119–1130.

Scharlemann, J.P.W., et al. (2014) Global soil carbon: understanding and managing the largest terrestrial carbon pool. *Carbon Manag* 5:81–91. https://doi.org/10.4155/cmt.13.77.

Schram, S.G. and Wallace, W.W. (1996) USDA global environmental change data assessment and integration project. *J Agric Food Inform* 3:39–47. https://doi.org/10.1300/j108v03n02_06.

Searchinger, T.D., et al. (2018) Assessing the efficiency of changes in land use for mitigating climate change. *Nature* 564:249–253. https://doi.org/10.1038/s41586-018-0757-z.

Sequeira, R.A., et al. (1994) Inclusion of plant structure and fiber quality into a distributed delay cotton model to improve management and optimize profit. *Ecol Model* 71(1):161–186. https://doi.org/10.1016/0304-3800(94)90080-9.

Silva, J.V. and Giller, K.E. (2021) Grand challenges for the 21st century: what crop models can and can't (yet) do. *J Agric Sci* 158:794–805. https://doi.org/10.1017/s0021859621000150.

Šimůnek, J. and Nimmo, J.R. (2005) Estimating soil hydraulic parameters from transient flow experiments in a centrifuge using parameter optimization technique. *Water Resour Res* 41. https://doi.org/10.1029/2004wr003379.

Šimůnek, J. and Suarez, D.L. (1993) Modeling of carbon dioxide transport and production in soil 1. model development. *Water Resour Res* 29:487–497.

Šimůnek, J., et al. (1992) *SWMS-2D Code for Simulating Water Flow and Solute Transport in Two-dimensional Variably Saturated Media: Version 1.1*. USDA.

Stockle, C.O. and Kemanian, A.R. (2020) Can crop models identify critical gaps in genetics, environment, and management interactions? *Front Plant Sci* 11:737. https://doi.org/10.3389/fpls.2020.00737.

Sun, W., et al. (2022) Effects of elevated CO2 and temperature on soybean growth and gas exchange rates: A modified GLYCIM model. *Agric For Meteorol* 312. https://doi.org/10.1016/j.agrformet.2021.108700.

Sun, W., et al. (2023a) Projected long-term climate trends reveal the critical role of vapor pressure deficit for soybean yields in the US Midwest. *Sci Total Environ* 878:162960. https://doi.org/10.1016/j.scitotenv.2023.162960.

Sun, W., et al. (2023b) Does drought stress eliminate the benefit of elevated CO2 on soybean yield? Using an improved model to link crop and soil water relations. *Agric For Meteorol* 343. https://doi.org/10.1016/j.agrformet.2023.109747.

Sun, W., et al. (2024) Simulating climate change effects on soil carbon dynamics in a soybean–maize ecosystem: Using improved CO2 emission and transport models. *Eur J Agron* 159. https://doi.org/10.1016/j.eja.2024.127226.

Tausz-Posch, S., et al. (2020) Elevated CO2 effects on crops: advances in understanding acclimation, nitrogen dynamics and interactions with drought and other organisms. *Plant Biol* 22:38–51. https://doi.org/10.1111/plb.12994.

Timlin, D.J. and Pachepsky Y.A. (1997) A modular soil and root process simulator. *Ecol Model* 94:67–80.

Timlin, D.J., et al. (2002a) Error analysis of soil temperature simulations using measured and estimated hourly weather data with 2DSOIL. *Agric Syst* 72:215–239. https://doi.org/10.1016/S0308-521X(01)00075-0.

Timlin, D.J., et al. (2002b) Experience with on-farm applications of GLYCIM/GUICS. Agricultural System Models in Field Research and Technology Transfer, CRC Press, Boca Raton. pp. 55–69.

Timlin, D.J., et al. (2023) CLASSIM: a relational database driven crop model interface. *Smart Agric Technol* 5. https://doi.org/10.1016/j.atech.2023.100281.

Timlin, D.J., et al. (2024) The role of crop simulation modeling in assessing potential climate change impacts. *Agrosyst Geosci Environ* 7. https://doi.org/10.1002/agg2.20453.

Van Der Velde, M., et al. (2012) Impacts of extreme weather on wheat and maize in France: evaluating regional crop simulations against observed data. *Climatic Change* 113:751–765. https://doi.org/10.1007/s10584-011-0368-2.

van Dijk, M., et al. (2021) A meta-analysis of projected global food demand and population at risk of hunger for the period 2010–2050. *Nat Food* 2:494–501. https://doi.org/10.1038/s43016-021-00322-9.

van Genuchten, M.T. (1980) A closed-form equation for predicting the hydraulic conductivity of unsaturated soils. *Soil Sci Soc Am J* 44:892–898.

van Oort, P.A., et al. (2015) Improved climate risk simulations for rice in arid environments. *PLoS One* 10:e0118114. https://doi.org/10.1371/journal.pone.0118114.

von Caemmerer, S. and Farquhar, G.D. (1981) some relationships between the biochemistry of photosynthesis and the gas exchange of leaves. *Planta* 153:376–387.

Vos, J. and Oyarzun, P.J. (1988) Water relations of potato leaves. II. Pressure-volume analysis and inferences about the constancy of the apoplastic fraction. *Ann Bot* 62:449–454.

Walker, D.W. and Van Loon, A.F. (2023) Droughts are coming on faster. *Science* 380:130–132. https://doi.org/10.1126/science.adh3097.

Wallach, D., et al. (2014) *Parameter Estimation with Classical Methods (Model Calibration), Working with Dynamic Crop Models (Second Edition): Methods, Tools and Examples for Agriculture and Environment*. Academic Press, Elsevier, London, UK. pp. 205–276.

Wallach, D., et al. (2016) Accounting for both parameter and model structure uncertainty in crop model predictions of phenology: a case study on rice. *Eur J Agron*. https://doi.org/10.1016/j.eja.2016.05.013.

Wallach, D., et al. (2021) The chaos in calibrating crop models: lessons learned from a multi-model calibration exercise. *Environ Model Softw* 145. https://doi.org/10.1016/j.envsoft.2021.105206.

Wang, Z., et al. (2020) Coupled model of surface runoff and surface-subsurface water movement. Advances in Water Resources 137. https://doi.org/10.1016/j.advwatres.2019.103499.

Wang, Z., et al. (2021a) Simulations of water and thermal dynamics for soil surfaces with residue mulch and surface runoff. *Water Resour Res* 57. https://doi.org/10.1029/2021wr030431.

Wang, Z., et al. (2021b) A diffusive model of maize root growth in MAIZSIM and its applications in ridge-furrow rainfall harvesting. *Agric Water Manag* 254. https://doi.org/10.1016/j.agwat.2021.106966.

Wang, Z., et al. (2022) Modeling vapor transfer in soil water and heat simulations: a modularized, partially-coupled approach. *J Hydrol* 608:127541. https://doi.org/10.1016/j.jhydrol.2022.127541.

Wang, Z., et al. (2025) Process-based vegetative growth model for cereal rye winter cover crop using object-oriented programming and linked-list data structure. *Comput Electron Agric* 231. https://doi.org/10.1016/j.compag.2025.109964.

Whisler, F.D., et al. (1986) Crop simulation models in agronomic systems. *Adv Agron* 40:141–208.

Xin Q., et al. (2016) Effects of the partitioning of diffuse and direct solar radiation on satellite-based modeling of crop gross primary production. *Int J Appl Earth Obs Geoinf* 50:51–63. https://doi.org/10.1016/j.jag.2016.03.002.

Xu, J., et al. (2021) What happens at night? Physiological mechanisms related to maintaining grain yield under high night temperature in rice. *Plant Cell Environ* 44:2245–2261. https://doi.org/10.1111/pce.14046.

Yang, Y., et al. (2009) Simulating canopy transpiration and photosynthesis of corn plants under contrasting water regimes using a coupled model. *Trans ASABE* 52:1011–1024.

Yin, X., et al. (2003) A flexible sigmoid function of determinate growth. *Ann Bot* 91:361–371. https://doi.org/10.1093/aob/mcg029.

Zhang, Y., et al. (2020) Adaptation in U.S. Corn Belt increases resistance to soil carbon loss with climate change. *Sci Rep* 10:13799. https://doi.org/10.1038/s41598-020-70819-z.

Zhou, G. and Wang, Q. (2018) A new nonlinear method for calculating growing degree days. *Sci Rep* 8:10149. https://doi.org/10.1038/s41598-018-28392-z.

Chapter 3

The Cycles Agroecosystem Model

Armen R. Kemanian, Yuning Shi, Charlie M. White, Felipe Montes, Rachel K. N. Rozum, Amanda R. Burton and Ele Saltmarsh, The Pennsylvania State University, USA; and Claudio O. Stockle, Washington State University, USA

1 Introduction

Cycles is a biophysically-based, multi-year, and multi-species agroecosystem modeling system (Kemanian et al., 2024). It has a structure that favors a generalized representation of elements and fluxes in the soil–plant–atmosphere continuum over taxonomic soil- and plant-specific nuances, as well as management interventions like tillage, fertilization, and irrigation. The structure is modular, so its components or the whole can be treated as objects. These objects can be instantiated to enable further development of physically distributed models like Cycles-L (Shi et al., 2023) or enable data assimilation at run-time in data-model workflows that require a flexible soft-coupling of modules (Gil et al., 2021).

The model is conceived as the coupling of the energy and water balance with the cycles of carbon (C) and nitrogen (N) (Figs. 1 and 2). Fluxes in and

http://dx.doi.org/10.19103/AS.2025.0155.04

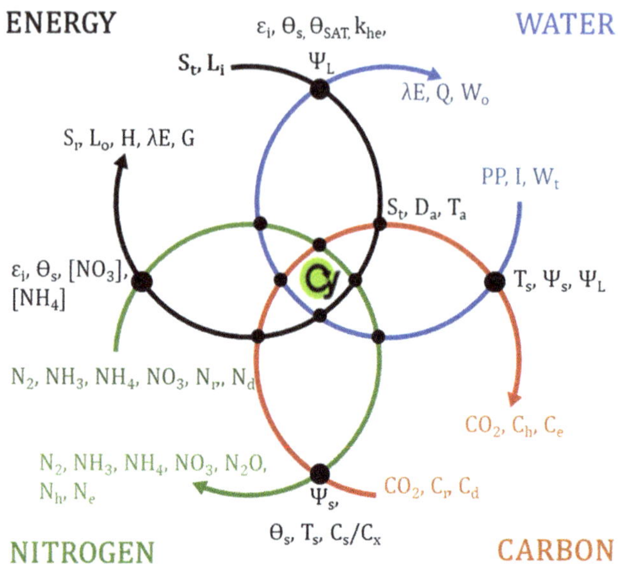

Figure 1 Schematics of the coupling of the energy, water, C, and N cycles, with fluxes in (arrow tail) and out (arrow point) for each component. Internal fluxes not represented. The nodes at the arrows' intersections represent transfers of information. Fertilization, tillage, and irrigation alter inputs and intensive properties; harvest alters output rates. Energy enters as shortwave (S_t) and thermal radiation (L_i) and is dissipated as reflected S_t (S_r), thermal outgoing radiation (L_o), sensible heat (H), latent heat (λE), or stored (G). Liquid water enters through precipitation (PP), irrigation (I), and capillary rise (W_t), and leaves via λE, runoff (Q), and soil percolation or lateral flow (W_o). Carbon enters through photosynthesis, soil ammendments (C_r), and deposition (C_d), and leaves via respiration, harvest (C_h), and erosion (C_e). D_a, T_a, T_s are air vapor deficit, air temperature, and soil temperature. ψ_S and ψ_L are soil and leaf water potential. θ_S is the volumetric soil water content. The ratio C_s/C_x is the soil carbon saturation ratio. Source: Figure adapted from Kemanian et al. (2024).

out of a pool define each differential equation. Intensive properties such as temperature, water potential, and N concentration in soil or plant tissue regulate fluxes. These fluxes are resolved with a forward Eulerian integration constrained by mass conservation, although a solver-based approach is under development. Cycles emerges from the fundamentals described in the landmark book *Soil Physics with Basic* by Campbell (1985), cropping system elements in CropSyst (Stöckle et al., 2014) and other models, and new developments to represent plant and soil processes. The latter include soil organic C and N saturation (Kemanian et al., 2005; Kemanian and Stöckle, 2010; White et al., 2014), plant growth based on stomata optimization theory (Cowan and Farquhar, 1977), and a model assembly that enables modeling annual and perennial polycultures explicitly (Pravia et al., 2019; McConnell et al., 2023).

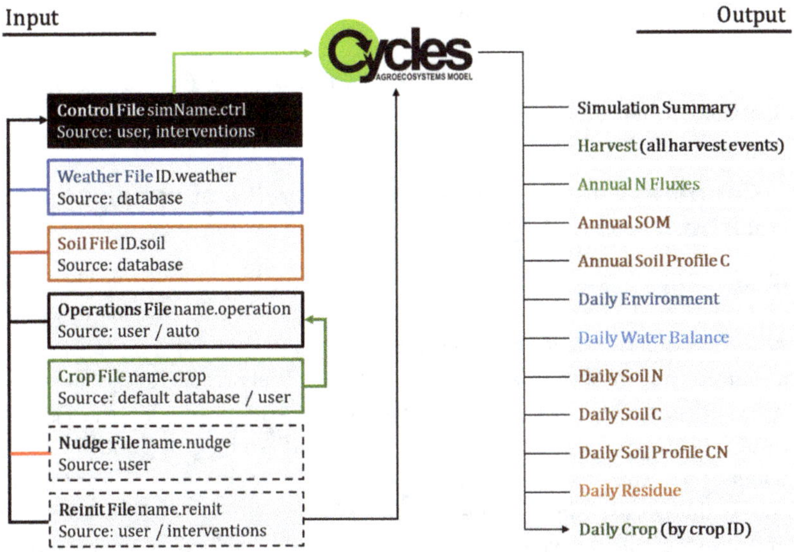

Figure 2 Input (left) and output (right) file structure of Cycles. The Cycles logo represents the executable. The control file in the upper left collects all the information needed to execute the model. The name of the file is the unique ID for an autogenerated folder to store outputs. Solid and dashed lines distinguish mandatory and optional input files. An external application is often used to generate input files from existing databases. The nudge file allows changing selected model parameters. The reinit or reinitialization file allows overwriting state variables at runtime (effectively breaking the mass and energy balance). Output files are printed on a fixed annual or daily time step or are event-based (harvest file). SOM stands for soil organic matter.

We introduce the model based on the physical compartments through which energy and mass flow in agroecosystems: soil, vegetation, and atmosphere. These compartments have storage capacity. In the 1D version of Cycles, a field is constrained by boundary conditions at the bottom of the soil profile and at a reference height in the atmosphere. In the 3D version of Cycles (Cycles-L), which includes a stream network represented by rectangular river segments and the terrain represented by a grid of triangular prisms, the boundary conditions are provided at the bedrock depth and at the watershed edge. Operations like irrigation, fertilization or tillage alter fluxes and system properties. While animals are not explicitly represented with their own compartment in Cycles, their effects via grazing and manure and urine deposition are functionally simulated. This chapter describes the main algorithms simulating carbon flows and water uptake, qualitatively describing the connections between the cycling of water, C, and N, an application of stomatal optimization theory to model carbon dioxide (CO_2) and soil C (C_s) saturation theory, combining theoretical developments and their practical implementation in the model. In addition, we present three applications that illustrate the model utility to explore crop traits in-silico, estimate nitrous oxide (N_2O) and C_s changes in almost 1 million fields in Iowa and the potential to grow maize (*Zea mays* L.), sorghum (*Sorghum bicolor* (L.) Moench), or a mixture of these two crops for silage production in the U.S. at the county level.

2 Coupling energy fluxes with the cycles of water, carbon, and nitrogen

Energy inputs are from solar radiation and atmospheric downward longwave radiation (Fig. 1). The incoming shortwave radiation is a forcing in the weather data. The atmospheric longwave radiation is estimated from air temperature and humidity as in Allen et al. (1998). The latent heat flux and soil heat storage are solved explicitly. The latent flux demand is estimated with the Penman-Monteith equation. Unlike models like SHAW (Flerchinger, 2000) or CUPID (Kustas et al., 2007), the temperature of each exposed surface contributing to the radiative exchange or heat transfer to the soil is not solved for each component explicitly, which can carry errors in the energy balance that are considered minimal. In addition, the net radiation implicit in the Penman-Monteith solution of the energy balance limits the potential latent heat transfer, somewhat protecting against large errors. However, important deviations can occur when the allocation of net radiation to soil, residues, and transpiring canopy is complex. Examples are row crops that create regular shading, winter crops covering a frozen ground (where transpiration is likely zero), and stony soils or other soils for which the root distribution is difficult to model.

In Cycles, the energy that is not dissipated as latent heat or stored in the soil is therefore assumed to be sensible heat. Energy capture via photosynthesis is not included in the energy balance. In Cycles-L (Shi et al., 2023), the energy balance is closed in every sub-daily time step in large part to account for differences in topography and aspect that are critical to simulate snow cover, snow melt, soil temperature, and water runoff.

Soil moisture is a key variable regulating multiple biophysical processes. The combination of soil water potential and leaf water potential controls both transpiration and photosynthesis, and therefore the C input to the system through plant growth and non-harvested residues. At the same time, soil moisture affects microbial activity and the decomposition of residues and soil organic matter. This dual role highlights the need to simulate soil moisture accurately in each soil layer not only to represent plant growth but also microbial respiration, as discussed insightfully by Manzoni and Katul (2014).

The complexity of these interactions increases when considering the N cycle. Nitrogen transformations involve diverse microbial processes with distinct enzymatic controls and sensitivities to substrate concentration and environmental conditions (e.g. Wallenstein et al., 2006). Oxic conditions in moist, well-drained soils favor nitrification, while hypoxic or anoxic conditions in wetter soils promote denitrification. In both pathways, a portion of N can be lost as nitrous oxide (N_2O), a potent greenhouse gas (e.g. Firestone, 1982; Saha et al., 2021). Soil moisture thus controls not only microbial and root oxygen demand but also oxygen diffusion into the soil and N_2O diffusion toward the surface. It also modulates nitrate transport, which typically occurs under wet conditions above field capacity. These processes are highly landscape-dependent (Saha et al., 2018; Weitzman et al., 2018; Shi et al., 2023). Their simulation requires both robust algorithms and accurate input data, particularly soil porosity and transport-relevant parameters. This illustrates the challenge of designing a modeling framework that is both broadly applicable and sufficiently detailed to capture key processes at meaningful spatial and temporal scales.

While it is not possible to display all equations in Cycles, a set of core formulations is presented here that reflects the model structure and philosophy. Before proceeding with these descriptions, we introduce the reader to the basic input and output structure of Cycles.

3 Input and output file structure

The input and output files provide the necessary information to run the model and reflect aspects of its internal structure. Required weather inputs include precipitation (PP, kg m^{-2} day^{-1}), maximum and minimum temperature (Tx and Tn, °C), solar radiation (St, MJ m^{-2} day^{-1}), maximum and minimum relative humidity (RHx and RHn, percent), and wind speed (WS, m s^{-1}). Relative humidity at Tn (RHx) and Tx (RHn) helps capture abrupt shifts in air moisture due to changing air masses or irrigation. These records also aid in identifying errors, since air humidity is often poorly measured. Gridded weather data has steadily improved. The reanalysis data from the North American Land Data Assimilation System Phase 2 (NLDAS-2; Xia et al., 2012) offers a reliable product beginning in 1979, with a global counterpart available from 2001 (GLDAS).

The soil file defines initial conditions by horizon or layer, specifying texture and other properties such as organic C and N. This information is often obtained from global databases like SoilGrids (Poggio et al., 2021). No soil attribute is static. For example, tillage can mix layers and alter texture and bulk density, thereby changing the thickness and hydraulic properties of those layers, which are dynamic.

The operations file (Fig. 2) contains fix-date and conditional field operations. These are tillage, fertilization, irrigation, planting, and harvest events associated to a year in a preset or autogenerated crop sequence. A preset crop sequence is grouped in a rotation of a known number of years. The autogenerated crop sequence is part of a special application named Cycles-A that uses a heuristic approach to build a crop sequence based on the available menu or crops for a region. Cycles-A is described in Kemanian et al. (2024).

The reinitialization file serves multiple purposes. It allows users to reset variables to define initial conditions for virtual experiments. It also supports the integration of models like Cycles into broader decision support systems rather than operating them as isolated units (Gil et al., 2021). The file enables soft coupling of models, such as importing water table depth from a hydrologic model into the 1D Cycles model. With increased access to live sensor data, it can reset state variables based on measurements, improving model relevance for decision support (Stöckle and Kemanian, 2020). However, it may break the mass balance if measurements are not synchronized with the time step of the model. For example, this occurs when soil moisture values are replaced with data from ground sensors or remote sensing (Kerr et al., 2001; Fu et al., 2024). Overall, the reinitialization file offers flexibility for model applications.

4 The crop growth module

Plant growth is simulated using the resource capture and resource use efficiency approach, where the two base resources are radiation and water. Plants capture radiation via the canopy and water mostly via the roots. These two captured resources are converted to potential growth by multiplying them by radiation and water use efficiency estimates, and the minimum of the two determines the growth rate, an approach used in Cropsyst for decades (Stöckle et al., 2014). Although not immediately apparent, this is a simplified version of Farquhar et al. (2001) photosynthesis model that, as shown below, conforms functionally but not strictly with stomata optimization theory (Cowan, 1977; Cowan and Farquhar, 1977; Lloyd, 1991).

In this framework, the crop growth rate G_B (g m^{-2} day^{-1} of dry matter) is the minimum of growth limited by radiation (G_R) or growth limited by transpiration (G_T):

$$G_B = \min(G_R, G_T) \tag{1}$$

$$G_R(\varepsilon_{ie}, S_t) = \varepsilon_R \varepsilon_{ie} S_t \tag{2}$$

$$G_T(D_a) = \frac{\varepsilon_W}{\sqrt{D_a}} T \tag{3}$$

The parameter ε_R (g MJ^{-1}) is the radiation use efficiency (Warren Wilson, 1967; Monteith, 1978), ε_i is the intercepted solar radiation (S_t). A conversion allows using the more appropriate PAR-based coefficient, and adjustments via biosynthesis cost can be embedded in the growth computation. Maximum ε_R

values for each species are available in the literature (e.g. Stöckle and Kemanian, 2009). Similarly, the parameter ε_W is the water use efficiency at a vapor pressure deficit (D_a, mol H_2O/mol air) of 0.01 mol/mol (or 1 kPa at sea level), and T is the transpiration rate (kg H_2O m^{-2} day^{-1}). Standard ε_W values are listed in Kemanian et al. (2024). The use of D_a instead of D_L, the leaf-to-air vapor pressure deficit, is considered acceptable for most situations. Both ε_R and ε_W should be reported for a reference atmospheric CO_2 concentration and scaled as CO_2 changes.

An interesting aspect of this approach is that it links water use efficiency (specifically transpiration use efficiency) to the inverse square root of vapor pressure deficit. This is consistent with empirical findings by Kemanian et al. (2005), who associated this with reduced c_i/c_a (ratio of leaf to air CO_2 concentration) due to stomatal regulation. Mechanistically, increased transpiration lowers stomatal conductance, limiting CO_2 diffusion and reducing c_i/c_a. Cowan and Farquhar (1977) derived a comparable relationship using optimization theory via the Lagrange multiplier. Though this theory is still debated (Hari et al., 1986; Katul et al., 2010; Buckley and Schymanski, 2014), it supports the modeling framework. Wang et al. (2020) proposed that such stomatal behavior co-evolved with vascular vulnerability, reinforcing its general applicability. For further discussion, see Bassiouni and Vico (2021). As modeling in this domain requires parameter estimation, the use of fewer parameters is preferable, and optimization theory offers a functional expression for ε_W compatible with the proposed growth model. A derivation of leaf-level water use efficiency follows, adopting the notation from Manzoni et al. (2011).

Representing photosynthesis by A (μmol CO_2 m^{-2} s^{-1}), leaf transpiration by T_L (mol H_2O m^{-2} s^{-1}), and conductance to CO_2 as g_c (μmol CO_2 m^{-2} s^{-1}):

$$\frac{\partial A}{\partial g_c} - \lambda \frac{\partial T_L}{\partial g_c} = 0. \tag{4}$$

The Lagrange multiplier λ (μmol CO_2/mol H_2O) is the ratio of the two derivatives. Transpiration can be expressed as the product of the combined (g_w) stomatal and boundary layer conductance to water vapor (g_w, g_{sw}, and g_{bw}, mol H_2O m^{-2} s^{-1}) and the leaf-to-air vapor pressure deficit (D_L, mol H_2O/mol air):

$$T_L = \left(\frac{g_{sw} g_{bw}}{g_{sw} + g_{bw}} \right) D_L = g_w D_L. \tag{5}$$

Similarly, photosynthesis can be expressed as the product of the combined stomatal and boundary layer conductance to CO_2 (g_c, g_{sc}, and g_{bc}, mol C_2O m^{-2} s^{-1}) and the leaf to air CO_2 molar fraction difference:

$$A = \left(\frac{g_{sc} g_{bc}}{g_{sc} + g_{bc}} \right) (c_a - c_i) = g_c (c_a - c_i). \tag{6}$$

A second expression for A based on the consumption rather than diffusion of CO_2 can be proposed and used later to eliminate c_i from Eq. [6]:

$$A = V_x \frac{c_i - \Gamma^*}{c_i + k_c\left(1 + c_o / k_o\right)} - R_d \approx k\left(c_i - \Gamma^*\right) - R_d. \tag{7}$$

This expression includes the unknown c_i and depends on enzyme kinetics where V_x is the maximum carboxylation rate and R_d is mitochondrial respiration (mol CO_2 m^{-2} s^{-1}), Γ^* is the CO_2 compensation point, and *k* is Lloyd's effective carboxylation efficiency (mol CO_2 m^{-2} s^{-1}) that by linearizing the response of *A* to c_i allows a simpler solution to the system of equations.

These two equations for A can be combined to eliminate c_i but cannot eliminate g_c. Instead, the problem can be reformulated as a function of the Lagrange multiplier by eliminating c_i to get a function for g_c that depends on λ, and by deriving then functions for *A* and for A/T_L. The use of these equations would require knowing the value of λ, but we are here only looking for the functional form. The solution for the A/T_L ratio is

$$\varepsilon_{WL} = \frac{A}{T_L} = \sqrt{\frac{\lambda\left(c_a - \text{``}^* - R_d / k\right)}{r_g}} \frac{1}{\sqrt{D_L}} \tag{8}$$

The term r_g in the denominator is the ratio g_{H2O}/g_{CO2}. The important result is the dependence of the leaf level water use efficiency, μ_{WL}, on $1/\sqrt{D_L}$. Without oversimplifying the complexities of transport in canopies, a similar proportionality may hold at the canopy scale, consistent with empirical findings in Kemanian et al. (2005). The exponent for D_L may vary from 0.5 (e.g. 0.4 to 0.6) but optimization theory suggests that photosynthesis and transpiration are balanced by plants in ways that minimize water loss penalties when D_L increases. A strictly inverse dependence would imply regulation solely by diffusive effects.

Mesophyll-level regulation of the H_2O and CO_2 exchange may be more complex than previously assumed, with c_i/c_a reductions potentially less pronounced than reported in the summary by Kemanian et al. (2005). Rather than stomatal closure, humidity in the substomatal cavity may fall below saturation (Jarvis and Slatyer, 1970; Marquez et al., 2024), limiting transpiration. Under this scenario, ε_W remains consistent with observed patterns not due to reduced c_i/c_a, but because D_a severely overestimates D_L. Further investigation is needed to reconcile lumped parameter and detailed canopy photosynthesis models with current insights into leaf and canopy gas exchange.

While detailed, hourly gas exchange and energy balance models can represent species- or cultivar-level responses, they require extensive parameterization. This demand can limit model applicability without guaranteeing improved outcomes. Moreover, interpretability may decline when model dependencies are built on layered assumptions (Monteith, 1978).

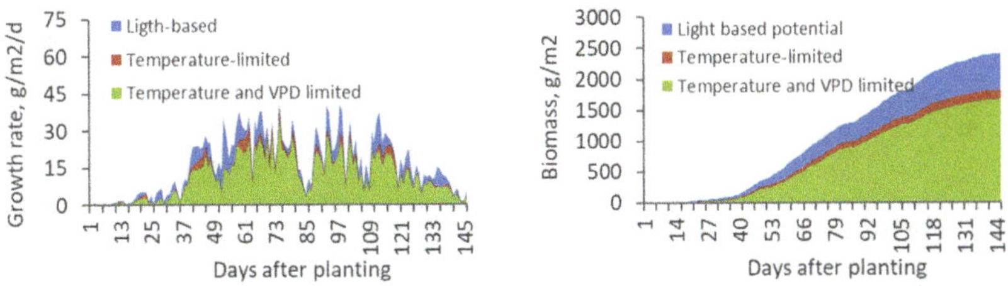

Figure 3 Light-limited, temperature-limited, and temperature- and vapor pressure-limited growth of maize in Centre Co PA for the year 1992. The left panel is the growth rate, and the second panel the cumulative growth. The model only used radiation, temperature and air humidity to illustrate the progressive limitation to growth by these factors, although the Cycles model operates as in Eq. [1]. The temperature limitation was calculated based on a daily temperature weighed 2/3 Tx and 1/3 Tn to emphasize daytime temperature and a beta function with minimum, optimum and maximum parameters of 0°C, 30°C, and 45°C. The vapor pressure deficit effect is the minimum of 1 and $1/D_a^{0.5}$. The radiation use efficiency is 1.6 g/MJ of solar radiation. The location is humid and the vapor pressure deficit effect limited.

In contrast, lumped parameter approaches offer parsimony while capturing key physiological constraints. Light availability, particularly its partitioning into beam and diffuse components, defines the upper photosynthetic limit (Sinclair et al., 1992). Actual photosynthesis is further reduced by temperature effects and CO_2 diffusion limitations, including stomatal responses to high evaporative demand. These collective effects, shown in Fig. 3, can be captured without complex models. Cumulative growth cannot exceed the green boundary.

5 Transpiration

The plant canopy functions as a variable conductance surface for vapor exchange. Although often assumed that the canopy substomatal cavity is saturated, evidence suggests that it can be unsaturated (Jarvis and Slatyer, 1970; Marquez et al., 2024). Nonetheless, vapor flow is still driven by a gradient of vapor concentration and a conductance, which remains challenging to model. A solution based on the leaf water potential was proposed 50 years ago by Campbell et al. (1974) and further detailed in Campbell (1985) and Stöckle and Jara (1998). The connection between vapor flow and liquid water flow is the leaf water potential. The gradient between soil water potential and leaf water potential drives liquid water flow. The leaf water potential in turn controls the stomatal conductance. This model, implemented in CropSyst, performs well with minimal parameters if root distribution and soil hydraulic properties are sufficiently characterized (Stöckle and Jara, 1998; Camargo and Kemanian, 2016). However, it is better parameterized using measurements of transpiration rather than measurements of leaf water potential (Camargo and Kemanian, 2016; Tardieu et al., 2017) for the model is more teleonomic than mechanistic.

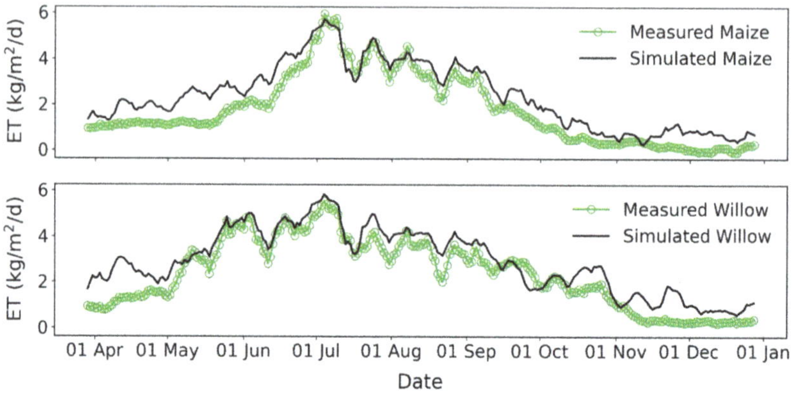

Figure 4 Measured and Cycles-modeled evapotranspiration (ET) for maize and willow in adjacent fields at Rockview in central Pennsylvania in 2014. Measurements were obtained with a pair of open path and close path eddy covariance flux towers in each field. Data presented as centered 7-day moving averages. Source: Reproduced from Kemanian et al. (2024).

Although leaf water potential does not directly control conductance, both respond to the same factors, making it an effective modeling proxy. Other modeling approaches may use different proxies.

We reproduce in Fig. 4 the measured and modeled ET of both maize and shrub willow (Kemanian et al., 2024). The measurements were obtained using flux towers. In willow, *T* dominates *ET* from May to October, owing to its rapid canopy closure and in maize *T* dominates ET from mid-June to September. The model simulates *ET*, and therefore *T*, remarkably well.

6 Crop parameters and genetic coefficients

Analogous to other crop models, most parameters used in Cycles to describe crops can be conceived as genetic coefficients reflecting the action of gene networks or individual alleles. These concepts have been formally developed by Hammer et al. (2002) and White and Hoogenboom (2003). From a physiological perspective, most plant traits are determined by the additive and interactive effects of genes and their alleles. The consideration of interactive effects is particularly challenging and could, in theory, be addressed through whole genome prediction, using mechanistic crop models as the link between genetic information and *trait* expression (Technow et al., 2015). This approach is appealing and methodologically transparent (Messina et al., 2018) but not without its limitations. At its core, it assumes that the model structure is sufficient to project genetic information into the phenotypic space. Mismatches between predictions and observations can offer important insights into trait genetic controls and inform model refinement.

An alternative approach involves genotype–environment associations, which use statistical links between allele frequencies in adapted genotypes and environmental variables. This approach surmises that variation at a given locus or loci reflects adaptation to local conditions and has been reviewed by Lasky et al. (2023). Rather than assembling and evaluating in-silico alleles with putative values assigned through crop coefficients (as in Messina et al., 2018), this approach can work in reverse by identifying allelic frequencies or combinations that reflect adaptation to distinct environments. A crop model can systematically characterize environmental variables, creating regression inputs that help detect alleles and allele frequencies suited to specific environmental stressors, as shown by McLaughlin et al. (2025) using Cycles. These authors then used such insights to assign landraces of barley, maize, rice, and sorghum to different geographies during a simulated 15-year environmental disruption following a soot-producing event such as nuclear war or volcanic eruption (Shi et al., 2025). Genotype–environment association and whole genome prediction can be complementary: the former can help narrow allele targets, while the latter can project them through models across broader environmental domains.

The crop parameters in Cycles are kept to a minimum but can represent relevant physiological variation. The phenology is controlled by thermal time using a beta function with an base, optimum, and maximum cardinal temperatures, so that temperatures above the optimum delay development in calendar time. Although these fundamentals are well understood (Parent and Tardieu, 2012), simulating with accuracy the phenological progression of a given genotype across environments remains challenging (e.g. Kumundini et al., 2014). Both photoperiod and vernalization affect development in Cycles. While the parameters are not presented here, the net effect of these two controls is to narrow down the window of flowering dates in spring or early summer in Mediterranean-like crops, even though the planting window in late summer and fall can be wide. This is important not only for winter grain crops like winter wheat, winter barley or winter canola but also for overwintering cover crops. In commercial conditions, cultivars or hybrids of any crop are already adapted to the target environment and a genotype-mismatch is unlikely except under sudden and drastic environmental change (e.g. McLaughlin et al., 2025).

The partitioning of biomass to roots and shoots is based on an allometric equation. The initial allocation to shoot exemplified in Table 1 for maize and sorghum is 0.45 kg/kg, and that proportion increases asymptotically to 0.95 kg/kg at maturity and is scaled based on the thermal time progression. The resulting allocation of biomass to roots follows the same pattern than the model proposed by Drouet and Pagès (2003). The harvest index (HI) is modeled as in Kemanian et al. (2007). The maximum harvest index (HI_x) is the asymptote of the relationship between fractional post anthesis phase growth (f_g, post anthesis to total growth ratio) and the HI, a notional asymptote as

Table 1 Crop parameters used in *Cycles*.

	Parameter	Units	Maize	Sorghum
Phenology	Thermal time to emergence	°C day^{-1}	65	65
	Flowering thermal time	°C day^{-1}	**1000**	**1250**
	Maturity thermal time	°C day^{-1}	1800	1800
	Base temperature for development	°C	6	6
	Optimum temperature for development	°C	28	28
	Maximum temperature for development	°C	46	46
Allocation	Initial partitioning to shoot	kg kg^{-1}	0.45	0.45
	Final partitioning to shoot	kg kg^{-1}	0.95	0.95
	Maximum harvest index	kg kg^{-1}	0.8	0.8
	Minimum harvest index	kg kg^{-1}	0.15	0.15
	Multiplier of slope harvest index	NA	1	1
Root	Maximum rooting depth	M	**1.8**	**1.7**
Growth	Radiation use efficiency	g Mg^{-1}	**2.2**	**2**
	Transpiration use efficiency	g kg^{-1}	**8.7**	**9.6**
Transpiration	Minimum temperature for transpiration	°C	**3**	**6**
	Threshold temperature for transpiration	°C	**15**	**18**
	K_C (crop coefficient)	NA	**1.1**	**1**
	Leaf water potential stress onset	J kg^{-1}	**−1100**	**−1400**
	Leaf water potential wilting point	J kg^{-1}	**−2000**	**−2300**
	Maximum transpiration	mm day^{-1}	**10**	**9**
Stress	Minimum temperature for cold damage	°C	**−5**	**−1**
	Threshold temperature for cold damage	°C	**3**	**6**
Nutrition	Maximum concentration of nitrogen	g g^{-1}	0.055	0.055
	Slope of the nitrogen dilution	NA	0.4	0.4

The crop parameters that differ between species are highlighted in bold font.

the maximum value of the fractional post-anthesis phase growth is < 1. The minimum harvest index (HI_o) is equivalent to the reserves accumulated up to anthesis that can be remobilized to the grain. The multiplier of the slope of the HI varies between 0 and 1 and adjusts the slope of the monotonical growth of the HI as fg increases. The default slope is (1 – HIo)/(HIx – HIo). A high value of HIx represents an increased sink capacity and can be easily used to model changes in HI due to breeding that increases either the grain number per unit of biomass or the potential size of each grain or both. This method to simulate HI has been verified independently (Campoy et al., 2020). A dynamic daily time-step version of this model is available in Cycles and uses these same parameters to estimate the potential growth of all grains (grain number × individual grain mass) per day and to allocate growth to grains accordingly.

The radiation and water use efficiency growth parameters were discussed in prior sections. The parameters of the model for transpiration are discussed in detail in Campbell (1985), Stöckle and Jara (1998), and Camargo and Kemanian (2016) but summarized here for completeness. Operationally, transpiration is calculated based on the evaporative demand and the soil water supply as limited by the plant conductance to liquid water flow and stomatal conductance. Both the leaf water potential at the onset of stomal closure and conductance to liquid water determine the maximum transpiration rate. In practice, the conductance to liquid water flow is calculated based on a set maximum transpiration, a model parameter. It allows the user to easily simulate the effect of genotypes with limited maximum transpiration as proposed by Sinclar et al. (2005). The FAO crop coefficient or K_c is a lumped parameter that represents variations in vapor conductance compared to the reference ET due to varying stomatal conductance or higher aerodynamic conductance; a more detailed discussion of the use of this coefficient and its limitations can be found in Kemanian et al. (2024). The maximum transpiration rate and the FAO crop coefficient are related, and furthermore they relate to both radiation and transpiration use efficiency (Kemanian et al., 2024).

Root depth is a user-defined parameter that defines the proclivity of a crop to reach a given depth when downward root growth is unimpeded. It is independent of the soil depth. The default fractional root distribution in Cycles is modeled using a normalized version of the maize root distribution described by Dwyer et al. (1996):

$$f_i = \frac{\int_{z_{bi}}^{z_{ti}} e^{-\omega z}}{\int_{0}^{z_d} e^{-\omega z}}$$

where f_i represents the fractional root activity in the layer i, z is depth, z_{ti} and z_{bi} represent the top and bottom of a rooted layer, and z_d the rooting depth. The maximum rooting depth is typically 2 m. The parameter α defines the rate of root distribution decay, with values ranging from top-heavy distributions ($\omega = 10$), to bottom-heavy distributions at ($\omega = -10$). This parameter is not open to the user. However, for the purpose of this book chapter, we altered this parameter to show how a shallow, even, or inverse root distribution (more root activity at the root depth progression front) affects water uptake. Results are shown in the applications section.

Cold damage is estimated as linear function between the two thresholds provided in Table 1 and based on the minimum temperature of the air only. If that temperature is that of the lower threshold, then all green leaves will be lost, with the fraction varying linearly when the minimum temperature approaches the threshold. Snow cover protects against cold damage. Finally, N demand is calculated using Greenwood et al. (1991) biomass dilution method. However, the N concentration of the new growth is calculated based on a crop growing unstressed, as this method lumps growth and development in the dilution curve. The parameters of the N dilution curve can depend on the ratio of leaf to stem and other parameters but for agronomic simulations the method provides a safe estimation of N demand.

7 The soil carbon balance and carbon saturation

In terrestrial systems that are not actively storing woody biomass, most of the CO_2 captured from the atmosphere is returned to it via autotrophic and heterotrophic respiration. Autotrophic respiration has been estimated to be *ca.* 50% of gross photosynthesis. In systems where C pools are at steady-state, heterotrophic respiration amounts to the other 50% (unless there is extraction of C via harvest or by other means). If soil C pools are accruing, the fraction of the photosynthates retained a C_s must be small, likely <0.2 kg/kg of C inputs (Mazzilli et al., 2014; Rasmussen et al., 1998). This makes modeling C_s cycling challenging (Jenkinson, 1989; Kemanian and Stockle, 2010). The N cycling is tightly coupled with that of C so modeling these two elements under the same framework requires accounting for the short-term mineralization and immobilization that often control mineral N pools as well as the more parsimonious changes in C_s (White et al., 2014).

The soil C model in Cycles is based on C-Farm (Kemanian and Stöckle, 2010). Rather than relying on multiple soil C pools that decay with first-order kinetics, it adopts the proposition of Hassink and Whitmore (1997), which posits a maximum and saturating mineral-associated organic C storage capacity (C_{sx}). This saturation framework yields valuable emergent properties in agroecosystem models. Most C_s consists of microbial detritus, as reflected in its C:N ratio of 5 to 15 by mass. The parameterization of C_{sx} and its influence on C_s dynamic is empiric and resembles an adsorption isotherm. Because it underpins Cycles, a simplified analytical version is presented here.

The organic C pools in Cycles are aboveground plant residue, roots, rhizodeposits, and manure, and C_s proper, which is separated into microbial biomass and non-microbial biomass organic C. For simplicity, we will develop the analytical solution lumping the microbial pool with organic matter. Ignoring the microbial biomass renders a good simulation of C fluxes but can smooth out mineral N oscillations caused by cycles of immobilization and mineralization (White et al., 2014). Without considering saturation, the simplest C_s model formulation is that from Hénin and Dupuis (1945) also used by Andriulo et al. (1999), with two pools, one for residue (C_r) and one for soil (C_s):

$$\frac{dC_s}{dt} = h_r C_r - k_s C_s \qquad ,(9)$$

The coefficient h_r is the retention of C originally in the residue as C_s (kg/kg). The coefficient k_s is the apparent C_s decomposition rate (1/time) (Jenkinson and Parry, 1989) and is a function of tillage, soil temperature, humidity, and any factor that alters microbial activity. The input C_r can be expressed in Mg ha^{-1} year^{-1} and assumed to be completely decomposed in 1 year. To include the concept of organic C saturation, Eq. [9] can be modified as follows:

$$\frac{dC_s}{dt} = h_x\left(1-\left(C_s/C_{sx}\right)^{\alpha}\right)C_r - k_x\left(C_s/C_{sx}\right)^{\beta}C_s. \qquad (10)$$

The coefficient h_x is now a maximum C retention coefficient that can be modulated by for example the soil clay content and clay type (as in Roth-C), and critically, by the saturation ratio C_s/C_{sx}. The higher the ratio the lower the retention rate, as shown experimentally by Pravia (2019). Analogously, k_s is replaced by a turnover rate that increases as the saturation ratio decreases. Priming is not integrated in this equation, but if so, the properties of the decomposing residue should alter k_s as shown by Mazzilli et al. (2014). The powers α and β are empiric adjustments.

A key component of this formulation is the saturation value of C_s, or C_{sx}. Lacking any other means to compute it, we use the equation provided by Hassink and Whitmore (1997), which makes C_{sx} a function of the clay fraction (f_{clay}, kg/kg) and multiply it by the soil layer bulk density (ρ_s, Mg/m^3) and thickness dz (m) to obtain:

$$C_{sx} = \rho_s dz\left(0.021 + 0.038 f_{clay}\right). \qquad (11)$$

The coefficients h_x and k_x are determined empirically. The highest value of h_x is approximately 0.18 kg/kg and is conceptualized as three microbial predation cycles with a maximum retention of 0.56 kg/kg per cycle (White et al., 2014). It is considered higher for manure (about 0.31 kg/kg), or two microbial predation cycles, as physical and chemical breakdown already occurred in the digestive tract of the animal. These figures are close to parameters summarized in Kemanian and Stockle (2010). When $C_s = C_{sx}$ then the retention is 0 and no accumulation of $C_s > C_{sx}$ can occur.

The apparent turnover rate k_S is reduced from the baseline of $k_x = 0.055$ kg kg^{-1} year^{-1} for undisturbed soils based on data in Andriulo et al. (1999) or increased by tillage (Balesdent et al., 2000). The disturbance rating for each tool as reported for USDA-NRCS is used to calculate the tillage factor (Kemanian and Stöckle, 2010). The important theoretical contribution is that the dependence of the turnover rate on $C_s^{1+\beta}$ with $\beta > 0$ makes the model nonlinear. Both the saturating behavior and the non-linear dependence on the current C_s of a soil layer are distinctive features of Cycles. The model algorithm includes the microbial pool and coupling with the N cycle (Kemanian et al., 2024).

To provide an intuitive understanding of the effects of saturation, Eq. [10] can be simplified setting $\alpha = \beta = 1$. If all residue decomposes in 1 year, the differential equation can be solved explicitly for C_s. Eq. [10] can be formulated as follows:

$$\frac{dC_s}{dt} = hC_r - \frac{hC_r}{C_{sx}}C_s - \frac{k}{C_{sx}}C_s^2 \tag{12}$$

The constants h and k stand for the effective values for these soils and do not affect the nature of the differential equation that can now be solved analytically. The solution to this equation introduced before in Kemanian et al. (2011) is

$$C_s(t) = \frac{hC_r}{2k}\left[\left(\frac{\Phi e^{\frac{\gamma t}{C_{sx}}} - 1}{\Phi e^{\frac{\gamma t}{C_{sx}}} + 1}\right)\sqrt{1 + \frac{4kC_{sx}}{hC_r}} - 1\right] \tag{13}$$

$$\gamma = hC_r\sqrt{1 + \frac{4kC_{sx}}{hC_r}}$$

$$\Phi = -\frac{2kC_s(t=0) + hC_r + \gamma}{2kC_s(t=0) + hC_r - \gamma}$$

The integration constant Φ depends on $C_s(t = 0)$. The steady-state solution for Eq. [10] is

$$C_s = \frac{hC_r}{2k}\left(\sqrt{1 + \frac{4kC_{sx}}{hC_r}} - 1\right) \tag{14}$$

The ratio hC_r/k is the steady state C_s for Eq. [9] when independent of the saturation ratio. In Eq. [14] C_s approaches C_{sx} as C_r increases. Linear models project a linear increase in C_s with C_r (as in Century or Roth-C), while the saturating model projects an asymptotic approach to C_{sx}. This distinction is important in scenarios like no-till where C_s accumulates in the topsoil causing stratification of soil properties. The equilibrium C_s is very sensitive to changes in C_{sx}, and estimating it correctly is critical for agroecosystem and earth system modeling. This is an area that requires careful research guided by a transparent modeling framework.

Cycles and similar models require careful initialization to reflect whether soil properties are in steady or transient states. Using uncurated data, such as

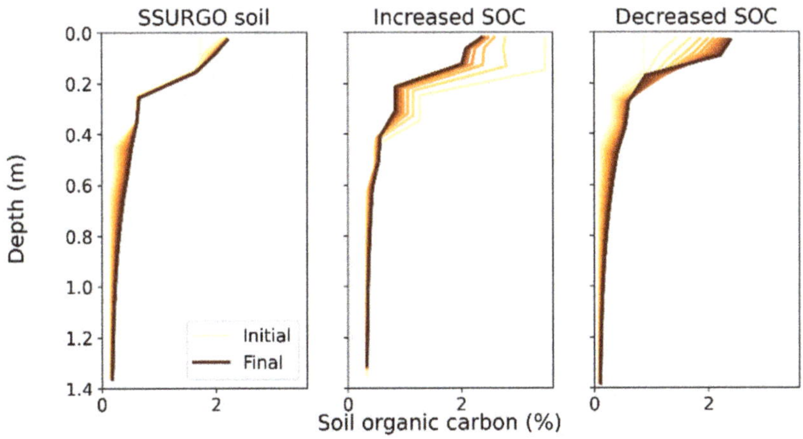

Figure 5 Soil carbon profile for Hagerstown soil from the SSURGO database, equilibrated with Cycles under a maize–soybean rotation. The left panel shows the original data. The middle and right panels start with C_s 50% higher or lower than SSURGO and track their return to steady-state C_s.

SSURGO-derived horizon averages, can result in major errors. In Cycles, users may reset soil conditions using observed data or allow the model to reach equilibrium before applying scenarios, as illustrated in Fig. 5. This process yields a stable C_s profile with depth and prevents confounding simulation drifts.

8 Model testing

Cycles performance has been tested for crop yield and N uptake, evapotranspiration and soil water depletion with depth (Kemanian et al., 2024), nitrous oxide emissions (Saha et al., 2017), and C_s dynamics in long-term experiments with tillage at Rothamsted, UK, and Pendleton, OR (Stockle and Kemanian et al., 2010), and under no-till (Pravia et al., 2019). In addition, as embedded in Cycles-L, the model has also been tested for watershed hydrology and N leaching (Shi et al., 2023). A summary of some of these tests is presented in Table 2.

To illustrate the versatility of the model, we present an additional evaluation of its yield prediction skill across states in the U.S. Midwest. Using a dataset with county-level soil information and daily weather from Xia et al. (2012), maize was simulated from 1980 to 2020. Yield first differences from the NASS database were compared with those from Cycles to detrend for technology changes. No calibration was performed. Soil depth was uniformly set at 2 m. Simulations were run without nutrient limitations.

Table 2 Summary evaluation of the Cycles model

Variable and other information	n	r^2	β_0	β_1	rRMSE	Ref.
	--	--	Mg/ha	Mg/Mg	--	
Multispecies perennial pasture yield, Uruguay	28	0.70	1.50	0.86		1
Sorghum yield, Uruguay	12	0.47	0.84	0.78		1
Soybean yield, Uruguay	13	0.47	1.70	0.72		1
Soybean yield, Rock Spring, Pennsylvania	10	0.66	0.50	1.04		2
Soybean yield, Landisville, Pennsylvania	12	0.90	1.40	0.86		2
Soybean yield, Passo Fundo, Brazil	8	0.71	-0.10	1.28		2
Soybean yield, Guarapuava, Brazil	9	0.73	2.80	0.66		2
Maize yield, NASS, first difference, Delaware	38	0.43	-0.09	1.10		3
Maize yield, NASS, first difference, Maryland	775	0.46	-0.05	1.10		3
Maize yield, NASS, first difference, New York	300	0.03	0.04	0.30		3
Maize yield, NASS, first difference, Pennsylvania	1053	0.51	-0.02	1.10		3
Maize yield, NASS, first difference, Virginia	1595	0.50	-0.02	1.00		3
Maize yield, NASS, first difference, West Virginia	209	0.29	-0.09	1.10		3
C_s, 20 years continuous cropping, Uruguay	20				0.13	1
C_s, 20 years crop–pasture rotation, Uruguay	20				0.14	1
C_s, 20 years perennial pasture, Uruguay	20				0.15	1

[1]Pravia et al. (2019) field-scale experiments; [2]Stefani-Fae (2019), plot-level experiments; [3]McConnell et al. (2023), NASS yield.
The column β0 and β1 represent intercept and slope of the 1:1 regression between modeled and observed records. The column rRMSE represents the relative root mean square error

The results, judged by the map of r^2 values, show that Cycles captured the environmental signal in yield especially well in the driest parts of the region, generally west of the 95th meridian and in counties less influenced by the water table (Fig. 6). In Iowa and southern Minnesota, where shallow water tables moderate drought effects, the model performance declined. This influence is better represented with models that consider topography and simulate water tables explicitly (Shi et al., 2023).

This can be better illustrated by focusing on three representative counties and using 1:1 plots and regression slopes for comparison. In Franklin County, Nebraska, and Clark County, Kansas, Cycles reproduced the environmental signal effectively. However, in Franklin County the response was apparently

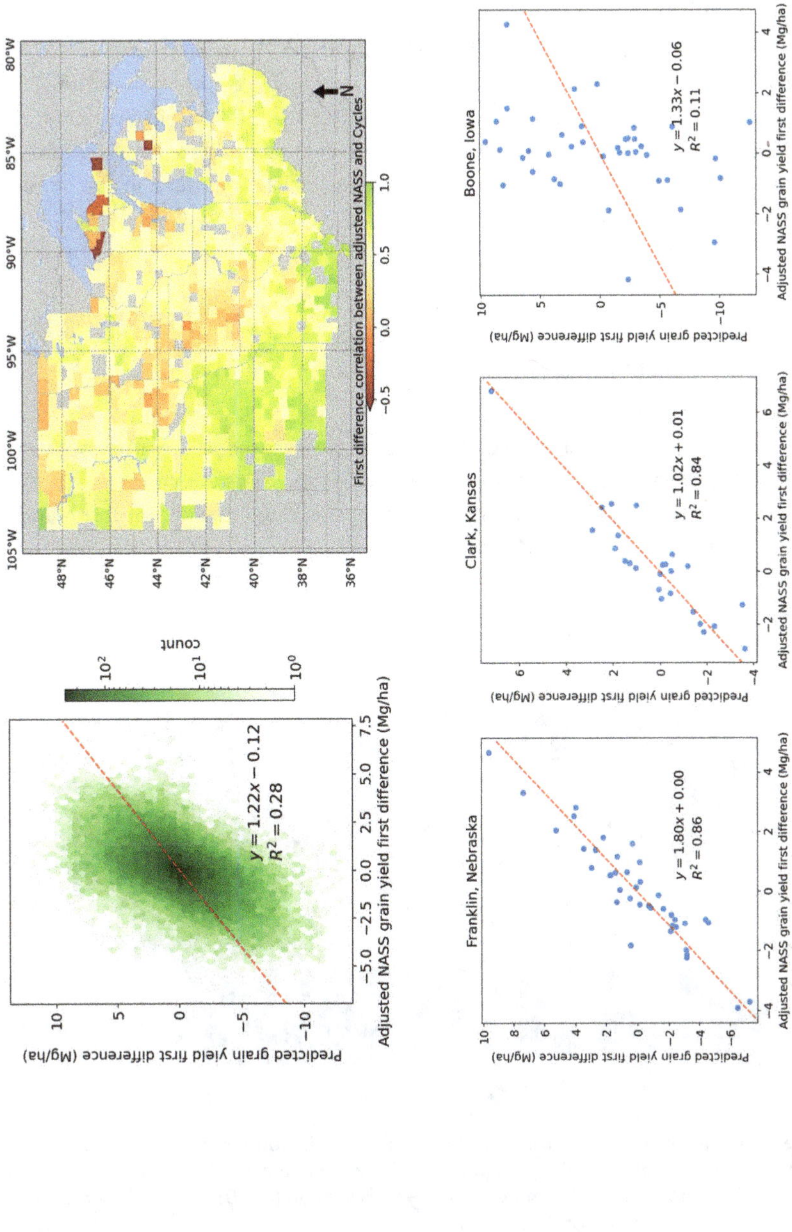

Figure 6 A comparison of Cycles-predicted and NASS-county level maize yield for 1980 to 2020 in the U.S. Midwest. The upper left panel shows the 1:1 plot for the yield's first difference. The upper right panel shows a map of the r^2 of the regression for each county. The lower panels show the 1:1 plot for the counties of Franklin (NE), Clark (KS), and Boone (IA). Simulations were run with N limitation.

overestimated (slope = 1.8 kg/kg) compared with Clark County (slope = 1.02 kg/kg). Without additional analysis, it is difficult to determine which result is more realistic. In Boone County, Iowa, where the water table is present and its dynamics was not accounted for, Cycles predicted yield variability up to 10 Mg/ha year-to-year compared to the subdued observed variability which remained below 4 Mg/ha (Fig. 7).

Given the limited and off the shelve inputs, the performance of the model is encouraging. Still, the results highlight limitations of mechanistic models in some settings. Their structure allows systematic interpretation, but they cannot and should not compensate for limited inputs or missing landscape features. In contrast, analysis with machine learning of a similar database in the same region, such as those by Hoffman et al. (2020), achieved r^2 values up to 0.86, even without soil data. These results underscore not only the power but also the limitations of process-based modeling.

9 Applications

We present three model applications. The first explores water uptake across maize root phenotypes. The second simulates maize–soybean rotations on over 871 000 Iowa fields, comparing till and no-till systems with a focus on soil carbon and nitrous oxide emissions. The third examines mixing sorghum and maize for silage production in the U.S.

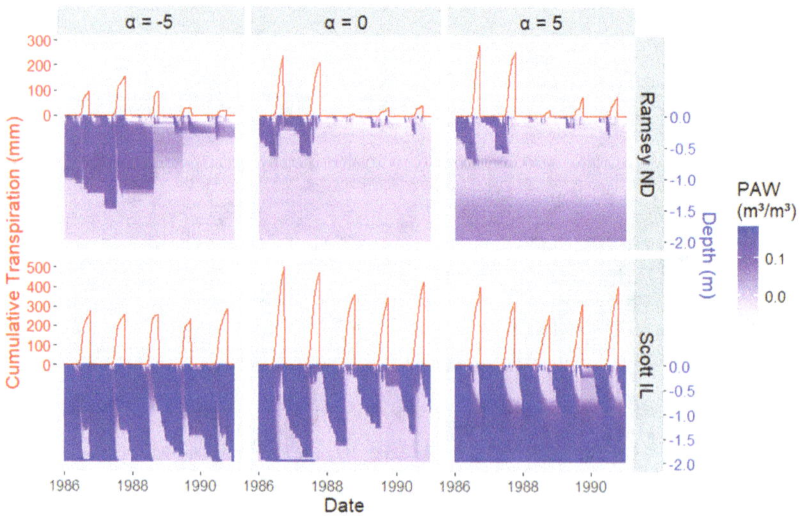

Figure 7 Cumulative season transpiration (red) and distribution of water with depth across the 2-m soil column (blue) between 1986–1991 in Ramsey County, ND and Scott County, IL, for moderately bottom-heavy (α = −5), balanced (α = 0) and moderately top-heavy (α = 5) exponential root distributions.

9.1 Case 1: Maize root distribution with depth affects water uptake

Root distribution of crops affects access to soil resources, particularly water. In Cycles, water uptake depends on root activity distribution, often surrogated with the relative root length density. Soil water availability is shaped by both precipitation regime and root distribution. In wet regions, plants access water from a consistently replenished topsoil and access to shallow or deep water has similar value. In regions that are drier or have erratic precipitation, predicting optimal root distribution is difficult. A bottom-heavy root system may lose surface recharge to evaporation, while a top-heavy root system may capture infiltrated water but miss deeper reserves. The ideal distribution may shift yearly. Experimentally testing this range is impractical, but models can offer valuable initial insights. If root architecture can be genetically tuned, what root distribution would be fitting for regions with contrasting hydric regimes?

We tested water uptake for ω values of 5, 0, and −5. An ω of 5 reflects surface-skewed root activity and shallower depth than typical crops (Fan et al., 2016). Conventional maize show ω values near 3–4 (Dwyer et al., 1996). Values from 0 to 3 resemble deeper-rooting species like alfalfa or taprooted crops (Fan et al., 2016). Negative ω values, which have no known annual crop counterparts, represent bottom-heavy distributions where uptake is dominated by the deepest roots, somewhat akin to apical dominance in water acquisition via aquaporins. The objective is to explore the magnitude of the changes in water uptake if designer roots are possible.

Simulations were conducted for a dry and a wet Midwestern U.S. county, Ramsey County, North Dakota, and Scott County, Illinois, over 1980–2020. Data from 1986 to 1991, spanning dry and wet years, are shown in Fig. 7. Dominant cropland soil textures were obtained from the Web Soil Survey. Weather inputs came from the grid nearest each county center using NLDAS-2 forcing data (Xia et al., 2012). Maize maturity and phenology were based on growing degree days (base temperature = 6°C), with planting windows constrained by soil temperature and moisture thresholds.

Over years of continuous cropping in drier Ramsey, ND, where water recharge is limited, root phenotypes showed distinct water depletion patterns: bottom-heavy phenotypes depleted the subsoil, top-heavy phenotypes depleted the topsoil, and balanced phenotypes depleted the entire soil profile (Fig. 7). The top-heavy phenotype performed well in the relatively wetter 1986 and 1987, relying on in-season precipitation. In dry 1988, all phenotypes had limited growth. The top-heavy phenotype could not access stored water while the balanced one had no water reserves, but the bottom-heavy type continued extracting water as the rooting front deepened. As a result, the bottom-heavy phenotype transpired more than the others. In 1989, moderate recharge did not restore subsoil moisture. The bottom-heavy phenotype, having depleted subsoil water, was unable to access shallow water and transpired little. The balanced phenotype slowly accessed infiltrated water but faced strong evaporative competition. The top-heavy phenotype increased transpiration by capturing incoming water. These year-to-year dynamics made the optimal ω variable. In contrast, in Scott IL where soil moisture is replenished throughout the profile, the balanced phenotype efficiently extracts water without depleting reserves or causing yield loss (Fig. 7).

An optimization algorithm could help identify the ideal root distribution for various hydric and soil conditions. Such results would offer breeding targets and prompt deeper research into the genetic and environmental regulation of root activity with depth.

9.2 Case 2: Tillage effects on soil carbon and nitrous oxide emission in Iowa

Increase interest in climate smart agriculture has renewed focus on modeling tools to estimate greenhouse gas emission changes resulting from management practices. When properly validated, such models can support payment schemes for emissions reductions using standardized and potentially low-cost methods. We present here some of the basic work needed for applying Cycles to this task using Iowa cropland, where we run simulations across more than 871 000 fields corresponding to the Common Land Units in Hunt et al., 2024.

Simulations used weather data from NLDAS (Xia et al., 2012), soil data from the Gridded National Soil Survey Geographic Database, field boundaries from a public CLU dataset, tile drainage presence (Nakagaki and Wieckzorek, 2016), and crop rotations from CropScape 2013–2016 (Han et al., 2012). Manure application was assigned using an algorithm based on the location and capacity of CAFOs with over 1000 animal units across Iowa (IDNR, 2016). Manure was distributed to nearby fields until stores were depleted (Fig. 8) and applied at 220 kg/ha of N for maize, with a 60 kg/ha of N credit following soybean. In manured fields and for simplicity, no additional fertilizer was applied apart from a 20 kg/ha of N starter. Applications occurred on April 22 and October 30.

To establish consistent baseline conditions, crop rotations were assigned using the 2013–2016 CropScape cropland data (Boryan et al., 2011). Rotations included continuous maize, continuous soybean, maize–soybean, and maize–maize–soybean sequences. Maize planting began April 25 followed by soybean planting a week later.

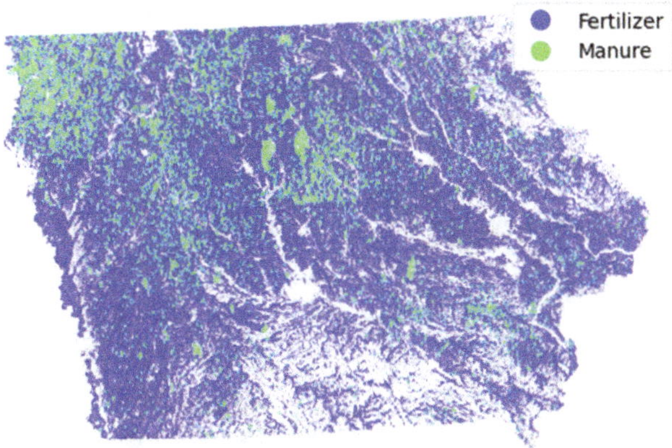

Figure 8 Manure distribution at field-scale for Iowa. Map based on CAFOs data from the Iowa Department of Natural Resources (2016), ASABE 2005 standards to calculate manure produced by each animal category, and Andersen and Pepple (2017) and Rotz (2004) to estimate nutrient losses during storage.

Yields, shown as quantiles in Fig. 9, are presented alongside simulated changes in N_2O–N emissions and C_s. Maize yields averaged 9.9 and 6.5 Mg/ha in the Eastern Iowa and Minnesota Drift Plains and Missouri Alluvial Plain ecoregions, respectively. Soybean yields were 3.4 and 2.3 Mg/ha in the same regions. The tiling pattern in Fig. reflects the NLDAS weather grid. Patterns in N_2O change are difficult to generalize, as emissions can swing widely with small changes in hydrology or residue placement (Saha et al., 2021). Still, in higher-yielding areas, no-till appears to reduce emissions. Changes in C_s are more consistent, with average gains below 0.5 Mg ha^{-1} $year^{-1}$. Yet these results must be interpreted cautiously, as the comparison reflects only the first 7 years after changing management. Early C_s gains often concentrate in the topsoil and can later reverse due to subsoil losses (Cai et al., 2022).

How well do the physical environment, Iowa's crop management, and biogeochemical processes align with model representation? That question can be reframed by asking modelers and users to rank desired improvements. These might include more precise regional hydrology and water table modeling (Shi et al., 2023), reliable manure and fertilizer management surveys to avoid reliance on inferred manuresheds (e.g. Preston et al., 2011; Fig. 8), and more comprehensive model testing, which demand data beyond that typically collected in field trials. Yet this example shows that it is possible to virtually represent all agricultural fields in a state or a country. That potential should drive the development of surveys, monitoring systems, and experiments matching the resolution with which models like Cycles can portray the agricultural landscape (Fig. 9).

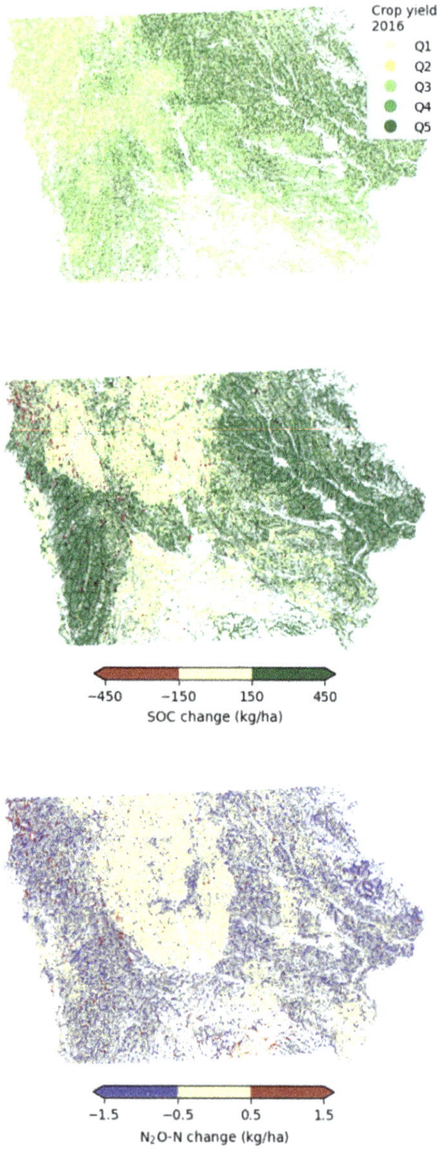

Figure 9 Crop yield (quantiles) in 2016 for a rotation managed with tillage (panel a), difference between no-till and tillage for N_2O-N in 2016 (panel b), and for soil organic carbon averaged from 2010 to 2016 (panel c). Each panel shows results for 871 000 individual fields.

9.3 Case 3: Mixing maize and sorghum for silage production in the United States

Maize silage is the backbone of dairy and ruminant production systems in the U.S., where it is grown in over 6.7 million acres (about 7% of the maize area). Despite steady maize yield increases over the past several decades, the crop remains sensitive to drought stress and high vapor pressure deficit during flowering (Lobell et al., 2014; Hoffman et al., 2020). Sorghum is a high-biomass crop that is commonly used for silage production in climates that are too hot and dry for maize, or where irrigation is limited, due to its favorable attributes in these suboptimal growing conditions (e.g. Sinclair et al., 2005; Laffray and Louguet, 1990). Burton et al. (2021) reported that silage polycultures comprising maize + sorghum yielded similar biomass to pure maize silage under rainfed conditions with adequate precipitation for both maize and sorghum growth. Mixing maize and sorghum in a polyculture may enable producers to stave off reductions in biomass during years with inadequate precipitation without compromising forage quality.

Investigating the use of maize + sorghum polycultures in a wide range of environments may be difficult as field studies are often costly and time consuming. We show here the use of Cycles to outline areas in the U.S. where maize + sorghum polycultures may produce comparably to maize monocultures and provide stability across years. These areas could serve as field sites to test polyculture performance.

We used Cycles to model rainfed maize and sorghum monocultures as well as a mixture of these crops in the cropland area of 3077 counties in the U.S. from 1980 to 2024. Only soils in areas with documented maize or sorghum were used in the simulation, as explained in Shi et al. (2025). The polyculture was simulated as a field comprising 50% maize and 50% sorghum randomly mixed. The crops were simulated in continuous sequence, a simplification that excludes rotational effects such as disease and pest pressures not represented in the model. Management was simplified by using automatic N, no irrigation, and no-till conditions. Planting was conditional on soil temperature, and both maize and sorghum were planted when soil at a depth of 0.1 m reached 11°C in spring, ensuring same-day planting for both species. The dominant soil series in each cropland and the corresponding NLDAS weather grid was used in each county.

Results are shown in relative terms to facilitate the comparison between the three alternatives (Fig. 10). To provide more quantitative information, overall, the model captured the east-west precipitation gradient east of the Rocky Mountains. For example, the average non-irrigated maize silage dry matter yield in Pennsylvania in 2013 was 18.8 Mg ha^{-1} and that of South Dakota was 8.6 Mg ha^{-1} while that of the model was 15–25 Mg ha^{-1} in Pennsylvania and

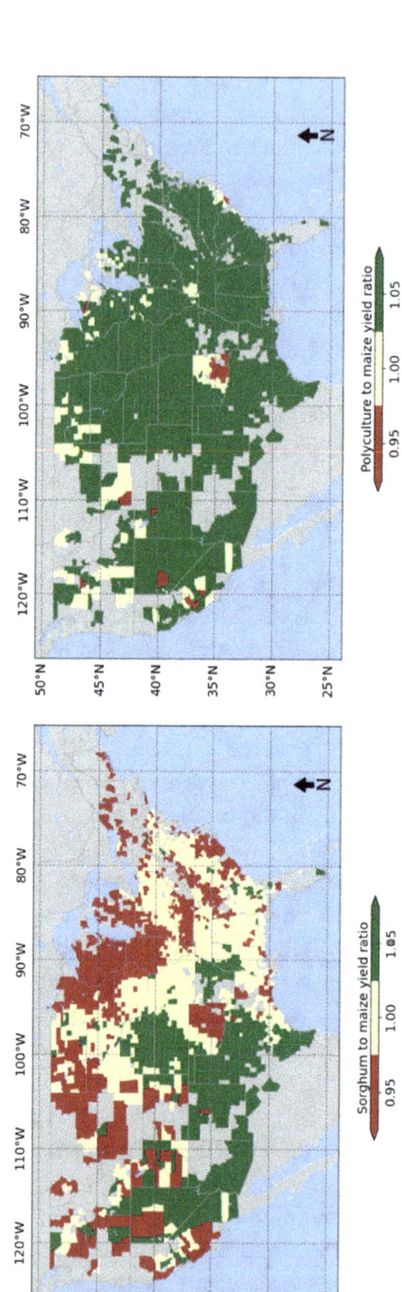

Figure 10 Ratio of sorghum and a polyculture of maize and sorghum silage yield to that of a monoculture of maize. Areas where either sorghum or the polyculture producing similar, higher, or lower biomass than maize are highlighted in yellow, green, and red, respectively. Simulations span from 1980 to 2024.

0–15 Mg ha^{-1} in South Dakota, with lower yields in the western part of the state (USDA NASS, 2020).

The results of the simulation show, as expected, that sorghum would outyield maize in a north south band around the 100° meridian, roughly corresponding to the U.S. sorghum belt (Fig. 10). It does so as well in areas of low rainfall in the west although actual production in those areas is marginal. The simulation results also show that the maize–sorghum polyculture would provide yield stability across years, since compared to maize, greater yields in sorghum were observed in the sorghum belt while polyculture yields were comparable to or slightly lower than maize (Fig. 10). Maize silage yields in the sorghum belt (both irrigated and non-irrigated) are around 8 Mg ha^{-1} (Bhattarai et al., 2019) and compare relatively well with the 5–15 Mg/ha simulated with Cycles here. In warm and arid rainfed conditions, when maize is low-yielding (<7 Mg ha^{-1}) or the availability of water for irrigation is uncertain, sorghum has already proved to be a successful alternative to maize silage monocultures in this area and is favored for its lower overall water use, equivalent silage quality, and lower production cost (Oliver et al., 2004; Bhattarai et al., 2019).

Using maize and sorghum polycultures for silage in the sorghum belt and beyond may stabilize yield and forage quality compared to maize monocultures. However, the added complexity of managing a mixture, like matching the phenology of both species, simultaneous planting of seeds of different sizes, and having a manageable weed control program are critical to realize the potential of this polyculture (Burton et al., 2021). Planting fractions of the area with each crop may provide an alternative to mixing. These results suggest that incorporating sorghum into maize silage systems may buffer against variable moisture conditions and possibly biotic stress, though practical hurdles in polyculture management remain.

10 Conclusion and future trends

In developing Cycles, our goal is to balance scientific rigor, practical utility, and flexible software design for agricultural modeling. In this chapter, we explored the theory and implementation of CO_2 capture through photosynthesis and its return via microbial respiration, emphasizing soil processes that control carbon turnover. Building a model sharpens our understanding of the processes of interest, an experience that we share with students through a simple motto: if we know something, we can model it. If we do not, modeling helps form hypotheses and design experiments to close those gaps. We also illustrated how models allow investigation of root phenotypes suited to hydric regimes in the Midwest and crop combinations that stabilize forage yield. These examples show that models like Cycles are not just repositories of knowledge but tools for inquiry.

The use of agroecosystem models for research and decision support is well established and will continue to grow in sophistication (Gil et al., 2021). Future development of Cycles includes a standardized 2D soil and canopy structure, a refined user version of the advanced 3D Cycles-L (Shi et al., 2023), integration of an animal production module, and nutrient modeling beyond nitrogen. Rather than a static endpoint, we see Cycles evolving as a scaffold for agricultural research and innovation in the U.S. and beyond.

11 Where to look for further information

Cycles is available in Github at:

- https://github.com/PSUmodeling/Cycles/releases.
- The manual is periodically updated and can be accessed here
- http://psumodeling.github.io/Cycles/.

12 References

Allen, R.G., et al., 1998. Crop evapotranspiration-Guidelines for computing crop water requirements-FAO Irrigation and drainage paper 56. *Fao, Rome, 300*(9), p.D05109.

Andersen, D.S. and Pepple, L.M., 2017. A county-level assessment of manure nutrient availability relative to crop nutrient capacity in Iowa: Spatial and temporal trends. *Transactions of the ASABE, 60*(5), pp.1669–1680.

Andriulo, A., et al., 1999. Modelling soil carbon dynamics with various cropping sequences on the rolling pampas. *Agronomie, 19*(5), pp.365–377.

Balesdent, J., et al., 2000. Relationship of soil organic matter dynamics to physical protection and tillage. *Soil & Tillage Res*, 53, pp.15–230.

Bassiouni, M. and Vico, G., 2021. Parsimony vs predictive and functional performance of three stomatal optimization principles in a big-leaf framework. *New Phytologist, 231*(2), pp.586–600.

Bhattarai, B., et al., 2019. Forage potential of pearl millet and forage sorghum alternatives to corn under the water-limiting conditions of the Texas high plains: A review. *Crop, Forage & Turfgrass Management*, *5*(1), pp.1–12.

Boryan, C., et al., 2011. Monitoring US agriculture: the US department of agriculture, national agricultural statistics service, cropland data layer program. *Geocarto International*, *26*, pp.341–358.

Buckley, T.N. and Schymanski, S.J., 2014. Stomatal optimisation in relation to atmospheric CO_2. *New Phytologist*, *201*(2), pp.372–377.

Burton, A.B., et al., 2021. Are polycultures for silage medleys or gallimaufries? *Agronomy Journal, 113*(6), pp.4969–4982. http://doi.org/10.1002/agj2.20602.

Cai, A., et al., 2022. Declines in soil carbon storage under no tillage can be alleviated in the long run. *Geoderma*, *425*, p.116028.

Camargo, G.G.T. and Kemanian, A.R., 2016. Six crop models differ in their simulation of water uptake. *Agricultural and Forest Meteorology*, *220*, pp.116–129.

Campbell, M.D., et al., 1974. A model describing soil–plant–water relations for potatoes. *American Journal of Potato Research*, *53*, 431–441.

Campbell, G.S. 1985. *Soil physics with BASIC: transport models for soil-plant systems.* (Vol. 14), Elsevier.

Campoy, J., et al., 2020. Estimation of harvest index in wheat crops using a remote sensing-based approach. *Field Crops Research*, *256*, p.107910.

Cowan I.R. and Farquhar G.D., 1977. Stomatal function in relation to leaf metabolism and environment. *Symposium of the Society for Experimental Biology*, *31*, pp.471–505.

Drouet, J.L. and Pagès, L., 2003. GRAAL: a model of GRowth, Architecture and carbon ALlocation during the vegetative phase of the whole maize plant: model description and parameterisation. *Ecological Modelling*, *165*(2–3), pp.147–173.

Dwyer, L.M., et al., 1996. Root mass distribution under conventional and conservation tillage. *Canadian Journal of Soil Science*, *76*(1), pp.23–28.

Fan, J., et al., 2016. Root distribution by depth for temperate agricultural crops. *Field Crops Research*, *189*, pp.68–74.

Farquhar, G.D., et al., 2001. Models of photosynthesis. *Plant physiology*, *125*(1), pp.42–45.

Firestone, M.K., 1982. Biological denitrification. *Nitrogen in Agricultural Soils*, *22*, pp.289–326.

Flerchinger, G.N., 2000. *The simultaneous heat and water (SHAW) model: technical documentation*. Northwest Watershed Research Center USDA Agricultural Research Service, Boise, Idaho.

Fu, Z., et al., 2024. Global critical soil moisture thresholds of plant water stress. *Nature Communications*, *15*(1), p.4826.

Gil, Y., et al., 2021. Artificial intelligence for modeling complex systems: taming the complexity of expert models to improve decision making. *ACM Transactions on Interactive Intelligent Systems*, *11*(2), pp.1–49.

Greenwood, D.J., et al., 1991. Growth rate and% N of field grown crops: theory and experiments. *Annals of Botany*, *67*(2), pp.181–190.

Hammer, G.L., et al., 2002. Future contributions of crop modelling–from heuristics and supporting decision making to understanding genetic regulation and aiding crop improvement. *European Journal of Agronomy*, *18*(1–2), pp.15–31.

Han, W., et al., 2012. CropScape: a Webservice based application for exploring and disseminating US conterminous geospatial cropland data products for decision support. *Computers and Electronics in Agriculture*, *84*, pp.111–123.

Hari, P., et al., 1986. Optimal control of gas exchange. *Tree physiology*, *2*(1-2-3), pp.169–175.

Hassink, J. and Whitmore, A.P., 1997. A model of the physical protection of organic matter in soils. *Soil Science Society of America Journal*, *61*(1), pp.131–139.

Hoffman, A., et al., 2020. The response of maize, sorghum, and soybean yield to growing-phase climate revealed with machine learning. *Environmental Research Letters*, *15*(9).

Hénin, S. and Dupuis, M., 1945. Essai de bilan de la matière organique du sol. *Annales Agronomiques*, *15*, pp.17–29.

Hunt, K.A., et al., 2024. Crop sequence boundaries using USDA National Agricultural Statistics Service historic cropland data layers. *Statistical Journal of the IAOS*, *40*(2), pp.237–246.

Iowa Department of Natural Resources (IDNR). 2016. Iowa geospatial data animal feeding operations. https://geodata.iowa.gov/documents/iowadnr::animal-feeding-operations-download/.

Jarvis, P.G. and Slatyer, R.O., 1970. The role of the mesophyll cell wall in leaf transpiration. *Planta*, *92*, pp.303–322.

Jenkinson D.S. and L.C. Parry, 1989. The nitrogen cycle in the Broadbalk wheat experiment: a model for the turnover of nitrogen through the soil microbial biomass. *Soil Biology and Biochemistry*, *21*, pp.535–541.

Katul, G., et al., 2010. A stomatal optimization theory to describe the effects of atmospheric CO2 on leaf photosynthesis and transpiration. *Annals of botany*, *105*(3), pp.431–442.

Kemanian, A.R. and Stockle, C.O., 2010. C-farm: a simple model to evaluate the carbon balance of soil profiles. *European Journal of Agronomy*, *32*(1), pp.22–29.

Kemanian, A.R., et al., 2005. Transpiration-use efficiency of barley. *Agricultural and Forest Meteorology*, *130*(1–2), pp.1–11.

Kemanian, A.R., et al., 2007. A simple method to estimate harvest index in grain crops. *Field Crops Research*, *103*(3), pp.208–216.

Kemanian, A.R., et al., 2011. Integrating soil carbon cycling with nitrogen and phosphorus in the watershed model SWAT: theory and model testing. *Ecological Modelling*, *222*(12), pp.1913–1921.

Kemanian, A.R. et al. 2024. The cycles agroecosystem model: fundamentals, testing, and applications. *Computers and Electronics in Agriculture*, *227*, 109510.

Kerr, Y.H., et al., 2001. Soil moisture retrieval from space: the Soil Moisture and Ocean Salinity (SMOS) mission. *IEEE Transactions on Geoscience and Remote Sensing*, *39*(8), pp.1729–1735.

Kumudini, S., et al., 2014. Predicting maize phenology: intercomparison of functions for developmental response to temperature. *Agronomy Journal, 106*(6), pp.2087–2097.

Kustas, W.P., et al., 2007. Utility of radiometric–aerodynamic temperature relations for heat flux estimation. *Boundary-Layer Meteorology*, *122*, pp.167–187.

Laffray, D. and Louguet, P., 1990. Stomatal responses and drought resistance. *Bulletin de la Société Botanique de France. Actualités Botaniques*, *137*(1), pp.47–60.

Lasky, J.R., et al., 2023. Genotype–environment associations to reveal the molecular basis of environmental adaptation. *The Plant Cell*, *35*(1), pp.125–138.

Lloyd, J. 1991. Modelling stomatal responses to environment in Macadamia integrifolia. *Functional Plant Biology*, *18*(6), pp.649–660.

Lobell, D.B., et al. (2014). Greater sensitivity to drought accompanies maize yield increase in the U.S. Midwest. *Science*, *344*, pp.516–519.

McLaughlin, C.M., et al., 2025. Maladaptation in cereal crop landraces following a soot-producing climate catastrophe. *Nature Communications*, *16*(1), pp.1–14.

Manzoni, S., et al., 2011. Optimizing stomatal conductance for maximum carbon gain under water stress: a meta-analysis across plant functional types and climates. *Functional Ecology*, *25*(3), pp.456–467.

Manzoni, S. and Katul, G., 2014. Invariant soil water potential at zero microbial respiration explained by hydrological discontinuity in dry soils. *Geophysical Research Letters*, *41*(20), pp.7151–7158.

Márquez, D.A., et al., 2024. Mesophyll airspace unsaturation drives C4 plant success under vapor pressure deficit stress. *Proceedings of the National Academy of Sciences*, *121*(39), p.e2402233121.

Mazzilli, S.R., et al., 2014. Priming of soil organic carbon decomposition induced by corn compared to soybean crops. *Soil Biology and Biochemistry*, *75*, pp.273–281.

McConnell, C.A., et al., 2023. Tradeoffs when interseeding cover crops into corn across the Chesapeake Bay watershed. *Agricultural Systems*, *209*, p.103684.

Messina, C.D., et al., 2018. Leveraging biological insight and environmental variation to improve phenotypic prediction: Integrating crop growth models (CGM) with whole genome prediction (WGP). *European Journal of Agronomy*, *100*, pp.151–162.

Monteith J.L., 1978. Models and measurements in crop climatology. In *Proceeding of the 11th International Society of Soil Science* (19–27 June 1978, Edmonton, Alberta, Canada.). Volume 3, 385–399.

Nakagaki, N. and Wieczorek, M.E., 2016. Estimates of subsurface tile drainage extent for 12 Midwest states, 2012. *US Geological Survey data release*. https://doi.org/10.5066/F7W37TDP.

Oliver, A.L., et al., 2004. Comparison of brown midrib-6 and-18 forage sorghum with conventional sorghum and corn silage in diets of lactating dairy cows. *Journal of dairy science*, *87*(3), pp.637–644.

Parent, B. and Tardieu, F., 2012. Temperature responses of developmental processes have not been affected by breeding in different ecological areas for 17 crop species. *New Phytologist*, *194*(3), pp.760–774.

Pravia, M.V., et al., 2019. Soil carbon saturation, productivity, and carbon and nitrogen cycling in crop-pasture rotations. *Agricultural Systems*, *171*, pp.13–22.

Preston, S.D., et al., 2011. Factors affecting stream nutrient loads: a synthesis of regional SPARROW model results for the continental United States 1. *JAWRA Journal of the American Water Resources Association*, *47*(5), pp.891–915.

Poggio, L., et al. 2021. SoilGrids 2.0: producing soil information for the globe with quantified spatial uncertainty. *SOIL*, *7*, pp.217–240.

Rasmussen, P.E., et al., 1998. Soil C and N changes under tillage and cropping systems in semi-arid Pacific Northwest agriculture. *Soil and Tillage Research*, *47*, pp.197–205.

Rotz, C.A., 2004. Management to reduce nitrogen losses in animal production. *Journal of animal science*, *82*(suppl_13), pp.E119–E137.

Saha, D., et al., 2017. Designing efficient nitrous oxide sampling strategies in agroecosystems using simulation models. *Atmospheric Environment*, *155*, pp.189–198.

Saha, D., et al., 2018. Lorenz curve and Gini coefficient reveal hot spots and hot moments for nitrous oxide emissions. *Journal of Geophysical Research: Biogeosciences*, *123*(1), pp.193–206.

Saha, D., et al., 2021. Organic fertility inputs synergistically increase denitrification-derived nitrous oxide emissions in agroecosystems. *Ecological Applications*, *31*(7), p.e02403.

Shi, Y., et al., 2023. Cycles-L: A coupled, 3-D, land surface, hydrologic, and agroecosystem landscape model. *Water Resources Research*, *59*(8), p.e2022WR033453.

Shi, Y., et al., 2025. Adapting agriculture to climate catastrophes: the nuclear winter case. *Environmental Research Letters*, *20*(6), p.064006.

Sinclair, T.R., et al., 1992. Variation in crop radiation-use efficiency with increased diffuse radiation. *Crop Science*, *32*(5), pp.1281–1284.

Sinclair, T.R., et al., 2005. Potential yield and water-use efficiency benefits in sorghum from limited maximum transpiration rate. *Functional Plant Biology*, *32*(10), pp.945–952.

Stefani-Faé, G., 2019. *Understanding Soybean yield limiting factors and the potential for agricultural intensification in the US and Brazil*. The Pennsylvania State University.

Stöckle, C.O. and Jara, J., 1998. Modeling transpiration and soil water content from a corn (*Zea maize* L.) field: 20 min vs. daytime integration step. *Agricultural and Forest Meteorology*, *92*(2), pp.119–130.

Stöckle, C.O. and Kemanian, A.R., 2009. Crop radiation capture and use efficiency: a framework for crop growth analysis. In V. O. Sadras and D. F. Calderini (eds.). *Crop physiology: applications for genetic improvement and agronomy* (pp. 145–170). Elsevier.

Stöckle, C.O. and Kemanian, A.R., 2020. Can crop models identify critical gaps in genetics, environment, and management interactions? *Frontiers in Plant Science*, *11*, p.737.

Stöckle, C.O., et al., 2014. CropSyst model evolution: From field to regional to global scales and from research to decision support systems. *Environmental Modelling & Software*, *62*, pp.361–369.

Tardieu, F., et al., 2017. Root water uptake and ideotypes of the root system: whole-plant controls matter. *Vadose Zone Journal*, *16*(9), p. vzj2017.05.0107.

Technow, F., et al., 2015. Integrating crop growth models with whole genome prediction through approximate Bayesian computation. *PLoS One*, *10*(6), p.e0130855.

Warren Wilson, J. 1967. *Ecological data on dry-matter production by plants and plant communities*. p. 77–123. In E.F. Bradley and O.T. Denmead (ed.) The collection and processing of field data. Interscience Publishers, New York.

Wallenstein, M.D., et al., 2006. Environmental controls on denitrifying communities and denitrification rates: insights from molecular methods. *Ecological applications*, *16*(6), pp.2143–2152.

Wang, Y., et al., 2020. A theoretical and empirical assessment of stomatal optimization modeling. *New Phytologist*, *227*(2), pp.311–325.

Weitzman, J.N. and Kaye, J.P., 2018. Nitrogen budget and topographic controls on nitrous oxide in a shale-based watershed. *Journal of Geophysical Research: Biogeosciences*, *123*(6), pp.1888–1908.

White, C.M., et al., 2014. Implications of carbon saturation model structures for simulated nitrogen mineralization dynamics. *Biogeosciences*, *11*(23), pp.6725–6738.

White, J.W. and Hoogenboom, G., 2003. Gene-based approaches to crop simulation: past experiences and future opportunities. *Agronomy Journal*, *95*(1), pp.52–64.

Xia, Y., et al., 2012. Continental-scale water and energy flux analysis and validation for the North American Land Data Assimilation System project phase 2 (NLDAS-2): 1. Intercomparison and application of model products. *Journal of Geophysical Research: Atmospheres*, *117*, D03109. https://doi.org/10.1029/2011JD016048

Chapter 4

Crop modeling with simple simulation models (SSM)

Afshin Soltani, Gorgan University of Agricultural Sciences and Natural Resources, Iran; and Thomas R. Sinclair, North Carolina State University, USA

1 Introduction

Ensuring global food security and protecting the environment at the same time is perhaps the single greatest scientific challenge currently facing humankind (Cassman, 2012). To meet this challenge, breeding crops for improved yield capacity, optimizing the use of land, water, and nutrient resources, and enhancing stress resilience and yield stability in response to unpredictable weather patterns will be required (Sinclair et al., 2020). To assess the potential benefit of the various approaches to increase crop productivity, we propose that crop modeling is an important component of such an evaluation.

Success in evaluating new ideas for crop production will almost certainly require local solutions due to Genotype × Environment × Management interactions. Crop models offer a valuable tool for evaluating Genotype × Environment × Management interactions if they capture the interactive

http://dx.doi.org/10.19103/AS.2025.0155.06

relationships among genetics, management practices, soil, and weather conditions that influence crop yield and quality (Sinclair et al., 2020). For example, the differences between two genotypes in a field experiment depend on the applied management and the environment (location and/or year) of the field test. Genotypic differences may be negligible under one management or environment but significantly different under another one. Thus, many field tests are indeed necessary. Conducting sufficient field experiments, however, to address geospatial responses is almost always impractical due to the extensive observations required to capture the temporal and geographical variability. Crop modeling that can approximate the complex interactions between plant growth and the environment as influenced by management will almost certainly be required to assess potential solutions leading to increased crop productivity (Sinclair et al., 2020).

Critically, for the full use of crop models as heuristic tools (Sinclair and Seligman, 1996), the models must be transparent and robust. One family of models that readily meets these criteria is the simple simulation models (SSM; Sinclair, 1986; Soltani and Sinclair, 2012a; Sinclair et al., 2020; these are accessible at: https://sites.google.com/view/ssm-crop-models). In the SSM approach, the first step toward transparency is defining objectives that allow for the identification of essential model features consistent with the objective, and also to help eliminate unnecessary elements in the model.

Another important factor contributing to model transparency is a model constructed to describe cropping systems at an explanatory level (defined temporally and spatially) that is only one level below the desired level of explanation (de Wit, 1970). This not only enhances transparency but also ensures logical and consistent fits within the model by avoiding mismatches among processes and relaxation times (Sinclair et al., 2020). The SSM models were constructed within the constraint of applying only a single explanatory level below the desired level of explanation.

A fundamental principle in the development of SSM models is to strive for simplicity without resorting to excessive reductionism or oversimplification. The number of equations and parameters used to describe each explanatory process needs to be kept to a minimum. To achieve this goal, conservative relationships and summary expressions are likely required at the explanatory level to build more comprehensible and transparent models. Such an approach enables users to grasp the underlying principles of the model without being overwhelmed by excessive detail (Soltani and Sinclair, 2012a).

In SSM models, all necessary parameters can be estimated from independent experiments, which avoids the critical problem of introducing empiricism in model application by adjusting parameter values based on simulation tests of the model, that is, 'model calibration'. In practice, model calibration is often based on the adjustment of parameters within the complex framework of the model so that model output matches a limited set of observations (Sinclair and Seligman, 1996). Since SSM does not require 'calibration', use of the model is not dependent on non-measurable or vague parameters as inputs (Soltani and Sinclair, 2012a). Nevertheless, before applying SSM to specific objectives, the robustness of SSM output must be documented over a range of conditions using all available data.

An *ad hoc* approach is employed in SSM modeling so that each modeling task is defined by specific objectives and the required model output. There is no specific software dedicated to SSM models. One common approach is the use of Visual Basic for Applications (VBA) in Excel to operate the model, with Excel worksheets serving as the interface for input and output operations. The code in this approach is open-access, allowing users the freedom to modify the model code and customize the output according to their objectives.

SSM models have been used in various applications worldwide. In nearly all applications, SSM has been used to do geospatial assessments of crop production in an array of locations over growing seasons. In this chapter, a brief description of the history of the expanding role of the SSM approach in crop assessments is reviewed. The overall guidelines for developing and using SSM models are discussed. Next, a short description of the functions and methods currently used in simulating various crop species and soil processes is presented. The genetic parameters and environmental input required by the SSM models are also provided. Furthermore, examples of geospatial applications of SSM models are described, and areas for future model development and application are identified.

2 The history of simple simulation models (SSM)

The original development of the basic form of SSM dates to the publication of a 'simple' soybean model by Sinclair (1986) (Fig. 1). This model was used to analyze water and nitrogen limitations in soybean production. In the model, the concept of radiation use efficiency (Sinclair and Horie, 1989) was employed to estimate daily dry mass production. Transpiration was simulated based on the linkage between dry mass production and transpiration (Tanner and Sinclair, 1983). Quantitative nitrogen limitations on yield formation and resulting leaf senescence were also included in the model simulations (Sinclair and de Wit, 1976). Grain yield formation was modeled based on a linear increase in harvest index (Spaeth and Sinclair, 1985). Responses of crop processes to soil water deficit were simulated by daily calculations of the fraction of total transpirable water currently retained in the soil (Sinclair and Ludlow, 1986).

SSM was subsequently applied to simulate crop species other than soybean. Sinclair et al. (1987) used the same model to simulate cowpea and black gram. A simple maize model was developed by Muchow et al. (1990)

Year	Potential Models	Water-limited models	Nitrogen-limited models
1986			Soybean (Sinclair, 1985)
1987			Tropical legumes (Sinclair et al., 1987)
:			
1990	Maize (Muchow et al., 1990)		
1991	Wheat (Amir and Sinclair, 1991)	Wheat (Amir and Sinclair, 1991); Maize (Muchow and Sinclair, 1991)	
1992			Wheat (Sinclair and Amir, 1992)
:			
1995		Peanut (Hammer et al., 1995)	Maize (Sinclair and Muchow, 1995)
:			
1997			Sorghum (Sinclair et al., 1997)
:			
1999		Chickpea (Soltani et al., 1999)	
:			
2003			Soybean (Sinclair et al., 2003)
:			
2005		Barley (Wahabi and Sinclair, 2005)	
:			
2011			Chickpea (Soltani and Sinclair, 2011)
2012			SSM for grain crops (Soltani and Sinclair, 2012a)
2013			SSM-Wheat (Soltani et al., 2013)
:			
2015			SSM-iLegume for grain legumes such as soybean, chickpea, lentil, peanut, dry bean (Ghanem et al., 2015; Soltani et al., 2016; Vadez et al., 2017)
2016			SSM-iCrop for grain crops like wheat, maize, soybean, and potato (Soltani and Sinclair, unpubl.)
:			
2020		SSM-iCrop2 (Soltani et al., 2020)	
2021			SSM-iCrop2 (Soltani et al., unpubl.)

Figure 1 Development of important SSM models over the past 35 years.

to analyze the effects of solar radiation and temperature on maize potential yield across different locations. Amir and Sinclair (1991a) developed a SSM model to study the effects of temperature and solar radiation on spring wheat growth and yield. This model was expanded later to incorporate water-deficit limitation (Amir and Sinclair, 1991b), as well as to assess nitrogen limitations

(Sinclair and Amir, 1992) on growth and yield of spring wheat. The model for potential maize yield was extended to simulate effects of water deficits (Muchow and Sinclair, 1991) and nitrogen supply on maize yields (Sinclair and Muchow, 1995). Early on, SSM models for peanut (Hammer et al., 1995), sorghum (Sinclair et al., 1997), and chickpea (Soltani et al., 1999) as listed in Fig. 1 were also developed.

During the period from 2010 to 2013, SSM models underwent a review and update. The term "SSM" was introduced for the first time, and a comprehensive description of SSM was presented by Soltani and Sinclair (2011; 2012a) and Soltani et al. (2013) (Fig. 1). In 2016, the SSM models for legumes (Soltani and Sinclair, 2011) and wheat (Soltani et al., 2013) were merged and updated into SSM-iCrop, which uses a single code to simulate crop development, growth and yield under potential, water-limited and nitrogen-limited conditions. The SSM-iCrop model currently simulates grain legumes, wheat, maize, and potato. Except for phenology, all other processes such as leaf area development and senescence, dry mass production, yield formation, soil water balance, and plant nitrogen budget are simulated with identical code. A multilayer soil is simulated and an estimate of N fixation in legumes is provided. This single, combined model aimed to facilitate model tracking and maintenance for different species. In 2020, a more universal, simplified version of SSM-iCrop, namely SSM-iCrop2, was developed and adapted to simulate over 30 species, including orchards and perennial forages (Soltani et al., 2020a). This model was originally focused on potential yield as well as water-limited conditions and did not include simulation of soil N balance and plant N budget. Model testing demonstrated model robustness in predicting crop yield and water use, making it suitable for assessing large-scale crop production across a wide number of plant species. SSM-iCrop2 has also been employed to simulate rangelands (Nehbandani et al., 2023) and the water balance of fish ponds (Soltani et al., 2019). Subsequently, in 2021, nitrogen sub-models for the soil and plants were incorporated into SSM-iCrop2 to enable simulations for nitrogen-limited conditions. Fig. 2 illustrates the current active updated SSM models used in education and research.

3 Basic principles underpinning SSM

The major advantages of SSM models are transparency and ease of use. Transparency in crop modeling refers to the ability of others to readily understand and comprehend a model, including its parameters, flow diagrams, and underlying code, even if the user was not involved in the development of the model. This transparency is particularly important for educational purposes as it facilitates the learning process. To prepare transparent, robust, and

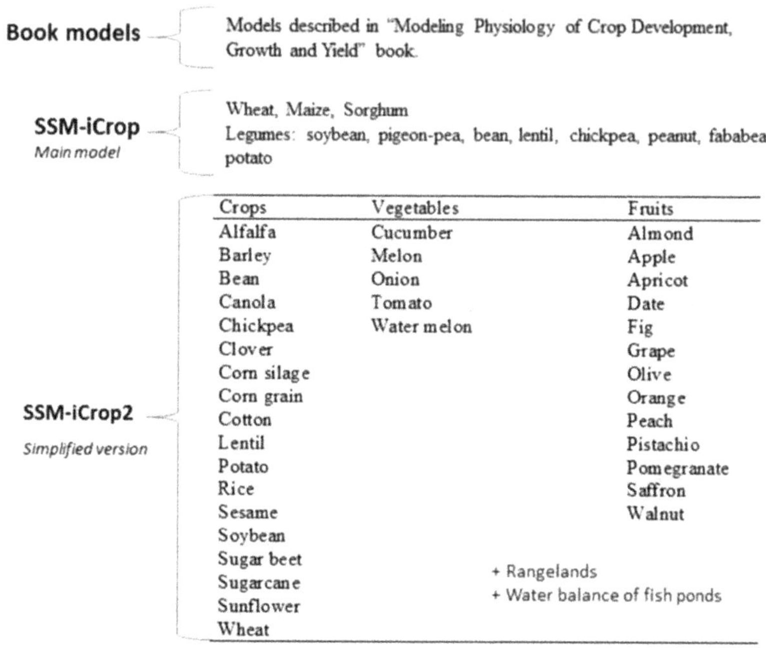

Figure 2 Current active, updated SSM models for education and research purposes.

easy-to-use simple simulation models, several basic guidelines within the SSM approach were followed. These guidelines are listed below:

1 The explanatory level (model functions and parameters) is only one level below the explainable level (model output).
2 Avoid reductionism beyond the explanatory level.
3 Conservative, mechanistic relations based on both theoretical derivation and experimental observation.
4 Model parameters that are properties of species and cultivars with clear meanings that can be directly measured in experiments. That is, model parameters are not estimated by the 'calibration' of the model.
5 Robustness tests are necessary before applying a model, and these tests should be conducted using all available experimental data.
6 Depending on objectives, the model should be tailored to specific problems, i.e. being *ad hoc*.
7 Models are open-access and fully documented, with codes and documents being consistent and easily understandable.
8 Models are easy to use in terms of programming and software.

3.1 Explanatory level one level below explainable level

A key objective of SSM is to describe cropping systems at one temporal and spatial level below the output of the model. That is, the output at a defined level is based on a description of activity at only one more complex level. This concept aligns with de Wit (1970) suggestion that the performance of a model at the explainable level is best described by the mechanisms at a single lower, explanatory level. In the context of SSM, the spatial explainable level is usually overall crop performance, while the lower explanatory level in the model is based on the behavior of the plant canopy. Similarly, the temporal explainable level is seasonal, while the explanatory level is typically daily. This approach of only two levels of hierarchy in the model not only enhances transparency but also ensures logical and consistent fits within the model by avoiding mismatches among processes and relaxation times (Sinclair et al., 2020).

3.2 Avoiding extensive reductionism

Simple models are more easily understood and interpreted by users. They facilitate clear communication of the model's structure, assumptions, and results, thereby enhancing transparency (Soltani and Sinclair, 2012a). Simplicity also promotes accessibility, making the model more inclusive for a wider range of users, including non-experts. A simple model aids in error detection and debugging processes, ensuring transparency by reducing the likelihood of hidden issues. It also supports model maintenance, updating, and reproducibility, further contributing to transparency. Therefore, a fundamental principle in the development of SSM models was to maintain simplicity without resorting to excessive reductionism (Soltani and Sinclair, 2012a).

The first step toward transparency through simplicity in SSM models is clarity in defining objectives. The complexity of a model originates from its original objective statement. A clearly defined, specific objective for each modeling task allows focus on key model features and helps eliminate unnecessary elements in the model.

Biological systems, including crops, are composed of numerous components and processes operating at various organizational levels. Unlike engineering models, biological models do not have clear specifications for all components. Further, the conceptual role and function of models in solving engineering problems do not directly translate to biological systems (de Wit, 1970). Thus, it must be recognized that it is impossible – and even inappropriate – to set an objective for a model to account for all factors that influence crop performance in every situation (Mayr, 1982; Pease and Bull, 1992).

Assumptions are necessary when simulating crops since assumptions are critical in facilitating the modeling process. Consequently, it is essential to remember that models are simplified descriptions of reality and are inherently incomplete and imperfect representations of reality. Even detailed crop models remain highly simplified representations of the actual system, and achieving a perfect one-to-one correspondence is unattainable. Adding complexity without extensive experimental justification may well increase the imperfections of the model (Sinclair and Seligman, 1996). Minimizing equations and parameters in achieving fully defined, specific objectives enhances transparency (Soltani and Sinclair, 2015). As Albert Einstein said, "If you can't explain it simply, you don't understand it well enough."

On the other hand, it is crucial to avoid developing models that are excessively simple and incomplete regarding vital aspects of the system. In such cases, applying the model may lead to an incomplete or even incorrect understanding of the system (Monteith, 1996). Again, Albert Einstein suggested, "Everything should be made as simple as possible, but not simpler." Monteith (1996) proposed resisting the tendency of creating complex models and instead advocated for models that are simple enough to be comprehensible by others (i.e. transparent) while being comprehensive in scope.

3.3 Using conservative summary equations

Incorporating conservative relationships and summary expressions can significantly enhance the transparency of a model (Sinclair and Seligman, 1996). These relationships should be based on well-established mechanistic understanding and can provide efficient and robust representations of empirical patterns. By incorporating such relationships where appropriate, the model becomes more comprehensible and transparent, enabling users to grasp the underlying principles without becoming overwhelmed by excessive detail (Sinclair et al., 2020).

Individual summary expressions, derived from both theoretical derivations and experimental observations, are employed in SSM to achieve transparency. For instance, crop mass accumulation is calculated based on canopy radiation use efficiency, which is sensitive to water and nitrogen availability. The theoretical dependence of radiation use efficiency showed it was directly dependent on leaf photosynthesis capacity (Sinclair and Horie, 1989). Experimental studies have provided experimental support for describing crop mass accumulation based on radiation use efficiency (Sinclair and Muchow, 1999). Also, theoretical derivation and experimental observations give a mechanistic representation of the impacts of water and nitrogen limitations on radiation use efficiency. Thus, radiation use efficiency serves as a documented parameter that can be appropriately used at the explanatory level.

Another example of a summary expression employed in SSM, which distinguishes the model, is the relationship between canopy transpiration and mass accumulation. The calculation of transpiration is particularly crucial in soil water balance calculations for rainfed crop simulations. Unfortunately, many models calculate water loss using the Penman-Monteith equation, which relies on uncertainty, empiricisms, and assumptions in describing its parameters. The alternative for calculating the transpiration rate used in SSM is the fundamental, well-established physiological linkage between transpiration and photosynthesis (Sinclair et al., 2020). Tanner and Sinclair (1983) derived a summary expression that demonstrates the close relationship between transpiration rate and photosynthesis rate at the canopy level. In SSM, daily canopy water loss based on transpiration is simply and directly calculated from daily mass accumulation (Tanner and Sinclair, 1983). None of the assumptions or calibrations needed in the application of the Penman-Monteith approach are necessary (Sinclair et al., 2020).

3.4 Parameters directly from measurement

Sometimes, models include parameter(s) that have no resolved physiological meaning but are still used to allow the matching of model output with observations. There is usually no way to estimate values for these parameters other than 'calibration' using the whole model to achieve observed results for a defined set of data. Transparency requires that crop parameters and state variables have clear meaning and that their values can be easily determined through simple, independent measurements or observations. It is crucial to avoid using non-measurable or vague parameters that require model calibration (Soltani and Sinclair, 2012a).

Attempting to calibrate the entire model to match relevant observations without a means to directly quantify these parameters greatly limits transparency and reduces the model to a complex framework that necessarily results in an empirical match with the 'calibration' data set. Unfortunately, there seems to be an increasing number of published papers that estimate parameter values in a model based on the 'calibration' of the entire model to match yield observations. For example, in the calibration approach, several crop parameters related to phenology, leaf development, and yield formation are estimated by running the model and adjusting parameters so that the model output matches with a calibration dataset. Such an approach allows numerous combinations of parameter estimates in the model to be obtained that might match model output with observations. Further, in some cases, parameters like plant phyllochron are estimated even when there are no observed leaf numbers within the available data.

A potential problem with calibration is that values for parameters can compensate for each other, and the derived parameter estimates may deviate substantially from the actual functioning of plants (Soltani and Sinclair, 2012a,

2015). C.T. de Wit, regarded as the father of crop modeling, aptly described this calibration approach as "going in circles chasing your own tail" (Sinclair and Seligman, 1996). In SSM models, all necessary parameters are estimable based on their definitions derived independently of the model from theory and/or experimental observations.

3.5 Robustness test using all data

When the objectives of model construction include model application to simulating real-world behavior, robustness needs to be tested based on predefined criteria in the objective-definition stage. Robustness is an evaluation of the ability of a model to simulate reality, and in the SSM approach, this statistical test is done using all available data (Sinclair et al., 2020). Use of all the available data contrasts with the approach sometimes required for calibrated models in which observations are divided into two subsets: one for "calibration" and another for "validation." Sometimes, these two sets come from one experiment and/or from one experimenter; hence, it cannot be said that the data sets are independent. Success in finding values for calibrated parameters that result in output matching a validation data set only indicates that the two data sets represent the same population of data (Soltani and Sinclair, 2012a). In other words, success with a calibrated model only confirms that the calibration dataset was adequate for empirically matching the validation dataset.

As an aside, the term "validation" in crop plant modeling is also inappropriate and incorrect for two major reasons. (1) Crop models involve multiple assumptions rather than a single hypothesis, and validation is typically associated with testing a single hypothesis where the result may allow the hypothesis to be rejected. To even consider scientific 'validation', a large number of tests are required under a range of conditions, usually by many investigators. (2) Validation requires comparing the output of a model with system properties that have been measured without error, whereas in crop modeling, the model's output is compared with measurements that inherently contain errors (Soltani and Sinclair, 2012a; Sinclair and Seligman, 1996).

Although, not all studies using a crop model require extensive robustness testing. If the purpose of the model is to integrate existing knowledge and information, high predictive capability may be less important compared to the need for identifying weaknesses in the conceptualization of hypotheses. In such cases, a research model is more concerned with identifying behavioral patterns rather than making precise quantitative predictions. The requirement for statistical tests of model robustness may not be as rigorous as using the model for real-world simulations (Sinclair and Seligman, 1996).

3.6 Being ad hoc

SSM modeling employs an *ad hoc* approach in crop modeling, recognizing that universal models are not feasible. Instead, crop models should be customized to address specific objectives (Sinclair and Seligman, 1996; Soltani and Sinclair, 2012a). In the SSM approach, existing models are viewed as valuable resources but must be evaluated in relation to the objectives of the current model construction. While it is important to consider successful approaches for efficiency, a new objective may necessitate the formulation of new hypotheses or significant modifications to an existing model (Sinclair and Seligman, 1996). Being open access and easy programming facilitates the use of SSM models for necessary modifications by users.

3.7 Open-access and fully documented code

Open access and full documentation are considered crucial for ensuring transparency in crop models (Soltani and Sinclair, 2015). It is essential to make models accessible to others, thereby enabling them to access and comprehend the model code and associated documents. This approach promotes transparency by allowing users to review, analyze, and understand the model independently. Moreover, it is vital for the model code and documents to be one and the same, ensuring that they are both understandable. The model code should be well-documented, providing detailed explanations and descriptions of its various components. This enables users to directly examine the code and develop a profound understanding of the model's internal mechanisms.

3.8 Easy programming and software

Another factor that enhances the transparency of a model and its application is programming and software. In SSM models, we use Visual Basic for Applications (VBA) in Excel, which is readily available on most computers, eliminating the need to learn and run new software. SSM models use Excel sheets for input and output (Soltani and Sinclair, 2012a). Since the code is open-access, users have the freedom to modify the model code and customize the output according to their needs. In the model code, parameters are separated from the main code and are read from an Excel sheet. The model code is organized into different sub-models, with each sub-model assigned to a specific canopy process, such as phenology, leaf area development, or dry mass production (Sinclair et al., 2020).

4 SSM critical functions

SSM models are available for simulating potential production, water-deficit production, and nitrogen-limited production (Fig. 1) (Soltani and Sinclair,

2012a). In the potential production models, it is assumed that there are no limitations in terms of water, nutrients, pests, diseases, and weed management. Therefore, crop growth and yield are primarily influenced by solar radiation, temperature, and atmospheric CO_2 concentration. In water-deficit models, the impacts of soil water deficit on various crop processes, including phenology, crop leaf area, and dry matter production, are taken into account. Consequently, the simulation of soil water balance becomes necessary for these models. In SSM models designed for nitrogen-limited conditions, both the plant nitrogen budget and the soil nitrogen balance are simulated.

Below is a brief overview description of the more critical functions of the SSM model. The functions used in the simpler version of the model, SSM-iCrop2, are presented. SSM-iCrop2 is designed to operate with limited and easily accessible input data, making it convenient for parameterization and application. Despite its simplicity, the model remains mechanistic, relying on functional responses of plant processes to managerial, genetic, and environmental factors. In Section 4.7, the more intricate SSM-iCrop is described and compared with SSM-iCrop2. For detailed descriptions of both models, readers are referred to Soltani and Sinclair (2011, 2012a), Soltani et al. (2013) and Soltani et al. (2020a).

4.1 Phenology

The selection of phenological stages of plant development to be simulated depends on the objectives of the model and the need to predict phenological stages to simulate other processes. Some of the simulated developmental stages for which parameters may be inputs to SSM include:

- Emergence from soil of field crop species (EMR);
- Beginning of leaf growth for trees and permanent forages;
- Beginning of root growth (BRG); equal to EMR in field crops;
- Beginning of effective seed growth or fruit growth (beginning linear harvest index increase, BSG);
- Beginning of leaf senescence (BLS); equal to BSG in field crops;
- Termination of root growth (TRG); equal to BSG in field crops;
- Termination of seed growth or fruit growth (termination linear harvest index increase, TSG);
- Physiological maturity (no increase in dry mass after this stage); equal to TSG in grain crops, but TSG is not the same as PM in tree plants because in some tree fruits like peach, growth continues after TSG; and
- Harvest in field crops and harvest or leaf fall in tree crops (HAR).

These stages are used to switch on or off other developmental and growth processes within the model. There can be links between phenological stages. It was assumed that bud burst in orchards and the start of the spring re-growth in perennial forages such as alfalfa occurs when cumulative temperature units from the first of January onwards reach or exceed a critical value (*ForReq*).

In SSM-iCrop2, the calculation of phenological development is based on the concept of temperature unit (Soltani and Sinclair, 2012a; Soltani et al., 2020a). The daily temperature unit (DTU, °C) is calculated assuming the temperature of the day is optimal for phenological development, and then it is corrected by multiplying scalar functions based on average daily temperature (*tempfun*) and soil water deficit (WSFDS):

$$\text{DTU} = (\text{TP1D} - \text{TBD}) \times \textit{tempfun} \times \text{WSFDS} \tag{1}$$

where TP1D and TBD are lower optimum and base temperature for development. Average daily temperature is calculated from the mean of the input minimum and maximum temperature. Schematics of example scalar functions used to calculate DTU are presented in Fig. 3 (Soltani and Sinclair, 2012a). The temperature response has three phases of ascending value at lower temperatures, plateau value at optimum temperature, and descending value at higher temperature.

Soil water deficit is based on the fraction of transpirable soil water (FTSW), which is tracked daily as described later. As shown in Fig. 1, the water deficit function is somewhat unique in that it allows the possibility for accelerated (WSFDS > 1) and retarded (WSFDS < 1) phenological development depending on soil water status (FTSW) (Soltani and Sinclair, 2011; 2012a).

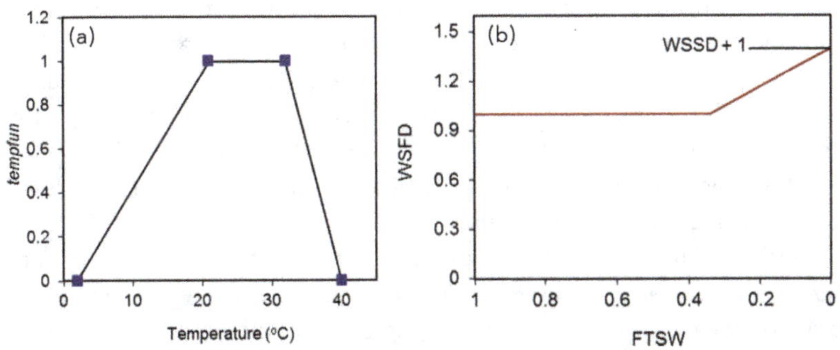

Figure 3 Example schematic representation of response of development rate to (a) average daily temperature and (b) fraction transpirable soil water (FTSW) (Soltani and Sinclair, 2012a). The temperature response is defined by the four cardinal temperatures indicated by squares in the figure. Response to FTSW is for a plant in which water deficit (FTSW < 0.35 in this example) accelerates its development rate. WSSD (water stress sensitivity threshold for development) is the maximum proportion in acceleration of development by water deficit.

The derived daily DTU is used to calculate cumulative temperature units as the season progresses. A phenological stage is predicted to occur when the cumulative temperature unit is equal to or exceeds the temperature unit required to reach each stage.

4.2 Leaf area development and senescence

Simulation of changes in leaf area index (LAI) in SSM models is initially calculated as a function of temperature unit, but then adjusted depending on soil water status. In many SSM models (e.g. Soltani and Sinclair, 2011), this is done by tracking the main stem leaf number using the phyllochron concept (the cumulative temperature between successive leaves) and then computing of potential plant leaf area expansion per day based on the main stem leaf number.

In SSM-iCrop2, however, due to the lack of required information for many plant species, including orchards, a more direct method for calculating leaf area was employed. In the more direct approach, a common exponential regression function was used to describe potential LAI expansion as a function of temperature units (Soltani et al., 2020a). This function needs a maximum plant leaf area under optimal conditions and two regression coefficients that can be estimated internally by the model if coordinates of two points on the function are provided from observation as input parameters.

In the SSM models, the potential increase in green leaf area index (GLAI, m^2 leaf area m^{-2} ground) each day is calculated by considering the difference in LAI between the current day and the previous day. The computed daily GLAI is constrained, however, at low soil water content (Soltani and Sinclair, 2012a).

Under non-limiting nitrogen conditions in which the model does not simulate plant and soil N budgets, accounting for leaf senescence requires specific functions for each crop species. These functions allow leaf senescence to be simulated linearly or nonlinearly by decreasing the leaf area index (LAI) to 0 from the beginning of leaf senescence (BLS) to harvest (HAR). One approach is simply to simulate a linear decrease in leaf area index to 0 from the beginning of leaf senescence to harvest maturity.

Under N-limited conditions, the amount of leaf area senesced per day (DLAI, $m^2\ m^{-2}\ d^{-1}$) is computed by dividing the amount of nitrogen mobilized from the leaves (XLN, g N m^{-2} ground) on that day by the difference between the specific leaf nitrogen of green leaves (SLNG, g N m^{-2} leaf) and senesced leaves (SLNS, g N m^{-2} leaf) (Sinclair et al., 2003; Soltani and Sinclair, 2012a):

$$DLAI = XLN/(SLNG - SLNS) \quad (2)$$

4.3 Crop mass production and partitioning

Daily crop mass increase is estimated by multiplying the daily incident photosynthetic active radiation (PAR, MJ m^{-2} d^{-1}) by the fraction of radiation intercepted by the leaf canopy (FINT) and daily crop radiation use efficiency (RUE, g MJ^{-1}) (Soltani and Sinclair, 2012a). The fraction of intercepted radiation is determined based on the crop leaf area index (LAI) and canopy extinction coefficient (KPAR):

$$FINT = 1 - \exp(-KPAR \times LAI) \quad (3)$$

The daily value of RUE is obtained as follows (Soltani and Sinclair, 2012a):

$$RUE = IRUE \times TCFRUE \times WSFG \times CO2RUE \quad (4)$$

where IRUE represents the potential RUE, and TCFRUE, WSFG, and CO2RUE are scalar factors (ranging from 0 to 1) that adjust IRUE for temperature, soil water deficit, and atmospheric CO_2 concentration, respectively.

Daily dry mass production is allocated among three sinks: leaves, "stems" (including stems, leaf sheaths, and ears/pods, excluding grains), and grains (Soltani and Sinclair, 2012a). First, dry mass is allocated between vegetative tissues and seeds/fruits, and then the distribution of vegetative dry mass between leaves and stems is determined. Assuming an essentially constant specific leaf area (SLA, m^2 g^{-1}), the dry matter allocated to leaves each day is calculated by multiplying the daily increase in LAI (GLAI) by the specific leaf area. Dry mass not allocated to leaves is accumulated in the stems. During seed growth, the growth rate of seeds/fruits and the formation of yield is based on empirical observations of linear increases in harvest index with time (dHI/dt) during seed growth (Soltani and Sinclair, 2011). The seed growth rate (SGR, g m^{-2} d^{-1}) is calculated based on the input slope of the linear increase in harvest index.

4.4 Plant nitrogen balance

The simulation of the plant nitrogen (N) budget differs between stages before and after seed growth. Before seed growth, the potential daily demand for N uptake (PNUP, g N m^{-2} d^{-1}) is calculated based on the daily increase in leaf area index (GLAI) and specific leaf N in green leaves (SLNG, g N m^{-2} leaf) plus the daily increase in "stem" dry matter (GST, g m^{-2} d^{-1}) and the N concentration in stems (SNCG, g g^{-1}):

$$PNUP = GLAI \times SLNG + GST \times SNCG \quad (5)$$

The values of SLNG and SNCG are input parameters of the model.

The actual daily N uptake (NUP) is calculated after accounting for several limitations on PNUP:

1 The value of NUP can be limited to a daily maximum amount when PNUP is greater than the input maximum value.
2 NUP (and N_2 fixation in legumes) is set to zero when there is no daily dry matter production by the crop.
3 NUP (and N_2 fixation in legumes) is sensitive to flooding, so NUP is set to zero when the soil is saturated.
4 NUP is limited to the amount of soil N available for crop uptake (SNAVL) when the initial calculation of NUP is greater than SNAVL.

In legume crops, biological N_2 fixation (BNF) is activated when the input cumulative temperature unit is exceeded. During vegetative growth BNF is calculated based on the demand for N by the crop. During seed growth, BNF is determined as a function of the vegetative mass during the seed filling period on days when photosynthate production is greater than that required by the growing seeds. Further, BNF is sensitive at all stages to soil water deficit, which is expressed as a scalar function of FTSW.

During periods when the rate of daily N uptake does not fully meet the N demand of the plants, a set of sequential actions are undertaken by the model (Soltani and Sinclair, 2011, 2012a): First, inadequate N uptake results in a decrease in stem N concentration until the time when it reaches a minimum value (SNCS). Under more severe N deficiency, when the stem N concentration reaches its minimum, the next response is simulated as continued stem growth at the minimum N concentration, but leaf area development is restricted by the amount of remaining N available for leaf development. Under extreme N limitation, when there is insufficient N uptake for stem growth, leaves are senesced as a source of remobilized N for stem growth. This means that leaf senescence is possible during vegetative growth under limited N accumulation. The N content of the senesced leaves (SLNS, g N m^{-2} leaf) is an input parameter to the model.

After the beginning of seed growth (BSG), seeds become the primary sink for N, and the daily N demand by the seeds is calculated as the product of the daily seed growth rate and the seed N concentration. However, in indeterminate crops such as peanuts, the daily increase in leaf area index (LAI) and "stem" dry matter, along with their corresponding N concentrations, are added to the daily crop N demand.

All the limitations described above for limiting N uptake during the vegetative stage also apply during seed growth. In addition, NUP (and N_2 fixation in legumes) is set to zero when the daily dry matter production by the crop does not exceed the seed growth rate (SGR), indicating a lack of assimilates to support NUP. When N uptake from the soil does not occur or is insufficient to meet the full requirements of the growing seeds, N is translocated from the leaves and the stems to the seeds. The fraction of N translocated from the leaves and the stems is determined based on the relative amount of translocatable N in each tissue. N mobilization from the leaves leads to leaf senescence, a reduction in PAR interception, and consequently, a decrease in daily crop mass production (Soltani and Sinclair, 2011, 2012a).

4.5 Soil water balance

In SSM-iCrop2 (Soltani et al., 2020a) and many other SSM models (e.g. Sinclair, 1986; Soltani and Sinclair, 2011), a two-zone soil budget is used for the calculation of soil water balance. The soil water balance for the two zones is calculated independently. One zone is the top layer of the soil profile, the depth of which is an input to the model, typically 15 to 50 cm deep. A critical feature of the top zone is accounting for evaporation from the soil surface, which is sensitive to the soil water content of this top zone. The addition of water to the soil is added directly to this top zone water content. The second zone includes the top zone along with the remaining soil depth occupied by roots. In this second zone, water is lost via soil evaporation and crop transpiration and added by precipitation and irrigation minus surface runoff. The depth of this second zone is increased daily if root extension is being simulated.

Available soil water in the top layer each day ($ATSW1_i$, mm) and in the total root zone ($ATSW_i$, mm) are obtained as follows:

$$ATSW1_i = ATSW1_{i-1} + RAIN + IRGW - DRAIN1 - RUNOF - SEVP - TR1 \quad (6)$$

$$ATSW_i = ATSW_{i-1} + RAIN + IRGW + EWAT - DRAIN - RUNOF - SEVP - TR \quad (7)$$

where RAIN is the precipitation, IRGW is the irrigation, DRAIN is the drainage from the zone, RUNOF is the run-off, and SEVP is the soil evaporation. The variable TR1 is the part of plant transpiration water removed from the top layer, and TR is the total crop transpiration. For the total root zone, EWAT is the additional water that may be added to the soil water budget as a result of root extension into deeper, wet soil. All these variables are expressed in millimeters of water.

Surface run-off is calculated using a modified curve number technique (Williams, 1991; Hoogenboom et al., 2019). This technique depends on the soil curve number, soil water content at the lower limit and saturation, actual soil water content in the first two layers, soil cover, and daily rainfall. Curve numbers are assigned to different soils based on their texture.

A critical output of the simulation of the soil water budget is the fraction of transpirable soil water (FTSW) calculated for the soil zone of rooting. The value of FTSW is calculated daily as the ratio of the actual amount of simulated transpirable soil water (ATSW, mm) in the rooting zone and the total transpirable soil water (TTSW, mm) that can be stored in this zone:

$$FTSW = ATSW/TTSW \quad (8)$$

Soil evaporation is determined using a two-stage soil evaporation model (Amir and Sinclair, 1991b; Soltani and Sinclair, 2011). Stage I evaporation occurs when abundant water is present in the top soil zone and the fraction of transpirable soil water (FTSW) of the total root zone exceeds 0.5. The potential evaporation rate is calculated using a modified Priestley and Taylor model (Ritchie, 1998). Stage II evaporation takes place when the water in the top layer is depleted or the FTSW of the total root zone is less than 0.5. In Stage II, the potential rate of soil evaporation is also computed as in Stage I, but it is significantly reduced based on the square root of the time elapsed since the start of Stage II. The calculation of soil evaporation reverts to Stage I only when rainfall or irrigation increases the soil water content of the top layer to 10 mm or greater.

The daily transpiration rate is calculated directly from the daily rate of crop mass production, based on the crop species transpiration efficiency coefficient and atmospheric vapor pressure deficit (Tanner and Sinclair, 1983; Soltani and Sinclair, 2012a). The calculation of daily vapor pressure deficit (VPD) was proposed by Tanner and Sinclair (1983) to be approximately 0.75 times the difference between saturated vapor pressure calculated from daily maximum and minimum temperatures.

Flooding can also hinder various physiological processes. On any day when the soil water in the soil reaches its saturation level, all water stress factors (WSFL, WSFG, and WSFD) can be set to 0 (Soltani and Sinclair, 2012a). Moreover, it is considered that a certain number of consecutive flooding days can lead to crop death.

4.6 Soil nitrogen balance

The iCrop2 version of SSM accounts for the soil nitrogen budget in the top soil zone, which is also the water reservoir for soil evaporation. The amount of N in the top layer (NSOL, g N m^{-2}) is recomputed each day as follows (Soltani and Sinclair, 2012a):

$$NSOL_i = NSOL_{i-1} + NMIN + NFERT - NVOL - NLEACH - NDNIT - NUP \quad (9)$$

where $NSOL_{i-1}$ is the soil nitrogen in the top soil zone on the previous day,

NMIN is the net daily mineralization amount (mineralization minus immobilization) of organic matter,

NFERT is the N amount derived if N fertilizer was applied on that day,

NVOL is the daily N volatilization amount,

NLEACH is the amount of N that leaches with soil percolation from the top zone,

NDNIT is the daily N denitrification amount when FTSW1 ≥ 1, and
NUP is the daily amount of crop N uptake.

The N concentration in the soil solution (NCON, g N g^{-1} water) is obtained using the equation:

$$NCON = NSOL/(WAT1 \times 1000) \quad (10)$$

where 1000 is a conversion factor from mm to g and WAT1 is the total water content (plant available and plant unavailable) of the top soil zone in mm. The amount of N available to the crop from the soil solution (SNAVL, g N m^{-2}) is then calculated as:

$$SNAVL = (NCON - 0.000001) \times ATSW1 \times 1000 \times FROOT1 \quad (11)$$

In Equation 11, it is assumed that no N is taken up by plants when NCON is less than 0.000001 g N g^{-1} water (equivalent to 1 mg N L^{-1}). The fraction of N available to the crop from the top soil zone is proportional to the fraction of the zone occupied by roots (FROOT1).

4.7 SSM-iCrop modifications

SSM-iCrop is an adaptation of the previous SSM models to allow more intricate features of the crop system to be simulated. SSM-iCrop is only available for grain crops and potato. The key differences between the more complex SSM-iCrop and the simpler SSM-iCrop2 are as follows:

- SSM-iCrop incorporates the effects of photoperiod and vernalization on phenological development.
- SSM-iCrop offers more complex options for dry mass distribution within the plant.
- Seed nitrogen concentration is allowed to vary during seed growth.
- Soil water balance is tracked in multiple soil layers instead of two soil zones.
- Soil nitrogen budget is calculated in several soil layers instead of in only the top soil layer.
- The slope of the linear increase in harvest index (dHI/dt) in some simulations is made sensitive to the pre-seed growth conditions that the crop experienced as represented by the total mass of the crop at the beginning of seed growth.

Details about the modifications incorporated into SSM-iCrop are presented in Soltani et al. (2013). An overview follows.

In SSM-iCrop, the calculation of phenological development is based on the concept of biological days (Soltani and Sinclair, 2012a). A biological day is defined as a day with optimal temperature, photoperiod, and moisture conditions for plant development. Each day from sowing until maturity, the biological day (BD) is calculated by multiplying temperature (*tempfun*), photoperiod (*ppfun*), vernalization (*verfun*), and water deficit (WSFDS) functions (factors) as follows:

$$BD = tempfun \times ppfun \times verfun \times WSFDS \quad (12)$$

tempfun and WSFDS are identical to that described for SSM-icrop2 (Section 4.1; Fig. 1). Schematics of the functions *ppfun* and *verfun* are presented in Fig. 2. Cumulative biological days are obtained by integrating the daily biological days. A phenological stage is predicted to occur when the cumulative biological days equal or exceed the biological days required to reach that stage. The biological day requirements for different phenological stages are inputs to the model.

The photoperiod function (ranging from 0 to 1) accounts for the effect of photoperiod on the development rate if the stage is sensitive to photoperiod. The daily photoperiod is used as the input for this function. The vernalization factor (ranging from 0 to 1) is determined based on the cumulative vernalization days experienced by the crop up to the current day (Fig. 2) (Soltani and Sinclair, 2012a). The model requires input values for the biological days when sensitivity to photoperiod and vernalization begins and ends. For stages that are not sensitive to photoperiod and vernalization, the values of the photoperiod and vernalization functions are set to 1.

The distribution of dry mass among specific plant components in SSM-iCrop is not based on a constant specific leaf area mass as in SSM-iCrop2 but rather on an observed biphasic pattern of leaf dry matter with increasing total plant dry matter from emergence until termination leaf production on the main stem (Fig. 3). In the initial phase (phase 1), when the total dry matter is relatively low, a higher proportion of dry matter is allocated to leaves, however, as the total dry matter increases, there is a shift to a higher portion of dry matter allocated to stems and less to leaves (phase 2). In grain crops, the leaf partitioning coefficients range from 0.5 to 0.7 for phase 1 and from 0.3 to 0.5 for phase 2. The breakpoint between phase 1 to phase 2 (WTOPL) commonly occurs at 150 to 200 g total dry mass per m^2.

In SSM-iCrop2 (Soltani et al., 2020a) and many other SSM models (e.g. Soltani and Sinclair, 2011; 2012a), seed N concentration is considered constant but in SSM-iCrop it is allowed to vary during seed growth (Manschadi et al., 2022). The initial estimation of N demand for seed growth is obtained by multiplying the seed growth rate by the maximum seed N concentration. This estimation is then updated with respect to N extractable from vegetative tissues plus the actual N uptake (NUP; uptake from the soil and/or N_2 fixation). All the limitations for NUP in SSM-iCrop2 during vegetative and seed growth also apply in SSM-iCrop. The daily extractable N from vegetative tissues is obtained by dividing N in vegetative tissues at beginning seed growth to the duration of seed filling. However, the minimum N demand for seed growth is limited to the seed growth rate and the minimum seed N concentration. A key consequence of invoking seed N demand based on minimum grain N concentration is that under low N conditions the approach accelerates N transfer to the grain and results in earlier crop maturity.

For the calculation of the soil water budget, SSM-iCrop simulates multiple layers in the soil (Soltani et al., 2013). The soil water content of each layer (WL, mm) on each day (i) is obtained using the following equation:

$$WL(L)_i = WL(L)_{i-1} + FLIN(L) - WU(L) - SE(L) - FLOUT(L) \quad (13)$$

where, L represents the layer number, FLIN is the water flux downward into the layer, WU is the amount of transpiration water extracted from the layer, SE is the amount of soil evaporation water extracted from the layer, and FLOUT is the downward water flux to the next lower layer. FLIN of a layer is set equal to the FLOUT of the layer above. For the top layer, FLIN is calculated as the sum of rainfall and irrigation, minus surface run-off on each day. The FLOUT from the user-specified layer (usually the deepest layer) is considered as deep drainage. Evaporated water is extracted from the layers (SE(L)) with a preference for the top layer, followed by the second layer, and so on. It is assumed that transpiration occurs from each layer with roots (WU(L)) in proportion to the water content of the layer. This loss decreases to zero as the FTSW in a layer approaches zero. Consequently, as a layer dries out, the crop derives less transpirational water from that layer.

Similar to the water budget, the nitrogen budget in SSM-iCrop is calculated for multiple layers in the soil. Layered soil simulations are used to simulate nitrogen movement in the soil and crop uptake. The top soil layer is especially important as the site of N fertilizer input and often a major location for denitrification. The successive concentration gradient with soil depth between soil layers is important in simulation of N transport down to lower layers in the soil.

4.8 Additional drought and temperature response features

SSM models possess two distinct features that make it suitable for assessing desirable crop traits under drought stress: (i) crop survival under severe water deficit and (ii) limited transpiration rate under elevated atmospheric vapor pressure deficit (VPD).

The mechanism of crop survival under drought stress has been elucidated by Sinclair (2000) and Ghanem et al. (2015) and is used to simulate crop termination under severe, prolonged drought. In this approach, the survival phase is characterized by stomatal 'closure. although leaf water loss is considered to continue to occur through the entire leaf epidermis based on a leaf epidermal conductance (EPCOND). As water is lost from the leaves during the survival phase, leaf relative water content decreases until it reaches lethal leaf relative water content (LTLRWC). As a result of leaves reaching lethal relative water content, leaf area is reduced. Plant death and termination of the simulation are defined when the leaf area index (LAI) decreases to less than 0.1.

To simulate sensitivity of transpiration rate to atmospheric VPD, these calculations in SSM are done on an hourly time step instead of a daily time step, following the approach described by Sinclair et al. (2005) and Guiguitant et al. (2017). Hourly solar radiation, temperature, and vapor pressure deficit values are estimated from daily input weather data. Whenever VPD exceeds a specific threshold (e.g., 2 kPa), water loss is limited to the rate calculated for the threshold VPD, resulting in soil water conservation under elevated VPD. A consequence of this trait, therefore, is an increase in daily transpiration efficiency due to the restriction of transpiration under high VPD conditions. Most importantly, however, is that soil water conservation under high VPD increases the amount of soil water available to the crop later in the growth cycle, which importantly includes the grain-filling period. The increased water availability later in the growing season allows for sustained crop and grain growth and an increased probability of greater grain yield. However, if a dry period later in the growing season does not occur to allow the crop to benefit from the conserved water, decreases in leaf gas exchange due to response to elevated VPD can result in a negative yield response (Soltani and Sinclair et al., 2012b).

Cold and heat events during flowering can damage pollination and seed set. This is incorporated in SSM-iCrop using the method proposed by Challinor et al. (2005) and Moriondo et al. (2011), but it has not been tested against experimental data. Maximum and minimum daily temperatures are used to compute stress scalar factors and then the factors are utilized to adjust the crop parameter 'linear increase in harvest index'.

5 Genetic parameter inputs to SSM

Table 1 presents the number of parameters required for the calculations in SSM-iCrop2 and SSM-iCrop. (Complete lists of the individual genetic parameter inputs for SSM-iCrop2 and SSM-iCrop, the current active SSM models, is presented in Appendix Tables A1 and A2). The total number of necessary parameters is 41 for SSM-iCrop2 and 55 for SSM-iCrop. There are also some optional parameters for possible specific processes if needed by the user. For instance, optional parameters can be incorporated to simulate heat and freezing effects on leaf area, and pollination and seed set (by adjusting the parameter for linear increase in harvest index).

Table 1 The number of parameters in SSM-iCrop2 and SSM-iCrop

	SSM-iCrop2		SSM-iCrop	
Component	Necessary	Optional	Necessary	Optional
Phenology	10		13	*
Leaf area changes	5	4[a]	8	6[a]
Dry matter production	6	1[b]	6	1[b]
Dry matter partitioning			6	
Yield formation	3		3	12[c]
Water relation	8		9	4[d]
Plant nitrogen	9		10	
Total	41[e]	5	55[e]	23

* Variable depending on the number of growth stages that user wants to simulate.
[a] To account for the effect of plant density on leaf area development and to simulate the damages of leaf area due to freezing (cold) and heat events.
[b] To simulate the effect of atmospheric CO_2 concentration.
[c] To adjust the slope of linear increase in harvest index for dry mass production before seed growth and the impacts of cold and heat events.
[d] To simulate sensitivity of transpiration rate to vapor pressure deficit trait and of crop survival to drought.
[e] Not all parameters are necessary for all species, for example, vernalization parameters.

The actual number of parameters required by the models is typically lower than listed in Table 1. For example, SSM-iCrop requires 13 parameters associated with phenology, but four of them are related to vernalization, which is not applicable for many grain species. Similarly, parameters such as the water-deficit stress threshold for biological nitrogen fixation and the biological days when biological nitrogen fixation is activated are only used for legumes. The same adjustments apply to SSM-iCrop2.

Soltani et al. (2020a) indicated that in SSM-iCrop2, the actual number of unique parameters for a crop species is around half of the number listed in Table 1. This is due to the interconnections between many parameters and the fact that certain parameters are not significant for certain species. For example, temperature units for the beginning of root growth and emergence are essentially the same for many species and input of unique values may not be necessary. Or, temperature units for the beginning of leaf senescence and the termination of root growth in grain crops can be set equal to the temperature unit for the beginning of seed growth. Many plant species share essentially equal values for the extinction coefficient of radiation in the leaf canopy.

In SSM models, crop parameters are not explicitly separated into species- and cultivar-related parameters. However, the number of parameters needed to distinguish between common cultivars is much lower than the total number of parameters. For example, in the study of Soltani and Sinclair (2015) only nine parameters were sufficient to differentiate among wheat cultivars, most of which were phenology parameters. In SSM-iCrop2, Soltani et al. (2020a) demonstrated that only three parameters were sufficient to distinguish among cultivars within a species.

Values for parameter inputs to SSM models to describe each genotype (species/cultivar) can be obtained from the literature (e.g. Soltani and Sinclair, 2012a) or derived directly from experimental observations as explained by Soltani and Sinclair (2012a). Soltani and Sinclair (2012a) also provided suggestions to identify parameter estimate(s) when doing simulations that may be causing discrepancies between observations and prediction from the model.

Probably the most challenging parameters to estimate are phenological parameters that describe the response of development rate to temperature, vernalization, photoperiod and water deficit. Fortunately, cardinal values for the temperature response are widely available in the literature and are fairly constant among cultivars of each species (Soltani and Sinclair, 2012a). However, if parameter values must be obtained for a new crop species or cultivar, it may be most convenient to undertake experiments in growth chambers under varying temperature. Villalobos and Ritchie (1992) described such experimentation. They evaluated the leaf appearance rate on sunflower plants subjected to differing temperatures and found cardinal temperatures. For determining the parameters of germination or emergence of species or cultivars, experiments can be done under different temperatures under laboratory conditions.

Development rates can be observed under field conditions by exploiting environmental variability and using regression analysis of the data to obtain cardinal temperatures. For example, Soltani et al. (2006a) used such an approach and obtained cardinal temperatures for chickpea development. The parameter that accounts for development rate acceleration due to water deficit (WSSD+1 in Fig. 1) can be estimated as the ratio of leaf temperature of water-deficit plants to a well-watered plants (Soltani and Sinclair, 2012a).

An appropriate way to estimate parameters related to photoperiod response is performing one-time, detailed growth-chamber experiments where the development rate of plants is evaluated under various photoperiods. These data can be analyzed to obtain the critical photoperiod and photoperiod sensitivity rate by the regression fitting of an appropriate function. For a good reference of such experimentation, refer to Ritchie (1991).

In the absence of growth chamber facilities, photoperiod parameters are obtained by fitting phenological data to functions from serially sown outdoor experiments (for example with one sowing date in every month of the year). Soltani et al. (2006b) used this approach to estimate cardinal temperatures and photoperiod parameters for chickpea. As noted earlier, estimating phenological parameters using model 'calibration' based on limited data is likely not valid because model parameters can compensate for each other, and the derived parameter estimates may deviate substantially from the actual functioning of plants.

Estimating parameters related to leaf area, crop mass production, mass partitioning in the plant, and plant nitrogen budget is rather straightforward. Parameters like phyllochron, specific leaf area, canopy extinction coefficient, radiation use efficiency, and nitrogen contents of plant organs can be obtained directly from observations in experiments under potential growth conditions with optimal water and nutrients conditions and free of insects, diseases, and weeds. In these experiments, observations and/or sampling can be done every 5 to 10 days during crop growing season. For details, refer to Soltani and Sinclair (2012a), and for a sample study in chickpea refer to Soltani et al. (2006c, 2006d, 2006e).

Dry-down experiments in pots are appropriate to estimate thresholds for process responses to soil volumetric water deficit (e.g. Soltani et al., 2000). In these experiments, plants are first grown in pots under well-watered conditions until the plant leaf area is sufficient to result in at least modest transpiration rates. Then, the pots are divided between two watering regimes, one optimal and one water deficit, which are imposed by simply watering and limited re-watering of pots, respectively. Daily measurements of plant leaf area and transpiration rate are performed. At the end of the dry-down of the water-deficit pots, relative or normalized rate of the plant process in the stressed plants is calculated and evaluated versus fraction transpirable soil water (FTSW). The response can be described using a 2-segment linear regression model to obtain thresholds for the decline of process activity as soil dries.

The differences among species and cultivars in SSM models lie in the parameter values (Table A1 and A2) used to define the processes. For example, difference between maize and wheat are described by differences in radiation use efficiency (RUE) and transpiration efficiency coefficient (TEC). The same is true for cultivars differences. For instance, spring wheat cultivars have less sensitivity to vernalization than winter cultivars and this difference is seen in parameters VDSAT and *vsen*.

6 Environmental inputs and model outputs

To simulate potential crop production with the SSM models, the following daily weather data are needed: minimum and maximum temperatures and daily solar radiation. If solar radiation data is unavailable, it can be estimated using sunshine hour data or temperature data, as explained by Soltani and Sinclair (2012a). Two simple programs for calculating solar radiation from sunshine hours or temperature can be found at "https://sites.google.com/view/ssm-crop-models".

In the case of simulating production when the crop may be subjected to water-stress conditions, it is necessary to input daily precipitation, and irrigation if applied. Similarly, if the impact of limited soil N is to be simulated, it is necessary to account for any N fertilization applied to the soil by inputting date(s) and amount of N application.

Soil inputs required by SSM for water-limited conditions include maximum rooting depth, soil water limits (saturation, drained upper limit, and lower limit), soil albedo, soil curve number for runoff calculation, and drainage factor for deep drainage calculation (Table 2). If the model is to be used for nitrogen-limited conditions, additional soil input parameters include soil bulk density, initial soil organic N, fraction of mineralizable organic matter, and initial soil soluble N (Table 2). SSM-iCrop2 employs a two-zone soil approach, while SSM-iCrop uses a multi-layered soil and requires specific soil inputs for each layer. Some of the soil inputs can be estimated from basic soil data, which may be available from soil surveys (Soltani and Sinclair, 2012a; Nehbandani et al., 2020).

A wide range of outputs is generated by SSM models. Output can be classified into daily and seasonal outputs. Daily outputs provide information on environmental conditions and the progress of crop development, growth, and yield. Some common output variables are: daily temperature, solar radiation and rainfall, daily and cumulative temperature unit or biological day,

Table 2 Soil input required by SSM models

Characteristic	Name
Water-related input	
Soil depth (mm)	SOLDEP
Soil albedo	SALB
Soil curve number	CN2
Soil water at saturation (mm mm^{-1})	SAT
Soil water at drained upper limit (mm mm^{-1})	DUL
Soil water at lower limit (mm mm^{-1})	LL
Soil water at air-dried (mm mm^{-1})	ADRY
Initial soil water content (mm)	iWL
Drainage factor	DRAINF
Nitrogen-related input	
Fraction coarse material in soil (%)	FG
Soil bulk density (g cm^{3})	BDL
Soil organic nitrogen (%)	NORG
Fraction soil organic nitrogen available for mineralization	FMIN
Initial amount of nitrogen in the layer (g N m^{-2})	iNSOL

LAI, fraction of intercepted PAR, daily dry matter production, seed growth rate, cumulative vegetative, grain and total dry mass, depth of root, daily and cumulative evapotranspiration, run-off, soil evaporation, plant transpiration, deep drainage, actual transpirable soil water, FTSW, daily and cumulative applied N fertilizer, N volatilization, mineralization and denitrification, and soil total and available N, daily rate of N accumulation and cumulative N in leaf, stem, and grain, and daily rate of N fixation in legumes.

Typical seasonal outputs are predicted days to phenological stages, such as days to emergence, beginning seed growth, and maturity. In tree fruits, the time of bud burst since the 1st of January is also predicted. Maximum achieved LAI, crop total dry mass, and yield are typical seasonal outputs related to crop growth and yield. Seasonal output related to soil water balance are soil water at sowing and harvest (expressed as amount and FTSW), the number and amount of irrigations if determined by the model, and cumulative rainfall, run-off, soil evaporation, plant transpiration, and deep drainage during the growing season. For plant and soil N balance, the model output includes nitrogen accumulated in leaf, stem, and grain, and the above-ground crop mass, total amount of biologically fixed N, cumulative N mineralization, volatilization, leaching, and denitrification are important seasonal outputs. Soil available N at sowing and harvest are also generated as seasonal output. Average environmental variables during main crop phases such as temperature, solar radiation, and rainfall are also generated as seasonal output. SSM-iCrop, the more intricate SSM model that simulates a multilayer soil, can output soil water and N in each layer at the beginning and the end of simulation for each season of crop growth.

7 Software and programming

There is no specific software dedicated to SSM models. Previously, SSM models were written in Basic programming languages such as GWBasic (e.g. Amir and Sinclair, 1991ab) and QBasic (e.g. Soltani et al., 1999). However, since 2010, Visual Basic for Applications (VBA) in MS Excel has been used. This choice was made because many users are already familiar with Excel, and it allows for the utilization of Excel's graphical capabilities without the need for extensive programming, thus avoiding the development of complex and cumbersome programs. Excel worksheets serve as the interface for input and output operations.

The main program (code) of SSM is organized into subprograms or sub-models, wherein initialization, rate calculations, and integration of each process are performed. SSM-iCrop2 employs the same codes for all plant species. In the case of SSM-iCrop, the same code structure is used to describe crop and soil processes for all crop species, except for phenology.

During the execution of SSM, the simulation sub-models are controlled by "administrative" sub-models responsible for inputting management and weather data, as well as performing daily and summary outputs. Excel sheets are used for input and output, and the model is implemented as a macro written in VBA within Excel. The SSM models can be downloaded from "https://sites.google.com/view/ssm-crop-models". Excel files provide separate sheets to define the location (weather), soil, management, and species/cultivar. These sheets serve as building blocks for constructing simulation scenarios or treatments. Users can prepare and store these blocks, then use them to create various scenarios or virtual experimental treatments. Therefore, the model can be readily configured to simulate multiple scenarios or virtual experimental setups.

8 Examples of SSM use

Depending on the objective for the use of a model, it is likely tests need to be done to assess the robustness of the model. SSM models has been tested and showed their robustness in many studies. Here, we refer to a few of them. As an introduction, SSM-iCrop2 has been tested for more than 30 important crops of Iran. The results (Fig. 4) showed the model was robust in prediction of crop yield and water use for this very wide range of species (Soltani et al., 2020a).

In a targeted test of robustness, Manschadi et al. (2022) evaluated the performance of SSM-iCrop in Austria for winter wheat subjected to four levels of soil nitrogen fertilization. The results (Fig. 5) confirmed the robustness of SSM-iCrop for predicting temporal dynamics of N uptake and distribution among plant tissue.

The original use of the SSM model by Sinclair (1986) was to examine the impact of water deficit on growth and yield of soybean. Since that time, SSM has been applied to a wide range of crop species (Fig. 7) and on all continents except Antarctica. Through a basic search on Google Scholar using the keywords "SSM", "crop" and "yield", we identified 55 papers published between 2011 and 2022. However, the actual number of papers using SSM is higher since not all published papers use the term "SSM," and only English papers were included in the count. The list of citations can be accessed on the SSM website at https://sites.google.com/view/ssm-crop-models.

Nearly all the more recent uses of SSM have been to do geospatial assessments of crop production. These applications of SSM can be broadly categorized into three main topics: (i) genetic modification of individual processes, (ii) assessment of management options, and (iii) analysis of crop-environment relationships.

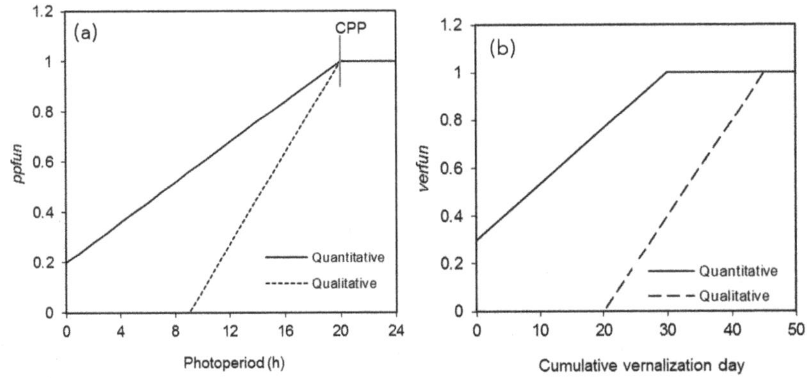

Figure 4 Schematic representation of response of development rate to photoperiod (a) and vernalization (b). Quantitative and qualitative responses to photoperiod and vernalization are presented for a long-day plant.

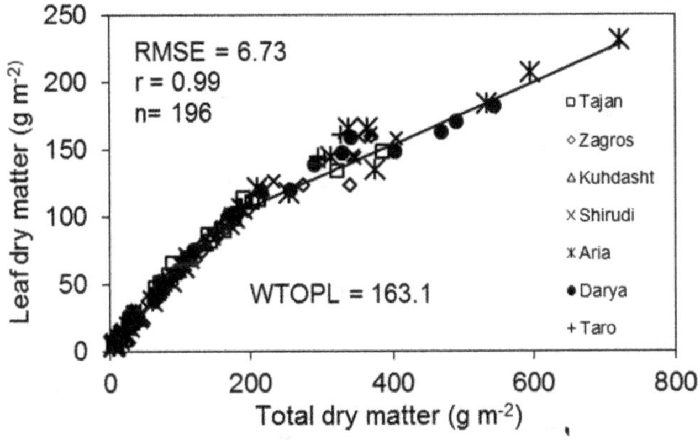

Figure 5 Relationship between cumulative leaf dry matter versus total dry matter for several wheat cultivars. WTOPL(g m^{-2}) is the total crop dry matter at the breakpoint. Data are from Bakhshandeh (2011) and Ghadiryan (2011).

8.1 Genetic alteration of individual processes

The application of the SSM models in the evaluation of desired plant genetic traits has a long history. Many of the studies have been to simulate the geospatial impact of genetic modifications under water-deficit conditions. A few examples of these studies include assessment of the impact of water-deficit thresholds on rainfed chickpea (Soltani et al., 2000), evaluation of drought plant traits to increase grain yield in maize under limited water supplies (Sinclair and Muchow, 2001), assessment of transpiration sensitivity to vapor pressure deficit in sorghum in Australia (Sinclair et al., 2005), assessment of soybean drought traits across the US (Sinclair et al., 2010), evaluation of the

VPD-limiting transpiration trait for lentil in South Asia (Guiguitant et al., 2017), identification of plant traits to increase winter wheat yield in semiarid and subhumid environments of the US (Sciarresi et al., 2019), and exploration of the importance of nocturnal transpiration traits in increasing wheat drought-tolerance in a Mediterranean environment (Schoppach et al., 2020). Below is a brief review of the study by Sadok et al. (2019) on changes in wheat yield by alteration of traits associated with plant sensitivity to water deficit.

Sadok et al. (2019) used SSM-Wheat to evaluate for Tunisia three types of water-saving traits recently identified in wheat: partial stomatal closure at high soil water content, an overall decrease in transpiration rate (the lower slope of transpiration rate versus VPD), and partial stomatal closure under elevated atmospheric VPD. To account for significant variation in seasonal precipitation across wheat-growing areas in Tunisia, a grid pattern of 29×29 km was employed for the simulations. Fig. 6 presents their findings specifically for the expression of a VPD threshold for imposition of a limited-transpiration rate under elevated VPD. By imposing a limited-transpiration rate under elevated

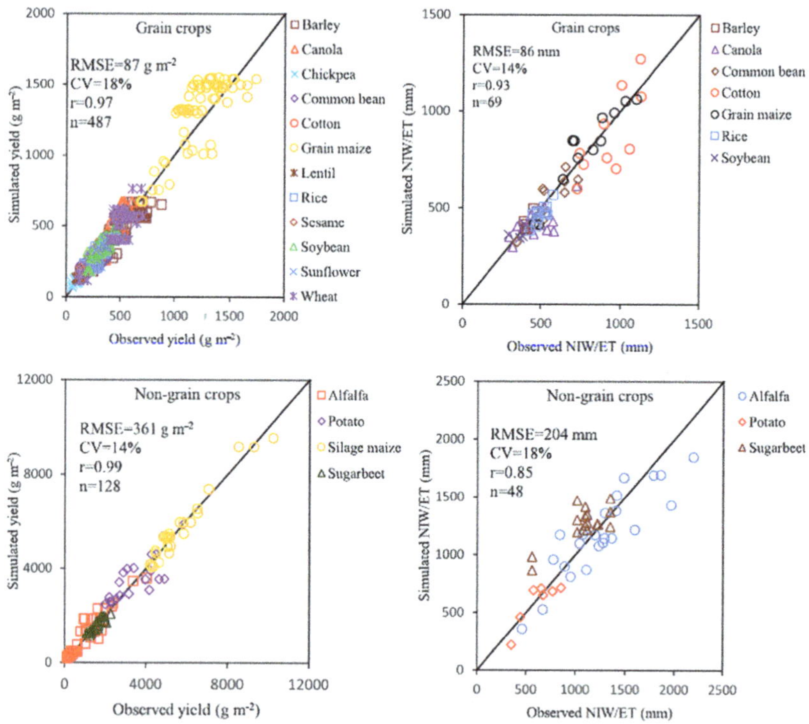

Figure 6 Simulated versus observed yields (fresh weight basis) and net irrigation water (NIW) or evapotranspiration (ET) in grain and non-grain crop species (Soltani et al., 2020a).

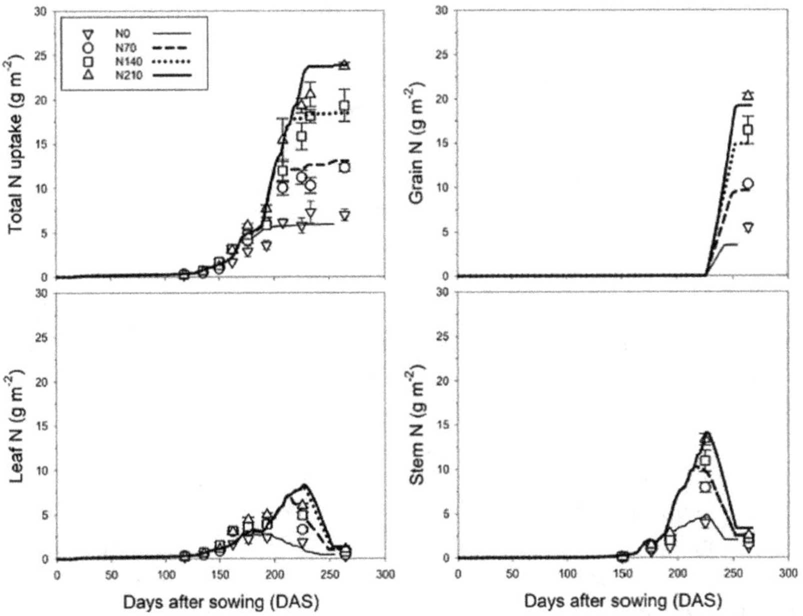

Figure 7 Simulated total and organ (grain, leaf, stem) N content (lines) compared to measured values (symbols) for the winter wheat cultivar Aurelius grown at 0 (N0), 70 (N70), 140 (N140), and 210 (N210) kg N ha^{-1} in the 2018/19 growing season. Vertical bars on the data indicate ± one standard error of the mean for measured values (Manschadi et al., 2022).

VPD, marginal yield increases were obtained with a 2 kPa VPD threshold. Even greater yield gains were achieved with a 1 kPa VPD threshold, particularly in the food-insecure central and southern regions of Tunisia. Furthermore, the implementation of water-saving measures through the limitation of transpiration at the lower VPD threshold (1 kPa) reduced yield *variability* by up to 10% in these areas. However, in contrast to regions that benefited from the trait, introduction of the limited-transpiration trait with a 1 kPa VPD threshold resulted in yield penalties of 5 to 10% in the northern, most productive region of Tunisia. These yield penalties stemmed from the inability of the crop to fully use available soil moisture for sustained gas exchange and growth. Therefore, the simulations clearly offered guidelines in making suggestions for the geographical regions in which cultivars with water-saving traits might be deployed or not deployed.

8.2 Evaluation of management options

SSM models have been used in various studies evaluating management options. Some examples include analysis of a straw-mulch system for wheat and barley in the Middle East (Sinclair and Amir, 1996; Shaaban et al., 2018), investigation of limited supplementary irrigation in chickpea (Soltani et al., 2001; Amiri et al., 2016), evaluation of crop rotation systems in Argentina (Salado-Navarro and Sinclair, 2009), assessment of irrigation management in beans in southern France (Marrou et al., 2014), examination of sowing date effects in wheat and barley in the Middle East (Schoppach et al., 2017; Shaaban et al., 2018), exploration of dormant seeding of rainfed lentils (Amiri and Deihimfard, 2018), and the analysis of yield loss caused by plant diseases in wheat (Bregaglio et al., 2021).

Vadez et al. (2017) used the SSM model to examine geospatially the influence of peanut plant density (40 plants m^{-2} vs. 20 plants m^{-2}) on yield. The simulations were done for both East and West Africa (Fig. 7). Contrary to the conventional view, their findings revealed that peanut yield in the West-Central Africa region, below latitude 12–13° N, was not limited by water-deficit conditions for either plant density. Similarly, in the East-Southern Africa region, peanut yield was unaffected by water deficiency except for the eastern fringes of Mozambique and Tanzania. As a consequence, simulation results showed that higher plant density resulted in increased average yield across nearly all of the simulated areas.

However, the simulation results for peanut by Vadez et al. (2017) showed there were geographical variations in the amount of yield increase obtained by increasing plant density from 20 plants m^{-2} to 40 plants m^{-2} (Fig. 7). Notably, significant yield increases were observed in southern West Africa and southwest East Africa, with eastern Zambia and Zimbabwe exhibiting especially high responsiveness to increased plant density. In these regions, the probability of yield increase was simulated to be at least 80 to 90% as a result of the increase in plant density, which was consistent with limited experimental reports (Vadez et al., 2017).

8.3 Assessing crop-environment relationships

An important application of SSM models has been their use in assessing crop-environment relationships. Examples include analysis of soybean yield in Argentina in response to weather variations among cropping seasons (Sinclair et al., 1992), analysis of water-limited yield in chickpeas in Iran (Soltani et al., 2016), assessment of meteorological limits to winter wheat productivity in the US southern Great Plains (Lollato et al., 2017), evaluation of prospects for increasing grain legume crop production in East Africa (Van Loon et al., 2018), characterization of the main cropping systems for chickpeas in India (Hajjarpoor et al., 2018), analysis of potential and water-limited yield and irrigation water demand for over 35 plant species across Iran (Soltani et al., 2020b, 2020c), exploration of the current status of barley yield in Iran (Alasti et al., 2022), and assessment of the impacts of climate change on crop yield, quality, and water use in various crop species (Sinclair and Rawlins, 1993; Sinclair and Seligman, 1995; Seligman and Sinclair, 1995; Amiri et al., 2021; Pourshirazi et al., 2022; Kafaie et al., 2023).

One example of the crop-environment application using the SSM model is a geospatial study of peanut yield across India by Hajjarpoor et al. (2021). The primary focus of the study was to comprehend the interactions between Genotype × Environment × Management (GEM) and use this understanding to propose a potential new environmental zonation for the crop. By evaluating the geographic distribution of simulated potential yields and comparing them with the actual yields, Hajjarpoor et al. (2021) identified seven geographically 'homogeneous production units. which exhibited distinct production-limiting constraints (Fig. 8). The identification of homogeneous production units

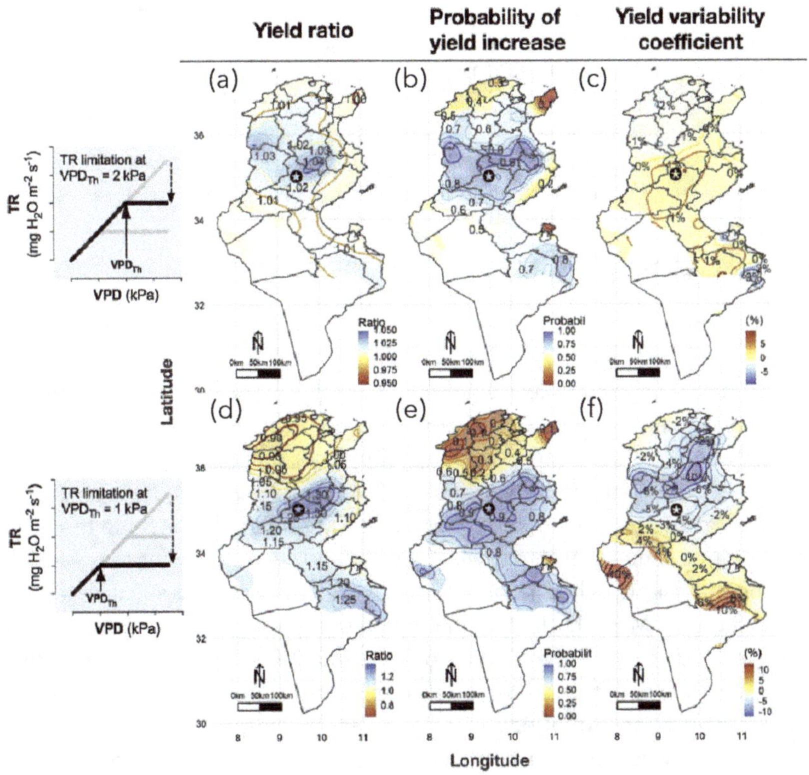

Figure 8 Yield ratios, probabilities for yield increase and yield variability coefficients resulting from deploying wheat genotypes exhibiting a limitation on transpiration rate (TR) response to increasing vapor pressure deficit (VPD) at VPD thresholds (VPD_{Th}) of 2 kPa (A–C), or 1.5 kPa (D–F (Sadok et al., 2019).

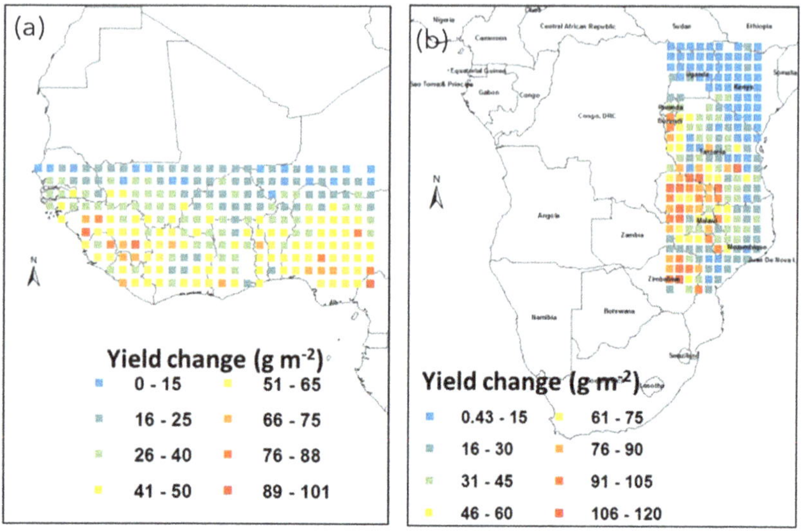

Figure 9 Peanut yield increase simulated in Africa by shifting plant density from 20 plants m^{-2} to 40 plants m^{-2} (Vadez et al., 2017).

resulted in significantly reduced variance in both actual and simulated yields when compared to the traditional geographical grouping (Fig. 9).

9 Future developments

The goal for the future is to maintain the SSM models and develop options that may prove necessary to expand the use of SSM within the guidelines established for these models. Ongoing efforts are being made to employ SSM-ICrop in simulating the impact of pests and diseases on crop growth and yield (R. Ferrise, personal communication), as well as simulating intercropping (A. Manschadi, personal communication). SSM-iCrop2 has already been used to simulate many crops including rangelands. Current studies include the use of the model to simulate crop growth and yield and water use in farmers' fields (A. Soltani et al., unpublished). Manschadi has used the model to simulate tea crop production (A. Manschadi, personal communication). An attempt is currently being explored to use the SSM model to simulate forests. Furthermore, there is an effort to incorporate sub-models into the existing model to simulate plant and soil phosphorus budgets.

The software for SSM continues to be Excel due to its simplicity and availability. However, for large-scale applications, SSM models may be implemented in *R* or Python. A Python version of SSM-iCrop2 is already accessible at: https://github.com/zackaryleady/Py_SSM_iCrop2

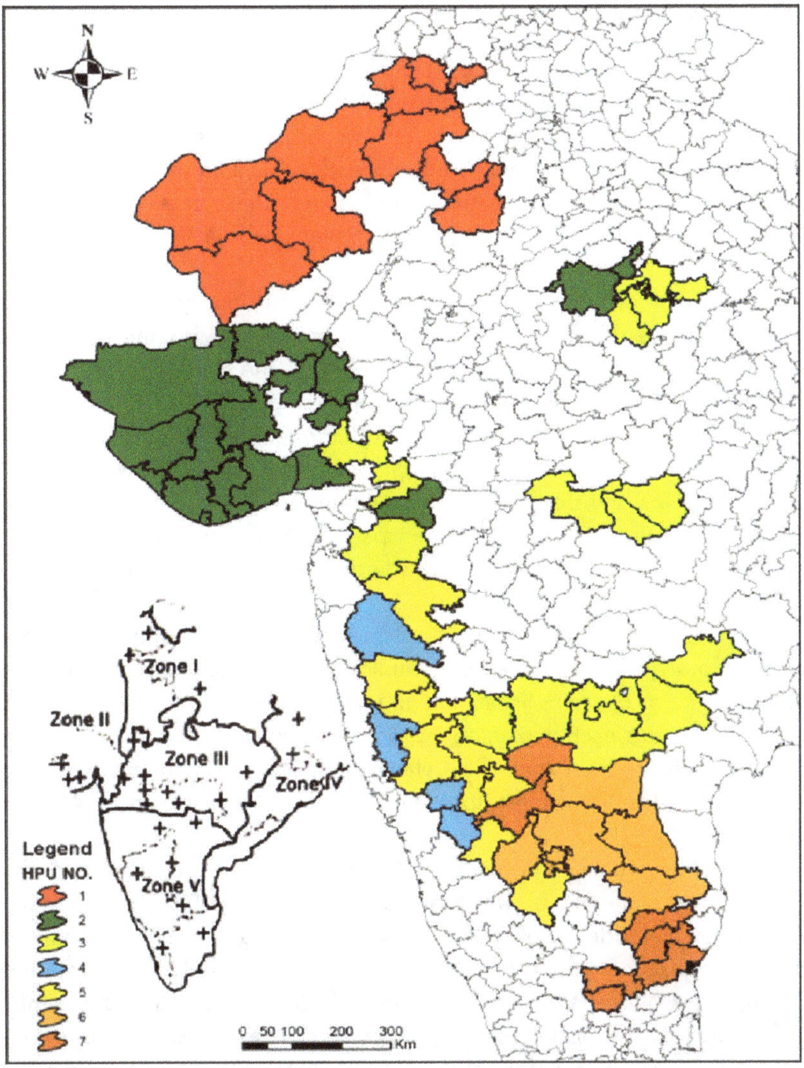

Figure 10 Results of the cluster analysis grouping primary groundnut producing districts into seven Homogenous Production Units (HPUs) (denoted by color), according to similarities in geo-biophysical characteristics, observed yield productions, and model output. Highlighted districts encompass more than 85 % of the groundnut production area in India. Inset shows the traditional groundnut zonation visualization, cited in Witcombe et al. (1998). Colors have no meaning other than to distinguish HPUs (Hajjarpoor et al., 2021).

10 Conclusions

An essential feature of crop models is their use as heuristic tools (Sinclair and Seligman, 1996) to understand and simulate crop development, growth, and yield under a range of conditions. To achieve successful heuristic investigations it is necessary that the models are both readily transparent and shown to be robust. One family of models that readily meets these criteria is the Simple Simulation Models (SSM; Sinclair, 1986; Soltani and Sinclair, 2012a; Sinclair et al., 2020).

In the SSM approach, the critical first step in using the models is a clear statement of the objective(s) of the model exercise to allow for the identification of essential model features in development of the SSM, and also to help eliminate unnecessary elements retained in the model to be used.

Another important factor leading to model transparency is a model constructed to describe a cropping system at an explanatory level that is only one level below the desired level of explanation. This approach to the crop model not only enhances model transparency but also ensures logical and consistent fits within the model by avoiding mismatches among processes and relaxation times (Sinclair et al., 2020).

Consequently, an overriding fundamental principle in the development of SSM models is to strive for simplicity without resorting to excessive reductionism or oversimplification. The number of equations and parameters used to describe each explanatory process needs to be kept to a minimum. To achieve this goal, conservative relationships and summary expressions are likely required at the explanatory level to build more comprehensible and transparent models.

In SSM models, all necessary parameters can be estimated from independent experiments, which avoids the critical problem of introducing empiricism in model application by adjusting parameter values based on simulation tests of the model, that is, 'model calibration'. In practice, model calibration is often based on adjustment of parameters within the complex framework of the model so that model output matches a limited set of observations. Since SSM does not require 'calibration', use of the model is not dependent on non-measurable or vague parameters as inputs.

SSM models have been used in various applications worldwide. In nearly all applications, SSM has been used to do geospatial assessments of crop production in an array of locations over growing seasons. Examples of geospatial applications of SSM models are described.

Another important factor leading to model transparency is a model constructed to describe a cropping system at an explanatory level that is only one level below the desired level of explanation. This approach to the crop model not only enhances model transparency but also ensures logical and consistent fits within the model by avoiding mismatches among processes and relaxation times (Sinclair et al., 2020).

Consequently, an overriding fundamental principle in the development of SSM models is to strive for simplicity without resorting to excessive reductionism or oversimplification. The number of equations and parameters used to describe each explanatory process needs to be kept to a minimum. To achieve this goal, conservative relationships and summary expressions are likely required at the explanatory level to build more comprehensible and transparent models.

In SSM models, all necessary parameters can be estimated from independent experiments, which avoids the critical problem of introducing empiricism in model application by adjusting parameter values based on simulation tests of the model, that is, 'model calibration'. In practice, model calibration is often based on adjustment of parameters within the complex framework of the model so that model output matches a limited set of observations. Since SSM does not require 'calibration', use of the model is not dependent on non-measurable or vague parameters as inputs.

11 Appendix: definition of necessary parameters

Table A1. Definition of necessary parameters in SSM-iCrop2 and their estimates for a wheat cultivar

Parameter	Name	Value
Phenology		
Base temperature for development (°C)	TBD	0
Lower optimum temperature for development (°C)	TP1D	25
Upper optimum temperature for development (°C)	TP2D	28
Ceiling temperature for development (°C)	TCD	40
Temperature unit from 1st January to bud burst or spring regrowth (°C)	ForReq	-
Temperature unit for emergence or beginning leaf growth (°C)	tuEMR	132
Temperature unit for beginning of seed or fruit growth (°C)	tuBSG	1620
Temperature unit for termination of seed or fruit growth (°C)	tuTSG	2172
Temperature unit for physiological maturity (end of dry mass accumulation) (°C)	tuPM	2170
Temperature unit for harvest or leaf fall (°C)	tuHAR	2400
Leaf area development and senescence		
Point #1 for normalized leaf area vs normalized temperature unit (x1, y1) *	X1, y1	(0.2, 0.06)
Point #2 for normalized leaf area vs normalized temperature unit (x1, y1) *	X2, y2	(0.5, 0.88)
Maximum expected leaf area index**	LAIMX	6.5
Temperature unit for beginning leaf senescence (°C)	tuBLS	1620

Parameter	Name	Value
Leaf senescence rate coefficient	SRATE	1
Dry mass accumulation		
Base temperature for dry matter production (°C)	TBRUE	0
Lower optimum temperature for dry matter production (°C)	TP1RUE	15
Upper optimum temperature for dry matter production (°C)	TP2RUE	22
Ceiling temperature for dry matter production (°C)	TCRUE	35
Extinction coefficient for photosynthetically active radiation	KPAR	0.65
Radiation use efficiency under optimal growth conditions (g MJ-1)	RUE	2.2
Yield formation		
Maximum harvest index/Liner increase in harvest index (g g-1 d-1)	HImax	0.5
Fraction of dry mass remobilizable from the vegetative tissue to the developing seeds/fruits (g g-1)	FRTRL	0.2
Grain conversion coefficient (g g^{-1})	GCC	1
Water relations		
Temperature unit for beginning root growth (°C)	tuBRG	132
Temperature unit for termination root growth(°C)	tuTRG	1620
Initial depth of roots at emergence or beginning leaf growth (mm)	iDEPORT	200
Maximum effective depth of water extraction from soil (mm)	MEED	1000
Transpiration efficiency coefficient (Pa)	TEC	5.8
FTSW*** threshold when dry matter production starts to decline	WSSG	0.3
FTSW threshold when leaf area development starts to decline	WSSL	0.4
A coefficient that specifies acceleration or retardation in development in response to water deficit	WSSD	0.4
Plant nitrogen budget		
Biological days from sowing to beginning nitrogen fixation	bdBNF	n.a.
Specific leaf area (m^2 g^{-1})	SLA	0.021
Specific leaf nitrogen in green leaves (target) (g N m^{-2})	SLNG	1.8
Specific leaf nitrogen in senesced leaves (minimum) (g N m^{-2})	SLNS	0.4
Stem nitrogen concentration in green stems (target) (g N g^{-1})	SNCG	0.019
Stem nitrogen concentration in senesced stems (minimum) (g N g^{-1})	SNCS	0.005
Grain nitrogen concentration	GNC	0.015
Maximum rate of nitrogen uptake (g N m^{-2} d^{-1})	MXNUP	0.6
FTSW threshold when nitrogen fixation starts to decline	WSSN	n.a.

*Can be replaced with estimates of related regression coefficients if they are available.
**Obtained as maximum plant leaf area under optimal condition (PLAMX: 186 cm^2 per plant) product by plant density (350 plant per m^2 for irrigated conditions).
***FTSW is fraction transpirable soil water.

Table A2. Definition of necessary parameters in SSM-iCrop and their estimates for a wheat cultivar

Parameter	Name	Value
Phenology		
Biological days from sowing to beginning response to vernalization	bdBRV	3.5
Biological days from sowing to termination response to vernalization	bdTRV	20
Biological days from sowing to beginning response to photoperiod	bdBRP	3.5
Biological days from sowing to termination response to photoperiod	bdTRP	20
Base temperature for development (°C)	TBD	0
Lower optimum temperature for development (°C)	TP1D	27.5
Upper optimum temperature for development (°C)	TP2D	27.5
Ceiling temperature for development (°C)	TCD	40
Number of vernalization days which saturates vernalization response	VDSAT	50
Vernalization sensitivity coefficient	vsen	0.00089
Critical photoperiod (h)	cpp	21
Photoperiod sensitivity coefficient	ppsen	0.001467
Biological days from physiological maturity to harvest maturity (Biological day)	bdPMHM	8
Leaf area development and senescence		
Biological days from sowing to beginning leaf growth (emergence)	bdBLG	3.5
Biological days from sowing to termination leaf growth on main-stem (booting)	bdTLM	27
Biological days from sowing to termination leaf growth on plant (beg. Seed growth)	bdTLP	43
Biological days from sowing to beginning leaf growth senescence (beg. Seed growth)	bdBLS	43
Phyllochron, the accumulated thermal time required for the successive leaves' appearance in the main stem (°C $leaf^{-1}$)	Phyl	95
A coefficient (constant) in the power relationship between plant leaf area and mainstem node number	PLACON	1
A coefficient (exponent) in the power relationship between plant leaf area and main stem node number (For 300 plants m^{-2})	PLAPOW	2.4336
Specific leaf area (m^2 g^{-1})	SLA	0.021
Dry mass accumulation		
Base temperature for dry matter production (°C)	TBRUE	0
Lower optimum temperature for dry matter production (°C)	TP1RUE	15
Upper optimum temperature for dry matter production (°C)	TP2RUE	22
Ceiling temperature for dry matter production (°C)	TCRUE	35
Extinction coefficient for photosynthetically active radiation	KPAR	0.65

Parameter	Name	Value
Radiation use efficiency under optimal growth conditions (g MJ^{-1})	IRUE	2.42
Dry mass partitioning		
Biological days from sowing to beginning seed growth	bdBSG	43
Biological days from sowing to termination seed growth	bdTSG	70
Partitioning coefficient to leaves during the main phase of leaf area development at lower levels of total crop mass (g g^{-1})	FLF1A	0.6
Partitioning coefficient to leaves during the main phase of leaf area development at higher levels of total crop mass (g g^{-1})	FLF1B	0.3
Total crop mass when leaf partitioning coefficient turns from FLF1A to FLF1B (g g^{-1})	WTOPL	160
Partitioning coefficient to leaves from termination leaf growth on mainstem to beginning seed growth (g g^{-1})	FLF2	0.1
Yield formation		
Fraction crop mass at the beginning of seed growth which is translocatable to grains (g g^{-1})	FRTRL	0.22
Grain conversion coefficient (g g^{-1})	GCF	1
The rate of linear increase in harvest index during effective grain filling period (g g^{-1} d^{-1})	PDHI	0.019
Water relations		
Biological days from sowing to beginning root growth	bdBRG	3.5
Biological days from sowing to termination termination growth	bdTRG	43
Initial depth of roots at emergence or beginning leaf growth (mm)	iDEPORT	200
Maximum effective depth of water extraction from soil (mm)	MEED	1000
Potential daily increase (growth) in root depth (mm d^{-1})	GRTDP	30
Transpiration efficiency coefficient (Pa)	TEC	5.8
FTSW* threshold when dry matter production starts to decline	WSSG	0.3
FTSW threshold when leaf area development starts to decline	WSSL	0.4
A coefficient that specifies acceleration or retardation in development in response to water deficit	WSSD	0.5
Plant nitrogen budget		
Biological days from sowing to beginning nitrogen fixation	bdBNF	n.a.
Specific leaf nitrogen in green leaves (target) (g N m^{-2})	SLNG	1.8
Specific leaf nitrogen in senesced leaves (minimum) (g N m^{-2})	SLNS	0.4
Stem nitrogen concentration in green stems (target) (g N g^{-1})	SNCG	0.019
Stem nitrogen concentration in senesced stems before beginning of seed growth (minimum) (g N g^{-1})	SNCS1	0.0095

Parameter	Name	Value
Stem nitrogen concentration in senesced stems after beginning of seed growth (minimum) (g N g^{-1})	SNCS2	0.0022
Grain nitrogen concentration (minimum)	GNCmin	0.016
Grain nitrogen concentration (maximum)	GNCmax	0.035
Maximum rate of nitrogen uptake (g N m^{-2} d^{-1})	MXNUP	0.6
FTSW threshold when nitrogen fixation starts to decline	WSSN	n.a.

*FTSW: fraction transpirable soil water.

12 References

Alasti, O., et al. (2022). Exploring the current status of barley yield and production gap of Iran. *European Journal of Agronomy*, 139, 126547.

Amir, J. & Sinclair, T.R. (1991a). A model of the temperature and solar radiation effects on spring wheat growth and yield. *Field Crops Research*, 28, 47–58.

Amir, J. & Sinclair, T.R. (1991b). A model of water limitation on spring wheat growth and yield. *Field Crops Research*, 29, 59–69.

Amiri, S.R., et al. (2016). A single supplementary irrigation can boost chickpea grain yield and water use efficiency in arid and semiarid conditions: a modeling study. *Agronomy Journal*, 108, 2406–2416. https://doi.org/10.2134/agronj2016.02.0087

Amiri, S.R., et al. (2021). Water use efficiency of chickpea agro-ecosystems will be boosted by positive effects of CO_2 and using suitable genotype × environment × management under climate change conditions. *Agricultural Water Management*, 252, 106928. https://doi.org/10.1016/j.agwat.2021.106928

Amiri, S.R. & Deihimfard, R. (2018). Can the dormant seeding of rainfed lentil improve productivity and water use efficiency in arid and semi-arid conditions? *Field Crops Research*, 227, 67–78. https://doi.org/10.1016/j.fcr.2018.07.014

Bakhshandeh, E. (2011). *Evaluation of allometric relationships in wheat cultivars of Golestan province*. Dissertation for M.Sc degree in Agronomy. Gorgan University of Agricultural Science and Natural Resources, 118p. (text in Persian with English abstract)

Bregaglio, S., et al. (2021). Comparing process-based wheat growth models in their simulation of yield losses caused by plant diseases. *Field Crops Research*, 265, 108108. https://doi.org/10.1016/j.fcr.2021.108108

Cassman, K.G. (2012). What do we need to know about global food security? *Global Food Security*, 2(1), 81–82.

Challinor, A.J., et al. (2005). Simulation of the impact of high-temperature stress on annual crop yields. *Agricultural and Forest Meteorology*, 135(1–4), 180–189.

de Wit, C.T. (1970). Dynamic concepts in biology. In Setlik, I. (Ed.), *Prediction and Measurement of Photosynthetic Activity* (pp. 17–23). Pudoc.

Ghadiryan, R. (2011). *Growth analysis of wheat cultivars of Golestan province*. Dissertation for M.Sc degree in Agronomy. Gorgan University of Agricultural Sciences and Natural Resource. (text in Persian with English abstract)

Ghanem, M.E., et al. (2015). Lentil variation in phenology and yield evaluated with a model. *Agronomy Journal*, 107, 1967–1977.

Guiguitant, J., et al. (2017). Relevance of limited-transpiration trait for lentil (Lens culinaris Medik.) in South Asia. *Field Crops Research*, 209, 96–107.

Hajjarpoor, J., et al. (2021). Environmental characterization and yield gap analysis to tackle genotype-by-environment-by-management interactions and map region-specific agronomic and breeding targets in groundnut. *Field Crops Research*, 267, 108160.

Hajjarpoor, A., et al. (2018). Characterization of the main chickpea cropping systems in India using yield gap analysis approach. *Field Crops Research*, 223, 93–104. https://doi.org/10.1016/j.fcr.2018.03.023

Hammer, G.L., et al. (1995). A peanut simulation model: I. Model development and testing. *Agronomy Journal*, 87, 1085–1093.

Hoogenboom, G., et al. (2019). The DSSAT crop modeling ecosystem. In Boote, K.J. (Ed.), *Advances in Crop Modeling for a Sustainable Agriculture* (pp. 173–216). Burleigh Dodds Science Publishing.

Kafaie Ghaeini, A., et al. (2023). Modifying sowing date as an adaptation strategy to climate change in Grain Maize (Zea mays L.) under mild-arid climates as simulated by the SSM-Maize model. *International Journal of Plant Production*, 17(3), 437–447.

Lollato, R.P., et al. (2017). Meteorological limits to winter wheat productivity in the US southern Great Plains. *Field Crops Research*, 203, 212–226. https://doi.org/10.1016/j.fcr.2016.12.014

Manschadi, A.M., et al. (2021). Full parameterisation matters for the best performance of crop models: inter-comparison of a simple and a detailed maize model. *International Journal of Plant Production*, 15, 61–78.

Manschadi, A.M., et al. (2022). Performance of the SSM-iCrop model for predicting growth and nitrogen dynamics in winter wheat. *European Journal of Agronomy*, 135, 126487. https://doi.org/10.1016/j.eja.2022.126487

Marrou, H., et al. (2014). Assessment of irrigation scenarios to improve performances of Lingot bean (Phaseolus vulgaris) in southwest France. *European Journal of Agronomy*, 59, 22–28. https://doi.org/10.1016/j.eja.2014.05.006

Mayr, E. (1982). *The growth of biological thought*. Harvard University Press.

Monteith, J.L. (1996). The quest for balance in crop modeling. *Agronomy Journal*, 88, 695–697.

Moriondo, M., et al. (2011). Climate change impact assessment: the role of climate extremes in crop yield simulation. *Climatic Change*, 104(3–4), 679–701.

Muchow, R.C. & Sinclair, T.R. (1991). Water deficit effects on maize yields modeled under current and 'greenhouse' climates. *Agronomy Journal*, 83, 1052–1059.

Muchow, R.C., et al. (1990). Temperature and solar radiation effects on potential maize yield across locations. *Agronomy Journal*, 82, 338–343.

Nehbandani, A., et al. (2020). Assessing HC27 soil database for modeling plant production. *International Journal of Plant Production*, 14, 679–687.

Nehbandani, A., et al. (2023). Estimating rangeland production in current and future conditions using SSM-iCrop2 model in Iran. *Agricultural Research*, 12(3), 346–355.

Pease, C.M. Bull, J.J. (1992). Is science logical? *BioScience*, 42, 293–298.

Pourshirazi, S., et al. (2022). Assessing the sensitivity of alfalfa yield potential to climate impact under future scenarios in Iran. *Environmental Science and Pollution Research*, 29(40), 61093–61106.

Ritchie, J.T. (1991) Wheat phasic development. In Hanks, R.J. and Ritchie, J.T. (Eds.), *Modeling Plant and Soil Systems* (pp. 31–54). Agronomy Monograph No. 31, American Society of Agronomy – Crop Science Society of America – Soil Science Society of America.

Ritchie, J.T. (1998). Soil water balance and plant water stress. In: Tsuji, G.Y., et al. (Eds.), *Understanding Options for Agricultural Production* (pp. 41–54). Kluwer Academic Publishers.

Sadok, W., et al. (2019). Wheat drought-tolerance to enhance food security in Tunisia, birthplace of the Arab Spring. *European Journal of Agronomy*, 107, 1–9. https://doi.org/10.1016/j.eja.2019.03.009

Salado-Navarro, L.R. & Sinclair, T.R. (2009). Crop rotations in Argentina: analysis of water balance and yield using crop models. *Agricultural Systems*, 102(1–3), 11–16.

Schoppach, R., et al. (2017). Yield comparison of simulated rainfed wheat and barley across Middle-East. *Agricultural Systems*, 153, 101–108. https://doi.org/10.1016/j.agsy.2016.12.017

Schoppach, R., et al. (2020). Sleep tight and wake-up early: nocturnal transpiration traits to increase wheat drought tolerance in a Mediterranean environment. *Functional Plant Biology*, 47(12), 1117–1127. https://doi.org/10.1071/fp20044

Seligman, N.G. & Sinclair, T.R. (1995). Global environment change and simulated forage quality of wheat II. Water and nitrogen stress. *Field Crops Research*, 40(1), 29–37.

Sciarresi, C., et al. (2019). Plant traits to increase winter wheat yield in semiarid and subhumid environments. *Agronomy Journal*, 111, 1–13. https://doi.org/10.2134/agronj2018.12.0766

Shaaban, A.S.A., et al. (2018). Sowing date and mulch to improve water use and yield of wheat and barley in the Middle East environment. *Agricultural Systems*, 165, 26–32. https://doi.org/10.1016/j.agsy.2018.05.011

Sinclair, T.R. (1986). Water and nitrogen limitations in soybean grain production. I. Model development. *Field Crops Research*, 15, 125–141.

Sinclair, T.R. (2000). Model analysis of plant traits leading to prolonged crop survival during severe drought. *Field Crops Research*, 68(3), 211–217.

Sinclair, T.R. & Amir, J. (1992). A model to assess nitrogen limitations on the growth and yield of spring wheat. *Field Crops Research*, 30, 1–11.

Sinclair, T.R. & Amir, J. (1996). Model analysis of a straw mulch system for continuous wheat in an arid climate. *Field Crops Research*, 47(1), 33–41.

Sinclair, T.R. & de Wit, C.T. (1976). Analysis of the carbon and nitrogen limitations to soybean yield. *Agronomy Journal*, 68, 319–324.

Sinclair, T.R. & Horie, T. (1989). Leaf nitrogen, photosynthesis, and crop radiation use efficiency: a review. *Crop Science*, 29, 90–98.

Sinclair, T.R. & Ludlow, M.M. (1986). Influence of soil water supply on the plant water balance of four tropical grain legumes. *Australian Journal of Plant Physiology*, 13, 329–341.

Sinclair, T.R. & Muchow, R.C. (1995). Effect of nitrogen supply on maize yield: I. Modeling physiological responses. *Agronomy Journal*, 87, 632–641.

Sinclair, T.R. & Muchow, R.C. (1999). Radiation use efficiency. *Advances in Agronomy*, 65, 215–265.

Sinclair, T.R. & Muchow, R.C. (2001). System analysis of plant traits to increase grain yield on limited water supplies. *Agronomy Journal*, 93, 263–270.

Sinclair, T.R. & Rawlins, S.L. (1993). Inter-seasonal variation in soybean and maize yields under global environmental change. *Agronomy Journal*, 85(2), 406–409.

Sinclair, T.R. & Seligman, N.G. (1995). Global environment change and simulated forage quality of wheat I. Nonstressed conditions. *Field Crops Research*, 40(1), 19–27.

Sinclair, J.R. & Seligman, N.G. (1996). Crop modeling: from infancy to maturity. *Agronomy Journal*, 88, 698–704.

Sinclair, T.R., et al. (1987). Field and model analysis of water deficits on carbon and nitrogen accumulation by soybean, cowpea and black gram. *Field Crops Research*, 17, 121–140.

Sinclair, T.R., et al. (1992). Soybean yield in Argentina in response to weather variation among cropping seasons. *Field Crops Research*, 30(1–2), 1–11.

Sinclair, T.R., et al. (1997). Model analysis of sorghum response to nitrogen in subtropical and tropical environments. *Agronomy Journal*, 89, 201–207.

Sinclair, T.R., et al. (2003). Modeling nitrogen accumulation and use by soybean. *Field Crops Research*, 81, 149–158.

Sinclair, T.R., et al. (2005). Potential yield and water-use efficiency benefits in sorghum from limited maximum transpiration rate. *Functional Plant Biology*, 32, 945–952.

Sinclair, T.R., et al. (2010). Assessment across the United States of the benefits of altered soybean drought traits. *Agronomy Journal*, 102, 475–482.

Sinclair, T.R., et al. (2020). Geospatial assessment for crop physiological and management improvements with examples using the Simple Simulation Model. *Crop Science*, 59, 1–9. doi:10.2135/cropsci2019.02.0093

Soltani, A. & Sinclair, T.R. (2011). A simple model for chickpea development, growth and yield. *Field Crops Research*, 124, 252–260.

Soltani, A. & Sinclair, T.R. (2012a). *Modeling Physiology of Crop Development, Growth and Yield*. CABI Publisher.

Soltani, A. & Sinclair, T.R. (2012b). Identifying plant traits to increase chickpea yield in water-limited environments. *Field Crops Research*, 133, 186–196.

Soltani, A. & Sinclair, T.R. (2015). A comparison of four wheat models with respect to robustness and transparency: simulation in a temperate, sub-humid environment. *Field Crops Research*, 175, 37–46.

Soltani, A., et al. (1999). A simple model for chickpea growth and yield. *Field Crops Research*, 62, 213–224.

Soltani, A., et al. (2001). A simulation study of chickpea crop response to limited irrigation in a semiarid environment. *Agricultural Water Management*, 49, 225–237.

Soltani, A., et al. (2013). SSM-wheat: a simulation model for wheat development, growth and yield. *International Journal of Plant Production*, 7, 711–740.

Soltani, A., et al. (2016). Analysis of chickpea yield gap and water-limited potential yield in Iran. *Field Crops Research*, 185, 21–30. https://doi.org/10.1016/j.fcr.2015.10.015

Soltani, A., et al. (2000). Thresholds for chickpea leaf expansion and transpiration response to soil water deficit. *Field Crops Research*, 68, 205–210.

Soltani, A., et al. (2006a). Modeling seedling emergence in chickpea as influenced by temperature and sowing depth. *Agricultural and Forest Meteorology*, 138, 156–167.

Soltani, A., et al. (2006b). Modeling chickpea growth and development: phenological development. *Field Crops Research*, 99, 1–13.

Soltani, A., et al. (2006c). Modeling chickpea growth and development: leaf production and senescence. *Field Crops Research*, 99, 14–23.

Soltani, A., et al. (2006d). Modeling biomass accumulation and partitioning in chickpea (Cicer arietinum L.). *Journal of Agronomy and Crop Science*, 192, 379–389.

Soltani, A., et al. (2006e). Modeling chickpea growth and development: nitrogen accumulation and use. *Field Crops Research*, 99, 24–34.

Soltani, A., et al. (2019). *Analysis of food security in Iran towards 2050 based on modeling water, land, food and environment nexus: prospects and required policies*. Gorgan University of Agricultural Sciences and Natural Resources, Gorgan, Iran and Agricultural Research, Education and Extension Organization, Tehran, Iran, Research Report, p. 129.

Soltani, A., et al. (2020a). SSM-iCrop2: a simple model for diverse crop species over large areas. *Agricultural Systems*, 182, 102855. doi:10.1016/j.agsy.2020.102855

Soltani, A., et al. (2020b). Modeling plant production at country level as affected by availability and productivity of land and water. *Agricultural Systems*, 183, 102859. doi:10.1016/j.agsy.2020.102859

Soltani, A., et al. (2020c). Future food self-sufficiency in Iran: a model-based analysis. *Global Food Security*, 24, 100351. doi:10.1016/j.gfs.2020.100351

Speath, S.C. & Sinclair, T.R. (1985). Linear increase in soybean harvest index during seed-filling. *Agronomy Journal*, 77, 207–211.

Tanner, C.B. & Sinclair, T.R. (1983). Efficient water use in crop production: research or re-search?. in Taylor, H.M., et al. (eds.) *Limitations to Efficient Water Use in Crop Production* (pp. 1–27). American Society of Agronomy.

Vadez, V., et al. (2017). Mapping water stress incidence and intensity, optimal plant populations, and cultivar duration for African groundnut productivity enhancement. *Frontier in Plant Sciences*, 8, 432.

Van Loon, M.P., et al. (2018). Prospect for increasing grain legume crop production in East Africa. *European Journal of Agronomy*, 101, 140–148.

Whitcombe J, Virk D, Farrington J. 1998. Seeds of Change. London Intermediate Technology

Williams, J.R. (1991). Runoff and soil erosion. In Hanks, R.J. & Ritchie, J.T. (Eds.), *Modeling Plant and Soil Systems* (pp. 439–456). Agronomy Monograph No. 31.

Chapter 5

The Systems Approach to Land Use Sustainability (SALUS) model

B. Basso, T. Tadiello, P. Sharma, L. Price and J.T. Ritchie, Michigan State University, USA

1 Introduction

The Systems Approach to Land Use Sustainability (SALUS) is a process-based model designed to simulate crop growth by accounting for genetic variation, and soil, water, and nutrient conditions under varying agronomic management and climate scenarios. SALUS can reproduce numerous management practices, including tillage types, planting methods, irrigation schedules, fertilization strategies, harvest operations, and residue handling (Basso et al., 2006). Operating on a daily time step, the model examines the interactions between the various components of an agricultural field system, with a focus on their impact on plant growth and yield, soil organic matter (SOM), water, nitrogen (N), and phosphorous (P) dynamics.

The components of SALUS have been built upon and modified from other major modeling sources that include the DSSAT suite of crop models (Jones et al., 2003), in particular CERES (Ritchie et al., 1998), as well as CENTURY (Parton et al., 1987). SALUS is a functional model used for prediction and control. It accounts for cultivar and management effects influenced by soil and climate conditions. The model does not account for the impacts of pests,

http://dx.doi.org/10.19103/AS.2025.0155.08

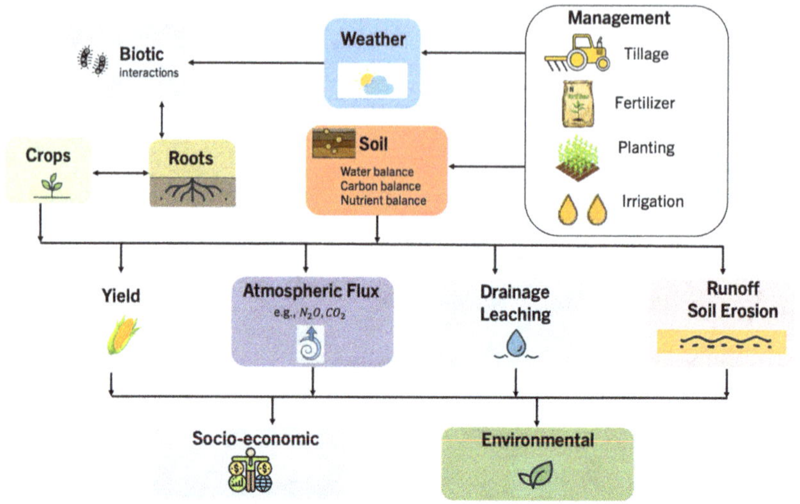

Figure 1 Overview of major SALUS (Systems Approach for Land Use Sustainability) model modules.

disease, but it can be coupled to external pest or risk modules. By prescribing the type and magnitude of these effects, external models or remote sensing observations can drive SALUS to reflect pest incidence or other damage processes. A diagram of the SALUS model is shown in Fig. 1. We provide a brief overview of some major components that include crop growth and development, SOM and nutrient cycling, and soil water balance and temperature, highlighting how simulation of varying crop genetics impacts these areas. More detailed information on model structure and parameters can be found in Basso et al. (2006, 2010, 2015), Dzotsi et al. (2013, 2015), and https://basso.ees.msu.edu/salus/index.html, as well as other sources noted throughout this chapter.

2 Modeling crop growth and development

SALUS crop growth components, derived from the CERES (Ritchie, 1998; Ritchie et al., 1989) and IBSNAT (Uehara and Tsuji, 1993) suite of crop production models were originally developed for single-year, monoculture simulations. The main external inputs required for simulating crop growth are genetic (variety-specific) coefficients and climate data (e.g. daily solar radiation, precipitation, and air temperature). SALUS can simulate crop growth using either a "simple" or "complex" module, which we refer to as SALUS-simple and SALUS-complex.

2.1 SALUS-simple

Based on modeling approaches from EPIC (Erosion Productivity Impact Calculator, Williams et al., 1989) and ALMANAC (Agricultural Land Management Alternatives with Numerical Assessment Criteria, Kiniry et al., 1992), SALUS-simple does not require extensive parameterization. For example, predetermined leaf area index (LAI) curves and growing degree days accumulation define the crop's phenological stages (Dzotsi et al., 2013). The simple model has been shown to capture fundamental growth dynamics while reducing extensive input data and calibration (e.g. Dzotsi et al., 2015). SALUS-simple determines the timing and duration of phenological phases based on accumulated heat units, measured in degree days above the crop-specific base temperature, as well as other temperature thresholds (e.g. see Fig. 2 for

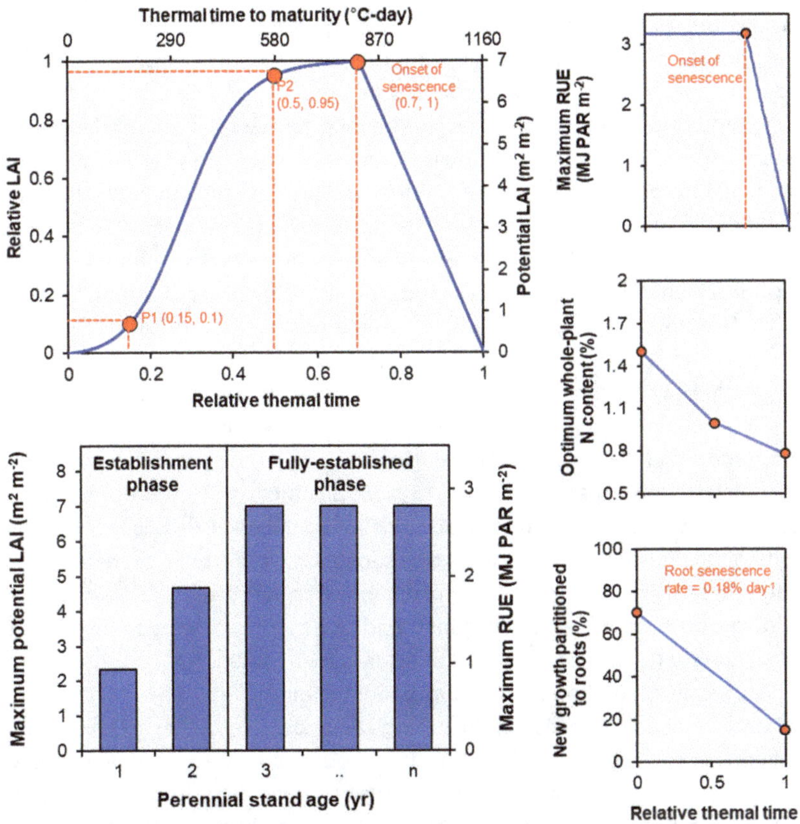

Figure 2 Summary of the main functional relationships in the simple model including perennial stand age for perennial crops from Supporting Information S1.2 of Martinez-Feria and Basso (2020).

perennial crops). When environmental factors such as water or nitrogen (N) are limited, the rate of biomass accumulation is reduced, representing plant stress and productivity decline. Root growth is modeled to extend deeper into the soil over time, with root allocated biomass decreasing over time. Root expansion is adjusted using a soil hospitality factor based on soil properties, as well as soil water and temperature conditions. The final grain yield is calculated based on the total aboveground biomass and the crop-specific harvest index. More information on the SALUS-simple crop model is given in Dzotsi et al. (2013).

2.2 SALUS-complex

SALUS-complex, based on the CERES suite of crop models, is characterized by the fractional progress toward the emergence of new leaves, a metric that allows leaf development to be quantified dynamically throughout the crop cycle, based on temperature, photoperiod, and crop-specific growth parameters. More detail is described in the genetic section below, but briefly, the phases include germination, emergence, vegetative growth until floral induction, reproductive development until crop ear growth is complete, grain filling until physiological maturity, and beyond to mark the end of the growing season. In SALUS-complex, the LAI is determined separately for the main shoot and tillers from the availability of photosynthates, leaf growth coefficients, and specific leaf weight. This adaptability enables the influence of stress factors on leaf growth to be simulated. Leaf senescence is modeled to account for the loss of photosynthetic capacity and the redistribution of nutrients within the plant.

In SALUS, net photosynthesis is modeled using the principle of radiation use efficiency (RUE). This concept assumes that daily biomass accumulation is directly proportional to the intercepted photosynthetically active radiation (IPAR), reflecting the central role of light capture in crop growth. Originally introduced by Monteith (1977), the RUE framework has proven to be a robust approach for linking solar radiation to plant productivity. However, subsequent studies have shown that the proportionality constant of RUE is not universal. Instead, it varies depending on factors such as the time scale of measurement (hourly, weekly, or seasonal), the form of biomass quantified (total aboveground, belowground, or individual organs), and the type of radiation used in the calculation (e.g. solar or photosynthetically active radiation).

Accurate estimation of LAI is fundamental for models based on IPAR, since LAI determines the fraction of incoming radiation intercepted by the canopy. Defined as the ratio of leaf area to the ground surface area covered, LAI is highly sensitive to plant population density, particularly in the early growth stages before the canopy closes. Thus, both the accuracy of LAI estimation and its dynamics over time are critical for reliable simulation of light interception and, ultimately, crop yield.

Plant development in SALUS is represented through two complementary processes: phasic development and morphological development. Phasic development describes the progression through distinct growth stages, along with the partitioning of assimilated biomass into different organs (leaves, stems, roots, reproductive structures). Genetic diversity within crop species allows for wide adaptation across global environments. For instance, wheat genotypes can be cultivated successfully from temperate Argentina (50° S) to northern Sweden (60° N), as well as in tropical environments in between. Morphological development, in contrast, refers to the growth and differentiation of specific plant organs throughout the life cycle. Both processes are strongly influenced by environmental conditions. Temperature is the dominant driver, with developmental progress quantified through growing degree-days (GDD) or thermal time. Cultivar characteristics and photoperiod sensitivity act as additional modulators of developmental rate and duration.

The balance between growth rate and growth duration is crucial in determining final yield. Record-breaking yields of annual crops are typically achieved in cooler environments where extended day length allows for prolonged growth. In contrast, warmer climates may compensate for shorter growing periods by allowing multiple cropping cycles per year, leading to comparable annual productivity. The most widely applied method to estimate crop development duration is through thermal time accumulation. Thermal time (td) is expressed as

$$td = \Sigma ni = 1(Ta - Tb)$$

where T_a is the daily mean temperature, T_b is the base temperature below which growth ceases, and n is the number of days. Daily mean temperature is typically approximated as the average of daily maximum and minimum values. Accurate simulation requires that temperature be measured near the plant's growing point, as air temperature alone may not reflect the microclimate experienced by developing tissues. For example, Ritchie and NeSmith (1991) found that calculating thermal time from air temperature overestimated leaf tip appearance and leaf number in the CERES-Maize model. Corrections required adjustments to the phyllochron – the interval between successive leaf tip appearances (Vinocur and Ritchie, 2001).

In addition to temperature, photoperiod sensitivity plays an important role in the development of many species. Short-day crops such as maize, sorghum, and soybean accelerate development under shorter day lengths, while long-day crops such as wheat and barley develop faster under extended photoperiods. Ritchie and NeSmith (1991) further demonstrated that photoperiod significantly influences both the number of leaves produced and the duration of vegetative growth stages in maize.

3 Incorporating genetics

Growth and development parameters establish the time framework for all physiological processes, determining how quickly a cultivar progresses from seed to maturity under various environmental conditions. The system starts with thermal time requirements, based on the degree-day accumulations that control key phenological transitions. These parameters serve as biological clocks, indicating when germination occurs, when seedlings emerge, when flowering begins, when grain filling starts, and when physiological maturity is reached. This thermal time method enables the model to reflect how different temperature patterns speed up or slow down crop development, offering the flexibility to simulate performance in various climate zones. A faster leaf appearance rate can lead to earlier light capture and potentially higher yields but may also increase vulnerability to late-season stress if resources become limiting. Grain and ear growth rates define the genotype-specific potential for dry matter accumulation into reproductive structures, while prolificacy captures the genetic tendency to produce multiple ears under favorable conditions. Conversely, the barrenness parameter acknowledges the harsh reality that stress conditions can prevent ear formation entirely, providing a mechanism to simulate yield loss under drought, nutrient deficiency, or other limiting factors. The transition from vegetative growth to reproductive development represents one of the most critical phases in crop production, and ear parameters capture the efficiency of this transition. These parameters recognize that initiating reproductive structures and successfully developing them are distinct processes, each with unique sensitivities to environmental conditions. Ear setting efficiency quantifies the proportion of initiated ears that successfully develop into harvestable structures. This parameter is particularly sensitive to plant population density, environmental stress, and genetic background, making it a crucial link between management decisions and reproductive success. High plant densities may reduce ear setting efficiency through increased competition, while stress during critical reproductive phases can lead to ear abortion. Cumulative IPAR thresholds for ear development links canopy development with reproductive initiation. The plant must accumulate a minimum amount of intercepted light energy before triggering ear formation, ensuring that reproductive investment occurs only when sufficient photosynthetic capacity exists to support it. This mechanistic approach moves beyond simple calendar-based development to capture the fundamental relationship between light capture and reproductive success. The final yield potential ultimately depends on kernel number and individual kernel weight. These parameters acknowledge that yield formation is not a single process but rather a series of sequential events, each with unique sensitivities and requirements. The framework begins with the genetic ceiling, the potential maximum kernel number per ear that defines the upper boundary of yield potential. However, realizing this potential depends on kernel setting efficiency, which determines what proportion of potential kernels survive the critical phases of pollination and early development. This parameter is exquisitely sensitive to stress conditions, particularly drought and nutrient limitations during the narrow window of kernel set. Once kernels are established, their filling becomes the final determinant of yield. The daily kernel fill rate at optimum temperature defines the potential for dry matter accumulation under non-limiting conditions, while the cumulative IPAR threshold for kernel development ensures that this filling process is constrained by the plant's actual photosynthetic capacity rather than proceeding independently. The inclusion of maximum kernel nitrogen concentration adds a quality dimension to yield formation, recognizing that kernel filling involves not just carbon accumulation but also nitrogen deposition for protein synthesis. This parameter links grain quality to nitrogen availability and demand, enabling the model to simulate trade-offs between yield quantity and quality under varying nitrogen management strategies.

The final category encompasses the cross-cutting physiological traits that enable crops to function across diverse environments. These parameters capture the fundamental processes of light interception and environmental sensitivity that underlie all crop performance. Light interception coefficients govern how effectively the canopy captures available solar radiation, implementing Beer's law to relate LAI to radiation capture efficiency. This seemingly technical parameter has profound implications for crop performance, as radiation interception drives photosynthesis and, ultimately, all crop growth and development processes. Environmental adaptation parameters acknowledge that crops are not equally suited to all environments. Cold tolerance coefficients capture genetic differences in the ability to maintain physiological processes under sub-optimal low temperatures, while photoperiod response parameters ensure accurate simulation of daylength sensitivity. The photoperiod framework includes both lower and upper daylength limits that trigger developmental responses, plus sensitivity coefficients that quantify how development rate changes with daylength variations. These photoperiod parameters are particularly crucial for capturing how genotypes adapted to specific latitudes behave when grown outside their traditional range – a scenario increasingly common as agriculture adapts to climate change and global food security challenges.

4 Soil carbon and nitrogen dynamics

The SOM and nitrogen (N) modules in SALUS are derived from the CENTURY model (Parton et al., 1987, 1988) and simulate the cycling of carbon (C) and N between organic and mineral pools in the soil. The SOM module (Fig. 3)

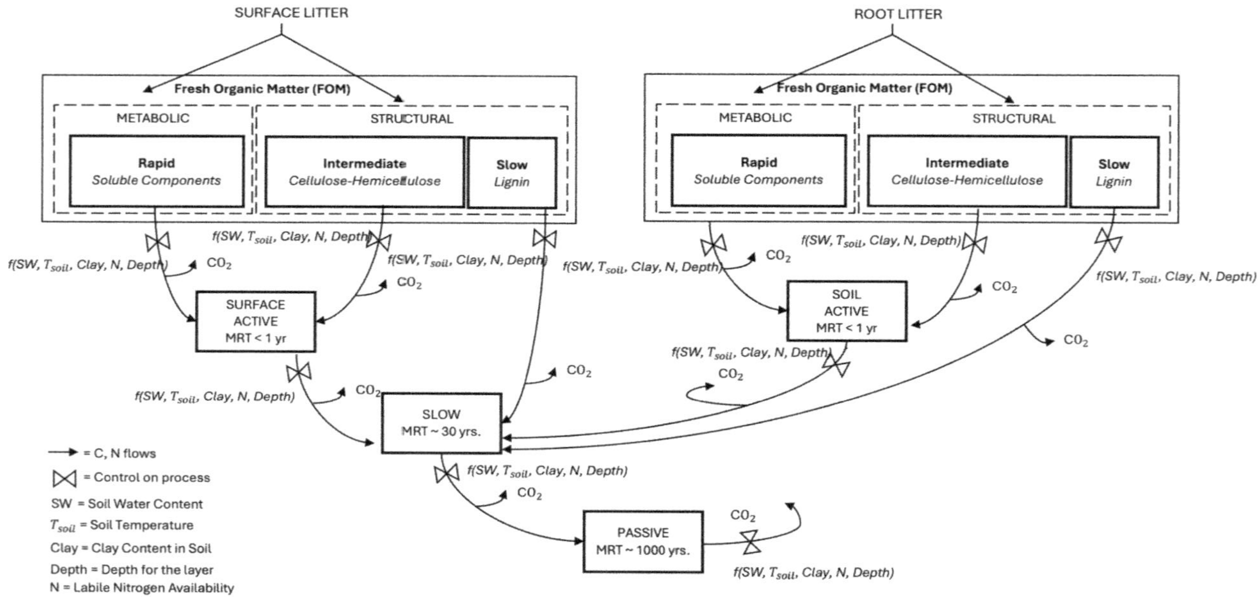

Figure 3 Conceptual diagram of the soil organic matter cycling processes in the SALUS process-based model. The figure represents the decomposition pathways of surface and root litter, categorized into metabolic (rapid soluble components) and structural (intermediate cellulose-hemicellulose) and slow (lignin) fractions. Fresh organic matter undergoes microbial transformation into active, slow, and passive pools, each with different mean residence times.

is divided into conceptual pools. Each C input or output follows first-order kinetics, characterized by a pool-specific decomposition rate and C:N ratio, and are adjusted by environmental factors, that include soil temperature, moisture, depth, and clay content. The inorganic N module (Fig. 4) is divided into ammonium (NH_4^+) and nitrate (NO_3^-) pools, while inorganic phosphorus (P) is partitioned into labile, active, and stable pools (not shown).

For the SOM module, the pools considered are fresh organic matter (FOM), microbial biomass (ACTIVE), humus (SLOW), and resistant (PASSIVE). The FOM pool is divided into fresh plant material from surface litter, such as aboveground crop residues incorporated into the soil, and dead root litter that decomposes in the soil. Each source of FOM is further divided into sub-pools based on residue quality and decomposition characteristics: soluble components (rapid, considered metabolic), and cellulose-hemicellulose (intermediate) and lignin (slow), both of which are considered Structural. The ACTIVE pool, which includes microbial biomass and highly labile by-products of microbial metabolism, undergoes rapid decomposition. The SLOW pool contains microbial by-products that are moderately resistant to further decomposition. Products of microbe metabolism that are extremely resistant to further breakdown reside in a PASSIVE pool. From each of these pools, a fraction of the decomposing SOM is lost as carbon dioxide (CO_2) from heterotrophic respiration (Parton et al., 1987, 1988). These SOM pools are differentiated by their mean residence time (MRT) and C:N ratios. The MRT can be interpreted as the average time required to completely renew the C in a pool, which is equal to the turnover time at steady state (Six and Jastrow, 2002; Rabbi et al., 2013). The approximate MRTs are ACTIVE (0.5–1 year; C/N = 8), SLOW (10–50 years; C:N = 11), and PASSIVE (1000–5000 years; C:N = 9).

The initialization of the FOM pools is user defined. The FOM content in the soil can vary rapidly depending on residue quality and prevailing environmental conditions (Elias et al., 2024). The user specifies the dry matter content of both surface and belowground residues from previous crops, while the C and N residues content is crop specific. Additionally, the user can specify whether surface residues have been incorporated after the harvest event (via choice of tillage practice) into the soil. For root residues, the amount of C and N distributed down the soil profile is determined using an exponential function, with C and N concentration decreasing with soil depth.

For the three belowground SOC pools, an "initial" organic carbon (OC, % or g/100 g of dry soil) value for each soil layer is required. This value can be taken from measured values in the field or other reliable sources (e.g. gSSURGO, 2020). This total amount of organic carbon (TOC) is distributed into the ACTIVE, SLOW, and PASSIVE pools. The fraction of active carbon in the soil is influenced by soil texture and is lower in sandy soils and higher in clay soils, but typically constituting between 2% and 5% of total SOC. The

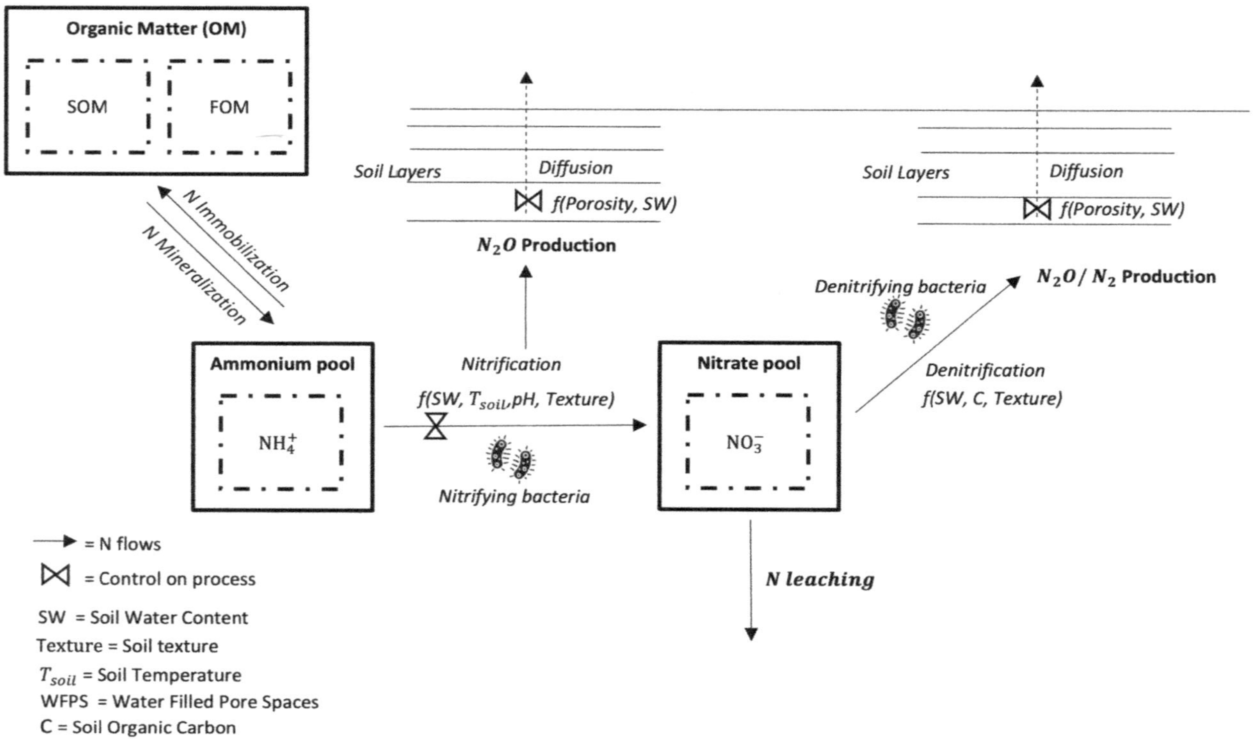

Figure 4 Conceptual representation of nitrogen cycling in the SALUS process-based model. The figure illustrates the conversion and movement of nitrogen through organic matter (OM) decomposition, into ammonium (NH_4^+) and nitrate (NO_3^-) pools. Nitrogen mineralization and immobilization regulate the conversion of soil organic matter into available nitrogen forms.

SLOW pool holds by default approximately 45% of TOC, while the PASSIVE pool remains relatively constant at around 50% of TOC, regardless of soil texture.

For each soil layer, the decomposition rate and amount of organic matter within the three pools is determined on a daily timestep, with the amount of C and N transferred between the pools determined by several environmental factors (e.g. soil temperature, texture, and water content), and management practices (e.g. tillage intensity).

As noted, the C:N of SOM pools are fixed values; however, the C:N of the FOM pools from crop residues and roots depends on the quality and composition of the residue. The movement of C between pools is accompanied by N movement to maintain the fixed C:N ratios, with N either released from organic matter through mineralization or immobilized into organic matter (Fig. 4).

During the mineralization process, released N is added to the NH_4^+ pool, while with immobilization, the N needed for decomposition is taken from NH_4^+ pool. Once mineralized, NH_4^+ undergoes nitrification (discussed below), a microbial process that converts it into NO_3^-. Both plants and microbes compete for their N needs from inorganic N pools, with insufficient N to maintain the decomposition required for fixed C:N, further reduction in the decomposition rates reduces microbial activity. Inorganic N produced during the process is susceptible to N loss via two pathways: in gaseous form from the soils surface and into the atmosphere or leaching through the soil profile into groundwater. A gas module simulates the production of the greenhouse gas (GHG) nitrous oxide (N_2O) and dinitrogen (N_2) gas from the microbial process of nitrification and denitrification in the soil. Nitrification, the oxidation of NH_4^+ via intermediates to NO_3^-, during which N_2O is released as a by-product, is estimated using nitrification rate – the fraction of NH_4^+ converted to NO_3^- each day, and is determined by the soil NH_4^+ concentration and modified by soil conditions, including temperature, water-filled pore spaces (WFPS), texture, and pH (Gilmour, 1984; Parton et al., 1994) in each soil layer. The amount of N_2O produced is proportional to nitrification rate. Denitrification is the reduction of NO_3^- via intermediates to N_2O and N_2. Denitrification is controlled by substrate C, soil temperature, moisture, and NO_3- availability. The module includes a water-filled pore space (WFPS) function (as a proxy for O_2 availability) to determine the fraction of N as N_2O emitted during denitrification, the remainder being emitted as N_2. The diffusion of N_2O between the soil incremental depths within the soil profile is also determined by considering soil porosity and air space (Shcherbak and Robertson, 2014). The solubility of N_2O in the soil water is temperature-dependent, and changes in N_2O concentration are tracked in each soil layer based on the production from nitrification and denitrification in that layer. Nitrate movement in the soil profile is dependent on water movement. Thus, nitrate concentrations are combined with modeled downward water drainage to provide estimates of nitrate leaching.

5 Soil water balances

The soil water balance module is derived from CERES models but with substantial revisions to the calculation of infiltration, drainage, evaporation, and runoff (Suleiman and Ritchie, 2003; Ritchie et al., 1998). The module accounts for soil surface runoff and evaporation, infiltration, saturated and unsaturated soil water flow, drainage, root water uptake, and soil evaporation and transpiration. Initially, precipitation is partitioned into rain or snow based on air temperature thresholds; if the air temperature is higher than 1°C, snow is considered to melt, as determined by a function that incorporates precipitation and snow cover. A time-to-ponding (TP) concept is used, where TP curves relate rainfall intensity to infiltration rate and define the point at which cumulative rainfall intensity exceeds the infiltration capacity of the soil (White et al. 1989), at which time water ponding occurs in micro-depressions in the soil surface. After ponding begins, infiltration is equal to the amount predicted by the TP curve if rainfall rate exceeds the infiltration capacity. When the rainfall rate is lower than the infiltration capacity, rainfall and any surface ponded water infiltrate into the soil until the ponded water is fully depleted. The key management-influenced parameter controlling the TP curve is the saturated hydraulic conductivity at the soil surface, which varies as a function of tillage, soil compaction, and surface residue amounts (Dadoun, 1993).

Evapotranspiration is modeled as the sum of soil evaporation and plant transpiration, both of which are first calculated as potential values before being adjusted for water deficit. Soil evaporation depends on radiation and aerodynamic factors, which are influenced by atmospheric pressure, temperature, and vapor pressure deficit. Similarly, plant transpiration is determined by LAIx, radiation, and CO_2 effects.

The module also accounts for surface water storage in superficial crop residues due to their ability to hold water. The model considers how precipitation interacts with residues, determining the amount of water stored in them before percolating into the soil. Evaporation from residues and ponded water reduces the potential soil evaporation accordingly.

Upward water movement within the soil profile is calculated by assessing the energy available for evaporation and the capacity for capillary rise, which is influenced by soil water potential, texture, and depth. The model defines coefficients for each soil layer to estimate the upward water flux.

6 Model inputs: soil, weather and agronomic data

To run a simulation, SALUS requires input data for three primary categories: soil properties, daily weather variables, and management characteristics. For ease of use, the model offers an Excel interface for essential input configuration, from which initial model files are generated. These files can be built upon if more "non-essential" data is available and helps reduce the time typically needed to translate and enter detailed field data, enabling users with less experience to quickly obtain functional simulations, while allowing more experienced users options for refinement to more precisely initialize environmental parameters for simulation.

6.1 Soil properties

The minimum set of soil inputs is listed below and is required per soil layer depth increment (see https://basso.ees.msu.edu/salus/tools/salus-layers.html), allowing flexibility for the user according to data available. The required soil properties include:

- soil texture: percentage of sand, silt, and clay;
- bulk density: expressed as Mg m^{-3};
- soil organic carbon content: percentage of SOC;
- soil water holding characteristics: lower limit (wilting point), upper limit (field capacity) of available water content, and saturation as volumetric content (m^3 m^{-3});
- saturated hydraulic conductivity: expressed as cm $hour^{-1}$;
- soil pH: soil acidity or alkalinity; and
- nitrogen concentrations: percentage of total nitrogen (%).

More detailed input data parameters include soil moisture as m^3 m^{-3}, and inorganic nitrogen, active inorganic phosphorus, labile phosphorus, slow-release phosphorus, and total phosphorus, expressed in g Mg^{-1}. Additionally, SALUS allows for the optional specification of initial residue amount in the form of FOM. These inputs, which include aboveground residues from previous crops, contribute to the soil carbon and nitrogen pools.

6.2 Weather data

SALUS requires daily weather data to simulate crop growth and soil processes. The critical weather variables are:

- maximum temperature (°C);
- minimum temperature (°C);

- precipitation (mm); and
- solar radiation (MJ m^{-2} day^{-1}).

Optional variables that can enhance model accuracy include wind speed, relative humidity, and dewpoint temperature. The preferred source of weather data is a local weather station with measured values. However, when local data are unavailable, alternative sources such as satellite-derived estimates or reanalysis datasets can be used.

6.3 Management

Management practices significantly impact soil properties, crop growth, and nutrient dynamics. For these reasons, SALUS requires detailed input data for each year, including the date of each management event. Management inputs include:

- crop type: users can select from pre-existing staple crops in SALUS. Specific cultivar coefficients can be refined to calibrate growth;
- planting: depth (cm), row spacing (cm), and population density (plants per square meter);
- fertilizer applications: type, quantity of nitrogen, phosphorus, and potassium (kg ha^{-1}), and method (e.g. surface application, incorporation, or injection);
- irrigation: amount (mm) and method (drip, sprinkler, or furrow);
- tillage operations: type and depth (cm);
- residue and manure applications: amount (kg ha^{-1}) and C and N contents and applied depth; and
- harvest: amount of grain harvested and amount of crop residue removed.

7 Case study: reducing nitrogen losses

Along with numerous outputs for crop and environmental variables, SALUS can be applied at multiple scales, including subfield, field, farm, and regional levels, and provides a robust framework for evaluating the biological and environmental impacts of specific genetic traits within cropping systems. Here we present case studies demonstrating how SALUS can help advance applications for innovative crop genetics. These examples highlight applications that improve soil health and the environment through for example reducing GHG emissions from increased SOC and reduced N_2O, and that are fundamental for approaches that seek to advance precision agriculture, regenerative management strategies, and the use of bioenergy feedstocks.

SALUS has been used to assess the impacts of crop breeding strategies aimed at reducing nitrous oxide (N_2O) emissions via improved on-farm nitrogen recycling, in the CERCA (Circular Economy that Reimagines Corn Agriculture) project (CERCA, 2025). This work targeted the development of early-season maize cultivars with lower nitrogen (N) and phosphorous (P) requirements, which SALUS simulated through adjustments to genetic parameters. These included enhanced cold tolerance, the base temperature threshold for accumulating thermal time and allowing developmental progression was lowered. This modified the timing of seedling germination and establishment, enabling earlier planting in the spring. Reductions in grain nutrient demand were incorporated by modifying parameters related to maximum kernel nitrogen concentration and the minimum permissible N and P concentrations in the plant. These adapted genetic configurations were integrated into SALUS simulations spanning diverse climatic zones and soil types across the Midwest U.S. Overall, the improved genetics increased yields, decreased N_2O emissions, and reduced N leaching when compared to the baseline cultivar (Fig. 5). Furthermore, SALUS facilitated the identification of optimal farm management practices tailored to improved genetic cultivars to maximize productivity while minimizing environmental impacts. This application of SALUS in genetic trait modeling enables the integration of genotype-specific performance data into agronomic decision-making, supporting plant breeders, agronomists, and policymakers in advancing agricultural sustainability and climate change adaptation.

SALUS simulations can be used to tailor crop management strategies within agricultural fields to optimize yield and reduce environmental damage. This can have particular benefit with variable input applications, such as nitrogen (N) fertilizer rate. Within-field spatial variability can be delineated by yield stability zones (YSZ), which classify sub-field areas based on historical yield performance that include stable areas that over time have high (HS), medium (MS), or low (LS) productivity, and areas with unstable (UN) yield (Maestrini and Basso, 2018). To achieve highly accurate simulation results, farm-specific cultivars can be calibrated in SALUS using historical yield data and growing degree day (GDD) information provided by the producer or phenological benchmarks sourced from state-level extension bulletins. For corn, this calibration involves adjusting the timing of key developmental stages, specifically the onset of grain fill, physiological maturity, and the duration of the grain fill period, to align with observed yield outcomes and phenological dates. To determine the optimal N fertilizer rates for each YSZ, scenarios with varying rates are simulated under historical weather conditions. Yield probabilities and responses to N inputs under different climatic scenarios determine whether a farmer's yield goal is attainable in the current growing season and provide estimates for the optimal N rate

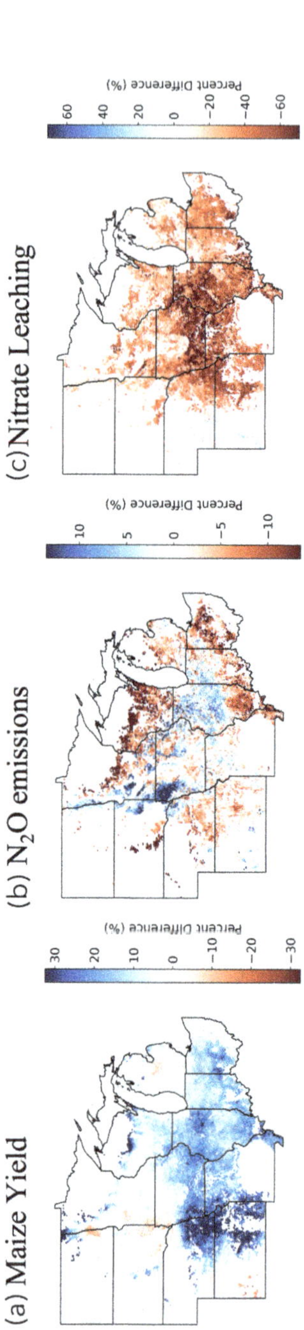

Figure 5 Percent difference from baseline cultivar to adapted CERCA cultivar with optimal management for maize grain yield (a), nitrous oxide emissions (b), and nitrate leaching (c).

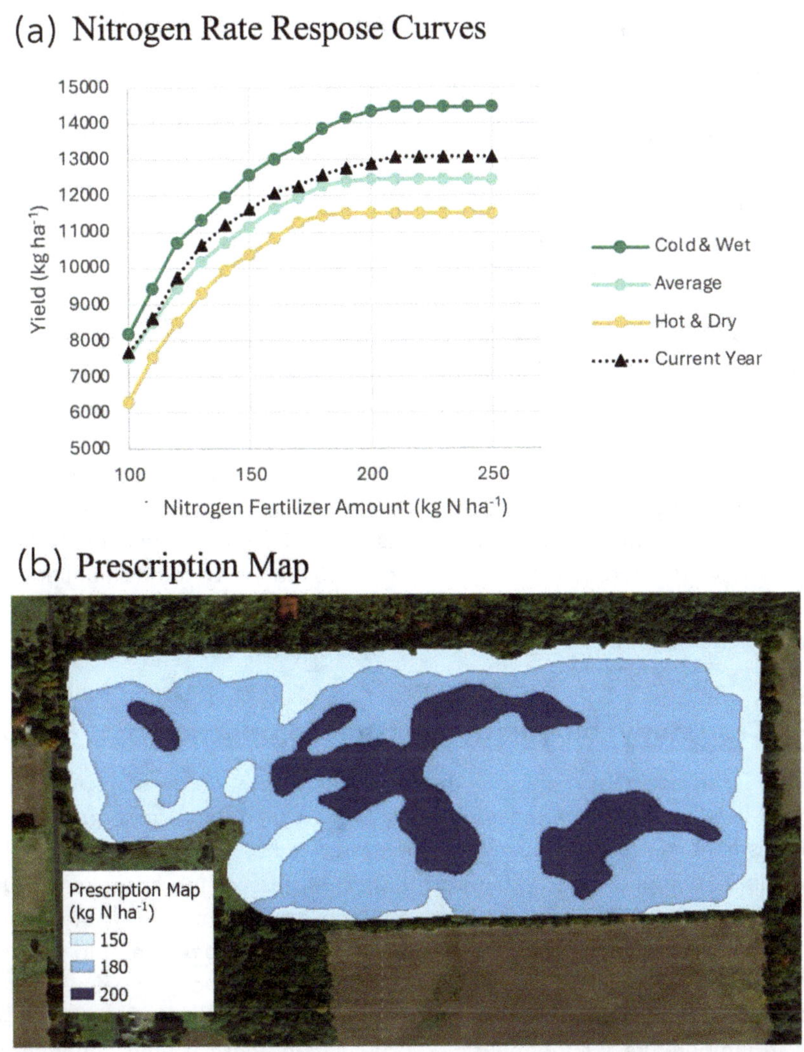

Figure 6 (a) Nitrogen rate response curves for the high and stable zone for multiple years representing various weather scenarios, cold and wet, average temperature and rainfall, and hot and dry conditions as well as the current growing season year. (b) Variable rate prescription map with optimal N fertilizer rates.

for each zone to maximize yields while reducing N waste (Fig. 6a). These zone-specific recommendations can be translated into a spatial prescription map (Fig. 6b). By adopting this precision approach, farmers can reduce N fertilizer input costs, increase profits, and reduce environmental N pollution (Basso et al., 2019).

8 Case study: regenerative management practices

SALUS can assess the impact of field management practices on a regional scale to investigate how crop productivity and soil health are affected. For example, the impact of regenerative management on soil organic carbon (SOC) was evaluated by Basso et al. (2025). Here, as an example, over a 42-year period at approximately 4 km spatial resolution, SALUS simulated the impact of no-till management compared with the business-as-usual (baseline) practice of conventional tillage over 46 million hectares of maize–soybean cropland across the U.S. Midwest. SALUS effectively captured the regional patterns in maize and soybean production by simulating representative genetic cultivars, providing confidence in its ability to simulate biomass accumulation, an important driver of SOC dynamics. The results show that no-till management consistently increased SOC content, whereas conventional tillage generally reduced it (Fig. 7). This tillage conversion had a pronounced effect in regions with high initial SOC levels, such as northern Iowa, southern Minnesota, and central Illinois, where substantial losses in SOC under conventional tillage transitioned to moderate gains under no-till. Overall, the findings highlight the potential for SALUS simulations to quantify the spatial and temporal effects of management decisions on SOC dynamics.

9 Case study: bioenergy feedstock production

SALUS can accurately simulate the key traits of bioenergy crops, and has been used to identity optimal locations to grow them through a quantitative assessment of both productivity and environmental impacts. For example, Martinez-Feria and Basso (2020) used SALUS to assess the changes in SOC under current and projected climate scenarios to determine the potential of switchgrass to increase SOC. A sensitivity analysis was performed to identify the cultivar parameters with the largest influence on above- and belowground biomass accumulation. Parameters exhibiting the highest influence included base temperature for thermal time accumulation, thermal time requirement to reach physiological maturity, timing of the onset of senescence, and maximum potential RUE. These parameters were then optimized and validated against measured biomass datasets to generate a high performing SALUS simulated cultivar that reliably predicted biomass accumulation across various site conditions with differing management practices, climates, and soil characteristics.

To assess large-scale bioenergy potential, simulations were extended to marginal lands, defined as areas unsuitable for food production due to soil limitations based on factors such as land capability class. Despite these limitations, marginal lands have the potential to sustain perennial grasses.

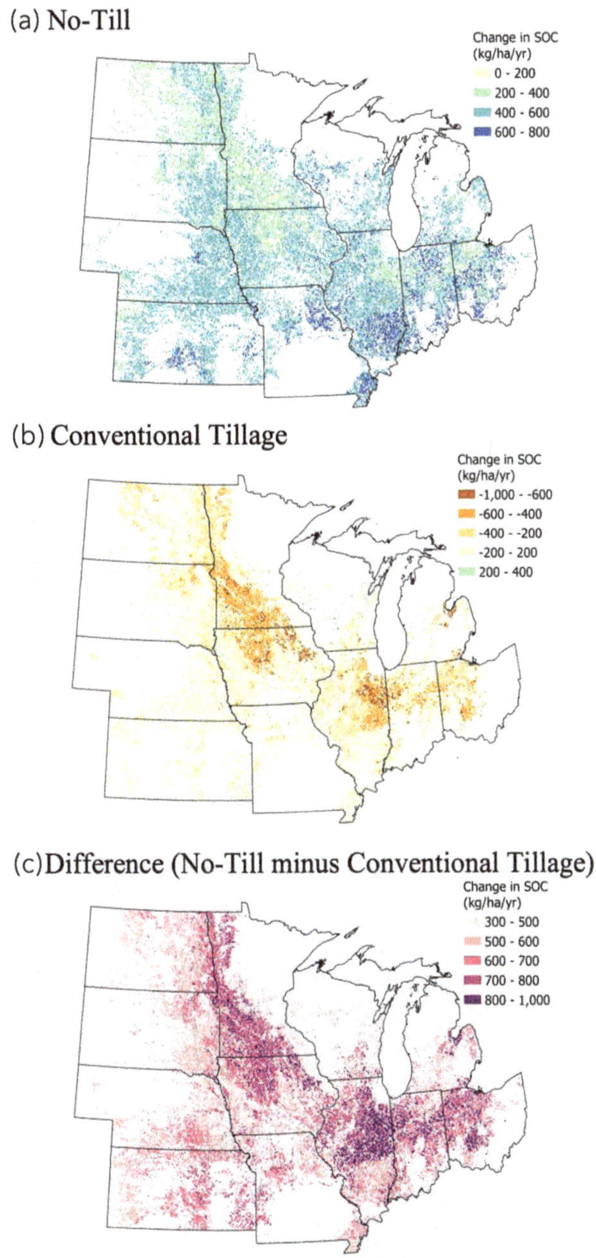

Figure 7 (a) Change in SOC (kg ha^{-1} $year^{-1}$) for no-till scenario, (b) conventional tillage scenario, and (c) the difference between the scenarios calculated as no-till minus conventional.

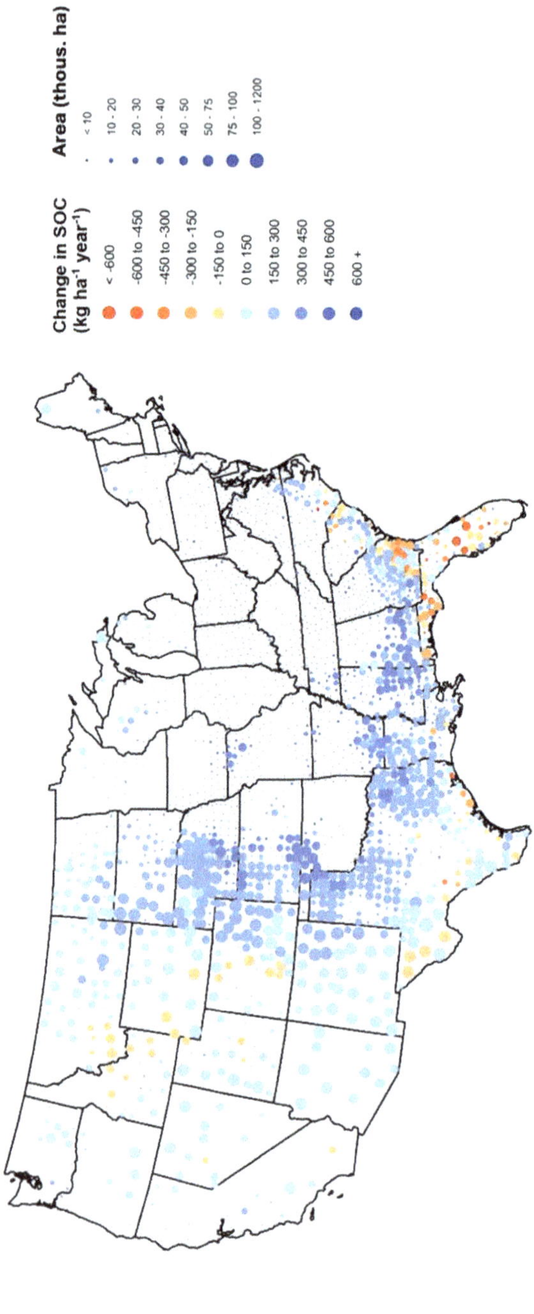

Figure 8 Change in soil organic carbon (kg ha^{-1} year^{-1}) resulting from switchgrass cultivation, with positive changes shown in shades of blue and negative changes in shades of red. The size of the points represents the total area of marginal land planted.

Accordingly, SALUS was used to quantify the productivity and changes in SOC. Spatial analysis of the results indicates that extensive regions within the U.S. Great Plains could achieve substantial carbon sequestration through switchgrass cultivation (Fig. 8). This application of SALUS provides critical insights that support the strategic establishment of biofuel production systems, enabling data-driven decisions for sustainable land use, carbon mitigation, and renewable energy development.

10 References

Basso, B., et al., 2006. Simulation of tillage systems impact on soil biophysical properties using the SALUS model. *Italian Journal of Agronomy* 1, 677–688. https://doi.org/10.4081/ija.2006.677

Basso, B., et al., 2010. Long-term wheat response to nitrogen in a rainfed Mediterranean environment: field data and simulation analysis. *European Journal of Agronomy* 33, 132–138. https://doi.org/10.1016/j.eja.2010.04.004

Basso, B., et al., 2015. Can impacts of climate change and agricultural adaptation strategies be accurately quantified if crop models are annually re-initialized? *PLoS One* 10, e0127333. https://doi.org/10.1371/journal.pone.0127333

Basso, B., et al., 2019. Yield stability analysis reveals sources of large-scale nitrogen loss from the US Midwest. *Scientific Reports* 9, 5774. https://doi.org/10.1038/s41598-019-42271-1

Basso, B., et al., 2025. A multi model ensemble reveals net climate benefits from regenerative practices in US Midwest croplands. *Scientific ReportsScientific Reports* 15, 24881. https://doi.org/10.1038/s41598-025-08419-y

Boote, 2003. Genetic coefficients in the CROPGRO–Soybean model: links to field performance and genomics. *Agronomy Journal* 95(1). https://acsess-onlinelibrary-wiley-com.proxy2.cl.msu.edu/doi/full/10.2134/agronj2003.3200 (accessed 9.22.25).

CERCA, 2025. Circular economy that reimagines corn agriculture (CERCA: On-Farm Nitrogen Recycling) (No. Project Number: 8062-21000-052-005-T).

Dadoun, F. A. 1993. *Modelling tillage effects on soil physical properties and maize (Zea mays, L.) development and growth*. Michigan State University, Michigan, USA. https://doi.org/10.25335/AZFG-MC33

Determining the Diffusivity of Nitrous Oxide in Soil Using in Situ Tracers - Shcherbak - 2014 - Soil Science Society of America Journal - Wiley Online Library, n.d. https://acsess.onlinelibrary.wiley.com/doi/abs/10.2136/sssaj2013.05.0181 (accessed 9.22.25).

Dzotsi, K.A., et al., 2013. Development, uncertainty and sensitivity analysis of the simple SALUS crop model in DSSAT. *Ecological Modelling* 260, 62–76. https://doi.org/10.1016/j.ecolmodel.2013.03.017

Dzotsi, K.A., et al., 2015. Parameter and uncertainty estimation for maize, peanut and cotton using the SALUS crop model. *Agricultural Systems* 135, 31–47. https://doi.org/10.1016/j.agsy.2014.12.003

Elias, D.M.O., et al., 2024. Microbial and mineral interactions decouple litter quality from soil organic matter formation. *Nature Communications* 15, 10063. https://doi.org/10.1038/s41467-024-54446-0

Gilmour, J.T., 1984. The effects of soil properties on nitrification and nitrification inhibition. *Soil Science Society of America Journal* 48, 1262–1266. https://doi.org/10.2136/sssaj1984.03615995004800060012x

gSSURGO, 2020. Gridded soil survey geographic (gSSURGO) database | Natural resources conservation service. https://www.nrcs.usda.gov/resources/data-and-reports/gridded-soil-survey-geographic-gssurgo-database (accessed 9.22.25).

Jones, J.W., et al., 2003. The DSSAT cropping system model. *European Journal of Agronomy, Modelling Cropping Systems: Science, Software and Applications* 18, 235–265. https://doi.org/10.1016/S1161-0301(02)00107-7

Kiniry, 1992. A general, process-oriented model for two competing plant species. *Transactions of the ASAE* 35(3), 801–810.

Maestrini, B. and Basso, B., 2018. Predicting spatial patterns of within-field crop yield variability. *Field Crops Research* 219, 106–112. https://doi.org/10.1016/j.fcr.2018.01.028

Martinez-Feria, R. and Basso, B., 2020. Predicting soil carbon changes in switchgrass grown on marginal lands under climate change and adaptation strategies. *GCB Bioenergy* 12, 742–755. https://doi.org/10.1111/gcbb.12726

Monteith, 1977. Climate and the efficiency of crop production in Britain | Philosophical Transactions of the Royal Society of London. B, Biological Sciences. https://royalsocietypublishing-org.proxy2.cl.msu.edu/doi/abs/10.1098/rstb.1977.0140 (accessed 9.10.25).

Parton, W.J., et al., 1987. Analysis of factors controlling soil organic matter levels in great plains grasslands. *Soil Science Society of America Journal* 51, 1173–1179. https://doi.org/10.2136/sssaj1987.03615995005100050015x

Parton, W.J., et al., 1988. Dynamics of C, N, P and S in grassland soils: a model. *Biogeochemistry* 5, 109–131. https://doi.org/10.1007/BF02180320

Parton, W.J., et al., 1994. A general model for soil organic matter dynamics: sensitivity to litter chemistry, texture and management, in: Bryant, R.B. and Arnold, R.W. (Eds.), *Soil science society of America*, SSSA Special Publications, Madison, WI, USA, pp. 147–167. https://doi.org/10.2136/sssaspecpub39.c9

Rabbi, S.M.F., et al., 2013. Mean residence time of soil organic carbon in aggregates under contrasting land uses based on radiocarbon measurements. *Radiocarbon* 55, 127–139. https://doi.org/10.2458/azu_js_rc.v55i1.16179

Ritchie, J.T., 1998. Soil water balance and plant water stress, in: Tsuji, G.Y., et al. (Eds.), *Understanding options for agricultural production*. Springer, Dordrecht, pp. 41–54. https://doi.org/10.1007/978-94-017-3624-4_3

Ritchie, J.T. and NeSmith, D.S. 1991. Temperature and crop development. In Hanks, R.J. and Ritchie, J.T. (Eds.), Modeling plant and soil systems: Agronomy Report 31, American Society of Agronomy (ASA), Madison (Wisconsin), USA.

Ritchie, J.T., et al., 1989. A user's guide to CERES Maize -V2.10. https://www.researchgate.net/publication/265205817_A_User's_Guide_to_CERES_Maize_-V210 (accessed 9.10.25).

Ritchie, J.T., et al., 1998. Cereal growth, development and yield, in: Tsuji, G.Y., et al. (Eds.),G.Y. Tsuji *Understanding options for agricultural production*. Springer, Dordrecht, pp. 79–98. https://doi.org/10.1007/978-94-017-3624-4_5

Ritchie, J.T., et al., 2009. Extension of an existing model for soil water evaporation and redistribution under high water content conditions. *Soil Science Society of America Journal* 73, 792–801. https://doi.org/10.2136/sssaj2007.0325

Shcherbak, I. and Robertson, G., 2014. Determining the diffusivity of nitrous oxide in soil using in situ tracers. *Soil Science Society of America Journal* 78, 79–88. https://doi.org/10.2136/sssaj2013.05.0181

Six and Jastrow, 2002. Organic matter turnover. ResearchGate. https://doi.org/10.1201/NOE0849338304.ch252

Suleiman, 2003. Modeling soil water redistribution during second-stage evaporation - Suleiman - 2003 - Soil Science Society of America Journal - Wiley Online Library. https://acsess-onlinelibrary-wiley-com.proxy2.cl.msu.edu/doi/full/10.2136/sssaj2003.3770 (accessed 9.16.25).

Uehara, G. and Tsuji, G.Y., 1993. The IBSNAT project, in: de Vries, F.P., et al. (Eds.), *Systems approaches for agricultural development: proceedings of the international symposium on systems approaches for agricultural development*, 2–6 December 1991, Bangkok, Thailand. Springer Netherlands, Dordrecht, pp. 505–513. https://doi.org/10.1007/978-94-011-2842-1_31

Vinocur, M.G. and Ritchie, J.T. 2001. Maize leaf development biases caused by air-apex temperature differences. *Agronomy Journal* 93: 767–772.

White, I., et al., 1989. Use and hydrological robustness of time-to-incipient-ponding. *Soil Science Society of America Journal* 53, 1343–1346. https://doi.org/10.2136/sssaj1989.03615995005300050007x

Williams, J.R., 1989. The EPIC crop growth model. in: Hargrove, W.L. (Ed.), Modeling of agricultural production: weather, soils, and crops. American Society of Agronomy, Madison, pp. 25–36.

Chapter 6

The denitrification-decomposition (DNDC) model

Kyle W. Proctor, Nilovna Chatterjee, Curtis Jones, Colin Hill, Gayathri Gopalakrishnan and William Salas, Regrow Ag, USA

1 Introduction

Agriculture is both extremely vulnerable to the impacts of climate change and a significant contributor to global greenhouse gas (GHG) emissions with nearly one-third of global GHG emissions coming from food production (Ritchie, 2021). At the same time, the industry is impacted by climate change in the form of rising temperatures (Malhi et al., 2021), increased variability and frequency of extreme weather events (Clarke et al., 2022), shifting agroecosystem boundaries (King et al., 2018), and an increased presence of invasive plants and pests (Skendžić et al., 2021). Furthermore, climate change has been shown to negatively impact crop yields, decrease the nutritional quality of major cereals, and lower livestock productivity (Ray et al., 2019).

It is critical to develop methods for improved measurement, modeling, and quantification of both the impacts of climate change on agriculture and the

http://dx.doi.org/10.19103/AS.2025.0155.09

GHG emissions that result from regenerative farming practices. Direct samples, whether crop yields, nutritional samples from cereals, samples of agricultural soils, or chamber measurements of methane (CH_4) and nitrous oxide (N_2O), are among the more accurate methods available. However, these samples are often difficult to collect, labor intensive, and expensive. Process-based models such as ApSIM (Keating et al., 2003), CENTURY (Parton, 1996), DAYCENT (Parton et al., 1998), and DNDC (Li et al., 1992) have been developed in the scientific community since the mid-1900s to enable an improved understanding of agroecosystems. These models describe the processes occurring in agricultural systems and use direct sample data to calibrate and validate the accuracy of model simulation results.

The denitrification decomposition model (DNDC) is a process-based, soil biogeochemical model designed to assess the impact of agricultural management practices on soil carbon (C) and nitrogen (N) dynamics (Li et al., 1992, 1994; Li, 2000). The model runs on a daily time step and is capable of simulating both aerobic and anaerobic soil conditions. The DNDC model has been applied across a wide range of agroecosystems globally, extensively validated, and studied in over 400 peer-reviewed publications. Many DNDC publications are focused on the calibration and/or validation of the model in specific contexts such as maize cultivation in Eastern Africa (Macharia et al., 2021), maize-winter wheat cropping systems in Northern China (Abdalla et al., 2020), and perennial grass production in Western Russia (Balashov et al., 2024). Other publications focus on improving or expanding the capabilities of the model to new cropping systems like boreal grasslands (Forster et al., 2022) or flooded rice paddies (Fumoto et al., 2008,Yin et al., 2020). Additionally, there are publications which integrate model results with other tools and data sources to explore broader contexts such as carbon emission inventories of swine farms (Li et al., 2021), life cycle assessments of precision agriculture technologies (Medel-Jiménez et al., 2024), and quantifications of the food-energy-water nexus in the U.S Pacific Northwest (Tabatabaie and Murthy, 2021). A more complete exploration of the wide-ranging uses of the DNDC model and the seminal papers can be found in "First 20 years of DNDC" (Gilhespy et al., 2014).

DNDC conducts a full accounting of C and N cycling by simulating the impacts of major ecological drivers (climate, soil, vegetation, and management) on soil climate conditions, plant growth, and decomposition (Fig. 1). The model is built using classical laws of physics, chemistry, and biology, as well as empirical equations generated from laboratory studies to parameterize specific biogeochemical processes.

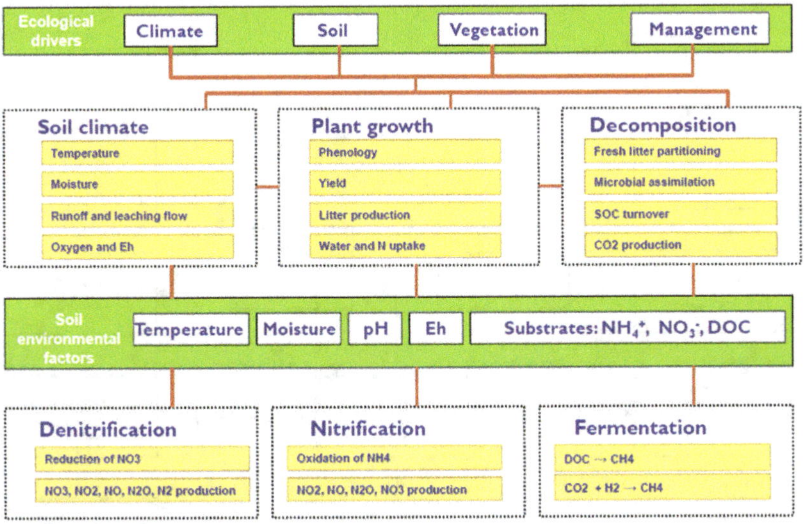

Figure 1 DNDC drivers (green boxes) and process-based sub-models (dashed boxes). Agricultural management practices directly modify the soil climate and plant growth sub-modules and indirectly alter organic matter decomposition and nutrient cycling (denitrification, nitrification, fermentation).

2 The history of DNDC

Multiple versions of DNDC have been created for research, education, and commercial purposes since the model's initial development (Fig. 2). The model was introduced by Dr. Changsheng Li in the early 1990s (Li et al., 1992) and the initial versions (v1.0 to v7.0) consisted of three main sub-modules: a soil-climate thermal flux sub-model, a decomposition sub-model for soil organic carbon (SOC) pools, and a denitrification model for nitrification/denitrification processes. Together, these sub-models simulate N_2O and N_2 emissions (Gilhespy et al., 2014).

An empirical plant growth model with subroutines for crop growth and land management practices was added in 1994 to better model biogeochemical cycling in agricultural soils (Li et al., 1994). Subsequent improvements to the model included the addition of sub-modules to predict methane (CH_4) and ammonia (NH_3) emissions from agricultural systems, particularly rice. Further modifications introduced several key crop algorithms from Crop-DNDC, resulting in the development of a phenological crop growth submodel (Zhang et al., 2002).

This framework was then used as the basis for the development of several research versions of DNDC for different countries and ecosystems, such as NZ-DNDC (Saggar et al., 2004) and UK-DNDC (Brown et al., 2002). Wetland-DNDC was developed for use in wetlands and other anaerobic systems and

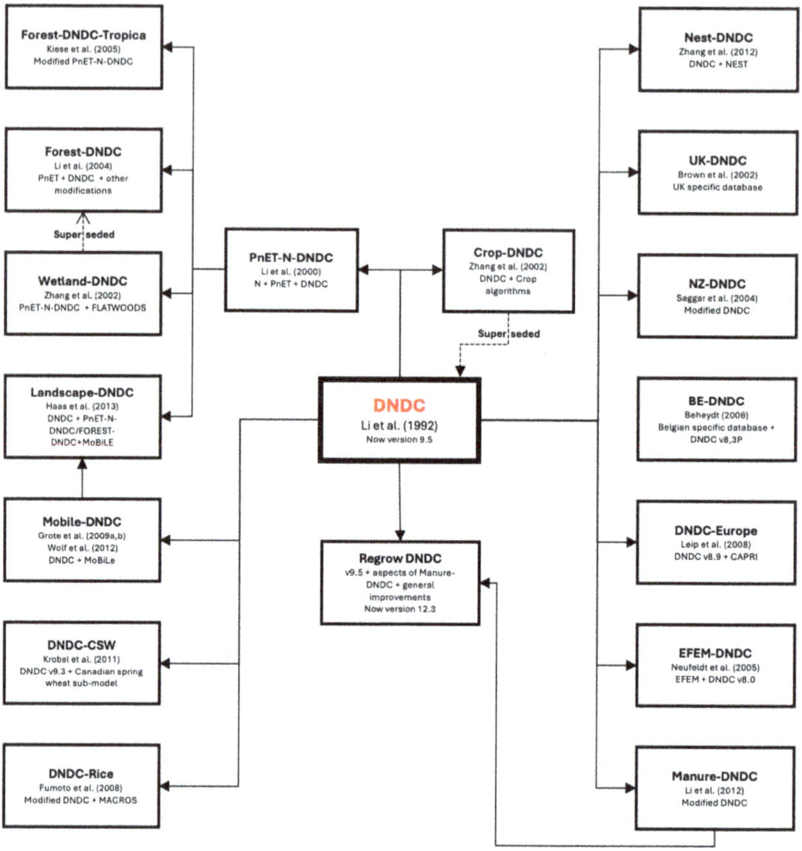

Figure 2 A schematic diagram of the different versions of DNDC (adapted from Gilhespy et al., 2014).

further refined the use of the Nernst and Michaelis-Menten equations (Li et al., 2004). Forest-DNDC extended the application of the model into upland and wetland forest ecosystems (Li et al., 2005). Manure-DNDC was developed to simulate livestock and grazing systems (Li et al., 2012). Customized versions of DNDC have been developed for the California Air Resources Board to support their Greenhouse Gas Emission Inventory Program (Deng et al., 2022, 2016) and for the Canadian government to support estimates of crop yield, GHG emissions, and water quality (Smith et al., 2013). Version 9.5 is the research version of DNDC that is publicly available and infrequently maintained by the University of New Hampshire (UNH). More information about how to access v9.5 can be found in Appendix 1. Although the various DNDC model versions build from a similar foundation, they represent different code bases and modeling approaches and are generally not compatible with each other or consistently maintained.

Regrow Ag has developed a streamlined and comprehensive version of the model, referred to hereafter as DNDC vRegrow. This version is frequently maintained and updated to keep pace with growing agroecosystem knowledge. It has been refined from the v9.5 foundation to create a more efficient, streamlined model suited for running simulations at scale. DNDC vRegrow improves fertigation modeling, enforces daily conservation of mass for C and N chemical species, and adds additional support for features that exist in v9.5 (precision fertilization, auto-fertilization, biomass and residue burning, biochar, event-based tile drainage, and dynamic erosion).

A major strength of DNDC vRegrow is that the code base has been overhauled to create more computationally efficient and well-engineered code. The base DNDC source code is still in C++, but it has undergone level 1 optimization to remove unneeded complexity, eliminate redundant operations, and generally improve computational resource management. Additionally, Regrow's model architecture includes various python-based services which simplify the process of creating a DNDC input file. For example, when a user specifies the location of a field, the services will automatically generate input files with the correct weather and soil inputs for that location. This greatly simplifies the process and enhances scalability. These improvements are further bolstered by more robust documentation and testing to reduce the occurrence of bugs. Although vRegrow is proprietary, Regrow periodically shares the source code and executable with UNH. More details about code access from UNH can be found in Appendix 1.

3 The crop growth model in DNDC

Plants play a critical role in regulating soil C, N, and water regimes in agroecosystems. They remove water, N, and other nutrients from soils and thus can alter several biogeochemical processes throughout their growth and development cycles. Additionally, plants produce biomass, including litter, as well as root exudates, and leguminous crops fix nitrogen, all of which can significantly impact soil health.

The first version of the plant growth sub-model in DNDC (Li et al., 1994) was an empirical model that simulated the effects of cumulative temperature, nitrogen uptake, and water stress on crop growth at a daily time step. In this version (v7.1), nitrogen uptake by plants is the key process connecting crop growth to climate and soil conditions. The daily N uptake rate is regulated by the crop potential maximum biomass, the crop C/N ratio, a generalized crop growth curve, and the availability of dissolved inorganic nitrogen. Total biomass carbon is allocated to the grain, straw/stalks, and roots based on empirical values of C/N ratios for the crop and the actual N uptake.

Later model versions like Crop-DNDC (Zhang et al., 2002) incorporated a phenological crop growth sub-model that simulated the impact of physiological processes (e.g. water and nitrogen stress) on crop growth at various stages of development. The phenological stages and stress factors influence carbon allocation and nitrogen demand, which, as described above, will impact plant growth and biogeochemical cycles.

To explicitly describe photosynthesis, respiration, and C allocation in crops, a generic crop model, Modules of Annual CROp Simulator (MACROS) (De Vries et al., 1989), was modified and integrated with DNDC v9.0 (Li et al., 2006) and subsequently used in DNDC v9.5 (Zhang and Niu, 2016). Here, net carbon gain was calculated by subtracting the respiration requirement from canopy photosynthesis. This assimilated carbon was allocated among plant organs using partitioning coefficients, which were defined as functions of the developmental stage. Zhang and Niu (2016) provide a more comprehensive review of the development of the DNDC plant growth sub-model over time. Figure 3 depicts the crop growth sub-model used in DNDC v9.5 and vRegrow.

The DNDC crop model is still actively being developed. DNDC vRegrow is a direct offshoot of DNDC v9.5, which also incorporates elements of Manure-DNDC (Li et al., 2012). In vRegrow, crop phenology is modeled using a standard thermal growing degree day (GDD) model with germination defined by a fixed crop-specific latency period from sowing to emergence. GDD accounting

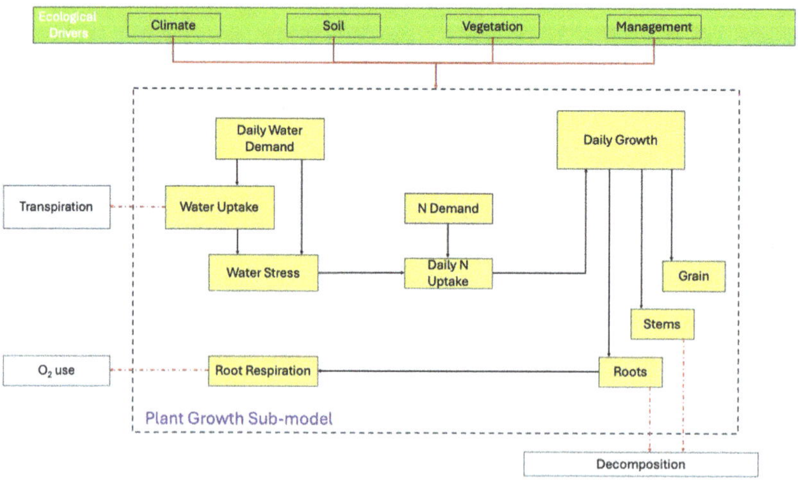

Figure 3 Representation of the crop growth sub-model in DNDC (adapted from Zhang and Niu (2016)). For context on how crop growth interacts with other sub-models, see Figure 1.

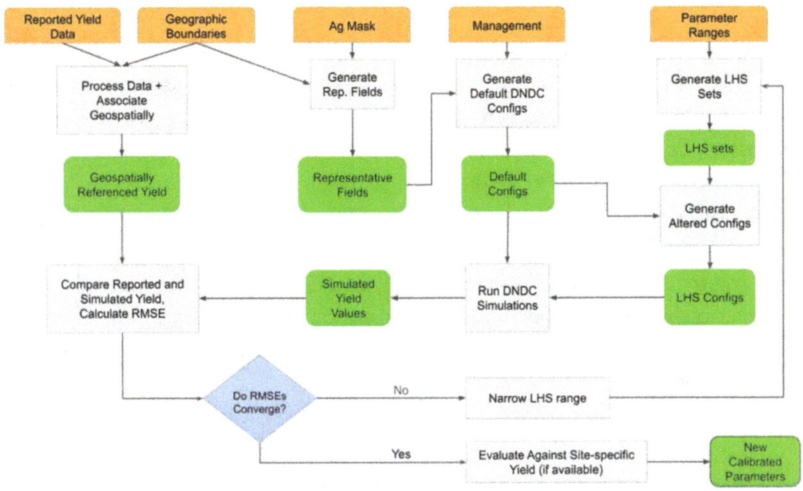

Figure 4 Simplified flowchart representation of the calibration process. LHS = Latin hypercube sampling, RMSE = root mean square error.

begins at emergence and accumulates on a daily basis according to the average daily temperature. The cumulative GDD requirement to progress from emergence to maturity is referred to as the total degree day requirement (TDD) (Fig. 4).

Perennial crops progress through phenological phases similarly but with a few key distinctions. Upon reaching maturation, the GDD value is reset to a lower crop-defined proportion of TDD. Following the reset, GDD accumulation resumes. Adjustments to GDD can also occur following cutting and grazing events. Winter dormancy can be initiated or terminated by prolonged exposure to cold and warm temperatures, respectively.

Crop growth simulation in DNDC vRegrow is conceptually driven foremost by nitrogen demand, which interacts with biomass accumulation via organ-specific carbon to nitrogen ratios (C:N). DNDC discretizes a crop into root, stem, leaf, and grain (reproductive) organs. The crop growth sub-model utilizes a crop-specific maximum potential biomass (max_biomass) input parameter that defines the ceiling for biomass accumulation under optimal, non-stressed conditions. The max_biomass value is translated into nitrogen demand according to organ-specific C:N ratios, which define the maximum seasonal nitrogen demand. Seasonal demand is then allocated to daily nitrogen demand according to the ratio of daily GDD accumulation to TDD. This daily nitrogen demand then influences the daily crop water demand. Water demand is primarily calculated with a crop-specific water use efficiency parameter that represents the variable efficiencies of crops to utilize water for biomass production.

Following determination of the key daily crop demands, actual crop uptake and biomass accumulation are calculated through limiting growth by stress factors, with water, nitrogen, and temperature serving as the key stressors. Water and nitrogen stresses are determined by constraining the daily crop demand by the availability of the respective resources in the soil profile. The soil nitrogen species made available for crop uptake are limited to nitrate and ammonium. Soil water availability is limited to plant available water, defined as the soil water exceeding the wilting point (~1500 kPa). Additionally, temperature stress is modeled following Yan and Hunt (1999), with calculations conducted distinctly depending on the phenological development of the crop and considering crop-specific optimum temperatures to penalize cold and heat stresses.

Finally, nitrogen uptake is converted into biomass accumulation by allocating nitrogen uptake among the four organs. Allocations are made among the vegetative organs as specified by crop-specific allocation parameters during the vegetative growth phase, which is defined by the first half of the TDD accumulation period for most crops. If organ-specific nitrogen allocation exceeds demand, a reserve concept is utilized to store excess nitrogen and make it available for use in other organs. Upon the onset of the reproductive phase, the grain or reproductive organ receives preferential allocation. Organ-specific biomass accumulation is then calculated according to the organ-specific C:N ratios.

While some models use a harvest index approach to allocate above-ground biomass between reproductive and vegetative biomass (Hay, 1995), DNDC instead accounts for biomass partitioning on a daily basis according to organ-specific growth allocations and C:N ratios as described in the previous paragraph. Biomass development of the different organs varies during the cropping season, depending on the current phenological stage. Ultimately, the yield of grain crops, roots and tubers, and perennial grasses are based on the biomasses of the grain, root, stem, and leaf organs, respectively. The model allows the user to specify multiple cutting events, including what fraction of each organ is harvested and what portion is left on the field as standing dead or returned to the soil surface as residue. A common annual grain crop would have a simple harvest of 100% of grain at the end of the cropping season, while perennial crops like alfalfa may use more complex cutting logic, including multiple harvests over the course of the season.

4 Use cases and running DNDC

Running DNDC simulations requires the user to provide information about the climatology, soil characteristics, and farmer management practices for the site of interest. Weather inputs are specified on a daily timestep. Only minimum and maximum air temperature and precipitation are required; however, the model also accepts solar radiation, wind speed, and humidity. If these parameters are not provided, they can be estimated.

Soil inputs are specified to define various physical (sand, silt, and clay percentages, hydraulic conductivity, field capacity, porosity, etc.) and chemical (pH, initial soil soc, nitrate, ammonium, etc.) soil characteristics. These parameters can either be obtained via in-situ soil sampling or from large soil databases, such as the US Geological Survey's Soil Survey Geographic Database (SSURGO) (Soil Survey Staff et al., n.d.) in the US or SoilGrids (Poggio et al., 2021) globally.

The number of soil layers which are simulated is dependent on the user-specified soil depth and the soil properties (specifically porosity). DNDC uses a simple tipping-bucket water balance approach where the soil moisture within each layer is impacted by factors such as precipitation, evaporation, and transpiration. When water input exceeds the capacity of the layer, it "tips", and water drains into the next layer. If "tipping" occurs in the bottom layer, this moisture is removed from the system and considered to be deep drainage.

The volume of water in the soil dictates whether portions of the soil experience aerobic or anaerobic conditions. One of the strengths of DNDC is its anaerobic-balloon modeling approach, which captures the dynamic nature of the anaerobic portion of the soil. In unsaturated/aerobic conditions, more oxidants are involved in oxidation reactions. In saturated conditions, oxygen content will decrease, causing the anaerobic balloon to swell, and more oxidants are involved in reductive reactions. This technique allows the model to simulate both aerobic and anaerobic reactions within the soil column simultaneously, better capturing true nitrification and denitrification cycles.

The user must also specify crop input parameters that define the crops planted in a given year, planting and harvest dates, crop residue management, and crop physiological and phenological parameters (Table 1). DNDC is well adapted to simulating complex crop rotations and can be used, for example, to track the impacts of growing a nitrogen-fixing legume in rotation with a nitrogen hungry grain and a winter cover crop.

DNDC v9.5 enhances usability with an extensive default library for 88 crops and 12 soil types (Appendix 2). These default parameters allow users to tailor crop parameters to real-world conditions, including temperature, water, and nitrogen needs, growth curves, biomass distribution, and maximum potential biomass. Continuous efforts are being made to expand and improve these defaults in DNDC vRegrow.

One of DNDC's greatest strengths is its ability to simulate the complex biogeochemical interactions that govern the nitrogen cycle and produce results that can be used to estimate Greenhouse Gas (GHG) emission/sequestration on a particular field and how those emissions vary with different inputs. Crop

Table 1 Main inputs to DNDC for crop parameters (adapted from the DNDC v9.5 User's Guide (Li, 2012))

Parameter	Description
Crop type	Select from the default list of crops in DNDC or create a "New crop" if not in the list
Fraction of leaf and stems left in the field after harvest	Fraction of above-ground crop biomass left on the field after harvest
Maximum biomass (kgC/ha)	Maximum biomass for leaves, grain, stem, and roots under optimum growing conditions
Biomass fraction	Grain, leaf, stem, and root biomass fractions at plant maturity
Biomass C/N ratio	C/N ratio of grain, leaves, stem, and roots
Total N demand (kg N/ha)	N required for the crop to reach maximum production
Total degree days (TDD) (°C)	Cumulative air temperature from emergence to maturity of plant
Water demand (g water/g dry matter)	Amount of water for the crop to produce 1 g of dry biomass
N fixation index	Ratio of total N content in plant/N content in plant taken from the soil for N-fixing crops like legumes. The default is 1 for non-leguminous crops
LAI adjustment factor	Factor to adjust default specific leaf weight
Optimum Temperature (°C)	Optimum crop growth temperature

demand for N is calculated based on the optimum daily crop growth and the plant C/N ratio. The actual N uptake by a crop could be limited by N or water availability during the growing season.

This actual N value is extracted from the available soil NO_3 and NH_4^+ pools in proportion to each pool's relative size, considering the depth of the plant roots. DNDC partitions the crop biomass production among the leaves, stems, roots, and grain based on the plant growth stage at each daily time step. After harvest, all root biomass remains in the soil profile, and a fraction of the above-ground crop residue also remains on top of, or in, the soil profile until the next tilling event. The crop residue incorporated into the soil will later be partitioned into three soil litter pools, namely "very labile," "labile," and "resistant" based on the residue's C/N ratio. The litter incorporation provides essential inputs for soil organic matter (SOM) storage and hence integrates the plant and soil into a biogeochemical system.

The plant growth portions of DNDC regulate the soil water, C, and N regimes by controlling the rates of water and N uptake from the soil and litter deposition into the soil. Ultimately determining many of the other biogeochemical reactions, including C sequestration and trace gas emissions.

The outputs of DNDC, especially the crop yield, soil organic carbon, nitrous oxide, and methane emissions, are often used by researchers and policymakers to make recommendations about ideal agricultural practices in a region. In recent years, the focus on voluntary carbon markets and attempts by organizations to reduce their carbon footprint have led to increased interest in process-based models and their ability to quantify GHG emissions and changes in emissions as a result of regenerative agriculture and other practices.

5 Spatial calibration approach

While DNDC has been used extensively to model GHG emissions at field sites around the world, one of the challenges associated with this and other process-based models is the need to calibrate and validate the model parameters to ensure accurate results. This process typically requires significant amounts of ground truth data; however, in the case of soil carbon values and emissions from nitrous oxide or methane, the data are not easily or widely available. As a result of the paucity of GHG data from field sites, researchers typically calibrate the model using crop yield data and other outputs that are more widely available (Deng et al., 2016; Li, 2000). Crop growth is a significant driver of the processes governing GHG emissions as described previously; hence, calibrating the crop growth model is often effective in improving the accuracy of model results.

However, the number of locales where DNDC can provide value is far greater than the number of areas with extensive peer-reviewed data available for calibration. To address this discrepancy, Regrow has developed a method for calibrating DNDC crop yield that requires relatively few calibration parameters and utilizes data that is either widely available or can be reasonably estimated. This approach is valuable for expanding the range of geographies where DNDC can be applied, and thus where the various benefits of the model, including quantifying reductions in GHG emissions, can be taken advantage of. In this context, calibration refers specifically to starting with an initial set of crop parameters and then adjusting those parameters to improve simulation accuracy and generally better represent that crop, in a particular location.

Regrow's current calibration approach leverages the fact that nearly all regions report some level of aggregated crop yield data. This significantly expands the regions where calibration is possible by reducing the reliance on sparse historical observational studies, which are often wholly absent in developing countries. This reported data serves as the measured value that simulated results can be compared against. Management data for the region of interest can be estimated by referencing local agricultural guides or via peer-reviewed literature. The yield calibration process described in this chapter focuses specifically on two parameters: (1) Total Degree Days to maturity (TDD) – a measure of the accumulated temperature (GDD) required for a particular crop to go from emergence to physiological maturity; and

(2) Maximum potential biomass (max_biomass) – a measure of the maximum possible biomass for a particular crop under ideal, non-stressed conditions. Daily GDD is calculated by taking the average daily temperature (the mean of the daily maximum and minimum temperatures) and subtracting a base temperature (the minimum temperature required for growth) (Equation 1). GDD accumulates over the growing season and is a widely used heuristic for estimating the developmental stage of a crop.

$$GDD = \frac{MaxTemp + MinTemp}{2} - BaseTemp \quad (1)$$

These two parameters were selected as the most impactful for crop yield based on an internal sensitivity analysis of the DNDC vRegrow model. The various parameters used in the crop growth model (Table 1) aim to represent the inherent variability in different crop cultivars. Although DNDC does not explicitly model these genetic differences, region-specific calibrations can capture growth trends in a particular area and serve as a proxy for those cultivar-specific differences.

This section of the chapter will present a case study of the calibration procedure, focusing on the Corn Belt in the US. To make clear the value of this approach, the section will compare simulation results for corn and soybean run using DNDC vRegrow with input parameters generated from three sources: (1) DNDC v9.5 default values, (2) input parameters generated using a data-intensive site-specific calibration which relies on peer-reviewed data, and (3) the best performing parameters from the spatial calibration approach. The U.S. Corn Belt was selected because it is a region that has a relatively high number of peer-reviewed publications that can be used to evaluate the accuracy of simulated yield results.

The following stages are used in implementing this approach: (1) Data acquisition and processing, (2) Generation of representative fields to ensure that the model is calibrated across a sufficiently diverse set of climate and soil conditions, (3) Simulation of the fields and calibration using a Latin Hypercube Sampling approach (LHS), and (4) Validation of the calibrated model against an independent dataset.

5.1 Data acquisition and pre-processing

The first step of the calibration process involves acquiring the relevant input management data for DNDC simulations and the reported yield data required for comparison with the results of DNDC simulations. Input management practices (planting and harvest dates, fertilizer type, timing, and application quantity, tillage type and timing) are generalized at the state

Table 2 Default management practices

State	Crops	Plant date	Harvest date	N quantity (kg N/ha)	Tillage
Illinois	Soybean	May 1	October 21	NA.	Conventional pre-plant, depth – 10 cm, No inversion
Indiana	Corn	May 15	October 9	179.62	Conventional pre-plant, depth – 10 cm, No inversion
Iowa	Corn, soybean	May 14 May 27	October 22 October 21	162.24 NA	Conventional pre-plant, depth – 10 cm, No inversion
Missouri	Corn	May 10	October 12	193.07	Conventional pre-plant, depth – 10 cm, No inversion
Nebraska	Corn Soybean	May 1 May 15	November 1 November 1	172.33 NA	Conventional pre-plant, depth – 10 cm, No inversion
South Dakota	Corn Soybean	June 13 May 31	October 1 October 19	150.39 NA	Conventional pre-plant, depth – 10 cm, No inversion

*Soybean crop is N fixing, so no additional N requirement.

level and are informed by the National Agricultural Statistics Service (NASS) (National Agricultural Statistics Service, 2024). Although farmer management practices vary widely, these values are intended to represent standard management practices for corn and soybean cultivation in the Corn Belt. This management data is outlined in Table 2. Daily weather data is acquired from the PRISM database (PRISM Climate Group) (Oregon State University, 2024). Soil texture data are acquired from the U.S. SSURGO Database (Soil Survey Staff et al., 2024).

The calibration exercise uses county level annual yield datasets for corn and soybean from NASS for six states in the US Corn Belt: Iowa (IA), Illinois (IL), Indiana (IN), Missouri (MO), Nebraska (NE), and South Dakota (SD) for the period from 2011 to 2021. This NASS data serves as the measured data for parameter calibration.

5.2 Generation of representative fields

Regrow employs a representative fields methodology to determine representative locations within a region by compiling information from geospatial agricultural land use masks (Ag masks) and soil texture maps, in addition to statistical sampling techniques. A total of 991 representative fields were generated across six states (IA, IN, IL, MO, NE, and SD) to capture inherent spatial variability. The number of samples generated per state is presented in Table 3 and was informed by the total agricultural land area in each region. The representative fields approach is intended to capture the variable environmental conditions across the landscape and thereby better reflect the

Table 3 Representative fields breakdown per state

State	# Sample fields
Iowa	226
Illinois	245
Indiana	143
Missouri	119
Nebraska	141
South Dakota	116
Total	991

heterogeneity of real-world farming scenarios. DNDC is a point-based model, and simulations are run at each of the 991 individual fields.

5.3 Simulation parameters and Latin hypercube sampling

As mentioned above, this calibration focuses exclusively on two input parameters that are known to have significant impacts on crop yield estimates: TDD and max_biomass. We initially established a parameter range for TDD and max_biomass by setting the bounds from 0.5 to 1.5 times their respective default values (Table 4). This broad spectrum was chosen to encompass a wide phenological range, ensuring that the model could accommodate crop variations across different geographies within the Corn Belt states.

This calibration relies on the LHS method to systematically explore the parameter space, selecting 24 distinct samples. This statistical sampling technique is known for its efficiency in covering the entire range of parameter values, ensuring that each sample is representative of the overall distribution without necessitating an exhaustive search (Wei-Liem Loh, 1996). The LHS approach thus allows for a computationally efficient yet comprehensive evaluation of the impact of varying TDD and max_biomass parameter values on crop yield estimates. All simulations are run using the DNDC vRegrow version of the model.

Following the initial round of simulations, an additional set of 24 samples was generated. The selection criteria for the subsequent round were based on the performance outcomes of the first set, specifically focusing on the top 5 parameter sets for each crop as quantified by the root mean squared error (RMSE) of the reported and simulated yields (Equation 2). A lower RMSE signifies closer alignment with observed data. This targeted approach allowed us to refine the parameter range further, honing in on the combinations that demonstrate the greatest potential for enhancing model accuracy. By iteratively adjusting the parameter space based on evidence from the initial simulations,

Table 4 Model NRMSEs using vRegrow default and spatial LHS calibrated TDD and MaxBiomass

	DNDC v9.5			Site-specific calibration			Spatial calibration			NRMSE Δ %
	TDD (∘C)	Max Biomass (kg DM C/ ha)	NRMSE (-)	TDD (∘C)	Max Biomass (kg DM C/ ha)	NRMSE (-)	TDD (∘C)	Max Biomass (kg DM C/ ha)	NRMSE (-)	(9.5 vs spatial calibration)
Corn	2550	10 309	0.413	2550	12 000	0.377	2360	11 080	0.381	−7.75
Soybean	1500	3512	0.538	1800	5000	0.419	2420	4910	0.344	−36.06

Table 5 Site-specific studies included in model validation

State	Crops	Study references	Number of sites/treatments
Indiana	Corn	Burzaco et al., 2013; Omonode et al., 2015	2
Iowa	Corn and soybean	Al-Kaisi et al., 2005; Iqbal et al., 2015; Kaspar et al., 2006; Poffenbarger et al., 2017; Parkin and Hatfield, 2010; Parkin and Kaspar, 2006	23
Missouri	Corn	Nash et al., 2012; Torralbo et al., 2022	2
Nebraska	Corn and soybean	Adviento-Borbe et al., 2007	2
South Dakota	Corn and soybean	SDBRREAP (Lehman, 2020); Clay et al., 2015	3

we effectively narrowed the focus to parameter sets with the largest influence on RMSE. The resulting most performant parameter sets are reviewed to ensure convergence in a particular region of the parameter space. The comparison of RMSEs for the default and calibrated parameter sets is shown in Table 5.

$$RMSE = \sqrt{\frac{1}{n}\sum_{i=1}^{n}(P_i - O_i)^2} \quad (2)$$

i observation in the dataset, O_i: Observed (or actual) value for the ith observation in the dataset, $(P_i - O_i)$: The squared difference between the predicted value P_i and the observed value O_i for the ith observation, n: the number of observations

5.4 Evaluation/validation of calibrated parameter sets

In new geographies with relatively few or no peer-reviewed site-specific studies to compare against, the calibration procedure would end at Stage 3. However, in the Corn Belt states, there is a rich catalog of peer-reviewed literature to draw from, and the calibrated parameter sets were evaluated at 22 sites across five states with each site providing multiple years of peer-reviewed yield data (Fig. 5, Table 5). At each location, a DNDC input config was generated, which accounts for the site-specific inputs as described by the related journal article. DNDC simulations were then run using the DNDCv9.5 defaults, input parameters from a site-specific calibration, and input parameters derived from the spatial calibration methodology. For each data point (2095), the measured yield value was compared with the simulated outputs from each of the input parameter sets and RMSEs were calculated. It is worth noting here that substantially more data points were available in Iowa than in other states,

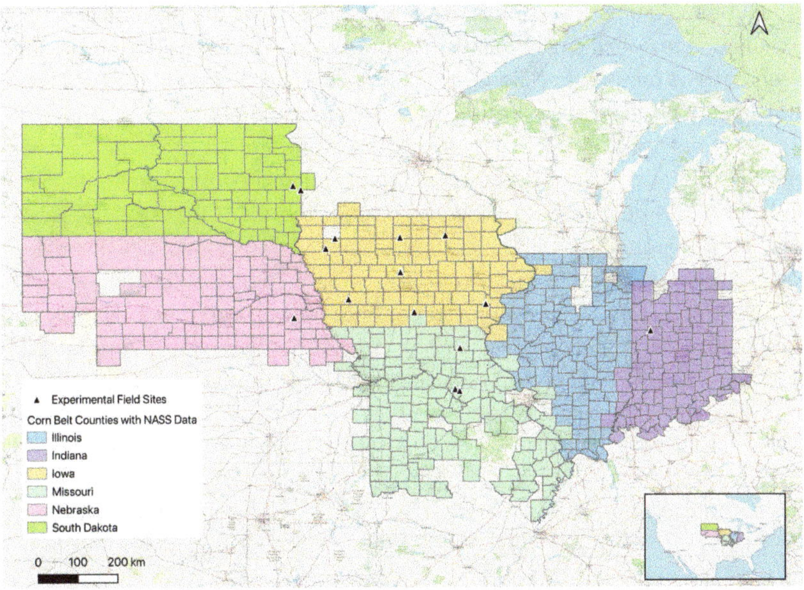

Figure 5 Locations of measured corn and soybean yield across the U.S. Corn Belt states and U.S. counties with NASS reported yield data.

and because each measured yield point is equally weighted in the evaluation, the analysis is somewhat biased towards representing Iowa.

6 Spatial calibration results

The calibration process involved over 47 000 DNDC simulations (991 field sites * 48 parameter sets). The grain biomass (kg DM/ha) from each simulation is compared against the reported county-level NASS yield (moisture corrected to kg DM/ha). Normalized Root Mean Square Errors (NRMSE) are calculated by dividing the RMSE by the average measured yield value. The NRMSE values from the calibrated parameter sets are compared against the default values used in DNDC v9.5 and parameters from a site-specific calibration. Table 4 shows the top-performing parameter sets across all six states. The LHS spatial calibration approach leads to a 7.75% and 36.06% reduction in NRMSE for corn and soybean, respectively, when compared with the DNDCv9.5 default parameters. The spatial LHS calibrated parameters perform similarly to the site-specific parameters for corn (~1% difference) and perform 18% better for soybean. This is encouraging as the LHS spatial calibration approach requires substantially less peer-reviewed data.

The most performant parameter set varies on a state basis with minimum state-specific NRMSEs ranging from 0.317 to 0.405 for corn and 0.288 to 0.446

for soybean (Tables 6 and 7). These findings highlight the possibility for more granular state-specific calibrations, although it is worth considering the risk of overfitting as parameter differences may be more a factor of input uncertainty than actual crop-specific differences from state to state.

As mentioned above, in many regions with limited data, the calibration effort would end after comparison with aggregated county-level data, as site-specific data is unavailable. However, within the Corn Belt, it is possible to evaluate the impact of this calibration when running site-specific simulations. Table 8 shows the NRMSEs when comparing simulation results using the DNDC v9.5 defaults, site-specific calibrated parameters, and the spatial LHS calibrated parameters with measured values of crop yield at the experimental sites. For corn, the improvement of the spatial calibrated parameter sets over DNDC v9.5

Table 6 Top performing spatial calibrated parameter sets by state for corn. DNDCv9.5 NRMSEs for each state are calculated using the default values shown in Table 4.

State	TDD (oC)	Max biomass (kg DM5862 C/ha)	NRMSE (-)	DNDCv9.5 NRMSE (-)	NRMSE Δ % (9.5 vs spatial calibration)
Iowa	2686	14423	0.405	0.526	−23.00
Illinois	2600	12753	0.343	0.373	−8.04
Indiana	2490	12645	0.377	0.422	−10.66
Missouri	2763	10821	0.320	0.319	0.31
Nebraska	2356	11082	0.336	0.339	−0.88
South Dakota	2490	12645	0.420	0.487	−13.76

Table 7 Top performing calibrated parameter sets by state for soybean. DNDCv9.5 NRMSEs for each state are calculated using the default values shown in Table 4.

State	TDD (oC)	Max biomass (kg DM5862 C/ha)	NRMSE (-)	DNDCv9.5 NRMSE (-)	NRMSE Δ % (9.5 vs calibrated)
Iowa	2536	5528	0.325	0.612	−46.90
Illinois	2468	5528	0.300	0.427	−29.74
Indiana	2458	5528	0.301	0.548	−45.07
Missouri	2512	2667	0.288	0.555	−48.11
Nebraska	2424	2909	0.380	0.635	−40.16
South Dakota	2204	4496	0.446	0.587	−24.02

Table 8 NRMSEs comparing measured vs modeled yield results for default, site-specific, and LHS spatial calibrated parameters

Crop	DNDC v9.5 defaults NRMSE (-)	Site-specific calibration NRMSE (-)	Spatial calibration NRMSE (-)
Corn	0.372	0.335	0.371
Soybean	0.453	0.442	0.336

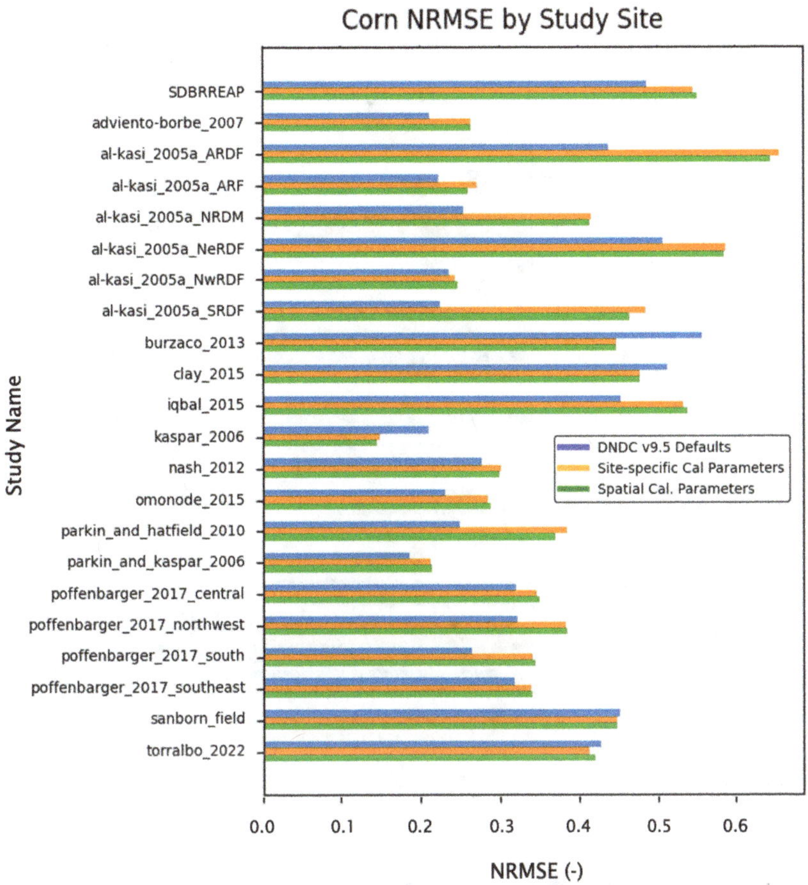

Figure 6 Comparison of measured vs simulated corn yield NRMSE using DNDC v9.5 defaults, site-specific calibration parameters, and spatial calibration parameters at 22 study sites. Specific study citations can be found in Table 4.

is minimal, and the site-specific calibrated parameter sets show a nearly 10% improvement. For soybean, the calibrated parameter set shows an over 25% reduction in NRMSE and is more performant than the site-specific parameters while using a less data-intensive methodology. Figures 6 and 7 show site-specific NRMSEs for each of the parameter sets.

7 Spatial calibration: conclusions

The calibration example described above demonstrates that there is clear potential to calibrate DNDC input parameters to improve yield estimation using widely available aggregated yield data and broad estimations about

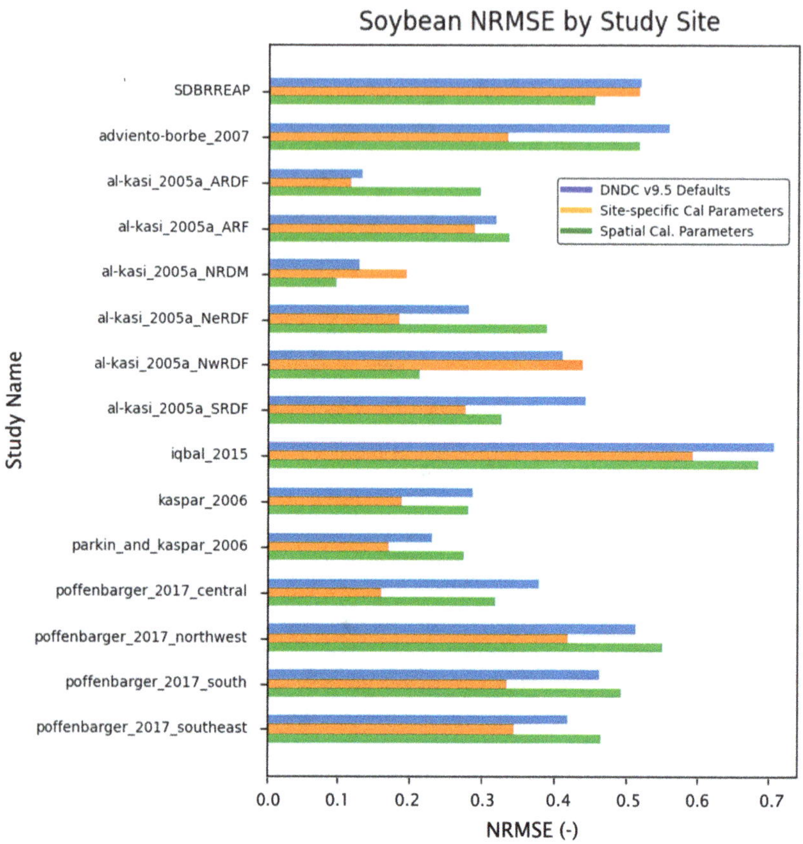

Figure 7 Comparison of measured vs simulated soybean yield NRMSE using DNDC v9.5 defaults, site-specific calibration parameters, and spatial calibration parameters at 15 study sites. Specific study citations can be found in Table 4.

management practices. The calibrated parameters produced with this approach reduce crop yield NRMSE when compared with DNDC v9.5 defaults for both corn and soybean. They were comparable to site-specific calibrated parameters, which were produced with a substantially more data-intensive method.

This approach is useful for expanding the usability of DNDC into geographies that have limited or no peer-reviewed studies. The method uses region-specific calibration as a proxy for the cultivar-specific attributes that vary in different regions. This approach may be most beneficial in developing nations which are more data sparse but are known to have substantially lower average yields than the Unied States.

Like many aspects of crop growth modeling, selecting the right calibration approach is fundamentally a matter of parsimony. In an ideal world, there would be rich sources of site and crop-specific empirical data to draw from for model calibration. Instead, we are faced with a world where agriculture is both affected by and is a substantial contributor to climate change and where carbon markets may be a major lever for influencing grower behavior and reducing carbon emissions. Biogeochemical crop growth simulation models like DNDC are a critical tool for informing these carbon markets; to make the best use of these tools, new methods must be developed to allow for calibration and implementation of the models in regions that are historically data sparse. Ultimately, the methodology described in this chapter enables the calibration of DNDC and potentially other process-based models at scale through the use of widely available data such as administrative yield reports and information from remote sensing.

8 Conclusion

As this chapter has highlighted, there is a great potential benefit from the strategic use of well-calibrated DNDC model parameters. DNDC has been widely used across the globe, and new publications and adaptations of the model are produced every year. Effective methods like the one described in this chapter can be used to further increase the impact and scalability of the model.

However, DNDC also has several limitations. Although DNDC is currently the gold standard for modeling soil carbon emissions, the model assumes a homogenous microbial population for both autotrophs and heterotrophs and thus may not simulate the timing of individual N_2O emission events very accurately (Smith et al., 2008). Furthermore, it does not simulate N_2O emissions in cold locations very well due to inaccuracies in the modeling of snow cover and freeze-thaw cycles. Research efforts to improve these aspects will further strengthen the model's value as a tool for simulating agroecosystems and informing carbon markets.

As a result of DNDC's focus on biogeochemical interactions, the soil moisture and cropping growth sub-models are not as advanced as other models which primarily focus on those aspects. DNDC is a 1D model, and the lateral flows of water and nutrients are represented in a relatively simplistic fashion. The model's ability to simulate soil moisture levels accurately is limited due to the lack of an unsaturated zone model and the use of a simplistic tipping-bucket model for water flow. On the crop growth side, DNDC's sub-model does not explicitly model the impact of genetics, carbohydrate accumulation, and photosynthetic rates on crop growth. Future researchers will need to deal with the challenge of improving DNDC's ability to model these aspects without drastically increasing the number of inputs required and reducing the accessibility that makes the model so powerful.

As highlighted in this chapter, the use of aggregated spatial datasets to facilitate calibration at scale enables the wider applicability of DNDC and other process-based models. Going forward, the continued development of techniques which incorporate multi-modal datasets, including existing ground-truth data, data from flux chambers and remotely sensed data, for calibration and validation can be effective in enabling model use at scale.

One of the challenges related to the use of the models such as DNDC in the scientific community is the necessity of maintaining the improvements in the codebase in a way that enables ease of use without significant loss of computational efficiency. Regrow will continue to maintain and update the codebase and share it with UNH; however, further collaboration and contributions from knowledgeable researchers in the field will be essential to enable the full range of model improvements and ensure that DNDC remains a great asset to the global community.

9 Where to look for further information

- UNH DNDC overview (https://www.dndc.sr.unh.edu/).
- Regrow's DNDC documents (https://www.regrow.ag/science-and-tech/dndc).
- First 20 years of DNDC by Gilhespy et al. 2014) (https://www.sciencedirect.com/science/article/pii/S0304380014004190).
- Review of DNDC crop growth sub-model by Zhang and Niu (2016) (https://www.sciencedirect.com/science/article/abs/pii/S0167880916303310?via%3Dihub).
- Brief history of agricultural systems modeling by Jones et al. 2017) (https://www.sciencedirect.com/science/article/pii/S0308521X16301585).
- Climate Action Reserve Soil Enrichment Protocol list of validated models including DNDC validation report (https://www.climateactionreserve.org/how/protocols/ncs/soil-enrichment/).

10 Appendix 1: Model access

An executable for DNDC Version 9.5 is available for download at this link https://www.dndc.sr.unh.edu/. The DNDC v9.5 download also includes 5 validation test cases and a "Crop_parameters.txt" file which contains the default parameters for 88 crops. A recreation of the v9.5 defaults can be seen in Appendix 2.

Regrow periodically shares the source code and executable for DNDC vRegrow with the University of New Hampshire; this information is available from UNH upon request. Note that the code available from UNH will differ from the most current commercial DNDC version.

11 Appendix 2: DNDC v9.5 defaults

Crop ID	Crop name	Max biomass C	Grain fraction	Leaf fraction	Stem fraction	Root fraction	Grain CN	Leaf CN	Stem CN	Root CN	Water demand	Optimum T	TDD	N fixation	Vascularity	Perennial crop
0	Fallow	0	0.1	0.25	0.25	0.4	1	1	1	1	0	20	0	1	0	0
1	Corn	10309	0.4	0.22	0.22	0.16	50	80	80	80	150	30	2550	1	0	0
2	Winter wheat	7610	0.41	0.21	0.21	0.17	40	95	95	95	200	22	1300	1	0	0
3	Soybean	3512	0.35	0.22	0.22	0.2	10	45	45	24	350	25	1500	2.5	0	0
4	Legume hay	11000	0.01	0.4	0.4	0.19	50	50	50	90	550	21	2500	2.5	0	0
5	Non-legume hay	11000	0.5	0.2	0.11	0.19	80	80	80	90	550	21	2500	2	0	0
6	Spring wheat	7800	0.4	0.22	0.22	0.16	50	80	80	80	300	22	1500	1	0	0
7	Sugarcane	17760	0.01	0.44	0.44	0.1	150	130	130	150	500	32	5000	1	0	0
8	Barley	8320	0.3	0.23	0.23	0.23	45	75	75	85	250	21	1300	1	0	0
9	Oats	10435	0.23	0.27	0.27	0.23	35	75	75	85	250	21	1650	1	0	0
10	Alfalfa	5614	0.5	0.05	0.05	0.4	10	10	10	35	300	23	2000	4	0	0
11	Annual grass	4444	0.01	0.22	0.22	0.54	33	33	33	50	300	21	2500	1.01	0	0
12	Perennial grass	9333	0.02	0.35	0.35	0.28	35	35	35	50	200	21	2000	1.5	0	1
13	Sorghum	6857	0.35	0.28	0.28	0.09	40	69	69	85	200	26	2600	1	0	0
14	Cotton	4500	0.32	0.26	0.26	0.16	10	45	45	75	400	25	2500	1	0	0
15	Rye	3048	0.28	0.23	0.23	0.25	20	50	50	50	250	25	2000	1	0	0
16	Vegetables	3692	0.65	0.15	0.15	0.05	30	30	30	50	500	25	1400	1	0	0
17	Papaya	20000	0.2	0.3	0.3	0.2	30	21.2	21.2	50	800	25	5000	1	0	0
18	Potato	8571	0.7	0.13	0.13	0.05	60	60	60	60	415	25	2100	1	0	0
19	Beet	11520	0.75	0.1	0.1	0.05	58	50	50	80	318	25	2550	1	0	0
20	Paddy rice	8238	0.41	0.23	0.24	0.12	45	85	85	85	508	25	2000	1.05	1	0
21	Banana	8000	0.3	0.3	0.3	0.1	45	37	37	37	456	25	5000	1	0	0

(Continued)

Crop ID	Crop name	Max biomass C	Grain fraction	Leaf fraction	Stem fraction	Root fraction	Grain CN	Leaf CN	Stem CN	Root CN	Water demand	Optimum T	TDD	N fixation	Vascularity	Perennial crop
22	Celery	4000	0.01	0.34	0.34	0.32	12	12	12	21	500	25	1300	1	0	0
23	Peanut	3757	0.35	0.23	0.23	0.18	20	40	40	40	300	25	2900	1	0	0
24	Upland rice	4444	0.45	0.24	0.24	0.07	35	55	55	40	400	25	2250	1	0	0
25	Rapeseeds	5634	0.23	0.34	0.34	0.08	20	75	75	95	300	25	1800	1	0	0
26	Tobacco	1920	0.45	0.24	0.24	0.07	15	15	15	40	700	25	3400	1	0	0
27	Millet	2129	0.3	0.25	0.25	0.2	20	40	40	50	331	25	1750	1	0	0
28	Sunflower	2401	0.3	0.25	0.25	0.2	10	45	45	50	495	25	1500	1	0	0
29	Beans	2933	0.36	0.23	0.23	0.18	9.2	45	45	55	300	28	1900	2	0	0
30	Deep water rice	3429	0.35	0.28	0.28	0.1	29	40	40	72	455	25	1500	1	1	0
31	Onion	9600	0.5	0.2	0.2	0.1	20	50	50	40	235	25	2550	1	0	0
32	Palm	280	0.2	0.25	0.25	0.3	10	30	30	50	250	25	3000	1	0	0
33	Strawberry	7309	0.5	0.1	0.1	0.3	20	30	30	40	800	25	1400	1	0	0
34	Lettuce	1428	0.64	0.08	0.08	0.2	11.5	20	20	30	800	25	1400	1	0	0
35	Artichoke	6882	0.5	0.1	0.1	0.3	20	30	30	40	800	25	1400	1	0	0
36	Flowers	1827	0.5	0.1	0.1	0.3	20	30	30	40	800	25	1400	1	0	0
37	Sprout	1827	0.5	0.1	0.1	0.3	20	30	30	40	800	25	1400	1	0	0
38	Berries	3200	0.5	0.1	0.1	0.3	20	30	30	40	800	25	1400	1	0	0
39	Truck crops	1428	0.64	0.08	0.08	0.2	10	20	20	30	800	25	1400	1	0	0
40	Fruit trees	3823	0.45	0.15	0.15	0.25	30	50	50	50	550	25	3000	1	0	0
41	Citrus	5381	0.5	0.13	0.13	0.25	30	50	50	50	550	25	3000	1	0	0
42	Grape	700	0.4	0.2	0.2	0.2	22	18.9	18.9	44	350	25	1600	1	0	0
43	Silage corn	11250	0.4	0.22	0.22	0.16	50	80	80	80	150	25	2500	1	0	0
44	Hops	1777	0.45	0.15	0.15	0.25	30	40	40	50	550	25	3000	1	0	0
45	Tomato	4613	0.36	0.22	0.22	0.2	26	26	26	45	900	25	1400	1	0	0
46	Rainfed rice	8238	0.41	0.27	0.27	0.05	45	85	85	85	508	25	2000	1.05	1	0

47	Cover crop	4000	0.01	0.4	0.4	0.19	15	25	25	30	300	25	1300	1.5	0	0
48	Safflower	6500	0.16	0.32	0.32	0.2	16	60	60	60	300	25	1000	1	0	0
49	Flax	1500	0.4	0.2	0.2	0.2	9	25	25	30	250	25	2300	1	0	0
50	Sedge	20000	0.1	0.3	0.3	0.3	50	61	61	44	800	25	3000	1	0	1
51	Cassava	9231	0.65	0.15	0.15	0.05	70	65	65	75	415	25	3100	1	0	0
52	Cattail	14000	0.01	0.28	0.28	0.44	66	66	66	40	550	25	3000	1	1	0
53	CA broccoli	4000	0.3	0.27	0.27	0.16	10	14.7	14.7	45	150	25	1800	1	0	0
54	Evergreens	4324	0.37	0.22	0.22	0.2	50	75	75	85	400	25	4000	1	0	0
55	Cabbage	2800	0.01	0.43	0.43	0.14	15	20	20	50	450	25	2500	1	0	0
56	Green onion	4000	0.05	0.4	0.4	0.15	10	10	10	15	400	25	2200	1	0	0
57	Mustard	4400	0.1	0.3	0.3	0.3	15	25	25	45	150	25	2200	1	0	0
58	Tule	10800	0.01	0.28	0.28	0.44	66	66	66	40	508	25	3000	1	1	1
59	Moss	4800	0.01	0.45	0.45	0.09	120	120	120	120	550	20	500	1	0	1
60	Radish	4081	0.75	0.13	0.13	0.01	19	45	45	44	508	15	1000	1	0	0
61	Shrub	2400	0.01	0.25	0.25	0.49	30	150	150	150	250	15	2000	1	0	1
62	Boreal sedge	1480	0.01	0.2	0.2	0.59	100	100	100	100	100	15	1200	1	1	1
63	Almond	3333	0.3	0.25	0.25	0.2	10	25	25	50	150	15	4000	1	0	1
64	Nut tree	8500	0.4	0.2	0.2	0.2	20	20	30	40	150	25	3000	1	0	0
65	Melon	10000	0.55	0.25	0.1	0.1	20	20	40	40	550	25	1500	1	0	0
66	Pasture hay	11000	0.5	0.2	0.11	0.19	80	80	80	90	550	21	2500	2	0	0
67	Small grain hay	10435	0.23	0.27	0.27	0.23	35	75	75	85	250	21	1650	1	0	0
68	Carrots	3692	0.65	0.15	0.15	0.05	30	30	30	50	500	25	1400	1	0	0
69	Peppers	3692	0.65	0.15	0.15	0.05	30	30	30	50	500	25	1400	1	0	0
70	Asparagus	3692	0.65	0.15	0.15	0.05	30	30	30	50	500	25	1400	1	0	0
71	Cauliflower	3692	0.65	0.15	0.15	0.05	30	30	30	50	500	25	1400	1	0	0
72	Artichokes	3692	0.65	0.15	0.15	0.05	30	30	30	50	500	25	1400	1	0	0

(Continued)

Crop ID	Crop name	Max biomass C	Grain fraction	Leaf fraction	Stem fraction	Root fraction	Grain CN	Leaf CN	Stem CN	Root CN	Water demand	Optimum T	TDD	N fixation	Vascularity	Perennial crop
73	Sweet potato	8571	0.7	0.13	0.13	0.05	60	60	60	60	415	25	2100	1	0	0
74	Beans green	2933	0.36	0.23	0.23	0.18	9.2	45	45	55	300	28	1900	2	0	0
75	COT	1175	0.32	0.25	0.26	0.16	10	45	45	75	400	25	3558	1	0	0
76	Olives	3823	0.45	0.15	0.15	0.25	30	50	50	50	550	25	3000	1	0	0
77	Plums	3823	0.45	0.15	0.15	0.25	30	50	50	50	550	25	3000	1	0	0
78	Cherries	3823	0.45	0.15	0.15	0.25	30	50	50	50	550	25	3000	1	0	0
79	Peach	3823	0.45	0.15	0.15	0.25	30	50	50	50	550	25	3000	1	0	0
80	Pears	3823	0.45	0.15	0.15	0.25	30	50	50	50	550	25	3000	1	0	0
81	Apples	3823	0.45	0.15	0.15	0.25	30	50	50	50	550	25	3000	1	0	0
82	Dates	3823	0.45	0.15	0.15	0.25	30	50	50	50	550	25	3000	1	0	0
83	Avocados	3823	0.45	0.15	0.15	0.25	30	50	50	50	550	25	3000	1	0	0
84	Apricots	3823	0.45	0.15	0.15	0.25	30	50	50	50	550	25	3000	1	0	0
85	Figs	3823	0.45	0.15	0.15	0.25	30	50	50	50	550	25	3000	1	0	0
86	Prunes	3823	0.45	0.15	0.15	0.25	30	50	50	50	550	25	3000	1	0	0
87	Lemons	5381	0.5	0.13	0.13	0.25	30	50	50	50	550	25	3000	1	0	0
88	Tea	1000	0.02	0.5	0.2	0.28	15	15	25	35	550	30	3000	1	0	1

12 References

Abdalla, M., et al. 2020. Calibration and validation of the DNDC model to estimate nitrous oxide emissions and crop productivity for a summer maize-winter wheat double cropping system in Hebei, China. *Environ. Pollut.* 262, 114199. https://doi.org/10.1016/j.envpol.2020.114199

Adviento-Borbe, M.A.A., et al. 2007. Soil greenhouse gas fluxes and global warming potential in four high-yielding maize systems. *Glob. Change Biol.* 13, 1972–1988. https://doi.org/10.1111/j.1365-2486.2007.01421.x

Al-Kaisi, M.M., et al. 2005. Soil carbon and nitrogen changes as influenced by tillage and cropping systems in some Iowa soils. *Agric. Ecosyst. Environ.* 105, 635–647. https://doi.org/10.1016/j.agee.2004.08.002

Balashov, E.V., et al. 2024. Assessing the impacts of climate change-induced variations in air temperature and precipitation on plant physiological and soil microbial processes with DNDC model. *Acta Hortic. Regiotect.* 27, 1–5. https://doi.org/10.2478/ahr-2024-0001

Brown, L., et al. 2002. Development and application of a mechanistic model to estimate emission of nitrous oxide from UK agriculture. *Atmos. Environ.* 36, 917–928. https://doi.org/10.1016/S1352-2310(01)00512-X

Burzaco, J.P., et al. 2013. Nitrous oxide emissions in Midwest US maize production vary widely with band-injected N fertilizer rates, timing and nitrapyrin presence. *Environ. Res. Lett.* 8, 035031. https://doi.org/10.1088/1748-9326/8/3/035031

Clarke, B., et al. 2022. Extreme weather impacts of climate change: an attribution perspective. *Environ. Res. Clim.* 1, 012001. https://doi.org/10.1088/2752-5295/ac6e7d

Clay, D.E., et al. 2015. Tillage and corn residue harvesting impact surface and subsurface carbon sequestration. *J. Environ. Qual.* 44, 803–809. https://doi.org/10.2134/jeq2014.07.0322

De Vries, et al. 1989. Simulation of ecophysiological processes of growth in several annual crops. *Int. Rice Res. Inst.* 29, 271.

Deng, J., et al. 2022. A decreasing trend of nitrous oxide emissions from california cropland from 2000 to 2015. *Earths Future* 10, e2021EF002526. https://doi.org/10.1029/2021EF002526

Deng, Q., et al. 2016. Assessing the impacts of tillage and fertilization management on nitrous oxide emissions in a cornfield using the DNDC model. *J. Geophys. Res. Biogeosci.* 121, 337–349. https://doi.org/10.1002/2015JG003239

Forster, D., et al. 2022. Simulating soil-plant-climate interactions and greenhouse gas exchange in boreal grasslands using the DNDC Model. *Land* 11, 1947. https://doi.org/10.3390/land11111947

Fumoto, T., et al. 2008. Revising a process-based biogeochemistry model (DNDC) to simulate methane emission from rice paddy fields under various residue management and fertilizer regimes. *Glob. Change Biol.* 14, 382–402. https://doi.org/10.1111/j.1365-2486.2007.01475.x

Gilhespy, S.L., et al. 2014. First 20 years of DNDC (DeNitrification decomposition): model evolution. *Ecol. Model.* 292, 51–62. https://doi.org/10.1016/j.ecolmodel.2014.09.004

Haas, E., et al. 2013. LandscapeDNDC: a process model for simulation of biosphere–atmosphere–hydrosphere exchange processes at site and regional scale. *Landsc. Ecol.* 28, 615–636. https://doi.org/10.1007/s10980-012-9772-x

Hay, R.K.M. 1995. Harvest index: a review of its use in plant breeding and crop physiology. *Ann. Appl. Biol.* 126, 197–216. https://doi.org/10.1111/j.1744-7348.1995.tb05015.x

Iqbal, J., et al. 2015. Does nitrogen fertilizer application rate to corn affect nitrous oxide emissions from the rotated soybean crop? *J. Environ. Qual.* 44, 711–719. https://doi.org/10.2134/jeq2014.09.0378

Jones, J.W., et al. 2017. Brief history of agricultural systems modeling. *Agric. Syst.* 155, 240–254. https://doi.org/10.1016/j.agsy.2016.05.014

Kaspar, T.C., et al. 2006. Examining changes in soil organic carbon with oat and rye cover crops using terrain covariates. *Soil Sci. Soc. Am. J.* 70, 1168–1177. https://doi.org/10.2136/sssaj2005.0095

Keating, B.A., et al. 2003. An overview of APSIM, a model designed for farming systems simulation. *Model. Crop. Syst. Sci. Softw. Appl.* 18, 267–288. https://doi.org/10.1016/S1161-0301(02)00108-9

Kiese, R., et al. 2005. Regional application of PnET-N-DNDC for estimating the N2O source strength of tropical rainforests in the Wet Tropics of Australia. *Glob. Change Biol.* 11, 128–144. https://doi.org/10.1111/j.1365-2486.2004.00873.x

King, M., et al. 2018. Northward shift of the agricultural climate zone under 21st-century global climate change. *Sci. Rep.* 8, 7904. https://doi.org/10.1038/s41598-018-26321-8

Kröbel, R., et al. 2011. Development and evaluation of a new Canadian spring wheat submodel for DNDC. *Can. J. Soil Sci.* 91, 503–520.

Lehman, M. 2020. *Residue removal study for greenhouse gas reduction through agricultural carbon enhancement network and resilient economic agricultural practices in Brookings, South Dakota*. U.S. Department of Agriculture - Agricultural Research Service.

Leip, A., et al. 2008. Linking an economic model for European agriculture with a mechanistic model to estimate nitrogen and carbon losses from arable soils in Europe. *Biogeosciences* 5, 73–94.

Li, C. 2012. User's Guide for the DNDC Model.

Li, C., et al. 1992. A model of nitrous oxide evolution from soil driven by rainfall events: 1. Model structure and sensitivity. *J. Geophys. Res. Atmospheres* 97, 9759–9776. https://doi.org/10.1029/92JD00509

Li, C., et al. 1994. Modeling carbon biogeochemistry in agricultural soils. *Glob. Biogeochem. Cycles* 8, 237–254.

Li, C., et al. 2004. Modeling impacts of management on carbon sequestration and trace gas emissions in forested wetland ecosystems. *Environ. Manage.* 33, S176–S186. https://doi.org/10.1007/s00267-003-9128-z

Li, C., et al. 2005. *Modeling carbon and nitrogen biogeochemistry in forest ecosystems.* 3rd International Nitrogen Conference, 893–898.

Li, C., et al. 2006. Modeling nitrate leaching with a biogeochemical model modified based on observations in a row-crop field in Iowa. *Ecol. Model.* 196, 116–130. https://doi.org/10.1016/j.ecolmodel.2006.02.007

Li, C., et al. 2012. Manure-DNDC: a biogeochemical process model for quantifying greenhouse gas and ammonia emissions from livestock manure systems. *Nutr. Cycl. Agroecosystems* 93, 163–200. https://doi.org/10.1007/s10705-012-9507-z

Li, C.S. 2000. Modeling trace gas emissions from agricultural ecosystems. *Nutr. Cycl. Agroecosystems* 58, 259–276. https://doi.org/10.1023/A:1009859006242

Li, Q., et al. 2021. Carbon emissions inventory of farm size pig husbandry combining Manure-DNDC model and IPCC coefficient methodology. *J. Clean. Prod.* 320, 128854.

Macharia, J.M., et al. 2021. Parameterization, calibration and validation of the DNDC model for carbon dioxide, nitrous oxide and maize crop performance estimation in East Africa. *Heliyon* 7, e06977. https://doi.org/10.1016/j.heliyon.2021.e06977

Malhi, G.S., et al. 2021. Impact of climate change on agriculture and its mitigation strategies: a review. *Sustainability* 13, 1318. https://doi.org/10.3390/su13031318

Medel-Jiménez, F., et al. 2024. Life cycle assessment of four different precision agriculture technologies and comparison with a conventional scheme. *J. Clean. Prod.* 434, 140198. https://doi.org/10.1016/j.jclepro.2023.140198

Nash, P.R., et al. 2012. Nitrous oxide emissions from claypan soils due to nitrogen fertilizer source and tillage/fertilizer placement practices. *Soil Sci. Soc. Am. J.* 76, 983–993. https://doi.org/10.2136/sssaj2011.0296

National Agricultural Statistics Service. 2024. Data and statistics.

Neufeldt, H., et al. 2006. Disaggregated greenhouse gas emission inventories from agriculture via a coupled economic-ecosystem model. *Agric. Ecosyst. Environ.* 112, 233–240.

Omonode, R.A., et al. 2015. Tillage and nitrogen rate effects on area- and yield-scaled nitrous oxide emissions from pre-plant anhydrous ammonia. *Agron. J.* 107, 605–614. https://doi.org/10.2134/agronj14.0440

Oregon State University. 2024. PRISM climate group.

Parkin, T.B. and Hatfield, J.L. 2010. Influence of nitrapyrin on N2O losses from soil receiving fall-applied anhydrous ammonia. *Agric. Ecosyst. Environ.* 136, 81–86. https://doi.org/10.1016/j.agee.2009.11.014

Parkin, T.B. and Kaspar, T.C. 2006. Nitrous oxide emissions from corn-soybean systems in the midwest. *J. Environ. Qual.* 35, 1496–1506. https://doi.org/10.2134/jeq2005.0183

Parton, W.J. 1996. The CENTURY model. in: Powlson, D.S., et al. (Eds.), *Evaluation of Soil Organic Matter Models*. Springer Berlin Heidelberg, pp. 283–291.

Parton, W.J., et al. 1998. DAYCENT and its land surface submodel: description and testing. *Glob. Planet. Change* 19, 35–48. https://doi.org/10.1016/S0921-8181(98)00040-X

Poffenbarger, H.J., et al. 2017. Maximum soil organic carbon storage in Midwest U.S. cropping systems when crops are optimally nitrogen-fertilized. *PLoS One* 12, e0172293. https://doi.org/10.1371/journal.pone.0172293

Poggio, L., et al. 2021. SoilGrids 2.0: producing soil information for the globe with quantified spatial uncertainty. *SOIL* 7, 217–240. https://doi.org/10.5194/soil-7-217-2021

Ray, D.K., et al. 2019. Climate change has likely already affected global food production. *PLoS One* 14, e0217148. https://doi.org/10.1371/journal.pone.0217148

Ritchie, H. 2021. *How much of global greenhouse gas emissions come from food?* Our World Data.

Saggar, S., et al. 2004. Modelling nitrous oxide emissions from dairy-grazed pastures. *Nutr. Cycl. Agroecosystems* 68, 243–255. https://doi.org/10.1023/B:FRES.0000019463.92440.a3

Skendžić, S., et al. 2021. The impact of climate change on agricultural insect pests. *Insects* 12, Article No. 440. https://doi.org/10.3390/insects12050440

Smith, W.N., et al. 2008. Evaluation of two process-based models to estimate soil N2O emissions in Eastern Canada. *Can. J. Soil Sci.* 88, 251–260. https://doi.org/10.4141/CJSS06030

Smith, W.N., et al. 2013. Assessing the effects of climate change on crop production and GHG emissions in Canada. *Agric. Ecosyst. Environ.* 179, 139–150. https://doi.org/10.1016/j.agee.2013.08.015

Soil Survey Staff, Natural Resources Conservation Service, United States Department of Agriculture, n.d. Web soil survey.

Tabatabaie, S.M.H. and Murthy, G.S. 2021. Development of an input-output model for food-energy-water nexus in the pacific northwest, USA. *Resour. Conserv. Recycl.* 168, 105267. https://doi.org/10.1016/j.resconrec.2020.105267

Torralbo, F., et al. 2022. Distinct enhanced efficiency urea fertilizers differentially influence ammonia volatilization losses and maize yield. *Plant Soil* 475, 551–563. https://doi.org/10.1007/s11104-022-05387-4

Wei-Liem, L. 1996. On Latin hypercube sampling. *Ann. Stat.* 24, 2058–2080. https://doi.org/10.1214/aos/1069362310

Yan, W. and Hunt, L.A. 1999. An equation for modelling the temperature response of plants using only the cardinal temperatures. *Ann. Bot.* 84, 607–614. https://doi.org/10.1006/anbo.1999.0955

Yin, S., et al. 2020. Carbon sequestration and emissions mitigation in paddy fields based on the DNDC model: a review. *Artif. Intell. Agric.* 4, 140–149. https://doi.org/10.1016/j.aiia.2020.07.002

Zhang, Y. and Niu, H. 2016. The development of the DNDC plant growth sub-model and the application of DNDC in agriculture: a review. *Agric. Ecosyst. Environ.* 230, 271–282. https://doi.org/10.1016/j.agee.2016.06.017

Zhang, Y., et al. 2002. A simulation model linking crop growth and soil biogeochemistry for sustainable agriculture. *Ecol. Model.* 151, 75–108. https://doi.org/10.1016/S0304-3800(01)00527-0

Chapter 7

The Agricultural Land Management Alternative with Numerical Assessment Criteria (ALMANAC) model

James R. Kiniry[1], USDA-ARS, USA

1 Introduction

Process-based simulation models, combining an easily parameterized plant model with readily available USDA-NRCS soils data and NOAA historical weather data, are valuable tools for a wide variety of management, environmental, and climate change scenarios. In addition, international soils and weather data are easily utilized for applications outside the U.S. and Mexico. Uses include:

- comparing crop production and grass/forage production on the same site;
- evaluating impacts of brush and undesirable plants on grassland productivity;
- quantifying the impacts of changes in plant communities on runoff and soil erosion;

1 Correspondence: jimkiniry54@gmail.com

http://dx.doi.org/10.19103/AS.2025.0155.10
Published by Burleigh Dodds Science Publishing Limited, 2025.

- optimizing which areas can to be used for biofuel production (both grasses and trees) to minimize impacts on overall crop production in an area;
- quantifying impacts of wetland destruction or wetland restoration from runoff from an area; and
- risk assessment with drought or flooding for production on croplands, pastures, and rangelands.

These and other similar applications require a robust model parameterized for pertinent crops, grasses, woody species, and wetland plants. The model must be capable of simulating competition between plant species for light, soil water, and soil nutrients. This type of process-based model is distinct from and has definite advantages over empirical models derived from data fitting (including those using artificial intelligence). The latter type suffers from lack of accuracy outside the range of fitted data as do any empirically derived equations. In contrast, process-based models, when they successfully quantify the various process involved, will not fail when simulating new scenarios. This becomes especially important when trying to simulate future climatic conditions not experienced before and when simulating impacts of severe drought and severe flooding.

The Agricultural Land Management Alternative with Numerical Assessment Criteria (ALMANAC) model is exactly that type of tool and has proven useful for all the above-mentioned applications. This chapter describes the basic processes simulated by the ALMANAC model and provides examples of several applications. It also describes the process of deriving parameters for various plant types. Finally, areas for future research to expand the use of this model are discussed.

2 The development of the Agricultural Land Management Alternative with Numerical Assessment Criteria (ALMANAC) model

Model development at the USDA's Texas Agricultural Experiment Station (at Temple in Texas, USA) has a long history. The models currently supported at Temple (ALMANAC, EPIC, APEX [Agriculture Policy/Environmental eXtender], and SWAT [Soil Water Assessment Tool]) are continuously used and updated and operate on spatial scales ranging from individual fields to river basins. These models have been used worldwide in many projects dealing with soil and water resources and environmental management.

The EPIC (Erosion Productivity Impact Calculator) crop growth model (Williams et al., 1989) was designed in a generic form to allow simulation of

Published by Burleigh Dodds Science Publishing Limited, 2025.

a large number of crops. Since the purpose of the original EPIC model was to assess the effects of erosion on soil productivity throughout the U.S., the crop model had to simulate a large variety of crops and grasses, and it had to be sensitive to changes in soil properties that affect the plants and nutrient supplies to plants. The EPIC crop model is the basis for the crop models contained in ALMANAC. Work on a special version of EPIC called ALMANAC, which featured a more advanced crop growth model that included plant competition, was first described by Kiniry et al. (1992). ALMANAC was developed first as a tool to assess the impacts of weeds on crop yields as atrazine was being removed from the market. The model was designed to quantify impacted crop yields in the US corn belt.

The ALMANAC model has been reviewed previously (Kiniry et al., 1992a) and is described at https://www.ars.usda.gov/plains-area/temple-tx/grassland-soil-and-water-research-laboratory/docs/193226. ALMANAC has been parameterized and validated for many crops (Baez-Gonzalez, et al., 2015, 2017, 2018; Meki et al., 2013, 2015; Xie et al., 2003) and grasses (Kiniry et al., 1992b, 2007, 2013, 2014, 2017). The model uses a more complex light competition system than in EPIC, using equations developed in The Netherlands by Spitter and Aerts (1983), as well as competition for water and nutrients. In addition, ALMANAC differs from EPIC in that it can simulate woody species over several years (MacDonald et al., 2008) and can simulate the bimodal growth curve of cool season grasses as exemplified by tall fescue (*Schedonorus arundinaceus* (Schreb.) Dumort) (Kiniry et al., 2018).

After development and testing with available crop – weed data sets (Kiniry et al., 1992) – the crop-yield simulation was validated at diverse sites in the U.S.(Kiniry et al., 1997; Kiniry & Bockholt, 1998; Xie et al., 2001). Subsequently, the model was extended to range simulation with multiple grass species at diverse sites (Kiniry et al., 2002). Parameters for common warm season grasses (Kiniry et al., 1999), improved grasses (Kiniry et al., 2007), and for mesquite (*Prosopis glandulosa* Torr. var. glandulosa) and juniper (*Juniperus virginiana* L.) (Kiniry, 1998) were developed to allow their simulation. The model has also been demonstrated as a valuable tool to assess switchgrass (*Panicum virgatum* L.) production for biofuels at diverse sites in the U.S. (Kiniry et al., 1996, 2005).

3 Plant growth processes simulated by ALMANAC

Plant growth simulation processes involve light interception, dry matter production, and biomass partitioned into plant parts (Kiniry et al., 2008a, 2011). Biomass is simulated via light interception and species-specific radiation use efficiency (RUE), which is the dry biomass produced per unit of intercepted light (Kiniry et al., 1999, 2007). Potential plant growth simulation depends on

RUE, LAI, and the light extinction coefficient (k) used to calculate the fraction of light intercepted by leaves (Kiniry, 1998). LAI has already been shown to be a promising indicator of coffee (*Coffea* species) yield, especially when combined with other data such as fertilizer use and precipitation (Taugourdeau et al., 2014).

Phenological development (the duration of various plant growth stages) defines the length of each growing season. Phenological development is simulated with a growing degree day (GDD) system, species-specific base temperature, and optimum growing temperature. The GDD sum (potential heat units or PHU) determines the duration of the growing season. Winter dormancy is simulated when day length gets sufficiently short in the fall. The DORMNT parameter defines this interval as the hours of the photoperiod that are near the minimum when plants are dormant.

The date of anthesis is simulated as a defined fraction of PHU (DLAI). Simulated leaf area growth is on a whole-canopy basis. The potential leaf area index (PotLAI) is defined for each species/ecotype/variety. The potential leaf area growth is defined as an 'S' curve over the growing season. Thus, LAI is simulated with this 0.0 to 1.0 'S' curve defined for each species. Daily dry matter production is simulated using the RUE. Potential dry matter produced is calculated from the amount of photosynthetically active radiation (PAR) intercepted by the leaf canopy on that day. The RUE value is species-specific and the units are g of dry matter per MJ of intercepted PAR. Plant part partitioning is also on a whole canopy basis. Partitioning between roots and shoots is defined by two parameters. Plants partition more of the total dry matter production into the roots initially. The fraction going to roots decreases as plants approach anthesis. Drought stress reduces the above-ground dry matter production more than the root dry matter production. Thus, drought stress changes the simulated root:shoot ratio.

Seed production is simulated with a harvest index (HI) approach. A species-specific HI parameter determines the fraction of the total above-ground plant weight in the seed at maturity relative to the total plant weight. Seeds begin growth after anthesis and seed growth stops at physiological maturity. Leaf area expansion and dry matter accumulation are reduced by environmental stresses. In addition, drought stress near anthesis can decrease this HI value. Water stress yield factor (WSYF) is the lowest HI value for a plant species under severe drought stress.

A variety of stresses are simulated each day. Leaf area growth and dry matter accumulation are constrained by the most severe stress each day. Leaf area growth is more sensitive to drought than by dry matter growth. PET is used to calculate drought stress. PET is calculated with weather variables. Plant available soil water is calculated based on rainfall, soil infiltration, and soil water-holding capacity. The model simulates a drought stress response if plant's available soil water is insufficient to meet the plant's demand (based on PET and LAI).

4 Assessing soil water and nutrient balances

Water and nutrient uptake are critical for yield and biomass simulation, particularly in water-limited conditions. The ALMANAC model calculates the effects of soil water on plant growth and yield with similar functions. Potential evaporation is calculated first, and then potential soil water evaporation and potential plant water transpiration are derived from potential evaporation and LAI. Based on the soil water supply and crop water demand, the water stress factor is estimated to decrease daily crop growth and yield. Potential evaporation is estimated by the Penman-Monteith method (Ritchie, 1972). Potential soil evaporation and plant transpiration were estimated by:

(1) $EP = E0(LAI/3)$
$0 < LAI < 3.0$

(2) $EP = E0$
$LAI > 3.0$

(3) $ES = \text{minimum of } (E0 \exp(-0.1BIO), E0\text{-}EP)$

Where EP and ES are potential plant transpiration and soil evaporation (mm), E0 is potential evaporation (mm), LAI is leaf area index, and BIO is the sum of the above-ground biomass and crop residue (Mg ha^{-1}).

Water stress factor (WSF) is the ratio of water use to water demand (potential plant transpiration) in ALMANAC, and water use (WU) is a function of plant extractable water and root depth.

4.1 Nutrient balance

The nutrient balance (N and P) also allows plants to acquire sufficient nutrients to meet the demands if adequate quantities are available in the current rooting zone. Nitrogen (N) and phosphorus (P) stresses reduce plant growth in the model. These stresses are simulated with a supply and demand approach. Plant nutrient uptake is calculated with three parameters defining how nutrient demand changes during the growing season. Optimum concentrations of N and P are defined early in plant development, near anthesis, and at physiological maturity for each species. These values define the potential nutrient uptake from the soil each day and if insufficient to meet the demand, ALMANAC simulates nutrient stress.

ALMANAC simulates temperature stress when the temperature is below the defined base temperature or above the defined optimum temperature. Plant height is simulated from the fraction of GDD relative to the physiological maturity value and a species-specific plant height parameter (CHT).

ALMANAC simulates competition between crops and weeds and simulates communities of plants, including woody plants competing with forages and native range sites. ALMANAC simulates competition for light, water, and nutrients. The fraction of incoming solar radiation intercepted by a leaf canopy (FIPAR) is:

$$(4)\quad \text{FIPAR} = 1.0 - \exp(-k * \text{LAI})$$

The light extinction coefficient (k) for Beer's law (Monsi and Saeki, 1953) is calculated as:

$$(5)\quad k = \text{logn}\,(1.0 - \text{FIPAR})/(-\text{LAI})$$

where logn = natural log of the number, and FIPAR = fraction of IPAR.

Values for k have been calculated for several plant species (Kiniry et al., 1992, 1999, 2017; Baez-Gonzalez et al., 2017). Simulation of LAI is critical for these equations describing light interception.

Simulated light competition (Spitters and Aerts, 1983) is computed as LAI*k using the LAI of each plant species in the mixture. These products (LAI*k) of each species are summed and then used in Beer's law to calculate the fraction of light interception by the whole plant community. This fraction for the whole plant community is divided among the competing species by weighted fractions. The weights accommodate for differences in plant heights and LAI*k of the species. Taller species and ones with higher LAI and higher k intercept more of the total light intercepted by the community. In addition, potential plant transpiration is simulated with the potential evapotranspiration and the total community LAI. Simulated water and nutrient competition use a balance sheet approach. After each plant species' intercepted light is computed as described above, each one's potential daily biomass growth is calculated as RUE multiplied by the intercepted PAR, assuming 45% of the total daily incident solar radiation is PAR (Meek et al., 1984; Monteith et al., 1965).

As part of the plant competition simulation, ALMANAC accounts for variability in root scavenging capacities between species through differences in the current rooting depth of each species. Potential rooting depths are derived from measurements reported in the literature (such as in Kiniry et al. (1999)). A deeper-rooted plant species may have access to soil water (and nutrients) not available to any competing shallow-rooted species. A deeper-rooted species can have adequate soil water and nutrients at the same time as when a shallow-rooted species is stressed.

5 Downloading soil and weather data for ALMANAC

The model uses USDA-NRCS soil data for the U.S. and soil data for the country of Mexico (Baez-Gonzalez et al., 2016, 2020). It uses solar, daily temperature, and rainfall data. It can also use international soils and weather data as described below and shown in Druille et al. (2020).

Within the U.S., soil data is taken from the USDA-NRCS (https://www.nrcs.usda.gov/wps/portal/nrcs/main/soils/survey/). This is publicly available, verified, and the most extensive soil database for the U.S. This data can be downloaded for any state, as described in the ALMANAC model documentation (https://www.ars.usda.gov/plains-area/temple-tx/-grassland-soil-and-water-research-laboratory/docs/193226/).

Soil data inputs are the texture, depth, and the number of rocks by soil layer. The model uses the values for saturation, drained upper limit, and lower limit for each soil layer. Soil organic matter impacts plant-available water and soil carbon balances in the model. Runoff is calculated using the traditional runoff curve number system. Runoff is simulated with the type of ground cover and slope.

Soil and weather data for Mexico are also available for the ALMANAC model (https://www.ars.usda.gov/plains-area/temple-tx/grassland-soil-and-water-research-laboratory/docs/almanacmex/). Accessing international soils data for other sites can be determined from available country data. Then the data can be manually input into the ALMANAC interface or database. First a site with a similar latitude and soil are input, then the soil data is augmented to match that from a country outside the U.S. or Mexico.

The input data is designed to make this model easily applied and readily available. The required daily weather inputs consist of solar radiation, wind speed, relative humidity, rainfall, and maximum and minimum temperature. Solar radiation can be approximated when unavailable for a given location derived from wind speed and relative humidity. The U.S. National Oceanic and Atmospheric Administration websites weather data are readily downloaded for the U.S. via the steps in the model documentation (see website link above). Weather in both U.S. and Mexico ALMANAC models are also available with the included weather generator. When needing observed weather that is unavailable in NOAA or internationally, this website from NASA provides all required weather inputs using satellite data from the latitude and longitude of the location (https://power.larc.nasa.gov/data-access-viewer/).

Published by Burleigh Dodds Science Publishing Limited, 2025.

6 Outputs from ALMANAC

The output list from ALMANAC is extensive. The tables show what was run as well as plant, soil, weather, and management parameters. Every year will begin by listing management, then the simulation details, followed by a summary. At the end of the output are the simulation averages. For details on what the output abbreviations mean, readers can refer to 'ALMANAC Output Abbreviation Key'. Below are tables of some of the key outputs. The complete list can be seen at ALMANAC Output Abbreviation Key of the link: https://www.ars.usda.gov/plains-area/temple-tx/grassland-soil-and-water-research-laboratory/docs/193226

Yearly Output Table

Variable	Description	Units
TMX	Maximum daily air temperature	°C
TMN	Minimum daily air temperature	°C
RAD	Solar radiation	MJ/m^2
RAIN	Precipitation	mm
UNO3	N uptake by the crop	kg/ha
PEP	Potential plant water evaporation	mm
Q	Surface runoff	mm
SSF	Lateral subsurface flow travel time	Days
PRK	Percolation below the root zone	mm
ET	Evapotranspiration	mm
EP	Plant evaporation (Transpiration)	mm
PET	Potential evaporation	mm
MUSL	Soil loss from water erosion using modified USLE (MUSLE)	t/ha
YW	Soil loss from wind erosion	t/ha
NFIX	N fixed by leguminous crops	kg/ha
UNO3	N uptake by the crop	kg/ha
SW	Total soil water in the profile	m/m
TNO3	Total NO_3 present in the soil profile	kg/ha
HU1	Heat units – average daily temperature minus base temperature of crop	°C
LAI1	Leaf area index	NA
RD	Root depth	m
RW1	Total root weight	t/ha
BIOM	Crop biomass (shoot + root)	t/ha
STRS	The type and number of days of stress by month for the three highest stress variables. Water=1, N=2, P=3, Temperature=4, Aeration=5. Example: 411107201 means that there were 11 days of temperature stress (is 411), 7 days of water stress (is 107), and 1 day of nitrogen stress (201).	NA

Published by Burleigh Dodds Science Publishing Limited, 2025.

Yearly Summary Table (Smaller table found at the end of every year. Shown for each crop simulated)

Variable	Description	Units
Shown as monthly intervals in the table:		
HU	Heat units – average daily temperature minus base temperature of crop	°C
LAI	Leaf area index	NA
RD	Root depth	m
RW	Total root weight	t/ha
BIOM	Crop biomass (shoot + root)	t/ha
CHT	Crop height	m
Shown as yearly values:		
(crop name) YLD	Harvested crop yield	t/ha
BIOM	Crop biomass (shoot + root)	t/ha
Stress Days	Days of crop stress due to water, N, P, temperature or aeration	Days

Summary Table

Average Monthly Values

Variable	Description	Units
C	Average water-erosion/crop-management factor	NA
MUSL	Soil loss from water erosion using modified USLE (MUSLE)	t/ha
YW	Soil loss from wind erosion	t/ha
RAIN	Precipitation	mm
DAYP	Number of days of rain per month	Days
PRK	Percolation below the root zone	mm
Q	Surface runoff	mm
ET	Evapotranspiration	mm
PET	Potential evaporation	mm

Average Annual Crop Yield Data (Shown for each crop simulated.)

Variable	Description	Units
CROP	Four-digit crop name	Text
YLD	Harvested crop yield	t/ha
BIOM	Crop biomass (shoot + root)	t/ha
HU	Heat units – average daily temperature minus base temperature of crop	°C
RD	Root depth	m

Average Stress Days (Shown for each crop simulated)

Variable	Description	Units
Biomass:		
WATER	Water stress on crop growth	Days
N	Nitrogen stress on crop growth	Days
P	Phosphorus stress on crop growth	Days
TEMP	Temperature stress on crop growth	Days
AIR	Aeration stress on crop growth	Days
Root growth:		
BD	Moist soil bulk density stress on root growth	t/m^3
ALSAT	Soil aluminum saturation stress on root growth	%
TEMP	Temperature stress on root growth	Days

7 ALMANAC applications

Publications describing the examples of applications given below as well as for other applications of the model are given in the 'publications' link in the website (https://www.ars.usda.gov/plains-area/temple-tx/grassland-soil-and-water-research-laboratory/people/james-kiniry/; checked 16 July, 2024).

7.1 Simulating different cultivars, hybrids, and ecotypes of crops and grasses

With ALMANAC, different groups within a species are simulated by changing a few key plant parameters. These include the potential LAI (DMLA), the duration of the growth period (defined by the degree days from planting or emergence to physiological maturity, PHU), the RUE, and, in the case of crops, the HI. In addition, if these plant groups differ in leaf angle, the extinction coefficient for Beer's law (k) is adjusted. These are better described in the following examples.

7.2 Simulating different switchgrass **(Panicum virgatum** *L.) types*

Switchgrass is a highly productive, native perennial warm season grass with great potential as a cellulosic biofuel. There are a wide variety of ecotypes that range from southern Canada to Texas. We first became involved in parameterizing switchgrass using 'Alamo' cultivar as part of a native grass field project (Kiniry et al., 1999). We used parameters derived in that study to successfully simulate several sites in Texas (Kiniry et al., 1996, 2005). We then began to develop parameters for other, more northerly cultivars (Kiniry et al., 2011, 2013). These parameters were used in ALMANAC to simulate biofuel production across a wide range of environments (Kiniry et al., 2008; Behrman et al., 2013; Kiniry et al., 2008a, 2008b; Kim et al., 2017; Aranguren et al., 2018; Kim et al., 2018, 2020; Rocateli et al., 2020; Proulx et al., 2021). The overall conclusion from all of these studies was that the ALMANAC model is a useful, realistic tool for predicting cellulosic biofuel production with switchgrass over a wide range of environments and a wide range of climatic conditions.

We used an extensive published data set (Casler et al., 2004) with 2 years of measured data at five sites of varying latitude to derive and verify switchgrass parameters. The four switchgrass types were Northern Upland (NU), Northern Lowland (NL), Southern Upland (SU), and Southern Lowland (SL). Parameters were adjusted to get reasonable simulated yields of each switchgrass type, as compared to the mean of the 2 years of measured yields. Using parameters developed for Alamo switchgrass in Texas (Kiniry et al., 1996, 1999, 2007), the only two parameters adjusted were the degree days to maturity (base 12°C) and the PotLAI. The light extinction coefficient for Beer's Law (Monsi and Saeki, 1953) was set *t* to −0.51, the average of the two means from Kiniry et al. (1999) and Kiniry et al. (2007). The values of PotLAI and degree days varied with latitude and switchgrass types (Table 1).

7.3 Maize simulations for different maturity groups at different locations

When we started working on simulating maize with ALMANAC, we had four main objectives. First, we wanted to be able to reasonably simulate adapted hybrids over a wide range of latitudes within the U.S. We developed the parameters and simulated sites from as far north as Minnesota and New York and as far south as Texas and Louisiana. We also wanted to be able to simulate at sites within the Corn belt. The successes with this effort were reported in Kinry et al. (1997). Next, we wanted to simulate diverse sites in Texas using reported yield trial data for various years. Successes in these efforts were reported by Kiniry and Bockholt (1998). Thirdly, we worked to simulate maize and sorghum (*Sorghum bicolor* (L.) Moench) yields under severe drought stress in central and southern Texas as reported for a very dry, very low-yielding year. The realistic accuracy of the model was reported by Xie et al. (2001). Finally, we wanted to test the model in extremely high-yielding conditions, with irrigated data from the High Plains of Texas. The model did an excellent job of simulation as reported by Kiniry et al. (2004). Thus the overall conclusion from these efforts was that once realistic parameters were developed for adapted hybrids at

Table 1 Data sets (Casler et al., 2004) to develop parameters for and validate the ALMANAC model for four switchgrass types switchgrass

Type:	SL	NL	SU	NU
Stillwater, OK (36°N)				
Degree days	1200	1100	1050	950
PotLAI	3.0	3.0	2.5	1.8
Msrd Yield	15.13	14.81	12.62	10.45
Sim. Yield	15.12	15.45	13.65	11.31
Sim./Msrd.	1.00	1.04	1.08	1.08
Overall mean Sim./Msrd = 1.05				
Manhattan, KS (39°N)				
Degree days	1200	1100	1050	950
Pot LAI	3.0	3.0	2.5	2.0
Msrd Yield	9.26	10.28	7.96	7.05
Sim. Yield	9.14	8.62	8.17	6.86
Sim./Msrd.	0.99	0.84	1.03	0.97
Overall mean Sim./Msrd = 0.96				
Mead, NE (41°N)				
Degree days	1100	1100	900	800
Pot LAI	3.0	3.0	2.5	2.0
Msrd Yield	17.46	20.93	14.99	12.71
Sim. Yield	16.84	17.75	15.30	12.92
Sim./Msrd.	0.96	0.85	1.03	1.03
Overall mean Sim./Msrd = 0.97				
Arlington, WI (43°N)				
Degree days	800	1100	800	800
Pot LAI	1.8	2.4	2.5	2.0
Msrd Yield	6.46	10.61	11.44	10.25
Sim. Yield	7.07	10.70	11.56	9.88
Sim./Msrd.	1.09	1.01	1.01	1.02
Overall mean Sim./Msrd = 1.02				
Spooner, WI (46°N)				
Degree days	540	630	900	800
Pot LAI	1.8	1.8	2.5	2.0
Msrd Yield	3.81	4.60	7.39	7.33
Sim. Yield	4.07	4.76	7.96	6.73
Sim./Msrd.	1.07	1.04	1.08	0.95
Overall mean Sim./Msrd = 1.03				

these sites, ALMANAC was an excellent simulation tool across latitudes and in high-yielding and low-yielding conditions.

Different hybrids and maturity types of maize are simulated by changing the values for potential LAI (DMLA) and degree days to maturity (PHU). This was done to simulate diverse sites across Texas (Kiniry and Bockholt, 1998) and across several latitudes in the U.S. (Kiniry et al., 1997).

7.4 Simulating grain sorghum

In another study on grain sorghum, the results demonstrated that the ALMANAC model is an appropriate tool to study the impact of climate warming on forage and grain sorghum yields (Druille et al., 2020). Its use in the Pampas Region of Argentina indicated that increases in mean annual temperature are projected to be positive for the forage yield and negative for the grain yield of sorghum. The variations detected in yields would be related not only to the direct effects of temperature on plant growth but also to indirect effects through changes in water and nitrogen availability. The use of simulation models in different agroecosystems worldwide to assess the effect of climate change on diverse crops and then food and forage production generates valuable information to make management decisions. Among others, the sowing dates, planted area, genotypes used, fertilization rates, and objectives in the genetic improvement projects can be modified.

7.5 Coffee and cocoa

A major hurdle restricting the development of any simulation model, but especially ones for tree crops, is the availability of seed yield data for model development and validation. Often, industry and private producers are reluctant to share such data. Thus, for this paper, we demonstrate the reasonableness of yield production for one region for coffee (Hawaii) and one location for cocoa (*Theobroma cacao*) (Ghana) (Kiniry et al., 2023b). The hope is that by demonstrating the capacity of the model to simulate such systems, those in the industry and researchers will adopt this simulation tool and apply it to their situations.

A major strength of the ALMANAC model is its ability to simulate competition between plant species (Kiniry et al., 1992, 1994; Debaeke et al., 1997). This becomes especially important for these two tropical tree species, as they are often grown in agroforestry conditions with companion trees much taller than the coffee or cocoa trees. ALMANAC accurately simulates light competition. It also simulates competition for water and nutrients among different plant species growing together. In this paper, we will demonstrate the seed yields of both tree species both as a monoculture and in an agroforestry setting.

For this project, coffee, cocoa, and a tropical overstory tree were simulated. The parameters for coffee came from parameters developed by J.G. Fernandez (personal communication). The parameters for cocoa were based on the coffee parameters but adapted according to supplementary information from Fati Aziz (personal communication). Lastly, a new aspect of the model developed for this study is a generic stand-in for tropical overstory trees. This was developed from parameters for carob trees (*Ceratonia siliqua*) by J.G. Fernandez, but our tropical tree was simulated as a non-legume. These trees take a few years to develop and they are evergreen; therefore, we set the total degree days (PHU) to values that allow development over 5 years for coffee and cocoa and 10 years for the overstory tropical tree. Once that number of years is reached, the LAI is assumed to be fairly stable and does not change except when trees are pruned and the leaves regrow.

ALMANAC plant parameters were calculated for physical descriptions, leaf area development, development rate response to temperature, radiation-use efficiency, and nitrogen and phosphorous concentrations in plant biomass. ALMANAC simulates the effects of stresses on plant biomass and LAI such as from nutrient deficiency, drought, and temperature (Kiniry et al., 1992). Plant parameter values and plant growth curves were optimized through ALMANAC application using the field data. Values for these curves were used to reproduce growth curves generated with measured field data of LAI. ALMANAC has simulated a wide range of species, including evergreen shrubs like creosote bush (*Larrea tridentata* (DC.) Cov.).

The ALMANAC model did an acceptable job simulating the yields of coffee. For coffee, the statewide mean coffee yield for Hawaii over 10 years was 0.983 Mg/ha. For 10 years of statewide yields, the average simulated over-measured ratio was also 0.983. While the correlation coefficient was low (0.554) due to the low variability in measured yields, the simulations show promise for future projects with more diverse simulated yields. For the 3 years of available yield data for the island of Kaua'i, the average simulated over-measured ratio was 1.022. Thus, for both types of data, the model was realistically close to the mean simulated values. For individual years, for the statewide yields, the simulated values were within 10% of the measured values for 8 of 10 years and within 5% of measured values in 6 of 10 years. For Kaua'i, all 3 years the simulated values were within 10% of the measured values. Looking graphically at the results, after the first year, the simulated values showed the same pattern as the measured ones, showing higher values in the higher measured years and lower values in the lower measured years (Fig. 1).

For cocoa, the ALMANAC model showed similar promising results. The simulated values were within 10% of the measured values for all 4 years.

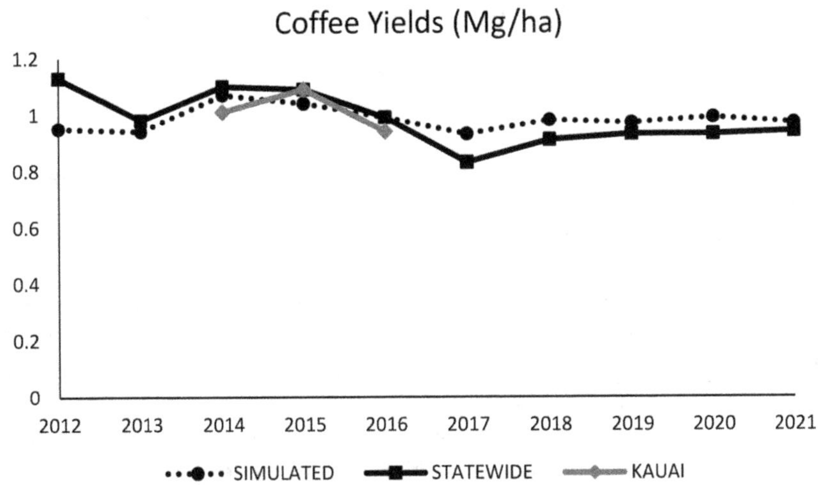

Figure 1 Coffee yields (in Hawai'i) measured and simulated with the ALMANAC model.

The simulated values were somewhat more stable over the years than the measured values, likely due to the high rainfall for this site. The measured values varied around the simulated values over the 4 years. Likewise, while the correlation coefficient was again low (0.577) due to the low variability in measured yields, the simulations also show promise for future projects with more diverse simulated yields. Thus, the overall conclusion was that the ALMANAC model shows great promise in simulating coffee and cocoa yields in tropical conditions (Fig. 2).

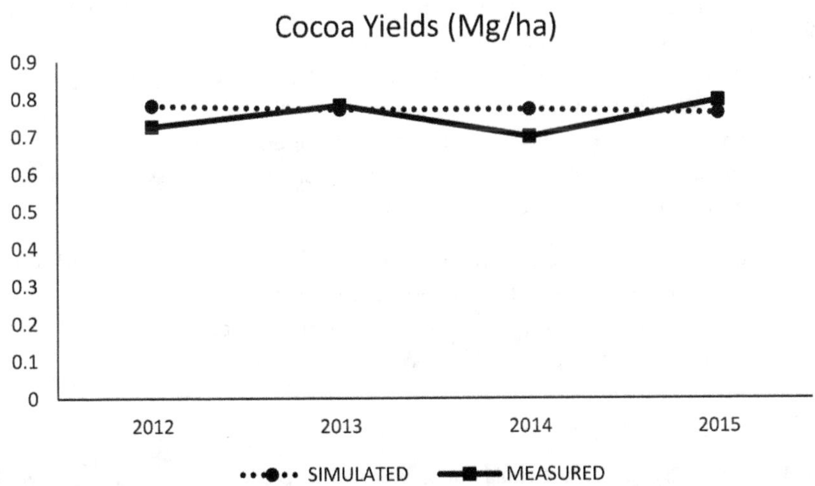

Figure 2 Cocoa yields (in Ghana) measured and simulated with the ALMANAC.

8 Derivation of plant parameters

To develop key plant parameters for the ALMANAC model, there needs to be field experimental plots established at the location of interest. Field experiments should be conducted for multiple years. Often, to avoid complications due to nutrient deficiency, plots are fertilized prior to planting as needed. The total aboveground biomass of all plants is harvested manually every 2 weeks throughout the growing season. At each harvest, the heights of all plants are measured. PAR is measured below the leaf canopy using an ACCUPAR LP-80 Ceptometer (Decagon Devices, Pullman, WA, USA) or a similar linear light bar. The PAR values are used to calculate the fraction of intercepted PAR (FIPAR). The light readings should be made between 10:00 and 14:00. Fresh total aboveground biomass weight, seeds, and dead leaves are measured. The leaf area (cm^2) for each plant sample is measured with a LI-3100 Area Meter (LI-COR Biosciences, Lincoln, NE, USA). Based on the leaf area, fresh weight, and sampling area, LAI of each plant sample is calculated. The light extinction coefficient (k) is calculated as the natural log of difference between 1 and FIPAR and then divided by LAI. Following these measurements, the fresh samples are dried at 66°C in a forced air-drying oven until dry weight stabilizes, after which, the dry weights of each sample are measured.

9 Conclusion and future trends

While the published results demonstrate that ALMANAC is a useful simulation tool for crops, grasses, and woody species, future work extending it to new varieties within species and new species will allow further extension of its applications. Work on simulating different genotypes within a plant species, such as described for eastern gamagrass (*Tripsacum dactyloides*) (Kiniry et al., 2023a), will allow more intensive analyses of genotype by environment interactions within a species. Likewise, as new crop hybrids are developed and become available, parameterizing them for ALMANAC will provide a means of optimizing where they are best adapted and their ability to sustain production in a wide range of environments and in a wide range of weather conditions.

The model and associated documentation are available at

- https://www.ars.usda.gov/plains-area/temple-tx/grassland-soil-and-water-research-laboratory/people/james-kiniry/; and
- https://www.ars.usda.gov/plains-area/temple-tx/grassland-soil-and-water-research-laboratory/docs/193226

10 References

Aranguren, M.F., et al. 2018. Simulation-optimization approach for the logistics network design of biomass co-firing with coal at power. *Plant Sustain.* 10(11), 4299.

Baez-Gonzalez, A.D., et al. 2015. Parameterization of ALMANAC crop simulation model for non-irrigated dry bean in semi-arid temperate areas in Mexico. *Interciencia*. 30, 185–189.

Baez-Gonzalez, A.D., et al. 2016. *ALMANACMEX. Agricultural Land Management Alternatives with Numerical Assessment Criteria Model (ALMANAC) with Mexican Interface; Version 1.0.18*. User's Manual. Special Publication No. 44; INFAP: Mexico City, Mexico.

Baez-Gonzalez, A., et al. 2017. Crop parameters for modeling sugarcane under rainfed conditions in Mexico. *Sustain*. 9. https://doi.org/10.3390/su9081337.

Baez-Gonzalez, A.D., et al. 2018. Potential impact of future climate change on sugarcane under dryland conditions in Mexico. *J. Agron. Crop Sci*. 204, 515–528.

Baez-Gonzalez, A.D., et al. 2020. High sowing densities in rainfed common beans (*Phaseolus vulgaris* L.) in Mexican semi-arid highlands under future climate change. *Agronomy* 10, 442.

Behrman, K.D., et al. 2013. Spatial forecasting of switchgrass productivity under current and future climate change scenarios. *Ecol. Appl.* 23(1), 73–85.

Debaeke, P., et al. 1997. Modelling crop:weed interactions in wheat with ALMANAC. *Weed Res*. 37, 325–341.

Druille, M., et al. 2020. Modeling climate warming impacts on grain and forage sorghum yields in Argentina. *Agronomy*. 10. https://doi.org/10.3390/agronomy10070964.

Kim, S., et al. 2017. Adaptation of C4 bioenergy crop species to various environments within the Southern Great Plains of U.S. *Sustain* 9, 89. https://doi.org/10.3390/su9010089.

Kim, S., et al. 2018. Two-phase simulation-based location-allocation optimization of biomass storage distribution. *Simul. Modell. Pract. Theory*. 86, 155–68.

Kim, S., et al. 2020. Simulated biomass, climate change impacts, and nitrogen management to achieve switchgrass biofuel production at diverse sites in U.S. *Agronomy* 10(4), 503–521.

Kiniry, J.R. 1994. Radiation-use efficiency and grain yield of maize competing with johnsongrass. *Agron. J*. 86, 554–557.

Kiniry, J.R. 1998. Biomass accumulation and radiation use efficiency of honey mesquite and eastern red cedar. *Biomass Bioenergy* 15(6), 467–473.

Kiniry, J.R. and Bockholt, A.J. 1998. Maize and sorghum simulation in diverse Texas environments. *Agron. J*. 90, 682–687.

Kiniry, J.R., et al. 1992a. A general, process-oriented model for two competing plant species. *Trans. ASAE* 35(3), 801–810.

Kiniry, J.R., et al. 1992b. Sunflower simulation using the EPIC and ALMANAC models. *Field Crops Res*. 30, 403–423.

Kiniry, J.R., et al. 1996. Simulating Alamo switchgrass with the ALMANAC model. *Agron. J*. 88, 602–606.

Kiniry, J.R., et al. 1997. Evaluation of two maize models for nine U.S. locations. *Agron. J*. 89(3), 421–426.

Kiniry, J.R., et al. 1999. Radiation use efficiency and leaf CO_2 exchange for diverse C_4 grasses. *Biomass Bioenergy* 17, 95–112.

Kiniry, J.R., et al. 2002. Simulating grass productivity on diverse range sites in Texas. *J. Soil Water Conserv*. 57, 144–150.

Kiniry, J.R., et al. 2004. Maize yield potential: Critical processes and simulation modeling in a high-yielding environment. *Agri. Syst.* 82, 45–56.

Kiniry, J.R., et al. 2005. Switchgrass simulation by the ALMANAC model at diverse sites in the southern U.S. *Biomass Bioenergy*. 29, 419–425.

Kiniry, J.R., et al. 2007. Coastal bermudagrass, Bahiagrass, and native range simulation at diverse sites in Texas. *J. Agron*. 99, 450–461. https://doi.org/10.2134/agronj2006.0119.

Kiniry, J.R., et al. 2008a. Switchgrass biomass simulation at diverse sites in the Northern Great Plains of the U.S. *BioEnergy Res*. 1(3–4), 259–264.

Kiniry, J.R., et al. 2008b. Biofuels and water use: comparison of maize and switchgrass and general perspectives. In Wright, J.H. and Evans, D.A. (eds.) *New Research on Biofuels*. Nova Science Publ, New York, NY, USA, pp. 17–30.

Kiniry, J.R., et al. 2012. Clash of the titans: comparing productivity via radiation use efficiency for two grass giants of the biofuel field. *BioEnergy Res*. 5(1), 41–48.

Kiniry, J.R., et al. 2013. Novel applications of ALMANAC: modelling a functional group, exotic warm-season perennial grasses. *Am. J. Exp. Agric*. 3, 631–650.

Kiniry, J.R., et al. 2014. Plant parameters for plant functional groups of western rangelands in enable process-based simulation modeling. *Am. J. Exp. Agric*., 4, 746–766.

Kiniry, J.R., et al. 2014. Modeling to evaluate and manage water and environmental sustainability of bioenergy crops in the United States. In *Practical Applications of Agricultural System Models to Optimize the Use of Limited Water*. Advances in Agricultural Systems Models 5. American Soc. of Agron, pp. 139–160.

Kiniry, J.R., et al. 2014. Plant parameters for plant functional groups of western rangelands to enable process based simulation modeling. *Am. J. of Exp. Agric*. 4(7), 746–766.

Kiniry, J.R., et al. 2017. Short duration, perennial grasses in low rainfall site in Montana deriving growth parameters and simulating with a process-based model. *Exp. Agric. Intl*. 15, 1–13.

Kiniry, J.R., et al. 2018. Simulating bimodal tall fescue growth with a degree-day-based process-oriented plant model. *Grass Forage Sci*. 73(2), 432–439.

Kiniry, J.R., et al. 2023a. Diverse eastern gamagrass ecotypes: general characteristics, ploidy levels, and biogeography. *Crop Science* 63, 3545–3556.

Kiniry, J.R., et al. 2023b. Tropical tree crop simulation with a process-based, daily timestep simulation model (ALMANAC): description of model adaptation and examples with coffee and cocoa simulations. Agron. 13(2), 580. https://doi.org/10.3390/agronomy13020580.

MacDonald, J.D., et al. 2008. A multi-species, process based vegetation simulation module to simulate successional forest regrowth after forest disturbance in daily time step hydrological transport models. *Environ. Eng. Sci*. 7, 127–143.

Meek, D.W., et al. 1984. A generalized relationship between photosynthetically active radiation and solar radiation. *J. Agron*. 76, 939–945.

Meki, M.N., et al. 2013. Energy sorghum biomass harvest tresholds and tillage effects on soil organic carbon and bulk density. *Ind. Crops and Prod*. 43, 172–182. https://doi.org/10.1016/j.indcrop.2012.07.033.

Meki, M.N., et al. 2015. Two-year growth cycle sugarcane crop parameter attributes and their application in modeling. *J. Agron*. 107(4), 1310–1320. https://doi.org/10.2134/agronj14.0588.

Monsi, M. and Saeki, T. 1953. The light factor in plant communities and its significance for dry matter production. *J. Jpn. Bot*. 14(1), 22–52.

Monteith, J.L. 1965. Light distribution and photosynthesis in field crops. *Ann. Bot*. 29, 17–37. https://doi.org/10.1093/oxfordjournals.aob.a083934

Proulx, R., et al. 2021. Improved ALMANAC simulations of upland switchgrass ecotypes in the northern US. *Agron. J.* 114. https://doi.org/10.1002/agj2.20970.

Ritchie, J.T. 1972. Model for predicting evaporation from a row crop with incomplete cover. *Agron. J.* 64(3), 329–332. https://doi.org/10.2134/agronj1972.00021962006400030010x.

Rocateli, A.C., et al. 2020. Simulating switchgrass biomass productivity using ALMANAC. I. Calibration of soil water. *Agron. J.* 112, 183–193. https://doi.org/10.1002/agj2.20054.

Spitters, C.J.T. and Aerts, R. 1983. Simulation of competition for light and water in crop-weed associations. *Asp. Appl. Biol.* 4, 467–483.

Taugourdeau, S., et al. 2014. Leaf area index as an indicator of ecosystem services and management practices: an application for coffee agroforestry. *Agric. Ecosyst. Environ.* 192, 19–37.

Williams, J.R., et al. 1989. The EPIC crop growth model. *Trans. ASAE* 32, 497–511.

Xie, Y., et al. 2001. Maize and sorghum simulations with CERES-Maize, SORKAM, and ALMANAC under water-limiting conditions. *Agron. J.* 93, 1148–1155.

Xie, Y., et al. 2003. The ALMANAC model's sensitivity to input variables. *Agric. Syst.* 78, 1–16.

Chapter 8

The USDA Root Zone Water Quality Model

Huihui Zhang, Liwang Ma and Jon D. Hanson, USDA-ARS, USA; and Zhiming Qi, McGill University, Canada

1 Introduction

The USDA Root Zone Water Quality Model (RZWQM) was initiated in the middle of the 1980s to provide detailed mechanistic process simulation in the root zone based on lessons learned from other USDA models such as CREAMS (Chemicals, Runoff and Erosion from Agricultural Management Systems), GLEAMS (Groundwater Loading Effects of Agricultural Management Systems), NTRM (Nitrogen-Tillage-Residue Management), Opus, and PRZM (Pesticide Root Zone Model) (Ahuja et al., 2000a). The first version of RZWQM was released in 1992, and a Windows user interface was developed in 1998.

http://dx.doi.org/10.19103/AS.2025.0155.25
Published by Burleigh Dodds Science Publishing Limited, 2025.

Initially, the model contains a generic plant module (GPM) that may be parameterized for any annual crop (Hanson, 2000). Yet, as of today, this generic crop module has only been used for maize, soybean, and winter wheat. In the early 2000s, the DSSAT (Decision Support System for Agrotechnology Transfer) cropping system models (CSM) were incorporated into RZWQM and renamed as RZWQM2 (Ma et al., 2005, 2006). However, RZWQM and RZWQM2 are used interchangeably in the literature. The HERMES crop model was added into RZWQM2 as another option a decade later (Malone et al., 2017; Sima et al., 2020; Ma et al., 2021) (Fig. 1). To improve surface energy balance simulation, the SHAW model was incorporated into RZWQM2 (Ma et al., 2012a, Fang et al., 2014; Qi et al., 2016). Since the DSSAT and HERMES crop models are covered in other chapters (Boote et al., 2025; Nendel et al., 2025), we only detail the generic plant module in RZWQM2 along with some example applications.

Besides the crop modules, RZWQM2 contained several other interrelated modules such as water balance, soil carbon, nitrogen cycle, surface energy

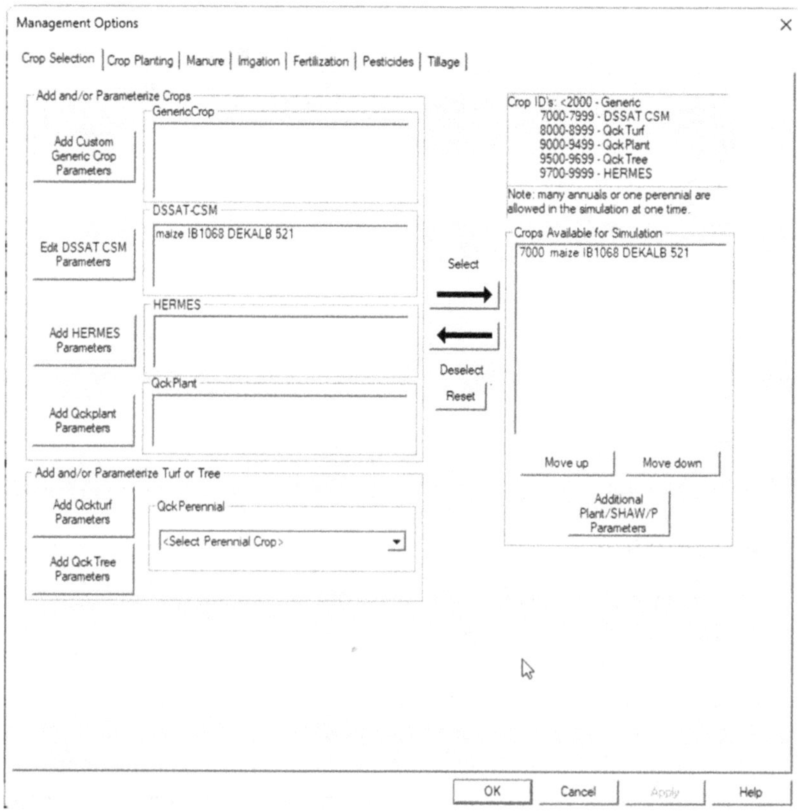

Figure 1 Plant module options in RZWQM2.

Published by Burleigh Dodds Science Publishing Limited, 2025.

balance, soil phosphorus, pesticides, chemical transport, and transformation (Fig. 2). The equilibrium chemistry module simulates soil cations and pH, but it has never been evaluated against experimental data. All the plant growth modules share the same water, nitrogen, and energy balance modules. Although the model runs on a daily time step, water balance is calculated on a sub-hourly basis by solving the Richards' equation, which contains plant water uptake as a sink term and soil potential evaporation (PE) as the upper boundary. Potential evapotranspiration is simulated by the Shuttleworth–Wallace equations with consideration of surface residue and partial canopy effects. Soil organic matter is divided into three organic matter pools (fast, intermediate, and slow) and two residue pools (past and slow) with their own degradation rates. Users may assign inter-pool transfer coefficients to control soil carbon mineralization rate. In addition, three microbial populations are simulated and used to mediate soil carbon and nitrogen cycling (Ahuja et al., 2000a; Hanson et al., 1998; Ma et al., 2000).

For solute transport, RZWQM does not solve the convection-dispersion equation. Instead, chemicals move with water flux and react with soil along the waterways via kinetic sorption and desorption. Soil temperature is simulated by solving the heat equation. Either the PRMS (Precipitation Runoff Modeling System) (Leavesley et al., 1983) or SHAW (Flerchinger and Saxton, 1989a, 1989b) may be used for snow dynamics. Agricultural management practices are simulated with the management module, including irrigation, fertilization, tillage, planting, harvesting, and pesticide applications. Rule-based irrigation and fertilization may be used for decision-making in field management.

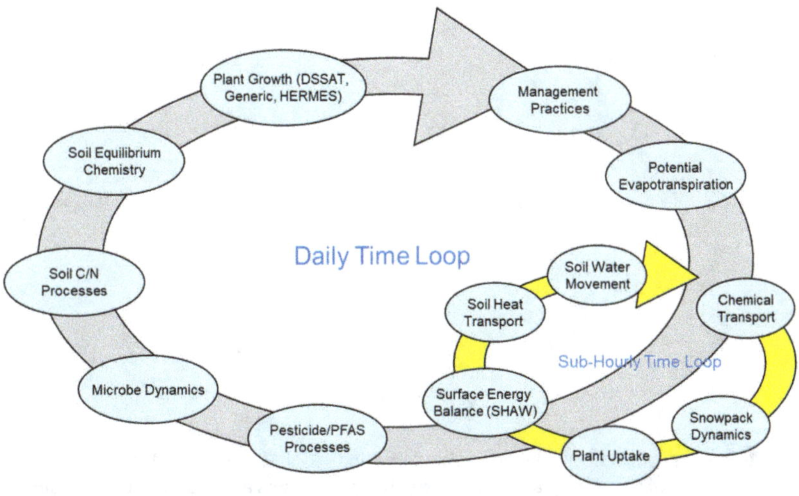

Figure 2 The processes and execution sequence of the RZWQM2 model.

RZWQM2 has a Windows user interface that facilitates data inputs and outputs. It also has an optimization algorithm, PEST (Parameter ESTimation), for model calibration of sensitive model parameters. In addition, the user interface can provide a confidence interval of simulation results based on distributions of input parameters due to spatial and temporal variability.

2 Model components: generic plant module

A major difference from other crop models is its simulation of plant population dynamics. A plant is divided into seven phenological growth stages: dormant seeds, germinating seeds, emerged plants, established plants (in the 4-leaf stage), plants in vegetative growth, plants in reproductive growth, and senescent plants (Fig. 3) (Hanson, 2000). Plant population dynamics are simulated using a modified Leslie-matrix model (Hanson and Olson, 1987). The Leslie matrix assumes that a population life history can be divided into a given number of discrete classes, with each class including a class-specific fecundity rate and a probability of surviving to the next age class. Individuals can move from lower to higher numbered classes if (1) a minimum amount of time has been spent in a class (genetically controlled), (2) the 'proper' growth stage has been realized, and (3) the environment is suitable for such a transition. Environmental conditions control the actual rate of movement between classes. At each daily time step, a given individual can move to the next class, die, or remain in its present class (Fig. 3).

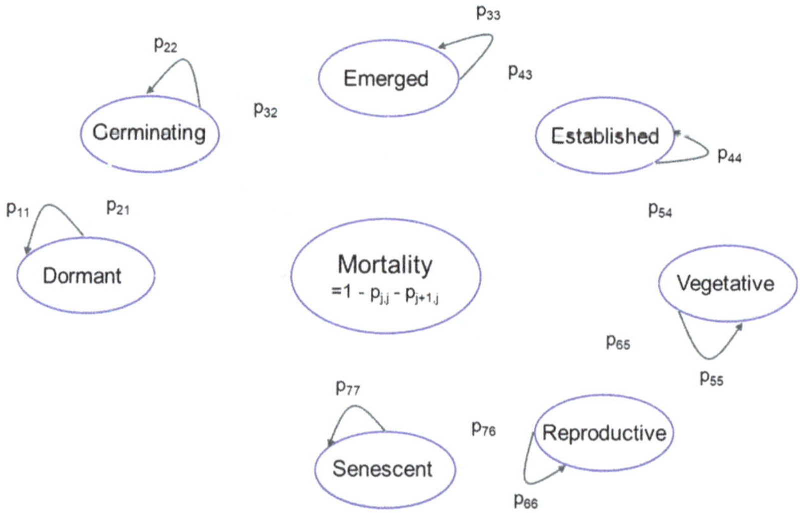

Figure 3 Crop phenological stages simulated in the GPM of RZWQM. The $p_{i,i}$ values are the probability of remaining at the *ith* stage. The $p_{i+1,i}$ values are probabilities transitioning from the *ith* to *i+1th* stage.

Published by Burleigh Dodds Science Publishing Limited, 2025.

Plant growth is simulated by carbon dioxide assimilation, carbon allocation, tissue respiration loss, plant mortality, root growth, water and nutrient uptake, and transpiration (Fig. 4). Currently, nitrogen is the only nutrient simulated. The desired output included information not only on plant growth but also on population development. Coupling a population development model and plant growth resulted in the GPM for RZWQM2.

2.1 Environmental fitness

Most of the crop models presently available use temperature as the primary driving variable for predicting plant growth and development. These models are controlled only by temperature and are used to predict germination rates, emergence, and other aspects of plant development. The most widely used model of this type uses the 'degree-day' concept.

In RZWQM, potential plant growth is modified by temperature, plant water, and nutrient stresses. An 'environmental fitness' (E_t^V), is defined as:

$$E_t^V = E_t^T \min\left(E_t^N, E_t^W\right) \tag{1}$$

where E_t^T, E_t^N, and E_t^W are the effects of temperature, nitrogen, and water on overall environmental fitness on day *t*. Each of these fitness parameters is determined on a scale from 0 (very stressful) to 1 (optimal for plant production).

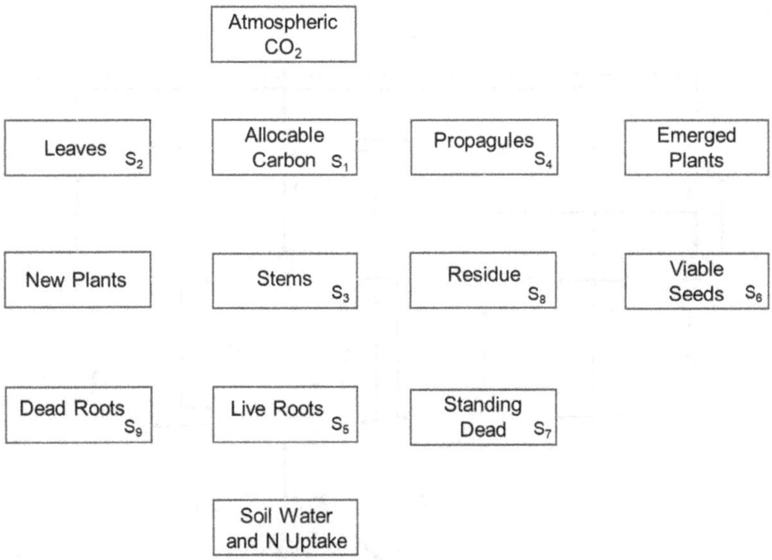

Figure 4 Plant growth and carbon allocation among individual components simulated in the GPM of RZWQM.

The effect of temperature on plant growth is simulated in a manner similar to that used by Hanson et al. (1988) for relating temperatures to various physiological processes. Within the same species, process rates can vary depending on whether soil (T_s, °C) or ambient (T_a, °C) temperature is used by the function in Eq. (2).

$$A_1 = \frac{T_{max} - T}{T_{max} - T_{opt}}, A_2 = \frac{T - T_{min}}{T_{opt} - T_{min}}, A_3 = \frac{T_{opt} - T_{min}}{T_{max} - T_{opt}} \tag{2}$$

Then, the effect of temperature on plant growth (E_t^T) is

$$E_t^T = (A_1 A_2^{A_3})^z \tag{3}$$

where T is the current temperature on day t (T_a or T_s), T_{max}, T_{opt}, and T_{min} (°C) are the maximum, optimal, and minimum temperature for plant growth. The shape parameter z is set at 1.328 (Detling et al., 1979).

Nitrogen affects the whole plant (E_t^N) and the plant photosynthetic mechanism (E_t^P). E_t^N is an estimate of the difference between actual nitrogen concentration (N) and the target nitrogen concentration (N_T) for the whole-plant (Fig. 5) and has the form

$$E_t^N = 1.0 - \frac{N_T - N}{N_T - N_L} \tag{4}$$

where N_L is the percentage nitrogen below which the plant will not grow. The target nitrogen concentration is defined as

$$N_T = (N_X e)^{-bGS} \tag{5}$$

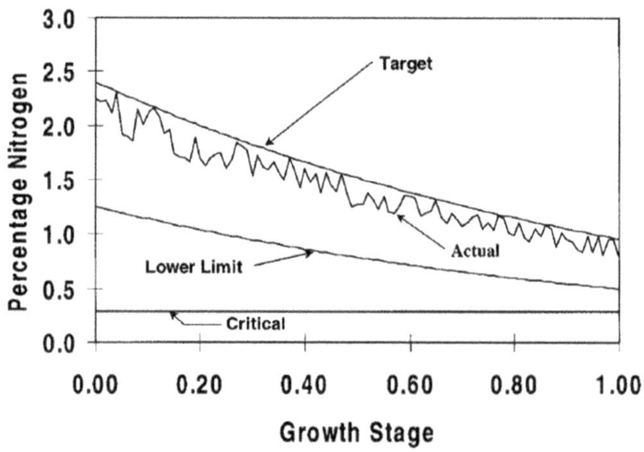

Figure 5 Relationships among targeted (N_T), actual (N), lower limit (N_L), and minimum or critical (N_N) whole-plant N concentrations simulated in GPM.

Published by Burleigh Dodds Science Publishing Limited, 2025.

where growth stage (GS) is the current growth stage at time *t*. The decay coefficient b is

$$b = -\ln\left(\frac{N_{GS=1}}{N_X}\right) \tag{6}$$

$$N_L = N_T \frac{N_N}{N_X e^{-b}} \tag{7}$$

where $N_{GS=1}$ is the percentage N in the plant when the growth stage is 1 and N_X is the maximum percentage N of the emerged plant. N_N is the minimum whole-plant percentage N (%).

For photosynthesis, a different N fitness (E_t^P) is used instead of (E_t^N), which is calculated from the distance between the present and the lower-bound percentage nitrogen in the leaf:

$$N_t^S = N_t^P - N^{min} \tag{8}$$

where N_t^P (%) is the current percentage leaf nitrogen and N^{min} (%) is the lower bound of percentage nitrogen in the leaf below which the plant is unable to assimilate carbon. So, the absolute fitness for nitrogen for photosynthetic activity (E_t^{Nit}) is defined as

$$E_t^{Nit} = \begin{cases} 1-\varepsilon N_t^P, N_t^P > N^{min} \\ 0, \quad Otherwise \end{cases} \tag{9}$$

where ε is the nitrogen-use efficiency coefficient. E_t^{Nit} is subsequently scaled between a lower limit l (e.g. 0.25) and 1.0 to improve overall model stability; therefore, the nitrogen fitness factor for photosynthesis is:

$$E_t^P = E_t^{Nit} + \lambda\left(l - E_t^{Nit}\right) \tag{10}$$

where l is a scaling value between 0 and 0.5.

The current water fitness of the crop (E_t^W) is defined as the ratio between the actual (“ $_a$, cm water day^{-1}) and potential transpiration (PT) rates of the crop (“ $_p$, cm water day^{-1}). It is bound within the generic plant model such that $0.15 \leq E_t^W \leq 1.0$. So,

$$E_t^W = 0.85\frac{\Gamma_a}{\Gamma_p} + 0.15 \tag{11}$$

2.2 Growth stage development

Plant development is measured in terms of GS, which is an index scaled between 0 (seeds) and 1 (a totally mature plant). Under perfect environmental conditions, a minimum amount of time is required for a plant to move from one phenological stage to the next. The rate of advancement between these stages is reduced according to the environmental fitness. If fitness approaches zero, the phenological development of the population and subsequent plant growth can actually terminate. Growth stage is defined as the stage reached by the dominant class at the development rate for the dominant class, reduced by the current environmental fitness

$$GS = \sum_{i=0}^{t} D_j E_i^V \tag{12}$$

where D_j (day^{-1}) is the inverse of the minimum time required to pass through the current average phenological stage under optimum environmental conditions. E_i^V is the environmental fitness at time *i*. Notice that GS is delayed by the past stresses experienced by the plant.

2.3 Plant biography

A deterministic matrix model is used to simulate plant biography in RZWQM. The Leslie matrix (Leslie, 1945) assumes a population life history can be divided into seven discrete age classes (Fig. 3). The transition probability matrix **P** for the RZWQM plant component has the form:

$$P = \begin{vmatrix} p_{11} & 0 & 0 & 0 & 0 & 0 & 0 \\ p_{21} & p_{22} & 0 & 0 & 0 & 0 & 0 \\ 0 & p_{32} & p_{33} & 0 & 0 & 0 & 0 \\ 0 & 0 & p_{43} & p_{44} & 0 & 0 & 0 \\ 0 & 0 & 0 & p_{54} & p_{55} & 0 & 0 \\ 0 & 0 & 0 & 0 & p_{65} & p_{66} & 0 \\ 0 & 0 & 0 & 0 & 0 & p_{76} & p_{77} \end{vmatrix} \tag{13}$$

The values of the **P** matrix are considered constant. Therefore, a new matrix **P′** must be constructed and updated each simulation day to account for the effect of environmental fitness on the progression of phenological development.

$$p'_{j+1,j} = p_{j+1,j} E_t^V \tag{14}$$

Then, the probability of staying in the present class is

$$p'_{j,j} = 1 - p_j^{Mort} - p'_{j+1,j} \tag{15}$$

and, by allowing environmental fitness (E_t^V) to control the development rate by keeping plants in the current stage. The mortality rate for each class (p_j^{Mort}) is constant.

Published by Burleigh Dodds Science Publishing Limited, 2025.

$$p_j^{Mort} = 1 - p_{j,j} - p_{j+1,j} \tag{16}$$

Three elements of **P**′ are calculated independent of E_t^V. Those include $p'_{1,1}$, $p'_{2,1}$, and $p'_{7,6}$. The probability that a dormant seed germinates during time-period t is dependent on the number of stress-free days (the sum over E_t^V), the current 5-day soil water potential, and the current 10-day mean temperature.

The new matrix **P**′ is then used to determine the number of plants in each class at time t+1 ($\mathbf{C_{t+1}}$), by left-multiplication, the vector $\mathbf{C_t}$ of population size for each class is updated at each time-step via by the matrix **P**′, so

$$C_{t+1} = P_t' C_t \tag{17}$$

2.4 Plant growth

The plant production component in RZWQM tracks carbon and nitrogen through the plant (Fig. 4). Initially, new plants are created in response to seed germination. Tissue from newly germinated plants is partitioned into leaves, stems, and roots as a basis for photosynthesis. Carbon dioxide is subsequently assimilated by the plant at a rate dependent on light intensity, canopy structure, and the current environmental fitness. Assimilated CO_2 is stored as allocable carbon, which is immediately partitioned between leaves, stems, roots, and reproductive organs based on the plant growth stage and growth demand of each tissue type. Viable seeds are produced from the propagule biomass pool. Leaf, stem, root, and propagule tissue can be sloughed throughout the growing season. The above-ground tissue passes through the herbaceous dead pool to the herbaceous residue. Root tissue is shed into the dead root pool. The herbaceous residue and dead root material are subsequently linked to the RZWQM soil organic matter module.

The partitioning algorithm used for RZWQM prioritizes carbon allocation as follows: survival of genes, reproduction, growth of existing organs, increase in the number of organs, and final storage. Also, environmental limitations may prevent some priorities from being addressed. The model contains an active mechanism to maintain the carbon balance of the plant following a high stress period. The model also assumes a functional equilibrium that allows growth of organs responsible for assimilating the element most limiting to production, thereby enhancing primary production during times of high environmental stress (e.g. if N is limiting, more roots grow, and if light is limiting, more leaves and stems grow). Photosynthesis is simulated by:

$$PN_t = \frac{\alpha RAD_t P_{max}}{\alpha RAD_t + P_{max}} * EAP * E_t^V \tag{18}$$

where PN_t (g C day^{-1}) is the net carbon assimilation rate, RAD_t (moles PAR m^{-2} day^{-1}) is the solar radiation incident at the top of the canopy, α is the light-use efficiency coefficient, and P_{max} (moles PAR m^{-2} day^{-1}) is the theoretical maximum net assimilation rate. EAP is a 0-1 aging factor (=A_p at propagule initiation and =A_s at seed initiation). The actual photosynthesis rate is reduced by the environmental fitness factor.

Whole-plant photorespiration rate (P_t^{Resp}, g C plant^{-1}) is calculated as a function of plant weight and current day photosynthetic rate. First, the temperature dependent respiration parameter (ξ) is determined from the respiration quotient of the plant (Q_{10}),

$$\xi = \frac{T_a\, \rho\left(Q_{10} - 1\right)}{10 - \rho\left(2\, Q_{10} + 1\right)} \tag{19}$$

where T_a (°C) is the average daily temperature and ρ is the respiration coefficient. Subsequently, whole-plant respiration is estimated as (McCree, 1970):

$$P_t^{Resp} = \phi PN_t + \xi P_t^{Bio} \tag{20}$$

where PN_t (g C day^{-1}) is the daily-assimilated carbon as previously calculated, P_t^{Bio} (g C day^{-1}) is the amount of carbon in the leaves and stems, and ϕ is the proportion of the photosynthate respired for general plant maintenance.

The difference between PN_t and P_t^{Resp} represents the allocable carbon for the plant. This carbon is distributed between the plant components based on a hierarchy of demand. Roots have first access to the available N. Subsequently, all remaining N is hierarchically allocated to the other plant organs. Propagule N demand is met first when the plants are in the reproductive stage. If N is still available after the requirements for roots and propagules are met, the remaining N is partitioned in proportion to the demand by the leaves and stems. Any remaining N is placed into a stored-N pool. The total amount of N in the storage pool is subtracted from the following day's N demand.

Proportion of carbon allocated to the roots (A_t^{Root}) is assumed to go from 0.5 to 0 as GS increases from 0 to 1 under optimal conditions, but was allowed to increase in relation to current water or nitrogen stress. The growth demand for the propagules is met at the appropriate growth rate (ψ, g C day^{-1}).

$$A_t^{Prop} = PN_t \psi \tag{21}$$

After accounting for the carbon allocated to the propagules and roots, the remaining allocable carbon is divided between leaves (A_t^{Leaf}) and stems (A_t^{Stem}) according to the growth stage of the plant. When the plant is in active growth, leaves receive up to 99.5% of the carbon (phenological stages 2 and 3). As the plant grows vegetatively and begins to flower, less of the carbon is allocated to the leaves (96% and 36% for plants in phenological stages 4 and 5, respectively). If the phenological stage is greater than 5, there is no leaf or stem growth. Propagule biomass is transferred to seed biomass (A_t^{Seed}) at a user-supplied rate. Finally, surface litter can accumulate by the breakdown of standing dead material (T_t^{Dead}, g C day^{-1}) and the death of propagule material (T_t^{Seed}, g C day^{-1}).

Up to 50% of the above-ground plant biomass can die during any given day of model simulation. Factors affecting the mortality rate include water stress, freezing temperatures, and tissue age. The actual contribution of each of these factors is determined by input parameters. Carbon lost through leaf (M_t^{Leaf}, g C day^{-1}) and stem (M_t^{Stem} , g C day^{-1}) mortality are passed to the standing dead biomass pool ($P_{7,t}^{C}$, g C day^{-1}). As roots die (M_t^{Root} , g C day^{-1}), the biomass is moved directly to the dead root pool ($P_{9,t}^{C}$, g C day^{-1}).

A dynamic fitness function is used to determine how new root material is distributed into the various soil layers. Roots seek to grow into areas that are the most fit as determined by soil layer temperature, moisture, aeration, and calcium and aluminum concentration. Algorithms are from the CERES-Maize model (Jones and Kiniry, 1986; Jones et al., 1991).

Nitrogen flows between the nitrogen pools ($P_{j,t}^{N} j = 2,9$) in proportion to the nitrogen content of the biomass that is being reallocated. New nitrogen is taken into the plant using passive uptake and active uptake algorithms. The amount of N that passively enters the plant (N_t^{Pass} ppm) is determined by the concentration of N in the soil water that enters the plant (transpiration). Soil water and, therefore, nitrogen are extracted by the soil layer in proportion to the root biomass present. If enough N is not brought into the plant with transpiration, then active N uptake is triggered. In a manner similar to the Michaelis-Menten substrate model, a rectangular hyperbola is used to determine the rate of active N uptake (N_t^{Act} , g C day^{-1})

$$N_t^{Act} = \mu_1 \frac{N_t^{Soil}}{\mu_2 + N_t^{Soil}} \quad (22)$$

where N_t^{Soil} (ppm) is the amount of N available in the soil layer, μ_1 is the maximum proportion of N that can be removed from the soil, and μ_2 the 1/2-maximum nitrogen-uptake amount for the species being simulated. The total amount of additional nitrogen available to the plant through uptake is the sum of N_t^{Pass} and N_t^{Act} .

Nitrogen fixation also brings nitrogen into the plant. At this time, nitrogen fixers are assumed to have met all nitrogen demands regardless of the amount of N in the soil without simulating N-fixature per se. Thus, N-fixature equals the difference between whole-plant demand and plant N uptake.

The model also estimates various plant dimensions by assuming the plant occupies a cylinder. The plant geometric characteristics include estimates of plant height, projected volume, aerial cover, and plant radius. In addition, the leaf area index (LAI) is calculated as a function of the current growth stage and leaf biomass using a specific leaf density (SLD).

3 Model components: the decision support system for agrotechnology transfer cropping system model crop modules

To provide users with other options for plant growth simulation, the DSSAT CSM plant modules were linked to RZWQM2 (Ma et al., 2005, 2006). Initially, CSM 3.5 was incorporated, and later it was upgraded to version 4.5. Recently, the forage, sweet corn, and sugarcane 4.8 were incorporated into RZWQM2. There is a total of 30 CSM crops in RZWQM2, i.e. maize, soybean, winter wheat, barley, millet (foxtail, pearl, proso), potato, sorghum, drybean, fababean, velvebean, brachiaria, cabbage, chickpea, cowpea, peanut, pepper, tomato, cotton, bahia, triticale, canola, sugarbeet, sunflower (oilcrop and cropgro versions), alfalfa, bermudagrass, sweet corn, and sugarcane. The selection of crop modules to be incorporated is primarily based on user demand. However, rice is not included because RZWQM2 does not simulate a water ponding soil environment. These crop modules have contributed to the majority of RZWQM2 applications in recent years. Details on DSSAT CSM are available in Jones et al. (2003) and Boote et al. (2025).

On a daily basis, RZWQM2's soil water and nutrient modules provide DSSAT CSM with soil water content, soil temperature, soil nutrients, and daily Evapotranspiration (ET) for plant growth simulations, and then receive LAI, rooting depth and distribution, biomass, yield, plant height, and crop residue from DSSAT CSM to update plant status at the end of the day. There are two potential ET routines in RZWQM2. One is the Shuttleworth-Wallace equation (Farahani and Bausch, 1995; Farahani and Ahuja, 1996), and the other is the FAO reference ET plus crop coefficient (Allen et al., 2005). The Nimah-Hanks equation or DSSAT water uptake routine may be used for plant water uptake from each soil layer and used as a sink term in solving the Richards' equation for water balance (Nimah and Hanks, 1973; Hanson et al., 1998; Ma et al., 2006), which would ensure mass balance in the soil system. The potential plant water uptake calculated from soil water content and root distribution in DSSAT CSM is the default method used to calculate plant water stress (Ma et al., 2005, 2006). However, users may also use other soil water stresses provided in the model (Kozak et al., 2006; Saseendran et al., 2014, 2015). N uptake from each soil layer is from DSSAT CSM.

To make it easier for DSSAT users, all DSSAT output files are preserved. Plant variables are also output in RZWQM2 format. All crop parameters can be calibrated with PEST (Ma et al., 2011). When a different water stress calculation method (not potential uptake/potential transpiration [PT]) is selected, it is recommended to recalibrate the DSSAT provided cultivar parameters using PEST. In addition, when PT is from the Shuttleworth-Wallace equation and water uptake is from the Nimah-Hanks equation in RZWQM2, it is inevitable to modify DSSAT provided soil parameters as well (Ma et al., 2006).

4 Model components: the HERMES crop module

HERMES is another generic crop model that can be parameterized for any crop. Crops that have been parameterized for are alfalfa, barley (spring and winter), wheat (spring and winter), bermudagrass, sugarbeet, sunflower, cowpea, mustard, maize, oilseed rape, oat, potato, soybean, winter rye, and turnip (Kersebaum 1995, 2007), which simulates different crops represented by external parameter sets. The model and its calibration have been described in more detail by Kersebaum (2011). Briefly, HERMES is a process-oriented functional model designed to simulate crop growth, water, and nitrogen dynamics for arable land. The model has been used under several conditions to calculate nitrogen fertilization recommendations at the field and sub-field scale (Kersebaum and Beblik, 2001; Kersebaum et al., 2003; Michalczyk et al., 2014) and for regional calculations of nitrogen pollution under different management options (Kersebaum, 2000; Kersebaum et al., 2003). It has also been applied in other regions such as the USA (Malone et al. 2017), Canada (Kersebaum et al., 2008), China (Michalczyk et al., 2020), Africa (Falconnier et al., 2020), and New Zealand (Dueri et al., 2022). The model has participated in several international model inter-comparison studies and has shown its capability to simulate crop growth, including climate effects for different crops (e.g. Asseng et al., 2025; Kollas et al., 2015; Falconnier et al., 2020; Nendel et al., 2023).

The crop growth module represents a generic approach, which is able to simulate different crops using external crop parameter files. The module is built on the basis of the SUCROS model (Van Keulen et al., 1982). Driven by global radiation and temperature, the daily net dry matter production by photosynthesis minus respiration is simulated considering the effect of atmospheric CO_2 (Kersebaum and Nendel, 2014). Dry matter production is partitioned among plant tissues (or organs) according to crop development stage, which is calculated from a thermal sum (°C days), modified for each stage by day length and vernalization effects, if applicable for a specific crop. Up to five different crop organs and ten development stages can be defined in the parameter file. Root dry matter is distributed over soil depth according to Gerwitz and Page (1974), with the rooting depth increasing exponentially with the modified thermal sum until a crop and soil-specific maximum is reached. Grain yield is estimated at harvest as a fraction of the weight of the storage organ.

Growth limitations are considered for water and nitrogen stress. Drought stress is simulated using the ratio of actual to PT. Sensitivity of the crop to this ratio can be adjusted in the parameter files for each development stage separately. Temporary limitation of soil oxygen by water logging is simulated through reducing transpiration and photosynthesis according to Supit et al. (1994), depending on the duration of limited air pore volume. For this, a critical percentage of air-filled pores has to be defined in the crop parameter file. Water and nitrogen uptake are simulated from PT and crop nitrogen status, depending on the simulated root distribution and water and nitrogen availability in different soil layers (Kersebaum, 1995). The concept of critical N concentration in plants as a function of crop developmental stage (Kersebaum and Beblik, 2001) or as a function of crop biomass (Greenwood et al., 1990) is applied to assess the impact of N shortage.

5 Model components: the simultaneous heat and water energy balance module

Simultaneous heat and water (SHAW) simulates canopy energy storage and energy balance (Flerchinger et al., 1989a, 1989b; 2012). This addition not only provides an accurate surface energy balance but also simulates canopy temperature to be used for irrigation purposes and plastic mulching effects (Ma et al., 2012a; Fang et al., 2014; Qi et al., 2016; Chu et al., 2022; Zhou et al., 2025). To maintain water balance in the system, PE from the Shuttleworth-Wallace equation is used as the upper boundary for the Richards' equation. Actual evaporation (AE) is obtained by solving the Richards' equation. Actual transpiration (AT) is from SHAW and used as a sink term in the Richards' equation (Chu et al., 2022). Since Actual evapotranspiration (AET) is only simulated between rainfall or irrigation events, SHAW is only called during water redistribution while solving the Richards' equation (Fang et al., 2014; Chu et al., 2022). Along with AE, soil water content, LAI, plant height, surface residue, and root distribution are also input into the SHAW module. In return, SHAW provides RZWQM2 with AT, ice content, frozen soil depth, canopy temperature, and other energy balance components (Fang et al., 2014; Chu et al., 2022) (Fig. 6). Different from the original SHAW model, rainfall interception by leaves and residue water content are not simulated in RZ-SHAW. Snow melting subroutine in SHAW is also used instead of the snow routine adapted from PRMS in RZWQM2. Another important feature in RZ-SHAW is the simulation of surface soil temperature under residue and/or canopy, rather than assuming it to be air temperature, which may improve plant phenology prediction when soil temperature is needed.

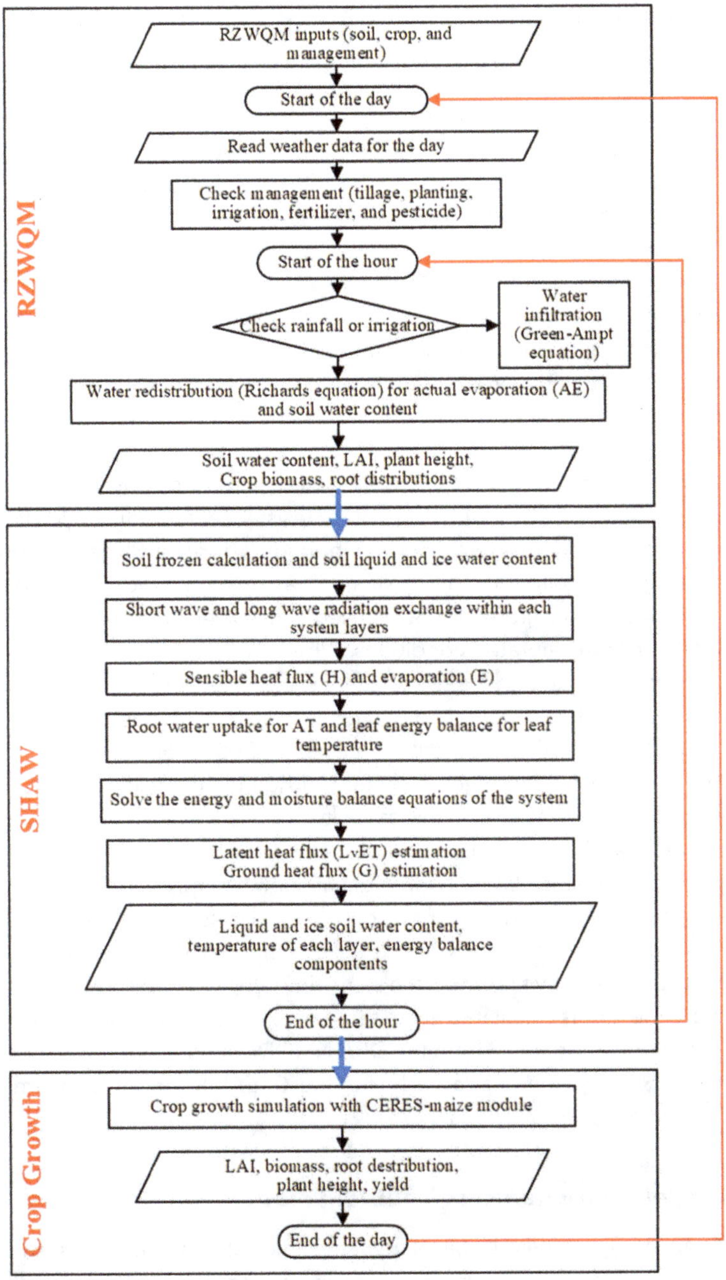

Figure 6 Flowchart between RZWQM2 and SHAW (Chu et al., 2022).

6 Model components: soil water and nutrient modules

6.1 Soil water module

RZWQM2 is a one-dimensional system model. It uses the Green-Ampt equation (Green and Ampt, 1911) for water infiltration during rainfall and irrigation events and solves the Richards' equation for water redistribution afterwards (Ahuja et al., 2000b; Nimah and Hanks, 1973). Soil horizons are converted to finer numerical layers to increase convergence of the Richards' equation. Since the Richards' equation is solved for soil water potential, the Brooks-Corey equations are used to convert water potential to volumetric soil water content, and vice versa. Subsurface tile flow and plant water uptake are treated as sink terms in solving the Richards' equation, whereas subsurface irrigation is treated as a source term.

Potential evapotranspiration is estimated from the Shuttleworth-Wallace equation and is divided into PE and PT based on canopy cover. Crop residue does not absorb water but serves as a resistance to reduce PE. Potential evaporation is used as the flux upper boundary for the Richards' equation. When soil water content cannot sustain PE, a constant head upper boundary is used instead. Unit gradient flow is used as a lower boundary for Richards' equation under unsaturated conditions. When there is a water table present, a constant leakage rate may be specified by users. Snow may accumulate on the surface of soil and/or residue and melt or sublimate. Snowmelt may run off the field or infiltrate into the soil profile (Ahuja et al., 2000b; Hanson et al., 1998).

Surface runoff is generated if rainfall or irrigation intensity exceeds the soil infiltration rate and macropore flow capacity. There are two types of macropores, planar cracks and cylindrical holes. The maximum flow capacity of macropores is estimated from Poiseuille's law with gravity flow (Ahuja et al., 2000b). Flowing water in macropores may be sorbed onto the walls as estimated by the lateral Green-Ampt equation. Plant water uptake is simulated by the Nimah-Hanks equation at sub-hourly time step, the same time step used to solve the Richards' equation. Tile flow is calculated at the same time step using Hooghoudt's steady state equation, as adapted from the DRAINMOD with a pseudo 2-dimensional approach. Soil erosion is adapted from the GLEAMS model (Leonard et al., 1987; Chen et al., 2017; DeMars et al., 2018).

6.2 Model components: soil nutrient module

RZWQM2 has a detailed soil N dynamic module that contains two crop residue pools, three soil organic matter pools, and three soil microbial pools (Shaffer et al., 2000). Crop residue may be incorporated into the soil via tillage or natural decomposition, and then be mineralized partially into CO_2 and NH_4, and the rest is converted to organic matter pools. Distinct rates of decomposition are assigned to each of the five organic matter pools along with inter-pool transfer coefficients between the pools. Users may adjust the inter-pool transfer coefficients to simulate soil organic mineralization rate under different conditions (Ma et al., 1998). Different from most soil nutrient modules, RZWQM2 simulates soil microbial dynamics, which mediates all degradation processes. Release of CO_2 from soil carbon degradation is accounted for in the model and used for the total soil C balance.

The C/N module also simulates nitrification, denitrification, NH_3 volatilization, CH_4 production, and greenhouse gas emission (N_2O, NxO). Plants can take up NO_3 and NH_4, depending on their respective concentrations in soil and plant type in the CSM and GPM, but only NO_3 is taken up in the HERMES model. Soil NH_4 is assumed to be immobile in soil, and NH_4-N can only be transported after nitrification into NO_3, which can move with water into the deep soil profile, in tile drainage, or runoff.

A phosphorus (P) component was recently developed and integrated into the RZWQM2 model, resulting in RZWQM2-P – the first modeling tool capable of simulating both dissolved reactive (DRP) and particulate P (PP) losses via tile drainage under both organic and inorganic P amendments (Sadhukhan et al., 2019). The structure and dynamics of soil P pools in the model are based on the EPIC model (Jones et al., 1984; Vadas et al., 2006), representing five key pools: stable inorganic, active inorganic, labile, fresh organic, and stable organic P (Fig. 7). A major advancement in RZWQM2-P is the addition of specialized surface P pools to represent fertilizer- and manure-derived phosphorus. Specifically, two pools are added for fertilizer P and four for manure P, based on the SurPhos model (Vadas, 2014), enhancing the model's ability to simulate P fates following different amendment strategies. Upon manure application, phosphorus is partitioned into water-extractable, stable inorganic, and organic forms. Tillage operations incorporate a portion of surface-applied manure P into the soil, influencing the distribution among internal soil P pools. The model simulates DRP leaching through the soil profile using the method proposed by Francesconi et al. (2016), while PP transport via tile drainage is modeled using a colloidal transport framework from Jarvis et al. (1999) and Larsson et al. (2007). For subsurface flow, RZWQM2-P applies a groundwater reservoir-based approach (Steenhuis et al., 1997), performing a daily mass balance that accounts for matrix and macropore flow of DRP, and macropore flow exclusively for PP. Recent model updates (Pan et al., 2023a, 2023b) have further refined soil P partitioning dynamics and the representation of tillage effects on P redistribution. A diagram to show the mass and information flow between the P module and RZWQM2 is shown in Fig. 8.

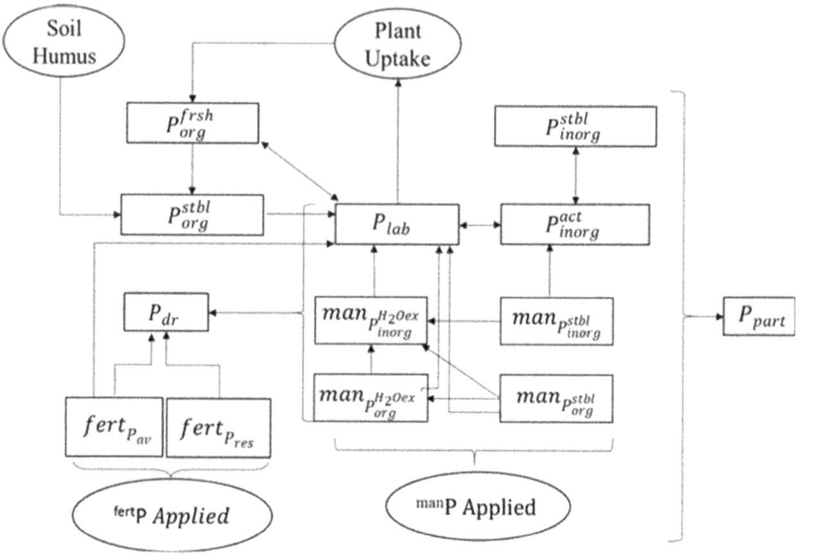

Figure 7 Model's P pools (from Sadhukhan et al., 2019). P_{org}^{frsh}, fresh organic P; P_{org}^{stbl}, stable organic P; P_{inorg}^{stbl}, stable inorganic P; P_{inorg}^{act}, active inorganic P; P_{lab}, labile P pool; ${}^{man}P_{inorg}^{H_2Oex}$, manure water extractable inorganic P; ${}^{man}P_{inorg}^{H_2Oex}$, manure water extractable organic P; ${}^{man}P_{inorg}^{stbl}$, manure stable inorganic P; ${}^{man}P_{org}^{stbl}$, manure stable organic P; ${}^{fert}P_{av}$, available fertilizer P; ${}^{fert}P_{res}$, residual fertilizer P; P_{dr}, dissolved reactive P; P_{part}, particulate P; ${}^{fert}P$, fertilizer P; ${}^{man}P$, manure P.

7 The Root Zone Water Quality Model scenario setup and model calibration

The science in RZWQM2 is coded in FORTRAN, and a user interface in C++ is developed to facilitate model setup. It also contains an NRCS Soil Survey Geographic Database for soils in the 50 US States (USDA-NRCS, 2024), a CLIGEN weather generator for the US states (Nicks et al., 1995), a pesticide database (Wauchope et al., 1992), and a collection of plant cultivar parameters from previous applications (Jones et al., 2003). Soil hydraulic properties may also be estimated from soil texture (Rawls et al., 1982). In addition, to facilitate model parameterization, the PEST algorithm is linked to the RZWQM2 Windows user interface (Fang et al., 2010a, 2019). Users may also conduct a Monte Carlo simulation of selected variables to find the best model parameters or generate confidence intervals due to spatial variability of model inputs. Rule-based irrigation and fertilization may be scheduled for decision-supporting purposes.

The model comes with a user manual to guide model setup and calibration. The Microsoft Windows user interface categorizes input parameters into soil, weather, plant, and management. Default parameters are available from

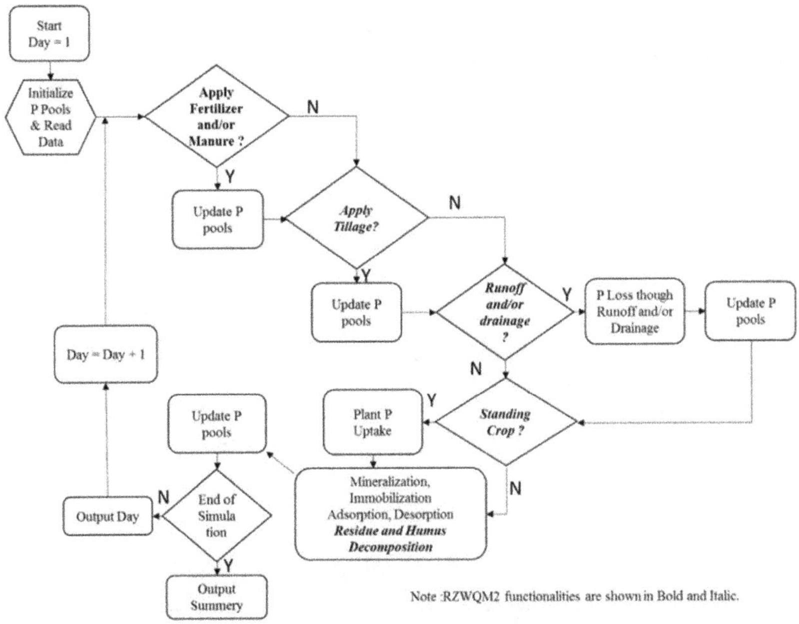

Figure 8 RZWQM2-P model's working algorithm and its dependencies on RZWQM2 (Sadhukhan et al., 2019).

databases and can be fine-tuned with PEST (Fig. 9). To use the PEST optimization feature, the model creates a file to compare with experimental results to be used in a user-defined PEST objective function. Users may assign weights for each experimental value (e.g. soil water content, LAI, biomass, yield) to be included in the objective function or have PEST determine the weights based on the magnitude of the experimental values so that all values contribute equally to the objective function. It is recommended that multiple treatments be included in PEST optimization (Ma et al., 2011).

When there are errors during a model run, they will be logged into the RZWQM2.LOG file. When DSSAT CSM is used, error messages are recorded in the WARNING.OUT or ERROR.OUT files. Since the majority of errors are made in weather and soil inputs, the user interface has built tools to check weather and soil data for consistency. It is also expected that users have knowledge of the model components they selected to be included. For example, if a user wants to use the DSSAT CSM modules, he or she should have some knowledge of the CSM model and data structure. If PEST is needed to calibrate some parameters, it is recommended to understand PEST and its assumptions. Documentation for each model component is available with installation. Users should go through all the soil and plant parameters on the interface or in the input files before engaging in random calibration.

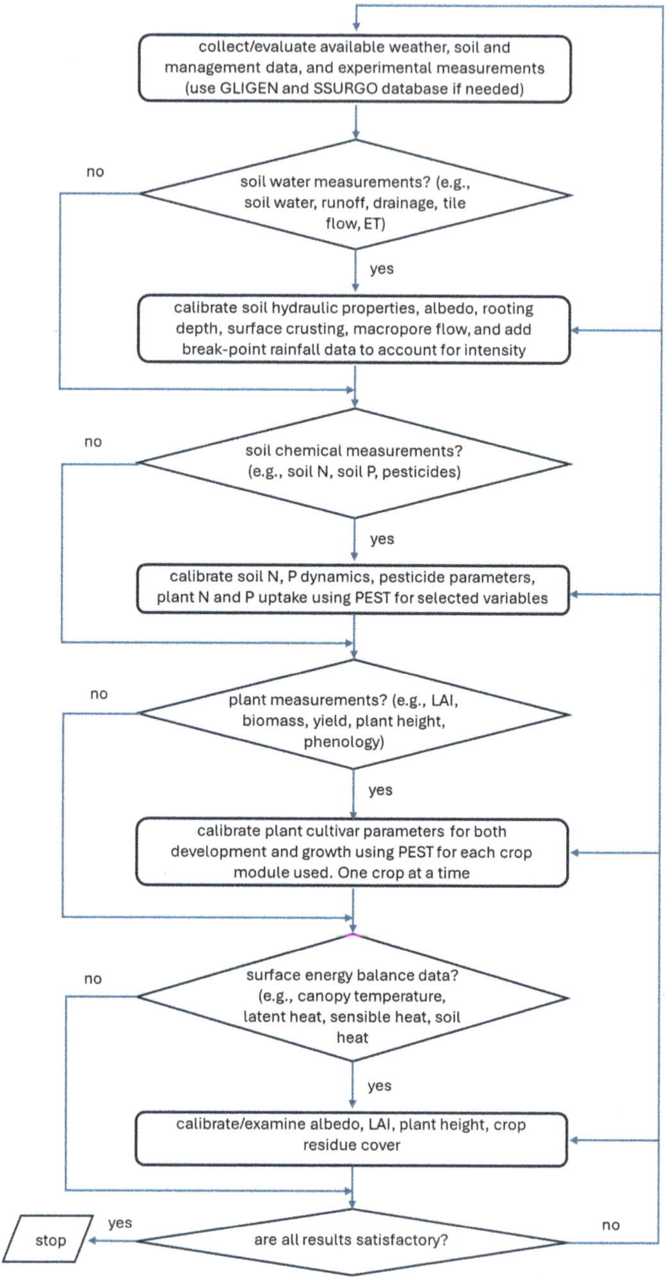

Figure 9 Iterative manual calibration procedure of RZWQM (adapted from Ma et al., 2003).

Published by Burleigh Dodds Science Publishing Limited, 2025.

Model calibration is the most difficult part of the application, even for experienced users. Since all model parameters are empirical and inherited G × E × M (genetic by environment by management) interactions, it is recommended to use multiple treatments to obtain transferable parameters. It is also critical to define an objective function that converges and accounts for all major contributing factors of interest, such as yield, biomass, and soil water. Selecting the most sensitive parameters and limiting the number of parameters will make PEST optimization converge faster. Another feature of the RZWQM2 interface is its ability to sample a group of parameters within a given range using the Latin Hypercube Sampling technique and then run each sampled parameter set. Users can then select a parameter set that meets the criteria set by the user.

Since most crop model applications end up in literature and seldom result in decision-making tools for producers, it has limited value by dividing experimental data into calibration and validation datasets, which may be logical and necessary for model developers, but not for model users. As a user, he or she should derive the best model parameter set for the whole data set, which is usually small and short-term, and then develop a decision tool with the model to address specific objectives of the study. Datasets from multiple locations may be desirable to obtain a parameter set that is transferable across soil and weather conditions.

RZWQM2 may run with minimum input data (Table 1) and use default soil, plant, weather, and pesticide databases. Model outputs are in text file format (Table 2), but the Windows user interface has the capability to view some results graphically (*.PLT files). If experimental data are provided, RZWQM2 will create a file to compare with simulation results in a text file called COMP2EXP.OUT. It is critical to review the RZWQM2.LOG and MANAGE.OUT file to make sure the model runs successfully, and management events are implemented correctly. Users should then check the daily mass balance files, especially MBLWAT.OUT and MBLNIT.OUT for water and nitrogen. Since RZWQM2 solves the Richards' equation for water redistribution at sub-hourly time steps, some users have reported water balance discrepancies on days with either very high or very low rainfall intensities. When this happens, it is suggested to smooth the breakpoint rainfall intensity using the tool in the interface so that they are limited within a reasonable range. Another solution is to smooth soil hydraulic properties between soil layers to avoid abrupt changes in soil texture, water holding capacity, and hydraulic conductivity. It is generally accepted that laboratory-measured soil hydraulic properties are not applicable in the field (Ma et al., 2012b).

Based on experiences collected from RZWQM2 users, it is suggested to calibrate soil water and soil nitrogen first before plant parameters (Ma et al., 2011, 2012c). It is best to exam annual water and N balance components first,

Table 1 Minimum data required to run RZWQM2. All files are managed by a Windows based user interface (adapted from Ahuja et al., 2000a)

Datafile	Minimum data required
Break-Point Rainfall (*.BRK)	Breakpoint rainfall data with a minimum of two pairs of rainfall amounts and times.
Daily or Hourly Meteorology (*.MET or *.HET)	Daily or hourly meteorology data (minimum and maximum air temperature, wind run, solar radiation, and relative humidity).
Snow file (*.SNO)	Snow parameters from PRMS. Users may need to pay special attention to the snowmelt partition between infiltration and runoff.
Site Description (RZWQM.DAT)	• Soil horizon delineation by depth. • Soil horizon physical properties: bulk density, particle size fractions for each horizon. Optional--soil horizon hydraulic properties: 330 or 100 cm-suction water content and saturated hydraulic conductivity if available for each horizon. • Estimate of dry mass and age of residue on the surface. • General pesticide data, such as common name, half-life, Koc, and dissipation pathway (this information can be found in the ARS pesticide database). • Specifying a crop from a database supplied with site-specific parameters. • Management selections and additions as needed.
Initial State (RZINIT. DAT)	• Initial soil moisture contents. • Initial soil pesticides. • Initial soil temperatures. • Initial soil chemistry (pH, CEC, and cations). • Initial nutrient model inputs (soil residue, humus, microbial populations, mineral NO_3-N, NH_4-N; use interface wizards to determine).
DSSAT Files (*.CUL, *.SPE, *.ECO, *.RZX)	Crop cultivar, species, and ecotype parameters and control files. *.RZX file contains rooting depth and paths for cultivar, species, and ecotype files.
GPM file (PLGEN.DAT)	Crop parameters calibrated on the MSEA projects.
HERMES file (HERMES.DAT)	Plant parameters from HERMES applications.
GLEAMS.DAT	Soil erosion parameters.
IPNAMES.DAT	File paths for input files, duration of model run, and climate modifiers.
EXPDATA.DAT	Experimental data inputs.

such as evaporation, transpiration, soil organic N mineralization, plant uptake, and water and N in runoff, leaching, and tile flow (Fig. 9). Users may not have measurements for all the system components, but they should have a rough idea from literature based on soil and weather conditions. If any of the annual variables are out of common knowledge for the site, users should look into

Table 2 RZWQM2 output files

Variable type	Description
Soil water	ACCWAT.OUT: Report accumulative soil water components in soil. MACRO.OUT: Report detailed macropore flow events.
Soil N	NUTRI.PLT: Record all N fates and can be plotted.
Soil P	PFLUX.OUT: P fluxes. PBALANCE_DAILY.OUT: Daily P in soil.
Soil pesticide	PEST1.PLT, PEST2.PLT, PEST3.PLT: Report pesticide fates in soils.
Plant (GPM)	PLANT0.PLT: Daily plant biomass and nitrogen. PLANT1.PLT: Nitrogen in plant tissue. PLANT2.PLT: Plant stress factors. PLANT3.PLT: Number of plants in each growth stage. PLANT4.PLT: Plant height and LAI. PLANT5.PLT: Plant nitrogen demand and supply. NUPTAK.PLT: Daily N uptake.
Plant (DSSAT)	OVERVIEW.OUT: Summary of plant growth. PLANTGRO.OUT: Detailed plant growth outputs. PLANTN.OUT: Report daily plant N in each tissue. Soil water, C, N, and temperature are also output in DSSAT format.
Energy balance (SHAW)	SHAW*.OUT: Report long-wave and short-wave radiation; canopy temperature, canopy moisture, ET, ice content, liquid water, leaf temperature, soil temperature, etc. EBALANCE.OUT: Energy component output, long- and short-wave radiation, sensible heat, latent heat, and soil heat flux.
Erosion	SEDIMENT.OUT: Report sediment and nutrient losses.
Experimental data comparison	EXPDATA.DAT: Record all experimental data to generate COMP2EXP.OUT for evaluating model performance and for PEST parameterization.
Management	MANAGE.OUT: Report implementation of management events.
Mass balance	MBLWAT.OUT: Report daily water balance. MBLNIT.OUT: Report daily nitrogen balance. MBLCARBON.OUT: Report daily soil carbon balance. MBLPHOS.OUT: Report phosphorus balance. MBLP1.OUT: Report mass balance for pesticide #1. MBLP2.OUT: Report mass balance for pesticide #2. MBLP3.OUT: Report mass balance for pesticide #3. MASSBAL.OUT: Report worse mass balance for each component above.
Plot files	DAILY.PLT: Contains scalar variables. (*.ANA contains all scalarvariables). LAYER.PLT: Contains vector variables with soil depth.
Run logging	RZWQM2.LOG: Report model execution and messages. PMODEL.LOG: Report run information related to P simulation. GLEAMS.LOG: Report run information related to soil erosion.

related parameters to correct the unusual simulation results. Once the water and N balance are reasonable, it is time to check phenology, LAI, biomass, and yield simulations. Again, users may not have all the measurements, but it is necessary to check the magnitude of these simulations to make sure they are within experimental ranges in the literature. This calibration is an iterative process as the plant and soil interact with each other. It is also desirable to calibrate RZWQM2 with multiple management treatments to account for the G x E x M interactions.

Goodness of model calibration is assessed with root mean square error (RMSE), coefficient of determination (r^2), relative biases, model efficiency, and the D-index (Ma et al., 2011). Some of the statistics are better descriptors for some variables but worse for others. Relative RMSE, defined as RMSE divided by the average measurement value, is preferable, which is also the one used in PEST optimization. However, it is still hard to define what criteria should be used to call a model successfully calibrated (Moriasi et al., 2007). Recently, an F-test has been proposed to compare model performance and should provide model users with truly testable statistics (Sima et al., 2018).

8 Applications: assessing the impact of agricultural practices

8.1 Model calibration

The Management Systems Evaluation Areas (MSEA) project was initiated in 1990 as part of the Midwest Water Quality Initiative to assess the impact of agricultural practices on water quality, enhance understanding of processes contributing to water contamination, and develop cost-effective strategies to reduce pesticide and nutrient runoff (Watts et al., 1999). The GPM was the only plant module available at that time. In 1991, the MSEA modeling group started to evaluate RZWQM, testing it with limited, commonly collected data. The evaluation of RZWQM covered several MSEA sites in the Midwest and a site in Colorado, influenced by specific weather, soil, crop, and management systems of each location. Soil water content, which was measured at all MSEA sites, was used to calibrate soil hydraulic properties, while soil N levels were both measured and simulated at most sites. Crop production was assessed in Iowa, Nebraska, Ohio, and Colorado. The MSEA project provided valuable insights that benefited both the RZWQM model and future experiments.

The group efforts of applying RZWQM in the MSEA projects showed that an interactive approach was needed to calibrate the generic plant module among soil water, available nitrogen, and plant production. Any errors in the first two components would be reflected by errors in the prediction of yield.

Therefore, the model developers recommended that the user calibrate the soil water component, then the nitrogen component, and finally the plant production component. At each stage of the calibration, the user repeated the preceding steps to make sure predictions in those steps were still reliable. The model user made certain all components of RZWQM were working properly before conducting evaluation experiments (Hanson et al., 1999; Hanson, 2000).

The model calibration was simplified by dividing the plant parameters into two general categories. 'Species-specific parameters' were those parameters set by the model developer and were held constant for each species, i.e. maize and soybean. Five parameters were found to give enough control to capture varietal differences among the MSEA sites. These were called the 'site-specific parameters.' Including maximum active nitrogen uptake (μ_1) (in Eq. 22), daily respiration as a portion of photosynthate (ϕ) (in Eq. 20), SLD, the age effect for plants in the propagule development stage (A_p), and the age effect for plants in the seed development stage (A_s) (in Eq. 18). Increases in μ_1 cause more nitrogen to be taken via active transport. Increasing ϕ causes a decrease in carbon allocated to plant tissue and therefore decreases production. Specific leaf density is a very sensitive plant parameter and should be kept within the bounds established for each crop. Decreasing SLD causes an increase in production. The age effect parameters A_p and A_s can be used to adjust the harvest index for the crop, i.e. the ratio between yield and total aboveground biomass. Increasing these parameters causes an increase in yield in proportion to the total aboveground biomass. The five plant parameters described previously were subsequently determined for each MSEA site (Table 3).

Aboveground biomass, yield, and leaf area index were calibrated to less than 15% for the MSEA project (Tables 4 and 5). Nitrogen in the aboveground biomass was more variable. This variable was not one for which the model was calibrated. As a result, errors of −17.65, −11.68, and 11.43% occurred for Missouri, Nebraska, and Colorado, respectively.

Table 3 Parameter estimation for field maize and summary statistics for the five site specific parameters of the GPM for the five primary MSEA sites (Hanson, 2000)

Parameter	IA	MN	MI	NE	OH	CO^d	CO^i
μ_1	1.5	1.5	2.0	1.5	2.5	1.5	1.5
ϕ	0.15	0.25	0.008	0.19	0.29	0.3	0.28
SLD	10.0	10.0	9.5	9.5	11.50	24.0	12.00
A_p	0.90	0.90	0.97	0.85	0.85	0.92	0.78
A_s	0.85	0.85	0.95	0.64	0.70	0.94	0.78

CO^d is dryland agriculture in Colorado and CO^i is irrigated agriculture in Colorado.

Table 4 Calibration of selected field maize indicator variables for the GPM (Hanson, 2000)

Site	Variable	Total biomass (kg ha^{-1})	Yield (kg ha^{-1})	Nitrogen uptake kg ha^{-1})	Leaf area index
Iowa	Obs.		9578		
	Pred.	18 026	9604	151	5.7
	D%		0.27%		
Minnesota	Obs.		7900	121	
	Pred.	13 389	7423	116	2.5
	D%		−6.04%	−4.13%	
Missouri	Obs.	14 583	7219	153	4.5
	Pred.	13 983	7339	126	4.6
	D%	−4.11%	1.66%	−17.65%	1.78%
Nebraska	Obs.	16 078	8200	214	4.9
	Pred.	16 457	8109	189	4.8
	D%	2.36%	−1.11%	−11.68%	−2.04%
Ohio	Obs.	15 636	9027	212	4.5
	Pred.	15 208	8293	193	4.6
	D%	−2.74%	−8.13%	−8.96%	2.22%
Colorado	Obs.	16 200	7900	175	
	Pred.	16 000	7650	195	4.4
	D%	−1.23%	−3.16%	11.43%	

Variables: Obs = observed value for indicator variable; Pred = GPM prediction of the observed indicator variable; D% = percentage difference between observed and predicted values.

Table 5 Calibration of selected soybean indicator variables for the GPM (Hanson, 2000)

Site	Variable1	Total biomass (kg ha^{-1})	Yield (kg ha^{-1})	Nitrogen uptake (kg ha^{-1})	Leaf area index	Nitrogen in profile (kg ha^{-1})
Minnesota	Obs.		1600	88		
	Pred.	4904	1638	88	1.5	
	D%		−2.32%	0%		
Missouri	Obs.		2868			
	Pred.	6524	2784	92		
	D%		2.93%			
Ohio	Obs.		3360			80
	Pred.	8172	3232	136		73
	D%		3.81%			8.75%

Variables: Obs = observed value for indicator variable; Pred = GPM prediction of the observed indicator variable; D% = percentage difference between observed and predicted values.

Published by Burleigh Dodds Science Publishing Limited, 2025.

8.2 Model validation

Each MSEA site provided data to be used during the validation phase. Data from the Colorado sites were also available. The Iowa and Minnesota sites provided data similar to that used for calibration, but in a subsequent year. The Missouri site provided data from a no-till and a reduced-till management system, even though the model was calibrated to a full-till system. Nebraska provided sites irrigated at 0.5 and 1.5 times the evapotranspiration rate. Ohio provided a 3-year sequence of data. The middle year was used for calibration. The Colorado

Table 6 Evaluation of selected field maize indicator variables for the GPM

Site	Variable	Total biomass (kg ha^{-1})	Yield (kg ha^{-1})	Nitrogen uptake (kg ha^{-1})
Iowa	Obs.	12 289	8760	136
	Pred.	14 457	8021	114
	D%	17.94%	−8.4%	−16.2%
Minnesota	Obs.		4938	69
	Pred.		5100	62
	D%		3.3%	−10.1%
Missouri (No-till)	Obs.	9374	5000	99
	Pred.	13 894	7274	125
	D%	48.2%	45.5%	26.3%
Missouri (Reduced till)	Obs.	11 971	6488	117
	Pred.	14 879	7047	132
	D%	24.3%	8.6%	12.8%
Nebraska (0.5 ET)	Obs.	15 692	8002	203
	Pred.	15 577	7541	163
	D%	−0.7%	−5.8%	−19.7%
Nebraska (1.5 ET)	Obs.	15 680	7997	212
	Pred.	16 484	8082	187
	D%	5.1%	1.1%	−11.8%
Ohio (3-years, 1st year)	Obs.	14 433	6595	183
	Pred.	12 493	5081	210
	D%	−13.4%	−23.0%	14.8%
Ohio (3-years, 3rd year)	Obs.	12 982	6868	
	Pred.	12 880	6651	
	D%	−0.8%	−3.2%	

Variables: Obs = observed value for indicator variable; Pred = GPM prediction of the observed indicator variable; D% = percentage difference between observed and predicted values.

data were collected from one center-pivot system, two dryland side slopes, one dryland summit position, and one toe slope position. Overall, the GPM performed adequately in the validation experiments for the MSEA sites (Table 6).

During validation, errors of more than 15% occurred for the total aboveground biomass for Iowa and for Missouri no-till and reduced-till sites. Yield errors greater than 15% occurred only for the Missouri no-till and Ohio (first year of a 3-year sequence) sites. These errors can be largely explained by the large simulation error in the soil nitrogen component. For Iowa and the Missouri reduced-till sites, nitrogen in the aboveground biomass was predicted with errors of −16.2% and 12.8%, respectively. The Ohio validation is interesting in that the model was calibrated to the second year of a 3-year run. Then, without knowing the initial conditions, the model was run for 3 years using the calibration parameters. Though an error of −23.0% occurred for yield in the first year, predicted aboveground biomass and yield errors were reduced to −0.8% and −3.2% by the third year.

Soybean data was available only for Ohio (3 years) and Missouri (both no-till and reduced-till). The pattern for soybeans was similar to the pattern for maize in Ohio. The first-year simulation underpredicted yield by −29.7% but had an error of only 2.6% by the third year. This again shows the ability of RZWQM to correct for the uncertainty of initial conditions. Soybean yield for no-till and reduced-till management systems in Missouri had errors of 9.8% and 48.5%, respectively.

Therefore, validation results for maize and soybean were good so long as the water and nitrogen dynamics were simulated properly. In cases where large discrepancies occurred in predicting nitrogen or water content in the soil profile, particularly for the Missouri no-till management system, larger deviations occurred in the predicted biomass and yield for the crops tested. The user must make certain that errors in simulating water and nutrient dynamics are minimal if good predictions of yield are desired. Later, another MESA site at Nashua, IA, was extensively studied with RZWQM2 with both GPM (Ma et al., 2007a) and DSSAT CSM (Saseendran et al., 2007). The results showed that both plant modules were able to simulate both maize and soybean production and N losses in tile flow in response to tillage, crop rotation (maize-soybean or maize-maize), and controlled drainage (Ahuja and Hatfield, 2007; Ma et al., 2007b).

9 Applications: irrigation management

Both GPM and DSSAT CSM in RZWQM2 have been used at two USDA-ARS sites in Colorado. One is in Akron and the other is in Greeley. RZWQM2 was used to evaluate crop irrigation responses to current and future climate (Nielsen et al., 2002; Ma et al., 2003, 2012a, 2017; Zhang et al., 2019, 2021), winter wheat-fallow system (Saseendran et al., 2005a), summer crops as alternative to summer fallow (canola, maize, millet) (Saseendran et al., 2010), planting window (Saseendran et al., 2005b), irrigation scheduling in real time (Zhang et al., 2025), surface energy balance under canopy and residue (Kozak et al., 2007; Qi et al., 2016), and N responses of wheat and maize (Saseendran et al., 2004).

One of the early studies with the GPM is that of Nielsen et al. (2002), who evaluated the performance of the GPM in RZWQM for soybean production in Colorado under various irrigation management in Akron, CO. It was calibrated using data from the wettest irrigation treatment and then was used to predict LAI, aboveground biomass, and yield for other irrigation treatments. Model calibration was focused on accurately predicting yield and ET, with attention to soil water content, LAI, biomass, and phenology. For the calibration data set, it correctly simulated the trends in LAI and aboveground biomass (Fig. 10). For the validation datasets, the model overpredicted ET except for the drip irrigation in 1986. Ironically, soybean yield was correctly simulated for those with over-predicted ET (Fig. 11). The reason for such a discrepancy for the

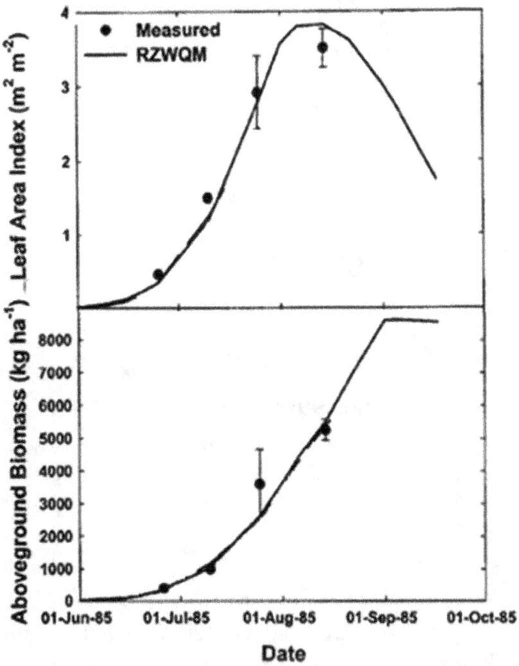

Figure 10 Measured and simulated leaf area index and aboveground biomass from Irrigation Level 4 of the 1985 line-source gradient irrigation data set. Bars are ± one standard deviation. From Nielsen et al. (2002).

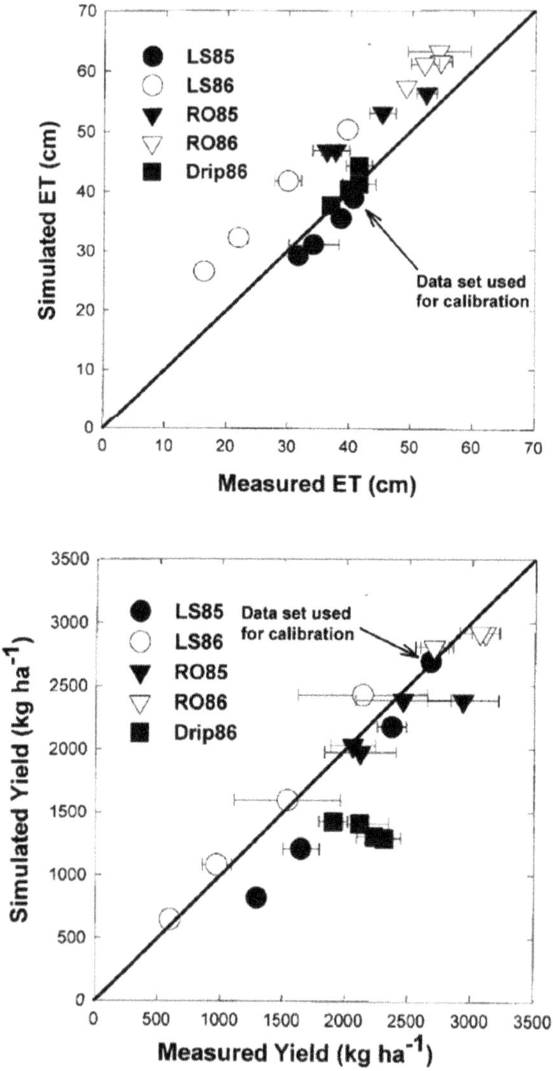

Figure 11 Measured vs. simulated soybean evapotranspiration (ET) and yield, Akron, CO. Bold diagonal line is 1:1 line. LS, line-source gradient irrigation experiment; RO, rainout shelter experiment; Drip, drip irrigation experiment. From Nielsen et al. (2002).

calibration datasets was due to overestimation of soil evaporation, which was also reported in another study in Colorado (Fang et al., 2010a).

In another study in Akron, CO, Ma et al. (2003) evaluated the GPM for maize responses to irrigation and found that the model was adequate in simulating LAI, biomass, and yield once it is calibrated for the wettest treatment based on the manual calibration procedure in Fig. 12. In this study, treatment #4 (wettest) in

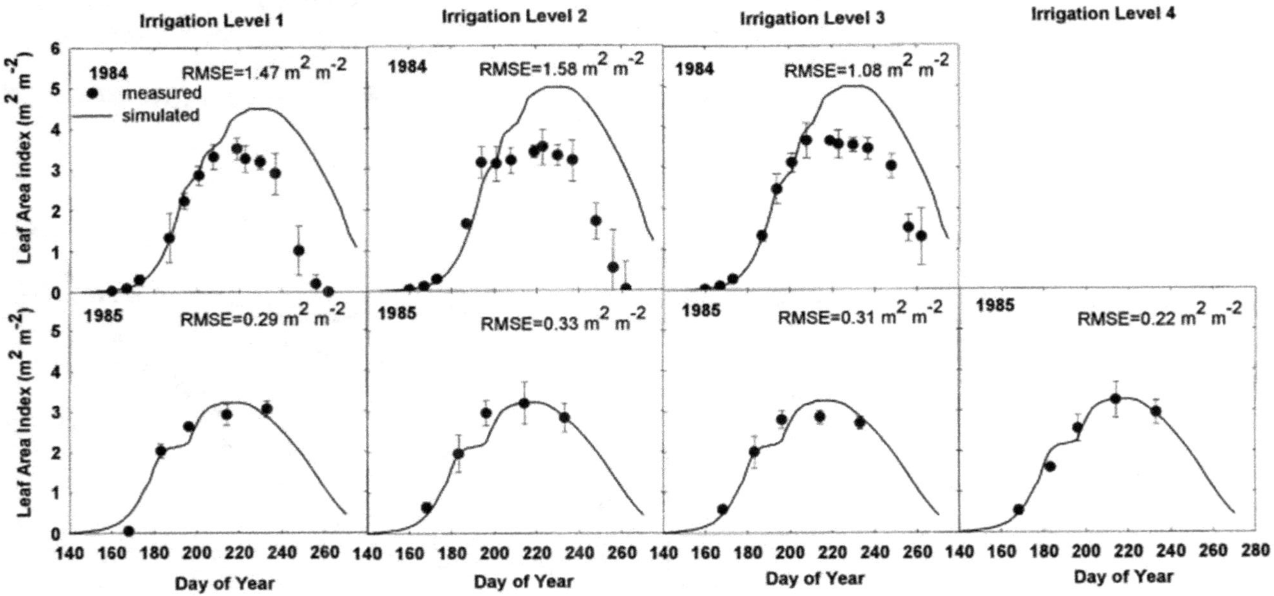

Figure 12 Measured and simulated maize LAI in 1984 and 1985. Vertical bars are standard deviation around the mean. The wettest irrigation treatment (level 4) in 1985 was used for model calibration. From Ma et al. (2003).

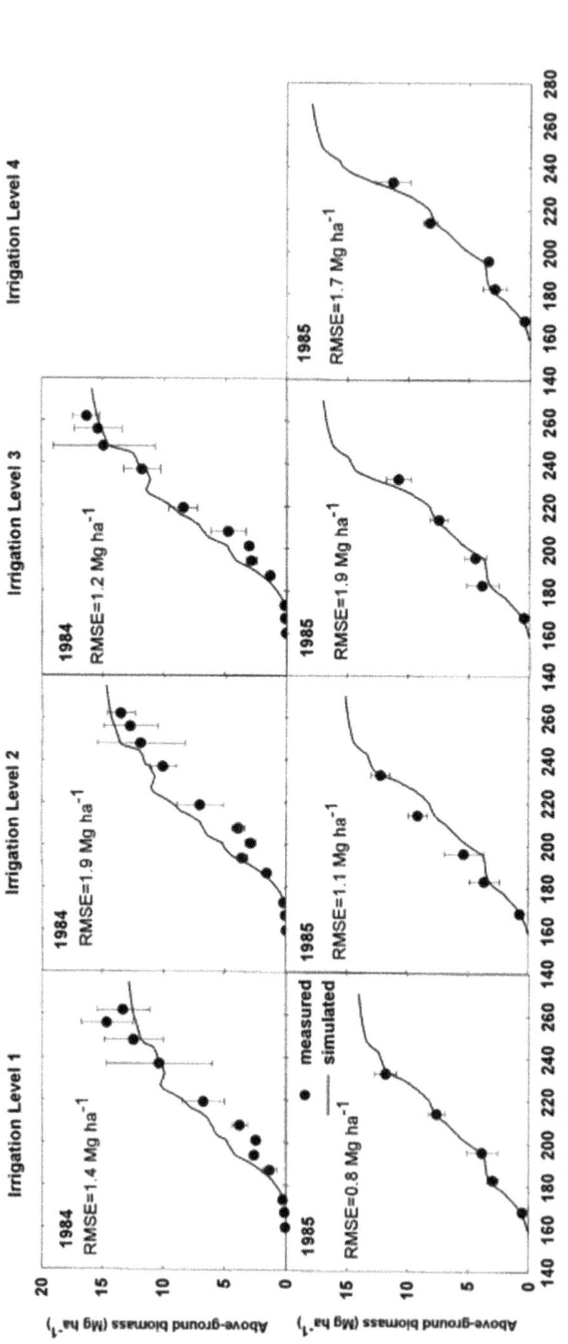

Figure 13 Measured and simulated maize biomass in 1984 and 1985. Vertical bars are standard deviation around the mean. The wettest irrigation treatment (level 4) in 1985 was used for model calibration. From Ma et al. (2003).

Published by Burleigh Dodds Science Publishing Limited, 2025.

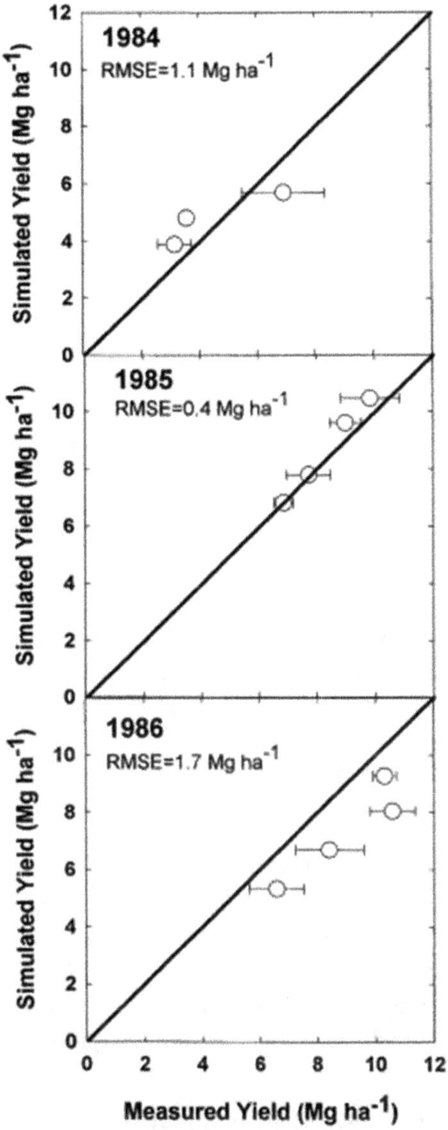

Figure 14 Measured and simulated maize yield in 1984, 1985, and 1986. Horizontal bars are standard deviations around the mean. From Ma et al. (2003).

1985 was used to calibrate the model and the rest was used for model evaluation. As shown in Figs. 12, 13, 14, the GPM simulated LAI, biomass, and yield well for all treatments, except for the overestimation of peak LAI in 1984. Crop ET was also well simulated in the 1985 calibration year (Fig. 15). Unfortunately, there were no soil water measurements in 1984 and 1986 to estimate ET. Thus,

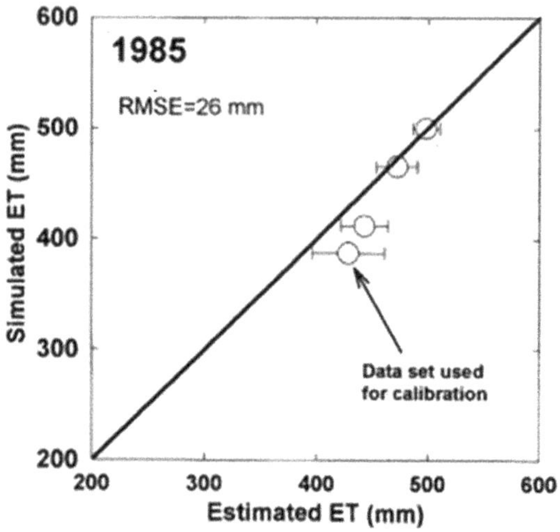

Figure 15 Measured and simulated maize ET in 1985 for all four irrigation levels. From Ma et al. (2003).

we could not see if overestimation of soil evaporation occurred for maize as observed in soybean (Nielsen et al., 2002). Leaf area index and biomass were not available in 1986 to further evaluate model performance.

In another study, using the DSSAT CSM-Maize in RZWQM2, Ma et al. (2012b) successfully simulated 3 years of irrigation treatments in Greeley, CO, from 2008 to 2010. Parameter estimation was used to calibrate the model with all six-treatment data from 2008 (Fig. 16). Model simulated transpiration was closely related to that estimated from FAO56 and crop coefficient (Fig. 17). Once the model was calibrated for all treatments in 2008, It was able to simulate maize production in 2009 and 2010 (Fig. 18). Fig. 18 also showed RZWQM2 simulated similar maize yield using either ET deficit irrigation rules or actual field irrigation amounts. This result indicated that RZWQM2 simulated water requirements were highly in agreement with actual irrigation amounts (Fig. 19), suggesting that RZWQM2 could be used as an irrigation decision support tool.

10 Applications: assessing cropping systems and other applications

RZWQM2 has been used worldwide to address many agronomic and environmental issues, especially in China, Canada, and Portugal (Yu et al., 2006; Hu et al., 2006; Fang et al., 2008; Cameira et al., 2007, 2014). It ranges from irrigation management (Fang et al., 2010b; Yu et al., 2006), nitrogen management (Hu et al., 2006; Fang et al., 2008; Cameira et al., 2007), surface energy balance (Fang et al., 2014; Yu et al., 2007; Chu et al., 2022; Zhou et al.,

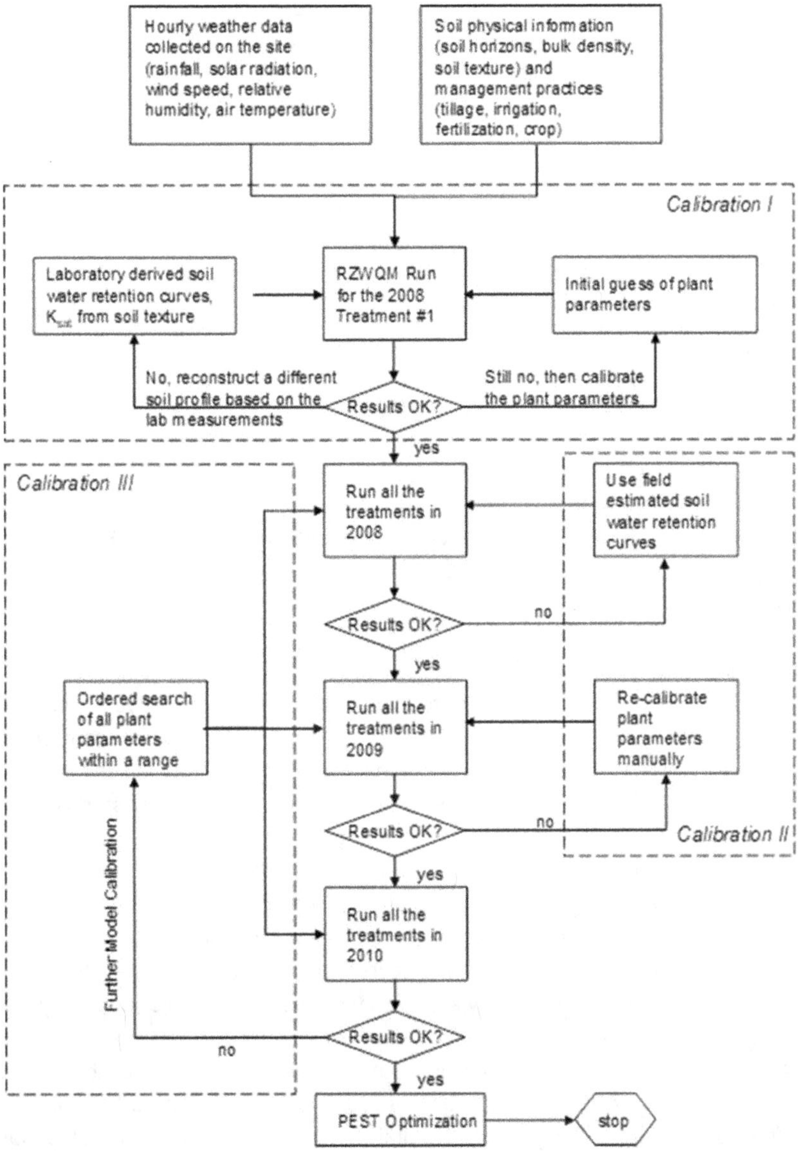

Figure 16 PEST calibration procedure for RZWQM2 in a Colorado study (adapted from Ma et al., 2012b).

2025), climate change impacts (Fang et al., 2013; Chen et al., 2019a, 2025), and decision support in real-time (Chen et al., 2019b, 2020).

In China, Yu et al. (2006) compared the GPM and DSSAT CERES modules (winter wheat and maize) in simulating wheat-maize double-cropping systems

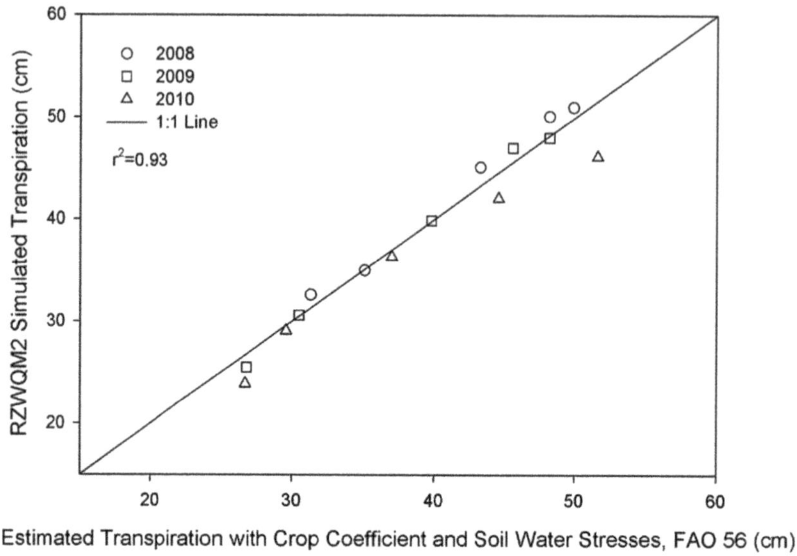

Figure 17 Simulated versus field estimated crop transpiration in 2008, 2009, and 2010. From Ma et al. (2012b).

in China. Both modules (RZWQM-G and RZWQM-C) were calibrated for the 1998–1999 crop season and validated with 1997–1998, 1990–2000, and 2000–2001 seasons because the 1997–1998 season had sparse measurements. There were no irrigation or nitrogen treatments. Both modules simulated biomass and yield adequately (Figs. 20, 21). Simulated soil water storage in the profile had slight differences between the two modules (Fig. 22), suggesting the interaction between the plant module and the soil module.

In Portugal, using GPM, Cameira et al. (2005, 2007) were able to simulate water and N balance in a maize field under two distinct agricultural settings. Both soil and crop parameters were calibrated manually with data from 1996 (silty loam soil) and 1997 (sandy soil) and validated from 1998 (both silty loam and sandy soils). Detailed calibration procedures are documented in Figs. 23, 24 for soil water, plant, and N components. This study was one of the few that reported root water and N uptake (Figs. 25, 26), demonstrating the capability of the GPM in simulating not only total water and N uptake, but also water extraction from each soil layer and N partitioning in plant organs. The GPM seemed to overpredict daily water uptake in the first soil layer (Fig. 25) and daily N uptake at later growth stages (Fig. 26) in the silty loam soil. However, overall water and N uptake were only slightly over-simulated (Cameira et al., 2005, 2007). It would be interesting to revisit the data using the DSSAT CSM-maize in RZWQM2, which has more crop-specific parameters, and use the PEST parameterization algorithm.

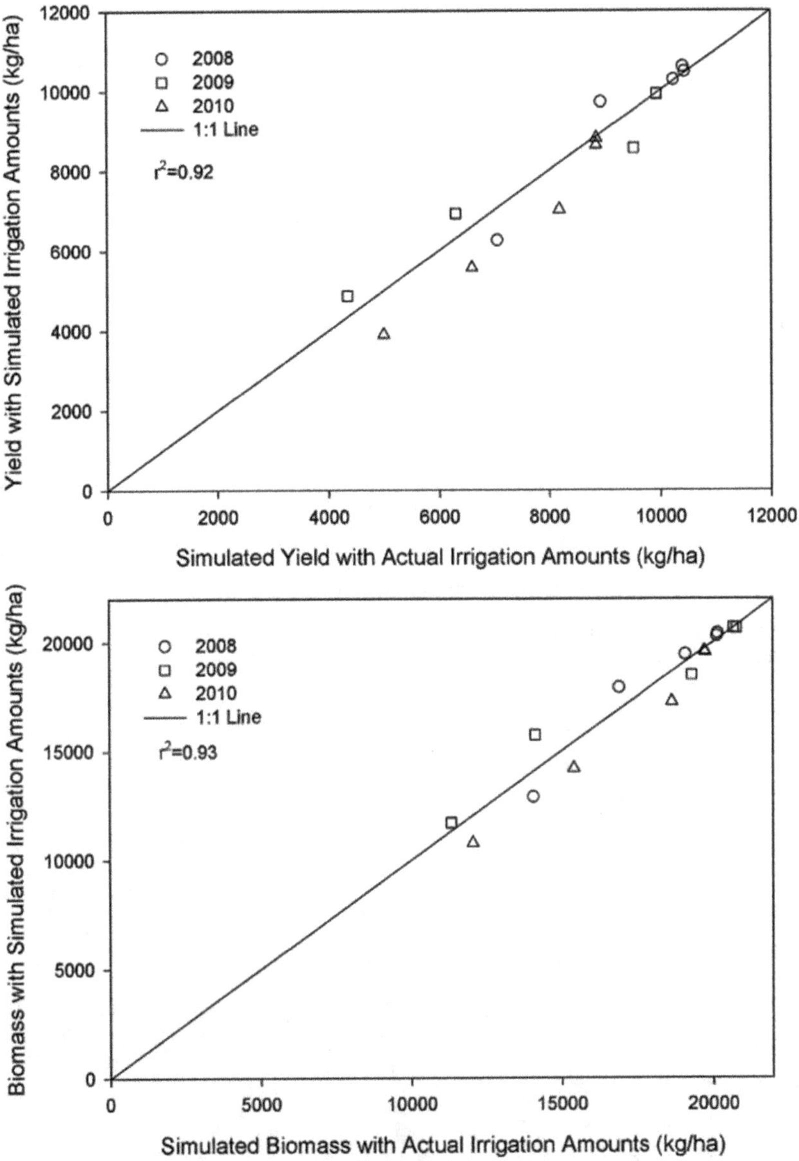

Figure 18 Simulated yield and biomass with simulated irrigation events (rule-based from experimental design) and actual irrigation events. From Ma et al. (2012b).

There have been various studies using RZWQM2 for surface energy balance, especially under plastic mulching (Zhou et al, 2025; Chu et al., 2022). Fang et al. (2014) evaluated surface energy balance with the SHAW and CSM modules in RZWQM2 for winter wheat and maize with data from 2003

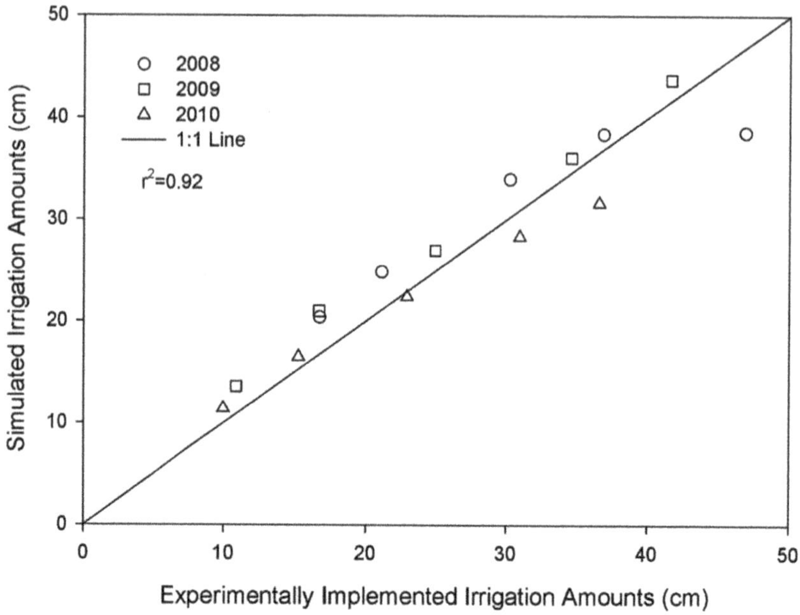

Figure 19 Irrigation amount simulated by RZWQM2 from rule-based irrigation design in 2008, 2009, and 2010. (Ma et al. 2012b).

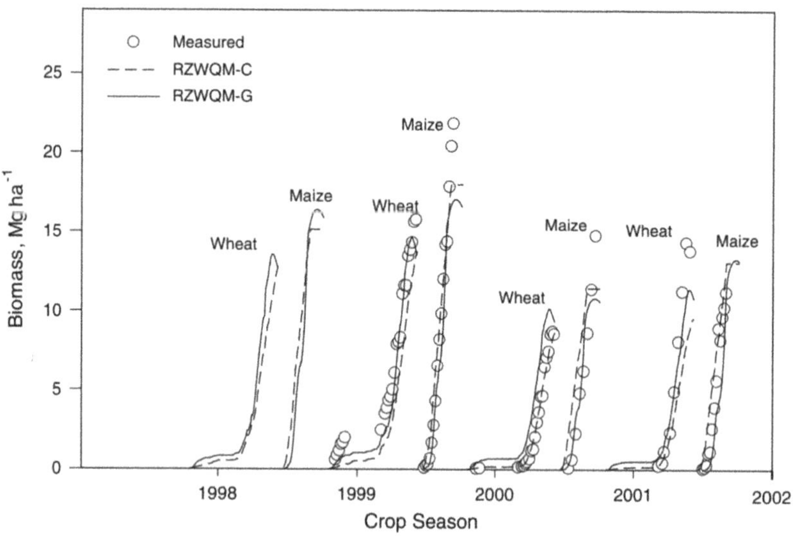

Figure 20 Comparison of measured (5-day intervals) and RZWQM-G and RZWQM-C model simulated crop biomass of winter wheat and maize from 1997–2001. No measurements were taken in 1997–1998 growing season. From Yu et al. (2006).

Published by Burleigh Dodds Science Publishing Limited, 2025.

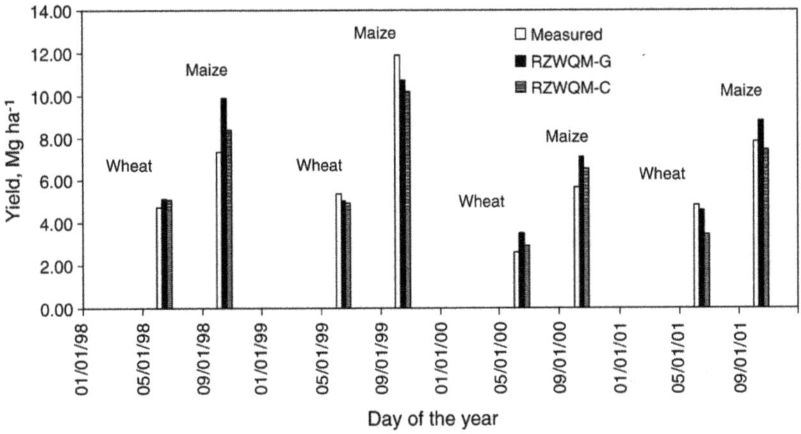

Figure 21 Comparison of measured and RZWQM-G (GPM) and RZWQM-C (DSSAT CSM) model simulated grain yield from 1997 to 2001. From Yu et al. (2006).

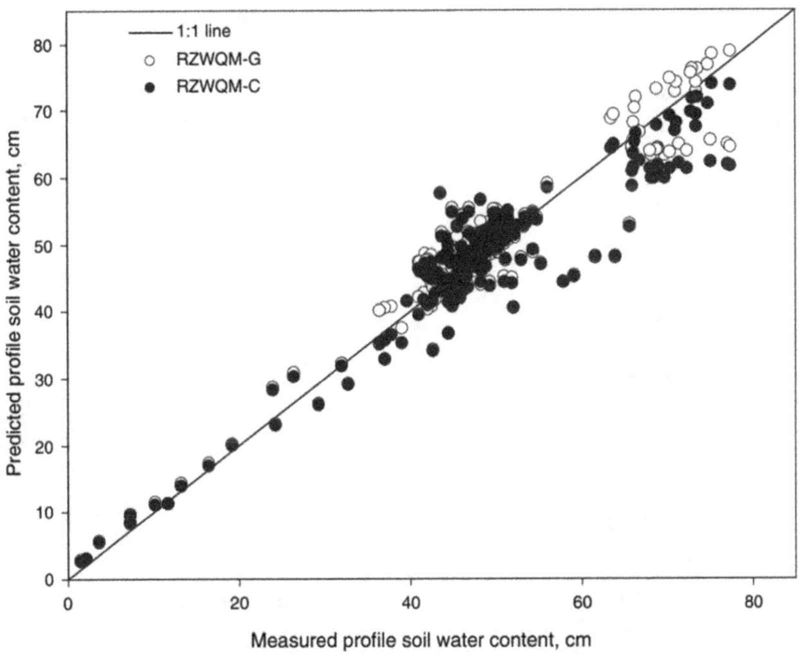

Figure 22 Comparison of measured and RZWQM-G (GPM) and RZWQM-C (DSSAT CSM) simulated total profile soil water from 1997–2001. Measurement soil profile depths varied between 10 and 220 cm. From Yu et al. (2006).

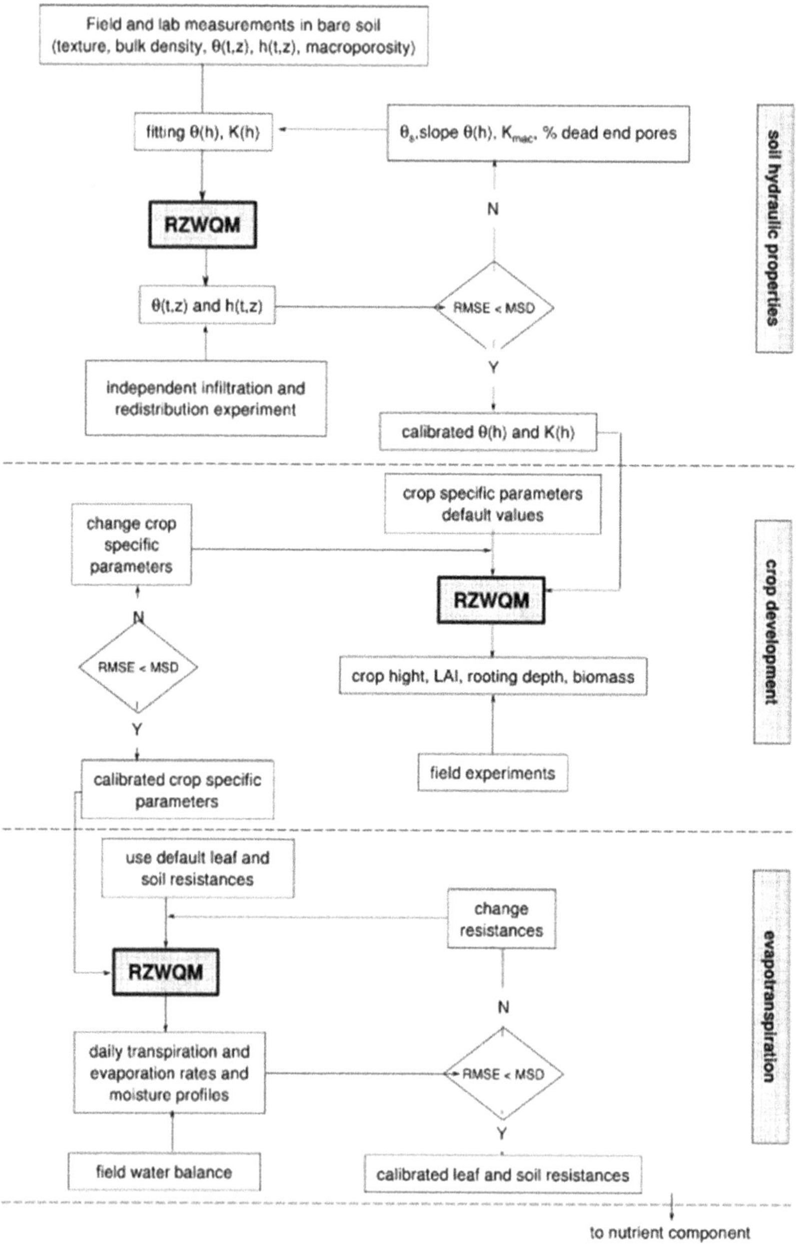

Figure 23 Calibration of RZWQM for maize production in Portugal. From Cameira et al. (2005).

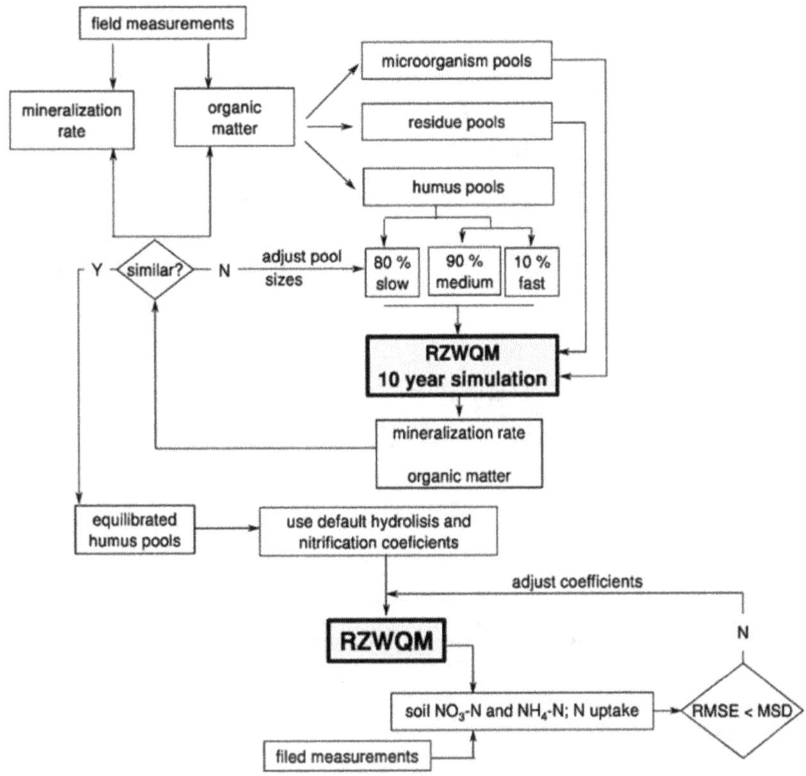

Figure 24 Flow chart describing the calibration of the nitrogen component of RZWQM. From Cameira et al. (2007).

to 2005 in a double-cropping system in China. The excellent simulation of LAI and canopy height (Fig. 27) contributed to a good simulation of energy components and canopy temperature for winter wheat (Fig. 28) and maize (Fig. 29). Similar results were reported for soybeans in an Iowa study in the US (Ma et al., 2012a).

11 Software availability and future development

The model is available for download at https://www.ars.usda.gov/research/software/download/?softwareid=412&modecode=30-12-30-25. The science code is programmed in Fortran and available from the senior author upon request. Questions on using RZWQM2 may be emailed to the senior author. ARS scientists will support the model in the near future.

It has been 33 years since its initial release in 1992, and it has gone through many milestones, such as incorporating the DSSAT CSM, HERMES,

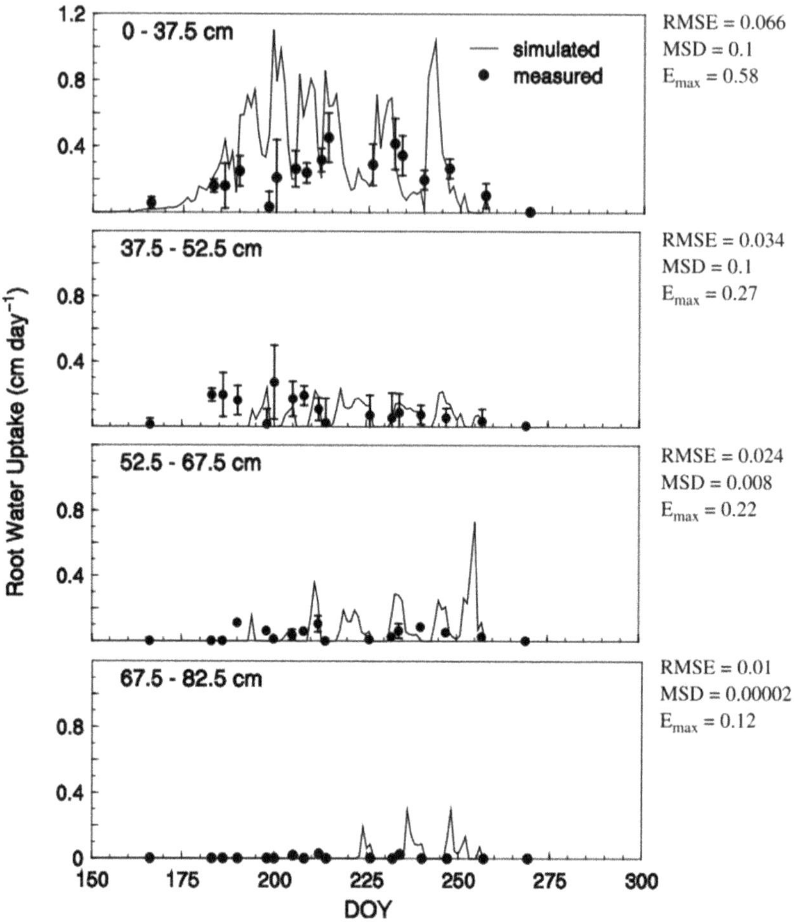

Figure 25 Simulated and measured root uptake by soil layer for the 1996 calibration year on the silty loam soil. From Cameira et al. (2005).

SHAW, GLEAMS' soil erosion, and EPIC's P module. The development of the Windows user interface in the late 1990s has greatly increased RZWQM2 usability. Recently, the PEST optimization algorithm has been extended to GPM and HERMES crop parameters. The DSSAT CSM in RZWQM2 was updated to version 4.5 for all crops, but some new additions (sugarcane, sweetcorn, and forage) are version 4.8. It is desirable to update all CSM modules to version 4.8 in the future because of the P uptake routine, but some known errors in 4.5 have been corrected by working with the DSSAT Team. If time permits, the DSSAT weed/disease module may be included. Another potential improvement is to update PEST to PEST++ to make model parameterization more robust. Future

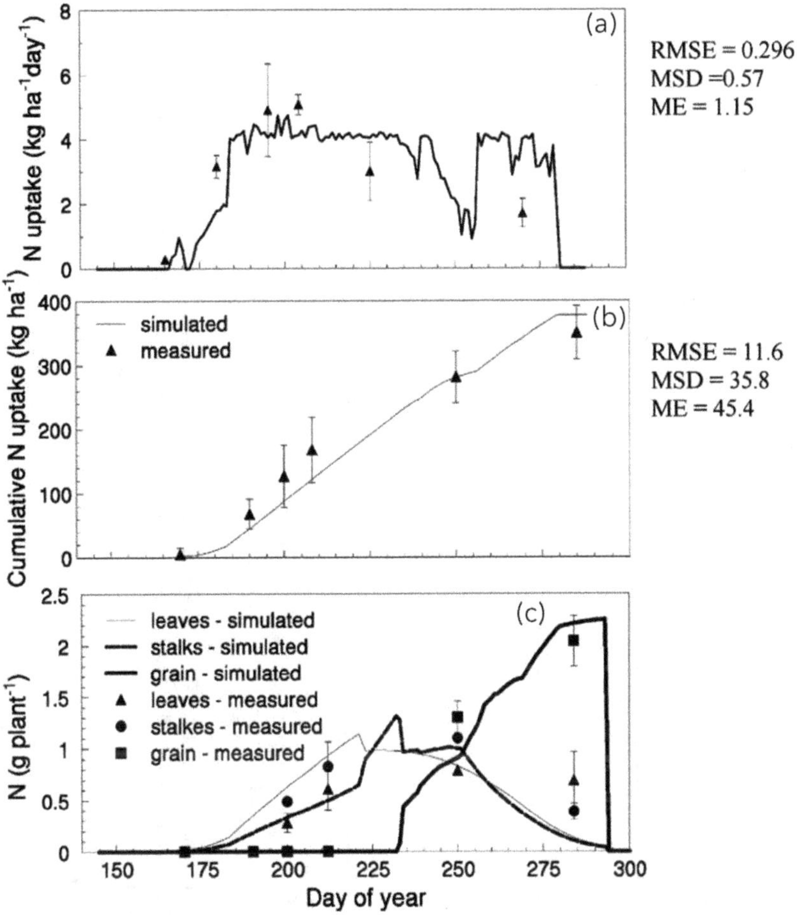

Figure 26 Nitrogen uptake daily (a) and accumulative (b), and N partitioning in plant components. Calibration results for the silty loam soil (1996). From Cameira et al. (2007).

development will be focused on improving current modules, rather than adding new ones.

12 Conclusion

RZWQM2 started as a system model to provide a detailed simulation of physical, chemical, and biological processes in the root zone to evaluate management effects on crop production and environmental quality. The three plant growth modules make the model attractive to users of various backgrounds. It is one of the few models that can simulate water balance, nutrient balance, pesticide fates, plant growth, and energy balance under a wide range of soil, weather, and

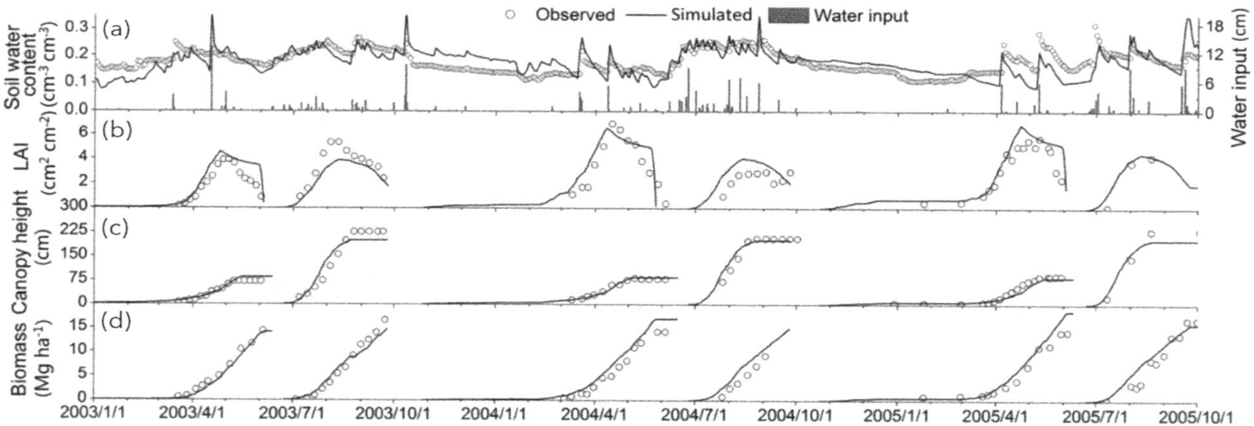

Figure 27 Comparisons between observed and simulated daily soil water content averaged from 10 cm and 30 cm depths, leaf area index, canopy height, and aboveground biomass by RZ-SHAW in the wheat (January–June) and maize (July–October) rotation system from 2003 to 2005 (Water input includes rainfall and irrigation). From Fang et al. (2014).

Published by Burleigh Dodds Science Publishing Limited, 2025.

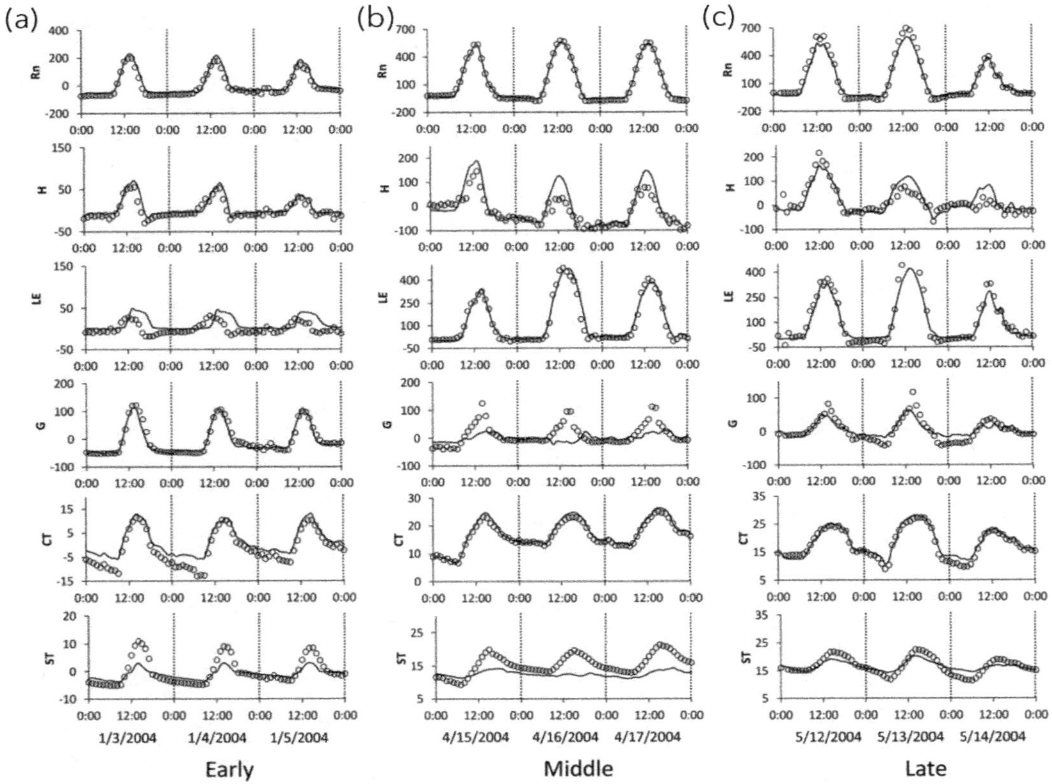

Figure 28 Observed and simulated diurnal net radiation (Rn, Wm^{-2}), sensible heat (H, Wm^{-2}), latent heat (LE, Wm^{-2}), ground heat flux (G, Wm^{-2}), crop canopy temperature (CT, °C) and soil temperature at 5 cm depth (ST, °C) by the hybrid RZ-SHAW during early (a), middle (b) and late (c) wheat seasons in 2004. From Fang et al. (2014).

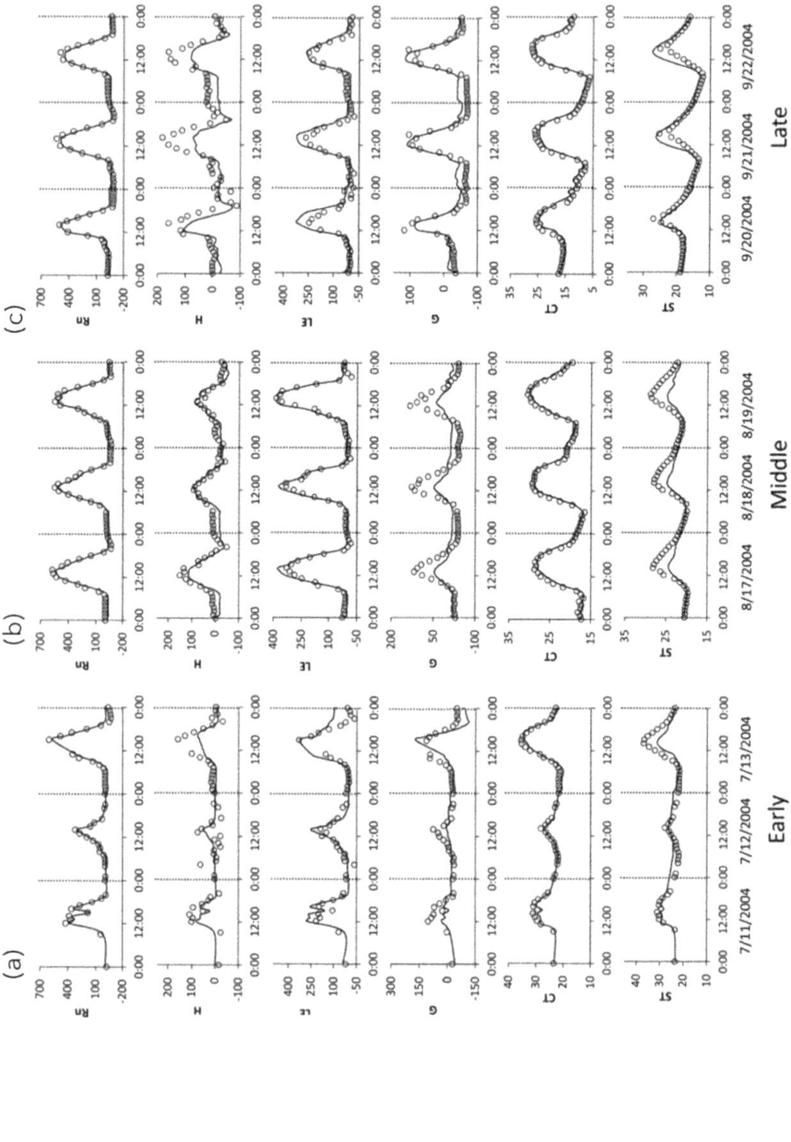

Figure 29 Observed and simulated diurnal net radiation (Rn, Wm^{-2}), sensible heat (H, Wm^{-2}), latent heat (LE, Wm^{-2}), ground heat flux (G, Wm^{-2}), crop canopy temperature (CT, °C) and soil temperature at 5 cm depth (ST, °C) by the hybrid RZ-SHAW during early (a), middle (b) and late (c) maize seasons in 2004. From Fang et al. (2014).

management conditions. Not only was it developed based on models before it (NTRM, GLEAMS, CREAMS, EPIC, and Opus), but it also consistently improved itself by adapting new science modules from existing models (DSSAT, HERMES, and SHAW). The Windows user interface has greatly expanded the RZWQM2 user base, especially after an automated parameterization algorithm, PEST, was integrated into the interface for calibration and uncertainty analysis. Recently, an effort has been made to include GPM and HERMES parameters in the PEST scheme for easy calibration. The inclusion of multiple plant growth modules in RZWQM2 to share the same soil, water, nutrient, and management environment should help crop modelers compare crop growth modules objectively and identify knowledge gaps in crop modeling.

As of today, RZWQM2 is being used to address many agronomic and environmental issues with close to 400 journal publications based on www .scopus.com. However, the majority of these applications are to explain the experiments themselves and are seldom extended to other locations or conditions or used for decision support (Chen et al., 2019b, 2020; Zhang et al., 2025). Even though PEST is integrated into the RZWQM2 interface to facilitate model calibration, it is still a formidable task for casual users. Parameter ESTimation has its limitations in defining an objective function that converges monotonically. Oscillation and nonconvergence are commonly encountered by users when too many parameters are included in the process or experimental data are sparse. What makes the situation worse is the lack of uniqueness of the parameters derived, and users cannot determine whether the parameters obtained are the best. They often must accept whatever parameters they get (manually, PEST, or in combination) as long as the simulation results appear reasonable, even if the parameters themselves may not be reasonable. More often than not, users do not verify all model outputs to ensure all processes are being simulated reasonably; instead, they tend to focus only on results for which experimental data are available or with which they are most familiar. (Ma et al., 2011).

With the retirement of model developers, most models will have to evolve into a modular form for others to adopt or improve. Otherwise, they will disappear from the modeling community. Since most crop models use old programming languages (e.g. Fortran), it is urgent to develop modules using newer languages in Python and R packages for new generations to learn and advance science in these models. Most importantly, these packages can be used to develop task-specific tools to solve individual problems, rather than a mega model to address all problems. Combining system models with real-time data (e.g. LAI or soil moisture from remote sensing, Zhang et al, 2025) will provide new opportunities for crop models.

13 References

Ahuja, L. R. and Ma, L. 2002. Parameterization of agricultural system models: current approaches and future needs. In: L. R. Ahuja, et al. (eds.), *Agricultural System Models in Field Research and Technology Transfer*. Lewis Publ., Boca Raton, FL, pp. 273–316.

Ahuja, L. R. and J. L. Hatfield. 2007. Integrating soil and crop research with system models in the Midwest USA: purpose and overview of the special issue. *Georderma*. 140: 217–222.

Ahuja, L. R., et al. 2000a. *The Root Zone Water Quality Model*. Water Resources Publications, Highlands Ranch, CO.

Ahuja, L. R., et al. 2000b. Water and chemical transport in soil matrix and macropores. In: L. R. Ahuja, et al. (eds.), *The Root Zone Water Quality Model*. Water Resources Publications, Highlands Ranch, CO, pp. 13–50.

Allen, R. G., et al. 2005. FAO-56 dual crop coefficient method for estimating evaporation from soil and application extensions. *J. Irrig. Drain. Eng*. 131: 2–13.

Asseng, S., et al. 2025. The DSSAT crop model ecosystem. In: G. Hoogenboom (ed.), *Current Crop Models: State-of-the-Art and Future Developments*. Burleigh Dodds Science Publishing, Cambridge, Chapter 1.

Boote, K. J., et al. 2025. The Decision Support System for Agrotechnology Transfer (DSSAT) crop modeling ecosystem: evolution and advances. In Hoogenboom, G. (Ed.), *Current crop models: State-of-the-art and future developments*. Burleigh Dodds Science Publishing, Cambridge, UK.

Cameira M. R., et al. 2005. Simulating the fate of water in field soil –crop environment. *J. Hydro*. 315: 1–24.

Cameira, M. R., et al. 2007. Simulating fate of nitrogen in field soil-crop environment in the Mediterranean Region. *Agric. Water Manag*. 90: 121–136.

Cameira, M. R., et al. 2014. Sustainability and environmental assessment of fertigation in an intensive olive grove under mediterranean conditions. *Agric. Water Manag*. 146: 346–360.

Chen, H., et al. 2017. Numerical simulation of agricultural sediment and pesticide runoff: RZWQM and PRZM comparison. *Hydrol. Processes*. 31: 2464–2476.

Chen, X., et al. 2019a. Simulating the impacts of climate change on cotton yield and water requirement using RZWQM2. *Agric. Water Manag*. 222: 231–241.

Chen, X., et al. 2019b. A model based real-time decision support system for irrigation scheduling to improve water productivity. *Agron*. 9: 686.

Chen, X., et al. 2020. Evaluation of a new irrigation decision support system in improving cotton yield and water productivity in an arid climate. *Agric. Water Manag*. 234: 106139.

Chen, X., et al. 2025. Potential deficit irrigation adaptation strategies under climate change for sustaining cotton production in hyper–arid areas. *Agric. Water Manag*. 312: 109417.

Chu, X., et al. 2022. Development of RZ-SHAW for simulating plastic mulch effects on soil water, soil temperature, and energy balance in a maize field. *Agric. Water Manag*. 269: 107666.

DeMars, C., et al. 2018. Integrating GLEAMS sedimentation into RZWQM for pesticide sorbed sediment runoff modeling. *Environ. Modell. Softw*. 109: 390–401.

Detling, J. K., et al. 1979. A simulation of *Bouteloua gracilis* dynamics on the North American shortgrass prairie. Oecologia. 38: 167–191.

Dueri, S., et al. 2022. Simulation of winter wheat response to variable sowing dates and densities in a high-yielding environment. *J. Exp. Bot.* 73(19): 6612–6627.

Falconnier, G. N., et al. 2020. Modelling climate change impacts on maize yields under low nitrogen input conditions in sub-Saharan Africa. *Glob Chang Biol.* 26: 5942–5964.

Fang, Q. X., et al. 2008. Modeling nitrogen and water management effects in a wheat-maize double-cropping system. *J. Environ. Qual.* 37(6): 2232–2242.

Fang, Q. X., et al. 2010a. Optimizing soil hydraulic parameters in RZWQM2 using automated calibration methods. *SSSAJ.* 74: 1897–1913.

Fang, Q. X., et al. 2010b. Irrigation strategies to improve the water use efficiency of wheat–maize double cropping systems in North China Plain. *Agric. Water Manag.* 97: 1165–1174.

Fang, Q. X., et al. 2013. Quantifying climate and management effects on regional N leaching and crop yield in the North China Plain. *J. Environ. Qual.* 42: 1466–1479.

Fang, Q. X., et al. 2014. Modelling evapotranspiration and energy balance in a wheat-maize cropping system using the revised RZ-SHAW model. *Agric. Forest Meteol.* 194: 218–229.

Fang, Q. X., et al. 2019. Uncertainty of model calibration using the PEST optimization algorithm. *Agron.* 9: 241.

Farahani, H. J. and Ahuja, L. R. 1996. Evapotranspiration modeling of partial canopy/residue-covered fields. *Trans. ASAE.* 39: 2051–2064.

Farahani, H. J. and Bausch, W. C. 1995. Performance of evapotranspiration models for maize - bare soil to closed canopy. *Trans. ASAE.* 38: 1049–1059.

Flerchinger, G. N. and Saxton, K. E. 1989a. Simultaneous heat and water model of a freezing snow-residue-soil system I. Theory and development. *Trans. ASAE.* 32: 565–571.

Flerchinger, G. N. and Saxton, K. E. 1989b. Simultaneous heat and water model of a freezing snow-residue-soil system II. Field Verification. *Trans. ASAE.* 32: 573–576.

Flerchinger, G. N., et al. 2012. Simultaneous Heat and Water Model: Model use, calibration, and validation. *Trans. ASAE.* 55(4): 1395–1411.

Francesconi, W., et al. 2016. Phosphorus modeling in tile drained agricultural systems using APEX. *J. Fertil. Pestic.* 7(1): 166.

Green, W. H. and Ampt, G. A. 1911. Studies on soil physics. 1. Flow of air and water through soil. *J. Agr. Sci.* 4: 1–24.

Gerwitz, A. and Page, E. R. 1974. An empirical mathematical model to describe plant root systems. *J. Appl. Ecol.* 11: 773–781.

Greenwood D. J., et al. 1990. Decline of percentage N of C3 and C4 crops with increasing plant mass. *Ann. Bot.* 66: 425–436.

Hanson, J. D., et al. 1998. RZWQM: Simulating the effects of management on water quality and crop production. *Agric. Systems.* 57: 161–195.

Hanson, J. D. 2000. Generic crop production. In: L. R. Ahuja, et al. (eds.), *The Root Zone Water Quality Model.* Water Resources Publications, Highlands Ranch, CO, pp. 81–118.

Hanson, J. D. and Olson, R. L. 1987. The simulation of plant biography. In: G. Fraiser and R. A. Evans (eds.), *Seed and Seedbed Ecology of Rangeland Plants*. USDA/ARS Misc. Publication, Tucson, AZ, pp. 110–116.

Hanson, J. D., et al. 1988. A multispecies model for rangeland plant communities. *Ecol. Model.* 44: 89–123.

Hanson, J. D., et al. 1999. Calibrating the root zone water quality model. *J. Agron.* 91: 171–177.

Hu, C., et al. 2006. Evaluating nitrogen and water management in a double cropping systems using RZWQM. *Vadose Zone J.* 5: 493–505.

Jarvis, N. J., et al. 1999. Modelling particle mobilization and leaching in macroporous soil. *Eur. J. Soil Sci.* 50(4): 621–632.

Jones, C. A. and Kiniry, J. R. 1986. *CERES-Maize: A Simulation Model of Maize Growth and Development*. Texas A&M Press, College Station, TX, 194 pp.

Jones, C. A., et al. 1984. A simplified soil and plant phosphorus model: I. Documentation. *Soil Sci. Soc. Am. J.* 48(4): 800–805.

Jones, C. A., et al. 1991. Simulation of root growth. In: J. Hanks and J. T. Ritchie (eds.), *Modeling Plant and Soil Systems*. ASA book series 31, Madison, WI, pp. 91–124.

Jones, J. W., et al. 2003. The DSSAT cropping system model. *Europ. J. Agron.* 18: 235–265.

Kersebaum, K. C. 1995. Application of a simple management model to simulate water and nitrogen dynamics. *Ecol. Mod.* 81: 145–156.

Kersebaum, K. C. 2000. Model-based evaluation of land use strategies in a nitrate polluted drinking water catchment in North Germany. In: R. Lal (ed.), *Integrated Watershed Management in the Global Ecosystem*. CRC Press, Boca Raton, FL, pp. 223–238.

Kersebaum, K. C. 2007. Modelling nitrogen dynamics in soil-crop systems with HERMES. *Nutr. Cycl. Agroecosys.* 77(1): 39–52.

Kersebaum, K. C. 2011. Special features of the HERMES model and additional procedures for parameterization, calibration, validation, and applications. In: L. R. Ahuja and L. Ma (eds.), *Advances in Agricultural Systems Modeling Series 2*, pp. 65–94.

Kersebaum, K. C. and Beblik, A. J. 2001. Performance of a nitrogen dynamics model applied to evaluate agricultural management practices. In: M. Shaffer, et al. (eds.), *Modeling Carbon and Nitrogen Dynamics for Soil Management*. CRC Press, Boca Raton, FL, pp. 551–571.

Kersebaum, K. C. and Nendel, C. 2014. Site-specific impacts of climate change on wheat production across regions of Germany using different CO2 response functions. *Europ. J. Agron.* 52: 22–32.

Kersebaum, K. C., et al. 2003. Site specific nitrogen fertilisation recommendations based on simulation. In: J. Stafford and A. Werner (eds.), *Precision Agriculture. Proc. 4th European Conference on Precision Agriculture*. Wageningen Academic Publishers, Wageningen, pp. 309–314.

Kersebaum, K. C., et al. 2008. Long term simulation of soil-crop interactions in semiarid southwestern Saskatchewan, Canada. *Europ. J. Agron.* 21: 1–12.

Kollas, C., et al. 2015. Crop rotation modelling – a European model intercomparison. *Eur. J. Agron.* 70: 98–111.

Kozak, J. A., et al. 2006. Evaluating various water stress calculations in RZWQM and RZ-SHAW for corn and soybean production. *J. Agron.* 98: 1146–1155.

Kozak, J. A., et al. 2007. Quantifying residue architecture effects on soil temperature and water. *Soil Tillage Res.* 95: 84–96.

Larsson, M. H., et al. 2007. A dual porosity model to quantify phosphorus losses from macroporous soils. *Ecol. Model.* 205(1–2): 123–134.

Leavesley, G. H., et al. 1983. c. USGS water-resour. *Investig. Rep.* 83–4238.

Leonard, R. A., et al. 1987. GLEAMS: groundwater loading effects of agricultural management systems. *Trans. ASAE.* 30: 1403–1418.

Published by Burleigh Dodds Science Publishing Limited, 2025.

Leslie, P. H. 1945. On the use of matrices in certain population mathematics. *Biometrika*. 33: 183–212.

Ma, L., et al. 1998. Manure management in an irrigated silage corn field: experiment and modeling. *Soil Sci. Soc. Am. J.* 62: 1006–1017.

Ma, L., et al. 2000. Integrating system modeling with field research in agriculture: applications of root zone water quality model (RZWQM). *Adv. Agron.* 71: 233–292.

Ma, L., et al. 2003. Evaluation of the RZWQM under various irrigation levels in Eastern Colorado. *Trans. ASAE*. 46: 39–49.

Ma, L., et al. 2005. Development and evaluation of RZWQM-CROPGRO hybrid model for soybean production. *J. Agron*. 97: 1172–1182.

Ma, L., et al. 2006. Evaluation of RZWQM-CERES-maize hybrid model for maize production. *Agric. Syst*. 87: 274–295.

Ma, L., et al. 2007a. RZWQM simulation of long-term crop production, water and nitrogen balances in Northeast Iowa. *Geoderma*. 140: 247–259.

Ma, L. et al. 2007b. RZWQM simulated effects of crop rotation, tillage, and controlled drainage on crop production and nitrate-N loss in drain flow. *Geoderma*. 140: 297–309.

Ma, L., et al. 2011. A Protocol for parameterization and calibration of RZWQM2 in field research. In: L. R. Ahuja and L. Ma, L. (eds.), *Methods of Introducing System Models into Agricultural Research*, SSSA Book Series, pp. 1–64.

Ma, L., et al. 2012a. Simulating the surface energy balance in a soybean canopy with SHAW and RZ-SHAW models. *Trans. ASABE*. 55(1): 175–179.

Ma, L., et al. 2012b. Simulating maize responses to irrigation water and crop ET based irrigation scheduling with RZWQM2. *Agric. Water Manag*. 103: 140–149.

Ma, L., et al. 2012c. Root zone water quality model (RZWQM2): model use, calibration and validation. *Trans. ASABE*. 55: 1425–1446.

Ma, L., et al. 2017. Modeling yield and biomass responses of maize cultivars to climate change under full and deficit irrigation. *Agric. Water Manag*. 180: 88–98.

Ma, L., et al. 2021. Simulated climate change effects on soybean production using two crop modules in RZWQM2. *J. Agron*. 113(2): 1349–1365.

Malone, R. W., et al. 2017. Winter rye as a cover crop reduces nitrate loss to subsurface drainage in central Iowa as simulated by HERMES. *Agric. Water Manage*. 184: 156–169.

McCree, K. J. 1970. An equation for the rate of respiration of white clover plants grown under controlled conditions. In: I. Setlik (ed.), *Prediction and Measurement of Photosynthetic Productivity*, pp. 221–229.

Michalczyk, A., et al. 2014. Model-based optimisation of nitrogen and water management for wheat–maize systems in the North China Plain. *Nutr. Cycl. Agroecosys*. 98: 203–222.

Michalczyk, A., et al. 2020. Quantifying nitrogen loss and water use via regionalization and multiple-year scenario simulations in the North China Plain. *J. Plant Nutr. Soil Sci*. 183: 718–733.

Moriasi, D. N., et al. 2007. Model evaluation guidelines for systematic quantification of accuracy in watershed simulation. *Trans. ASABE*. 50: 885–900.

Natural Resources Conservation Service (NRCS). 2024. Soil survey geographic (SSURGO) database. https://www.nrcs.usda.gov/resources/data-and-reports/soil-survey-geographic-database-ssurgo

Nendel, C., et al. 2025. The MONICA and HERMES crop model. In: G. Hoogenboom (ed.), *Current Crop Models: State-of-the-Art and Future Developments*. Burleigh Dodds Science Publishing, Cambridge, Chapter 19.

Nendel, C., et al. 2023. Future area expansion outweighs increasing drought risk for soybean in Europe. *Glob Chang Biol*. 29: 1340–1358.

Nicks, A. D., et al. 1995. *WEPP Model Documentation, Chapter 2: Weather Generator*. USDA-ARS, Durant. http://ars.usda.gov/SP2UserFiles/ad_hoc/36021500WEPP/chap2.pdf

Nielsen, D. C., et al. 2002. Simulating soybean water stress effects with RZWQM and CROPGRO models. *J. Agron*. 94: 1234–1243.

Nimah, M. and R. J. Hanks. 1973. Model for estimating soil water-plant-atmospheric interrelation: I. Description and sensitivity. *SSSA Proc*. 37(4): 522–527.

Pan, P., et al. 2023a. Modeling phosphorus losses to subsurface drainage under tillage and compost management. *Soil Till. Res*. 227: 105587.

Pan, P., et al. 2023b. Modification of the RZWQM2-P model to simulate labile and total phosphorus in an irrigated and manure-amended cropland soil. *Comput. Electron. Agric*. 206: 107672.

Qi, Z., et al. 2016. Simulating maize production, water and surface energy balance, and canopy temperature, and water stress under full and deficit irrigation. *Trans. ASABE*. 59: 623–633.

Rawls, W. J., et al. 1982. Estimation of soil water properties. *Trans. ASAE*. 25: 1316–1320, 1328.

Sadhukhan, D., et al. 2019. Development and evaluation of a phosphorus (P) module in RZWQM2 for phosphorus management in agricultural fields. *Environ. Syst. Res*. 113: 48–58.

Saseendran, S. A., et al. 2004. Modeling nitrogen management effects on winter wheat production using RZWQM and CERES-wheat. *J. Agron*. 96: 615–630.

Saseendran, S. A., et al. 2005a. Effectiveness of RZWQM for simulating alternative Great Plains cropping systems. *J. Agron*. 97: 1183–1193.

Saseendran, S. A., et al. 2005b. Simulating planting date effects on corn production using RZWQM and CERES-Maize models. *J. Agron*. 97: 58–71.

Saseendran, S. A., et al. 2007. Simulating management effects on crop production, tile drainage, and water quality using RZWQM-DSSAT. *Geoderma*. 140: 260–271.

Saseendran, S. A., et al. 2010. Simulating alternative dryland rotational cropping systems in the central great plains with RZWQM2. *J. Agron*. 102: 1521–1534.

Saseendran, S. A., et al. 2014. Nimah-Hanks approach for computing water uptake and water stress factors in RZWQM-CERES-maize model. *J. Agron*. 106: 81–94.

Saseendran, S. A., et al. 2015. Quantification of crop water stress factors in limited irrigation experiments. *Agric. Sys*. 137: 191–205.

Shaffer, M. J., et al. 2000. Nutrient chemistry processes – OMNI. In: L. R. Ahuja, et al. (eds.), *The Root Zone Water Quality Model*. Water Resources Publications LLC, Highlands Ranch, CO, pp. 119–144.

Sima, M. W., et al. 2020. Field and model assessments of irrigated soybean responses to increased air temperature. *J. Agron*. 112(6): 4849–4860.

Sima, N. Q., et al. 2018. A modified F-test for evaluating model performance by considering both experimental and simulation uncertainties. *Environ. Modell. Softw*. 104: 236–248.

Steenhuis, T. S., et al. 1997. A simple model for predicting solute concentration in agricultural tile lines shortly after application. *Hydrol. Earth Syst. Sci*. 1(4): 823–833.

Supit, I., et al. (eds.). 1994. *System Description of the WOFOST 6.0 Crop Simulation Model Implemented in CGMS. Vol. 1: Theory and Algorithms.* EC Publication EUR 15956, Luxemburg.

Vadas, P. A. 2014. *Surface Phosphorus and Runoff Model, Theoretical Documentation Version 1.0.* USDairy Forage Research Center, USDA.

Vadas, P. A., et al. 2006. Modeling phosphorus transfer between labile and nonlabile soil pools: updating the EPIC model. *Soil Sci. Soc. Am. J.* 70: 736–743.

Van Keulen, H., et al. 1982. A summary model for crop growth. In: F. W. T. Penning de Vries and H. H. van Laar (eds.), *Simulation of Plant Growth and Crop Production*, PUDOC, Wageningen, pp. 87–97.

Wallor, E., et al. 2018. The response of process-based agro-ecosystem models to within-field variability in site conditions. *Field Crops Res.* 228: 1–19.

Watts, D. G., et al. 1999. Background of the MSEA-RZWQM modeling project. *J. Agron.* 91: 169–170.

Wauchope, R. D., et al. 1992. The SCS/ARS/CES Pesticide properties database for environmental decision-making. *Rev. Environ. Cont. Toxicol.* 123: 1–164.

Yu, Q., et al. 2006. Modeling a wheat-maize double cropping system in China using two plant growth modules with RZWQM. *Agric. Systems.* 89: 457–477.

Yu, Q., et al. 2007. Simulation of energy balance and water and heat transfer by SHAW and RZ-SHAW. *Trans. ASABE.* 50(5): 1507–1516.

Zhang, H., et al. 2019. Response of maize yield components to growth stage-based deficit irrigation. *J. Agron.* 111: 3244–3252.

Zhang, H., et al. 2021. Modelling maize production under growth stage-based deficit irrigation management with RZQWM2. *Agric. Water Manag.* 248: 106767.

Zhang, H., et al. 2025. Comparisons of different irrigation scheduling methods on maize growth, yield, and water productivity in Northern Colorado. *Agric. Water Manag.* (Submitted and under review).

Zhou, L., et al. 2025. Simulating the effects of mulched drip irrigation on soil water balance and corn production using a modified RZ-SHAW model. *Field Crops Res.* 322: 109712.

Part 2

Europe

Chapter 9

The STICS soil-crop model: a generic model to simulate the functioning of agroecosystems

Benjamin Dumont and Mathieu Delandmeter, University of Liège, Belgium; Marie-Odile Bancal, Université Paris-Saclay, INRAE, AgroParisTech, UMR EcoSys, France; Nicolas Beaudoin and Joël Léonard, INRAE, UMR Transfrontalière BioEcoAgro, France; Samuel Buis, INRAE, Avignon Université, UMR EMMAH, France; Julie Caubel, EcoClimaSol and INRAE, US Agroclim, France; Hugues Clivot, Université de Reims Champagne-Ardenne, INRAE, FARE, UMR A 614, France; Julie Constantin, Université de Toulouse, INRAE, UMR AGIR, France; Antoine Couëdel and Mathilde de Freitas, CIRAD, University of Montpellier, UPR AIDA, France; Audrey Deheinzelin, INRAE, US Agroclim, France; Gatien N. Falconnier, CIRAD, University of Montpellier, UPR AIDA, France and CIRAD, UPR AIDA International Maize and Wheat Improvement Centre (CIMMYT) and University of Zimbabwe, Zimbabwe; Fabien Ferchaud, UMR Eco&Sols, University of Montpellier, CIRAD, INRAE, IRD, Institut Agro Montpellier, France; Iñaki García de Cortázar-Atauri, Marie Launay, Patrice Lecharpentier and Dominique Ripoche, INRAE, US Agroclim, France; Anne-Isabelle Graux, Agrocampus Ouest, INRAE, UMR PEGASE, France; Guillaume Jego, Agriculture and Agri-Food Canada, Canada; Arthur Lenoir, University of Liège, Belgium; Gaetan Louarn, INRAE UR P3F, France; Alain Mollier, INRAE, Bordeaux Sciences Agro, UMR1391 ISPA, France; Céline Schoving, Université de Toulouse, INRAE, UMR AGIR, France; Mounir Seghouani, INRAE, Bordeaux Sciences Agro, UMR1391 ISPA and CIRAD, University of Montpellier, UPR Recyclage et Risque, France; Loïc Strullu, Strullu Company, France; Remi Vezy, CIRAD, University of Montpellier, CNRS, INRAE, IRD, UMR AMAP, France; Tiphaine Vidal, Université Paris-Saclay, INRAE, UR BIOGER, France; Eric Justes[1], CIRAD, France; and Jean-Louis Durand[1], INRAE UR P3F, France*

1 Authors share senior co-authorship. *Corresponding author e-mail: benjamin.dumont@uliege.be

http://dx.doi.org/10.19103/AS.2025.0155.11

1 Introduction

The STICS model was developed in the mid-1990s by a group of INRA crop scientists from diverse research units in France. This acronym stands for "multidisciplinary standard crops simulator" (*Simulateur mulTIdisciplinaire pour les Cultures Standard*, in French). Behind STICS lies a story based on a collective project that owes a lot to the first coordinator of the STICS "adventure", the late Nadine Brisson. The STICS model was born from the project ECOSPACE – "heterogeneity of cultivated environments", an INRAE's (*Institut national de recherche pour l'agriculture l'alimentation et l'environnement, France*) AIP program (*Action Incitative Programmée*), which provided an opportunity to combine the expertises of three former departments of Bioclimatology, Agronomy and Soil Sciences. At that time, there were already agronomic models available at the international level with different orientations: photosynthetic production models (e.g. SUCROS), biogeochemical models (e.g. PASTIS, CENTURY) and more generalist soil-crop models (e.g. CERES, EPIC, APSIM, CROPSYS). While some of these models were in the process of being adapted to the French context, a bunch of French researchers sought to merge the most recent crop science knowledge in a new agroecosystem model: STICS.

The first version of STICS resulted from the merging of three dynamic models, namely GOA – a plant growth model (Ruget et al., 1994), BYM – a water balance model (Brisson et al., 1992) and LIXIM – a model dedicated to simulate or to calculate, using measured data, the soil water and nitrogen balances (Mary et al., 1999). As many other models presented in this book, STICS is a daily time-step crop model with input variables relating to atmosphere, soil, crop and system management. Its output variables relate to crop growth and yield elaboration – both in terms of quantity and quality – and to the environmental changes related to the agrosystem-water, -nitrogen and -carbon balances. STICS simulates the response of the soil–crop system over one crop cycle but also allows, as few other models (Basso et al., 2018), to simulate complex rotations of annual and perennial crops over several crop cycles, including fallow periods between crops.

STICS was originally applied to simulate wheat and maize crop growth (Brisson et al., 1998, 2002). The model then evolved, driven by new and very diverse issues, such as:

- estimation of regional production potential;
- water and nitrogen requirements of crops;
- effect of agricultural practices on N losses;
- precision agriculture, including remote sensing data assimilation;
- scenarios of French forage production monitoring and projections;
- climate change impacts and adaptation; and
- effects of cover crops on succeeding main crops.

In recent years, the topics covered by the model have expanded to include the evaluation of agroecological cropping systems, intercropping of bispecific mixtures and companion crops, perennial crops (i.a. for bioenergy), greenhouse gas emissions and long-term soil carbon sequestration.

Early in its infancy, the idea emerged that STICS should not be a fixed model but rather an interactive modeling platform, able to evolve according to the needs and interests of researchers. Since then, STICS has always been developed collaboratively at the intersection of scientific and practical concerns. The genesis and first evolution of the model were made possible by the exceptional leadership of Nadine Brisson and her ability to establish collaborations with researchers of diverse disciplines and affiliations, as well as technical experts. After Nadine's passing in 2011, the STICS Project Team (SPT) took over the maintenance and development of the model.

Since the beginning of its development, STICS was designed to meet four main criteria (Brisson et al., 1998):

- ***Balance*** in the complexity and description of the processes involved in the soil-plant-atmosphere continuum. This balance aims at having a model that can be applied in a wide variety of agricultural contexts.
- ***Genericity*** of the algorithms used to describe the plant functioning using non-specific eco-physiological concepts in order to have a single model able to simulate the growth of a wide variety of plants.
- ***Robustness*** of its algorithms and parameter setting to provide realistic results in a wide range of agri-environmental conditions.
- ***Simplicity*** of the input data required to run the model as well as model parameters readily accessible with small sensitivity to changes in scale. This simplicity in the use of the model was designed to facilitate any operational use of the STICS model in the real conditions of the agricultural sector.

The design of a dynamic and functional model with a strong agri-environmental focus has led to the emergence of a fifth quality: ***scalability***. This is illustrated by the development of new algorithms over time to simulate perennial cropping or intercropping systems (Brisson et al., 2004). This scalability also resulted in the model being used at different spatial scales, from the plot scale to the macro-regional scale.

2 The STICS model: overview

The equations, algorithms and formalisms behind the STICS model are fully described in a recent dedicated book (Beaudoin et al., 2023). This section will describe the main concepts behind the model. We invite the readers to have a look at the book (freely available as ebook and pdf formats) for more details. The term "formalism/formalization" will be preferably used in this chapter to refer to equations, algorithms or conceptualization of real-life processes as implemented in the model.

The STICS model is a deterministic functional process-based soil-crop model that allows to describe the energy, water, C and N dynamic balances across the crop cycle and during fallow periods (Fig. 1). The STICS model allows for the simulation of annual crops, intercropping of two species and perennial crops (i.e. pasture, vineyard, miscanthus), at the cropping scale or over multiple cropping seasons (crop rotations).

Using a 1D description of the agrosystem (pseudo 3D for some specific crops and for intercropping) within given boundaries, STICS simulates the processes occurring in a cropping system. The model allows users to simulate the dynamic evolution of a large range of agronomic and environmental variables of interest at daily time-step, over the soil profile and across the whole plant cycle (or even multiple cycles).

The upper boundary of the simulated system is delimited by the low layer of the atmosphere (2 m height), where standard weather variables are characterized (radiation, minimum and maximum temperatures, rainfall, reference evapotranspiration (ETP) and possibly wind and humidity). The lower boundary of the simulated system corresponds to the soil/sub-soil interface where water and nutrients are leached out of the crop rooting system. The soil compartments discretization depends on the soil parameterization and on the simulated process. Soil properties are defined at the pedologic layer; the discretization of the simulated processes ranges from centimetric (elementary layer) for water and solute contents and transfers to decimetric (pedologic layer) for waterlogging.

Phenology, which refers to plant development, governs the crop cycle time period. Depending on plant type, crop development is driven either by (i) a thermal index (degree-days), (ii) a photo-thermal index which also takes photoperiod into account or (iii) a vernal-photo-thermal index which also takes

Figure 1 Schematic representation of the organization of the STICS model. It represents the main inputs, outputs and modules of the model, following Beaudoin et al. (2023) – adapted from Lenoir (2024).

vernalization into account. The crop development module drives the kinetics of other processes, such as the leaf area index (lai) or the root growth, and defines the harvested organ filling phase (Brisson et al., 1998).

Foliar apparatus setting relies on the concept of a simple "big leaf" model; it can also be simulated using a pseudo 3D light capture submodel (mandatory for intercropping). Crop growth is driven by energy interception by the leaf apparatus and results in plant carbon accumulation (de Wit, 1978). Solar radiation absorbed by the foliage is transformed into aboveground biomass (concept of radiation use efficiency). Biomass is allocated to the different crop organs, and some fraction can be remobilised to the harvested organs when they become sinks (Fig. 2). Crop organs are defined as plant compartments to which generic ecological functions are assigned. Here, the distinction between the structural and temporary pool of carbohydrates is a key point. Only the structural aspect is assigned to the leaves, stems, fruits, perennial reserves and roots. Conversely, the temporary reserves are not physically located in the crop organs, to allow for keeping the model genericity for all crops. The storage organs and harvested organs can be grains, fruits, tubers or even stems. The STICS V10 (last version published in 2023) considers the C and N fluxes to and from temporary and perennial storage organs.

The crop nitrogen uptake depends upon (i) carbon accumulation in the crop and (ii) mineral nitrogen availability in the soil. Biological nitrogen fixation of N_2 fixed from the atmosphere can be accounted for in the case of legumes.

The model simulates the water and nitrogen element balances in soil to assess their availability for plants and potential environmental losses. Their

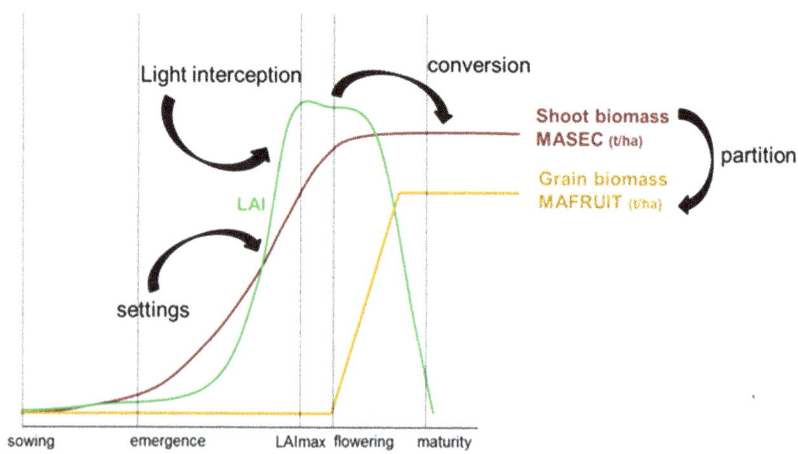

Figure 2 Schematic representation of the main processes related to crop growth (*MASEC* and *MAFRUIT* are STICS output variable names) – From Artru (2017).

descriptions are based on the classic compartmental approach, which defines different pools in the system, their evolution and their relationships. The functions encompass physical, physico-chemical and biochemical processes.

Computed from water and nitrogen balance modules, stress indices will allow to simulate reduction of crop growth. Other abiotic stresses, such as thermal (frost or high temperatures), trophic or waterlogging stresses, can also be taken into account as limiting factors of growth and yield formation. The response function to environmental constraints can also integrate global enzymatic activities, such as nitrate uptake. Biotic stress is not simulated *per se*, however, it can be accounted for through coupling of STICS with other models (e.g. Caubel et al., 2012).

The strong effect of crop management on the dynamics of the soil-crop-microclimate systems is also given particular attention. Indeed, crop specificities influence both ecophysiology and crop management (e.g. accounting for the impact of fertilizers or the various management of forage such as grazing or cutting, etc.). The links between the main routines and formalisms of the model are presented in Fig. 3.

Finally, at the daily scale, the STICS model can be considered as a functional or process-based model and not a pure mechanistic model (see next section). At the crop cycle scale, since it deals with the main interactions between the agrosystem components, STICS can be considered as a mechanistic model. Moreover, it allows to design and simulate the emergent properties at the crop cycle and crop rotation scales.

The processes implemented in the model are described in a way which can be qualified as functional and process-based. It aims to favor a simple integration of all the processes interacting in the soil-crop system. The most iconic illustrations of this conceptual choice in STICS are as follows:

- The interception efficiency of the total solar radiation uses Beer's law (1852), as an exponential response of the shadowing effect of leaves, in the case of a homogeneous canopy;
- The conversion of photosynthetically active radiation (PAR) in aerial biomass is based on the concept of radiation use efficiency (RUE), assuming a specific potential linear response of biomass accumulation, at a stable radiation level (Monteith, 1972);
- The nitrogen uptake and accumulation in the plant is based on the concept of N dilution curve in the aerial biomass of the crop, both for isolated plants or dense canopies crop (Lemaire and Gastal, 1997);
- Fruit filling includes the concept of dynamic harvest index, as proposed by Spaeth and Sinclair (1985), for crop species with determinate development;
- Based on the hypothesis of infinite hydric conductivity, the downward soil water storage uses the tipping bucket concept at the elementary

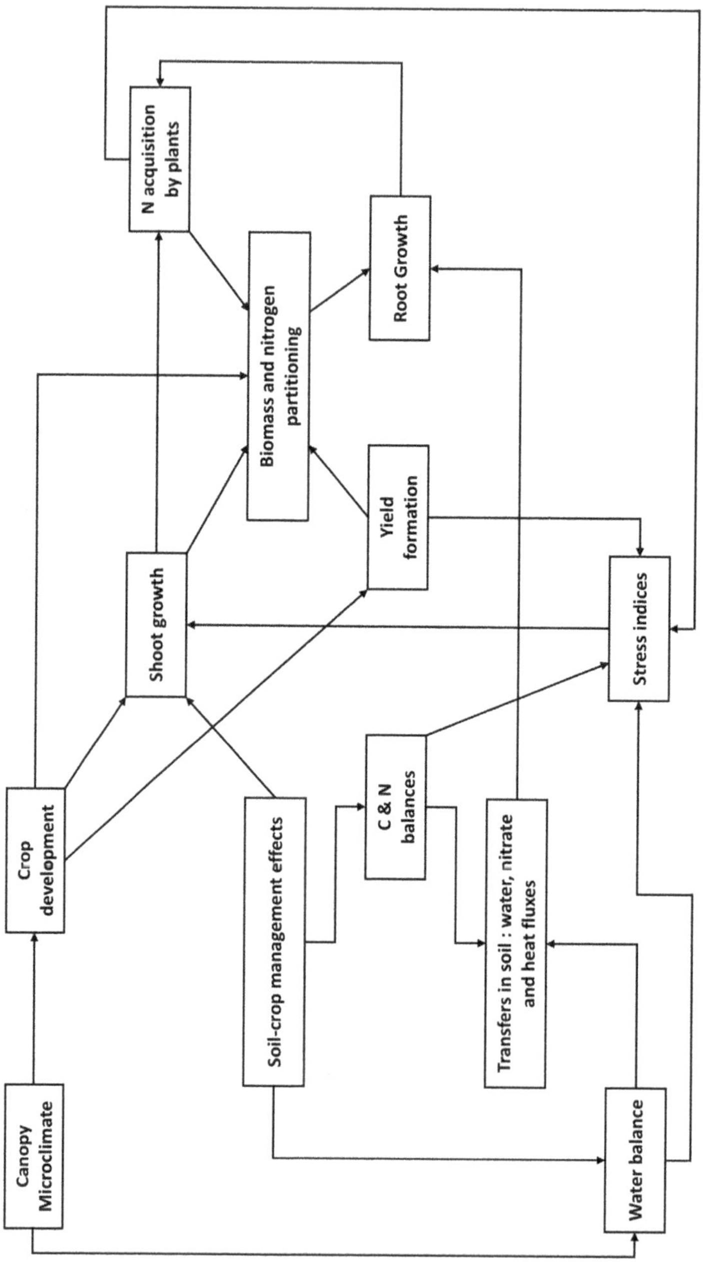

Figure 3 Main functional links between STICS model routines and formalisms – From Beaudoin et al. (2023)

layer infiltration scale, which is known to converge toward the convective-dispersive equations of Richard-Darcy law (Van Der Ploeg et al., 1995), when resolution is small enough (~1 cm – Mary et al., 1999);

- Nitrate content in the soil solution is simulated using a mixing cell model (Mary et al., 1999); and
- Soil C and N mineralization are simulated using a compartmental approach, with only three pools of organic matter, similar to the AMG model (Clivot et al., 2019). The compartmental approach allows an independent parameterisation linked to a residue typology (Nicolardot et al., 2001), which uses the C/N ratio of residues to drive their mineralization and humification rates.

These processes are selected by default when users download the standard STICS version. When the basic hypothesis of implemented processes is not met, an alternative process can usually be chosen, involving a more complex conceptual approach. For instance, for a heterogeneous canopy (e.g. row crops such as vineyard or cereal-legume intercropping systems), the simple 'big leaf' model associated with the Beer's law and the Priestley-Taylor (1972) or Penman-Monteith (Monteith, 1965, 1972) ETP functions can be replaced with a more complex energy balance based on crop simplified architecture and a resistive ETP approach (Wallace, 1995).

3 Accounting for genetic differences at species and cultivar levels

Globally speaking, the genetic dimension of plant production is implicit within STICS. Yet, a distinction is made at the inter- and intra-specific levels. Indeed, some parameters are considered as species-dependent – i.e. common to the different cultivars of a same species – while others are considered as cultivar-dependent. The number and list of specific or cultivar parameters vary in function of options activated for the simulation of a given crop or the need to be more or less accurate on specific processes (e.g. root growth).

Overall, parameters involved in the equations of physical processes (e.g. base temperature, etc.) or trophic relationships (e.g. radiation use efficiency, nitrate affinity in the rhizosphere, etc.) are considered species dependent. Parameters involved in most of the processes known to be genetically variable are ascribed to cultivar particularities. They are linked to the different processes simulated by the model: crop phenology, leaves growth, radiation interception, biomass partitioning, yield components and elaboration, root front growth, sensitivity to thermal and water stress.

It is important to note that not all parameters have the same genetic variability. Specific and cultivars parameters' nominal ranges – i.e. boundaries within which parameters may vary – were determined by expert knowledge (or within known physical boundaries).

The list of parameters is not exhaustive and could evolve, with future model releases, in function of the user's needs and knowledge. In the future, while the STICS model is a functional process-based model, we are confident that it still could help define theoretical ideotypes, through dedicated sensitivity analysis and the exploration of crop response across the parameters known to be genetically variable.

4 STICS simulated crop species

The STICS model can be downloaded with a set of pre-parameterized crops. STICS allows for parameters to be defined at specific and cultivar levels. The files containing the set of plant parameters can be qualitatively distinguished and include crop species ranging from officially validated to prototypes:

- ***Officially validated*** crop means the plant file is currently released after being tested in routine against a large dataset consolidated over time by the core team behind the STICS model development. Such sets of parameters were partly parameterised based on a literature review and went through a calibration phase against a specific database before being tested and validated against an independent database. Each new release of STICS version is documented following a specific procedure regarding this aspect. Some of the species are awaiting for publication or for greater documentation of their validation phase, prior to official release.
- ***Prototype*** crop refers to a set of parameters whose performance quality is uncertain for two main reasons: calibration was performed using either a limited dataset or an old STICS version. In these cases, old parameterized crops were adapted to be compliant with the current model version, including the newest development, but users should proceed with caution as the set of parameters was not validated under new model formalization. Users running a prototype parameterisation are invited to test it and, when possible, improve it through proper calibration/validation procedure. They are invited to share back the results of improvements – and ideally their dataset – so that they can be spread within the STICS community – or be included in the official validation procedure.

Figure 4 shows the diversity of plant files that are considered within STICS. Note that some species, such as turmeric and strawberry crops, which are perennial in botanical terms, can be parameterized as annuals based upon

Category	Herbaceous gramineous	Herbaceous leguminous	Market gardening	Fruit trees
Perennial crops	Fescue **Grass** **Miscanthus** **Timothy** Sugarcane	Alfalfa *White clover **	Strawberry	**Grapevine** Banana
	Cereals	**Leguminous**	**Vegetables**	**Oilseeds**
Annual crops	Winter barley Spring barley Spring barley IC **Corn** **Winter wheat** **Durum wheat** Sorghum Rice *Triticale ** *Finger-millet **	Soybean **Spring pea** Spring pea IC *Winter faba bean ** *Pigeon pea **	Lettuce Tomato **Sugarbeet** Potato Turmeric	Flax **Winter rapeseed** Sunflower
Catch crop	**Italian ryegrass CC** **Bristle oat CC**	**Vetch CC** **Crimson clover CC**		**Mustard CC**
Bare soil				

Figure 4 List of currently parameterized plant species. The species given in bold have been officially validated by the team maintaining STICS; those in regular font are yet considered as prototypes considering the recent model developments, while those in italic marked with a "*" are awaiting for greater documentation prior to their official release. "IC" refers to plant files that require specific parameterization when used to simulate intercropping, while "CC" refers to plant files that have been parameterised to simulate cover crops.

how farmers manage them. Additionally, the parameterization of triticale, white clover, finger-millet, pigeon pea and winter faba bean are waiting for documentation.

4.1 Unlocking simulations of perennials with undetermined growth and fruits: grapevine simulations

In 2006, the STICS crop model was adapted to grapevines (García de Cortázar-Atauri, 2006). The objective was to add a perennial crop to the list of existing crops in STICS. A preliminary analysis allowed to identify the necessary integrations and modifications in the existing formalisms to allow representing the functioning of the grapevine, including different grape varieties or management systems.

Among the different modifications, new formalisms were introduced to calculate budbreak, inspired from the BRIN model (García de Cortázar-Atauri et al., 2009a, 2009b). This model has the particularity to calculate bud break taking into account the dormancy period, allowing for a more accurate physiological representation. The set of parameters representing the other main phenological stages of the grapevines (flowering and veraison) were also calibrated. The grape ripening process was completely revised to account for water and carbon dynamics and their asynchrony. A new mathematical description of berry carbon growth, in line with the existing literature and confirmed with three years of experimental data, was introduced (García de Cortázar-Atauri et al., 2009a, 2009b).

Special focus was given to the management of carbon and nitrogen reserves. The relationships between canopy structure, viticultural practices and radiation interception were examined in detail. The model was adapted to account for specific technical management (such as planting density). The nitrogen demand of the vine was described to consider low planting density and rapid nitrogen stress variations. The nitrogen dilution curve, typically used for annual crops, was defined to allow for the vine's nitrogen demand to be managed (Celette and Gary, 2013). The control between sinks strength within indeterminate crops was solved to allow the simulation of leaf and fruit competition. This point was very important in order to avoid an imbalance between vegetative and reproductive biomass. The effects of photoperiod and vine vigor on leaf senescence were included, questioning the genetic nature of leaf senescence (previously defined by a lifespan in STICS). It allowed to simulate the critical period between harvest and leaf fall, which is crucial for the plant's reserve buildup. In addition, new pruning techniques (such as single or double Guyot or Royat cordon) were proposed to manage leaf density and canopy structure.

Finally, measuring the water balance of grapevine systems and soils was a major challenge, grapevine being usually cultivated on soils with a high density of stones and having a deep root system. To overcome these limitations, the calculation of the leaf water potential, a variable commonly measured in viticultural experiments. The integration allowed retro-adjustment of soil parameters in the model, some of which remain difficult to measure in the field, such as rooting depth.

To ensure robustness, the model was validated against experiments conducted in a wide range of pedoclimatic contexts (seven vineyards) and with a great diversity of varietal conditions (eight grape varieties), over two years. García de Cortázar-Atauri (2006) and Fraga et al. (2016) successfully applied the model to explore climate change impacts in France and Europe, respectively. These studies revealed different regional variations in phenological stages, harvest dates, vegetative biomass and water stress sensed by the plant. Work is currently underway to adapt the vineyard to the new version of the model, which represents the different reserve compartments more explicitly and takes advantage of the latest improvements introduced to the model in recent years (see Strullu et al., 2014, 2015, 2020).

4.2 Grassland simulation: relying on the concept of functional types

Multispecies grassland vegetation can be described simply by the formation of groups of species with similar functions and sharing common biological characteristics known as functional traits. Cruz et al. (2010) have proposed a classification of these grasslands into four and then six functional groups, grouping together grass species with similar growth strategies (capture or conservation of resources), phenology (early or late) and capacity to accumulate biomass. Although this functional classification was only established on aerial functional traits, it seems that species with a conservative strategy have a deep, coarse root system and species with an acquisitive strategy have a shallow, fine root system (Fort et al., 2013). These groups form in response to environmental and management factors and are found in more or less fertile environments and under more or less intense pasture management conditions.

The functional diversity of grasslands has been taken into account in some multispecific grassland simulation models by using this functional classification of grasslands and parameterizing the models with the value of functional traits describing groups of species, which often correspond directly or indirectly to model parameters (Duru et al., 2010; Jouven et al., 2006). These models, therefore, use a functional representation of grasslands assuming that grasslands can be modeled as one or a combination of these groups.

This approach is currently used to parameterize the STICS model for some of the functional groups in the Cruz et al. (2010) classification corresponding to rather intensive productive grasslands (groups A and B). This is done by entering each functional group as a "cultivar" in the STICS sense and also by modifying the value of some of the model's plant parameters according to the functional group under consideration (Graux et al., 2024). This approach has several advantages in terms of modeling. Parameterization of grassland functional groups rather than grass species present in grasslands greatly simplifies model parameterization. Taking into account the role of functional diversity of multispecific grasslands in models can help to better simulate changes in forage production and soil organic matter, which are key issues for grasslands in the context of climate change. The approach also has its limitations and is not easily applicable in the absence of grassland data to verify that the development and growth dynamics of the functional group under consideration are simulated correctly.

5 Current applications: accounting for Genotype x Environment x Management (GxExM) interactions

5.1 *Sensitivity analysis of cultivar parameters in different environments: an alfalfa case study*

In an attempt to parameterize different cultivars of alfalfa, data were collected by INRAE Lusignan's genetic team. Results revealed that the cultivars were characterized by variable biomass production during the whole growing season and notably at spring and autumn harvests. Such results led to hypothesize that the observed variability could be explained by a difference of sensitivity of the cultivars to the photoperiod.

The STICS model was used to conduct a sensitivity analysis of the Alfalfa aboveground biomass production to two parameters linked to the photoperiod sensitivity of the crop (namely, *sensiphot* and *phobase* parameters). A tool called OptimiSTICS, allowing to conduct sensitivity analysis and perform optimization of parameters value, was used to perform the study. The tool is described in Buis et al. (2011) (OptimiSTICS is not maintained anymore and has since be replaced by CroptimizR – See Buis et al. (2023))

Variation of the parameter *sensiphot*, which corresponds to the sensitivity to the photoperiod *sensu stricto*, was allowed between a value of 0 (very sensitive) to 0.6 (less sensitive), while the parameter *phobase*, which corresponds to the base photoperiod of the crop in hours, ranged from 10 to 14 h. Those thresholds were defined following the analysis of Major et al. (1991). Furthermore, in order to isolate the effect of photoperiod sensitivity on aboveground biomass production from other stresses, the sensitivity analysis was performed with a deactivation of water or N stresses. The sensitivity analysis was performed over two French sites with different latitude and climatic conditions: Avignon (43°57′N) and Fagnières (48°57′N). The crop was sown on July 28 and harvested a first time on November 10. During the second year of growth, four cuts were applied, namely on: June 11 (DOY 527), July 23 (DOY 569), September 03 (DOY 611) and finally October 15 (DOY 653 – "Day of the Year" are here cumulated over the two growing seasons, starting from January 1 of the sowing year).

Results revealed that aboveground biomass harvested in autumn was very sensitive to these two parameters, with yield variation of more than 57% in September and over 83% in October, with a more pronounced effect at the lower latitude. Biomass production in spring appeared less sensitive to those parameters with a yield variation of 28%, with no effect of latitude (Fig. 5). Biomass production appeared only sensitive to the *phobase* parameter in spring (Fig. 5a) while it was sensitive to both parameters in autumn (Fig. 5b), due to the model formalization.

In view of these results, measurements of biomass production made in autumn (September and November) were used to estimate the parameters

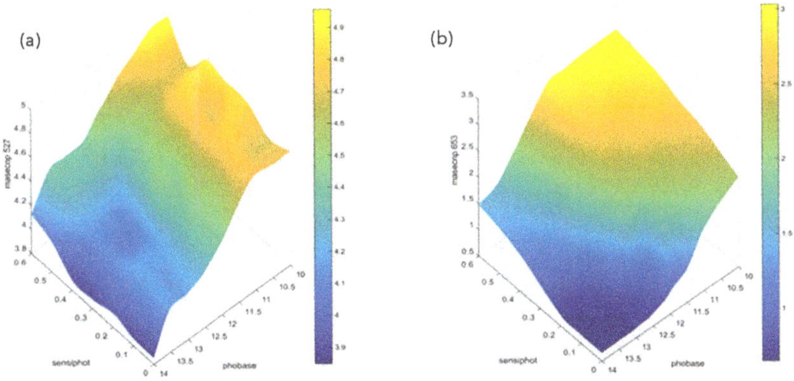

Figure 5 Sensitivity analysis of the aboveground biomass production in t DM ha^{-1} in spring (a), at DOY 527 (masecnp527) or in autumn (b) at DOY 653 (masecnp653) for various combinations of the parameters related to photoperiod base hour (*phobase*) and crop sensitivity to photoperiod (*sensiphot*).

sensiphot and *phobase* for the different cultivars studied at Lusignan (Table 1). The results of parameter estimation were consistent with the empirical knowledge; for example, Anik – which is a cultivar known as very sensitive to photoperiod – had the highest base photoperiod with 13.7 h and the highest sensitivity with 0.08. On the contrary, cultivar Magali, which is a mediterranean cultivar, had a base photoperiod of 11 hours and a sensitivity of 0.50.

However, even if the model was able to reproduce with accuracy the biomass production of some cultivars, it still tended to overestimate or underestimate biomass for others, meaning that more parameters should be calibrated to simulate with accuracy the large diversity existing within Alfalfa cultivars. To do this, sufficient data are needed to allow an estimation of all the parameters considered as cultivar-dependent within the STICS model.

5.2 Assessing ExM interaction to improve wheat N fertilization at the Walloon territorial scale

Recently, the STICS model has been used at the subfield scale to develop strategic approaches to N fertilization that precisely optimize agronomic and environmental objectives (Lenoir, 2024). The procedure was based on the concept of management zone delineation. Inspired by the work of Basso et al. (2019), Lenoir et al. (2024) proposed a methodology to advance Blackmore's delineation of management zones (Blackmore, 2000) and to transpose it to the use of Sentinel-2 images. This work served as a basis to deploy the methodology at the territorial scale of Wallonia, southern part of Belgium (Fig. 6). It led to the successful characterization of uniform zones of management at the sub-field

Table 1 Alfalfa cultivars for which photoperiod-related parameters (*phobase* [h] and *sensiphot* [-]) were optimised.

Cultivar/Genotype	Sensiphot [-]	Phobase [h]
Villa	0.53	13.2
Monte	0.29	13.1
Pancrudo	0.35	13.0
D15	0.51	11.3
Pool3	0.47	11.1
Pool5	0.48	11.5
Flamande	0.51	11.2
Provence	0.51	11.4
Lucon	0.52	11.5
Anik	0.08	13.7
Magali	0.50	11.0
Maron	0.54	13.1
Malzeville	0.54	13.0
Krasnostkaya	0.54	13.1
Europe	0.51	11.4
Coerulea	0.29	13.1
Romanica	0.17	12.6
Quasifalcata	0.15	12.9
Gabes	0.13	10.6
Tunetana	0.51	11.4
Aragon	0.50	11.4
Mediterraneo	0.44	11.6
Tierra	0.51	11.5
Ampurdan	0.47	11.2

scale, differentiating between zones in field with more frequent higher-than-average yield potential (HS zones), with more frequent lower-than-average yield potential (LS zones), and unstable zones within which yields could strongly shift from year to year (U zones). Only fields having at least three occurrences in the crop rotation of a winter cereal crop (winter wheat, winter barley, spelt or triticale) over the period 2016–2021 were considered, as required by the delineation method.

The territorial map of delineated fields was crossed with the soil profiles contained within the Aardewerk database (Tavernier and Maréchal, 1958, Legrain et al. 2011). As a result, a total of 10 fields whose soil profiles intersected at least two different management zones were selected to carry out the modeling analysis.

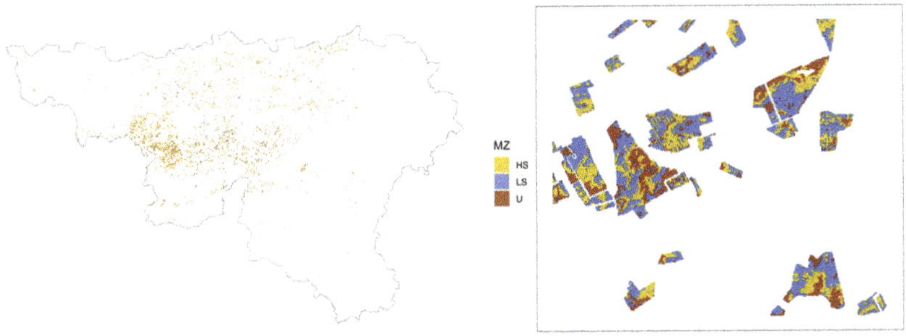

Figure 6 Spatial distribution of the satellite-based delineation conducted at the scale of Wallonia, Southern region of Belgium (left) and zoom on a group of fields where management zones were delineated using advanced Blackmore approach (right). MZ stands for "management zones", HS stands for "high and stable yielding zones", LS stands for "low and stable yielding zones", U stands for "unstable zones". Extracted from Lenoir PhD thesis (2024).

The STICS model was then used to perform simulations to capture the influence of climate (over 26 years) and its interaction with soils under varying N fertilization schemes. Following the study of Dumont et al. (2015a, 2015b, 2016), N fertilization schemes were simulated on each soil profile to concomitantly assess their agronomic and environmental performances and determine the most adapted management of nitrogen regarding the potential of each zone. A total of 10 fertilization schemes were carried out, ranging from 0 to 270 kg $.ha^{-1}$, split into three equal applications of N at critical growth stages, namely "tillering", "stem elongation" and "flag leaf".

Complementary, the STICS model was parameterized as follows:

- Soils properties from the Aardewerk database were used to compute pedotransfer functions and parameterize the model;
- Climatic conditions were retrieved from the Agromet dataset using an API request (https://www.agromet.be/fr/pages/home/), over the period 1995–2021; and
- Plant files were calibrated from a local experiment dataset dedicated to N fertilization (Dumont et al., 2014).

Yields simulated under the different N fertilization schemes were represented as a function of N left in the soil after harvest (Fig. 7). Overall, increasing fertilization resulted in higher yields and higher nitrate levels in each management zone. Yield differences between HS and LS zones were consistent with the features of the management zones, with higher yields in HS than LS. However, yield differences between the two zones under similar fertilizations did not systematically indicate an opportunity to spatially differentiate N management

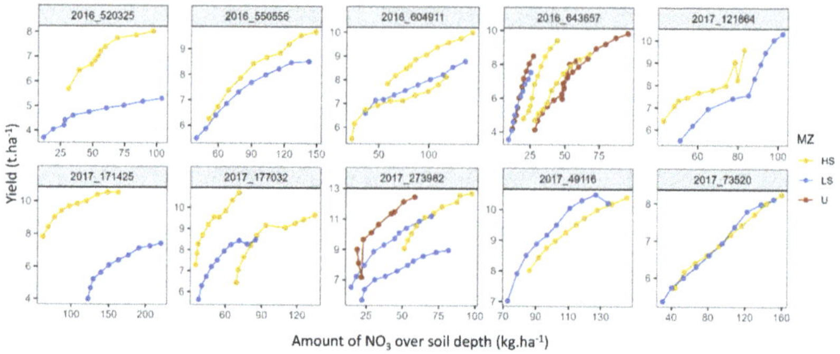

Figure 7 Median yields obtained in HS (yellow), LS (blue) and U (red) zones as a function of soil nitrate content (NO_3) simulated at harvest. Points from left to right correspond to increasing nitrogen fertilizer inputs. Graphic titles correspond to the first year of remotely sensed wheat data and soil index in the database.

at the subfield scale. Also, N left in the soil after harvest in HS and LS highlighted an inconsistent pattern among management zones, contradicting the idea of a higher efficiency in HS than LS zones. While the results proved interesting to advance fertilization redesign in management zones, further analysis should explore the effect of varying N supplies at the different growth stages on each zone.

5.3 Using STICS to better understand, assess and optimize GxExM interactions: example with soybean

Positioning the cycle of an adapted soybean genotype in a given environment is a strategy often suggested in Europe to avoid water and heat stress during the most sensitive phases. Schoving et al. (2022) evaluated different sowing strategies (from very early to late) for contrasting soybean genotypes (Maturity Group 000 to II) in environments with different water constraints in southwestern France, resulting in 157 contrasted situations. This assessment was carried out by combining statistical analysis of experimental data and modeling. GxExM interactions were highlighted by the statistical analysis, but an explanation of early sowing performance in certain sites and years was needed. STICS crop model was used to (i) complete experimental data with phenology key stages and growth dynamics, (ii) calculate stress dynamics (e.g. water stress indicator), (iii) virtually evaluate the effect of avoiding water stress by sowing soybean early in three contrasting situations in south-west France, two soil available water contents, with maturity group I soybean and under 28 historical years.

The output variables produced by STICS were used to calculate ecoclimatic and agrophysiological indicators for three phenological phases between crop emergence and harvest (VE-R1, R1-R5, R5-harvest) for the 157 simulations

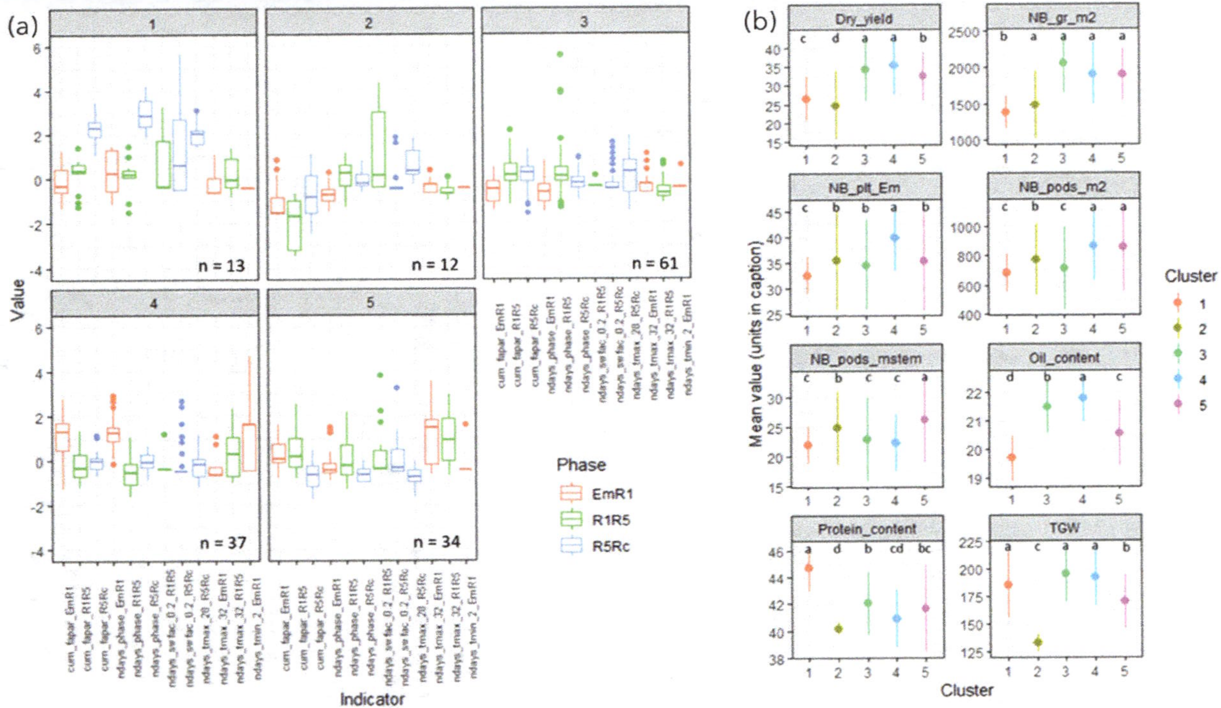

Figure 8 Graphical representation of (a) the centered-reduced values of the indicators characterizing each class for the different phases of the crop cycle: EmR1 (emergence to first flower), R1R5 (first flower to beginning of grain filling), and R5Rc (beginning of grain filling to harvest) (*n* = number of USMs per class), and (b) the mean observations in the field and their standard deviation for each class. The data originate from field trials conducted in the south of France (Mauguio, Béziers, Mondonville, Rivières, En Crambade, and Auzeville-Tolosane), spanning Mediterranean to altered Mediterranean climates from 2010 to 2018. The effect of the classes on the observations is highly significant at the $p < 0.001$ threshold. Means with a common letter are not significantly different at $p < 0.01$ according to the Student Newman Keuls test. Indicators are calculated per phase as follows: cum_fapar = cumulated fraction of intercepted photosynthetically active radiation; ndays_phase = phenophase duration in days; ndays_swfac_0.2 = number of days with stomatal water stress below 0.2 (1 = no stress; 0 = high stress); ndays_tmax_28 = number of days with max temperature above 28°C; ndays_tmax_32 = number of days with max temperature above 32°C; ndays_tmin_2 = number of days with min temperature below 2°C. Additional variables include: Dry_yield: dry grain yield (q/ha), NB_gr_m2: number of grains per m^2, NB_plt_Em: number of plants per m^2, NB_pods_m2: number of pods per m^2, NB_pods_mstem: number of pods on the main stem, Oil_content: oil content in the grain (%), Protein_content: protein content in the grain (%), TGW: thousand grain weight (g).

corresponding to one site-year-genotype-sowing date-water management. A *pls* regression was used to select the indicators with the greatest weight in explaining grain yield. Reduced centered values of these indicators were used to classify the treatments simulated into five environments (ward method then *k*-means) which were then characterized by the mean field observations (e.g. yield, oil content, number of plants) of the corresponding experiments (Fig. 8). The poor performance of soybean in classes 1 and 2 was mainly explained by major water stress from flowering onwards and high temperatures at the end of the cycle.

Secondly, a detailed analysis of the dynamics of water stress indicator, aerial biomass and N_2 fixation was used to validate the differences in performance between early sowing and conventional sowing. The water stress (intensity and duration) in the flowering-harvest phase was partially avoided in the case of early sowing. This avoidance was also demonstrated for an early genotype compared to a late genotype for the same sowing date.

Finally, the scenarios simulated on three sites across the oceanic to Mediterranean climatic gradient showed the success of early sowing for maturity group I soybean. Its efficacy clearly increases with water stress level in the flowering-harvest phase in conventional sowing. This study could be completed on earlier genotypes or by simulating the impact of early sowing date under future climatic conditions. This study demonstrated the interest of STICS to simulate environmental variables and stress dynamics to better understand GxExM interactions, identify the best genotype for a given environment and design more adapted crop management and/or ideotypes.

6 The contribution of STICS to the international Agriculture Model Intercomparison and Improvement Project (AgMIP)

The international Agriculture Model Intercomparison and Improvement Project (AgMIP, Rosenzweig et al., 2013 - https://agmip.org/) is an initiative launched by the NASA (National Aeronautics and Space Administration) and the University of Florida following the line of CMIP (Coupled Model Intercomparison Project), the scientific tool of IPCC (Intergovernmental Panel on Climate Change). Modeling groups across the world were asked to join and analyze the source of uncertainty in the assessment of climate change impacts on yields of main cultivated crops.

STICS was soon involved in many exercises such as the ones on wheat (Asseng et al., 2019), maize (Bassu et al., 2014), rice (Confalioneri et al., 2016), soybean (Kothari et al., 2021), calibration (Wallach et al., 2023), evapotranspiration (Kimball et al., 2019), low input farming systems (Couedel et al., 2024) or on soils and crop rotations (Basso et al., 2018). The exercises gathered between 16 and more than 40 models, depending on the study. In a cases, the first step consisted in breaking down the yield estimate's uncertainty into different components, such as parameter values, parameterization and input data. Detailed datasets from different regions were used for that purpose and the adaptation of model parameters to the cultivar use was a challenge.

In STICS, the phenology parameters such as temperature or photoperiod thresholds, as well as freezing and drought stress coefficients were modified for each situation to improve the model calibration. From those calibrated values, projections using simulated projected climatic time series were computed. Finally, the variability between models was compared to the uncertainties linked to climate input. In general, STICS performed relatively well for predicting yields or nitrogen use by crops. Results on leaf area or water consumption were less accurate.

In a second step, several intercomparison exercises aimed at isolating some modules of models to compare their specific contribution to simulated uncertainties. For instance, this was done for wheat to further study the response of phenology to temperature or the response of biomass and yield to the air concentration in CO_2. Globally, STICS was relatively reliable in simulating experimental wheat or maize data gained in Free Air Carbon Enrichment (FACE) experiments. However, like most other models, the water x CO_2 interaction was very difficult to simulate using the available inputs data. In maize, in particular, the high sensitivity of maize yields to water deficiency at the silking stage appeared challenging to reproduce. The main difficulty lies in the variations in phenological development and the occurrence of the sensitive stages within the time course of soil water deficits. One of the main issues remaining to address consists of converting this knowledge into specific calibration of parameters controlling the formalisms such as implemented in the current models versions. This remains a topic to be addressed more systematically in AgMIP.

Furthermore, in general, crop models are validated using well-spread genetic material, while their ability to reproduce newly released or projected genetics remains highly uncertain. Given the magnitude of the genetic progress achieved over the last 80 years, important efforts should be put to better consider new genetic material into study aiming at simulating projected yields and assessing the use of resources. To do this, it would be desirable for geneticists to use the model to allow parameterization of the various varieties with automated methods based on plant traits and phenotyping information.

Recently, the very rich routines built around STICS in the R environment that are used for data handling, multi-simulations, sensitivity analysis, parameter optimization and data reporting (with automatic graphs construction), were adopted by AgMIP as a common tool for model calibration in different exercises, especially for wheat. The corresponding packages initially

STICS team can be used with any crop model, provided a …n is built to interface it.

…ent STICS model improvements: crop-related …atures

… Improving root system simulation

Originally in the STICS crop model, the root system development was simulated in two steps: (i) first, the root front growth in soil depth was simulated, (ii) then the growth in root length density by soil layers was accounted for (Beaudoin et al., 2023). At the infancy of STICS, the root system was considered only as an exchange interface for water and nitrogen uptake, and its growth was simulated independently from the shoot growth. Recently, this conceptualization has been enhanced with the addition of several formalisms in order to simulate the major roles that the root system plays, notably in soil C-N storage (Wichern et al., 2008) or in response to mineral availability in the soils (Wedin and Tilman, 1990). Until recently, STICS only calculated the amount of recycled C-N at the end of the crop cycle, which allowed simulation of C-N storage for annual crops but jeopardized the model relevance for perennial crops. Fortunately, recent works improved the root description as described in the STICS V10 version, provided appropriate options are activated.

First, roots are described in terms of front depth and density profile, with a potential growth function of degree-days. Their growth begins now, starting at germination (for sown plants), at planting or at the bud break stage for perennial crops, and stops at a given stage defined by the user (end of leaves growth, flowering or senescence); this first step remained unchanged in STICS V10. Second, actual growth depends on the soil environment (bulk density, humidity, etc.) and is linked to daily biomass production. Third, when growing, the root system uses a part of the plant's nitrogen uptake for its own demand, which allows it to simulate the evolution of root N content. Fourth, two root classes are simulated in order to simulate the root system turn-over, which is mandatory for simulation of perennial crops like miscanthus (Strullu et al., 2015). These classes differ by their diameter, root tissue density and lifespan, following the principles of the ArchiSimple model (Pagès et al., 2014).

An example of the simulation of *Miscanthus* x *giganteus* root system growth and turn-over is given in Fig. 9. Simulated root C and N inputs to the soil vary between years, with annual inputs varying between 1.5 t C ha^{-1} y^{-1} and 3.2 t C ha^{-1} y^{-1} for C and between 21 kg N ha^{-1} y^{-1} and 73 kg N ha^{-1} y^{-1} for N when the crop is fully established. Further evaluation of the model will be done in the near future to improve parameterization of root turn-over to reproduce soil

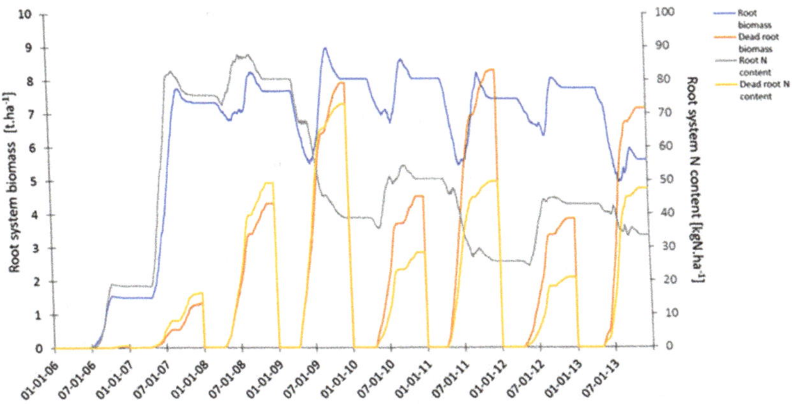

Figure 9 Simulation of *Miscanthus* root biomass and N content and its turn-over in the entire soil profile (until 210 cm depth) from rhizome planting in 2006 until the end of 2013, at Estrées-Mons in Picardie region, France.

C-N storage under perennial crops like *Miscanthus* x *giganteus* or Switchgrass, thanks to newly acquired observed data on root growth or C-N stocks in the soil.

7.2 Allocation of C and N to perennial and non-perennial organs

Under the pressure of climate change, the increasing interests for the use of perennial crops as forage, as a means for bioenergy production or second generation biofuels production, as well as a mitigation lever of climate change impacts by soil C sequestration, reinforced the need for developments of the STICS model to better simulate perennial crops over the long term.

Prior model versions could not accurately simulate biomass or nitrogen dynamics within perennial or non-perennial organs, and the remobilization between those organs was not taken into account. As a consequence, the model was not able to simulate long-term perennial crop functioning, including C and N recycling from one year over the other.

The STICS modules were substantially modified to account for the allocation of C and N assimilates in the plant. *Miscanthus* x *giganteus* was used as a reference crop for the development of new equations (Strullu et al., 2014, 2015). Ultimately, novel equations were tested and adapted, when necessary, for the simulation of other perennial crops like *Medicago sativa* L. (Strullu et al., 2020) or grapevine. Table 2 provides the results of model evaluation on independent data for biomass partitioning between perennial and non-perennial organs, as well as biomass partitioning between stems and leaves. For each crop, datasets used for the model evaluation included different climatic

Table 2 Evaluation of model ability to simulate biomass partitioning between organs for different perennial crops. *masecnp* = non-perennial organs biomass (t.ha^{-1}); *maperenne* = perennial organs biomass (t.ha^{-1}); *mafeuilverte* = structural part of green leaves biomass (t.ha^{-1}); *matigestruc* = structural part of stems biomass (t.ha^{-1}). All details related to the datasets used for model validation, including pedoclimatic conditions and management itineraries, can be found in Strullu et al. (2014, 2015, 2020 and 2023).

Variable	masecnp	maperenne	mafeuilverte	matigestruc
Miscanthus				
Mean obs.	15.0	15.1	2.4	11.6
Mean sim.	15.5	14.5	1.9	11.4
R^2	0.81	0.51	0.73	0.84
rRMSE	22	28	41	24
EF	0.80	0.24	0.60	0.83
IPQ	Very good	Good	Good	Very good
Alfalfa				
Mean obs.	2.6	4.8	0.7	1.3
Mean sim.	2.4	4.8	0.7	1.0
R^2	0.77	0.65	0.54	0.70
rRMSE	33.8	20.0	52.9	56.6
EF	0.75	0.45	0.52	0.61
IPQ	Very good	Good	Acceptable	Acceptable
Grapewine				
Mean obs.	2.1	–	0.9	0.9
Mean sim.	2.5	–	1.1	1.0
R^2	0.88	–	0.56	0.73
rRMSE	38	–	43	42
EF	0.80	–	0.45	0.61
IPQ	Very good	–	Acceptable	Good

conditions, various soil types (in terms of texture, water availability and organic matter content), variable durations (from 4 to 20 years for Miscanthus, from one to three of four growing seasons, respectively for Alfalfa and grapevine), as well as different crop management practices (i.e. harvest date, irrigation and fertilization for Miscanthus; cutting rate, irrigation and fertilization for Alfalfa; pruning and fertilization for grapevine). For Miscanthus, only one cultivar was parameterized, *Miscanthus* x *giganteus*. For Alfalfa, although different cultivars were used in the different experimentations, due to a lack of data to allow a finer parameterization, we chose to parameterize only one virtual cultivar corresponding to the Flemish gene pool. Finally, for grapevine, three cultivars were parameterized: Pinot Noir, Chardonnay and Meunier. For more information about each specific dataset used, please refer to the references

(Strullu et al., 2014 and 2015 for Miscanthus; Strullu et al., 2020 for alfalfa and Strullu et al., 2023 for grapevine).

These studies allowed to verify the genericity of the novel equations across different species and ensure a proper simulation of biomass and nitrogen partitioning between perennial and non-perennial organs, as well as between different vegetative organs (stems, leaves and temporary reserves). The recent parameterization of grapevine simulation with such novel equations permitted to evaluate and improve the simulation of biomass and N allocation to the reproductive organs (Strullu et al., 2023).

Finally, another advantage in simulating perennial organs biomass and N content of perennial organs is the model's ability to reproduce the rear effect of perennial crop destruction (like Alfalfa) on mineral N availability in the soil, which is mandatory to simulate cropping systems integrating such perennial crops (Strullu et al., 2020).

7.3 Ability to simulate intercropping

STICS was adapted 20 years ago for the simulation of intercropping (Brisson et al., 2004). It stimulates the competition between two species for light, water and nitrogen. The competition for light is computed either simply as a Beer-Lambert inspired light extinction computation adapted for intercrops in the case of homogenous canopies, or using a more complex energy transfer module between a main crop and an understory crop that takes into account the extinction coefficient of the plants, the row spacing and the shape of the canopy, as a pseudo 3D submodel (Brisson et al., 2004). STICS calculates the water and nitrogen uptake in intercropping, taking into account the distribution of the roots of each of the two cultivated species along the soil profile. Water demand is calculated with a daily time step using the resistance analogy to account for the microclimatic effect between the two species. The model has been proven to perform reasonably well in temperate (Launay et al., 2009, Shili-Touzi et al., 2010), Mediterranean (Kherif et al., 2022) and tropical conditions (Traoré et al., 2022).

In a recent study, STICS has been proven to well reproduce cereal and legume intercrops competition and complementarity effects in contrasting tropical conditions (de Freitas, 2023). STICS well reproduced the partial Land Equivalent Ratio (pLER) of cereals (maize, millet and sorghum) and legumes (cowpea and pigeon pea) but couldn't reproduce the facilitation effects when observed pLER were higher than one (Fig. 10). Some other discrepancies were identified in several cases. Corre-Hellou et al. (2007, 2009) had difficulties in calculating light competition related to insufficient simulation of plant height, while Shili-Touzi et al. (2010) found a tendency to overestimate N uptake by a red fescue when intercropped with winter wheat.

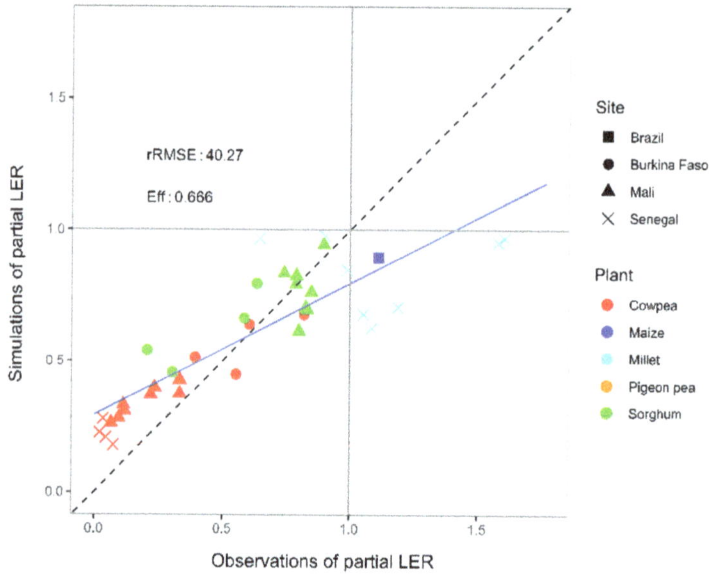

Figure 10 STICS simulations of partial Land Equivalent Ratio (LER) of grain yield for tropical sites in several cereals and legumes intercrops - From de Freitas (2023).

To overcome these limitations, Vezy et al. (2023) proposed new equations and algorithms that improve the simulation of important processes such as development phases, light interception, plant growth, nitrogen and water balance and yield formation in intercropping. The model was found consistent and generic under a large range of bispecific spring and winter intercrops, including Faba Bean-Wheat, Pea-Barley, Soybean-Sunflower, and Wheat-Pea mixtures (nRMSE = 25% for maximum leaf area index, 23% for shoot biomass at harvest, and 18% for grain yield). Several tropical intercropping systems are also currently being evaluated using these new STICS formalisms (de Freitas, 2023, de Freitas et al., 2025).

8 Recent STICS model improvements: soil-related processes

8.1 Mineralization of soil organic matter

The mineralization of humified soil organic matter (SOM) into mineral N (NH_4^+ and NO_3–) and CO_2 in STICS is assumed to occur in an upper soil layer (*profhum*) considered biologically active and to be negligible below that layer. The humified organic matter is divided into (i) an active pool that is fed by the decomposition of fresh organic residues and affected by the mineralisation process, and (ii) a stable pool considered to be inert at the century scale. In

STICS, the SOM mineralisation depends on the amount of active soil organic N and C and on an actual mineralisation rate constant. This latter depends on soil temperature and moisture and on a potential mineralization rate constant (*Khum*), which depends on soil properties. The considered soil properties affecting *Khum* were updated from the STICS V9.0 version according to the study on N mineralization in bare fallow (BF) soils published in Clivot et al. (2017). *Khum* thus integrates the influences of clay and $CaCO_3$ contents (already considered in previous versions) and also the additional effects of soil pH and C:N ratio of organic matter (within *profhum* layer).

Integrating these new mineralization formalisms improved the model performances for simulating soil mineral N contents (and particularly for near-surface layers) under BF soils in 566 simulation units from the STICS evaluation database. This improved model version has also demonstrated its ability to predict satisfactorily long-term soil organic carbon (SOC) dynamics across BF sites in Europe (Farina et al., 2021). Ensuring the validity of model simulations under BF conditions is crucial to ensure a robust and independent parameterization of the model soil submodel. It is important to avoid situations where plants interact, sometimes strongly, with N, C and water cycles in soils and for which inadequate model parameterization would potentially allow to compensate for errors between soil and plant submodels.

The improved prediction of long-term SOC dynamics were then also confirmed in diverse arable cropping systems, not only for STICS but also for the more simple AMG model (Clivot et al., 2019, 2020), both models sharing a common representation of the influence of soil properties on SOM mineralization.

8.2 N_2O emissions

In recent years, the description of N_2O emissions has been improved on the basis of the concepts described in Bessou et al. (2010). Nitrification and denitrification processes, and the N_2O emissions associated with each process, are simulated separately. Coupling of nitrification and denitrification exists through the production of nitrate by nitrification, which then serves as substrate for denitrification and N_2O emissions associated with this pathway. Processes are described using a functional approach, which is similar to that used in other models such as DayCent (Del Grosso et al., 2001) or Ecosse (Smith et al., 2010).

The nitrification rate is considered to be proportional to NH_4^+ content, which proved to be a good approximation over the typical range of ammonium content in soil. It is affected by temperature, increasing until an optimum rate and then decreasing again (Benoit et al., 2015). Water-filled pore space (WFPS) strongly influences the rate of nitrification as its intensity increases until field capacity is reached, then decreases because of the decline of soil aeration (Khalil et al., 2004). pH values decreasing below neutrality progressively constrain nitrification before stopping it at a pH of 4. N_2O emission associated with the nitrification pathway is calculated as a variable fraction of the nitrification rate. That fraction remains low for WFPS values below 60% (< 0.3%) but strongly increases with WFPS afterwards to reach a maximum of 2.5% of nitrified N in the topsoil at 100% WFPS, as a consequence of the resulting decline in oxygen availability (Khalil et al., 2004).

Denitrification is calculated as the product of a soil-dependent potential rate and functions expressing the effects of nitrate concentration, soil temperature and soil water content (Brisson et al., 2004). The denitrification rate increases with NO_3– content following Michaelis-Menten kinetics (Bessou et al., 2010), and it increases with temperature over most of the typical range of temperature as the optimum rate is 47°C (Benoit et al., 2015). As in the NEMIS model (Hénault and Germon, 2000), the denitrification rate is null when WFPS is below a threshold whose default value is 62% and then increases exponentially with WFPS. N_2O emission from denitrification is calculated as a variable fraction of the denitrification rate, which mainly depends on pH and WFPS. Acid pH strongly inhibits N_2O reduction to N_2 (Rochester, 2003): for pH values below 6, denitrification has N_2O as the main end-product. Finally, soil water content values close to saturation favor N_2O reduction to N_2. This is due to the development of anoxia (Vieten et al., 2008) and to the increase in the residence time of N_2O into the soil associated with lower gaseous diffusion rates. This effect is taken into account through a linear decrease of the denitrification end-product ratio between the threshold WFPS for the onset of denitrification and maximal soil saturation (Bessou et al., 2010).

One main area for future development of the N_2O emission model is to better couple the simulation of denitrification to carbon availability in the soil (Chaves et al., 2024). This would avoid using a denitrification potential parameter, which remains difficult to measure while at the same time dealing better with the temporal dynamics of denitrification, in particular regarding the influence of practices which influence carbon availability in the soil.

9 Recent STICS model improvements: new modules and tools

9.1 Considering grazing impacts on pasture simulations through an internal module

A grazing module was developed within STICS to account for animal returns to the soil in grazed grasslands (Graux et al., 2020) with a parameterization available for dairy and suckling cows. This grazing module is a simplified but useful representation of animal returns during a grazing period in which animals are assumed to be fed with grazed grass, without concentrates' supply. Grassland intake is simulated as a cutting event, with all of the herbage removed being assumed to feed a fluctuating number of animals that have a uniform and fixed dry matter intake. Animal feces and urine are represented by an application of cattle liquid manure and urea, respectively, to the soil surface on the day of herbage defoliation. This module is based on known relationships between animal N returns during grazing, animal herbage dry matter intake, and herbage N content, all of which respond to N availability (including fertilization). The decomposition of animal feces uses existing STICS equations for simulating the mineralization of organic residues with grazing-related parameter values. Urea return is represented as an application of mineral fertilizer, of which up to 15% can be volatilised, taken up by plants, nitrified, leached, denitrified and immobilized in soil organic matter.

In STICS, animal stocking density and the length of each grazing period are thus model outputs instead of inputs, but it is possible to calculate grazing days. This simplification makes it difficult to compare predictions to observations, as simulated herbage growth may differ from reality, and simulated grazing days may not reflect the stocking density or grazing period length observed. Yet, based on this methodology, Delandmeter et al. (2024) successfully simulated the response of Brazilian integrated crop and livestock systems over 18 years and assessed their potential to mitigate climate change impacts.

9.2 Performing multi-simulation, sensitivity analysis and parameter optimization with external tools

In recent years, new applications of the STICS model have emerged, necessitating large-scale simulations and finer control over its inputs and outputs. This requires the use of software tools that enable efficient, flexible and reproducible management of the model and its data via scripts.

For many years, an existing graphical interface has allowed users to define the model's input files, initiate simulations and display corresponding results. Supplementary tools were also developed in Matlab to facilitate the simulation of large experimental designs and model calibration by coupling with optimization methods available in that language (Wallach et al., 2011). New software tools have been developed since 2018 under the SticsRPacks project to drive the STICS model from R, aiming to facilitate and extend its usage. The objective was to create open-source, modular, well-documented tools using a programming language accessible to a wide audience.

SticsRPacks comprises four R packages. Two of them, SticsRFiles (Lecharpentier et al., 2024a) and SticsOnR (Lecharpentier et al., 2024b), are designed to interface with the STICS model from R. SticsRFiles provides functions for creating and managing STICS inputs, as well as importing and pre-formatting its simulated outputs. SticsOnR facilitates running STICS simulations from R, optionally distributed over different CPUs.

The remaining two packages, CroptimizR (Buis et al., 2023) and CroPlotR (Vezy et al., 2024), aim to simplify the implementation, comparison and automation of parameter estimation and evaluation processes for crop models. They offer a generic interface for connecting crop models with contemporary parameter estimation algorithms and for generating customized plots and statistics for evaluation purposes. These packages have been so far integrated as standard tools in STICS, APSIM NextGen and SiriusQuality2 and have been utilized with over 12 different crop models to implement and compare calibration protocols within the AgMIP international project (Wallach et al., 2023).

All SticsRPacks packages are open-source and freely available on Github under the LGPL license. They adhere to a strict development workflow, including shared concepts, enforced code style and automated tests for both their functions (unit tests) and workflows (functional tests). Detailed documentation, including a tutorial and examples of application on several crop models for the generic packages, is provided. Interested users can refer to each package *via* the link to the dedicated Github pages.

10 Areas for further model enhancement

10.1 Phosphorus stress

Crop phosphorus (P) nutrition is one of the key sustainability challenges of the 21st century and especially in low inputs agriculture (Cordell and White, 2014). Crop models are relevant tools to study and manage phosphorus in agroecosystems. However, P modeling has lagged behind nitrogen and carbon modeling. A major reason for this delay is the difficulty in developing a semi-mechanistic model that adequately predicts the temporal evolution of soil P availability and crop uptake (Das et al., 2019), as well as the P nutritional status of crops and the feedback of P deficiency on crop growth. Based on the FUSSIM-P model (Mollier et al., 2008), a new phosphorus module is currently being developed and will soon be integrated into the STICS model. This P module allows the simulation of the P availability in the soil, the P demand of the crop, the P uptake by the roots and the feedback effects of P deficiency on crop growth.

Plant-available soil P in the ploughed layer is calculated daily for each 1 cm soil layer (z). It has been evaluated by a functional and process-based approach, which consists in determining together the orthophosphate ions (oPions) concentration (C_P, mg P L^{-1}) and the oPions quantity (Q_w) in solution, and the amounts of diffusive oPions bound to soils (P_r, mg P kg^{-1} soil) that equilibrate the solution over time (t in minutes) (Morel et al. 2000).

Daily plant P demand is determined by plant growth and biomass production. The aerial part demand is based on the concept of dilution curves as presented for N by Lemaire et al. (2008) and extended to P (Lemaire et al., 2019; Morel et al., 2021). The root P demand is calculated on the basis of a constant optimal root P content.

The root P uptake submodule uses the formalisms of the FUSSIM-P root P uptake module (Mollier et al. 2008) adapted to the one-dimensional soil description of STICS. The microscopic model for nutrient uptake developed by De Willigen and Van Noordwijk (1987) is used for each soil layer according to root length density and plant available soil P. Phosphorus transport within the soil solution around the root is driven by mass flow and diffusion. The maximum P uptake rate is equal to the maximum possible transport rate (by diffusion and mass flow) to the root, i.e. the root acts as a zero sink. The maximum nutrient uptake rate per unit root area is derived from the steady-state approximation of the concentration profile around the root for the zero-sink condition (De Willigen and Van Noordwijk, 1994). Root P uptake is determined by crop P demand. The actual root P uptake may or may not meet the crop P demand. It is assumed that the actual P uptake rate is equal to the required P uptake rate as long as it is less than the maximum P uptake rate, otherwise, it is equal to the maximum P uptake rate. Total P uptake by the whole root system is the sum of P uptake from all soil layers.

The Phosphorus Nutrition Index (PNI) is calculated as a function of P uptake and is used to apply feedback to crop growth. The main processes affected by P deficiency are plant development, root elongation and leaf production and senescence.

A first version of the STICS P model was evaluated using a dataset from four field trials with deficient to excessive plant-available soil P at different sites in mainland France (Seghouani et al., 2025). Briefly, the trials consisted of fertilizing maize with a mineral fertilizer at three application rates (P0, P1, P2), corresponding to no P added or P added at one or two times the crop yield. Model predictions of both crop P uptake and plant growth indicators (yield, LAI and biomass) were evaluated. The model has shown great ability in predicting P uptake both dynamically and at the end of the growing season for the whole data set (EF > 0.75). The model has satisfactory predictions of crop biomass accumulation (EF > 0.5) and leaf area index. Considering each fertilization level separately, the evaluation showed that the model had predicted the fertilized treatments better than the unfertilised ones (EF of 0.73, 0.75 and 0.64 for P2, P1 and P0, respectively). However, the evaluation of the latter remains satisfactory for both P uptake and plant growth.

The good performance of the model is promising as it shows that the model is sufficiently robust to simulate maize P uptake over a range of soil P availability under contrasting temperate climatic conditions. Despite the relative simplicity of the model, which does not include all rhizosphere mechanisms, it appears to perform well even at low P levels under the conditions considered. However, the fact that the model predicted the fertilized treatment better than the unfertilized treatment raises the question of model performance under conditions of more severe P deficiency or even lower soil P availability status (e.g. in tropical soils). Further validation on different crop species and soil and climatic conditions is therefore needed.

10.2 Towards biotic stress simulation: the case of airborne fungal diseases

Although the STICS model does not directly include the impacts of diseases on crop growth, it can be coupled with disease models thanks to simulation outputs such as phenology, LAI, biomass and canopy microclimate variables, including hourly reconstitution of temperatures and humidity.

Previous works (Caubel et al., 2012; Caubel et al., 2017) have coupled STICS with a generic disease model called MILA (MILA is not an acronym, just the regular model name). MILA-STICS simulates the development of pathogenic fungi and their impacts on plant organs and crop growth for a range of polycyclic airborne fungal diseases. MILA's conceptual design is organized around several connected modules representing the processes involved in the dynamics of an epidemic, namely:

- dispersal of spores and interception by plant organs;
- infection by intercepted spores;
- latency of deposited spores;
- secondary inoculum production by infectious lesions; and
- lesion and spore lifespan.

MILA runs at a daily time step with feedback to and from STICS' daily loop (Fig. 11). Each day, cohorts of spores and lesions are created and/or transferred from one state to the next (available, dispersed, infectious, etc.) depending on crop phenology, canopy microclimate and pathogen parameters, which all increment a transition variable until it reaches a given threshold. The components of the relevant organs of the plant are updated depending on status (healthy, diseased or dead tissues), and a severity index is provided as an output. In the case of leaves, severity is a percentage of diseased LAI which is removed from the photosynthetic area used in STICS. Hence the simulations from MILA are directly connected to plant function processes handled in STICS.

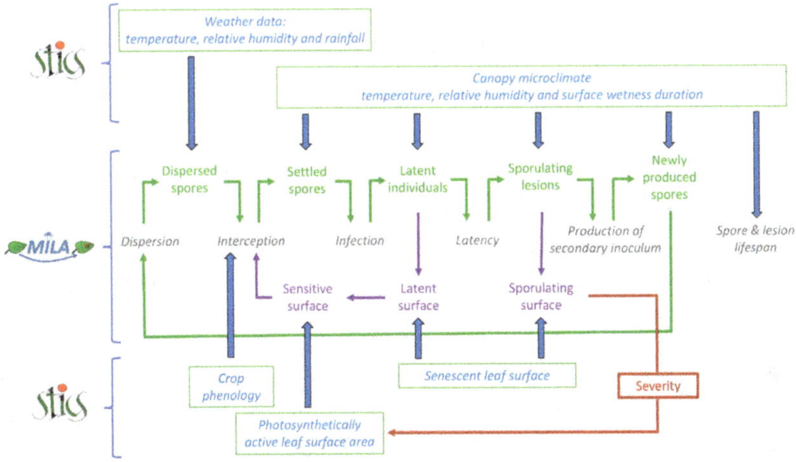

Figure 11 Schematic representations of the coupling between STICS and MILA models (from Caubel et al., 2012; Caubel et al., 2017).

Until the creation of the coupled model MILA-STICS, little work had been conducted on modeling the relationships between climate, soils, plants, their microclimate and diseases in an integrated, mechanistic manner while remaining generic enough to be applied to a number of various diseases and crops. Several works have used the MILA-STICS coupled model to study disease-crop systems such as wheat with brown rust and vine with mildew (Caubel et al., 2012).

Since STICS can also run long-term simulations, climate change studies are possible. For example, results have already shown that changing weather patterns affect the phenology of both pathogens and crops as well as the microclimate of the canopy, thereby desynchronizing the developmental phases and increasing damage to crop tissues (Caubel et al., 2017). Current research projects include a case study on yellow rust and the adaption of MILA to STICS intercropping.

10.3 Pasture simulations: the need to better account for C sequestration

The capacity of grasslands to mitigate climate change by storing carbon in soils is well known but difficult to quantify because it depends on soil and climate conditions, nature of the vegetation and grassland management. Modeling the evolution of soil organic carbon (SOC) stocks requires that models are sufficiently robust and accurate in their prediction. The ability of the STICS model to simulate SOC stocks has already been evaluated in arable cropping systems, with satisfactory results (Autret et al., 2020; Yin et al., 2020; Clivot et al., 2020), but only few have explored this aspect under rotations that include temporary grassland (Autret et al., 2020), or under permanent grassland soils. Work in this area has been proposed by Graux et al. (2023, 2024).

A new research version of the model derived from STICS V10 was calibrated and used to evaluate SOC simulation using data from three long-term (14–27 years) French experimental trials, located in contrasted pedoclimatic conditions in France (Lusignan, Kerbernez and Theix sites), and offering data for arable crops alone, arable crops alternating with temporary grassland and permanent grassland. These sites include several treatments with different grassland duration, different levels of N fertilization, and mown and grazed situations. Some grassland shoot parameters were adjusted based on the literature to better represent the productive grassland functional groups present in the experimental trials. Then, part of the data from the Lusignan trial was used to calibrate parameters involved in simulating the allocations of biomass to shoots and roots and the roots' lifespan and N demand. The final grassland parameterisation was then evaluated against observations in the remaining treatments in Lusignan and in the Kerbernez and Theix trials. We evaluated the model's performance by calculating statistical criteria and comparing the root variables with the orders of magnitude known from the literature (Freschet et al. 2017; Legay et al. 2016; Mokany et al. 2006).

With the final grassland parameterization, SOC dynamics at Lusignan and Kerbernez trials are well simulated under both temporary and permanent grassland (Graux et al., 2024). It also makes it possible to accurately simulate root C and N in grasslands, variables that were not previously simulated. At Theix, the SOC stock evolution under permanent grasslands is less well simulated, with a tendency towards destocking where observations indicate storage or slight destocking. However, SOC observations for this site were only available for a rather short period (7 years) and with a high spatial variability, even at the start of the simulation. In general, the order of magnitude of the grass biomass, root/shoot ratio, root N content and root C/N ratio is in line with the expected values according to literature and experimental data. Evaluation of the STICS model shows that it is able to reproduce observed trends in SOC evolution in rotations, including temporary grassland and in permanent grassland soils.

10.4 A generic external module to simulate a complete CO_2 balance from model outputs: illustration with STICS

Taking the most of cropland's potential to mitigate climate change requires a deep understanding of (i) soil-plant CO_2 fluxes and (ii) how agricultural management and environmental variables affect them. Originally, STICS does not explicitly compute all CO_2 fluxes within the agroecosystem. STICS computes Heterotrophic Respiration (HR), i.e. the carbon losses from the decomposition by microorganisms of soil organic matter. However, it uses the concept of radiation use efficiency which combines photosynthesis and respiration to directly convert photosynthetically active radiation in aerial biomass. An external module was recently developed around STICS (Delandmeter et al., 2023) to explicitly compute (i) Gross Primary Productivity (GPP), i.e. carbon fixed by plants, and (ii) Autotrophic Respiration (AR), i.e. carbon lost by internal plant metabolisms for plant growth and maintenance. With these additions, the whole Ecosystem Respiration (RECO) and the Net Ecosystem Exchange (NEE) can be computed, enabling to assess the balance between photosynthesis and respiration.

The methodology is extensively described in Delandmeter et al. (2023). In brief, GPP is computed from the daily change in all plant carbon pools, i.e. the net primary productivity (NPP), to which we subtract AR. This latter is itself computed by differentiating, on the one hand, maintenance respiration (MR), derived from plant nitrogen concentration and plant biomass, and, on the other hand, growth respiration (GR), which is estimated as a fixed fraction of the available carbon for growth (GPP-MR, where GPP is here the one of the previous day). RECO is logically the sum of AR and HR; AR is the sum of MR and GR, and NEE is the sum of GPP and RECO. The generic approach is transferable to most soil-crop process-based models.

Based on the data collected on a field experiment situated in Lonzée (Belgium), which is part of the ICOS (Integrated Carbon Observation System)

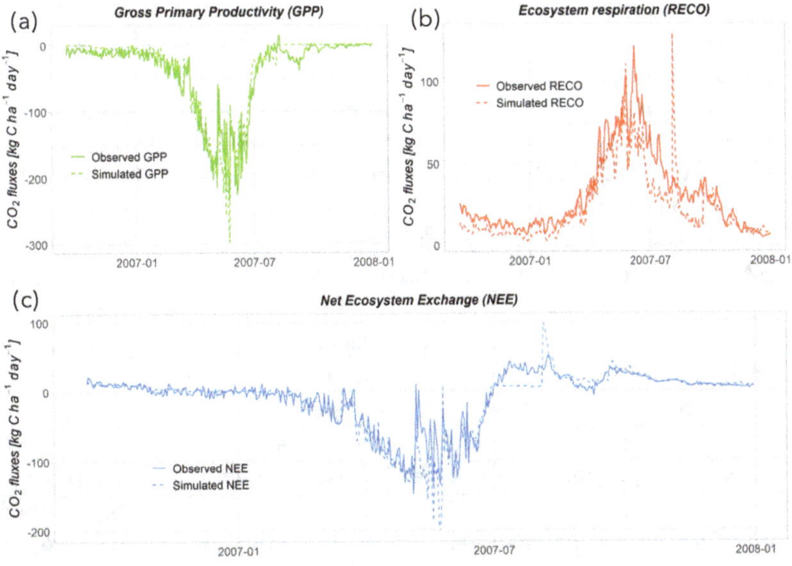

Figure 12 Dynamics of the CO_2 fluxes related to NEE (c), GPP (a) and RECO (b), observed and simulated, for a winter wheat cropping season at the BE-LON ICOS site of Lonzée (Belgium) - From Delandmeter et al., 2023.

international network, the methodology was calibrated and validated over a 16-year crop sequence, comprising winter wheat, sugar beet, maize, potato and cover crops. It globally proved to be able to simulate accurately daily CO_2 fluxes, as illustrated on Fig. 12. The model also allows to evaluate GPP and RECO budgets accurately, with the contribution of each of their components. Yet cumulated NEE fluxes are overestimated, due to slight underestimation of respiration and slight overestimation of photosynthesis, indicating that more precise estimations of heterotrophic respiration are required to obtain reliable net C budgets.

The STICS outputs integrated in the C budget model also demonstrated its ability to successfully capture the variability of CO_2 fluxes due to various environmental factors, allowing to estimate the CO_2 fluxes associated to different crop rotations under different pedo-climatic contexts and/or management strategies. We believe that it opens the door to better exploration of the climate change impacts on cropping systems and their potential of mitigation of climate change due to CO_2 fluxes. Yet the whole methodology would benefit from further validation on contrasted agricultural systems (including e.g. grasslands) and/or under various pedoclimatic contexts, such as the ones offered by ICOS sites.

11 Conclusion

Throughout this chapter, constructed around a general presentation and numerous illustrations of STICS model uses, the reader would have the opportunity to discover the STICS soil-crop model abilities and performances. This was also a great occasion to present an overview of the wide domain of validity of the model, which has demonstrated its potential for a wide range of agronomic and environmental applications. Since the beginning of its story in the early 90s, STICS has been able to remain a generic and robust model that allows to simulate the functioning of a wide range of agroecosystems, both in temperate and tropical environments.

As highlighted in this chapter, and particularly in the last two sections, many efforts were, and are still, put to keep the model up-to-date and in a permanent state of evolution towards better representation of cropping systems, allowing to analyze novel research and applied questions. This is made possible thanks to a strong collaboration between the team of engineers and researchers associated to (re)design and enhance the model and a large community of users with complementary expertise. The users' community is well-organized and freely opened to any contributors willing to improve STICS. We believe that STICS will be valuable to help support the design of innovative cropping systems, meeting tomorrow's future challenges of an ever-changing world, in particular for the necessary ecological transition of agriculture and to sustain the adaptation (and mitigation) to climate change.

12 Where to look for further information

The model's conceptual framework, the equations underlying each formalization, as well as case studies and instructions for use are extensively detailed in Beaudoin et al. (2023), which is available as a free ebook (epub or PDF) at https://stics.inrae.fr/eng/resources/documentation or at https://doi.org/10.35690/978-2-7592-3679-4.

The STICS model and its graphical user interface, JavaSTICS, are freely accessible at https://stics.inrae.fr/eng/download under the CeCILL-B license. Compiled versions are available for Windows, Mac and Linux users. Comprehensive documentation is provided to guide users through the installation process and the use of the interface. The source code of the STICS model is also available under the CeCILL-C license and can be downloaded from the same link. The full text of the CeCILL-B and CeCILL-C licenses can be found at: https://cecill.info/licences/Licence_CeCILL-B_V1-en.html and https://cecill.info/licences/Licence_CeCILL-C_V1-en.html.

Additionally, the SticsRPacks R packages are open-source and come with detailed documentation. These packages are freely available on GitHub under the LGPL license at https://github.com/SticsRPacks. Links to each specific pack out of the four R packs can be found directly in the reference section: Lecharpentier et al. (2024a) for SticsRFiles, Lecharpentier et al. (2024b) for SticsOnR, Buis et al. (2023) for CroptimizR and Vezy et al. (2024) for CroPlotR.

13 Acknowledgements

The authors would like to thank the INRAE AgroEcoSystem department for providing financial support to the STICS Project Team. The authors thank all physiologists and modelers whose work and publications enabled the modeling of plant processes. The authors would like to thank all the people and institutes involved in acquiring the experimental data used for the development of the STICS model. The authors also thank the F.R.S.-FNRS (Belgian Fund for Scientific Research), which awarded a Research Fellow grant (number 44221) to M. Delandmeter.

14 References

Artru, S. 2017. *Impact of spatio-temporal shade on crop growth and productivity, perspectives for temperate agroforestry*. Doctoral thesis, ULiège - Université de Liège. ORBi-University of Liège. https://hdl.handle.net/2268/211276

Asseng, S., et al. 2019. Climate change impact and adaptation for wheat protein. *Global Change Biology*, 25(1), 155–173. https://doi.org/10.1111/gcb.14481

Autret, B., et al. 2020. Long-term modelling of crop yield, nitrogen losses and GHG balance in organic cropping systems. *Science of the Total Environment*, 710, 134597. https://doi.org/10.1016/j.scitotenv.2019.134597

Basso, B., et al. 2018. Soil organic carbon and nitrogen feedbacks on crop yields under climate change. *Agricultural & Environmental Letters*, 3, 180026.

Basso, B., et al. 2019. Yield stability analysis reveals sources of large-scale nitrogen loss from the US Midwest. *Scientific Reports*, 9, 5774. https://doi.org/10.1038/s41598-019-42271-1

Bassu, S., et al. 2014. How do various maize crop models vary in their responses to climate change factors? *Global Change Biology*, 20(7), 2301–2320. https://doi.org/10.1111/gcb.12520

Beaudoin, N., et al. 2023. Stics Soil-Crop Model: Conceptual framework, equations and uses. Versailles, éditions Quae, 516p. https://doi.org/10.35690/978-2-7592-3679-4

Beer, A. 1852. Ann. *Physik (Poggendorf)*, 86, 78.

Benoit, M., et al. 2015. Temperature dependence of nitrous oxide production of a luvisolic soil in batch experiments. *Process Biochemistry*, 50, 79–85. https://doi.org/10.1016/j.procbio.2014.10.013

Bessou, C., et al. 2010. Modelling soil compaction impacts on nitrous oxide emissions in arable fields. *European Journal of Soil Science*, 61, 348–363. https://doi.org/10.1111/j.1365-2389.2010.01243.x

Blackmore, S. 2000. The interpretation of trends from multiple yield maps. *Computers and Electronics in Agriculture*, 26, 37–51. https://doi.org/10.1016/S0168-1699(99)00075-7

Brisson, N., et al. 1992. Agrometeorological soil water balance for crop simulation models. *Agricultural and Forest Meteorology*, 59, 267–287. https://doi.org/10.1016/0168-1923(92)90097-N

Brisson, N., et al. 1998. STICS: A generic model for the simulation of crops and their water and nitrogen balances. I. Theory and parameterization applied to wheat and corn. *Agronomie*, 18, 311–346. https://doi.org/10.1051/agro:19980501

Brisson, N., et al. 2002. STICS: A generic model for simulating crops and their water and nitrogen balances. II. Model validation for wheat and maize. *Agronomie*, 22, 69–92. https://doi.org/10.1051/agro:2001005

Brisson, N., et al. 2004. Adaptation of the crop model STICS to intercropping. Theoretical basis and parameterisation. *Agronomie*, 24, 409–421. https://doi.org/10.1051/agro:2004031

Buis, S., et al. 2011. The STICS crop model and associated software for analysis, parameterization, and evaluation. In L.R. Ahuja and L. Ma (Eds.), *Methods of Introducing System Models into Agricultural Research*, Volume 2, Chapter 14. https://doi.org/10.2134/advagricsystmodel2.c14

Buis, S., et al. 2023. CroptimizR: A Package to Estimate Parameters of Crop Models. Zenodo. https://doi.org/10.5281/zenodo.4066451, https://github.com/SticsRPacks/CroptimizR.

Caubel, J., et al. 2012. Generic response functions to simulate climate-based processes in models for the development of airborne fungal crop pathogens. *Ecological Modelling*, 242, 92–104. https://doi.org/10.1016/j.ecolmodel.2012.05.012

Caubel, J., et al. 2017. Climate change effects on leaf rust of wheat: implementing a coupled crop-disease model in a French regional application. *European Journal of Agronomy*, 90, 53–66. https://doi.org/10.1016/j.eja.2017.07.004

Celette, F. and Gary, C. 2013. Dynamics of water and nitrogen stress along the grapevine cycle as affected by cover cropping. *European Journal of Agronomy*, 45, 142–152. https://doi.org/10.1016/j.eja.2012.10.001

Chaves, B., et al. 2024. Modelling decomposition of crop residue mulches and the associated N2O emissions in a no-till system in southern Brazil. *Soil and Tillage Research*, 240, 106065. https://doi.org/10.1016/j.still.2024.106065

Clivot, H., et al. 2017. Quantifying in situ and modeling net nitrogen mineralization from soil organic matter in arable cropping systems. *Soil Biology and Biochemistry*, 111, 44–59. https://doi.org/10.1016/j.soilbio.2017.03.010

Clivot, H., et al. 2019. Modeling soil organic carbon evolution in long-term arable experiments with AMG model. *Environmental Modelling & Software*, 118, 99–113. https://doi.org/10.1016/j.envsoft.2019.04.004

Clivot, H., et al. 2020. Simulating soil organic carbon dynamics in long-term bare fallow and arable experiments with STICS model. In Book of Abstracts. Montpellier (France), pp. 70–71. https://hal.inrae.fr/hal-02737949

Confalonieri, R., et al. 2016. A taxonomy-based approach to shed light on the babel of mathematical models for rice simulation. *Environmental Modelling & Software*, 85, 332–341.

Cordell, D. and White, S. 2014. Life's Bottleneck: sustaining the World's phosphorus for a food secure future. *Annual Review of Environment and Resources*, 39, 161–188. https://doi.org/10.1146/annurev-environ-010213-113300

Corre-Hellou, G., et al. 2007. Effect of root depth penetration on soil nitrogen competitive interactions and dry mat- ter production in pea–barley intercrops given different soil nitro- gen supplies. *Field Crops Research*, 103:76–85. https://doi.org/10.1016/j.fcr.2007.04.008

Corre-Hellou, G., et al. 2009. Adaptation of the STICS intercrop model to simulate crop growth and N accumula- tion in pea–barley intercrops. *Field Crops Research*, 113:72–81. https://doi.org/10.1016/j.fcr.2009.04.007

Couëdel, A., et al. 2024. Long-term soil organic carbon and crop yield feedbacks differ between 16 soil-crop models in sub-Saharan Africa. *European Journal of Agronomy*, 155, 127109. https://doi.org/10.1016/j.eja.2024.127109

Cruz, P., et al. 2010. Functional typology of perennial forage grasses : a classification based on several characteristics. *Fourrages*, 201, 11–17.

Das, B., et al. 2019. Soil phosphorus modeling for modern agriculture requires balance of science and practicality: a perspective. *Journal of Environmental Quality*, 48, 1281–1294. https://doi.org/10/gf9x6d

De Freitas, M. 2023. Intercropping cereals and legumes to stabilise yield in the tropics: evaluation of the STICS soil-crop model to simulate bi-specific intercrops. Montpellier: AgroParisTech, 57 p. Mémoire d'ingénieur: Produire et innover dans les systèmes techniques végétaux (PISTv): AgroParisTech.

De Freitas, M., et al. 2025. Associations céréale-légumineuse pour stabiliser les rendements en régions tropicales: évaluation du modèle de culture STICS. *Innovations Agronomiques*, 99, 133-146. DOI: https://doi.org/10.17180/ciag-2025-vol99-art11.

Delandmeter, M., et al. 2023. A comprehensive analysis of CO2 exchanges in agro-ecosystems based on a generic soil-crop model-derived methodology. *Agricultural and Forest Meteorology*, 340, 109621. https://doi.org/10.1016/j.agrformet.2023.109621

Delandmeter, M., et al. 2024. Integrated crop and livestock systems increase both climate change adaptation and mitigation capacities. *Science of the Total Environment*, 912, 169061. https://doi.org/10.1016/j.scitotenv.2023.169061

Del Grosso, S.J., et al. 2001. Simulated interaction of carbon dynamics and nitrogen trace gas fluxes using the DAYCENT model. In: M. Schaffer, et al. (Eds.), *Modeling Carbon and Nitrogen Dynamics for Soil Management*. CRC Press, Boca Raton, Florida, pp. 303–332.

De Willigen, P., et al. 1987. *Roots, Plant Production and Nutrient Use Efficiency* (PhD). Wageningen Agricultural University, Wageningen, The Netherlands.

De Willigen, P. and Van Noordwijk, M. 1994. Mass flow and diffusion of nutrients to a root with constant or zero-sink uptake. II. Zero-sink uptake. *Soil Science*, 157, 171–175.

de Wit C.T. (Ed.). 1978. *Simulation of Assimilation, Respiration and Transpiration of Crops*. Centre for Agricultural Publishing and Documentation, Simulation monographs. Pudoc, Wageningen, The Netherlands.

Dumont, B., et al. 2014. Parameter optimisation of the STICS crop model, with an accelerated formal MCMC approach. *Environmental Modelling & Software*, 52, 121–135. https://doi.org/10.1016/j.envsoft.2013.10.022

Dumont, B., et al. 2015a. Climatic risk assessment to improve nitrogen fertilisation recommendations: a strategic crop model-based approach. *European Journal of Agronomy*, 65, 10–17. https://doi.org/10.1016/j.eja.2015.01.003

Dumont, B., et al. 2015b. Systematic analysis of site-specific yield distributions resulting from nitrogen management and climatic variability interactions. Precision Agric 16, 361–384. https://doi.org/10.1007/s11119-014-9380-7

Dumont, B., et al. 2016. Assessing and modeling economic and environmental impact of wheat nitrogen management in Belgium. *Environmental Modelling & Software*, 79, 184–196. https://doi.org/10.1016/j.envsoft.2016.02.015

Duru, M., et al. 2010. Herb'sim, a model for a rational management of grass production and grass utilization. *Fourrages*, 201, 37–46.

Farina, R., et al. 2021. Ensemble modelling, uncertainty and robust predictions of organic carbon in long-term bare-fallow soils. *Global Change Biology*, *27*(4), 904–928. https://doi.org/10.1111/gcb.15441

Fort, F., et al. 2013. Root and leaf functional trait relations in Poaceae species: implications of differing resource-acquisition strategies. *Journal of Plant Ecology*, 6, 211–219. https://doi.org/10.1093/jpe/rts034

Fraga, H., et al. 2016. Modelling climate change impacts on viticultural yield, phenology and stress conditions in Europe. *Global Change Biology*, 22(11), 3774–3788. https://doi.org/10.1111/gcb.13382

Freschet, G.T., et al. 2017. Climate, soil and plant functional types as driv-ers of global fine-root trait variation. *Journal of Ecology*, 105, 1182–1196. https://doi.org/10.1111/1365-2745.12769

García de Cortázar-Atauri, I., 2006. *Adaptation du modèle STICS à la vigne (Vitis vinifera L.): Utilisation dans le cadre d'une étude d'impact du changement climatique à l'échelle de la France*. Doctoral thesis. École nationale supérieure agronomique de Montpellier. http://www.theses.fr/2006ENSA0030

García de Cortázar-Atauri, I., et al. 2009a. Performance of several models for predicting budburst date of grapevine (Vitis vinifera L.). *International Journal of Biometeorology*, 53, 317–326. https://doi.org/10.1007/s00484-009-0217-4

García de Cortázar-Atauri, I., et al. 2009b. Asynchronous dynamics of grapevine ("Vitis vinifera") maturation: experimental study for a modelling approach. *Journal International des Sciences de la Vigne et du Vin*, 43, 83–97. https://doi.org/10.20870/oeno-one.2009.43.2.801

Graux, A.-I., et al. 2020. High-resolution assessment of French grassland dry matter and nitrogen yields. *European Journal of Agronomy*, 112, 125952. https://doi.org/10.1016/j.eja.2019.125952

Graux, A.I., et al. 2023. STICS ability to simulate long-term soil organic matter dynamics in crop-grassland rotations. XIII Séminaire STICS, Bordeaux, 13–16 Novembre 2023. https://hal.inrae.fr/hal-04637939

Graux, A.I., et al. 2024. Calibration and evaluation of the STICS model for long-term simulation of soil organic carbon dynamics in grassland soils (in prep.)

Hénault, C. and Germon, J.C. 2000. NEMIS, a predictive model of denitrification on the field scale. *European Journal of Soil Science*, 51(2), 257–270. https://doi.org/10.1046/j.1365-2389.2000.00314.x

Jouven, M., et al. 2006. Model predicting dynamics of biomass, structure and digestibility of herbage in managed permanent pastures. 1. Model description. *Grass and Forage Science*, 61, 112–124. https://doi.org/10.1111/j.1365-2494.2006.00515.x

Khalil K., et al. 2004. Nitrous oxide production by nitrification and denitrification in soil aggregates as affected by O2 concentration. *Soil Biology and Biochemistry*, 36, 687–699. https://doi.org/10.1016/j.soilbio.2004.01.004

Kherif, O., et al. 2022. The first calibration and evaluation of the STICS soil-crop model on chickpea-based intercropping system under Mediterranean conditions. *European Journal of Agronomy*, 133, 126449. https://doi.org/10.1016/j.eja.2021.126449

Kimball, B.A., et al. 2019. Simulation of maize evapotranspiration: An inter-comparison among 29 maize models. *Agricultural and Forest Meteorology*, 271, 264–284. https://doi.org/10.1016/j.agrformet.2019.02.037

Kothari, K., et al. 2021. Are soybean models ready for climate change food impact assessments? *European Journal of Agronomy*, 135, 126482. https://doi.org/10.1016/j.eja.2022.126482

Launay, M., et al. 2009. Exploring options for managing strategies for pea-barley intercropping using a modeling approach. *European Journal of Agronomy*, 31, 85–98. https://doi.org/10.1016/j.eja.2009.04.002

Lecharpentier, P., et al. 2024a. SticsRFiles: Read and Modify 'STICS' Input/Output Files. https://doi.org/10.5281/zenodo.4443206, https://cran.r-project.org/web/packages/SticsRFiles/index.html .

Lecharpentier, P., et al. 2024b. SticsOnR: Manage STICS Simulations Running the Executable or JavaStics. https://doi.org/10.5281/zenodo.4443130, https://github.com/SticsRPacks/SticsOnR.

Legay, N., et al. 2016. Influence of plant traits, soil microbial properties, and abiotic pa-rameters on nitrogen turnover of grassland ecosystems. *Ecosphere*, 7, e01448. https://doi.org/10.1002/ecs2.1448

Legrain, X., et al. 2011. Cartographie des sols en Belgique: aperçu historique et présentation des travaux actuels de valorisation et de révision de la Carte Numérique des Sols de Wallonie. *Biotechnologie, Agronomie, Société et Environnement*, 15(S2), 647–656.

Lemaire, G. and Gastal, F. 1997. N uptake and distribution in plant canopies. In: G. Lemaire (Ed.), *Diagnosis of the Nitrogen Status in Crops*. Springer, Berlin, Heidelberg, pp. 3–43. https://doi.org/10.1007/978-3-642-60684-7_1

Lemaire, G., et al. 2008. Diagnosis tool for plant and crop N status in vegetative stage theory and practices for crop N management. *European Journal of Agronomy*, 28, 614–624. https://doi.org/10.1016/j.eja.2008.01.005

Lemaire, G., et al. 2019. Allometric approach to crop nutrition and implications for crop diagnosis and phenotyping: a review. *Agronomy for Sustainable Development*, 39, 27. https://doi.org/10.1007/s13593-019-0570-6

Lenoir, A. 2024. *Opportunité des outils numériques pour optimiser la fertilisation azotée du blé à l'échelle d'un territoire*. Doctoral thesis, ULiège - Université de Liège. ORBi-University of Liège. https://hdl.handle.net/2268/318782

Lenoir, A., et al. 2024. Advancing Blackmore's methodology to delineate management zones from Sentinel 2 images. *Precision Agriculture*, 25, 1360–1385. https://doi.org/10.1007/s11119-024-10115-2.

Major, D., et al. 1991. Photoperiod response characteristics of alfalfa (Medicago sativa L.) cultivars. *Canadian Journal of Plant Science*, 71(1), 87–93. https://doi.org/10.4141/cjps91-010

Mary, B., et al. 1999. Calculation of nitrogen mineralization and leaching in fallow soil using a simple dynamic model. *European Journal of Soil Science*, 50(4), 549–566. https://doi.org/10.1016/0168-1923(92)90097-N

Mokany, K.I, et al. 2006. Critical analysis of root : shoot ratios in terrestrial biomes. *Global Change Biology*, 12(1), 84–96. https://doi.org/10.1111/j.1365-2486.2005.001043.x.

Mollier, A., et al. 2008. A two dimensional simulation model of phosphorus uptake including crop growth and P response. *Ecological Modelling*, 210, 453–464. https://doi.org/10/bkqvbc

Monteith, J.L. 1965. *Evaporation and Environment*. Cambridge University Press, Cambridge, pp. 205–234.

Monteith, J.L., 1972. Solar radiation and productivity in tropical ecosystems. *The Journal of Applied Ecology*, 9, 747–766. https://doi.org/10.2307/2401901

Morel, C., et al. 2000. Transfer of phosphate ions between soil and solution: Perspectives in soil testing. *Journal of Environmental Quality*, 29, 50–59. ttps://doi.org/10.2134/jeq2000.00472425002900010007x

Morel, C., et al. 2021. Calibration of maize phosphorus status by plant-available soil P assessed by common and process-based approaches. Is it soil-specific or not? *European Journal of Agronomy*, 122, 126174. https://doi.org/10.1016/j.eja.2020.126174

Nicolardot, B., et al. 2001. Simulation of C and N mineralisation during crop residue decomposition: a simple dynamic model based on the C:N ratio of the residues. *Plant and Soil*, 228, 83–103. https://doi.org/10.1023/A:1004813801728

Pagès, L., et al. 2014. Calibration and evaluation of ArchiSimple, a simple model of root system architecture. *Ecological Modelling*, 290 76–84. https://doi.org/10.1016/j.ecolmodel.2013.11.014.

Priestley, C.H.B. and Taylor, R.J. 1972. On the assessment of surface heat flux and evaporation using large-scale parameters. *Monthly Weather Review*, 100, 81–92. https://doi.org/10.1175/1520-0493(1972)100%3C0081:OTAOSH%3E2.3.CO;2

Rochester, I.J. 2003. Estimating nitrous oxide emissions from flood-irrigated alkaline grey clays. Austr. J. Soil Res., 41 (2). https://doi.org/10.1071/SR02068

Rosenzweig, C., et al. 2013. The agricultural model intercomparison and improvement project (AgMIP): protocols and pilot studies. *Agricultural and Forest Meteorology*, 170 166–182. https://doi.org/10.1016/j.agrformet.2012.09.011

Ruget, F., et al. 1994. Principes generaux du systeme GOA. Le module mais. Application a l'estimation des potentialités agroclimatiques dans la Communauté Européenne. *Délimitation de zones aux potentialités agricoles équivalentes en France et dans la Communauté pour optimiser les réseaux d'expérimentation des variétés*, INRA. (hal-02845413)

Schoving, C., et al. 2022. Combining multi-environmental trials and crop simulation to understand soybean response to early sowings under contrasting water conditions. *European Journal of Agronomy*, 133, 126439. https://doi.org/10.1016/j.eja.2021.126439

Seghouani, M., et al. 2025. Simulating phosphorus dynamics between the soil and the crop with the STICS model: Formalization and multi-site evaluation on maize in temperate area. *European Journal of Agronomy*, 164, 127475.

Shili-Touzi, I., et al. 2010. Does inter- cropping winter wheat (Triticum aestivum) with red fescue (Fes- tuca rubra) as a cover crop improve agronomic and environmental performance? A modeling approach. *Field Crops Research*, 116:218–229. https://doi.org/10.1016/j.fcr.2009.11.007

Smith, J., et al. 2010 Estimating changes in Scottish soil carbon stocks using ECOSSE. I. Model description and uncertainties. *Climate Research*, 45, 179–192. https://doi.org/10.3354/cr00899

Spaeth, S.C. and Sinclair, T.R. 1985. Linear increase in soybean harvest index during seed-filling. *Agronomy Journal*, 77, 207–211. https://doi.org/10.2134/agronj1985.00021962007700020008x

Strullu, L., et al. 2014. Simulation of biomass and nitrogen dynamics in perennial organs and shoots of Miscanthus * giganteus using the STICS model. *BioEnergy Research*, 7, 1253–1269. https://doi.org/10.1007/s12155-014-9462-4

Strullu, L., et al. 2015. Multisite yield gap analysis of miscanthus × giganteus using the STICS model. *BioEnergy Research*, 8, 1735–1749. https://doi.org/10.1007/s12155-015-9625-y

Strullu, L., et al. 2020. Simulation using the STICS model of C&N dynamics in alfalfa from sowing to crop destruction. *European Journal of Agronomy*, 112, 125948. https://doi.org/10.1016/j.eja.2019.125948

Strullu, L., et al. 2023. Improvement of grapevine yield simulation in Champagne with the STICS model. 13th Seminar of the STICS users, November 13–16, Bordeaux, France. https://stics.inrae.fr/content/download/4009/47243?version=1

Tavernier, R. and Maréchal, R. 1958. Carte des Associations de Sols de la Belgique. *Pédologie*, 8, 134–182.

Traoré, A., et al. 2022. Modeling sorghum-cowpea intercropping for a site in the savannah zone of Mali: strengths and weaknesses of the Stics model. *Field Crops Research*, 285, 109180. https://doi.org/10.1016/j.fcr.2022.108581

Van Der Ploeg, R.R., et al. 1995. Late fall site-specific soil nitrate upper limits for groundwater protection purposes. *Journal of Environmental Quality*, 24, 725–733. https://doi.org/10.2134/jeq1995.00472425002400040025x

Vezy, R., et al. 2023. Modeling soil - plant functioning of intercrops using comprehensive and generic formalisms implemented in the STICS model. *Agronomy for Sustainable Development*, 43(5). https://doi.org/10.1007/s13593-023-00917-5

Vezy, R., et al. 2024. CroPlotR: A Package to Analyze Crop Model Simulations Outputs with Plots and Statistics. https://doi.org/10.5281/zenodo.4442330, https://github.com/SticsRPacks/CroPlotR

Vieten, B., et al. 2008. The fate of N2O consumed in soils. *Biogeosciences*, 5, 129–132. https://doi.org/10.5194/bg-5-129-2008

Wallace, J.S. 1995. Towards a coupled light partitioning and transpiration model for use in intercrops and agroforestry. In: SH. inoquet and P. Cruz P (Eds.), *Ecophysiology of Tropical Intercropping*. INRA Editions, Versailles, France, pp 153–162.

Wallach, D., et al. 2011. A package of parameter estimation methods and implementation for the STICS crop-soil model. *Environmental Modelling & Software*, 26(4), 386–394. https://doi.org/10.1016/j.envsoft.2010.09.004

Wallach, D., et al. 2023. Proposal and extensive test of a calibration protocol for crop phenology models. *Agronomy for Sustainable Development*, 43(4), 46. https://doi.org/10.1007/s13593-023-00900-0

Wedin, D.A. and Tilman, D. 1990. Species effects on nitrogen cycling: a test with perennial grasses. *Oecologia*, 84, 433–441.

Wichern, F., et al. 2008. Nitrogen rhizodeposition in agricultural crops: methods, estimates and future prospects. *Soil Biology and Biochemistry*, 40(1), 30–48. https://doi.org/10.1016/j.soilbio.2007.08.010

Yin, X., et al. 2020. Long-Term Modelling of Soil N Mineralization and N Fate Using STICS in a 34-Year Crop Rotation Experiment. *Geoderma*, 357, 113956. https://doi.org/10.1016/j.geoderma.2019.113956.

Chapter 10

The Simulator of crop trait Assembly, MAnagement Response and Adaptation crop model

Myriam Adam[1], UMR AGAP Institut and Université de Montpellier, Institut Agro, France and University of Battambang, Cambodia; Gregory Beurier, Lauriane Rouan and Michael Dingkuhn, CIRAD, UMR AGAP Institut and Université de Montpellier, CIRAD, INRAE, Institut Agro, France; Florian Larue, Université de Montpellier, France; and Bertrand Muller, CIRAD, UMR AGAP Institut and Université de Montpellier, Institut Agro, France and CIRAD, Madagascar

1 Introduction

SAMARA (Simulator of crop trait Assembly, MAnagement Response and Adaptation) is a deterministic model for cereal crops that operates on a daily time step. It was mostly used for simulating rice and sorghum growth and development. It simulates the growth and development of cereals at the canopy level, considering limitations from radiation, water, and CO_2. The model uses daily inputs such as solar radiation, air temperature, potential evapotranspiration (FAO), and precipitation. SAMARA captures source–sink relationships in cereals, including rice, with detailed morphological aspects.

1 Contact : myriam.adam@cirad.fr

http://dx.doi.org/10.19103/AS.2025.0155.13

This is based on phenotypic plasticity due to competition for assimilates within the plant (Kumar et al., 2016, 2017).

What sets SAMARA apart from most agronomic models are:

1. its detailed representation of plant morphology, including the size and timing of organ cohorts like leaves, tillers, and internodes, and their senescence; and
2. its approach to biomass allocation, which is driven by competition for a single assimilate pool among organs. This means that growth can be limited either by the supply or by the demand for assimilates.

The model's principle of internal competition, Ic = supply/demand at the plant level, is adapted from Ecomeristem (Luquet et al., 2012, 2016), a Functional Structure Plant Model.

The SAMARA model was initially implemented on the EcoTrop platform (programmed in Delphi, Kouressy et al., 2008) with the following features:

- a database for observed and simulated data;
- crop, environment, and management data;
- a library of simulation modules constituting different models;
- simulation scenarios executable in bulk, e.g. sensitivity analyses;
- a simulation engine; and
- a user interface with graphical and numerical output.

The SAMARA model has undergone significant restructuring and optimization through a recoding process in C++, which ensures cross-platform operability without the need for external library dependencies. This revision focuses on enhancing computational efficiency by implementing near-native C++ practices and optimizing the simulator's execution performance. While the simulator itself operates sequentially (since SAMARA's small execution blocks do not benefit from parallelization), the tools developed around SAMARA – such as those for parameter estimation and simulations of Target Population of Environments – are designed to run in parallel. This approach leverages parallel processing where it is most impactful, enabling rapid parameter estimation and extensive-scale simulations.

The model's integration features are expanded through various interfaces, including direct command-line execution, a Qt graphical user interface for C++ that facilitates chart visualization and direct parameter modifications, and an R package that encapsulates the C++ core, thereby widening the model's applicability. Data input, including parameters and meteorological records, can be managed through several methods: plain text files, SQL databases (such as Access or SQLite), or directly from R dataframes. The model's output is versatile, providing either comprehensive time-series arrays of all simulated variables or selective subsets, thus meeting a range of analytical needs.

2 Simulating crop plasticity: trait responses to environmental and management factors

SAMARA was developed to explore plant type, i.e. ideotype, concepts under various weather, soil, and management conditions (Adam et al., 2014). It allows for the evaluation of the adaptive and agronomic value of different traits that interest breeders, both individually and in combination. The model's strength lies in its ability to simulate physiological interactions among traits and their responses to environmental and management factors. Consequently, SAMARA is a valuable tool for pre-breeding research, including the characterization of target populations of environments (TPE) and the development of *in-silico* ideotypes.

However, for purposes such as agronomic decision support or assessing yield potential, SAMARA is somewhat over-parameterized with respect to genotype characteristics. Despite efforts to differentiate between frequently and rarely used parameters, the sheer number of parameters remains large, and calibrating some is challenging because their values are not directly measurable. An optimization tool for the most-challenging parameters is currently being developed. As a result, SAMARA also serves as a useful educational tool for crop physiologists and agronomists. Experimenting with parameter values helps users understand how plant adjustment processes (e.g. tillering vs. leaf size), trait trade-offs (e.g. plant height vs. harvest index), and cultural practice trade-offs (e.g. plant spacing vs. varietal type) influence agronomic outcomes. Table 1 presents the most important genetic parameters that can be calibrated, with the range value for rice and sorghum, the two cereals modeled in SAMARA. These parameters highlighted the importance of morphological details integrated within SAMARA.

SAMARA distinguishes itself from other plot-scale crop models through its approach to assimilate partitioning among sinks, with greater detail on morphology and phenology. Unlike many crop models, SAMARA considers not only the carbon assimilation (source or supply) but also the demand for growth and respiration from the accumulated sink capacity of various organs on a given day during vegetative and reproductive stages. Since organ potential size and number, such as leaf appearance, tillering rates, and panicle size, can be genetically limited or vary in response to resources and stresses, the demand may sometimes fall short of the supply. This specific feature of SAMARA enables to represent the plasticity of the plant in terms of organ size and numbers in response to environmental conditions.

The model uses the state variable Ic (Index of Internal Competition = supply/demand) to assess the daily source–sink balance and provide feedback

Table 1 Key genetic parameters of interest

Crop parameter	Definition	Unit	Value range	Default sorghum	Default rice
Tbase, Topt1, Topt2, Tlim	Cardinal temperatures used for calculating thermal time by broken-stick poly-linear model	°C	–	10, 25, 35, 40	10, 24, 36, 44
SDJBVP	Thermal duration from germination to panicle initiation (PI) in the absence of day length effects	°Cd	200–1000	400	500
SDJRPR	Thermal duration from PI to flowering (period of internode elongation and panicle development)	°Cd	200–600	400	380
SDJMatu1	Thermal duration from flowering to end of grain filling	°Cd	200–600	400	380
SDJMatu2	Thermal duration from end of grain filling to grain maturity	°Cd	0-300	50	30
PPSens	PP sensitivity, important variable. Range 0.3–0.6 is PP sensitive, and sensitivitydisappears toward values of 0.7–1. Dingkuhn et al. 2008		0–1	0.4	0.95
Phyllo	Phyllochron (duration from one leaf appearance to the next). The first three leaves already formed in the embryo appear faster. After PI phyllochron increases (set by user).	°Cd	20–100	50	70
LeafLengthMax	Potential length of the longest leaf blade. Actual length increases with phytomer rank and is reduced if Ic1.	mm	200–1200	800	450
CoeffLeafWLRatio	Leaf blade width as a fraction of length	mm mm^{-1}	0.02–0.2	0.08	0.025
CoeffLeafDeath	Coefficient for leaf senescence response to assimilate supply/demand (Ic). Senescence is proportional to [1-min(Ic, 1)]		0–0.1	0.03	0.07
RankLongestLeaf	Phytomer rank of longest leaf on main stem, usually the leaf preceding the flag leaf in rice		6–20	12	10

InternodeLengthMax	Potential average length of elongated internode (actual length can be smaller if Ic1)	mm	0–500	200	80
CoeffInternodeMass	Maximal structural mass of internode per mm length g mm^{-1}	0–0.1	0.01	0.001	
TilAbility	Coefficient for response of tiller outgrowth to Ic, provided Ic IcTillering. At TilAbility = 0 tiller production is suppressed. Tiller outgrowth rate is capped by fn(current tiller number per plant) to account for available tiller buds.		0–1	0.2	0.7
CoeffPanicleMass	Sets growth rate of structural parts of panicle between PI and flowering, subject to limitation by resource availability and genetic size limitation of panicle (PanStructMassMax). Most important parameter for harvest index yield potential		0–0.3	0.1	0.005
PanStructMassMax	Upper limit of individual panicle mass (structural parts including peduncle). Multiplying this para with CoeffPanSinkPop and dividing it by PoidsSecGrain gives max grain number/panicle. This is a varietal feature usually known (yield components)	g	0–10	3	0.3
CoeffPanSinkPop	Sets the grain mass (yield) that can be produced per structural mass of panicle including peduncle. Should be species feature, don't use it to calibrate for varieties	fraction	0–1	0.5	0.5

to morphogenetic and physiological processes. This includes reserve storage and mobilization, tiller initiation or senescence, leaf size and senescence, internode elongation, root growth, and pre-floral panicle dimensioning. Carbon demand for root growth, influenced by factors such as soil depth and plant spacing, affects how effectively assimilates are used. If there are insufficient sinks or storage, assimilates may not be fully utilized, leading to feedback inhibition of photosynthesis and reduced radiation use efficiency (RUE). Conversely, excessive sink development, such as profuse tillering, can cause excessive senescence and also reduce RUE. Such crowding effects are difficult to simulate with crop models that have fixed partitioning and senescence patterns.

This concept, adapted from the EcoMeristem model (Luquet et al., 2012), diverges from the traditional 'prescriptive' partitioning approach, which assumes a constant RUE in the absence of physiological stresses. SAMARA, in contrast, allows for the simulation of phenotypic plasticity, capturing both adaptive and non-adaptive traits, and enables the exploration of various adaptation strategies, such as aggressive or conservative resource utilization. According to SAMARA, a genotype's growth and yield potential are determined not only by its potential photosynthetic rate and light interception (source) but also by the dynamics and resource responsiveness of its morphogenetic processes (demand). The model incorporates factors like water resources, drought, flooding, and thermal stresses but does not account for mineral nutrition.

SAMARA has a simple water balance derived from the SARRAH upland crop model, adapted with additional functionalities to flooded-irrigated and rainfed-lowland systems and bunded plots catching ponded water. A surface soil reservoir (e.g. 100 mm) is used to calculate soil evaporation. A second soil layer has variable depth because it ends at the root front, which progresses during crop development. Root front progression extends the second layer according to development-phase dependent rates set by the user, unless it is (1) stopped as it reaches the current wetting front (a phenomenon happening after a dry season as the rains fill up the soil gradually) or (2) when it reaches the limit of the user-defined soil depth. Soil reservoirs fill up from top to bottom, with no capillary rise simulated. Soil shrinkage and swelling are also not considered. Deep drainage is limited by the soil parameter PercolationMax (mm/day). Each reservoir contains four functional water reservoirs:

- the non-evaporable part, water content of air-dry soil;
- the evaporable but non-transpirable part, fixed by the wilting point;
- the transpirable but non-drainable part located between wilting point and field capacity; and
- the drainable part in a saturated soil (located in macropores that normally are filled with air) - this reservoir is available for transpiration unless

the genotype is declared as highly susceptible to water logging (crop parameter WaterLoggingSens is set to 1).

An additional water reservoir is located outside the soil, generated by situations where water supply has filled up all soil reservoirs and the supply continues to exceed PercolationMax. Ponding only occurs if the parameter BundHeight is superior to 0, otherwise the water runs off. Filling-up of bunded plots can lead to spill-over runoff as well.

3 Simulating crop plasticity: source and sink relations

The index of internal competition (Ic) is a state variable representing the ratio of total fresh assimilate supply to total demand at the whole-plant level, adjusted for maintenance. This supply does not account for carbon reserves stored in stems, leaf sheaths, and internodes, nor does it include recycled materials from organ senescence, which only occurs when Ic < 1. The demand component encompasses the following:

- the incremental, potential growth of expanding leaves and sheaths on all tillers, based on their potential size, which is determined by leaf rank and the proportion of growth to be achieved within a single day. This assumes that full expansion occurs within one phyllochron.
- the incremental, potential growth of all elongating internodes, on the same basis;
- the incremental, potential, aggregate assimilate required for grain filling, calculated based on the fraction of thermal time elapsed during the day relative to the total maturity phase. Grain filling is assumed to proceed linearly throughout the entire maturity phase;
- the demand of the root system for growth, determined by the day's advancement of the root front (including proportional lateral expansion, constrained by the soil area allocated to the plant in a canopy context), the current root dry weight per soil volume, and the potential root dry weight per soil volume defined by a parameter. Root partitioning is managed by a crop parameter called RootPartitMax (typically set to 1.0), which indicates that root partitioning cannot exceed shoot partitioning; and
- the user can declare the stem reserve storage compartment as being an active sink, by setting the parameter CoeffReserveSink > 0. The unfilled part of the sink compartment then contributes to overall demand. Otherwise, the storage compartment acts only as a spill-over pool (passive sink).

Ic serves as a proxy for hypothetical sugar signaling that influences several key processes: organ initiation in meristems or tiller buds, reduction in organ size under resource constraints, and the triggering of tiller outgrowth and leaf and tiller senescence. Tiller formation is permitted when Ic > 0.5, as tiller growth can occur even with a mild shortage of assimilates. The extent of tiller outgrowth is proportional to (Ic −0.5) and is also influenced by the parameter TilAbility. The critical Ic value for tillering can be adjusted as needed using the parameter IcTillering.

Ic has the potential to influence various plant processes, including development rate (with slower progression to flowering under assimilate shortages), leaf appearance rate, photosynthesis, respiration, and the growth of specific organs. While these feedback functions are not currently implemented, they can be easily added if needed. Additionally, water stress, represented by the variable Cstr, can optionally affect the development rate by extending the thermal time. Ic-driven feedback mechanisms are powerful for adjusting plant growth and development according to resource availability. However, this complexity makes the model challenging to parameterize. Since Ic impacts all development and growth processes, adjusting one parameter often necessitates adjustments to many others.

4 Input variables to simulate different environments

New simulations are established by adjusting various factors, including plot conditions, weather and soil, location specifics, altitude and coordinates, agricultural practices such as sowing dates and irrigation management, and crop variety. All external input data is sourced from text files. Weather files contain daily data on solar radiation, air temperature, potential evapotranspiration (FAO), and precipitation. For soil information, we need to specify the thickness (mm) of the surface layer and the depth of the second layer. Additionally, we define key hydrologic properties, including soil volumetric water content ($m^3.m^3$) at the wilting point, field capacity, and saturation. Two parameters detail runoff potential with PourRuiss, which indicates the percentage of rainfall that runs off above a specified daily threshold (SeuilRuiss in mm). A daily percolation rate (PercolationMax, mm) triggered deep drainage if there is standing water above bunded plots or filled macropores. Soil texture information is not required.

In terms of management practices, SAMARA offers various options, including transplanting versus direct seeding, flexible water management strategies (such as alternate wetting and drying and deficit irrigation), and mulching. The transplanting parameter indicates whether sowing is done through direct seeding (value of 0) or transplanting (value of 1). The sowing date and density are specified using two parameters with the same names. Additionally, transplanting of seedlings is permitted only when the BundHeight parameter is greater than 0, indicating bunded fields that can retain ponded water. Users can set the seedbed duration and population, as well as the intensity of transplanting shock during the first approximately seven days post-transplanting, indicated by CoeffTransplantingShock (ranging from 0 to 1). This shock impacts both assimilation and development rates, potentially extending the seed-to-seed crop duration if CoeffTransplantingShock is greater than 0. Transplanting shock is classified as a cultural parameter rather than a crop parameter, as it is influenced by management factors such as competition in the seedbed, the quality of seedling removal, and shading and watering during transport. The depth of the root system after transplanting varies based on the farmer's practices, and the Transplanting Depth cultural parameter specifies the appropriate value in millimeters.

SAMARA allows for the programming of individual irrigation events for upland crops, as well as irrigation options for aquatic rice cultivation (when BundHeight is greater than 0). If irrigation is not selected (irrigation = 0), the bunded plots may experience varying conditions between flooding and dryland. Generally, opting for irrigation significantly increases yields by reducing runoff and enhancing water retention on the plot, unless PercolationMax is excessively high, as in very sandy soils or during periods of little rainfall. A unique feature enables automatic drainage in bunded plots to prevent submergence, which is particularly crucial for young seedlings or very high bunds. In this mode, it is assumed that farmers will implement surface drainage to ensure that no more than 50% of the plant canopy is submerged.

When the IrrigAuto parameter is set to 1, the automatic irrigation is executed daily to maintain plot water levels up to a specified percentage of BundHeight each morning. The IrriAutoTarget parameter, which ranges from 0 to 1, indicates the target fraction of BundHeight. This feature should typically be paired with life-saving drainage activation, as farmers aiming to irrigate will want to avoid drowning their seedlings. The IrrigAuto mode also includes an initial irrigation flush before planting to saturate the soil. For transplanted crops, an initial irrigation flush is applied prior to transplanting as well. In this case, the water used for the seedling nursery is not included in the overall crop water balance metrics (such as CumIrrig, CumWaterUsed, and WUE), since the nursery occupies a very small area.

The parameter FtswIrrig (ranging from 0.2 to 2) determines the soil moisture level at which automatic irrigation is activated. To maintain flooding conditions, this value should be set to 2. If irrigation is intended to commence when the soil water content drops below field capacity, it should be set to 1. Setting the value below 1 allows for a water deficit to develop before irrigation resumes, facilitating the easy implementation of alternate wetting and drying (AWD). Additionally, there is a new feature that allows for programming a drought period for crops under automatic irrigation. The duration of the non-irrigation period is defined by the parameters IrrigAutoStop and IrrigAutoResume (measured in days after sowing).

5 Recent model developments

5.1 Photosynthesis adapted to e-CO_2 response

The local source–sink ratio (LSSR) source–sink proxy parameter is defined as the ratio of flag leaf surface (cm) over the grain (or spikelet) number of the corresponding panicle (Fabre et al., 2020). Fabre et al. 2020 considered that LSSR is negatively correlated with the e-CO_2 effect on photosynthesis and grain yield, hence LSSR can be predictive of e-CO_2 response. The crop model SAMARA can predict this effect of genotypic source-sink ratio on yield response to e-CO_2. The model's canopy net photosynthesis simulation is simple, as it depends only on intercepted radiation, temperature, plant water status (function of water available in the soil), and atmospheric CO_2 via a curvilinear model, with respiration subtracted *a posteriori*. However, growth and yield can be source- (photosynthesis) or sink-limited (current C demand for organ growth), reversibly buffered by a reserve compartment. Consequently, under severe sink limitations, not all simulated photosynthate may be utilized, leading to some assimilates being classified as "assimilate not used." This forfeited assimilate can serve as a rough indicator of feedback inhibition of photosynthesis.

We illustrate this concept, using the rice variety IR72 in the dry-season climate of the PRRI station in the Philippines in 2013 (Kumar et al., 2016). The environment is irrigated (flooded) and warm-tropical with high solar radiation (dry season, rainless). IR72 is a standard semidwarf, high-yielding, indica variety yielding 8–9 t/ha under the PRRI dry season conditions with a harvest index of about 0.50–0.55. From the simulation outputs, LSSR was calculated by dividing the simulated last-leaf area by the simulated spikelet number per panicle (Table 1). SAMARA makes no distinction between main culms and tillers, thus all flag leaves and panicles are considered identical. Under the 2015 ambient CO_2 level (400 ppm), the simulated yield of IR72 was high but normal for this site and type of variety (Table 2). About 18% of assimilates were 'not used,' indicating sink-limited growth during part of the crop cycle. Under e-CO_2 conditions (600 ppm), grain yield increased by 26% and biomass by 27%. This was mainly due to increased tillering and increased panicle number, whereas panicle size changed little. The LSSR also changed very little. However, the 'assimilates not used' value, accrued over the entire crop cycle, increased by 70%, indicating much greater sink limitation under e-CO_2 conditions.

In order to modify source-sink relationships (and vary the LSSR), two parameters were adjusted:

Table 2 Simulated crop performance for IR72 in DS2013 at PRRI site in the Philippines at two CO_2 levels

	Observed	400 ppm CO_2	600 ppm CO_2	% change
Aboveground biomass (t/ha)	16.25	15.3	19.3	+27
Grain yield, 0%MC (t/ha)	8.9	8.1	10.2	+26
Panicles/hill (20 x 20cm spacing)	16.8	18.0	21.4	+19
Spikelets per panicle	92	89.4	89.9	+1
LSSR	–	0.411	0.408	–1
Assimilates not used (t/ha)	–	3.55	6.04	+70
Assimilates not used (%)	–	18	23	–

Model calibration as in Kumar et al. (2016).
LSSR, local source-sink ratio.

- Source: LeafLengthMax (350–550mm, proportionally affecting leaf width via an allometric factor).
- Sink: CoefPanicleMass (0.0050–0.0075, affecting structural growth rate before flowering).

Variations of these two crop parameters enable to vary the potential leaf area (LA) and potential panicle size, representing a population of virtual genotypes, differing from IR72 in two traits. Variation of each parameter alone has a non-linear effect on grain yield (Fig. 1), both at 400 and 600 ppm CO_2. Figure 1 shows that the crop parameter variations caused grain yield variations between 6 t/ha and 8.5 t/ha under ambient CO_2 (400 ppm) and between 8.5 t/ha and 11 t/ha under e-CO_2 (600 ppm). Simulated LSSR varied between 0.25 and 0.62 and was not affected by CO_2 level. There was no uniform effect of LSSR on grain yield because the effect depended on whether sink (panicle) or source (leaf) traits were varied. As posed at the beginning of our section, we could show that LSSR is negatively correlated with the e-CO_2 effect on grain yield: for a small LLSR the difference in yield between the two different CO_2 concentration (400ppm vs. 600ppm) was higher than for a higher value of LLSR (Fig. 1). These results support the fact that SAMARA can simulate the effect of higher CO2 concentration on grain yield, by using a pool 'assimilates not used' and having the Ic parameter to adapt the plant response to environmental conditions, through the plasticity of the different organs.

5.2 SAMARA adapted to West African sorghum varieties

To simulate sorghum grain yield accurately across a wide range of conditions in the drylands of West Africa, we set up a network of crop trials to gather a complete dataset for model calibration and evaluation. This network of trials

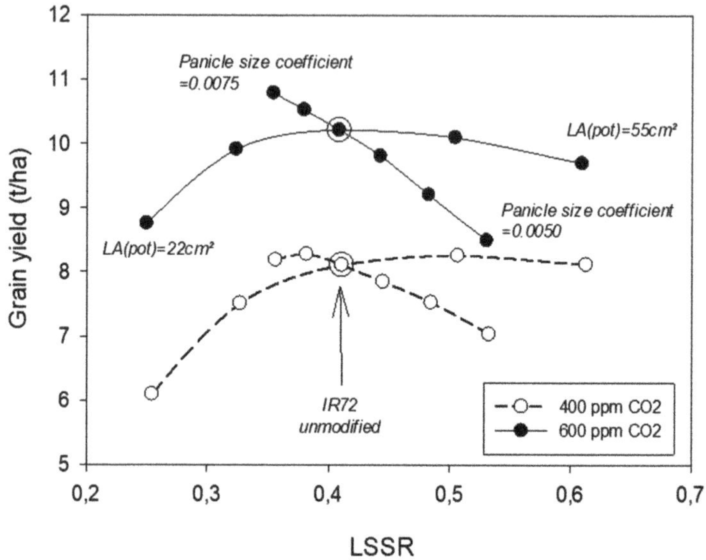

Figure 1 Changes in simulated grain yield of IR72 in response to source and sink crop parameter values under two CO_2 levels. Predictions are plotted against simulated local source–sink ratio.

comprised 23 trials carried out between 2013 and 2016 in Senegal, Mali and Burkina Faso, under optimal and suboptimal conditions (Adam et al., 2015). Data were collected for 10 West African sorghum varieties contrasted in terms of architecture (crop morphology), phenology and yield potential (biomass and grain). The choice of these 10 varieties is representative of the existing sorghum diversity (and their associated plant trait plasticity) and enables to represent the existing 'genetic diversity' in the region (Table 2). Looking at this diversity, we aimed at capturing sorghum plant traits' variance within the SAMARA crop model. Within the 23 trials of the dataset (Table 3), we selected 9 trials for calibration: three trials per country considered at the most optimal growing conditions (i.e. no water and nutrient stress and minimal pest and disease effects), and with different sowing dates to capture the photoperiod sensitivity of studied West African varieties. The 14 remaining trials were used for evaluation purposes.

Estimated model parameters led to a satisfactory evaluation of crop development (leaves number), plant height, LAI and aboveground biomass (AGB) for all varieties at calibration and evaluation sites (example for LAI in Fig. 2). Independently of the variety and site, the leaf number observed varied from about six leaves at 30 days after sowing to a maximum of 36, and we had an average MAE of four leaves for the calibration and evaluation sites. With respect to plant height, it varied from about 44 cm at 30 days after sowing to

Table 3 Characteristics of the studied sorghum genotypes

Genotype name	Type*	Photo-sensitivity	Cycle length	Purpose	Plant height	Potential grain yield	Origin*
621B	C	No	90–115 days	Grain	=< 2 m	2.5–3 t/ha	SN
CSM63E	G	Low	=< 90 days	Biomass	>3 m	2 t/ha	MLI
FADDA	GH	Moderate	90–115 days	Dual	From 2 m to 3 m	4.5 t/ha	MLI
GRINKAN	C	No	=< 90 days	Dual	=< 2 m	4 t/ha	MLI
IS15401	GC	High	>115 days	Biomass	>3 m	2 t/ha	CMR
Local (F2-20)	C	Low	90–115 days	Grain	From 2 m to 3 m	3– 5.3t/ha	SN
NIELENI	CH	Low	90–115 days	Dual	From 2 m to 3 m	4 t/ha	MLI
PABLO	GH	Moderate	90–115 days	Biomass	>3 m	4 t/ha	MLI
SK5912	C	High	>115 days	Biomass	=< 2 m	2.5–3.5 t/ha	NGR
SOUMBA	C	Low	90–115 days	Dual	From 2 m to 3 m	2.5 t/ha	MLI

*G (Guinea), C (Caudatum), GH (Guinea Hybrid), CH (caudatum hybrid), GC (Guinea Caudatum) MLI (Mali), CMR (Cameroun). NGR (Nigeria), SN (Senegal).

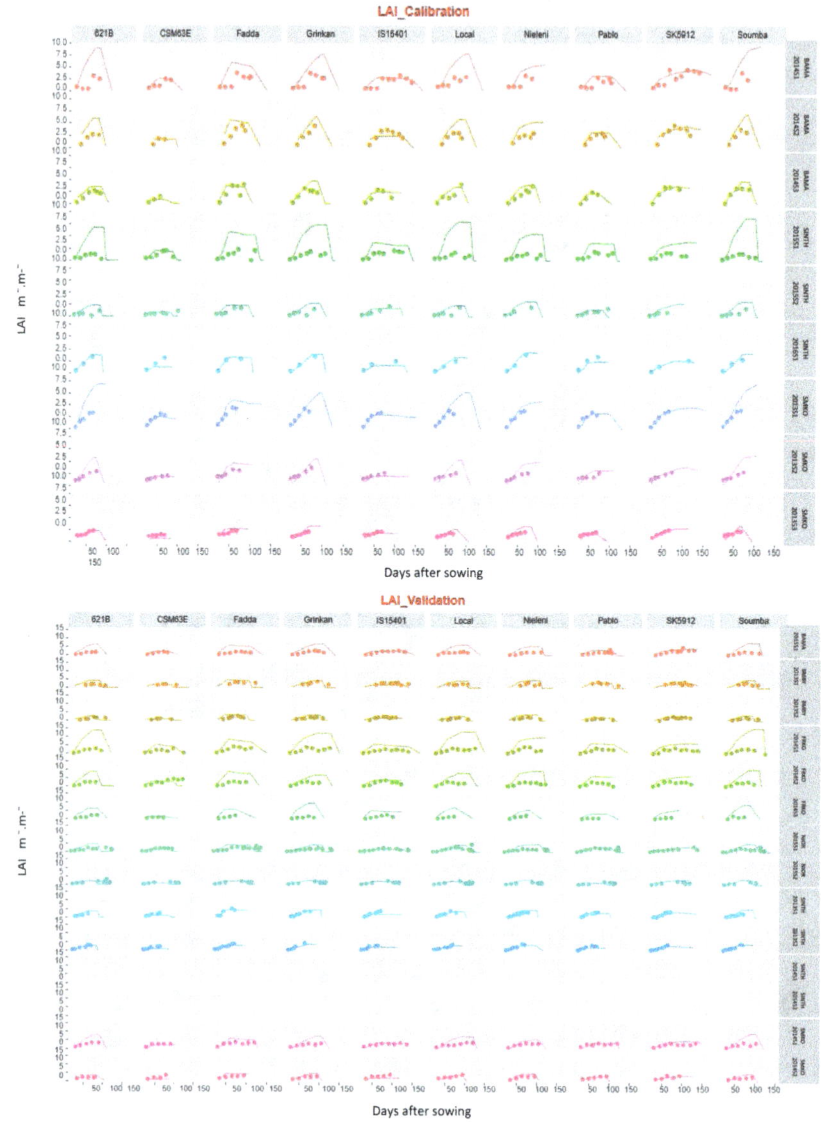

Figure 2 Evaluation of LAI dynamics for all varieties at 9 calibration sites and 14 evaluation sites. Abbreviations of the sites are indicated in Table 4.

a maximum plant height of 46 cm, with the model giving an average Mean of Absolute value of Errors (MAE) of 51 cm and 58 cm, for the calibration and evaluation sites, respectively. The observed LAI varied from 0.05 to a maximum of 4.82 and had an average MAE of 1.27 and 1.78 for the calibration and evaluation sites, respectively. The AGB exhibited a minimum value of 9 kg/ha at

Table 4 Trials' overview with abbreviations used in figures, the grey-shaded cells indicate the sites chosen for calibration

		2013	2014	2015	2016
SN	Bambey (2)	Two sowing dates (BMBY2013S1; BMBY2013S2)			
	Nioro (2)			Two sowing dates (NIOR2015S1; NIOR2015S2)	
	Sinthiou (7)	Two sowing dates (SINTH2013S1; SINTH2013S2)	Two sowing dates (SINTH2014S1; SINTH2014S2)	Two sowing dates (SINTH2015S1; SINTH2015S2)	One sowing date (SINTH2016S1)
ML	Samanko (5)	Two sowing dates (SMKO2013S1; SMKO2013S2; SMKO2013S3)	Two sowing dates (SMKO2014S1; SMKO2014S2)		
BF	Bama (4)		Three sowing dates (BAMA2014S1; BAMA014S2; BAMA2014S3)	1 sowing date (BAMA2015S1)	
	Farako Ba (3)		Three sowing dates (FRKO2014S1; FRKO014S2; FRKO2014S3)		

SN: Senegal, ML: Mali, BF: Burkina Faso.

30 DAS and a maximum of 28 780 kg/ha at flowering with an average MAE of 3280 kg/ha and 4507 kg/ha for the calibration and evaluation sites, respectively. Grain yield and AGB at harvest were satisfactorily predicted (Fig. 3). The main differences observed in the evaluation set of grain yield are mostly due to stresses not simulated by SAMARA modeled, and that we experienced in the experiment trials that are not at optimal conditions in West Africa region (i.e. grain moldiness, phosphorus deficiency, and bird attack).

6 Current applications emphasizing G x E interactions

6.1 Drought stress pattern characterization in sorghum in West Africa

After calibration of SAMARA on 10 West African sorghum varieties across 23 sites located on a rainfall gradient in Senegal, Mali, and Burkina Faso (Section 5.2), long-term simulations across the dryland area of West Africa highlighted a high variability of sorghum yield across isohyets and varieties (Fig. 4a). Average

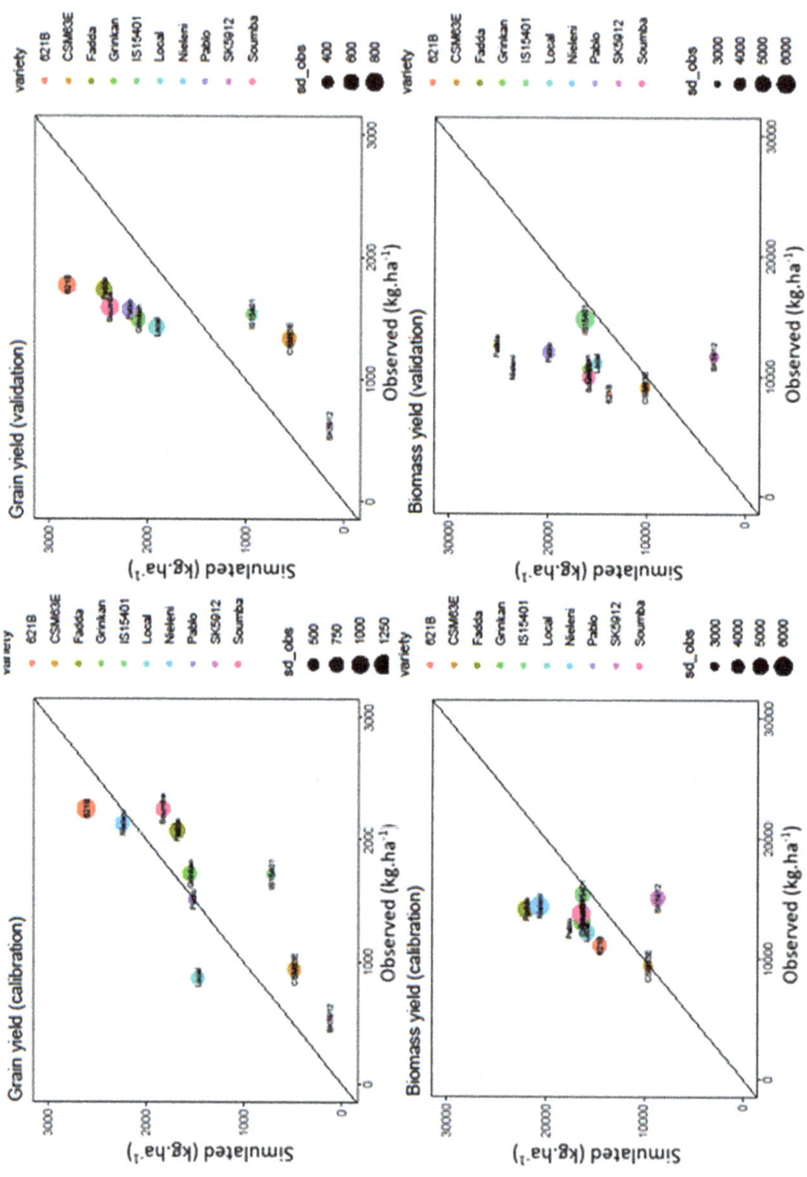

Figure 3 Comparison of observed and SAMARA-simulated grain yield and aboveground biomass for calibration and validation dataset. The black line is the 1:1 line, 9 calibration sites were used for calibration and the 14 remaining trials were used for evaluation (see Table 4).

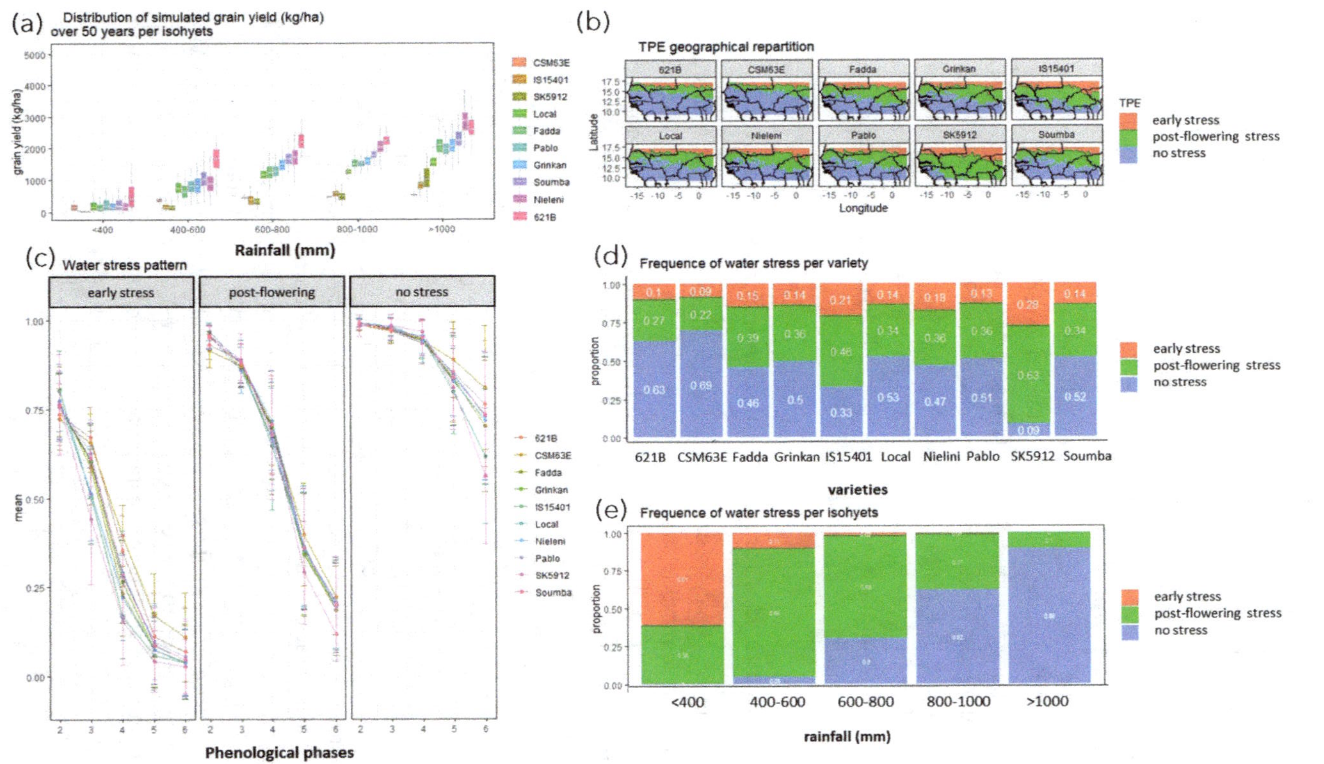

Figure 4 Main outputs of the target population environment characterization. (a) Distribution of the simulated grain yield in the different isohyets, ordered driest to wettest. (b) Geographical repartition of the target population environment according to variety. (c) Water-stress index throughout the crop cycle, represented as a function of development phase for three target population environment and frequencies of each target population environment according to (d) variety and (e) isohyets.

yield ranged between 0.2 t/ha (in isohyet under 400 mm, for IS15401 and SK5912) and 3.8 t/ha (for Nieleni in the isohyet receiving more than 1000 mm). Using a cluster analysis, three major water-deficit patterns were identified (Fig 4b and 4c): the first target population environment (TPE1) was characterized by a severe water stress starting at the beginning of plant growth till maturity; the second target population environment (TPE2) was characterized by post-flowering stress, starting late in the vegetative period until the end of grain filling; and the third target population environment (TPE3) comprised situations where plants were not limited by water or experienced only short-term water deficit. Figure 4a-b-d shows that some varieties were more affected by water stress (in lower rainfall isohyets) than others. We can note that CMS63E appears as the most water stress tolerant, while SK5912 and is15401, varieties more adapted to the southern latitude, did experience higher water stress. As expected, the more rainfall, the least water stress. However, even for the locations with the highest isohyets, there is still 10% of the situation that the crop experienced a stress (Fig. 4e).

6.2 Scenario analyses for rice adaptation in Madagascar

Through sensitivity analysis of key parameters, the model SAMARA was used to test different adaptation strategies in terms of rice varieties or management practices for rice cropping systems in Madagascar. We tested the effect of (1) date of transplanting, (2) sowing date, and (3) varieties more or less sensitive to photoperiod not only on grain yield but also on some other key rice plant traits (i.e. crop cycle, sterility tolerance, and yield components). Grain yield, AGB, and the number of panicles per plant are very sensitive to plant age at transplanting, even more so for short-cycle variety: the longer-cycle varieties are more tolerant to late transplanting (Fig. 5a). The early variety has a higher harvest index and a lower biomass. It appears that a transplantion at about 30 days in the nursery would optimize grain yield, independently of the rice variety, after this duration, grain yield will dramatically decrease. The second strategy shows that a good window for sowing is around the beginning of December, then there's a risk of sterility due to cold. An aggravating factor is that late sowing lengthens the cycle, accentuating the risk of sterility. Finally, in Fig. 5c we illustrate that for the photo-insensitive variety, the ripening date shifts roughly in proportion to the sowing date. In contrast, the photoperiodic variety reaches maturity at an almost constant date for a certain range of sowing dates (October to December). In other words, it varies the length of its cycle to make its maturity date less variable. For late sowing dates, the arrival of cold extends the cycle for both types of variety.

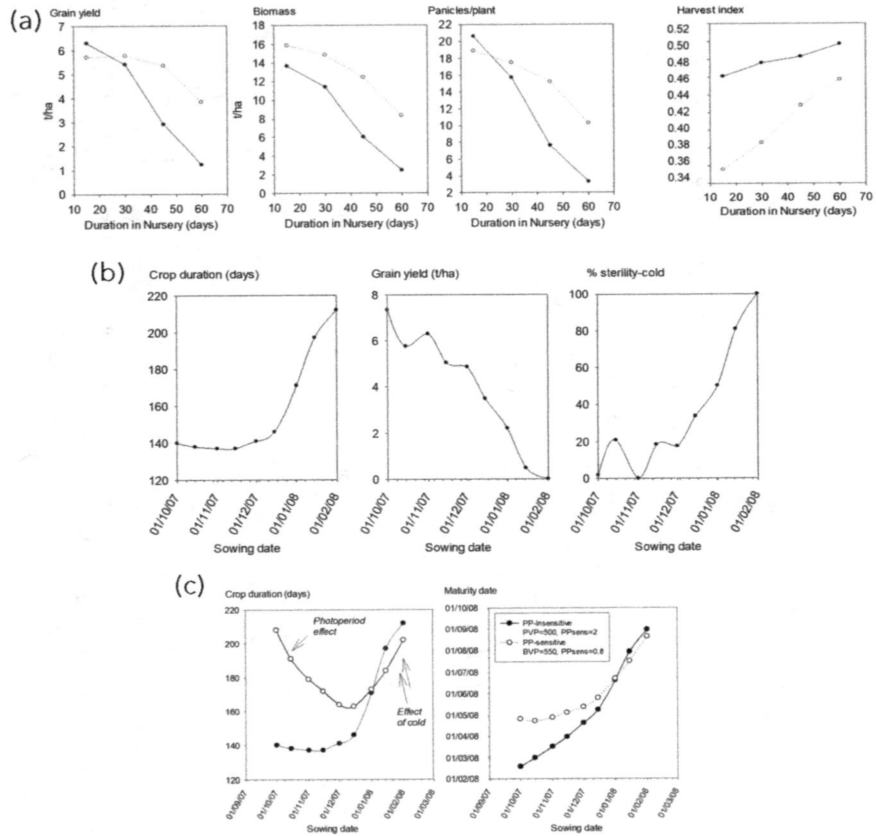

Figure 5 Climatic risks in the uplands of Madagascar: the effect of nursery duration, sowing schedule, and choice of variety type (PP: photoperiod; BVP: basic vegetative phase; PPsens: photoperiod sensitivity index). (a) Effect of duration in nursery on grain yield, biomass, panicles per plant, and harvest index for two different rice varieties (in full line short cycle variety, in dashed line long cycle varieties). (b) Effect of sowing dates on crop cycle, grain yield, and percentage of sterility due to cold temperature. (c) Effect of photoperiod sensitivity on crop duration and maturity date.

7 Conclusion and future trends

SAMARA is a research-oriented crop model primarily used for pre-breeding applications rather than for widespread use. Its strength lies in its ability to simulate plant plasticity in response to environmental conditions, as demonstrated by its modeling of rice responses to elevated CO_2 levels. Another key application is characterizing drought stress and assessing the adaptability of rice or sorghum varieties to various drought scenarios. Note that SAMARA does not simulate nutrient limitations.

For drought characterization applications, we are currently working on integrating SAMARA into a pipeline that will enable regional applications by linking grid data with satellite-derived information. Additionally, future developments may include new approaches to model temperature stress effects on sorghum growth and development (Berger et al., 2023), as well as shading tolerance for rice in agroforestry and agro-voltaic systems (Perez et al., 2024).

Model availability: The SAMARA model's source code is accessible at https://github.com/GBeurier/rSamara. A comprehensive tutorial, including compiled binaries and an R package, is available at https://github.com/GBeurier/Samara_Tutorial. Additionally, the Qt frontend source code can be found at https://github.com/PAMDeveloper/QtSamara.

8 References

Adam, M., et al. 2014. La conception d'idéotypes : de la pratique à la théorie.

Adam, M., et al. 2015. *Network of experiments to phenotype contrasted sorghum and to model its adaptability in West African environments* [P40]. Presented at the Climate-Smart Agriculture 2015: Global Science Conference, CIRAD, Montpellier, France, pp. 129–129.

Berger, A., et al. 2023. Plant response to a late heat stress can be modified by an earlier one. A case study on sorghum grain production. In: *Sorghum in the 21th century: Resiliency and Sustainability in the Face of Climate Change*. Book of Abstracts. CIRAD, Kansas State University, Sorghum International Development, IRD, CERAAS. Montpellier: CIRAD-Kansas State University, Résumé, p. 211. Sorghum in the 21st Century, Global Sorghum Conference, Montpellier, France, 5 June 2023/9 June 2023.

Dingkuhn, M., et al. 2008. A Model of Sorghum Photoperiodism Using the Concept of Threshold-Lowering during Prolonged Appetence. *European Journal of Agronomy* 28, no. 2 (February 1, 2008): 74–89. https://doi.org/10.1016/j.eja.2007.05.005.

Fabre, D., et al. 2020. Genotypic variation in source and sink traits affects the response of photosynthesis and growth to elevated atmospheric CO_2. *Plant Cell & Environment* 43, 579–593. https://doi.org/10.1111/pce.13693

Kouressy, M., et al. 2008. Adaptation to diverse semi-arid environments of sorghum genotypes having different plant type and sensitivity to photoperiod. *Agricultural and Forest Meteorology* 148, 357–371. https://doi.org/10.1016/j.agrformet.2007.09.009

Kumar, U., et al. 2016. Compensatory phenotypic plasticity in irrigated rice: Sequential formation of yield components and simulation with SAMARA model. *Field Crops Research* 193, 164–177. https://doi.org/10.1016/j.fcr.2016.04.036

Kumar, U., et al. 2017. Analysis and simulation of phenotypic plasticity for traits contributing to yield potential in twelve rice genotypes. *Field Crops Research* 202, 94–107. https://doi.org/10.1016/j.fcr.2016.04.037

Luquet, D., et al. 2012. Developmental dynamics and early growth vigour in rice 2. Modelling genetic diversity using ecomeristem. *Journal of Agronomy and Crop Science* 198, 385–398. https://doi.org/10.1111/j.1439-037X.2012.00527.x

Luquet, D., et al. 2016. Heuristic exploration of theoretical margins for improving adaptation of rice through crop-model assisted phenotyping. In: Yin, X. and Struik, P.C. (Eds.), *Crop systems biology*. Springer International Publishing, Cham, pp. 105–127.

Perez, R.P.A., et al. 2024. Combining modelling and experimental approaches to assess the feasibility of developing rice-oil palm agroforestry system. *Journal of Experimental Botany* 75(13), 4074–4092.

Chapter 11

The Genotype-by-Environment interaction on CROp growth Simulator model

Xinyou Yin, Wageningen University & Research, The Netherlands

1 Introduction

In Wageningen, since the pioneering work of de Wit (1965), many crop models for predicting the productivity of both arable and greenhouse crops have been available. This section of the chapter only covers the models used for predicting the growth and yield of arable crops.

http://dx.doi.org/10.19103/AS.2025.0155.14

Benchmark Wageningen models, where photosynthesis and respiration were explicitly described, often had a name ended with CROp growth Simulator (CROS). The first model developed was a simple model labelled ELCROS – 'ELementary CROp growth Simulator' (de Wit et al. 1970). Next, a comprehensive model labelled BACROS (BAsic CROp growth Simulator) was developed that predicts canopy photosynthesis and transpiration with the time step of an hour or shorter (de Wit et al. 1978). Later, simplified crop models, such as SUCROS (Simple and Universal CROp growth Simulator, van Keulen et al. 1982) and MACROS (Modules of an Annual CROp Simulator, Penning de Vries et al. 1989), were derived for general applications and for crop simulation training programs.

Crop models were often distinguished according to three levels: potential, water-limited, and nutrient (mainly nitrogen)-limited production (Penning de Vries et al. 1989). Models for the potential production level, such as SUCROS1 (Goudriaan and van Laar 1994), assume that crop yields are determined by weather variables (temperature and radiation).

When water stress occurs, the ratio between actual and potential transpiration rates is used as a stress indicator within the model, serving as the basis for estimating photosynthesis. This approach has led to the development of crop models for water-limited production, such as SUCROS2 (van Laar et al. 1997). A similar, though less mechanistic, approach has been applied to model crop production where nitrogen stress occurs.

By incorporating crop-specific processes, models in the SUCROS-family have also evolved into crop-specific versions, e.g. ORYZA models for rice (Kropff et al. 1994; Bouman et al. 2001).

Temporal and spatial up-scaling of photosynthesis and transpiration is a relatively sophisticated component in the SUCROS models, whereas other parts of models (such as assimilate partitioning over growing organs) are over-simplified, often using empirically derived coefficients. This means that SUCROS models are less effective in predicting interactive effects of various environmental variables on growth processes (especially with assimilate partitioning).

General information about the history of crop model development in Wageningen has been described by Bouman et al. (1996) and van Ittersum et al. (2003). The models WOFOST and a simpler model LINTUL, which use the radiation use efficiency (RUE) concept, also contributed to the history.

2 Improving crop models: algorithms, structure, and input parameters

Traditionally, crop models have been widely used to study and predict crop performance in response to environmental conditions and management practices. As crop science research has shifted from focusing on broad adaptation to targeting specific genotypes to particular management practices and local environments, a central question has emerged on how to deal with genotype–environment–management interactions (Kropff and Struik 2002; Peng et al. 2020).

Genotypic impacts on crop performance (especially in the context of plant breeding, where a large number of genotypes are involved) receive little attention in crop modelling.

Physiological parameters (also termed 'genetic coefficients') are used as inputs to models, which reflects genotypic differences at the process level, making crop models seem to be capable of dealing with crop responses to genotypic strategies. White and Hoogenboom (1996) reported that values of genetic coefficients in a DSSAT crop model could be estimated from allelic variations of several known genes across various cultivars.

Yin et al. (2000a) identified molecular markers for physiological parameters of a SUCROS-family model measured across a large number of genotypes in a segregating genetic population and examined model performance in predicting yield differences among these genotypes. The SUCROS-type model did not perform sufficiently well. This was partly owing to inadequacy in defining model structure, and partly to inaccuracy in measuring model-input parameters under field conditions. Yield progress in major crops during the Green Revolution largely resulted from improvement in harvest index; so, assimilate partitioning among growing organs must differ among genotypes. A major difficulty in using many of the earlier Wageningen models is their reliance on frequent destructive sampling of field plants to measure coefficients of assimilate partitioning among the organs. In fact, the ease to parameterize a model is a prerequisite if the model is applied to any studies where a number of crops or genotypes are involved. In order to become a useful tool for addressing G × E interactions, models have to be improved in both input parameters and model structure (Yin et al. 2004; Boote et al. 2013).

2.1 Improvements in model algorithms and structure

Firstly, improvements of crop models can be achieved by upgrading algorithms, for both the component processes and linking model components.

Among individual processes, photosynthesis and transpiration have been subjected to study most extensively. This is not surprising because unlike many other processes, modelling of photosynthesis and transpiration involves scaling from leaf level to canopy level, owing to spatial heterogeneity of radiation in a canopy. These two processes are strongly co-regulated, via the stomata, by environmental factors. For these reasons, numerical quantification of photosynthesis and transpiration has often accounted for an overwhelming portion of algorithms in crop models, such as BACROS and SUCROS.

An alternative is to develop analytical algorithms (e.g. de Pury and Farquhar 1997) for these two processes. One criterion to evaluate model components is whether they are capable of predicting interactive responses to relevant environmental factors. For example, if a crop model is used for assessing crop production in response to global change, interactions of CO_2 and other environmental factors on photosynthesis (e.g. Peng et al. 2020) or on RUE (van Oijen 2004) need to be well quantified. A change of optimum temperature for photosynthesis with levels of ambient CO_2 and irradiance is well described by the model of Farquhar et al. (1980), but this model is only occasionally incorporated into a crop/grass model (Rodriguez et al. 1999).

Crop growth relies on the functional balancing of a number of contrasting components (e.g. shoots vs roots, sources vs sinks) and processes (e.g. carbon vs nitrogen metabolisms, assimilation vs dissimilation, biosynthesis vs senescence). In addition, crop yield formation is associated with many feedback features in response to environmental perturbations. For example, more nitrogen uptake results in higher leaf nitrogen and leaf photosynthesis, which results in more growth and more nitrogen uptake (positive feedback). However, nitrogen uptake is an active process requiring energy to support; so, more nitrogen uptake is accompanied by more respiration and less growth, which results in less nitrogen uptake (negative feedback). Another indication of the (negative) feedback is apparent down-regulation of photosynthesis under conditions of high carbohydrate status.

Models have to be structured to embody the physiological causes that drive crop dynamics and produce emergent consequences. The model concept that demarcates crop production into potential, water-limited, or nitrogen-limited levels has facilitated model development by focusing on one major factor at a time but may not help to model individual processes that may interact. For example, leaves with high nitrogen levels transpire more than low-nitrogen leaves, because of the coupling of photosynthesis, stomatal conductance and transpiration (Wong et al. 1985). These issues are all important for predicting G × E interactions under a wide range of conditions, whereby major environmental factors (e.g. nitrogen and water) may be co-limiting or become limiting in tandem. The whole crop systems dynamics appear to have been insufficiently captured by many earlier crop models.

2.2 Improving model input parameters

The following can be considered in order to improve model-input parameters.

Firstly, input parameters are best set to be determined at the individual plant level. Random errors in input traits of current models are largely caused by field sampling (Yin et al. 2000a). The need for destructive samplings of field crop plants to determine key input parameters is a major limitation in using many crop models in breeding not only because of the required amount of work but also the limited amount of available material. For example, the extinction coefficient of light in a crop canopy is a widely used input trait. However, its measurement is laborious, especially involving sampling leaf materials of various canopy heights. The leaf angle would be its obvious alternative, as the extinction coefficient can be estimated from leaf angle (Goudriaan 1988).

Secondly, input parameters of current models may vary depending on the environment. For example, temperature sum is a common input parameter to predict crop phenological events, such as flowering. However, evidence from more careful analysis shows that temperature sum requirement for plants to reach flowering is not a constant but depends on (photo)thermal conditions that plants experienced (Yin 1999). Model parameters, known as 'genetic coefficients', are better designed as environment-independent whenever possible.

Thirdly, in addition to phenological traits (e.g. flowering or maturity times), morphological traits (e.g. plant height and seed size under the optimum condition) are better suited than physiological traits (e.g. coefficients of carbon or biomass partitioning among organs) as model input parameters, this is because morphological traits are more readily and precisely measured. Phenotyping physiological traits usually involve a number of steps, and small random errors at each step may aggregate to a significant level. Also, like phenological traits, there is ample information readily available about the inheritance of morphological traits scored by breeders for various crop species. The use of these traits as model input parameters would facilitate the conversion to gene-based models, thereby allowing a flexibility of physiological parameters being potentially grounded in gene-level understanding. Proper algorithms are needed for translating these morphological inputs into the required model coefficients.

3 The Genotype-by-Environment interaction on CROp growth Simulator model

Based on the previous discussions on improved model structure and input parameters, the GECROS model was developed (for version 1.0, see Yin and van Laar 2005). The model predicts crop biomass and yield as affected by daily weather variables (radiation, temperature, wind speed, and vapour pressure) and available amount of soil, water, and nitrogen (Fig. 1a). The model runs on a daily time step, but with subroutines for photosynthesis, transpiration, and phenology computed on shorter time steps.

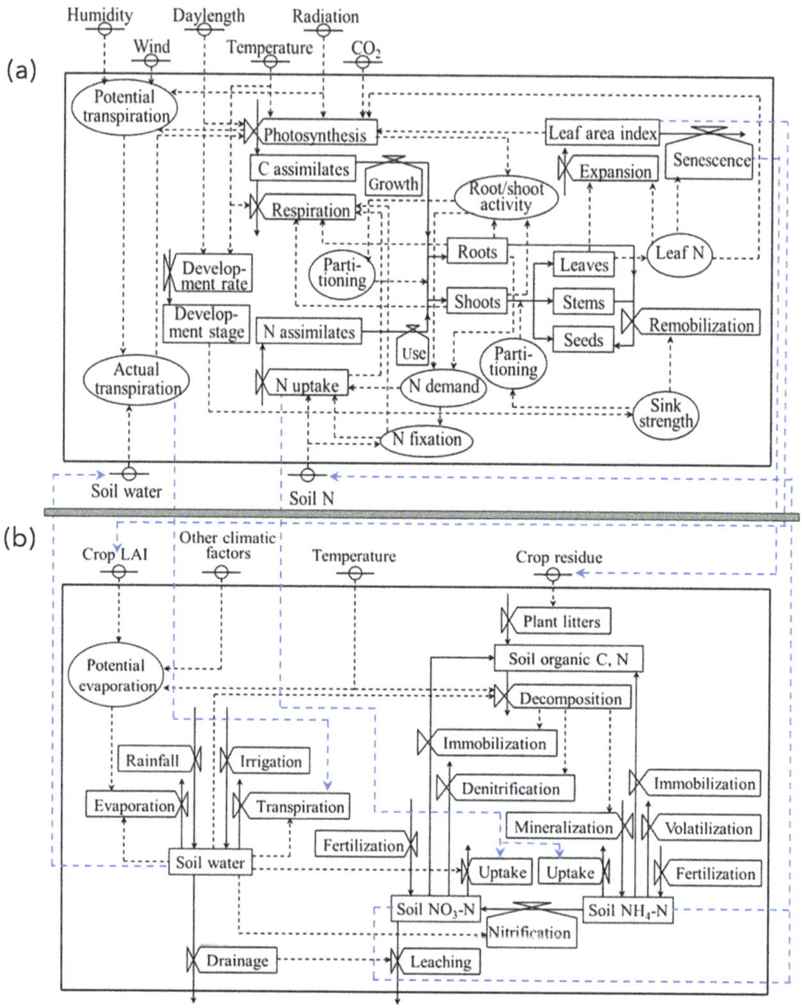

Figure 1 Relational diagram of the GECROS crop model (a, the part above the thick grey line) and its connected soil model (b, the part below the line). The processes and their interactions modelled are shown inside two big rectangle boxes (one for the crop model, and one for the soil model), while environmental input variables to each model are indicated by small crossed circles outside the boxes. The connections between crop and soil models are indicated by blue dashed arrows. The crop model can also run with user-defined scenarios for daily water and N available for crop uptake, by choosing to decouple it from the soil model. The soil model is a simplified version of the Roth-C/N model described by Coleman et al. (1999), where ten (sub)layers of the 150-cm soil depth were reduced to two dynamically varying layers – the top-rooted layer and the layer below, and water and mineral N in the rooted layer are assumed to be available for crop uptake. Indicative values of soil model input parameters for three broad soil types (sand, loam and clay) were given by Yin and van Laar (2005).

Since its first release, GECROS has been updated a few times, mainly to improve its photosynthesis algorithms: version 2.0 as used by Yin and Struik (2010), version 3.0 (Yin 2013; Gu et al. 2014a), and version 4.0 (Yin and Struik 2017). GECROS is implemented in the FST (FORTRAN Simulation Translator) language (van Kraalingen et al. 2003), but its 4.0 version in Python was used by Wang et al. (2024). It is important to note that the GECROS model is not for public use. However, the model can be made available for academic purposes upon reasonable request.

The model represents crop functions and interactive responses of contrasting processes, or components to environmental variables, thereby embodying physiological mechanisms that drive crop dynamics and generate emergent feedback features. The contrasting components primarily emphasize carbon (C) vs nitrogen (N) interactions but also include root vs shoot, source activity vs sink capacity, and growth vs senescence relationships. While it was previously discussed elsewhere (e.g. Priesack and Gayler 2009), here I outline model algorithms in modelling C-assimilation and growth processes, nitrogen uptake, root–shoot relation, sink demand, senescence, and phenology.

4 Model algorithms: photosynthesis, transpiration, respiration, and senescence

4.1 Stomatal conductance, leaf photosynthesis, and transpiration, and their scaling-up to canopies

GECROS (v.1.0) calculates instantaneous leaf net photosynthesis (*A*) from the C_3-photosynthesis model of Farquhar et al. (1980), which assumes the intercellular:ambient [CO_2] ratio only as a function of leaf-to-air vapour pressure deficit. This approach was also applied to C_4 crops, with additional assumptions (i) to set internal [CO_2] to an arbitrarily high value, and (ii) to consider the extra ATP consumption by the CO_2-concentrating mechanism. In a later version (v.2.0), *A* was calculated from the analytical algorithms that are based on the model of Farquhar et al. (1980) for C_3 photosynthesis, and its equivalent for C_4 photosynthesis (von Caemmerer and Furbank 1999), coupled with a model for stomatal conductance g_s and temperature- and leaf N-dependent mesophyll conductance g_m (see Yin and Struik 2009). The analytical cubic polynomials (see Yin and Struik 2009) simultaneously solve g_s, internal [CO_2] level, and leaf photosynthesis rate (*A*) for a given temperature. The obtained g_s was used in the Penman-Monteith equation (Monteith 1973) to model leaf transpiration and temperature through surface energy balance, accounting for factors such as elevated [CO_2]. Leaf temperature was then used for re-calculating leaf photosynthesis and transpiration. The effect of leaf N content on photosynthesis, g_s and transpiration, is reflected by the effect of leaf

N on parameters (i.e. V_{cmax} – maximum capacity of Rubisco activity, and J_{max} – maximum rate of electron transport) of the photosynthesis model. This coupled approach allows the decrease of g_s at elevated [CO_2] to be well predicted. A further development of GECROS (v.3.0; Yin 2013; Gu et al. 2014a) allowed an option that g_m varies in proportion with g_s, given reports that g_m may resemble g_s in response to various environmental variables (e.g. Flexas et al. 2008; Yin et al. 2009).

Spatial extension from leaf to canopy photosynthesis and transpiration was established using the analytical sun/shade model of de Pury and Farquhar (1997). Temporal extension from instantaneous rates to daily total was performed using the five-point Gaussian integration (Goudriaan 1986) to account for (a)symmetric diurnal course of radiation and temperature, to which photosynthesis and transpiration respond nonlinearly. These approaches for spatial and temporal extensions apply to the case in the absence of water stress.

For simplicity, the diurnal course of available water is assumed to follow that of radiation. In the presence of water stress (i.e. water availability does not satisfy the requirement for potential transpiration), the available water is partitioned between sunlit and shaded leaves, according to the relative share of their potential transpiration to obtain their instantaneous actual transpiration. The actual transpiration is transformed into the actual level of g_s using a reversed Penman-Monteith equation, and the actual g_s was then used as input to the analytical quadratic model (Yin and Struik 2017; Liu et al. 2025), to estimate the instantaneous actual photosynthesis of the sunlit and shaded leaves. The Gaussian integration is again used to obtain the daily total of the actual photosynthesis.

4.2 Crop respiration and growth efficiency

Crop respiration was modelled, based on the framework of Cannell and Thornley (2000) that recognises individual relationships between respiration and each process it supports. In this framework, component processes are differentiated: symbiotic N_2 fixation (relevant to legumes), root nitrogen uptake, nitrate reduction, other ion uptake, phloem loading, residual maintenance component, and biosynthesis.

Most of these processes can be well quantified and the default carbon costs used are 6 g C (g N fixed)$^{-1}$, 0.17 g C (g ammonium-N uptake)$^{-1}$, 0.34 g C (g nitrate-N uptake)$^{-1}$, 1.71 g C (g nitrate-N reduction)$^{-1}$, 0.06 g C (g mineral uptake)$^{-1}$, and 0.06 g C (g C loaded from shoot to root)$^{-1}$. However, the residual maintenance respiration is less quantifiable, but suggested to be related to crop N content. The efficiency of biosynthesis (growth) is obtained from the chemical compositions (lipids, lignins, organic acids, minerals, proteins, and carbohydrates) of plant material, based on carbon fraction and glucose requirement of these chemical components (Penning de Vries et al. 1989).

4.3 Leaf area index and senescence

A common method for modelling senescence is to use empirical, developmental-stage-dependent leaf-turnover coefficients (e.g. Penning de Vries et al. 1989; Goudriaan and van Laar 1994). These coefficients differ across crops, and across environments (e.g. N supplies), requiring periodical samplings or specific calibration to determine.

Yin et al. (2000b) derived an equation for describing leaf area index (L) in relation to the amount of canopy leaf-nitrogen (N) as

$$L = \frac{1}{k_n}\log_e\left(1 + k_n N / n_b\right) \quad (1)$$

where k_n is the coefficient for leaf-N extinction in the canopy, and n_b is a base N content at or below which leaf photosynthetic rate at a saturating light is nil. A broad use of equation (1) is to describe the generic logarithmic relation between L and N for the full canopy, in which the leaf N at the bottom of the canopy is as low as n_b. However, equation (1) can also be used to generate the relationship between L and N for young canopy, where $L \leq 1$, which showed that the relationship is virtually linear (Yin et al. 2003b).

A more important use of equation (1) is to engender a simple robust method to predict the onset and quantity of leaf senescence. The value calculated by equation (1) can be designated as the N-limited leaf area index (L_N). As in other Wageningen crop models, leaf area index has been calculated from a specific leaf area constant (s_{la}) and C partitioned to or biomass accumulated in leaves, denoted here as L_C. The rate of decrease in leaf area index owing to senescence can be formulated as $[L_C - \min(L_C, L_N)]/t_c$, where t_c is the time constant, which can be equal to, or more than, the time step for simulation (Yin et al. 2000b). This approach is generic and avoids the use of time-dependent, empirical leaf-turnover coefficients. The input parameter n_b is required in a crop model for simulating leaf photosynthesis, and the other parameter k_n may be calculated theoretically from light extinction coefficient (Yin et al. 2003b). The approach agrees with a coherent biological picture of leaf senescence in relation to the decreasing amount of N in the bottom canopy, especially during seed growth. A similar approach was used in GECROS for describing root senescence, based on root N content. The approach can also predict, to a large extent, high temperature-induced senescence as high temperature accelerates seed filling that requires a fast remobilization of leaf and root N, thereby causing fast senescence. However, any direct impact of extreme high temperature on senescence needs further modelling studies.

5 Model algorithms: nitrogen uptake and assimilate partitioning

5.1 Nitrogen uptake

As usual, N uptake is the minimum of crop N demand and soil N supply; the latter is provided by a soil model (Fig. 1b). Crop N demand is commonly based on the critical N concentration in aboveground biomass in the course of the growth cycle reported for various crops or similar empirical equations.

The main equation to describe crop N demand (N_{dem}) in GECROS is (Yin et al. 2003b)

$$N_{dem} = C_R\sigma_N = C_R\sigma_C^2 / (d\sigma_C / d\kappa) \tag{2}$$

where C_R is the amount of C in roots. This was based on the analysis of Hilbert (1990) for balanced growth conditions that achieving the optimum plant N:C ratio for a maximized relative C gain requires that relative root activity for N uptake (σ_N) and relative shoot activity for C assimilation (σ_C) be balanced as defined by equation (2), where $d\sigma_C/d\kappa$ is the first-order derivative of σ_C with respect to κ, the N:C ratio in the whole plant. The value of σ_C is defined as $\sigma_C = (\Delta C / \Delta t) / C_S$, where C_S is the amount of C in shoots; $\Delta C/\Delta t$ is daily crop net C gain (which can be calculated from the abovementioned models for daily canopy photosynthesis and crop respiration). Because the quantity $d\sigma_C/d\kappa$ in equation (2) cannot be analytically calculated given the sophistication of estimating σ_C, it can be numerically calculated by $d\sigma_C / d\kappa = [\sigma_{C(\kappa+\Delta\kappa)} - \sigma_{C(\kappa)}] / \Delta\kappa$, where $\Delta\kappa$ is a small increment of κ; $\sigma_{C(\kappa)}$ and $\sigma_{C(\kappa+\Delta\kappa)}$ are relative shoot activities when plant N:C ratio is κ and ($\kappa+\Delta\kappa$), respectively. This generic method is less trivial than the 'critical N concentration' approach and can reflect that root N uptake may be related to crop photosynthetic activity (Triboi and Triboi-Blondel 2002). The method can produce the commonly observed sigmoid shape of cumulative N uptake over a growing season (Yin and van Laar 2005).

5.2 Partitioning between root and shoot

In other Wageningen crop models, partitioning of newly formed assimilates between root and shoot is simply fixed as an empirical function of development stage (e.g. Penning de Vries et al. 1989; Goudriaan and van Laar 1994; Fig. 2a). Derivation of these partitioning functions for various crops/genotypes needs substantial experimental data. More importantly, this approach does not account for the plasticity of root–shoot ratios in response to environmental changes, as reported by Poorter and Nagel (2000).

Yin and Schapendonk (2004) presented an equation for the partitioning of C between shoots and roots. The equation is derived from the classical

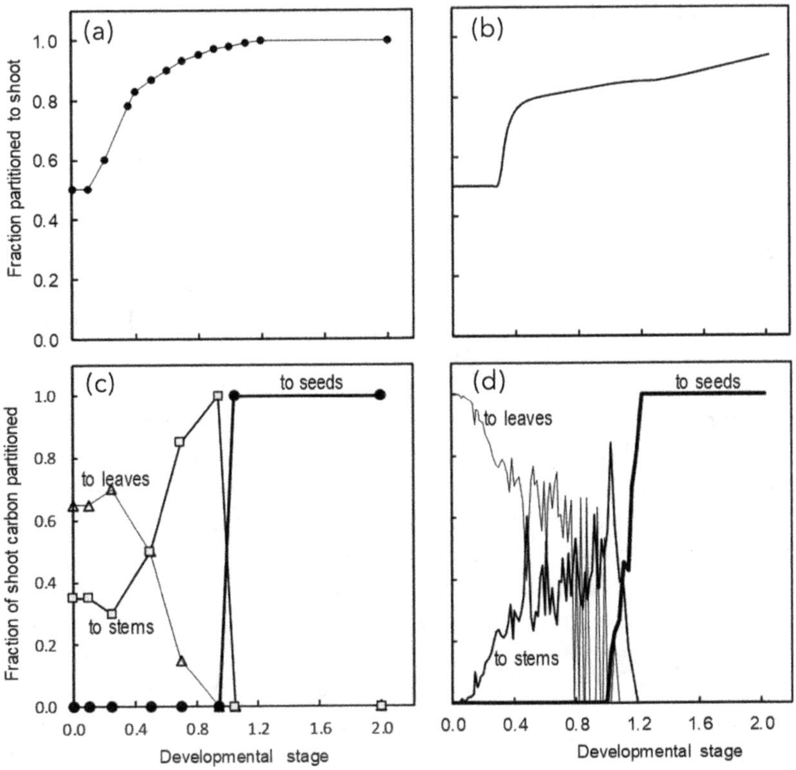

Figure 2 Fractions of newly formed carbon partitioned to growing organs defined by the SUCROS model (Goudriaan and van Laar 1994) (a, c) vs the typical patterns of the partitioning modelled by the GECROS model (b, d), for a determinate crop. In both models, partitioning was made first between roots and shoots (a, b) and then within-shoot among leaves, stems and seeds (c, d). In SUCROS, partitioning coefficients are linearly interpolated from a table in relation to development stage, where values in the table (indicated as symbols in panels a and c) would need frequent samplings of experimental plant materials to observe if they are set as genotype specific. In contrast, in GECROS, root–shoot partitioning is modelled according to equation (3), while within-shoot partitioning to stems and seeds is calculated according to equation (4). Any change in partitioning to stems is determined by genotype-specific maximum plant height (H_{max}, which influences w_{max} in equation 4 for stems), and that to seeds is determined by modelled sink demand (which also depends on genotype-specific single-grain weight, s_W). Partitioning to leaves is calculated according to equation (5), and as can be seen from panel (d), it may fluctuate for some days when carbon and nitrogen limits leaf area index alternately. This framework of modelling partitioning also allows the fraction of newly formed carbon going to reserves to be calculated daily (see equation (6)), rather than using a fixed fraction of stem weight being remobilized to seeds as used in the SUCROS model. In short, based on simple equations (3–6) with a few morphological parameters (like H_{max} and s_W) as input (Table 2), GECROS provides a coherent biological picture about assimilate partitioning and remobilization, without relying on heavy field sampling of plant materials as needed for describing genotype-specific partitioning and remobilization dynamics in most other Wageningen crop models. GECROS, Genotype-by-Environment interaction on CROp growth Simulator; SUCROS, Simple and Universal CROp growth Simulator.

root–shoot functional balance theory (Brouwer 1983), with an incorporation of the mechanism that plants control root–shoot partitioning in order to maximize their relative C gain (Hilbert 1990). The fraction of the newly assimilated C partitioned to the shoot ($\lambda_{C,S}$), calculated by

$$\lambda_{C,S} = \frac{1}{1+(\eta / \sigma_C)d\sigma_C / d\kappa} \quad (3)$$

where η is the N:C ratio in newly formed biomass, $d\sigma_C/d\kappa$ and σ_C are the same as defined for equation (2). A similar equation was derived for estimating the fraction of the newly obtained N partitioned to the shoot (Yin and Schapendonk 2004). Equation (3) is simple and has no parameters to estimate as all the inputs can be calculated elsewhere in a crop model. The equation, strictly speaking, holds for steady-state growth conditions (Yin and Schapendonk 2004). van der Werf et al. (1993) showed evidence that partitioning functions, if related to plant N status, held for both steady-state and general conditions. When incorporated into a crop model, equation (3) produces a pattern of root-shoot partitioning similar to the observed ones (Fig. 1a,b), yet addresses the plasticity of root:shoot ratios in response to environmental conditions (Yin and Schapendonk 2004). The simulated effects of light, nutrients, and water on the partitioning agreed with the experimentally observed; the simulated effect of CO_2 was not consistent, which seemed to agree with the meta-result of Poorter and Nagel (2000) that a doubled $[CO_2]$ did not significantly affect the average allocation of C to roots or shoots.

5.3 Sink demand, within-shoot assimilate partitioning, and remobilization

Assimilates distributed to the shoot need to be further modelled for partitioning among the shoot organs, and again, empirical partitioning coefficients are fixed in relation to development stage in SUCROS (e.g. Goudriaan and van Laar 1994; see Fig. 2c). In GECROS, the strength of growing organs (stems, grains) as sinks of available C assimilates determines the partitioning; any surplus assimilate goes to the pool of reserves which can be remobilized later if there is assimilate deficit, as described below.

Yin et al. (2003a) developed an equation for describing any asymmetric sigmoid pattern of determinate growth. By setting the initial value of a sink organ at the start of its growth as zero, the equation uses three parameters w_{max}, θ_e and θ_m (where w_{max} is the maximum quantity, which is reached at the end of growth θ_e; θ_m is the stage at which the growth rate is maximum). The differential form of the equation, when multiplied by the daily development rate (d^{-1}), can describe the daily demand of a sink (e.g. seed or stem) for C assimilates, Δw, as (Yin and van Laar 2005):

$$\Delta w = \omega w_{\text{max}} \frac{(2\theta_e - \theta_m)(\theta_e - \theta)}{\theta_e(\theta_e - \theta_m)^2} \left(\frac{\theta}{\theta_e} \right)^{\theta_m/(\theta_e - \theta_m)} \tag{4}$$

where ω is the development rate at stage θ. The commonly used classical (like Logistic) growth equations have an asymptotic form, meaning that an expected w_{max} can never be achieved, even if source activity is not limiting. In contrast, equation (4) ensures that w_{max} is achieved exactly at θ_e. The model uses only three parameters (i.e. w_{max}, θ_m and θ_e) to define any asymmetric sigmoid pattern that would need at least four parameters to define with classical growth equations. Moreover, θ_e for a given sink organ can be defined, so its timing can be predicted by the phenology part of the crop model. These features allow equation (4) to be better suited for being embedded in crop models to simulate sink demand than classical growth equations and thus to simulate the partitioning to stems and to seeds (Fig. 2d).

The fraction of new shoot carbon partitioned to the leaf ($\lambda_{C,leaf}$) is determined, depending on whether it is carbon or nitrogen that limits the growth of canopy:

$$\lambda_{C,\text{leaf}} = \begin{cases} 0 & L_C \geq L_N \\ 1 - \lambda_{C,\text{seed}} - \lambda_{C,\text{stem}} & L_C < L_N \end{cases} \tag{5}$$

where L_C and L_N are carbon-limited and nitrogen-limited leaf area index, respectively (see earlier), and $\lambda_{C,stem}$ and $\lambda_{C,seed}$ are fractions partitioned to stems and seeds, respectively, based on their stage-dependent demands for carbon as quantified by equation (4). The fraction of carbon partitioned to leaf, predicted by equation (5), can fluctuate for some days in the middle growth phase (Fig. 2d). The fraction of new shoot carbon going to the shoot reserve pool ($\lambda_{C,Sres}$) is then calculated by

$$\lambda_{C,\text{Sres}} = 1 - \lambda_{C,\text{seed}} - \lambda_{C,\text{stem}} - \lambda_{C,\text{leaf}} \tag{6}$$

GECROS also assumes the existence of the root carbon reserve pool. The fraction of new the root carbon going to root reserve pool is either 0 or 1, depending on whether it is carbon or nitrogen that limits the growth of structural root biomass. If the new shoot carbon is less than the demand for carbon by seed filling, the carbon in reserve pools, if any, is remobilized to fill the shortfall, for which shoot and root carbon reserves are remobilized impartially.

6 Model algorithms: phenological development

Developmental stage (θ) is set to be 0 at seedling emergence, 1 at the onset of seed filling, and 2 at maturity. Daily development rate (ω), which has a unit of day^{-1}, is calculated separately for pre-seed filling (vegetative growth) period and for the seed-filling (reproductive growth) period. For the vegetative period, ω is calculated by (Yin et al. 1997; Nakagawa et al. 2005):

$$\omega = \begin{cases} g(T)/m_V & \theta \le \theta_1 \text{ or } \ge \theta_2 \\ g(T)h(D_{lp})/m_V & \theta_1 < \theta < \theta_2 \end{cases} \tag{7}$$

where m_V is the minimum number of days for the vegetative period when the photothermal environment is at the optimum, θ_1 and θ_2 are the development stages for the onset and the end of photoperiod-sensitive phase, respectively, D_{lp} is photoperiodic daylength.

The temperature-effect function, $g(T)$, is a bell-shaped function (Yin et al. 1995) as

$$g(T) = \left[\left(\frac{T_c - T}{T_c - T_o}\right)\left(\frac{T - T_b}{T_o - T_b}\right)^{\frac{T_o - T_b}{T_c - T_o}}\right]^{c_t} \tag{8}$$

where c_t is temperature response curvature coefficient, T_b, T_o, and T_c are the base, the optimum, and the ceiling temperature for development [i.e. $g(T) = 0$ if $T \le T_b$ or $\ge T_c$]. As temperature is diurnally fluctuating under field conditions, $g(T)$ is estimated on a hourly basis and hourly $g(T)$ values are averaged for the daily value. The hourly temperature is estimated from daily maximum and minimum temperature by a sine function, assuming the daily maximum occurs at 14:00 each day. In case of the lack of data for determining c_t by curve fitting, c_t can be assumed to be equal to 1.

$h(D_{lp})$ in equation (7) has a value between 0 and 1 with $h(D_{lp}) = 1 - p_{sen}(D_{lp} - M_{op})$, where M_{op} is the maximum optimum photoperiod for a short-day crop (about 11 h) or the minimum optimum photoperiod for a long-day crop (about 18 h); p_{sen} is the photoperiod-sensitivity parameter, being positive for short-day crops and negative for long-day crops. A zero value of p_{sen} characterises absolutely insensitive cultivars. Genotypic difference within a crop in response to photoperiod is assumed not in M_{op} but mainly in p_{sen}, θ_1 and θ_2 (Yin et al. 1997).

Development rate for seed filling phase is calculated dependent on temperature only, using genotypic coefficient m_R, the minimum number of days for the seed-filling period when temperature is at its optimum. The effect of temperature on development rate of this period is also based on equation (8), but with a restriction that T is set to T_o if $T > T_o$ to abolish the decline of development rate at high temperatures. This restriction is incorporated to account for the accelerated seed filling when plants are exposed to high temperatures.

7 Recent model improvements

Like most crop models, GECROS follows the principle that a plant grows as much as light, water, and nutrient resources would potentially enable, and thus, net photosynthesis (source) is the driving process. However, both vegetative and reproductive organs (sinks) frequently do not commensurate with the potential source, and growth is uncoupled from photosynthesis when drought and heat stress occur (Muller et al. 2011). For these cases, crop yield does not depend much on the availability of photosynthates, but more on the sink size, such as the number of spikelets (Dingkuhn et al. 2020; Li et al. 2025). However, accurately simulating the number of fertile spikelets (sink) across diverse environments with a minimum set of parameters is challenging.

In earlier versions, the number of seeds to be filled (the number of spikelets m^{-2} for cereal crops like rice) is modelled to be proportional to the amounts of either carbon or nitrogen (depending on which is more limiting) accumulated until flowering. Gu et al. (2014b) showed that this approach overestimated yield of rice genotypes under drought, probably under stress in general. It is known that when stress occurs during the flowering phase, the percentage of filled spikelets depends more on the panicle temperature during the flowering time window in a day, ~8:00–13:00 h (Julia and Dingkuhn, 2013; Li et al. 2025). Therefore, the model was modified to account for the direct effect of panicle temperature on sink size (GECROS v5.0 as described in Kadam et al. 2019) in such a way that the potential number of spikelets formed depends on photosynthates, while their survival depends more directly on the tissue temperature (Julia and Dingkuhn 2013). Kadam et al. (2019) considered the time window in days for determining the potential number of spikelets, and hours within a day and tissue transpirational cooling for determing spikelet survival. The simulation of panicle temperature was done using the same algorithms in GECROS for simulating leaf-surface energy balance, based on a coupled conductance–photosynthesis–transpiration routine, whereby panicles were treated as a photosynthesizing organ and its conductance was calculated using a leaf conductance model. Since the panicle temperature during the flowering time window is the most crucial factor in determining the spikelet sterility (Julia and Dingkuhn, 2013; Li et al. 2025), simulation of instantaneous photosynthesis and transpiration to daily total was changed from the five-point Gaussian integration (see earlier text) to computation of 24 times from sunrise to sunset. A factor for reduction induced by any high panicle temperature at flowering hours was introduced to simulate the actual spikelet fertility under stress, based on the linear relationship between sterility and panicle temperature reported by Julia and Dingkuhn (2013) for rice.

While this extension of the model was mainly implemented for rice, the principles may be applicable to other crops as well. However, in contrast to the 'source' processes like photosynthesis that is generic across species, 'sink' processes like spikelet formation, especially their response to environmental variables, which are highly specific to species, cultivars, and environmental factors *per se*, are more difficult to model (see later discussion) especially when applied to predicting genotypic differences (which are often subtle among genotypes in a genetic population).

8 Model applications: genotype-by-environment (G × E) interactions

When running the model, it is necessary to first specify crop management practices, which include crop planting date, amounts, and timing of (NH_4^+- and NO_3^--) nitrogen-fertiliser application, and amounts and timing of irrigation. These later input parameters join the simulation of soil nitrogen and water balance (Fig. 1b). As described above, GECROS is a generic crop model, as

- it simulates potential, water-limited or/and N-limited growth and productivity;
- it predicts not only biomass yield, but also protein contents of end products (amounts of N accumulated in grains is converted to protein content by a factor 6.25); and
- it suits various crops: C_3 and C_4 crops; legume and non-legume crops; short-day and long-day crops; determinate and indeterminate crops; and field and greenhouse crops.

Running the model requires specifying crop types and defining some crop-specific parameter values (Yin and van Laar 2005). Indicative values of these parameters are given in Table 1 for several crops.

Genetic differences among genotypes within a crop species are defined by genotype-specific input parameters, or 'genetic coefficients'. The model was designed in such a way that most of these genetic coefficients can be directly determined from measurements, without recourse to an optimization that may require to conduct many experiments. Such an optimization procedure may not suit for parameterization of sophisticated crop models like GECROS that have different time steps for different sub-processes. Main genetic coefficients used in the model are listed in Table 2, broadly according to phenological, morphological, and physiological categories.

Many genetic coefficients correspond to those that breeders usually score in their selection and test trials. This may facilitate the parameterization of GECROS from existing variety test trials and the transformation of insights from the model into options that breeders may use. Morphology-related traits were chosen, because they are more readily measured than those physiological equivalents (such as assimilate partitioning coefficients and light extinction coefficients) used in other Wageningen crop models. However, they should not be considered as only genetic coefficients nor should they all be considered in every study to indicate genotypic differences. Maximum rooting depth (D_{max}) is an important extra parameter when examining genetic differences under the conditions where drought prevails during the growing season. Similarly, daily maximum N uptake (n_{maxUp}) is important for examining genotypic differences in nitrogen use. Phenological parameters θ_1, θ_2, and p_{sen} would be important, only if the model is specifically used to model genetic differences in responding to photoperiod. Below I show a few examples, in which (some of) the above parameters with an additional trait 'total crop nitrogen uptake' (N_{max}; but see later for discussion on this) were determined to predict yield differences among the genotypes in real genetic populations.

The GECROS model was used to analyze differences in tuber yield of potato (*Solanum tuberosum* L.) in five cultivars covering a wide range of maturity types and 100 individuals of a diploid F1 population segregating for maturity type (Khan et al. 2014). Six field experiments were conducted, in which contrasting nitrogen availabilities were created to represent six environments. Values of five genetic coefficients (m_V, m_R, H_{max}, n_{SO}, and N_{max}) were estimated and calibrated. Variation among the 100 F1 genotypes was as wide as, or slightly wider than, that among the five contrasting cultivars for any of the five parameter values. For the 100 F1 genotypes, the model accounted for 86% of genotypic differences in across-environment average tuber yield, and 89% of environmental difference in across-genotype average yield. However, the percentage in the genotypic differences in yield for a given experiment accounted for by the model ranged from 2% to 65%. Model sensitivity analysis (by fixing genetic coefficients, one at a time, to their across-genotype average value) identified N_{max} (i.e. total crop N uptake) and tuber N concentration (n_{SO}) as key components affecting tuber yield for all six experiments. Genotypes with higher N_{max} and lower n_{SO} exhibited higher tuber dry matter yield. The development of potato ideotypes for any specific environments should prioritize optimizing N-related traits.

Yin et al. (2000a) showed the performance of a SUCROS-family crop model when incorporated with molecular markers in predicting differences of genotypes in a genetic population of barley. In such an analysis, modelled genetic coefficients are considered as standard quantitative traits, subject to standard QTL (quantitative trait loci) analysis to identify associated molecular markers. Using GECROS, Gu et al. (2014b) designed markers and virtual ideotypes for improving yields of rice (*Oryza sativa*) under drought stress. Genotypic variations for seven genetic coefficients (s_W, n_{SO}, H_{max}, m_V, m_R, s_{la}, and N_{max}) were identified for a bi-parental rice population of 94 introgression lines (ILs) derived from two parents differing in drought tolerance (Fig. 3), and with these parameter values as input, the model predicted the variations of yield and biomass of these ILs under well-watered and drought conditions (Fig. 4). Next, loci underlying these variations of the model-input parameters

Table 1 Indicative values of crop-specific parameters of the GECROS (Genotype-by-Environment interaction on CROp growth Simulator) model

Parameter	Wheat	Barley	Rice	Maize	Sorghum	Soybean	Pea	Potato	Sunflower	Rape	Sugar beet
$Y_{G,V}$	0.81	0.81	0.81	0.81	0.81	0.75	0.75	0.81	0.81	0.81	0.81
$f_{c,V}$	0.48	0.48	0.43	0.48	0.48	0.48	0.48	0.48	0.48	0.48	0.48
ε_g	0.25	0.25	0.25	0.25	0.25	0.35	0.35	0.25	0.45	0.45	0.25
f_{lip}	0.02	0.01	0.02	0.05	0.04	0.20	0.02	0.00	0.29	0.48	0.00
f_{lig}	0.06	0.04	0.12	0.11	0.12	0.06	0.06	0.03	0.05	0.05	0.05
f_{oac}	0.02	0.02	0.01	0.01	0.02	0.05	0.04	0.05	0.03	0.02	0.04
f_{min}	0.02	0.04	0.01	0.01	0.02	0.04	0.03	0.05	0.03	0.04	0.04
f_{pro}	0.14	0.09	0.08	0.10	0.12	0.38	0.27	0.09	0.20	0.23	0.05
f_{car}	0.74	0.80	0.76	0.72	0.68	0.27	0.58	0.78	0.40	0.18	0.82
T_b	0	0	8	8	8	8	0	0	8	0	0
T_o	25	25	30	30	30	30	25	25	30	25	25
T_c	37	37	42	42	42	42	37	37	42	37	37
w	0.01	0.01	0.01	0.05	0.05	0.025	0.025	0.025	0.10	0.025	0.08
ρ	460	450	450	570	560	350	345	170	270	285	150
D_{max}	130	130	80	145	145	140	100	100	145	130	120
n_b	0.35	0.30	0.30	0.25	0.25	0.80	0.60	0.35	0.30	0.30	0.25
n_{cri0}	0.050	0.050	0.050	0.050	0.050	0.060	0.055	0.050	0.050	0.050	0.050
n_{Rmin}	0.005	0.005	0.005	0.005	0.005	0.010	0.007	0.005	0.005	0.005	0.005

n_{Smin}	0.010	0.010	0.010	0.008	0.008	0.020	0.015	0.010	0.010	0.010	0.010
s_{la}	0.028	0.031	0.023	0.022	0.025	0.025	0.033	0.033	0.025	0.030	0.020
E_{Jmax}	48270	30200	88380	70890		97020	53580		67100		

Symbol definitions (units, where dw = dry weight of biomass):
$Y_{G,V}$ vegetative-organ (leaf, stem, root) growth efficiency (g C g^{-1}C)
$f_{c,V}$ fraction of carbon in vegetative-organ biomass (g C g^{-1}dw)
ε_g germination efficiency, i.e. dry weight of seedling to per gram dry seed (g dw g^{-1}dw)
f_{lip} fraction of lipids in seed biomass (g g^{-1})
f_{lig} fraction of lignins in seed biomass (g g^{-1})
f_{oac} fraction of organic acids in seed biomass (g g^{-1})
f_{min} fraction of minerals in seed biomass (g g^{-1})
f_{pro} fraction of proteins in seed biomass (g g^{-1}); = $6.25n_{SO}$ where n_{SO} is a genotype-specific parameter (see Table 2)
f_{car} fraction of carbohydrates in seeds biomass (g g^{-1}); = $1 - f_{lip} - f_{lig} - f_{oac} - f_{min} - f_{pro}$
T_b base temperature for phenological development (°C)
T_o optimum temperature for phenological development (°C)
T_c ceiling temperature for phenological development (°C)
w leaf blade width (m)
ρ proportion factor between stem biomass and plant height (g dw m^{-1})
D_{max} maximum rooting depth (cm)
n_b minimum leaf nitrogen for photosynthesis (g N m^{-2} leaf)
n_{cri0} initial critical shoot nitrogen concentration (g N g^{-1} dw)
n_{Rmin} minimum nitrogen concentration in root (g N g^{-1} dw)
n_{Smin} minimum nitrogen concentration in stem (g N g^{-1} dw)
s_{la} specific leaf area constant for new leaves (m^2 leaf g^{-1} leaf)
E_{Jmax} activation energy (J mol^{-1}) for maximum electron transport rate J_{max} used in the model of Farquhar et al. (1980).
Source: Adapted from Yin and van Laar (2005).

Table 2 List of genotype-specific model-input parameters (or called 'genetic coefficients') of the GECROS (Genotype-by-Environment interaction on CROp growth Simulator) model, belonging to three broad categories[a]

Category	Parameter[b]	Definition	Unit
Phenological	$\boldsymbol{m}_{\mathbf{V}}$	Minimum days for vegetative growth (entire pre-seed filling) phase	d
	$\boldsymbol{m}_{\mathbf{R}}$	Minimum days for reproductive growth (seed-filling) phase	d
	p_{sen}	Photoperiod sensitivity	$hour^{-1}$
	θ_1	Development stage for the onset of the photoperiod-sensitive phase (with $0 \leq \theta_1 < \theta_2$)	–
	θ_2	Development stage for end of the photoperiod-sensitive phase (with $\theta_1 < \theta_2 \leq 1$)	–
Morphological	$\boldsymbol{H}_{\mathbf{max}}$	Maximum plant height	m
	D_{max}	Maximum rooting depth	cm
	s_{la}	Specific leaf area constant for newly growing leaves	m^2 leaf (g dry leaf)$^{-1}$
	$\boldsymbol{s}_{\mathbf{W}}$	Single average seed (grain, or tuber) weight	g
Physiological	$\boldsymbol{n}_{\mathbf{SO}}$	Nitrogen concentration in seed organs (grains, or tubers)	g N (g dry seed)$^{-1}$
	n_{maxUp}	Maximum daily nitrogen uptake	g N (m^2 ground)$^{-1}$ d^{-1}
	N_{max}[c]	Maximum nitrogen uptake	g N (m^2 ground)$^{-1}$
	b_{fer}[d]	Baseline spikelet fertility	%
	F_{h1}[d]	Onset of flowering (anthesis) time of a day	Hours (after sunrise)
	F_{h2}[d]	End of flowering (anthesis) time of a day	Hours (after sunrise)
	f_{erSen}[d]	Fertility sensitivity to flower-organ temperature	°C^{-1}

[a]In principle, all parameter values refer to those obtained (or estimated) from the control experimental treatment in the absence of abiotic stress; this list does not include photosynthesis parameters relevant when simulating impacts of natural variations in these parameters on crop yield (but see Gu et al. 2014a).

[b]Five parameters in **bold** are always needed in general, while others may be included, depending on the context and objective of the study (see the text); so, some of them are also listed as crop-specific parameters in Table 1.

[c]This is not meant as an input parameter (in fact, an output trait) in the GECROS model but is listed here to indicate that it is sometimes needed for reliable simulation of genotypic differences (see the text).

[d]These parameters are required only when simulating the effect of heat or/and drought on floret fertility (as implemented in the version 5.0 of GECROS; see the text).

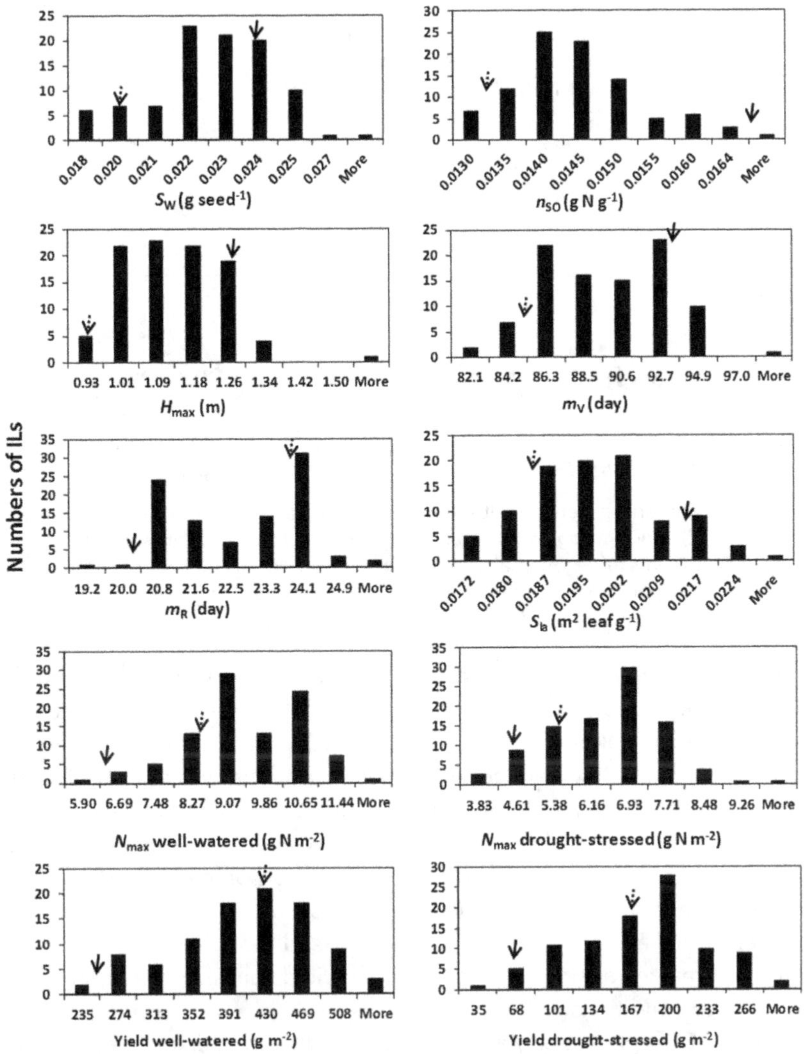

Figure 3 Frequency distribution of seven model parameters and yield in the population of introgression lines of rice (after Gu et al. 2014b). Arrows show values for the two parents (full arrow for 'Haogelao' and dotted arrow for 'Shennong265').

were mapped. Marker-based values of ILs for each of these parameters were estimated from additive allele effects of the loci and were fed to the model to simulate yields of the ILs grown under well-watered and drought conditions and to design virtual ideotypes for those conditions. Genetic variation in N_{max} had the most significant effect on yield; five other parameters also significantly influenced yield, but seed weight and leaf photosynthesis

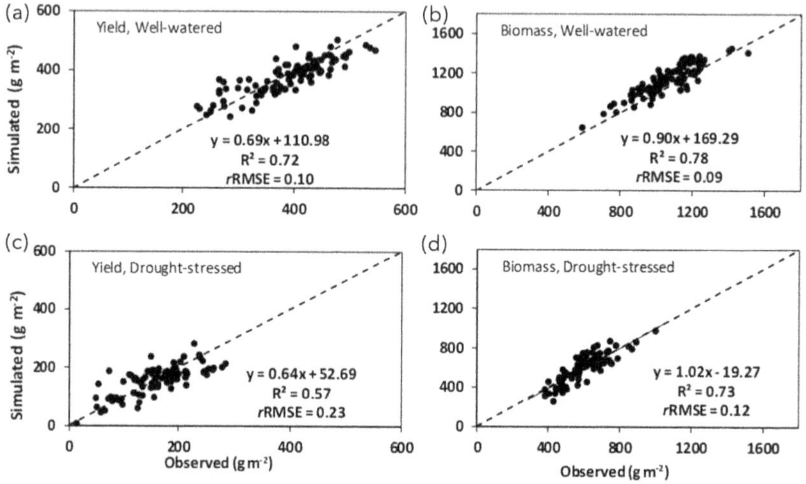

Figure 4 Comparison between observed and simulated values of grain yield (a, c) and biomass (b, d) of rice introgression lines for both well-watered (a, b) and drought-stressed (c, d) environments (after Gu et al. 2014b). The simulation was based on the phenotypic values of seven parameters as shown in Fig. 3. The diagonal line is the 1:1 line.

did not. Using the marker-based parameter values, GECROS also simulated yield variation among 251 recombinant inbred lines derived from the same parents. The model-based dissection approach detected more markers than the analysis using only yield *per se*. Model-based sensitivity analysis (by fixing markers, one at a time, i.e. by assuming all ILs carry identical alleles at the marker locus) ranked all markers for their importance in determining yield differences among the ILs. Virtual ideotypes based on markers identified by modelling had 10–36% more yield than those based on markers for yield *per se*. This analysis shows that GECROS-based modelling can be used as a tool for marker design for improved rice yields and for ideotyping under contrasting conditions.

The GECROS model was later used to identify markers for improving rice yield under well-watered (control) and water-deficit conditions based on genetic differences exhibited in a GWAS (genome-wide association study) panel (Kadam et al. 2019). Eight genetic coefficients (m_V, m_R, p_{sen}, H_{max}, n_{SO}, s_W, b_{fer}, and N_{max}) were measured from the control in one season (2013) for 267 *indica* rice genotypes. The model accounted for 58% of yield variation among genotypes under control and 40% under water-deficit conditions of the same year (Fig. 5a) but accounted for less for the independent year 2014 (Fig. 5b). Using 213 randomly selected genotypes as a training set, 90 SNPs (single-nucleotide polymorphisms) were identified using GWAS, explaining 42–77% of genetic-coefficient variation. SNPs-based coefficient values estimated from the additive

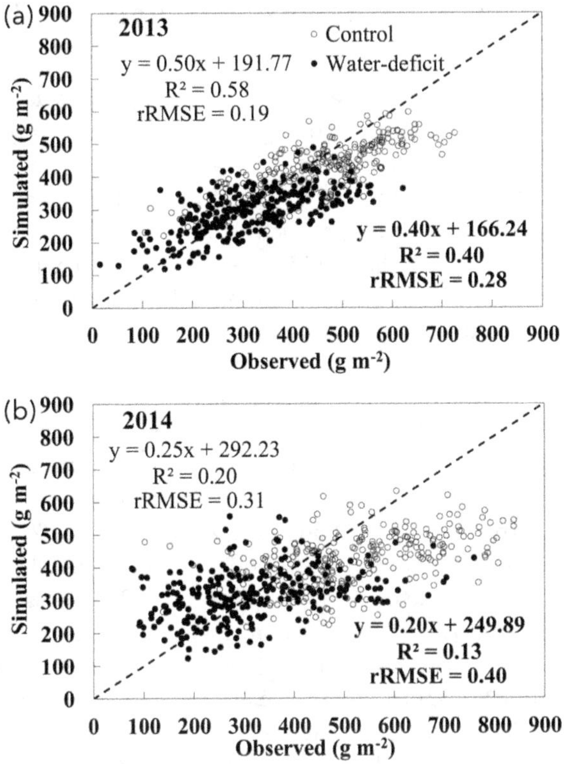

Figure 5 Relationship between simulated and observed values of grain yield in 267 genotypes of a rice genome-wide association study population under control (open symbols, statistical indicators in non-bold) and water-deficit (filled symbols, statistical indicators in bold) conditions in the 2013 experiment (a) and the 2014 experiment (b) (after Kadam et al. 2019). The diagonal line is the 1:1 line. rRMSE = relative Root Mean Squared Error.

loci effects were fed into the model. For the training set, the SNPs-based model accounted for 37% (control) and 29% (water-deficit) of yield variation, less than the 78% explained by a statistical genomic-prediction model for the control treatment. Both models failed in predicting yields of the 54 testing genotypes. However, compared with the genomic-prediction model, the SNPs-based crop model was advantageous when simulating yields under either control or water-stress conditions in an independent season. GECROS-based sensitivity analysis ranked the SNP loci for their relative importance in accounting for yield variation, and the rank differed greatly between control- and water-deficit environments. It was shown that compared with the genomic-prediction model, GECROS has a potential to use single-environment information for predicting phenotypes under different environments.

9 Model applications: improving photosynthesis

Yield progress in major crops during the Green Revolution largely resulted from introducing the dwarfing genes that allow crops to have higher harvest index and become resistant to lodging, even with high-N input. Given that the improvement of harvest index may have reached the plateau, improving photosynthesis to increase total biomass production has been proposed as a new avenue of improving crop productivity (e.g. Long et al. 2006; Zhu et al. 2022; Stirbet et al. 2024). This could be achieved either through exploiting natural variation of photosynthetic parameters or by synthetic biology approach to engineer for improved photosynthesis.

Gu et al. (2014a) examined the extent to which natural variation in photosynthetic biochemical parameters can contribute to increasing rice productivity in Asia. Using GECROS, they analysed the impact of genetic variation in leaf photosynthetic rate (A) on crop biomass production based on the QTL for various biochemical photosynthetic components within their rice IL population (Gu et al. 2012). The genetic variation in A of 25%, created by 20–50% variation in biochemical parameters, can be scaled up almost equally to crop level, resulting in a potential increase in biomass of 28–29% across different years and Asian locations. Using the simulation results, multiple regression analysis showed that the genetic variation in κ_{2LL} (efficiency of converting incident light into linear electron transport) contributed most to the variation in total biomass, followed by g_s (stomatal conductance) and g_m (mesophyll conductance). This was probably because the genetic variation in A of this rice population resulted not only from Rubisco-limited photosynthesis but also from electron transport-limited photosynthesis; as a result, photosynthetic rates could be improved more from increasing both light-saturated and light-limited leaves in the canopy. This simulation study demonstrates the potential of improving rice productivity by mining the natural variation in existing germplasm, especially the variation in parameters determining light-limited photosynthesis. Genetic variation of these parameters still needs more experimentation to investigate as photosynthesis under light limitation is generally conservative.

Various synthetic-biology routes to enhance leaf photosynthesis have been proposed to improve crop productivity (e.g. Long et al. 2006). In order to assess their potential contribution to crop productivity under realistic field conditions, Yin and Struik (2017) used 31-year weather data in three Asian rice-producing locations to run the model GECROS for rice in tropical, subtropical, and temperate environments, and evaluate the following routes: (1) improving mesophyll conductance (g_m), (2) improving Rubisco specificity ($S_{c/o}$), (3) improving both g_m and $S_{c/o}$, (4) introducing C_4 biochemistry, (5) introducing C_4 anatomy that effectively mimic C_4 CO_2-concentrating mechanism (CCM) for minimizing CO_2 leakage, (6) engineering the complete C_4 mechanism, (7) engineering cyanobacterial bicarbonate transporters, (8) engineering a more elaborate cyanobacterial CCM with the carboxysome in the chloroplast, and (9) a mechanism that combines the low ATP cost of cyanobacterial CCM and the C_4 high photosynthetic capacity per unit leaf nitrogen. All routes improved crop mass production, but benefits from Routes 1, 2, and 7 were ≤10%. Benefits were higher in the presence than in the absence of drought, and under the present climate than for the climate predicted for 2050. The route combinations gave larger effects than the sum of the effects of the single routes but only Route 9 could bring an advantage of ≥50% under any environmental conditions. This analysis showed that to supercharge crop productivity, exploring a combination of routes in improving CCM, photosynthetic capacity, and quantum efficiency is required. Moreover, manipulating photosynthesis was simulated to have unwanted secondary effects on other traits at the crop level, e.g., inducing faster senescence if nutrient uptake was not increased (Yin and Struik 2017). Previous predictions by photosynthesis biologists, who only used a photosynthesis model that could not identify this type of feedback mechanisms, might have overestimated the beneficial effect of the routes to engineer for high photosynthesis on crop productivity.

Overall, the results from both Gu et al. (2014a) and Yin and Struik (2017) using GECROS showed the importance of light-limited photosynthesis in determining canopy and crop productivity for rice in Asia. These results were supported by Harbinson and Yin (2023) using the same model but for wheat, potato, and maize in Europe, as well as by simulation studies using other crop models for sorghum in Australia (Wu et al. 2019) and for soybean in US (He and Matthews 2023). These simulation studies showed that increases in electron transport parameter values had a predominant effect on the light-use efficiency for sub-saturating irradiances while improvements in other parameters produced smaller increases. Yin et al. (2022) further showed that the potential of improving these photosynthetic traits can be maximally realized when combined with improved crop N status.

10 Model applications: assessing agroecosystem functioning in response to climate change

The GECROS crop model (v.1.0) was included in the Expert-N software package for simulating plant growth and matter fluxes in agroecosystems, which is particularly useful for simulations and projections at the regional or national scale (Priesack et al. 2006). The full GECROS crop model (v.1.0) was incorporated with some algorithms for root growth in the soil from the CERES model, resulting in the DANUBIA model (Lenz-Wiedemann et al. 2010). The model was evaluated against various field measurements of sugar beet, spring barley, maize, winter wheat, and potato crops and was shown to be a suitable tool for assessing the consequences of global change on biomass production and water and nitrogen demand, when taking into account the complex interplay of water, carbon, and nitrogen fluxes in agro-ecosystems. They also showed that the model was used for the assessment at regional scales (Krimly et al. 2016). Also, integrated GECROS was integrated into a land surface model (Ingwersen et al. 2018; Warrach-Sagi et al. 2022) or into an atmospheric mixed-layer model (Combe et al. 2015).

The GECROS model was used to simulate the impact of elevated atmospheric [CO_2] (Biernath et al. 2013; Baslam et al. 2020). Using data from a FACE (Free-Air CO_2 Enrichment) field experiment of Kimball et al. (1995) for wheat, Yin and Struik (2010) evaluated GECROS (v.2.0) in estimating the impact of stimulation of yield by elevated [CO_2]. Like most of other crop models Tubiello and Ewert (2002) earlier evaluated using the same data set, GECROS correctly showed that the 'CO_2-fertilization' effect was larger under drought than non-stress conditions. However, in contrast to other crop models that overestimated yield responses to elevated [CO_2], GECROS did not overestimate this response even when standard parameters for instantaneous leaf photosynthesis and other growth processes of wheat were used.

Yin and Struik (2010) attributed the performance of GECROS without overestimation to its ability in generating the photosynthetic acclimation to elevated [CO_2]. The photosynthetic acclimation refers to the observation that initial enhancement of photosynthesis by elevated [CO_2] cannot be sustained over a longer term, and was explained by biochemical mechanisms like Rubisco downregulation (e.g. Stitt and Krapp 1999). GECROS does not incorporate the experimentally observed results as algorithms, nor does it model any fine-scale biochemical mechanism. Instead, relying on modelled crop-level C-N interactions, the model gave rise to this acclimation as emergent properties at both leaf and canopy levels. At the leaf level, the photosynthetic advantage of plants grown under elevated [CO_2] decreased with age as a result of a more strongly decreased leaf N content, compared to those under ambient [CO_2]. At the canopy level, an additional mechanism contributes – C determined LAI was initially higher, but leaves senesced faster under elevated than ambient [CO_2] for a given canopy N content. As a result, LAI eventually became smaller, compared with the canopy at ambient [CO_2] (Yin and Struik 2010).

Previous models' overestimations of the stimulation of elevated [CO_2] under field conditions was attributed to model parameterization for [CO_2] response from data of enclosure (e.g. growth chamber) experiments, leading to the suggestion of needs for a FACE experimentation to parameterize crop models (e.g. Ainsworth et al. 2008). The performance of GECROS suggested that it does translate input information at the single-organ level in a short-time scale (e.g. photosynthetic rates measured in an enclosure on single leaves in μmol m^{-2} s^{-1}) to crop performance in a continuously changing field environment, not necessarily relying on expensive FACE experimental facilities for parameterization (Yin 2013). Afterall, Allen et al. (2020) questioned whether FACE is a best experimental set-up for showing the effect of elevated [CO_2].

11 Model applications: evaluating agrivoltaic systems to optimize land use for energy production

Among renewable energies, solar photovoltaics is the fastest growing power generating technology. A system combining soil-grown crops with photovoltaic panels installed above the crop are referred as agrivoltaic systems. To evaluate to what extent crop yield was affected by such systems, GECROS was revised to couple with a radiation and shading model (Amaducci et al. 2018). The simulation was conducted using a 40-years' climate input at Piacenza, Italy, with a rainfed maize crop. Simulated agrivoltaic configurations consisted of two panel densities and two sun-tracking setups, and simulations with full irrigation and full light were considered as control.

Global radiation under the agrivoltaic system was more reduced by solar panel area: 29.5% and 13.4%, respectively, for double-density and single-density than by panel management: 23.2% and 20.0% for sun-tracking and static panels. The land-equivalent-ratio value of agrivoltaic systems as a special *intercropping* system was calculated being always above 1, indicating that any of the agrivoltaic scenario tested was more advantageous than the separate production of maize for biogas and electric energy from ground-mounted photovoltaics systems. The ratio increased with panel density, and it was higher with sun tracking than with static panels.

One surprising result was that simulated maize yield did not decrease but increased, compared to control, for many cases. The model showed that reduced radiation under agrivoltaic systems affected mean soil temperature, evapotranspiration, and soil water balance, which provided more favourable conditions for plant growth afterwards. Consequently, the lowest average grain yield was obtained under full light and the yield advantage under agrivoltaic systems was proportional to drought stress. Thus, a partial shading under such systems may increase yield, as it reduces water consumption by evapotranspiration during the summer and under drought conditions. The systems may be especially advantageous for shade-tolerant crops. Overall, energy production coupled with crop production under agrivoltaic systems can generate increases in economic value compared to conventional agriculture.

12 Model applications: supporting smart nitrogen management

Modern farming for improved sustainability may benefit from supports by crop modelling in combination with modern *in situ* crop sensing technologies. Wang et al. (2024) proposed a N-management optimization method, in which the numerical 'bisection algorithm' is combined with the GECROS modelling and assimilating the data from hyperspectral remote sensing. Data collected from a field experiment of rice with six N treatments (each with four times of topdressing) were used to illustrate the methods, where the first two N topdressings were applied to the experiment while the last two topdressing were optimized. The two methods were compared with a reference method – farmer practice optimized by the yield response curve (FP_{opt}). Crop yields using N applications from the reference method were also simulated by GECROS. Compared with FP_{opt}, the recommendation by the new method decreased the sum of the last two topdressings by 61.2%, with little yield penalty (the recommended N topdressing decreased the simulated yield by only 1.2%) yet increased profit by 4.4%, compared with FP_{opt}. While using crop modelling was only shown to be used for optimizing in-season N application, the combined crop modelling-remote sensing method further benefited from assimilating data from *in situ* remote sensing information into the crop modelling framework. The result of this analysis is yet to be validated experimentally, but it highlights the potential of combining GECROS modelling and sensing technology to guide smart fertilizer management, thereby providing more effective support to in-season adaptive management and improving fertiliser use efficiency and sustainability of crop production.

13 Conclusion and future trends

Few crop models have been confronted directly against simulating G × E interactions observed when a large number of genotypes are involved (a GWAS panel and bi-parental or other populations). As discussed above, when applied for a genetic population, the moderate performance of GECROS in predicting genotypic differences in grain yield still relied on using N_{max} as input (Figs. 4-5). This could be partly attributed to the facts that (i) noise in data is high as yield is measured from plants in a small plot (1–2 m^2 ground per replicate per genotype) when involving a GWAS panel or other genetic populations; (ii) the model is generic, not made specific for a particular crop; and (iii) the model uses a minimum set of *measurable* parameters for expressing genotypic differences. Nevertheless, this also suggests that root N capture is a major component to be improved, in addition to earlier discussed sink size, among other model components.

The general principle in GECROS for describing crop N demand was quantified in equation (2). However, actual crop N uptake also depends on N available from rooted soil layers and whether roots in these layers can effectively take up the available N. Compared with weather variables that are in the form of recorded weather data and thus have less uncertainty, the edaphic variables (soil N and water supply) need to be simulated (Fig. 1b), which is a major uncertainty for any crop model projection. However, the 'hidden half' that is in belowground may be an important determinant of genotypic differences observed for aboveground traits. For example, genetic differences in photosynthetic parameters are often associated with differences in leaf N content (e.g. Gu et al. 2012), which may be owing to differences in root N-uptake ability. Like plant height, rooting depth is quantified in GECROS as a state variable, and its rate is simply estimated by the differential form of a sigmoid growth function with the same form as equation (4), assuming a crop- or genotype-specific maximum rooting depth (D_{max}) that is reached around the onset of seed filling. Effects of any abiotic stress on rooting depth are accounted for by multiplying the rate equation with a simulated stress factor. Such simple methods may not suffice as the profile of root N-capture ability and other important characteristics are not accounted for. Given that genetic potential exploiting aboveground traits can be best realized in combination with improved crop N status (Yin et al. 2022), there is a need to improve simulation of genotypic difference in root N uptake.

An alternative method would be to bypass physiological modelling of the traits (like N_{max}) that are highly uncertain but retrieve them from *in situ* real-time remote sensing in combination with statistical methods (e.g. Banerjee et al. 2020). For this instance, assimilating spatially and temporally referenced data or those predicted from machine learning into model simulation will enable crop models to be more accurate (Wang et al. 2025a; 2025b). A major question is to what extent these sensing or imaging technologies in combination with advanced statistics and even artificial intelligence can detect (subtle) differences among genotypes in a genetic/breeding population (Chang-Brahim et al. 2024).

This chapter summarizes the GECROS model from both theoretical and practical applications' perspectives. Model equations were developed, based on existing understandings, to achieve both biological rigor and model simplicity. As discussed in this chapter, GECROS as a result of integrating these equations can give rise to unexpected results, possibly new understandings of unknowns. Crop models are not only for practical prediction but more as a research tool for data synthesis, system design, and (model-)assisted heuristics. To that end, crop modelling should complement experimentation, not merely in a way that experiments generate data for parameterization and validation of the model, but more in that modelling and experimentation interact in an iterative manner to co-contribute the unravelling of biological insights (Yin et al. 2021). Given that data-based methods may overtake crop models in terms of predictabilities, unique heuristic roles of crop modelling should further be explored for unravelling complex genotype-to-phenotype relationships.

14 Acknowledgements

Many people contributed, either directly or indirectly to GECROS-model development and applications. I would especially thank J. Goudriaan, M.J. Kropff, A.H.C.M. Schapendonk, P.C. Struik, H.H. van Laar, and M. van Oijen, as well as later Prof. P. Stam, for their years-long inspiration, discussion and collaboration, and J. Gu, N. Kadam, M.S. Khan, and D. Wang for their PhD research that was partially discussed in this chapter.

15 References

Ainsworth, E.A., et al., 2008. FACE-ing the facts: inconsistencies and interdependence among field, chamber and modelling [CO_2] impacts on crop yield and food supply. *New Phytologist* 179: 5–9.

Allen, L.H., et al., 2020. Sunlit, controlled-environment chambers are essential for comparing plant responses to various climates. *Agronomy Journal* 112: 4531–4549.

Amaducci, S., et al, 2018. Agrivoltaic systems to optimise land use for electric energy production. *Applied Energy* 220: 545–561.

Banerjee, B.P., et al., 2020. High-throughput phenotyping using digital and hyperspectral imaging-derived biomarkers for genotypic nitrogen response. *Journal of Experimental Botany* 71: 4604–4615.

Baslam, M., et al., 2020. Photosynthesis in a changing global climate: Scaling up and scaling down in crops. *Frontiers in Plant Science* 11: 583899. https://doi.org/10.3389/fpls.2020.00882.

Biernath, C., et al., 2013. Modeling acclimation of leaf photosynthesis to atmospheric CO_2 enrichment. *European Journal of Agronomy* 48: 74–87.

Boote, K.J., et al., 2013. Putting mechanisms into crop production models. *Plant, Cell and Environment* 36: 1658–1672.

Bouman, B.A.M., et al., 1996. The 'school of de wit' crop growth simulation models: a pedigree and historical overview. *Agricultural Systems* 52: 171–198.

Bouman, B.A.M., et al., 2001. *ORYZA2000: modeling lowland rice*. International Rice Research Institute, Los Baños, Philippines.

Brouwer, R., 1983. Functional equilibrium: sense or nonsense? *Netherlands Journal of Agricultural Science* 31: 335–348.

Cannell, M.G.R. and Thornley, J.H.M., 2000. Modelling the components of plant respiration: some guiding principles. *Annals of Botany* 85: 45–54.

Chang-Brahim, I., et al., 2024. Reviewing the essential roles of remote phenotyping, GWAS and explainable AI in practical marker-assisted selection for drought-tolerant winter wheat breeding. *Frontier Plant Science, Sec. Plant Breeding* 15: 13199380. https://doi.org/10.3389/fpls.2024.13199380.

Coleman, K., et al., 1999. Detailed report of the individual partner, IACR-Rothamsted. p.16–20. In: *First Annual Report of the Modelling Agro-ecosystems under Global Environmental Change project* (EU contract ENV4-CT97-0693).

Combe, M., et al., 2015. Two perspectives on the coupled carbon, water and energy exchange in the planetary boundary layer. *Biogeosciences* 12: 103–123.

de Pury, D.G.G. and Farquhar, G.D., 1997. Simple scaling of photosynthesis from leaves to canopies without the errors of big-leaf models. *Plant, Cell and Environment* 20: 537–557.

de Wit, C.T., 1965. *Photosynthesis of leaf canopies*. Agricultural Research Report 663, Wageningen, The Netherlands.

de Wit, C.T. et al., 1978. *Simulation of assimilation, respiration and transpiration of crops*. Pudoc, Wageningen.

de Wit, C.T., et al., 1970. The simulation of photosynthetic systems. In: Setlik, I. (ed.) *Prediction and measurement pf photosynthetic productivity. Proceedings International Biological Program/Plant Production Technical Meeting Trebon*. Pudoc, Wageningen, pp. 47–70.

Dingkuhn, M., et al., 2020. The case for improving crop carbon sink strength or plasticity for a CO_2-rich future. *Current Opinion in Plant Biology* 56: 259–272.

Farquhar, G.D., et al., 1980. A biochemical model of photosynthetic CO_2 assimilation in leaves of C_3 species. *Planta* 149: 78–90.

Flexas, J., et al., 2008. Mesophyll conductance to CO_2: current knowledge and future prospects. *Plant, Cell and Environment* 31: 602–621.

Goudriaan, J., 1986. A simple and fast numerical method for the computation of daily totals of crop photosynthesis. *Agricultural and Forest Meteorology* 38: 249–254.

Goudriaan, J., 1988. The bare bones of leaf-angle distribution in radiation models for canopy photosynthesis and energy exchange. *Agricultural and Forest Meteorology* 43: 155–169.

Goudriaan, J. and van Laar, H.H., 1994. *Modelling potential crop growth processes*. Kluwer Academic Publishers, Dordrecht, The Netherlands.

Gu, J., et al., 2012. Physiological basis of genetic variation in leaf photosynthesis among rice (*Oryza sativa* L.) introgression lines under drought and well-watered conditions. *Journal of Experimental Botany* 63: 5137–5153.

Gu, J., et al., 2014a. Can exploiting natural genetic variation in leaf photosynthesis contribute to increasing rice productivity? A simulation analysis. *Plant, Cell and Environment* 37: 22–34.

Gu, J., et al., 2014b. Linking ecophysiological modelling with quantitative genetics to support marker-assisted crop design for improved rice (*Oryza sativa* L.) yields under drought stress. *Annals of Botany* 114: 499–511.

Harbinson, J. and Yin, X., 2023. Modelling the impact of improved photosynthetic properties on crop performance in Europe. *Food and Energy Security* 12: e402. https://doi.org/10.1002/fes3.402.

He, Y. and Matthews, M.L., 2023. Seasonal climate conditions impact the effectiveness of improving photosynthesis to increase soybean yield. *Field Crops Research* 296: 108907.

Hilbert, D.W., 1990. Optimization of plant root:shoot ratios and internal nitrogen concentration. *Annals of Botany* 66: 91–99.

Ingwersen, J., et al., 2018. Coupling the land surface model Noah-MP with the generic crop growth model Gecros: model description, calibration and validation. *Agricultural and Forest Meteorology* 262: 322–339.

Julia, C. and Dingkuhn, M. 2013. Predicting temperature induced sterility of rice spikelets requires simulation of crop-generated microclimate. *European Journal of Agronomy* 49: 50–60.

Kadam, N., et al., 2019. Incorporating genome-wide association into eco-physiological simulation to identify markers for improving rice yields. *Journal of Experimental Botany* 70: 2575–2586.

Khan, M.S., et al, 2014. An ecophysiological model analysis of yield differences within a set of contrasting cultivars and an F1 segregating population of potato (*Solanum tuberosum* L.) grown under diverse environments. *Ecological Modelling* 290: 146–154.

Kimball, B.A., et al., 1995. Productivity and water use of wheat under free-air CO2 enrichment. *Global Change Biology* 1: 429–442.

Krimly, T., et al., 2016. Effects of future climate changes on yields, land use and agricultural incomes. In: Mauser, W. and Prasch, M. (eds) *Regional assessment of global change impacts*. Springer, Cham, pp. 609–614. https://doi.org/10.1007/978-3-319-16751-0_70.

Kropff, M.J. and Struik, P.C., 2002. Developments in crop ecology. *NJAS-Wageningen Journal of Life Sciences* 50: 223–237.

Kropff, M.J., et al., 1994. *ORYZA1: an ecophysiological model for irrigated rice production*. Wageningen Agricultutal University, Wageningen and International Rice Research Institute, Los Baños, Philippines.

Lenz-Wiedemann, V.I.S., et al., 2010. Development and test of a crop growth model for application within a Global Change decision support system. *Ecological Modelling* 221: 314–329.

Li, H., et al., 2025. Panicle temperature explains contrasting yield responses of rice genotypes to elevated CO_2 and increased temperature in T-FACE environments. *Journal of Experimental Botany* 76: 3611-3626. https://doi.org/10.1093/jxb/eraf170

Liu, Q., et al., 2025. Estimating stomatal conductance under drought: Parameterizing a phenomenological model and evaluating roles of the energy balance equation. *Ecological Modelling* 505: 111133. https://doi.org/10.1016/j.ecolmodel.2025.111133.

Long, S.P., et al., 2006. Can improvement in photosynthesis increase crop yields? *Plant, Cell and Environment* 29: 315–330.

Monteith, J.L. 1973. *Principles of environmental physics*. Edward Arnold, London, 241pp.

Muller, B., et al., 2011. Water deficits uncouple growth from photosynthesis, increase C content, and modify the relationships between C and growth in sink organs. *Journal of Experimental Botany* 62: 1715–1729.

Nakagawa, H., et al., 2005. Flowering response of rice to photoperiod and temperature: a QTL analysis using a phenological model. *Theoretical and Applied Genetics* 110: 778–786.

Peng, B., et al., 2020. Towards a multiscale crop modelling framework for climate change adaptation assessment. *Nature Plants* 6: 338–348.

Penning de Vries, F.W.T., et al., 1989. *Simulation of ecophysiological processes of growth in several annual crops*. IRRI, Los Banos and Pudoc, Wageningen.

Poorter, H. and Nagel, O., 2000. The role of biomass allocation in the growth response of plants to different levels of light, CO_2, nutrients and water: a quantitative review. *Australian Journal of Plant Physiology* 27: 595–607.

Priesack, E. and Gayler, S., 2009. Agricultural crop models: Concepts of resource acquisition and assimilate partitioning. In: Lüttge, U., et al. (eds) *Progress in Botany*. Springer-Verlag, Berlin Heidelberg, pp. 195–222.

Priesack, E., et al., 2006. The impact of crop growth sub-model choice on the simulation of water and nitrogen balances. *Nutrient Cycling in Agroecosystems* 75: 1–13.

Rodriguez, D., et al., 1999. LINGRA-CC: a sink-source model to simulate the impact of climate change and management on grassland productivity. *New Phytologist* 144: 359–368.

Stirbet, A., et al., 2024. From leaf to multiscale models of photosynthesis: applications and challenges for crop improvement. *Photosynthesis Research*. Advance online publication. https://doi.org/10.1007/s11120-024-01083-9.

Stitt, M. and Krapp, A., 1999. The interaction between elevated carbon dioxide and nitrogen nutrition: the physiological and molecular background. *Plant, Cell and Environment* 22: 583–621.

Triboi, E. and Triboi-Blondel, A.-M., 2002. Productivity and grain or seed composition: a new approach to an old problem. *European Journal of Agronomy* 16: 163–186.

Tubiello, F.N. and Ewert, F. 2002. Simulating the effects of elevated CO_2 on crops: approaches and applications for climate change. *European Journal of Agronomy* 18: 57–74.

van Ittersum, M.K., et al., 2003. On approaches and applications of the Wageningen crop models. *European Journal of Agronomy* 18: 201–234.

van Keulen, H., et al., 1982. A summary model for crop growth. In: Penning de Vries, F.W.T. and van Laar, H.H. (eds) *Simulation of plant growth and crop production*. Simulation Monographs, Pudoc, Wageningen, pp. 87–97.

van Kraalingen, D.W.G., et al., 2003. The FORTRAN simulation translator, a simulation language. *European Journal of Agronomy* 18: 259–361.

van Laar, H.H., et al., 1997. *SUCROS97: Simulation of crop growth for potential and water-limited production situations, as applied to spring wheat*. Report series No. 14, Quantitative Approaches in Systems Analysis, Wageningen.

van der Werf, A., et al., 1993. Allocation of carbon and nitrogen as a function of the internal nitrogen status of a plant: Modelling allocation under non-steady-state situations. *Plant and Soil* 155/156: 183–186.

van Oijen, M., et al., 2004. Simple equations for dynamic models of the effects of CO_2 and O_3 on light use efficiency and growth of crops. *Ecological Modelling* 179: 39–60.

von Caemmerer, S. and Furbank, R.T., 1999. Modeling C_4 photosynthesis. In: Sage R.F. and Monson R.K. (eds) *C_4 plant biology*. Academic Press, Toronto, pp. 173–211.

Wang, D., et al., 2024. Developing remote sensing- and crop model-based methods to optimize nitrogen management in rice fields. *Computers and Electronics in Agriculture* 220: 108899.

Wang, D., et al., 2025a. Enhancing crop growth forecasting by incorporating estimated uncertainties for time series hyperspectral data and crop model GECROS simulations into Ensemble Kalman Filter. *Computers and Electronics in Agriculture*: in press, 239: (Article 111057).

Wang, D., et al., 2025b. Estimating leaf and canopy nitrogen contents in major field crops across the growing season from hyperspectral images using nonparametric regression. *Computers and Electronics in Agriculture* 233: 110147.

Warrach-Sagi, K., et al., 2022. Noah-MP with the generic crop growth model Gecros in the WRF model: Effects of dynamic crop growth on land-atmosphere interaction. *JGR Atmospheres* 127: e2022JD036518.

White, J.W. and Hoogenboom, G., 1996. Simulating effects of genes for physiological traits in a process-oriented crop model. *Agronomy Journal* 88: 416–422.

Wong, S.C., et al. 1985. Leaf conductance in relation to rate of CO2 assimilation. I. Influence of nitrogen nutrition, phosphorous nutrition, photon flux density and ambient partial pressure of CO2 during ontogeny. *Plant Physiology* 78: 821–825.

Wu, A., et al., 2019. Quantifying impacts of enhancing photosynthesis on crop yield. *Nature Plants* 5: 380–388.

Yin, X. 1999. A critical appraisal of thermal time approach for predicting crop development. *Acta Agronomica Sinica* 25: 474–482 (in Chinese with English Abstract).

Yin, X., 2013. Improving ecophysiological simulation models to predict the impact of elevated atmospheric CO_2 concentration on crop productivity. *Annals of Botany* 112: 465–475.

Yin, X. and Schapendonk, A.H.C.M., 2004. Simulating the partitioning of biomass and nitrogen between root and shoot in crop and grass plants. *NJAS-Wageningen Journal of Life Sciences* 51: 407–426.

Yin, X. and Struik, P.C., 2009. C_3 and C_4 photosynthesis models: An overview from the perspective of crop modelling. *NJAS-Wageningen Journal of Life Sciences* 57: 27–38.

Yin, X. and Struik, P.C., 2010. Modelling the crop: from system dynamics to systems biology. *Journal of Experimental Botany* 61: 2171–2183.

Yin, X. and Struik, P.C., 2017. Can increased leaf photosynthesis be converted into higher crop mass production? A simulation study for rice using the crop model GECROS. *Journal of Experimental Botany* 68: 2345–2360.

Yin, X. and van Laar, H.H., 2005. *Crop systems dynamics: an ecophysiological simulation model for genotype-by-environment interactions*. Wageningen Academic Publishers, Wageningen, The Netherlands, 155pp.

Yin, X., et al., 1995. A nonlinear model for crop development as a function of temperature. *Agricultural and Forest Meteorology* 77: 1–16.

Yin, X., et al., 1997. A model for photothermal responses of flowering in rice. II. Model evaluation. *Field Crops Research* 51: 201–211.

Yin, X., et al., 2000a. Coupling estimated effects of QTLs for physiological traits to a crop growth model: predicting yield variation among recombinant inbred lines in barley. *Heredity* 85: 539–549.

Yin, X., et al., 2000b. A generic equation for nitrogen-limited leaf area index and its application in crop growth models for predicting leaf senescence. *Annals of Botany* 85: 579–585.

Yin, X., et al., 2003a. A flexible sigmoid function of determinate growth. *Annals of Botany* 91: 361–371 (erratum in 91: 753, 2003).

Yin, X., et al., 2003b. Some quantitative relationships between leaf area index and canopy nitrogen content and distribution. *Annals of Botany* 91: 893–903.

Yin, X., et al., 2004. Role of crop physiology in predicting gene-to-phenotype relationships. *Trends in Plant Science* 9: 426–432.

Yin, X., et al., 2009. Using combined measurements of gas exchange and chlorophyll fluorescence to estimate parameters of a biochemical C_3 photosynthesis model: a critical appraisal and a new integrated approach applied to leaves in a wheat (*Triticum aestivum*) canopy. *Plant, Cell and Environment* 32: 448–464.

Yin, X., et al., 2021. Loomis review: on the needs for combining physiological principles and mathematics to improve crop models. *Field Crops Research* 271: 108254. https://doi.org/10.1016/j.fcr.2021.108254.

Yin, X., et al., 2022. A model-guided holistic review of exploiting natural variation of photosynthesis traits in crop improvement. *Journal of Experimental Botany* 73: 3173–3188. https://doi.org/10.1093/jxb/erac109.

Zhu, X., et al., 2022. Improving photosynthetic efficiency through multidisciplinary efforts: the next frontier of photosynthesis research. *Frontiers in Plant Science (section Photosynthesis and Photobiology)* 13: 967203.

Chapter 12

The WOFOST (WOrld FOod STudies) cropping system model

Allard de Wit, Hendrik Boogaard, Herman Berghuijs and Martin K. van Ittersum, Wageningen University, The Netherlands

1 Introduction

Stakeholders of current agricultural systems require accurate and up to date information for decision-making at all levels. On the one hand, up-to-date information on the growth and development of crops is crucial for managing practical aspects of farming and agricultural value chains such as local farm and crop management, storage and logistics, and trading and contract management. On the other hand, policymakers require insight into the short- and long-term impact of their policies on production, socio-economic conditions, food security and the availability of land, water and energy resources. The latter is becoming increasingly important in particular due to the impact of climate change on agriculture.

Cropping systems modelling has been recognized as a mature technology that can assist with decision-making aspects in many of the societal relevant application domains outlined above. For this purpose, cropping system models that have been developed over the last decades, including DSSAT (Hoogenboom et al., 2019; Jones et al., 2003), Apsim (Holzworth et al., 2014),

http://dx.doi.org/10.19103/AS.2025.0155.15

STICS (Brisson et al., 2003) and EPIC (Wang et al., 2012), have been implemented in decision support systems for a range of applications (Holzworth et al., 2015). In Wageningen, the WOrld FOod STudies (WOFOST) model (van Diepen et al., 1988; de Wit et al., 2019; Berghuijs et al., 2024) has been in continuous development over the past four decades and is applied for diverse applications related to farmer support, yield gap analysis, regional crop monitoring, yield forecasting and climate change impact analysis.

Development of crop models in Wageningen started already in the 1960s with the work carried out by the group of C.T. de Wit. Their work focused on the understanding of growth and development of crops based on the underlying processing such as phenology and plant photosynthesis (Wit, 1965). Within the Wageningen 'School of De Wit', many algorithms were developed leading to a number of crop models including BACROS, SWHEAT and later SUCROS, which have been described by Bouman et al. (1996) and van Ittersum et al. (2003). The WOFOST cropping system model was also developed within the 'School of De Wit' sharing much of its biophysical core with the SUCROS model (Laar et al., 1997). Similar to SUCROS, WOFOST simulates many different crops with a single set of algorithms; only the parametrization changes between crops.

Most of the crop models developed in Wageningen were developed for scientific purposes. Instead, WOFOST was developed as a more operational model having clear version control, standardized input/output files and proper documentation. The first WOFOST version that was released and documented was WOFOST version 4.1 (Diepen et al., 1988), earlier versions were undocumented and no information is available anymore. Version 6.0 was released in 1994 (Supit et al., 1994) based on major efforts to integrate WOFOST into the European Crop Growth Monitoring System (Meyer-Roux and Vossen, 1994). The availability of a model implementation with standardized input/output also allowed the development of a graphical user interface: the WOFOST Control Centre integrated with WOFOST 7.1 and released in 1998 (Boogaard et al., 2014, 1998).

Today, several implementations of the WOFOST model are maintained by Wageningen University & Research which differ in their details but are all rigorously tested to yield the same results given the same inputs. For this purpose, a large reference set is used to ensure that the model results are identical between different implementations. This reference set is available online,[1] which allows other researchers to test their implementation of WOFOST for correctness as well.

First, the Python implementation of WOFOST embedded in the Python Crop Simulation Environment (PCSE) is focused on research and education (de Wit and Berghuis, 2025). This implementation is versatile, easy to modify

1 https://github.com/ajwdewit/pcse/tree/master/tests.

and can interface with various data sources. The versatility allows to test ideas quickly, run sensitivity analysis or calibrations against datasets and explore scenarios related to management practices. Moreover, it integrates very well with today's scientific software stack where Python is the dominant language. For example, the use of PCSE/WOFOST in combination with Jupyter notebooks has proven useful in teaching crop modelling principles and allowing students to explore strategies for climate change impact mitigation.

Second, for operational applications that require running WOFOST over large number of locations, soil types or scenarios, we implemented WOFOST in Java as part of the WISS (Wageningen Integrated Systems Simulator) framework (van Kraalingen et al., 2020). This implementation is particularly suited for applications that deal with big data. For example, it was demonstrated that WISS/WOFOST can be used in so-called Virtual Research Environments (Knapen et al., 2020) and can be efficiently combined with Apache/Spark dataframes to run on distributed systems (Knapen et al., 2025).

Third, WOFOST is implemented in C# under the NMODCOM framework (Evert and Lamaker, 2007). This implementation is targeting services for precision agriculture. For example, it can integrate information from farm management information systems and it can assimilate field-specific observations from drones or satellites to adjust the model during the growth cycle (Been et al., 2023; Evert et al., 2021, 2024).

Finally, WOFOST is implemented in the Soil-WAter-Plant (SWAP) model (Heinen et al., 2024), which combines the crop simulations of WOFOST with a highly detailed soil water balance that can take into account the impact of shallow groundwater, impeding layers and soil profile inhomogeneity. Moreover, the SWAP model offers the simulation of solute transport and impact of saline conditions on crop growth, and it provides advanced approaches for root water uptake and the impact of oxygen shortage on root functioning.

The success of the WOFOST cropping system model is reflected in the variety of applications that exist today. One of the flagship applications of the model is its embedding within the MARS Crop Yield Forecasting System (van der Velde et al., 2019). This system has been operational for more than 25 years and is important for monitoring agricultural production in Europe and other areas in the world. Other applications are the implementation within the Watervision Agriculture framework, which is used to quantify the effects of hydrology and climate on agricultural crop production in the Netherlands (Broeke et al., 2019). Moreover, also the private sector is increasingly using the WOFOST model for crop advisories and predicting yield and phenology (Baron and Uitdewilligen, 2023). Finally, WOFOST is one of the crop models implemented within the Global Yield Gap Atlas for estimating the gap between current and potential production in order to estimate the scope to increase production on current crop areas through intensification (van Ittersum et al., 2016).

Despite the success of the WOFOST model, there have been relatively few improvements in the underlying algorithms, and the biophysical core of WOFOST has essentially stayed the same since the early 2000s. Some important improvements were implemented in WOFOST 7.2 such as the impact of vernalization on phenological development (Ceglar et al., 2019) and the possibility to simulate crop rotations instead of a single growth cycle. However, these improvements were relatively minor. Reasons for this relative standstill in model development was a shift in focus from scientific development towards application of the model as well as reduction in funding for experimentation and field trials.

With the release of WOFOST version 7.3 and 8.1 in May 2024, a major improvement in the simulation capabilities of WOFOST has been made (Berghuijs et al., 2024). WOFOST 7.3 extends the model with biomass reallocation and a new multi-layered soil water balance, while WOFOST 8.1 also includes the nitrogen dynamics in the crop and the soil. In the next chapter, we will describe improvements in WOFOST 7.3 and 8.1 from a conceptual point of view and highlight the new simulation capabilities that were gained from it. We will highlight the use of WOFOST for simulating G × E × M interactions as well as improvements that will be implemented in the next release.

2 Recent developments in WOFOST

An important characteristic of 'Wageningen School' models is that they are tied to a hierarchy in growth and production factors (van Ittersum and Rabbinge, 1997). These growth factors are organized in three hierarchical levels of crop growth: potential production, limited production and actual production. Each of these growth levels adds a set of growth factors. At the potential production level, growth is defined by temperature, radiation, atmospheric CO_2 concentration, planting date and crop features. At the limited production level, limitations in water and nutrient availability can have a limiting impact on the production. Finally, at the actual production level, there are reducing factors that can play a role such as the impact of weeds, pest and disease, pollutants, etc.

An important limitation of WOFOST 7.2 was that WOFOST could simulate potential and limited production with water limitations but could not deal with the impact of limitations in nutrients. This limitation was somewhat alleviated by the integration of the QUEFTS approach (Janssen et al., 1990; Pushpalatha and Byju, 2022) within the WOFOST Control Centre (Boogaard et al., 2014). However, QUEFTS is basically a post-processing of WOFOST results and not a dynamic simulation model. Therefore it cannot simulate the impact of timing of nutrient applications nor can it deal with the dynamic behaviour of nutrients in the soil due to processes like de-nitrification, leaching and adhesion. Instead such processes are lumped together in the QUEFTS approach through a single 'recovery fraction' which empirically reduces the amount of nutrients that are actually available for the plant during the growing season.

A second limitation of WOFOST 7.2 was the simplicity of its soil water balance. This water balance used a dual layer concept where the upper layer consisted of the root zone that gradually increased with depth as a result of root growth. This water balance was originally developed for the application of WOFOST within the MARS project in the early 1990s. At that time, very limited information on soil characteristics was available at European scale. Therefore, the soil water balance was kept very simple with a minimum number of parameters. However, with increasing availability of better soil information, this model has become outdated. Moreover, it suffers from conceptual problems such as instantaneous wetting of the entire root zone (due to the single-layer root zone) and the inability to simulate the impact of impeding layers in the soil as only one rooted soil layer is defined in terms of soil physical characteristics.

The model improvements in WOFOST 7.3 and 8.1 have therefore focused on addressing these limitations. A new soil water balance was developed and integrated in WOFOST 7.3, which uses a multi-layer approach based on physical principles of soil moisture flow while still retaining the ability to run on daily time steps. In addition, nitrogen stress was integrated into WOFOST 8.1 by simulating and coupling leaf nitrogen content to the maximum assimilation rate for individual leaves. Moreover, to estimate availability of nitrogen in the soil, a new soil nitrogen/carbon model was integrated as well. The WOFOST versions 7.2, 7.3 and 8.1 will co-exist for the foreseeable future allowing users to select a model version that matches their requirements in terms of complexity and crop production level.

In the next sections, we will describe the new modules of WOFOST from a conceptual point of view. A detailed description of the nitrogen models can be found in Berghuijs et al. (2024) and a description of the soil water balance in Rappoldt et al. (2012).

3 Modelling soil water, carbon and nitrogen balances

3.1 A multi-layer soil water balance

The new soil water balance which is available in WOFOST 7.3 and 8.1 holds the middle ground between the very simple approach of the original model and the highly detailed complex approach that is used by the SWAP model. It also avoids the requirement of variable time steps used by SWAP, which complicates the computational procedures considerably.

The model separates the soil into a number of layers that can be defined by the user based on the structure of the soil for which the simulation is applied.

However, the thickness of each layer should be at least 10–15 cm. Thinner layers would require smaller time steps than one day to simulate realistically, since rain storms will fill up the top layer very quickly leading to surface runoff because the model cannot handle the infiltration of the rainfall in a single timestep (a day) over multiple soil layers. Moreover, the maximum rootable depth of the crops should coincide with a layer boundary in order to avoid that roots directly tap water from the part of the layer below the rootable depth.

The new soil water balance uses the same soil physical information as the SWAP model. This consists of the soil moisture content and soil hydraulic conductivity as a function of the soil matric suction (pF). But instead of solving the Richards equation to estimate the flow of moisture through the soil layers, it uses the Matric Flux Potential (MFP) (Pinheiro et al., 2018; Shaykewich and Stroosnijder, 1977). Mathematically, the MFP is known as a Kirchhoff transformation, and this concept is used also in similar problems related to heat flow in inhomogeneous materials (Bonani and Ghione, 1995). Rappoldt et al. (2012) provide a description of the approach, which is repeated below.

The matric flux potential ϕ at a certain value of matric suction head ψ and hydraulic conductivity K as a function of ψ is defined as

$$\phi(\psi) = \int_{-\infty}^{\psi} K(\psi)\,d\psi \tag{1}$$

where ψ is the variable of integration ranging from $-\infty$ to the actual ψ for which the potential is calculated. At a given soil status, the quantity φ exists at each point, very much like a soil moisture content or the matric suction itself. The MFP approach is attractive because in a stationary state, the stationary flow is found directly from the derivative of φ according to

$$flow = -\frac{d\phi}{dz} \tag{2}$$

where z is the depth coordinate in the soil. Especially under dry conditions, the use of a gradient in matric flux potential avoids the problem of a huge gradient in matric suction in combination with a gradient in conductivity. The gradient in φ 'automatically' accounts for both factors, matric suction and conductivity.

The matric flux potential φ, however, does not account for gravity. Hence, a stationary flow may be calculated as $-d\varphi/dz$ only if gravity is small compared to the differences in matric head, which is usually the case under dry circumstances. Another problem is that the matric flux potential φ is discontinuous at the interface of two soil layers with different characteristics. The reason is that a difference in φ is not actually driving the flow. Although $\varphi(\psi)$ can be calculated everywhere, $d\varphi/dz$ is equal to the stationary flow only between soil layers with equal properties. Between layers with different soil properties, there is a jump in the matric flux potential (and only the matric suction ψ itself is continuous). In such cases, the model will iterate to find a pF value at the boundary of the two layers from the requirement that the flows at both sides of the layer boundary must be the same. Finally, a limitation of the MFP concept is that it does not allow to consider the impact of hysteresis as a result of drying and wetting cycles of the soil.

To estimate the flow of water between the layers, the model computes two flow rates (Fig. 1):

1 A 'dry flow', which accounts for the large gradient in matric suction under dry conditions but neglects gravity. This flow can be upward or downward depending on the distribution of matric suction in the soil profile.

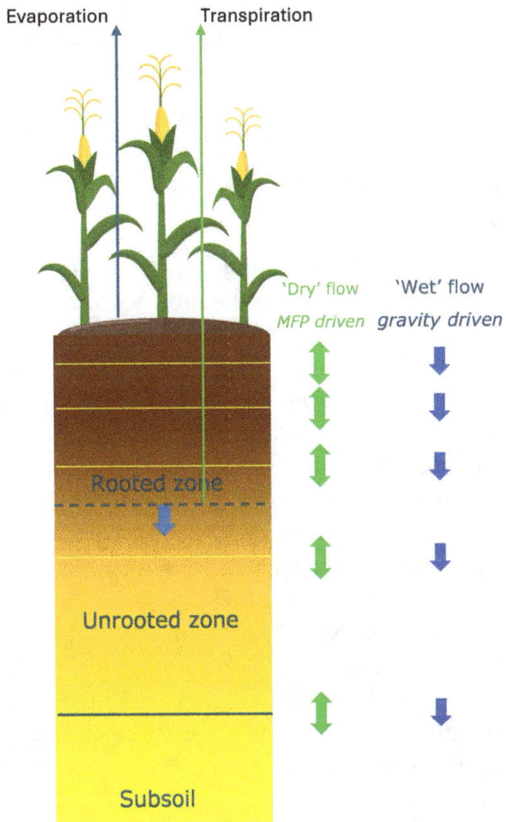

Figure 1 Schematic representation of the multilayer water balance based on the Matric Flux Potential (MFP) concept.

2 A 'wet flow', which takes into account gravity only and which will dominate under wet conditions. This flow can be downward only.

The final flow is computed as the maximum values of dry and wet flow. Moreover, there are limits applied on the flow rates in order to avoid over-saturation or water content below wilting point. Finally, the impact of soil evaporation and transpiration are taken into account. Evaporation from the soil is assumed to happen from the soil surface only and impacts the upper layer directly although additional water can be transported upwards through an increase in MFP gradient. Transpiration can be derived from the entire rooted soil profile.

Similar to the simple water balance, the reduction in transpiration rate due to drought is simulated with a macroscopic Feddes approach (Heinen et al., 2024). However, the transpiration reduction factor is now derived from a layer-weighted average reduction factor instead of from the average root-zone soil moisture content as in the old water balance (since there was only one layer). This implies that a dry layer somewhere in the profile will lead to drought stress even when other (wetter) layers would be able to compensate for the lack of water. In the current version, compensatory mechanisms are not implemented.

3.2 The Soil Nitrogen model for Organic and MIneral Nitrogen (SNOMIN)

The purpose of the soil carbon and nitrogen balance is to estimate the amount of nitrogen available in the soil during the growth cycle of the crop as well as between crop cycles in a rotation. The new soil carbon and nitrogen module is abbreviated as SNOMIN (Soil Nitrogen model for Organic and MIneral Nitrogen). It takes into account the initial amount of inorganic nitrogen, the input from inorganic nitrogen fertilizers, the amount of nitrogen becoming available from organic matter in the soil and organic amendments (manure, plant residues). The model is subdivided into two sub-models, one for inorganic nitrogen and one for organic nitrogen.

The sub-model used by SNOMIN for organic nitrogen is largely based on the Janssen model (Janssen, 1984) for the simulation of organic matter decomposition and on the NDICEA model (van der Burgt et al., 2006), MINIP-C and MINIP-N models (Heinen and Willigen, 2005) for the simulation of organic carbon and nitrogen dissimilation. The main difference between the sub-model and these other models is that SNOMIN distinguishes different soil layers, which allows coupling it to the previously described multilayered soil water balance.

The change of the state of each individual amendment in each soil layer is simulated separately from the moment that it is applied until the end of the

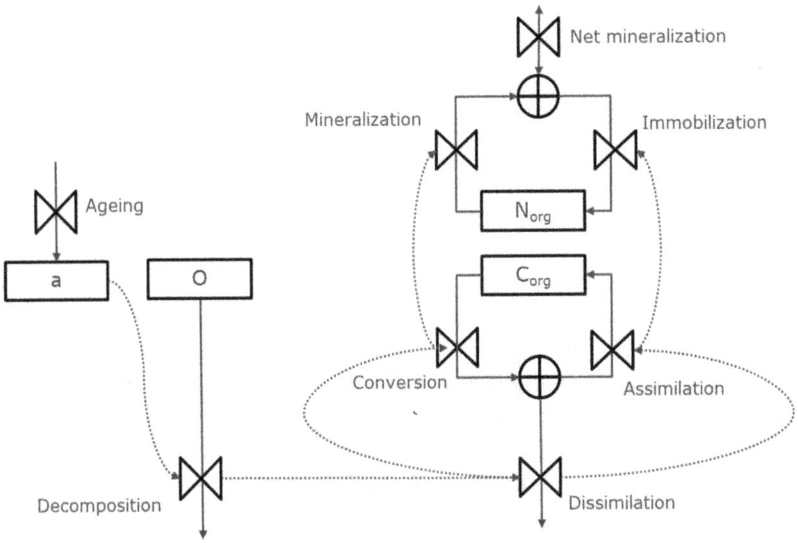

Figure 2 Forrester diagram of the state variables of a single amendment and the processes that affect them. State variables for apparent age (a), organic matter amount (O), organic C amount (C_{org}) and organic N amount (N_{org}) are represented by blocks. The rate variables of individual processes are represented by valves.

simulation (Fig. 2). The state of an amendment is described by a set of state variables. These are its amounts of organic matter, organic C and organic N, and its apparent age. The latter variable is a measure for the ease at which an amendment is decomposed. The apparent age of an amendment increases once it is applied and a higher apparent age results in a lower relative decomposition rate. For each amendment, values have to be provided for the application date, application depth, fresh weight of applied amendment, mass fraction of organic matter, the C:N mass ratio of this organic matter and mass fractions of nitrogen in the form of ammonium and nitrate in the amendment. Values of initial ages are amendment specific have been determined for various types of manure, slurry, plant residues and other organic materials (Janssen, 1984; Groenendijk et al., 2016). The state of the soil organic matter that was already in a soil layer at the start of a simulation is described by the same state variables as those of an amendment. In order to simulate the state of this initial organic matter in a soil layer, values have to be provided for the soil layer thickness, soil bulk density, the mass fraction of organic matter, the C:N ratio and the initial age.

From the application date onwards, organic matter will decompose following the Janssen model (Janssen, 1984). The relative decomposition rate decreases exponentially over time, because the apparent age increases.

SNOMIN assumes a fixed ratio between organic C and organic matter in an amendment. Therefore, the rate at which organic C is lost to the atmosphere in the form of CO_2 (i.e. dissimilation rate) is proportional to the decomposition rate of organic material. The dissimilation rate of organic C is the net result of the process in which organic C is decomposed (conversion) and the process in which the decomposed C is taken up by the microbial biomass (assimilation). The conversion rate of organic C is proportional to the rate in which organic N is released in the environment in the form of NH_4^+-N (gross mineralization). Since the microbial biomass is assumed to keep its C:N ratio constant, it will take up NH_4^+-N from the soil layer (immobilization). The net mineralization rate from an amendment is the difference between the gross mineralization rate and the immobilization rate. The total net mineralization rate in a soil layer is positive if the sum of the gross mineralization rates of all amendments and from the initial soil organic matter in that layer is higher than the immobilization rates. Otherwise, the net mineralization rate is negative and there is net removal of NH_4^+-N from the layer (net immobilization) (Fig. 3).

The sub-model for inorganic N calculates the dynamics of the amounts of N in the form of ammonium (NH_4^+-N) and nitrate (NO_3^--N) in each soil layer. It is largely based on the module from SWAP to WOFOST (Groenendijk et al., 2016) for inorganic nitrogen with, again, the important difference that the soil inorganic sub-module distinguishes separate soil layers. Sources for NH_4^+-N in a soil layer are mineralization, atmospheric deposition, fertilizer application and NH_4^+-N inflow from adjacent soil layers. Sinks are root uptake, outflow of NH_4^+-N to adjacent soil layers and conversion into NO_3-N (nitrification). Sources for NO_3^--N are atmospheric deposition, fertilizer application, nitrification and NO_3^- inflow from adjacent layers. Sinks are root uptake, outflow of NO_3^--N to adjacent layers and denitrification. See Fig. 4 for a complete overview of SNOMIN for all soil layers.

It should finally be noted that SNOMIN is not a derivative of RothC (Giongo et al., 2020; Jenkinson et al., 1997), unlike the soil organic C and N modules used by many cropping systems models in Wageningen like GECROS (Yin and Laar, 2005) SWAP-WOFOST (Heinen et al., 2024) and TIPSTAR (Jansen, 2008). RothC and its derivatives lump all plant residues and other types of organic matters in a limited number of organic matter, organic C and organic N pools. SNOMIN is conceptually different as it simulates the state of the soil organic matter and all individual amendments and their mineralization and immobilization rates separately. This approach prevents the need of subdividing the organic matter in pools that cannot be determined from measurements.

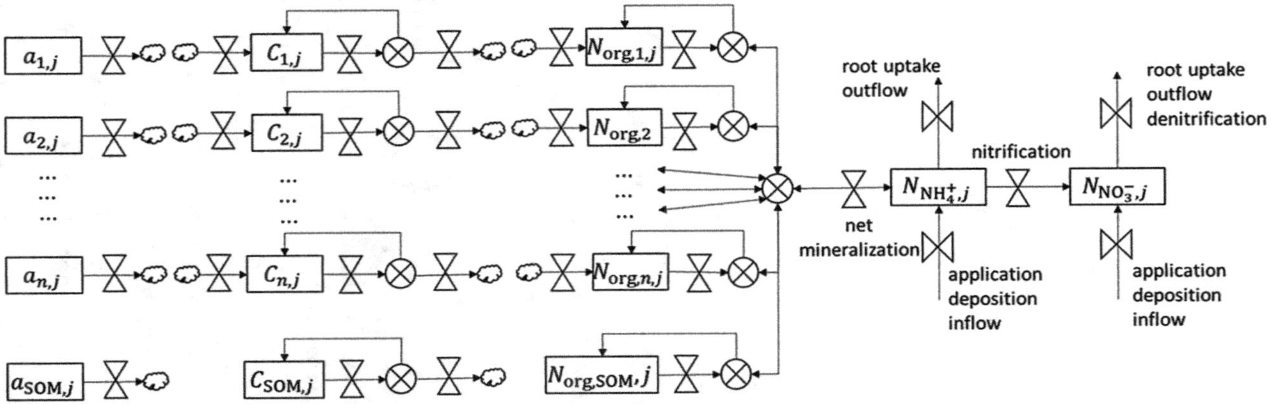

Figure 3 Schematic overview of all state variables for the amendments, initial soil organic matter (SOM), and amounts of inorganic N and processes that affect them. The state of individual amendments in a soil layer *j* and, also, of the initial SOM are described by their apparent age (a), organic C amount and organic N amount. The state of the inorganic N in layer *j* is described by the amounts of NH_4^+-N (N_{NH4^+}) and NO_3^--N (N_{NO3-}). The difference between gross rates of mineralization and immobilization determines the overall net mineralization rate which can be either positive (net release of NH_4^+-N) or negative (net immobilization).

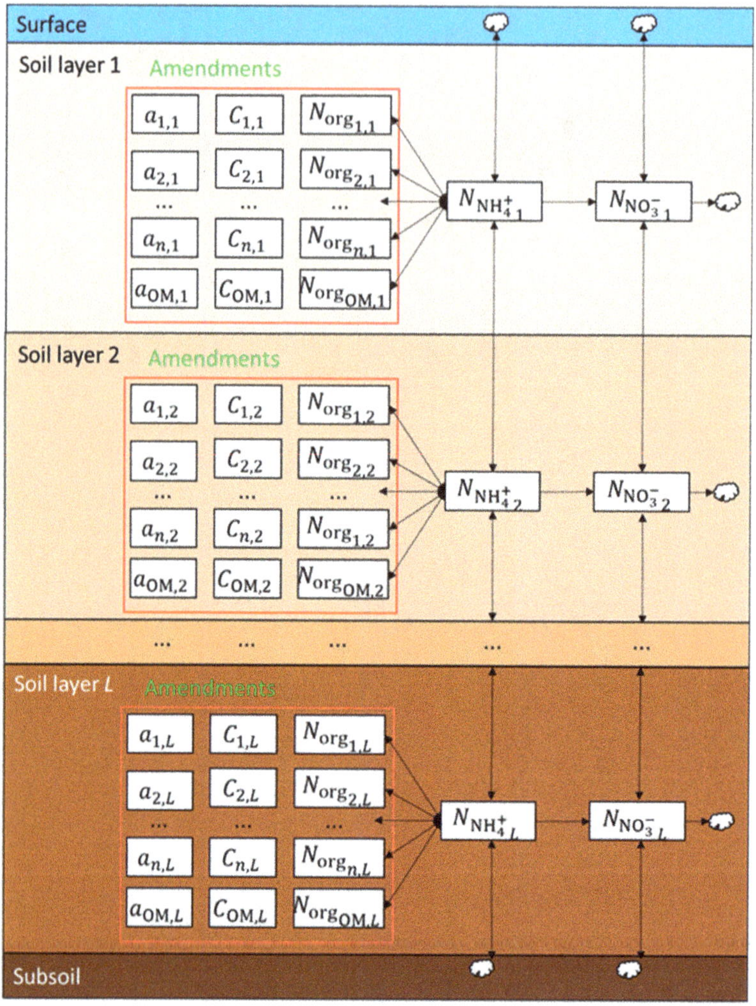

Figure 4 Overview of all state variables of SNOMIN in all soil layers.

3.3 New crop processes implemented in WOFOST 8.1: nitrogen dynamics and reallocation of dry matter

3.3.1 Integrating N-dynamics in WOFOST

Crop N dynamics in WOFOST 8.1 are described by the amount of N in grains, leaves, stems and roots. For most crops, N uptake is solely from the soil but in case of legume crops, N can also be taken up from N_2 fixation which is assumed a fixed fraction of total N demand. Sinks of N in the vegetative organs (i.e. roots, leaves, and stems) consist of N loss due to senescence and

N translocation to the storage organs. For storage organs, N translocation is thus a source. WOFOST calculates the daily N demand of each organ, i.e. the amount of N uptake to maintain the maximum N concentration in each organ. If N demand is both lower than the crop's maximum daily N uptake rate (a crop parameter) and the amount of N available for uptake (from soil and, in case of legumes, N_2 fixation), then crop N uptake equals N demand. If there is not enough N available to fulfil N demand for crop uptake, then crop uptake equals the remaining amount of available N. N is further distributed over the different organs after crop uptake. Translocation of N occurs if N demand of the storage organs cannot be met by root uptake and N_2 fixation only. This usually happens because root N uptake is assumed to halt after a given development stage in WOFOST. N is then transferred from the vegetative organs to the storage organs until either the grain N demand is fulfilled or exceeds the maximum daily translocation rate.

3.3.2 Simulating the impact of N limitation

WOFOST 7.2 had a crude and empirical way of dealing with N stress that may happen in most crops during their growth cycle. Figure 5a shows the shape of the AMAXTB parameter for wheat in WOFOST 7.2. This parameter describes the leaf-level maximum CO_2 assimilation rate as a function of crop development stage. The decline in AMAX from development stage 1.3 onward is to mimic the decrease in maximum assimilation rate as a result of declining leaf N concentration due to aging of the leaves and translocation of N from vegetative to reproductive organs. Although this approach is rather empirical, it does demonstrate that there is an elegant way to implement N stress in WOFOST by relating leaf N concentration to maximum assimilation rate.

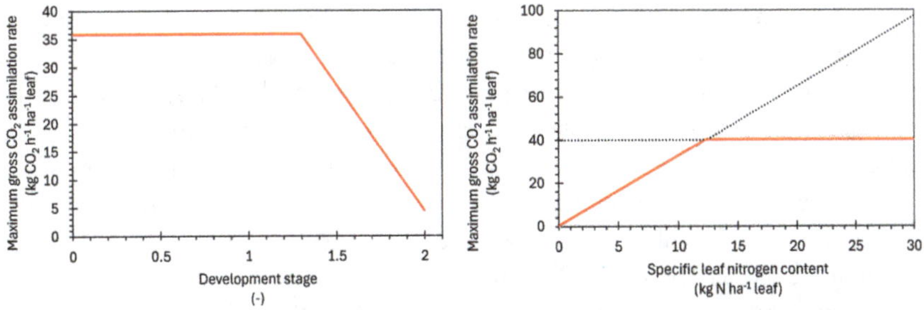

Figure 5 Left panel: The leaf-level maximum CO_2 assimilation rate as a function of crop development stage for wheat in WOFOST 7.2. Right panel: Shape of the relationship between specific leaf nitrogen content and leaf-level assimilation rate. AMAX increases linearly with leaf N content until the reference AMAX is reached (in this example, the value 40).

WOFOST 8.1 discards the tabular AMAXTB parameter and replaces this with a single AMAX value at a CO_2 reference condition and optimal temperature. This AMAX is used as the maximum value of a linear relationship between AMAX and leaf N content (Peng et al., 1995; Van Keulen and Seligman, 1987) (Fig. 5b). This approach is similar to the approach used by ORYZA2000 (Bouman and van Laar, 2006). Initially an increase in leaf N concentration will quickly increase the CO_2 assimilation rate until the reference AMAX is reached. Hereafter, increasing leaf N will not lead to higher assimilation rate. The leaf level CO_2 assimilation derived from the leaf N concentration then propagates through the assimilation routines which integrate instantaneous assimilation rates over the crop canopy depth and during the day. The assimilation routines also assume a canopy N gradient to take into account that leaf N content generally declines with canopy depth. WOFOST thus does not use the concept of nitrogen nutrition index (Lemaire et al., 1989) to account for N stress on CO_2 assimilation, which is often used by other models such as LINTUL3 (Shibu et al., 2010), STICS (Brisson et al., 2003), APSIM (Holzworth et al., 2014) and DSSAT-CERES (Jones et al., 2003).

Other areas where N stress has an impact in WOFOST 8.1 is on the relative growth rate of the leaf area index during the juvenile stage and on the relative death rates of leaves. See Berghuijs et al. (2024) for specific details on the mathematical implementation of these effects.

3.3.3 Reallocation of dry matter

Reallocation of biomass of stems and leaves can be an important source of biomass for the storage organs depending on the crop type and variety. The impact of biomass reallocation is particularly important when the assimilation rate during the reproductive stage is low due to stress factors (such as drought) and a large part of the assimilates required for grain filling has to be obtained from the reserves in stems and leaves. WOFOST 7.3 and 8.1 were therefore extended with a module for reallocation of dry matter from aboveground vegetative organs to storage organs (Berghuijs et al., 2023). This module calculates the amount of reallocatable dry matter in the stem and leaves as a fraction of the stem and leaf dry matter once a certain development stage is reached. From this development stage until maturity, a fixed fraction of the initial amount of reallocatable dry matter is translocated daily to the grains assuming a fixed reallocation efficiency.

Currently, the parameters for biomass reallocation have been calibrated for wheat (Berghuijs et al., 2024) and several potato varieties only. For all other crops, the fraction of reallocatable dry matter has currently been set to zero which essentially disables reallocation.

4 Simulating variety differences and G × E × M interactions with WOFOST

4.1 Modelling approach for different crops

WOFOST simulates a large diversity of crops and varieties with a single set of algorithms. Therefore, differences between crops and between crop varieties are expressed as different model parameter values rather than differences in the algorithms or model structure. The advantage of this approach is that WOFOST has a relatively small code base of around 5300 lines of code to simulate 23 different crop types. Instead, the DSSAT model suite uses around 155 000 lines[2] to simulate an equivalent number of crops. It is obvious that a smaller code base has an advantage in terms of maintenance, and it also explains why there are many implementations of WOFOST developed by organizations outside of Wageningen (de Wit et al., 2019).

The disadvantage of this approach is that certain traits of specific crops are more difficult to simulate with WOFOST. The clearest example of this limitation is the model for phenological development in WOFOST. This was originally developed for cereals, and it forces all crops into the same development scale based on a dimensionless variable 'DVS' which is 0 at emergence, 1 at the start of storage organ development (anthesis in cereals) and 2 at crop maturity. The result is that differences between crops are expressed by the partitioning patterns (which are a function of DVS) while a crop-specific model could use a dedicated phenological development model to steer partitioning between plant organs.

4.2 Expressing variety differences

The WOFOST 7.3 model requires 62 parameters to describe a crop and variety, while WOFOST 8.1 requires 79 parameters due to the inclusion of crop nitrogen dynamics. Default values are available for 23 crops and for most crops several varieties are specified as well (Table 2 in Supplementary material). Fortunately, only a limited number of parameters are relevant for expressing differences between varieties. In many cases, WOFOST can be adapted for different varieties by adapting the temperature sums for the phenological stages (TSUMEM, TSUM1, TSUM2) by using only observations of crop phenology (sowing, emergence, flowering and maturity). This is often sufficient when the model has to be adapted to a new but otherwise similar variety growing under the same agro-environmental conditions.

2 Computed as the number of lines of Fortran code in the Plant/ directory of the DSSAT source tree for the same crops as WOFOST. Analysis carried out on 14 February 2025.

When a variety is from a very different agro-environmental region or when parameters are strongly outdated (for example based on a decades old variety) more parameters may need to be estimated which requires detailed field experimental data. For example, ten Den et al. (2022) calibrated the WOFOST model for six different potato cultivars based on 2 years of experimental data from modern potato cultivars. The existing WOFOST parameters for potato were based on old cultivars from the 1980s, and therefore 19 parameters were updated. However, only six of those parameters were found to be variety specific, which included the temperature sum from emergence to tuber initiation (TSUM1), the relative growth rate for leaf area (RGRLAI), leaf span of leaves (SPAN) and the partitioning coefficients (FLTB, FOTB, FSTB).

For spatial applications such as the European MARS forecasting system, it is important to realize that crop varieties and sowing dates gradually change based on the gradient in climatic conditions. Therefore, it is often required to modify the phenological parameters of WOFOST for different agro-ecological zones over Europe. The parameters for each zone are estimated from local observations of crop phenology. The exact varieties for such observations are usually not known; however, they are assumed to have similar phenological characteristics because we assume that farmers grow varieties that are adapted to the local climatic conditions and sowing dates. The spatial distribution of sowing dates was estimated using climatic rules based on temperature and calibrated on observations of sowing dates. The general procedure for calibration and checking consistency of results is described in Wit et al. (2019), while a specific application of this approach for winter-wheat can be found in Ceglar et al. (2019).

A description of model parameters, guidance on parameter selection for calibration and sensitivity analysis as well as requirements on field experimental data can be found in the Appendix I to IV of the 'Gentle introduction to WOFOST' document available from the WOFOST website.[3] This document describes the parameters for all the major processes implemented in WOFOST and divides them into 'Free', 'Additional' and 'Static' parameters. The first category can be freely calibrated as they are known to vary considerably between varieties. The second category contains parameters that can be calibrated when the calibration of the first category does not yield good results and when sufficient observations are available. Finally, there is a category of parameters that are relatively static and should not be touched unless very specific measurements are available to estimate their value.

3 http://www.wofost.wur.nl.

4.3 Simulating G × E × M interactions

WOFOST can be used to simulate the impact of genotype (variety), environment (weather and soil) and crop management. Hajjarpoor et al. (2022) provide an overview of the ways that process-based crop simulation models can help in understanding G × E × M interactions and below we provide an overview of which processes in WOFOST are responsive to the environment (E) and crop management (M). As discussed above, the differences between genotypes (varieties) in WOFOST are expressed as variety-specific parameters rather than differences in algorithm or model structure. Therefore, it is relatively straightforward to describe the environmental responses to the processes in WOFOST as these are quite similar between crops and varieties.

4.3.1 Atmospheric CO_2 concentration

In WOFOST 7.3 and 8.1, the atmospheric CO_2 concentration has an impact on the shape of the radiation-photosynthesis relationship by modifying the maximum assimilation rate (AMAX) and the initial light use efficiency (EFF). WOFOST uses empirical modification factors which assume a 90% increase for AMAX and an 11% increase for EFF for C3 crops when the CO_2 concentration increases from 360 ppm to 720 ppm. Moreover, for C3 crops a 10% decrease in crop transpiration is assumed at 720 ppm CO_2 concentration. For C4 crops, no impact of atmospheric CO_2 is assumed on the assimilation rate but the crop transpiration is assumed to decrease by 26% at 720 ppm. All modification factors assume a linear impact between 360 ppm and 720 ppm on AMAX, EFF and crop transpiration.

4.3.2 Temperature

Temperature regulates many processes in WOFOST including phenology, assimilation, respiration and leaf development. In general, higher temperatures will generally shorten the growth cycle, decrease assimilation rate, increase respiration and cause early canopy senescence. More specifically, higher temperatures accelerate phenological development and thus shorten the growth cycle for many crops but this can be modified by photoperiod and vernalization.

Assimilation is assumed to be optimal within a temperature window taken from the day-time average temperature. Temperature values above or below this window will reduce the assimilation rate. Moreover, WOFOST uses a 7-day average of the daily minimum temperature to simulate growth reduction due to low night-time temperatures. Respiration is simulated by an exponential function that predicts a doubling of respiration with every 10 degree increase in temperature. Increases in temperature will therefore quickly reduce net available assimilates.

Finally, leaf development is driven by temperature in the initial phase through a relative growth rate and in the senescing phase due to leaf death, which is determined by a temperature sum. WOFOST has no algorithms that determine the impact of high temperature extremes, for example, to simulate the impact on leaf death or flower abortion. Similarly the impact of low temperatures on crop death is not taken into account either.

4.3.3 Radiation

Radiation has an impact on crop phenology through day length, on the computed assimilation rate as well as on the reference evapotranspiration. For crops that are sensitive to photoperiod, the development rate can be reduced if the day length is below (long day) or above (short day) the critical day length. However, in WOFOST, this is only applied during the vegetative phase (DVS < 1).

Higher radiation levels will initially lead to a higher assimilation rate but this will quickly level off due to the plateauing of the radiationphotosynthesis relationship used in WOFOST. When radiation levels increase further, this will not lead to an increase in assimilation, but it will increase the reference evapotranspiration. This may lead to the counter-intuitive effect that with increasing radiation, potential production is stable but water-limited production may decrease due to the increased water use and increased drought impact. Day length has an impact on the assimilation rate. With a longer day length assimilation rate increases because plants experience a lower average radiation but for a longer time period.

4.3.4 Vapour pressure and windspeed

Vapour pressure and windspeed influence the calculation of the reference evapotranspiration but have no other impact on WOFOST itself. For example, WOFOST does not implement impact of lodging due to wind or the impact of closing of plant stomata with a high vapour pressure deficit (Grossiord et al., 2020).

4.3.5 Rainfall and snow

Rainfall has no direct impact on the crop simulation but has a major impact through the crop–soil interface and the impact of drought or oxygen stress. It should be noted that WOFOST does not simulate the impact of extreme rainfall events which may damage the plant canopy or decrease yield due to the shattering of grains from the pods or leaf damage. The impact of snow is not simulated by WOFOST. All precipitation is assumed to be rainfall with no impediments due to frozen soil or buffering of water in the snow pack.

4.3.6 Soil characteristics

Soil characteristics can have a major impact on the water-limited and nitrogen-limited production simulated by the WOFOST model. First of all, the soil water-holding capacity is an important characteristic which influences the water-limited simulation results. The water-holding capacity is determined by two aspects: first the soil rootable depth and second the amount of water per unit of soil that can be held. The latter is roughly determined by the difference between field capacity and wilting point. Additionally, soil properties like drainage and the existence of impeding layers can influence water availability.

The impact of the soil on nitrogen availability in WOFOST is mainly determined by its organic matter content and composition, which can provide nitrogen through mineralization. Moreover, the soil hydraulic characteristics strongly determine the prospect of leaching of N to deeper soil layers where it becomes unavailable for root uptake.

4.4 Crop management

WOFOST is highly flexible in its definition of crop management. The management for a given crop or crop rotation is defined by an agromanagement structure which consists of a series of campaigns. Each campaign has a starting date and contains one crop calendar or none in case of fallow land, zero or more timed events and zero or more state events. The crop calendar specifies the timing of the crop, while the timed and state events can be used to specify management actions that are either dependent on time (a specific date) or a certain model state variable such as crop development stage. Crop calendars and event definitions are only valid for the campaign in which they are defined.

Due to this structure, WOFOST is able to simulate a large number of management options including:

- impact of sowing and harvest dates including harvest at fixed dates or harvest determined by crop maturity;
- impact of fallow periods;
- the application of various irrigation regimes either determined by fixed dates or by automatic scheduling when soil moisture levels drop below a predefined value (state events); and
- the application of organic and inorganic fertilizers and their influence on nitrogen availability for the crop. Fertilization events can happen at fixed dates or be scheduled connected to certain development stages.

Because crops in a rotation are simulated as a continued sequence, the model can also simulate the impact of carry-over effects between crop cycles. This is particularly important to understand the effect of fallow periods on soil water availability (Adil et al., 2022) and the impact of organic manure on the following crop cycles. In a similar way, WOFOST can also simulate the impact of cover crops and green manures since the SNOMIN model can treat them as organic amendments and estimate the release of nitrogen from their decay.

4.5 Linking genetics with WOFOST parameters

Linking genetics with crop models has been regarded as a pathway to connect genotypes with phenotypes in order to support the development of crop varieties that are high yielding in certain target environments (Wang et al., 2019; Yin et al., 2018). Understanding the link between specific genes and crop model parameters would allow to capture G × E × M interactions. In the ideal situation, the crop model could directly predict the impact of environment and management on a given genotype.

So far, for WOFOST there have been no attempts to link model parameters to genotypes and to predict model parameter values from specific genes (or combination of genes) of a given variety. Currently, WOFOST parameters are derived from classical field experiments or field observations and many of the parameters in WOFOST were designed to be measurable from such data. It is also questionable whether the approach used in WOFOST is suitable for building such links as its crop physiology may not be detailed enough.

5 Accuracy of WOFOST simulation results

The accuracy of the results of a model such as WOFOST depends on many factors including the level of calibration for a given crop/variety and the accuracy of model inputs such as initial conditions, agromanagement, soil type and weather data. Moreover, WOFOST has been designed to simulate crop growth and development for theoretical production levels. Therefore comparisons with observations which do not strictly apply to those production levels (such as measurements form farmers' fields) usually show that WOFOST overestimates yield and total biomass (which is the expected behaviour).

Although results for WOFOST have been reported for many studies, there has never been a systematic attempt to collect information about the accuracy of WOFOST predictions. Such effort is also complicated by the use of different accuracy metrics which are sometimes hard to compare. Nevertheless, we collected some accuracy metrics for different crops and regions for which WOFOST was applied (Table 1).

Table 1 Statements on the accuracy of WOFOST output in predicting the harvestable product (grain, root or tuber yield) for an independent dataset for different crops and regions

Crop	Statement on accuracy of model results[1]	Region	Version	Source
Wheat	RMSE Between 0.62 and 1.353 ton/ha for the 2014 season, 1.3 ton/ha and 4.7 ton/ha for the 2015 season (Table S12)	The Netherlands	8.1	(Berghuijs et al., 2024)
Wheat	RMSE for the simulated yield under irrigated conditions ranged from 11% to 23%, for water-limited conditions from 22% to 28%.	China	7.1	(Huang et al., 2017)
Potato	Difference between 0.3% and 8.0% of observed yields (Table 6)	The Netherlands	7.1	(ten Den et al., 2022)
Potato	RMSE between 4.5% and 25.8% of dry matter yield over 7 cultivars (Table 6)	India	7.1	(Dua et al., 2018)
Barley	RMSE of around 1300 kg/ha (Fig. 3)	Europe	7.1	(Rötter et al., 2012)
Barley	RMSE of 1.81 ton/ha (figure 3 Suppl. Mat.)	Czech republic	7.1	(Pohanková et al., 2018)
Maize	RMSE of 6.58% and 9.5% of observed yields (Table 5)	The Netherlands	7.2	(Eweys et al., 2017)
Maize	Difference of 0.1 ton/ha and 0.7 ton/ha with observed yields (Table 7)	Ethiopia	7.1	(Kassie et al., 2014)
Rice	Difference between −8.25% and +2.23% of observed yields	Punjab, India	7.1	(Mukherjee et al., 2011)
Rice	RMSE of 1232 kg ha·1 (ca. 17% of the observed mean) and similar to the mean absolute error (MAE) of 1110 kg ha^{-1}	Philippines, Indonesia, Vietnam	7.1	(Cabrera et al., 1998)
Soybean	RMSE of 1.61 ton/ha for observed seed yield (Fig. 6).	Nebraska, Indiana, Iowa (US)		(Setiyono et al., 2010)
Sugar beet	RMSE of 0.2–0.5 ton/ha in predicting root dry matter	The Netherlands, UK	7.2	Berghuis 2025, unpublished
Sugar cane	RMSE of 5.35–5.96 ton/ha of total dry weight (Fig. 34)	Brazil	7.1	(Scarpare, 2011)
Sunflower	RMSE of 0.79 ton/ha over all treatments	Bari, Italy	7.1	(Pohanková et al., 2018)

[1]References to tables and figures point to the respective sources not to tables or figures in this manuscript.

Given the variability in accuracy values, it is hard to draw general conclusions. However, based on our experience and reported values in the literature, WOFOST can be expected to be accurate to within 10–15% for the harvestable product yield. This is on the condition that the input data are accurate and observed values are not influenced by pests, diseases or weeds. This is also supported by the well-designed studies of Ten Den et al. (2022) for potato and partially for Berghuijs et al. (2024). Also for 'outliers' like the 5.35–5.96 ton/ha RMSE for sugar cane (Table 1), it should be noted that these refer to total biomass values of 50–60 ton/ha and are therefore still within 10–15% range. WOFOST is expected to be more accurate for total aboveground biomass where errors are usually smaller than 10%. Total biomass is often easier to predict as it is less affected by uncertainty in the partitioning of assimilates between plant organs.

Having accurate model predictions is important for applications that rely on absolute estimates of yield and total biomass. For many other applications, a correct prediction of the relative interannual and spatial variability in yield and biomass can be much more important than the absolute value. For example, the MARS Crop Yield Forecasting system uses a statistical approach to derive yield predictions that takes care of technological trends in yield time-series and systematic differences in yield levels between regions due to socio-economic conditions (De Wit et al., 2010). In such a system, the absolute yields predicted by WOFOST are less important because they are corrected by the statistical approach. Instead, the yield prediction is driven for a large part by the interannual variability in WOFOST results (Lecerf et al., 2019). The latter is driven by variability in weather and agromanagement such as varying sowing dates.

6 Further enhancements

An important improvement that will be included in the upcoming 8.2 release (to be expected in 2026) is the ability to simulate C4 crops. Currently C4 crops have been disabled in the crop parameter files for WOFOST 8.1 because the relationship between maximum assimilation rate and leaf nitrogen content was only developed and evaluated for C3 crops. Current applications (such as the MARS system) do not rely on WOFOST 8.1 yet due to this limitation but use the 7.3 version which allows for simulation of C4 crops. Therefore effort will spent to develop such a relationship and evaluate the simulation result using experimental data. We expect this to be a modest effort given that considerable literature exists on the relationship between leaf N content and assimilation rate and/or radiation use efficiency (Muchow and Sinclair, 1994; Sinclair, 1991; Sinclair and Horie, 1989).

The current implementation of the multi-layer water balance can take into account upward movement of water through the pressure gradient in the soil, but it cannot take into account the impact of capillary rise due to shallow groundwater tables. For many low-lying areas such as the Netherlands with clay soils and groundwater tables at less than 2 meter depth, this can be an important source of water for crops (Kroes et al., 2018). Rappoldt et al. (2012) implemented an approach for taking into account capillary rise from shallow ground tables. This approach has not been implemented in the water balance in WOFOST 8.1 but will be included in the next release.

An important drawback of the more detailed soil components for water, nitrogen and carbon that are included with WOFOST 8.1 is the complexity of the parameterization of such components. On the one hand, this is due to the complex data structures that are required, while on the other hand, the data to parameterize them is often hard to find. We are developing specific data providers that can support users with parameterizing these components by querying publicly available APIs. For Dutch soils, the BOFEK soil physical units map can be queried using a web API providing information on Dutch soil types and their physical properties (Heinen et al., 2022). At a global scale, the SoilGrids database provides global information on soils and soil properties (including uncertainties) that can be used to derive soil physical properties from its web API (Poggio et al., 2021). Currently, examples of deriving soil properties from such web API are available as jupyter notebooks[4] but these will be integrated as data providers within the model framework.

In a similar fashion, data providers will be developed that can retrieve key agronomic observations that are collected by the AgroSTAC initiative.[5] Such observations will facilitate calibrating important model parameters based on observations of phenology, leaf area index and biomass. Making such datasets available makes it easier to adjust model parameters for new locations. This applies not only to WOFOST but also to many other cropping system models that need calibration of parameters to adjust the model to local crop varieties.

7 Obtaining the WOFOST model and documentation

General information about the WOFOST model can be found on the WOFOST page from Wageningen University & Research (WUR): https://wur.nl/wofost. This provides links to many resources including model implementations and versions, documentation and references to papers where WOFOST is applied to different crops and use cases.

4 https://github.com/ajwdewit/pcse_notebooks.
5 https://agrostac.org/en.

The easiest way to obtain the most recent versions of WOFOST (7.2, 7.3 and 8.1) is to use the WOFOST implementation integrated in the PCSE:

- PCSE can be installed from the Python Package Index and installation procedures, user guide and code documentation can be found at https://pcse.readthedocs.io.
- Source code for PCSE can be found at https://github.com/ajwdewit/pcse.
- Jupyter notebooks demonstrating many aspects of cropping system modelling with WOFOST can be found at: https://github.com/ajwdewit/pcse_notebooks.

An alternative implementation of WOFOST is available in the SWAP model where WOFOST is coupled to the highly detailed soil water balance of SWAP. Documentation, binary executables and Fortran90 source code of SWAP are available from http://swap.wur.nl. Other implementations of WOFOST that are maintained by WUR are not available publicly.

Finally, documentation on WOFOST independent from any model implementation is available:

- A description of algorithms can be found at https://wofost.readthedocs.io.
- A general introduction and user guide can be found at https://www.wur.nl/en/show/A-gentle-introduction-to-WOFOST.htm.

8 Conclusion

The new WOFOST 8.1 release provides a major step forward in the simulation capabilities of WOFOST by including the nitrogen dynamics in the crop and coupling it to a more capable soil module for simulating the dynamics of water, nitrogen and carbon. Moreover, the model has the ability to simulate crop rotations over longer periods and the possibility for defining complex agromanagement that can be easily extended with new management options. With these new capabilities WOFOST has evolved from a crop model which mainly focused on describing the growth and development of a single arable crop, towards a cropping system model that can simulate the state of the crop and soil and the impact of management over multiple cropping cycles. Similar to (Keating and Thorburn, 2018), the soil actually has a central position in WOFOST 8.1 and crops are added and removed according to the cropping cycle thereby altering the soil as a result of crop growth and management.

Although the new model components were evaluated thoroughly by (Berghuijs et al., 2024) more work is required. The approach developed for soil water, carbon and nitrogen in WOFOST 8.1 must be evaluated for more case studies with varying weather conditions, crop types, soils and agromanagement regimes. This will help to build experience and confidence in the algorithms implemented in WOFOST 8.1 but will also point out deficiencies and uncertainties in the modelling approach which require further improvements.

WOFOST 8.1 and the PCSE modelling framework were developed in Python and are available under the European Union Public License which grants broad permissions for re-use and integration. The model is available from public package repositories (such as PyPI) which makes installing and running the model relatively easy. The documentation includes the algorithm theoretical background, user guide and reference guide as well as introductory document that explains the model for novice users that start from scratch. Moreover, many examples of model use are available as jupyter notebooks that help users to start with many topics. We believe that both the thematic expansion of the model and the quality of other resources available for WOFOST ensure that the model remains cutting-edge, relevant and useful for a future generation of researchers.

9 Supplementary material

Overview of crops and varieties

Table 2 Overview of crop and varieties available for WOFOST and the origin of the parameterization

Crop	Variety	Origin
Barley	Spring_barley_301	1
Cassava	Cassava_VanHeemst_1988	2
Chickpea	Chickpea_VanHeemst_1988	2
Cotton	Cotton_VanHeemst_1988	2
Cowpea	Cowpea_VanHeemst_1988	2
Fababean	Faba_bean_801	1
Groundnut	Groundnut_VanHeemst_1988	2
Maize	Maize_VanHeemst_1988	2
	Grain_maize_201	1
	Grain_maize_202	1
	Grain_maize_203	1
	Grain_maize_204	1
	Grain_maize_205	1
	Fodder_maize_nl	8
Millet	Millet_VanHeemst_1988	2
Mungbean	Mungbean_VanHeemst_1988	2
Pigeonpea	Pigeonpea_VanHeemst_1988	2

(*Continued*)

Table 2 (*Continued*)

Crop	Variety	Origin
Potato	Potato_701	1
	Potato_702	1
	Potato_703	1
	Potato_704	1
	Innovator	3
	Fontane	3
	Markies	3
	Premiere	3
	Festien	3
Rapeseed	Oilseed_rape_1001	1
	Oilseed_rape_1002	1
	Oilseed_rape_1003	1
	Oilseed_rape_1004	1
Rice	Rice_501	1
	Rice_HYV_IR8	4
	Rice_IR64616H_DS	4
	Rice_IR64616H_WS	4
	Rice_IR64	4
	Rice_IR72	4
	Rice_IR72_DS	4
	Rice_IR72_WS	4
	Rice_IR8A	4
Sorghum	Sorghum_VanHeemst_1988	2
Soybean	Soybean_901	1
	Soybean_902	1
	Soybean_903	1
	Soybean_904	1
	Soybean_905	1
	Soybean_906	1
	Soybean_VanHeemst_1988	2
Sugarbeet	Sugarbeet_601	1
	Sugarbeet_602	1
	Sugarbeet_603	1
	Sugarbeet_604	1
Sugarcane	Sugarcane_VanHeemst_1988	2
Ssunflower	Sunflower_1101	1
Sweetpotato	Sweetpotato_VanHeemst_1988	2

(*Continued*)

Table 2 (*Continued*)

Crop	Variety	Origin
Tobacco	Tobacco_VanHeemst_1988	2
Wheat	Winter_wheat_101	1
	Winter_wheat_102	1
	Winter_wheat_103	1
	Winter_wheat_104	1
	Winter_wheat_105	1
	Winter_wheat_106	1
	Winter_wheat_107	1
	Bermude	5
	Apache	5
	Arminda	6
	Kanzler	6
	Bussard	6
	Ramiro	6
	Greif	6
	Julius	6
	Ritmo	6
	Tabasco	6
Seed_onion	onion_agriadapt	7

1 MARS project (Boons-Prins et al., 1993).
2 Monograph on crop parameters (Heemst, 1988).
3 Study on modern potato cultivars (ten Den et al., 2022).
4 Various studies at the CGIAR-IRRI institute (Philippines).
5 (Wallach et al., 2021).
6 (Berghuijs et al., 2024).
7 (Wolf et al., 2010).
8 (Eweys et al., 2017).

10 References

Adil, M., et al., 2022. Effects of fallow management practices on soil water, crop yield and water use efficiency in winter wheat monoculture system: a meta-analysis. Front. Plant Sci. 13, 825309. https://doi.org/10.3389/fpls.2022.825309

Baron, F.-J. and Uitdewilligen, L., 2023. Dacom Precision Agriculture System, in: Cammarano, D., et al. (Eds.), *Precision Agriculture: Modelling, Progress in Precision Agriculture*. Springer International Publishing, Cham, pp. 263–268. https://doi.org/10.1007/978-3-031-15258-0_15

Been, T.H., et al., 2023. Akkerweb and Farmmaps: Development of Open Service Platforms for Precision Agriculture, in: Cammarano, D., et al. (Eds.), *Precision Agriculture: Modelling*. Springer International Publishing, Cham, pp. 269–293. https://doi.org/10.1007/978-3-031-15258-0_16

Berghuijs, H.N.C., et al., 2023. Catching-up with genetic progress: simulation of potential production for modern wheat cultivars in the Netherlands. Field Crops Res. 296, 108891. https://doi.org/10.1016/j.fcr.2023.108891

Berghuijs, H.N.C., et al., 2024. Expanding the WOFOST crop model to explore options for sustainable nitrogen management: a study for winter wheat in the Netherlands. *Eur. J. Agron*. 154, Article 127099. https://doi.org/10.1016/j.eja.2024.127099

Bonani, F. and Ghione, G., 1995. On the application of the Kirchhoff transformation to the steady-state thermal analysis of semiconductor devices with temperature-dependent and piecewise inhomogeneous thermal conductivity. *Solid-State Electron*. 38, 1409–1412. https://doi.org/10.1016/0038-1101(94)00255-E

Boogaard, H.L., et al., 1998. WOFOST 7.1; user's guide for the WOFOST 7.1 crop growth simulation model and WOFOST Control Center 1.5. SC-DLO.

Boogaard, H.L., et al., 2014. WOFOST CONTROL CENTRE 2.1; User's guide for the WOFOST CONTROL CENTRE 2.1 and the crop growth simulation model WOFOST 7.1. 7. Wagening. Neth. Alterra.

Boons-Prins, E.R., et al., 1993. Crop-specific simulation parameters for yield forecasting across the European Community. CABO-DLO [etc.].

Bouman, B.A.M. and van Laar, H.H., 2006. Description and evaluation of the rice growth model ORYZA2000 under nitrogen-limited conditions. *Agric. Syst*. 87, 249–273. https://doi.org/10.1016/j.agsy.2004.09.011

Bouman, B.A.M., et al., 1996. The 'School of de Wit' crop growth simulation models: a pedigree and historical overview. *Agric. Syst*. 52, 171–198. https://doi.org/10.1016/0308-521X(96)00011-X

Brisson, N., et al., 2003. An overview of the crop model stics. *Eur. J. Agron*. 18, 309–332. https://doi.org/10.1016/S1161-0301(02)00110-7

Broeke, M.J.D.H., et al., 2019. Quantitative land evaluation implemented in Dutch water management. *Geoderma* 338, 536–545. https://doi.org/10.1016/j.geoderma.2018.11.002

Cabrera, J., et al., 1998. Preliminary Results of Crop Model Development and Evaluation for Rice, in: Roetter, R.P., et al. (Eds.), A Systems Approach to Analysing Land-Use Options for Sustainable Rural Development in South and Southeast Asia (IRRI Discussion Paper No. 28). International Rice Research Institute, Manila, pp. 40–57.

Ceglar, A., et al., 2019. Improving WOFOST model to simulate winter wheat phenology in Europe: Evaluation and effects on yield. *Agric. Syst*. 168, 168–180.

de Wit, A. and Berghuis, H., 2025. PCSE: the Python crop simulation environment.

de Wit, A., et al., 2010. Using ERA-INTERIM for regional crop yield forecasting in Europe. *Clim. Res*. 44, 41–53.

de Wit, A., et al., 2019. 25 years of the WOFOST cropping systems model. *Agric. Syst*. 168, 154–167. https://doi.org/10.1016/j.agsy.2018.06.018

de Wit, C.T., 1965. *Photosynthesis of leaf canopies* (No. 663). Pudoc, Wageningen.

Dua, V.K., et al., 2018. Calibration and validation of WOFOST model for seven potato (Solanum tuberosum) cultivars in India. Indian J. Agron. 63, 357–365.

Evert, F.K. and Lamaker, E.J.J., 2007. The MODCOM Framework for Component-Based Simulation, in: *Farming Systems Design 2007, International Symposium on Methodologies on Integrated Analysis on Farm Production Systems. Presented at the International Symposium on Methodologies on Integrated Analysis on Farm Production Systems*, Goliardica Pavese, pp. 223–224.

Evert, F.K. et al., 2021. A Digital Twin For Arable Crops and for Grass, in: *Proceedings of the 16th International Conference on Precision Agriculture. Presented at the The 16th International Conference on Precision Agriculture*, Manhattan, Kansas, US.

Evert, F.K., et al., 2024. Operational Yield Forecasting and Crop Management with a Digital Twin, in: *Synergies for a Resilient Future: From Knowledge to Action. Presented at the 18th Congress of the European Society for Agronomy*, Rennes, France.

Eweys, O.A., et al., 2017. Integrating WOFOST and Noah LSM for modeling maize production and soil moisture with sensitivity analysis, in the east of The Netherlands. *Field Crops Res.* 210, 147–161. https://doi.org/10.1016/j.fcr.2017.06.004

Giongo, V., et al., 2020. Optimizing multifunctional agroecosystems in irrigated dryland agriculture to restore soil carbon – Experiments and modelling. *Sci. Total Environ.* 725, 138072. https://doi.org/10.1016/j.scitotenv.2020.138072

Groenendijk, P., et al. 2016. Simulation of nitrogen-limited crop growth with SWAP/WOFOST: process descriptions and user manual. https://www.cabidigitallibrary.org/doi/full/10.5555/20173088584

Grossiord, C., et al., 2020. Plant responses to rising vapor pressure deficit. *New Phytol.* 226, 1550–1566. https://doi.org/10.1111/nph.16485

Hajjarpoor, A., et al., 2022. How process-based modeling can help plant breeding deal with G x E x M interactions. *Field Crops Res.* 283, 108554. https://doi.org/10.1016/j.fcr.2022.108554

Heemst, H.D.J., 1988. *Plant data values required for simple crop growth simulation models: review and bibliography, Simulation reports CABO-TT.* Centre for Agrobiological Research (CABO): Department of Theoretical Production Ecology, Agricultural University, Wageningen.

Heinen, M. and Willigen, P. de, 2005. *Vergelijking van de organische stofmodellen MOTOR en MINIP*. Alterra, Wageningen.

Heinen, M., et al., 2022. The Dutch soil physical units map: BOFEK. *Geoderma* 427, 116123. https://doi.org/10.1016/j.geoderma.2022.116123

Heinen, M., et al., 2024. SWAP 50 years: advances in modelling soil-water-atmosphere-plant interactions. *Agric. Water Manag.* 298, 108883. https://doi.org/10.1016/j.agwat.2024.108883

Holzworth, D.P., et al., 2014. APSIM – Evolution towards a new generation of agricultural systems simulation. *Environ. Model. Softw.* 62, 327–350. https://doi.org/10.1016/j.envsoft.2014.07.009

Holzworth, D.P., et al., 2015. Agricultural production systems modelling and software: current status and future prospects. *Environ. Model. Softw.* 72, 276–286. https://doi.org/10.1016/j.envsoft.2014.12.013

Hoogenboom, G., et al., 2019. The DSSAT Crop Modeling Ecosystem, in: Boote, K. (Ed.), *Advances in Crop Modelling for a Sustainable Agriculture*. Burleigh Dodds Science Publishing, Cambridge.

Huang, J., et al., 2017. Dynamic simulation of growth process of winter wheat in main production areas of China based on WOFOST model. https://landbouwwagennld.library.ingentaconnect.com/content/tcsae/tcsae/2017/00000033/00000010/art00029# (accessed 8.22.25).

Jansen, D.M., 2008. *Beschrijving van TIPSTAR: hét simulatiemodel voor groei en productie van zetmeelaardappelen* (No. 547). Plant Research International, Wageningen.

Janssen, B.H., 1984. A Simple Method for Calculating Decomposition and Accumulation of 'Young' Soil Organic Matter, in: Tinsley, J. and Darbyshire, J.F. (Eds.), *Biological*

Processes and Soil Fertility. Springer Netherlands, Dordrecht, pp. 297–304. https://doi.org/10.1007/978-94-009-6101-2_26

Janssen, B.H., et al., 1990. A system for quantitative evaluation of the fertility of tropical soils (QUEFTS). *Geoderma* 46, 299–318. https://doi.org/10.1016/0016-7061(90)90021-Z

Jenkinson, D.S., et al., 1997. The turnover of organic carbon and nitrogen in soil. *Philos. Trans. R. Soc. Lond. B. Biol. Sci.* 329, 361–368. https://doi.org/10.1098/rstb.1990.0177

Jones, J.W., et al., 2003. The DSSAT cropping system model. *Eur. J. Agron.* 18, 235–265. https://doi.org/10.1016/S1161-0301(02)00107-7

Kassie, B.T., et al., 2014. Climate-induced yield variability and yield gaps of maize (*Zea mays L.*) in the Central Rift Valley of Ethiopia. *Field Crops Res.* 160, 41–53. https://doi.org/10.1016/j.fcr.2014.02.010

Keating, B.A. and Thorburn, P.J., 2018. Modelling crops and cropping systems—evolving purpose, practice and prospects. *Eur. J. Agron.* 100, 163–176. https://doi.org/10.1016/j.eja.2018.04.007

Knapen, M.J.R., et al., 2020. AGINFRA PLUS: Running Crop Simulations on the D4Science Distributed e-Infrastructure, in: Athanasiadis, I.N., et al. (Eds.), *Environmental Software Systems*. Data Science in Action. Springer International Publishing, Cham, pp. 81–89. https://doi.org/10.1007/978-3-030-39815-6_8

Knapen, R., et al., 2025. Efficient and scalable crop growth simulations using standard big data and distributed computing technologies. *Comput. Electron. Agric.* 236, 110392. https://doi.org/10.1016/j.compag.2025.110392

Kroes, J., et al., 2018. Impact of capillary rise and recirculation on simulated crop yields. *Hydrol. Earth Syst. Sci.* 22, 2937–2952. https://doi.org/10.5194/hess-22-2937-2018

Laar, H.H., et al., 1997. *SUCROS97: Simulation of Crop Growth for Potential and Water-Limited Production Situations. As Applied to Spring Wheat*. AB-DLO, Wageningen/Haren.

Lecerf, R., et al., 2019. Assessing the information in crop model and meteorological indicators to forecast crop yield over *Europe. Agric. Syst.* 168, 191–202. https://doi.org/10.1016/j.agsy.2018.03.002

Lemaire, G., et al., 1989. Analysis of the effect of N nutrition on dry matter yield of a sward by reference to potential yield and optimum N content. *Presented at the Proceedings of the XVI International Grassland Congress*, 4–11 October 1989, Nice, France., Association Française pour la Production Fourragère, Centre National de Recherche Agronomique, pp. 179–180.

Meyer-Roux, J. and Vossen, P., 1994. The first phase of the MARS project, 1988–1993: overview, methods and results, in: *Proceedings of the Conference on the MARS Project: Overview and Perspectives*. Commission of the European Communities, Luxembourg, pp. 33–85.

Muchow, R.C. and Sinclair, T.R., 1994. Nitrogen Response Of Leaf Photosynthesis And Canopy Radiation Use Efficiency In Field-grown Maize And Sorghum. *Crop Sci.* 34, cropsci1994.0011183X003400030022x. https://doi.org/10.2135/cropsci1994.0011183X003400030022x

Mukherjee, J., et al., 2011. Comparative evaluation of WOFOST and ORYZA2000 models in simulating growth and development of rice (Oryza sativa L.) in Punjab. *J. Agrometeorol.* 13, 86–91. https://doi.org/10.54386/jam.v13i2.1347

Peng, S., et al., 1995. Relationship between leaf photosynthesis and nitrogen content of field-grown rice in tropics. *Crop Sci.* 35, cropsci1995.0011183X003500060018x. https://doi.org/10.2135/cropsci1995.0011183X003500060018x

Pinheiro, E.A.R., et al., 2018. A matric flux potential approach to assess plant water availability in two climate zones in Brazil. *Vadose Zone J.* 17, 160083. https://doi.org/10.2136/vzj2016.09.0083

Poggio, L., et al., 2021. SoilGrids 2.0: producing soil information for the globe with quantified spatial uncertainty. *SOIL* 7, 217–240. https://doi.org/10.5194/soil-7-217-2021

Pohanková, E., et al., 2018. Estimating the water use efficiency of spring barley using crop models. *J. Agric. Sci.* 156, 628–644. https://doi.org/10.1017/S0021859618000060

Pushpalatha, R. and Byju, G., 2022. QUEFTS model, a tool for site-specific nutrient management of crops: a review. *Commun. Soil Sci. Plant Anal.* 53, 2339–2352. https://doi.org/10.1080/00103624.2022.2071924

Rappoldt, C., et al., 2012. Extension of the WOFOST soil water submodel (No. 6), EcoCurves report.

Rötter, R.P., et al., 2012. Simulation of spring barley yield in different climatic zones of Northern and Central Europe: a comparison of nine crop models. *Field Crops Res.* 133, 23–36. https://doi.org/10.1016/j.fcr.2012.03.016

Scarpare, F.V., 2011. *Simulação do crescimento da cana-de-açucar pelo modelo agrohidrológico SWAP/WOFOST* (Doutorado em Física do Ambiente Agrícola). Universidade de São Paulo, Piracicaba. https://doi.org/10.11606/T.11.2011.tde-12092011-172246

Setiyono, T.D., et al., 2010. Simulation of soybean growth and yield in near-optimal growth conditions. *Field Crops Res.* 119, 161–174. https://doi.org/10.1016/j.fcr.2010.07.007

Shaykewich, C.F. and Stroosnijder, L., 1977. The concept of matric flux potential applied to simulation of evaporation from soil. *Neth. J. Agric. Sci.* 25, 63–82. https://doi.org/10.18174/njas.v25i2.17133

Shibu, M.E., et al., 2010. LINTUL3, a simulation model for nitrogen-limited situations: application to rice. *Eur. J. Agron.* 32, 255–271. https://doi.org/10.1016/j.eja.2010.01.003

Sinclair, T.R., 1991. Canopy Carbon Assimilation and Crop Radiation-Use Efficiency Dependence on Leaf Nitrogen Content, in: *Modeling Crop Photosynthesis–from Biochemistry to Canopy*. John Wiley & Sons, Ltd, Chichester, pp. 95–107. https://doi.org/10.2135/cssaspecpub19.c6

Sinclair, T.R. and Horie, T., 1989. Leaf nitrogen, photosynthesis, and crop radiation use efficiency: a review. *Crop Sci.* 29, cropsci1989.0011183X002900010023x. https://doi.org/10.2135/cropsci1989.0011183X002900010023x

Supit, I., et al., 1994. System description of the WOFOST 6.0 crop simulation model implemented in CGMS. Vol 1, Theory and algorithms, Agriculture series. European Commission.

ten Den, T., et al., 2022. Modelling potential potato yields: accounting for experimental differences in modern cultivars. *Eur. J. Agron.* 137, 126510. https://doi.org/10.1016/j.eja.2022.126510

van der Burgt, G.J.H.M., et al., 2006. The NDICEA model, a tool to improve nitrogen use efficiency in cropping systems. *Nutr. Cycl. Agroecosys.* 74, 275–294. https://doi.org/10.1007/s10705-006-9004-3

van der Velde, M., et al., 2019. The European crop monitoring and yield forecasting system: celebrating 25 years of JRC MARS Bulletins. *Agric. Syst.* 168, 56–57. https://doi.org/10.1016/j.agsy.2018.10.003

van Diepen, C.A., et al., 1988. *CWFS Crop Growth Simulation Model WOFOST Documentation: Version 4.1, Staff Working Paper / Centre for World Food Studies*. Centre for World Food Studies, Amsterdam.

van Ittersum, M.K., et al., 2003. On approaches and applications of the Wageningen crop models. *Eur. J. Agron.* 18, 201–234. https://doi.org/10.1016/S1161-0301(02)00106-5

van Ittersum, M.K., et al., 2016. Can sub-Saharan Africa feed itself? *Proc. Natl. Acad. Sci.* 113, 14964–14969. https://doi.org/10.1073/pnas.1610359113

Van Keulen, H. and Seligman, N., 1987. *Plant growth processes*. Simul. Water Use Nitrogen Growth Spring Wheat Crop PUDOC Wagening, Netherlands.

van Kraalingen, D.W.G., et al., 2020. WISS a Java Continuous Simulation Framework for Agro-Ecological Modelling, in: Athanasiadis, I.N., et al. (Eds.), *Environmental Software Systems. Data Science in Action*. Springer International Publishing, Cham, pp. 242–248. https://doi.org/10.1007/978-3-030-39815-6_23

Wallach, D., et al., 2021. How well do crop modeling groups predict wheat phenology, given calibration data from the target population? *Eur. J. Agron.* 124, 126195. https://doi.org/10.1016/j.eja.2020.126195

Wang, E., et al., 2019. Improving process-based crop models to better capture genotype×environment×management interactions. *J. Exp. Bot.* 70, 2389–2401. https://doi.org/10.1093/jxb/erz092

Wang, X., et al., 2012. EPIC and APEX: model use, calibration, and validation. *Trans. ASABE* 55, 1447–1462. https://doi.org/10.13031/2013.42253

Wolf, J., et al, 2010. *Methodologies for Analyzing Future Farming Systems and Climate Change Impacts in Flevoland as Applied within the AgriAdapt Project, AgriAdapt Project Report*. Wageningen UR, Wageningen.

Yin, X. and Laar, H.H., 2005. *Crop Systems Dynamics: An Ecophysiological Simulation Model for Genotype-by-Environment Interactions*. Wageningen Academic Publishers, Wageningen.

Yin, X., et al., 2018. Bringing genetics and biochemistry to crop modelling, and vice versa. *Eur. J. Agron.* 100, 132–140. https://doi.org/10.1016/j.eja.2018.02.005

Chapter 13

The Light INTerception and UtiLisation family of crop models

A.G.T. Schut, H.N.C. Berghuijs, A.J.W. de Wit and M.K van Ittersum, Wageningen University and Research, The Netherlands

1 Introduction

Summary models are simple, yet are built on solid understanding of underlying processes, balancing simplicity and robustness (Monteith, 1996). The LINTUL (Light INTerception and UtiLisation) models are such summary models in which many detailed processes are left out and captured by some simpler concepts. The LINTUL approach has been used extensively in teaching as its simplicity allows students to study and understand key processes and interactions between crops and their environment in a quantitative manner. At the same time, it has also been used frequently for research purposes, either for studies where a summary model fits the purpose of the study or in cases where less detailed crop physiological observations are available. In such cases, the LINTUL approach is attractive because important model parameters can be easily derived from observations.

http://dx.doi.org/10.19103/AS.2025.0155.16

Models based on the LINTUL approach are part of the Wageningen pedigree of models and have been launched after 1–2 decades of work on more comprehensive types of models including daily photosynthesis at leaf level and various types of respiration (Bouman et al., 1996; van Ittersum et al., 2003). The development of these more detailed models and the insights derived from it allowed the development of the simpler light use efficiency (LUE) concept of daily biomass accumulation.

LINTUL models use the linear relationship between dry matter production and the amount of radiation intercepted by the crop canopy (Monteith, 1977), grown under well-watered conditions and ample nutrient supply, in the absence of pests, diseases and weeds. This relationship sets a finite limit on yield potential (Sinclair, 1994), which can be modelled without going into detailed descriptions of the processes of photosynthesis and respiration, which are modelled explicitly in more comprehensive models such as SUCROS (van Laar et al., 1997), ORYZA (Bouman et al., 2001), DSSAT (Hoogenboom et al., 2023) and derivates (Jones et al., 2003; van Ittersum et al., 2003).

Spitters and Schapendonk (1990) introduced the concept of LUE in Wageningen-based crop growth models. The name 'LINTUL' was first used during the release of the LINTUL POTATO model (Kooman and Haverkort, 1995), which applied this approach. LINTUL has been developed for a range of crops, using crop-specific LUEs (often also named radiation use efficiencies (RUEs); see e.g. Connor et al., 2011) and basically the same model structure. To account for water-limited or even nutrient-limited conditions, the LUE component must be linked to a soil water component and a nutrient limitation component. These are not LINTUL specific, but similar to soil water and nutrient limitation components used with, e.g., SUCROS or ORYZA.

The concept of RUE captures the chain of processes from intercepted photons to the formation of structural carbohydrates. A proportion of intercepted photons in the photosynthetically active range is absorbed by the chloroplasts and triggers the photosystems to produce energy. This energy is used to produce carbohydrates from CO_2 and water, which are partly converted into structural carbohydrates. These processes are sensitive not only to the amount of radiation, temperature, CO_2 concentration of the ambient air, water and nutrient stress but also to age of the leaf and phenological stage of the crop. In LINTUL, the RUE is a parameter that reflects the average conversion of intercepted photosynthetically active radiation (PAR) to dry matter, including roots, under non-limiting conditions over a longer period of time. This parameter needs to be measured in the field when the ground is fully covered but before anthesis. Under these conditions, the RUE is surprisingly similar across latitudes (Sinclair and Muchow, 1999) and varies little between crops and cultivars, as all crops grow roots and mostly leaves and stems with similar characteristics in the early growth stage. From anthesis (or, e.g. tuber of bulb initiation in tuber and root crops such as potato, cassava or sugar beet) onwards, biomass is also allocated to grains, tubers and leaf senescence, and re-allocation of dry matter from leaves and stems to the grains and tubers becomes important.

In this chapter we further introduce the LINTUL family of models. We first discuss in more detail versions of LINTUL developed for different production situations (i.e. potential, water-limited and nutrient-limited conditions). We then provide an overview of different LINTUL applications and of calibrated and tested parameter sets for different crops. Next, we give a short description of how LINTUL models are used in Wageningen courses and we end with reflections and the way forward.

2 Development of Light INTerception and UtiLisation (LINTUL) crop growth models for different production situations

Different production situations can be simulated, where crops grow at potential (P), water-limited (WL), N limited (WL-N) and NPK limited (WL-NPK) growth rates. These production situations include additional components (compared to the potential production version) to describe evapotranspiration and soil water dynamics, nutrient demand and uptake and account for stress on crop growth rates.

The first versions of LINTUL were developed for potato and spring wheat for P and WL production situations. The N-limited model (LINTUL3) was first developed for paddy rice (Shibu, 2010). These concepts were used as starting point for PCSE–LINTUL3 (Berghuijs et al., 2023) and LINTUL–Cassava–NPK (Adiele et al., 2022) that account for WL-N and WL-NPK production situations. The PCSE–LINTUL3 model includes functions to adapt the RUE based on ambient CO_2 concentrations (Berghuijs et al., 2023), which were also employed in the 2023 versions of LINTUL 1 and 2 that are made available for R users.

These models use a single parameter value for RUE that does not vary with crop development stage (DVS). Wolf (2012) developed LINTUL4 where the RUE varied with DVS to better capture the effect of leaf aging and temperature on RUE. In the conceptual LINTUL5 model, the concepts of LINTUL3 to describe N limitations were also used to describe N, P and K limitations, while LINTUL 6 also included crop rotations. These latter two models are experimental and are not well tested on independent datasets, so fall out of the scope of this chapter and therefore will not be discussed in detail here.

Standard parameter sets are available for a selected set of crops. For most crops, suitable RUE values are available in literature that can be corrected for ambient CO_2 concentration and converted into the reference RUE parameter values (see below). The critical crop development parameters (see below) are similar to WOFOST and can be derived from calibrated parameter sets, available for a large number of crops and calibrated for a wide geographical range.

3 Modelling potential crop production

The LINTUL models use the linear relationship between the amount of intercepted radiation by the crop (Monteith, 1977) and the total dry matter to calculate the daily rate of dry matter production. First, they calculate the fraction of the daily total amount of global radiation that it intercepted (f_{int}(t) at time *t*) by the canopy using Beer's law as

$$f_{int}(t) = 1 - e^{-K \cdot LAI(t)} \qquad \text{[eq. 1]}$$

where *K* is the light extinction coefficient of the canopy (m^2 ground m^{-2} leaf) and *LAI*(t) is the leaf area index (m^2 leaf m^{-2} ground) at day *t*. Secondly, they calculate the total dry matter production rate. Since plants can only use PAR for dry matter production, the daily total dry matter production rate (R_{Wtot}: g DM m^{-2} ground day^{-1}) under potential growth conditions is calculated as

$$R_{Wtot}(t) = f_{PAR} \cdot f_{int}(t) \cdot f_{CO2} \cdot RUE\big(DVS(t)\big) \cdot R_{tot}(t) \qquad \text{[eq. 2]}$$

where f_{PAR} is the fraction of PAR in the global radiation. RUE(DVS(*t*)) is the RUE (g DM MJ^{-1} PAR) as function of the DVS at time *t*. DVS is a state variable of the LINTUL models, and its calculation is explained in more detail in the next section. $R_{tot}(t)$ is the total amount of global radiation at day *t* (MJ PAR m^{-2} ground day^{-1}). In most versions of the model, the RUE is a parameter that does not vary with DVS. Wolf (2012) accounted for leaf aging effects on RUE in LINTUL4, by adjusting the value as function of DVS Further, in LINTUL4 the RUE is also reduced when minimum temperatures or daily average temperatures are sub-optimal as described in Wang et al. (2017). In LINTUL4, the recent PCSE–LINTUL3 model (Berghuijs et al., 2023) and the 2023 updates of LINTUL 1 and 2 in R the effect of the changing ambient CO_2 concentrations on RUE are accounted for by a correction factor f_{CO_2}. In versions of LINTUL that do not incorporate a CO_2 effect on dry matter production, $f_{CO_2} = 1$.

Each day, newly produced dry matter is partitioned to different organs. The LINTUL models distinguish leaves, stems, roots and storage organs. The newly produced dry matter is allocated to each organ at day *t* according to the so-called partitioning functions which depend on the DVS (Rappoldt and van Kraalingen, 1990). These tabular functions determine the partitioning fractions to each organ by linear interpolation between different combinations of known values of DVS or temperature sum and partitioning fraction, as illustrated in Fig. 1.

Under potential growth conditions, the daily dry matter growth of each organ *o* ($R_{W_o}(t)$: g DM m^{-2} ground day^{-1}) is calculated as

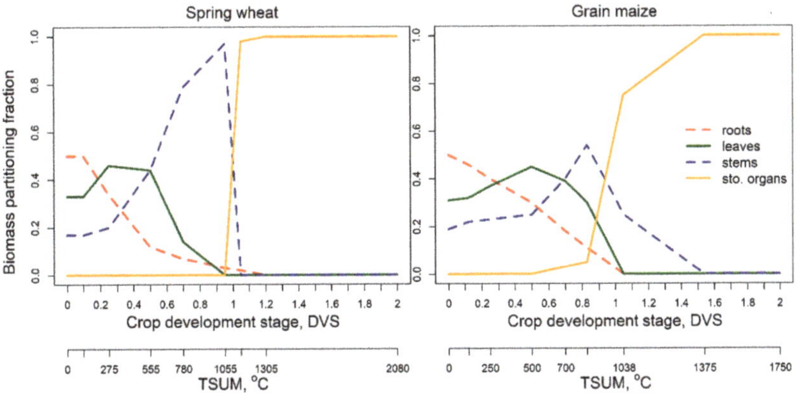

Figure 1. Allocation of dry matter to organs as function of temperature sum or development stage for spring wheat and grain maize.

$$R_{W_o}(t) = f_{\text{part,o}}(DVS(t)) \cdot f_{\text{PAR}} \cdot f_{\text{int}}(t) \cdot f_{\text{CO}_2} \cdot RUE(DVS(t)) \cdot R_{\text{tot}}(t) \qquad \text{[eq. 3]}$$

where $f_{\text{part,o}}$ is the partitioning fraction of newly produced dry matter to the storage organs at a given DVS. In contrast to cereals, the RUE and therefore the potato growth rates strongly depend on temperature. The effect of temperature differs between the four potato development phases that are defined (Kooman and Haverkort, 1995).

Equations 2 and 3 show that both the RUE and the partitioning fractions depend on DVS. A crop passes a number of DVSs throughout its life cycle. The different DVSs that are distinguished differ between different models in the LINTUL family (Table 1), but all of them, at least, distinguish between the DVSs emergence, anthesis (or start of tuber initiation in root and tuber crops) and physiological maturity. The rate of change of the DVS is called development rate. The development rate can be determined as the inverse of time that it takes for the crop to develop to one DVS, **s**, to the next DVS, **s + 1**. Since there is a strong linear correlation between the daily development rate and the daily average temperature in spring crops (van Keulen and Seligman, 1987), the development rate can be calculated as

$$\frac{dDVS(t)}{dt} = \max\left(0, \frac{T_{av}(t) - T_b}{TSUM_{s+1} - TSUM_s}\right) \qquad \text{[eq. 4]}$$

where $T_{av}(t)$ is average daily temperature (°C) at day *t*, based on the average of minimum and maximum temperature. T_b is the base temperature (°C); the daily average temperature below which there is no development. $TSUM_s$ and $TSUM_{s+1}$ are the respective cumulative amounts of heat, expressed in °C day, that the crop needs to accumulate to reach stage *s + 1* while it is in state *s*.

Table 1 Overview of Light INTerception and UtiLisation implementations under potential, and water-limited combined with and nitrogen or nitrogen-, phosphorus- and potassium-limited production situations

Crop type	Production situation	Software	CO_2	Comments	Reference	Repository
Summer grains, maize, Amaranth	P, WL	FST, FSE	No	–	Farré et al., 2000; Gimplinger and Kaul, 2009	Models.pps.wur.nl
Summer grains	P, WL	R, MS Excel	Yes	Teaching		Models.pps.wur.nl
Potato	P, WL	FSE	Yes	LINTUL-POTATO	Kooman and Haverkort, 1995; van Haren and Haverkort, 1998; Haverkort et al., 2013	
Winter oilseed rape	P, WL	FSE	No	Vernalisation	Habekotté, 1996	Models.pps.wur.nl
Sugar beet	P, WL	FSE	No		De Koeijer and van der Werf, 1995	
Grass	P, WL	FSE	No	Source-sinkLINGRA, LINGRA_CC, GrazeGrow	Schapendonk et al., 1998; Barrett et al., 2005; Rodriguez et al., 1999	
Rice	P, WL, WL-N	FSE	No	–	Shibu et al., 2010	Models.pps.wur.nl
Cassava	P, WL, WL-NPK	R	Yes	Dormancy	Ezui et al., 2018; Adiele et al., 2020, 2021b, 2022	Models.pps.wur.nl
Banana	P, WL	FSE, R	No		Taulya, 2015; Arqués Viver, 2022	
Winter grains, rice	P, WL, WL-N	PCSE	Yes	Vernalisation	Berghuijs et al., 2023	wofost.readthedocs.io/en /7.2 github.com/ajwdewit/ pcse/tree/develop_lintul3
Various	P, WL, WL-N	FSE	No	LINTUL 4	Wolf, 2012	Models pps.wur.nl
Various	P, WL, WL-N	FSE	Yes	LINTUL 4-VSHT	Wolf, 2012	Models pps.wur.nl
Various	P, WL, WL-NPK	FSE	No	LINTUL 5	Wolf, 2012	Models pps.wur.nl
Various	P, WL, WL-N	FSE	Yes	LINTUL 6	Wolf, 2012	Models pps.wur.nl

FSE, Fortran Simulation Environment; FST, Fortran Simulation Translator; LINTUL, Light INTerception and UtiLisation; P, potential; WL, water-limited; WL-N, water-limited along with nitrogen; WL-NPK, nitrogen-, phosphorus-, and potassium-limited production.

The relationship between development rate and average daily temperature is more complicated in winter crops, because of the effects of vernalisation and the photoperiodicity on the development rate. Also for potato, the photoperiod affects crop development rates (Kooman and Haverkort, 1995). Vernalisation is the process in which the development towards anthesis in winter crops are only induced after the crop has been exposed by prolonged low temperatures. Photoperiodicity of the development rate is the process in which the development towards anthesis is only induced after the daylength exceeds a critical value. Some LINTUL versions (Table 1) take one or both of these processes that only occur before anthesis into account for winter crops by extending equation 4 as following (Habekotté, 1996; Ceglar et al., 2019):

$$\frac{dDVS(t)}{dt} = V(t) \cdot P(t) \cdot \max\left(0, \frac{T_{av}(t) - T_b}{TSUM_{s+1} - TSUM_s}\right) \qquad \text{[eq. 5]}$$

where $V(t)$ is the vernalisation factor and indicates to which extend the vernalisation requirement of a winter crop has been satisfied. Its value varies between 0 (maximum inhibition owing to insufficient vernalisation) and 1 (no inhibition) owing to vernalisation, because the vernalisation requirement is fulfilled. Similarly, $P(t)$ is the photoperiod factor that varies between 0 (daylength too short for development) and 1 (daylength does not inhibit development, as the daylength exceeds the optimum daylength). The methodology to calculate $V(t)$ and $P(t)$ varies between different LINTUL models. The state variable V is integrated using a computed vernalisation rate which depends on daily average temperature (Fig. 2).

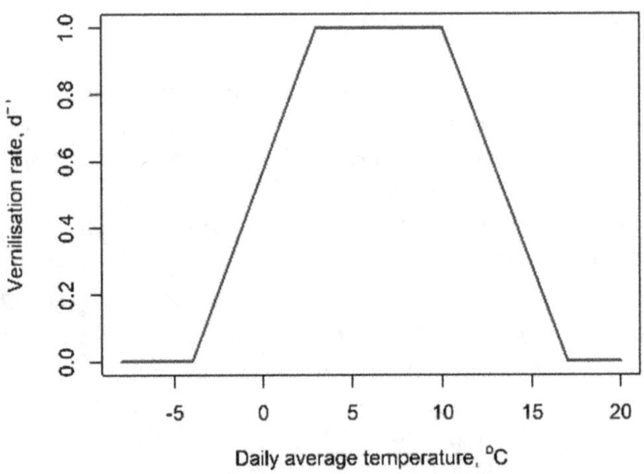

Figure 2. The vernalisation rate as function of daily average temperature.

The reallocation of assimilates within the crop, e.g. from stems to storage organs, is not modelled explicitly in most versions of LINTUL. Yet, to compensate for this, in LINTUL, the partitioning of dry matter to storage organs does start before anthesis, even when such storage organs are not yet present in reality. This is done by setting the parameter for allocation to storage organs to zero when e.g. DVS equals 0.9 and to 1.0 when DVS equals 1.1 (see Fig. 1). The linear interpolation of the partitioning fractions between these two DVSs then results in dry matter accumulation that starts just before anthesis. Differences in reallocation of dry matter to storage organs between crops and cultivars can be captured by calibrating these parameter values. Reallocation to storage organs before DVS 1 is no longer needed in LINTUL3, as implemented in the Python Crop Simulation Environment (PCSE-LINTUL3). In this model, reallocation is described by three parameters: to define the fraction of dry matter in the stem that can be reallocated to storage organs; to define the DVS when reallocation is possible; and to quantify the reallocation rate with a relative reallocation rate. See Berghuijs et al. (2023) for further details.

4 Modelling water-limited production

In the water-limited production situation, crop growth rates are reduced by the ratio between actual transpiration rate $T_{act}(t)$ and potential transpiration rate $T_{crop}(t)$:

$$TRANRF(t) = \frac{T_{act}(t)}{T_{crop}(t)} \qquad \text{[eq. 6]}$$

The $TRANRF$ is the transpiration reduction factor with values between 0 (no growth owing to drought) and 1 (non-water-limited growth). This reduction factor also affects the increase in root depth in the early stages of the season and the partitioning of dry matter to roots. Dry matter partitioning changes in favour of root growth during the vegetative phase when $TRANRF(t)$ falls below 0.5 (Farré et al., 2000). Except for PCSE implementations (that use the Penman-Monteith equation), the potential transpiration is determined by the Penman equation (Penman, 1948). Penman's equation computes the evapotranspiration in kg(H_2O) m^{-2}, equivalent to millimetres under normal conditions, which is attributed to evaporation or transpiration as function of the fraction of light absorbed by crop and soil (albedo) and LAI.

$$E_{soil}(t) = e^{(-0.5 \times LAI(t))} \times \left(\frac{\Delta(t)}{\Delta(t)+\gamma} \cdot \frac{R_{net\_soil}(t)}{\lambda} + \frac{\gamma}{\Delta(t)+\gamma} \cdot \frac{E_{air}(t)}{\lambda} \right) \qquad \text{[eq. 7]}$$

$$T_{crop}(t)=\max\left(0,\left(1-e^{(-0.5\times LAI(t))}\right)\cdot\left(\frac{\Delta(t)}{\Delta(t)+\gamma}\cdot\frac{R_{net\_canopy}(t)}{\lambda}+\frac{\gamma}{\Delta(t)+\gamma}\cdot\frac{E_{air}(t)}{\lambda}\right)\right)$$

[eq. 8]

where Δ (*t*) represents the first derivative value at a given temperature of the relationships between air temperature and saturated vapour pressure (hPa °C^{-1}); γ refers to the adiabatic psychrometer coefficient that indicates what the change in air temperature is when air vapour pressure changes (hPa °C^{-1}); λ the heat needed for vaporisation of liquid water (MJ kg(H_2O)$^{-1}$). The values for $R_{net\_soil}(t)$ and $R_{net\_canopy}(t)$ indicate how much net energy is available for evapotranspiration, accounting for differences in albedo between crop and soil (MJ m^{-2} day^{-1}). The $E_{air}(t)$ refers to the evaporative demand from the air, defined by the difference between saturated and actual vapour pressure (i.e. the actual vapour pressure deficit) and wind speed. The R_{int} is the daily intercepted amount of rainfall (in mm day^{-1}) and accounts for the reduction in transpiration after rainfall.

Following the macroscopic concept of Feddes (Heinen et al., 2024), the *TRANRF* has a value of 1 when actual soil water content is favourable and declines linearly when the soil water content is below a critical water content ($\theta_{cr}(t)$) in case of water shortage and above a high water content (θ_{wet}, when soil pF values are about 0.5 equivalent to 3.2 hPa) in case of a surplus of water, see Fig. 3. The latter reflects a shortage of oxygen that occurs in very wet soils for most plant species, except for species with aerenchyma such as rice, cattail and reed.

The critical water content ($\theta_{cr}(t)$) on day *t* lies between the water content at field capacity (θ_{fc}, at a pF value of 2.0, equivalent to 100 hPa) and wilting point (θ_{wp}, at a pF value of 4.2, equivalent to 15 848 hPa). It is assumed that θ_{cr} is affected by the drought tolerance of the crop: a very drought-tolerant crop will unrestrictedly extract water from the soil to a lower water content as compared to a drought-sensitive crop. The critical water content is calculated by

$$\theta_{cr}(t)=\theta_{wp}+\frac{T_{crop}(t)}{T_{crop}(t)+TRANCO}\cdot(\theta_{fc}-\theta_{wp})$$ [eq. 9]

where T_{crop} is the potential transpiration rate of the crop as computed with the Penman equation, and *TRANCO* is the transpiration coefficient, which is a measure of the drought tolerance of the crop. From equation 19, it is clear that the critical water content will be higher when the value for TRANCO is lower for the same value of T_{crop}. This means that a crop with a lower TRANCO will more quickly respond to drought conditions, i.e. is more sensitive to drought than a crop with a higher TRANCO.

The ratio $\frac{T_{\text{crop}}(t)}{T_{\text{crop}}(t)+TRANCO}$ varies between 0 (at an infinite drought tolerance, where $\theta_{cr}(t) = \theta_{wp}$) and 1 (if the crop would be extremely sensitive to drought, resulting in $\theta_{cr}(t) = \theta_{fc}$). By including the dynamic potential transpiration rate in the calculation of $\theta_{cr}(t)$, we recognize that on days with a high potential transpiration rate it is more difficult for even a drought-tolerant crop to take up enough water from the soil and to maintain its turgor against the large demand, compared to days with a low potential transpiration rate.

Transpiration completely ceases at wilting point, θ_{wp}. Actual evaporation ($E_{act}(t)$), however, can continue until the air-dry water content in the soil (θ_{ad}, at a pF value of 5.0, equivalent to 100 000 hPa) is reached according to Fig. 3. This evaporation reduction factor, EVAPRF(t) is used to compute the actual evaporation $E_{act}(t)$ from the potential evaporation E_{soil}:

$$E_{\text{act}}(t) = EVAPRF(t) \cdot E_{\text{soil}}(t) \qquad \text{[eq. 10]}$$

Most LINTUL versions use a simple tipping bucket approach to keep track of soil water, and therefore, actual soil water contents. This tipping bucket approach requires few parameters: it is implicitly assumed that the rooted zone is homogeneous; it does not account for capillary rise; and does not include lateral water flows. Consequently, it is not suited for sloping terrain or areas with high groundwater tables where capillary rise can be substantial. The size of the bucket depends on the rooting depth, which increases from an initial depth to

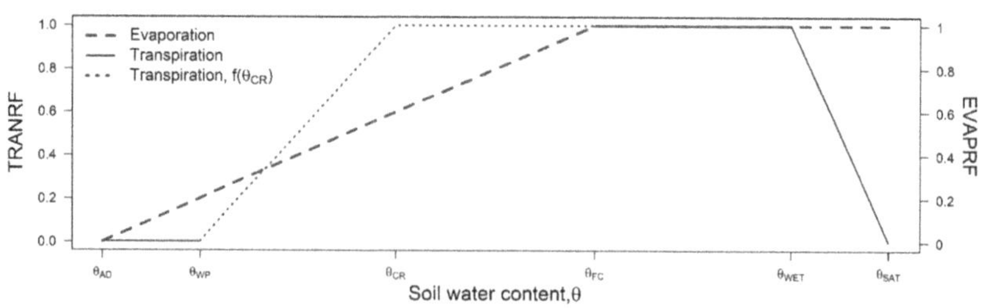

Figure 3. The transpiration (TRANRF, primary y-axis) and evaporation (EVAPRF, secondary y-axis) reduction factors as a function of water content (θ, x-axis) between wilting point (θ_{wp}) and completely saturated soil (θ_{sat}). The evaporation is reduced gradually when water content in the soil decreases from field capacity to air dry (θ_{ad}). This $E_{red-evap}$ only depends on θ and soil water contents at θ_{fc} and θ_{ad}. When water contents are $<\theta_{cr}$ or $>\theta_{wet}$, transpiration reduction takes place. The θ_{cr} can vary between θ_{fc} and θ_{wp}, depending on weather conditions and crop characteristics. The red dashed line displays the reduction of $T_{red-trans}$ if the soil water content is below the critical soil water content. The blue dashed line displays the reduction of the evaporation if the soil moisture content is below field capacity.

a maximum value with a fixed rate, at least when growth is not terminated by drought. Both values are parameters that need to be set by the model user.

The soil water balance (kg H_2O m^{-2} day^{-1} or mm day^{-1}) is computed with

$$\frac{dw(t)}{dt} = Rain(t) + Irr.(t) + Exploration(t) - R_{int}(t) - runoff(t) - drainage(t) - T_{act}(t) - E_{act}(t)$$

[eq. 11]

Inputs of water into the soil is from effective rain – rain minus intercepted rain by leaves, which depends on LAI (Rain(*t*) – $R_{int}(t)$). The irrigation (Irr.(*t*)) term refers to the kg of water m^{-2} day^{-1} supplied by irrigation. The exploration term refers to increasing rooting depth and increasing volume of the tipping buck where new soil and additional water is added to the bucket. Next to actual transpiration ($T_{act}(t)$), water is lost from the soil by actual soil evaporation ($E_{act}(t)$), drainage and runoff. Drainage only occurs when soil water contents are above field capacity and cannot exceed the maximum drainage capacity defined by the user (as a parameter value). Runoff occurs when daily soil water inputs exceed remaining storage capacity ($\theta_{sat} - \theta(t)$) and drainage.

5 Modelling nitrogen-limited production

The photosynthetic capacity of leaves strongly depends on the nitrogen concentration of leaves. Even in the absence of nitrogen deficiencies, the older leaves low in the canopy have a lower nitrogen concentration owing to the translocation of nitrogen to other parts of the plant. This is an adaptation mechanism to lower radiation levels low in the canopy (Van Keulen and Seligman, 1987). Various Wageningen school crop growth models outside the LINTUL family, like ORYZA (Bouman et al., 2001), NWHEAT (Groot, 1987; Groot and de Willigen, 1991), NPOTATO (Wolf, 2002) and WOFOST 8.1 (Berghuijs et al., 2024) consider this explicitly. They do so by simulating the amount of N in the leaves over time and calculating the maximum leaf photosynthesis rate, as related to the leaf N concentration (van Keulen and Seligman, 1987; Peng et al., 1995; Bouman et al., 2001).

The LINTUL models do not include detailed processes for leaf photosynthesis. Instead, they use the nitrogen nutrition index (NNI) approach to reduce crop growth rates under N-deficient conditions (Shibu et al., 2010). In principle, the NNI approach requires information about the minimum, maximum and critical N concentration of aboveground biomass that is sufficient to ensure maximum crop growth rates, captured with typical dilution curves for N concentration at crop level, as crop development and growth proceeds (Greenwood et al., 1990; Lemaire, 1997; Makowski et al., 2020).

N-limited growth was first introduced in the LINTUL models during the release of LINTUL3 (Shibu et al., 2010). For this purpose, LINTUL was extended with state variables that describe the N status of the crop and the soil. The soil N state consists of a single variable for the amount of N available for uptake in the soil (g N ha^{-1} ground). The crop N state variables are the amounts of N in the various crop organs (g N ha^{-1} ground). During each time step, LINTUL3 calculates the N demand of the vegetative organs, determined by the amount of N needed to reach the maximum N concentration in each of its vegetative organs (Fig. 4).

If there is enough available N in the soil, the crop takes up the amount of N that the vegetative organs need to reach their maximum N concentration. If not, the crop takes up all available N. Subsequently, the N taken up by the crop is distributed over the vegetative organs based on their relative contribution to the total N demand of the vegetative organs. If the N concentrations of the vegetative organs drop below the critical concentrations owing to N deficiencies, growth reduction owing to N stress can occur. In LINTUL3, the

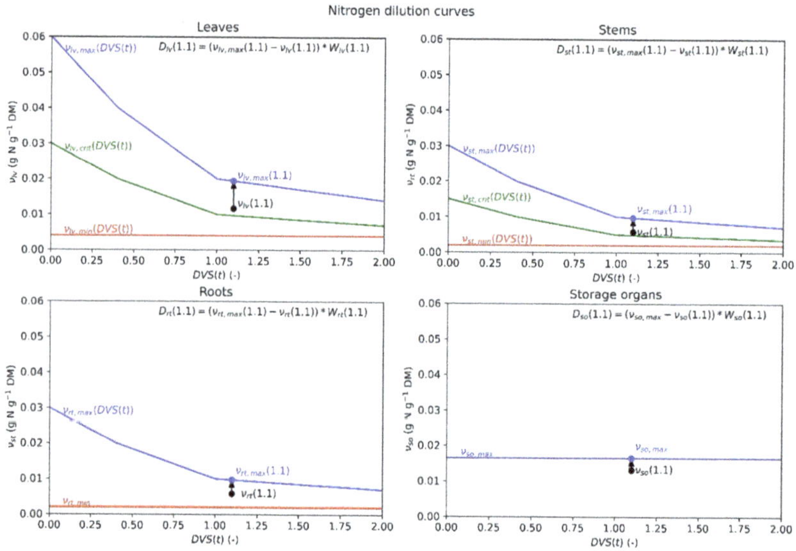

Figure 4 Nitrogen dilution curves for each organ for spring wheat in the FST version of LINTUL-3. For each organ (o), nitrogen dilution curves are defined as concentrations (kg N kg^{-1} DM) as a function of development stage. The N demand of each individual organ (D_o) is calculated as the amount of N that needs to be added to reach the maximum N concentration $\nu_{o,max}$ of that organ. As an example, the calculation of the N demands at DVS 1.1 for particular N organ concentrations ν_o are shown. The critical N concentration curves $\nu_{lv,crit}$ and $\nu_{st,crit}$ are the critical N concentrations of green organs (stems, leaves). If the actual N concentration of these organs drop below their critical N concentration, N stress occurs. The N concentrations of vegetative organs (leaves, stems, roots) cannot drop below their minimum N concentrations $\nu_{o,min}$. Vegetative organs fulfil their N demand by N uptake from the soil, while storage organs fulfil their demand by N translocation from vegetative organs.

degree of N stress is entirely determined by the N concentration (g N g^{-1} DM) of the green parts (green leaves lv and stems st) of the crop. The maximum N concentration of the green parts of the crop is calculated as

$$\nu_{max}(t) = \frac{\nu_{max,lv}\left(DVS(t)\right)\cdot\left(W_{lv}(t) + f_{max,stlv}\cdot W_{st}(t)\right)}{W_{lv}(t) + W_{st}(t)} \quad \text{[eq. 12]}$$

where $\nu_{max,lv}\left(DVS(t)\right)$ is a tabular function that calculates the maximum N concentration of the leaves for a given DVS. $f_{max,stlv}$ is the ratio of the maximum N stem concentration to the maximum N leaf concentration, which is assumed to remain constant during the season.

Crops are capable of taking up more N than necessary to achieve potential or water-limited production (Ågren, 1988; Justes et al., 1994). This is sometimes referred to as luxury N consumption (Gastal et al., 2015). Therefore, the crop production is not immediately affected if the N demand can no longer be met. However, if the N demand cannot be met for a longer period, the crop N concentration of the green parts of the crop (leaves, stems) drops below a critical N concentration and the dry matter production is reduced, compared to potential or water-limited production (Ågren, 1988; Greenwood et al., 1990). LINTUL3 calculates the critical N concentration of the green parts as

$$\nu_{crit}(t) = f_{crit}\cdot\nu_{max}(t) \quad \text{[eq. 13]}$$

where f_{crit} is the fraction of the critical and the maximum N concentration of the green parts of the crop. Once the N concentration of the green parts of the crops reaches its critical N concentrations, the dry matter production rate reduces relative to the potential or water limited dry matter production rate. It starts to experience N stress. Once the green organs only contain structural N, the N concentration of these green parts is at their minimum N concentration and dry matter production is no longer possible (Ågren, 1988). The minimum N concentration of the green parts of the crop is calculated as

$$\nu_{min}(t) = \frac{\nu_{min,lv}\cdot W_{lv}(t) + \nu_{min,st}\cdot W_{st}(t)}{W_{lv}(t) + W_{st}(t)} \quad \text{[eq. 14]}$$

LINTUL3 quantifies the degree of N-stress by the NNI. It is calculated as

$$\nu(t) = \frac{N_{lv}(t) + N_{st}(t)}{W_{lv}(t) + W_{st}(t)} \quad \text{[eq. 15]}$$

$$NNI(t) = \min\left(1, \frac{\nu(t) - \nu_{min}(t)}{\nu_{crit}(t) - \nu_{min}(t)}\right) \quad \text{[eq. 16]}$$

Where $N_{lv}(t)$ and $N_{st}(t)$ are the amounts of N in the leaves and stems. $v(t)$ is the N concentration at time t. $NNI(t)$ varies between 0 and 1. If $NNI(t)$ = 1, the dry matter production rate equals potential or water limited production. If $NNI(t)$ = 0, there is no dry matter production possible. Nitrogen stress occurs once the $NNI(t)$ drops below 1. This stress affects crop growth in in various ways. First, N deficiencies can result in a reduction of the daily rate of dry matter production relative to the potential dry matter production rate (Eq. 2), if the potential growth reduction owing to N stress is more severe than owing to water stress:

$$R_{Wtot}(t) = \begin{cases} f_{PAR} \cdot f_{int}(t) \cdot f_{CO2} \cdot RUE(DVS(t)) \cdot R_{tot}(t) \cdot e^{-r_{NLUE} \cdot NNI(t)}, e^{-r_{NLUE} \cdot NNI(t)} < TRANRF \\ f_{PAR} \cdot f_{int}(t) \cdot f_{CO2} \cdot RUE(DVS(t)) \cdot R_{tot}(t) \cdot TRANRF, e^{-r_{NLUE} \cdot NNI(t)} \geq TRANRF \end{cases}$$

[eq. 17]

where r_{NLUE} is a parameter that links the dry matter growth reduction to the $NNI(t)$. Additionally, N stress also results in a reduction of the growth of the leaf area index in juvenile growth stages; decreases the specific leaf area index; and can reduce the fraction of newly produced dry matter that is partitioned to the leaves and it enhances leaf mortality.

For the reproductive organs (i.e. the storage organs), a separate N demand is calculated. In LINTUL, storage organs satisfy their entire N demand (Fig. 4) by N translocation from vegetative organs rather than by uptake from the soil. If there is enough non-structural N available in the crop organs to satisfy the N demand of the storage organs, this demand is fulfilled. Otherwise, all available non-structural N is translocated, and the vegetative organs reach their minimum N concentration. A result of translocation, N is removed from the leaves and the stems, which lowers the N concentration of the green parts. If N translocation results in a combined leaf and stem concentration below the critical N concentration, the NNI drops below 1 and the growth of the storage organs is reduced. In PCSE–LINTUL3, this effect is reinforced by an additional parameter called DVSNLT, to limit N uptake (NLT) from the soil once the DVS of the crop is larger than DVSNLT.

6 Modelling nitrogen, potassium and phosphorus co-limitation

A similar approach based on critical dilution curves was developed for N, P and K in cassava (Adiele et al., 2021a). Following the equations for NNI above, nutrition indices for P (PNI) and K (KNI) are computed accordingly. In the LINTUL–cassava–NPK model (Adiele et al., 2022), the critical NPK concentrations are defined as a fraction of the maximum concentration (FR_{MAX}) with default parameter value 0.8:

$$crit_{N,P,K}(t) = \min_{N,P,K}(t) + FR_{MAX} \cdot \left(max_{N,P,K}(t) - min_{N,P,K}(t)\right) \quad \text{[eq. 18]}$$

Where $crit_{N,P,K}(t)$ is the concentration of either N, P or K in leaves, stems and storage organs on day *t*. Decreasing crop growth rate by a simple multiplication of NNI, PNI and KNI reduces crop growth rates strongly when there is moderate stress for all three nutrients, or if one nutrient is strongly limiting. This is not consistent with observations. If one nutrient is limiting, concentrations of the other nutrients also decrease. For example, when compared to the full treatment, the treatment of a full N and P dose and a reduced dose of K had lower N, P and K concentrations in all crop organs results in a NNI, PNI and KNI of about 0.8 (Adiele et al., 2021a). Multiplication of these three indices would result in a value of 0.51, which would translate in a 50% growth rate reduction, while crop biomass was only 20% lower (Adiele et al., 2022). Therefore, the three nutrition indices for N, P and K are combined into a NPKI index:

$$NPKI(t) = \begin{cases} \dfrac{(MM_{NI}+1)\cdot NNI(t)\cdot PNI(t)\cdot KNI(t)}{MM_{NI}+NNI(t)\cdot PNI(t)\cdot KNI(t)} & \text{when } MM_{NI} \le K_{max} \\[2ex] \dfrac{(MM_{NI}+1)\cdot\left(1-NNI(t)\cdot PNI(t)\cdot KNI(t)\right)}{MM_{NI}+\left(1-NNI(t)\cdot PNI(t)\cdot KNI(t)\right)} & \text{when } MM_{NI} > K_{max} \end{cases} \quad \text{[eq. 19]}$$

The values for the Michaelis-Menten (or Monod) parameter (MM_{NI}) and K_{max} were calibrated and determined below what NPKI value growth reduction occurs. The MM_{NI} value determines the curvature of the relationship between ($NNI(t)\cdot PNI(t)\cdot KNI(t)$) and *NPKI(t)*, while the K_{max} determines if the curve is above or below the 1:1 for a specific value of MM_{NI} (Fig. 5).

This approach is flexible, i.e. it allows for adjusting the combined impact of nutrient deficiency on crop growth depending on the situation. For example, growth rate reductions as a function of ($NNI(t)\cdot PNI(t)\cdot KNI(t)$) can range from nearly linear to strongly curvilinear. This approach worked well for cassava (Adiele et al., 2022).

Concentrations of N, P and K in crop organs are only near their maximum values when all nutrients are abundantly available. If one or more nutrients are limiting, uptake of other nutrients is also reduced and concentrations in crop organs are well below their maximum, as observed by Adiele et al. (2021a). To capture this behaviour, the nutrient equivalent approach as developed by Janssen (2011) was adopted. Both nutrient demand (NED) and soil supply

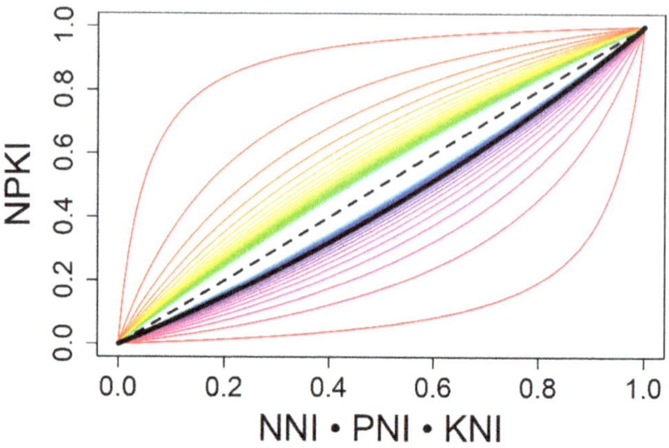

Figure 5. The NPKI as function of the product of NNI, PNI and KNI for MM_{NI} values ranging from 0 to 8, where the value for K_{max} was set at 4.15. MM_{NI} values < K_{max} produced lines from red, yellow green to cyan above and values > K_{max} lines from cyan, blue, purple to red below the 1:1 line. The striped line indicates the 1:1 line, the thick solid black line reflects the curve for Cassava using the calibrated parameter value for MM_{NI} of 5.86

(NES) were expressed in nutrient equivalents (NE) based on actual (Act) and maximum (Max) N, P and K amounts (g m^{-2}) in the crop:

$$NED(t) = \left(Max_N(t) - Act_N(t)\right) + \frac{Max_P(t) - Act_P(t)}{Max_P(t)} \cdot Max_N(t) + \frac{Max_K(t) - Act_K(t)}{Max_K(t)} \cdot Max_N(t)$$

[eq. 20]

For soil supply, the actual plant available amounts (A, g m^{-2}) were used:

$$NES(t) = A_N(t) + \frac{A_P(t)}{Max_P(t)} \cdot Max_N(t) + \frac{A_K(t)}{Max_K(t)} \cdot Max_N(t) \quad \text{[eq. 21]}$$

Based on these NE values, the uptake rate for NE were determined:

$$\frac{dNE_{UP}}{dt}(t) = RNE \cdot \min\left(NES(t), NED(t)\right) \quad \text{[eq. 22]}$$

Here, RNE (day^{-1}) is the highest recorded relative nutrient equivalent uptake rate in experiments.

These daily NE uptakes were converted to actual daily uptakes of N, P and K, using ratios of maximum contents in the crop. This approach limits the uptake of a nutrient that is in excess in the soil when other nutrients are deficient. It resulted in reasonably accurate total uptakes of N, P and K (Adiele et al., 2022).

The nutrient supply from the soil was derived from crop N, P and K uptakes in control and omission treatments. Uptakes in omission plots are nearly always higher than in control plots: plants not only dilute a single-deficient element more strongly but may also strategize to increase uptake that differs for N, P and K. A stronger dilution will also result in more growth per kilogram of uptake, hence more resources are available for additional root growth and more soil exploration. Such detailed processes are not modelled explicitly. Fertilizer N, P and K are applied in events and take time to dissolve in soil water and become available to plants with some delay that strongly differs between nutrients. The potential recovery fraction of applied N, P and K that is available for uptake differs between nutrients and depends not only on soil conditions, such as pH, soil water contents and soil nutrient contents, but also on management (placed, banded or broadcasted application, covered/not covered etc.). These processes are not modelled, but a simple approach was used based on a potential recovery and a typical release pattern of both soil and fertilizer N, P and K. This is obviously a major simplification of reality that needs further improvement. More explicit approaches are used in other models, where a large number of pools are needed to capture dynamics in soil N, P and K. N and P both have organic and inorganic components interacting with soil conditions. K is not part of soil organic matter yet can be exchanged with the cation exchange complex and clay lattices, hence plant availability depends also on the concentration of other cations in the soil.

7 LINTUL-derived models

The overarching characteristic of the LINTUL family of models is that they are all based on the RUE concept. Differences in the model simulation approach among implementations and production levels can be subtle as well as more fundamental. A wide range of LINTUL models has been described in literature (Table 1). The model code of various versions with additional information can be found on the repositories as indicated in Table 1. These models are provided under an EUPL-1.2. license that allows users to download, use and adapt these models free of charge. This chapter provides only a brief overview, as e.g., both SIMPLACE (Enders et al., 2023) and MULTICROP (Yetik et al., 2023) also use LINTUL components.

The LINTUL models have been implemented in a range of software environments, including Fortran Simulation Environment (FSE) (van Kraalingen, 1995) and related Fortran Simulation Translator (FST) (van Kraalingen et al., 2003). The latter is relatively simple and more suitable for teaching purposes than FSE. Currently, also R and Python versions[1] are available and used by many students. These environments offer a range of integration methods. The LAI growth rate early in the growing season, when LAI < 0.75 and TSUM < 300, uses

1 https://pcse.readthedocs.io

an exponential function. The calculated growth rates of LAI are very sensitive to the integration method and the integration errors made. Therefore, all LINTUL models with this exponential function should be solved by an Euler integration with a time-step of 1 day.

Despite the success of the LINTUL concept, a problematic aspect of the LINTUL model family is a lack of standardisation and consistency between different LINTUL production levels and the absence of a clear model versioning scheme. For example, the exact implementation of the crop growth processes between potential production (LINTUL1) and water-limited production (LINTUL2) differs slightly at the code level. This is more prominent for nutrient-limited production levels (LINTUL3, LINTUL4) with considerable differences in the way growth limitations and phenology have been implemented. One of the practical problems is that parameter sets are currently not exchangeable between LINTUL production levels so, e.g., a parameter set for a LINTUL3 model may require adaptations for LINTUL2 and LINTUL1.

Other aspects to consider are that the grassland models LINGRA, and LINTUL-Cassava, as well as the potato-specific version LINTUL-POTATO, are essentially quite different from the other LINTUL models. The LINGRA model uses source-sink concepts and implements detailed growth and tillering processes that are specific for grass crops (Schapendonk et al., 1998), which are not relevant for LINTUL models that simulate other crop species. LINTUL-Cassava includes a description of tuber dormancy (Ezui et al., 2018), which is specific for cassava and allows the crop to survive longer dry periods. Similarly, LINTUL-POTATO includes detailed functions to quantify the impacts of photoperiod and temperature on potato physiology, which is absent in other LINTUL models. The LINTUL models have not been developed specifically to evaluate or analyse differences between genotypes. Owing to the described differences in concepts between crops, the LINTUL models vary in their capacity to capture differences between cultivars within a crop type. Important differences in light interception capacity or crop phenology can be quantified by changing key parameters. For many crops, the required phenological parameters are the same for LINTUL and WOFOST, and parameter values can be derived from existing WOFOST parameter sets.

8 Use of LINTUL models in teaching

LINTUL is used in current bachelor and master courses of the plant sciences curriculum of Wageningen University. The LINTUL concepts are taught in a few lectures, while students explore scenarios and evaluate how changes in parameters impact leaf area indices, crop growth rates and yields in a series of hands-on practical sessions.

The advantage of LINTUL is that the list of processes and parameters is short, which allows students to fully understand what happens in a short period of time. First, students learn the key concepts related to a potential production situation and use these to understand how differences in radiation, ambient CO_2 concentrations and temperature affect crop growth and crop development rates during the growing season, and how this translates into differences in biomass and yield at full maturity. Second, the water-limited production situation is used to study and explore how temperature, radiation and vapour pressure deficits determine water demand by the crop. Students learn about how the interaction between crops and soil determines growth rate reduction.

The relatively simple concepts provide excellent means to show how early growth rate reductions affect root, stem and leaf biomass, and consequently leaf area index, transpiration and evaporation rates later in the season. This is a first step towards a better insight in the importance of the timing of drought stress and how this affects harvest index values and water use efficiency. LINTUL simulations are particularly useful for students to explore such interactions in a range of different soil–climate contexts.

9 Conclusion and future improvements

Summary models, such as LINTUL, are not only valuable for didactical purposes but also provide robust simulation capability for a wide range of crop species, environments and applications. For instance, LINTUL models were able to reproduce measured dry matter growth and yields of rice grown at various locations in Asia (Shibu et al., 2010), various winter wheat cultivars grown in the Netherlands in different time periods (1982–1984 and 2013–2015) (Berghuijs et al., 2023), maize grown in North-eastern Spain under different irrigation regimes (Farré et al., 2000), and casava grown in Nigeria under various N, P and K fertilisation regimes (Adiele et al., 2022). The core of LINTUL has not changed much over all these years and provides a robust basis. The approach to keep it as simple as possible and use Occam's razor forces modellers to focus on the most important processes and leave out less important ones. Naturally, this also means that there is a somewhat larger gap between the model and reality, leaving less room to adjust parameter values to better fit the model to observations. For students new to crop growth modelling, the LINTUL approach enables them to explore and understand how a field crop responds to its environment and how various main factors interact, without getting distracted by all minor or detailed processes. For example, temperature is an important factor that determines the LAI growth rate early in the season (when the LAI is below 0.75) and the crop development rate. Temperature affects both assimilation and respiration processes, yet with a small net effect in the normal temperature range. Therefore, these more detailed processes were left out and temperature does not appear in the crop growth rate equation (eq. 17).

One of the challenges is the harmonisation of the different models that apply the LINTUL approach in order to obtain a more consistent suite of LINTUL-based models that can be used for education as well as research purposes. Harmonizing LINTUL-based models will provide many benefits, such as (1) the possibility to harmonize the sets of crop parameters that can be used for different production levels; (2) a reduction in the amount of code for model implementation; (3) more consistent documentation regarding processes, variables and model parameters; and (4) making it easier to interface the model with other submodules particularly those related to soil water, nutrients and carbon. An improved implementation and harmonisation would also address one of the weaker aspects of the current LINTUL implementations which is the simple tipping bucket water balance which does not allow to account for groundwater or lateral flows. In the PCSE implementation, various other soil modules can be chosen that can better describe layered soils with a matrix potential approach to quantify upward and downward water fluxes, broadening the scope of the model. Some weaker aspects related to phenology have also been addressed in the recent PCSE versions (Berghuijs et al., 2023). Effects of extreme temperatures are likely poorly captured, although adjustment of RUE based on daily average temperatures is possible in LINTUL4. However, effects of more extreme temperatures on crop growth and phenological development rates have not been addressed fully and are also out of scope as other models are better suited for explorative studies related to e.g. climate change.

Providing better interfaces between LINTUL-based models and models for carbon, water and nutrients in the soil also provides opportunities for simulating advanced crop management strategies for estimating nutrient requirements. In many parts of the world, co-limitations between various nutrients occur. A better understanding of the interactions between these co-limitations and drought is needed to explore how crop management can be improved and to evaluate how much nutrients are removed from the field in more or less dry years. The recently developed LINTUL-Cassava-NPK model (Adiele et al., 2022) provides insights in interactions between nutrients. It is one of the first crop growth models to account for N, P and K limitations under field conditions. Its nutrient equivalents approach deviates from those used for e.g. LINTUL5 and suggests that plants balance rather than maximize uptake.

Finally, one of the attractive features of the LINTUL-approach is that the canopy light interception computed by the model can be easily replaced by external estimates. Satellites routinely make observations of canopy reflectance over agricultural fields and operational algorithms exist to derive the fraction of absorbed PAR (fAPAR) from canopy reflectance (Weiss et al., 2020). This satellite-derived fAPAR can be used to replace the internal calculation of light interception based on Equation 1. Time-series of satellite fAPAR estimates can therefore drive a model based on the LINTUL approach and replace a simulated estimate of light interception with an external observation. The use of fAPAR reduces the number of model parameters and uncertainties related to estimating LAI, and can capture real-time stresses normally not captured by LINTUL, for example pests and diseases that affect ground cover, crop chlorophyll content or leaf vigour. This approach has already been implemented in operational products that combine a LINTUL approach with satellite fAPAR estimates to monitor crop productivity over large areas (Chevuru et al., 2023).

10 Where to look for further information

Interested readers can find a short history and background of the Wageningen crop growth models in Bouman et al. (1996). The books of van Keulen and Seligman (1987), Penning de Vries et al. (1989) and Soltani and Sinclair (2012), together with articles of van Ittersum and Rabbinge (1997) and van Ittersum et al. (2003) provide good starting points to concepts and developments in field crop models. The model repository models.pps.wur.nl provides source code and background information for selected models, including a series of short introduction videos for LINTUL 1 and 2.

11 References

Adiele, J.G., et al., 2020. Towards closing cassava yield gap in West Africa: agronomic efficiency and storage root yield responses to NPK fertilizers. *Field Crops Research* 253, 107820.

Adiele, J.G., et al., 2021a. Dynamics of N-P-K demand and uptake in cassava. *Agronomy for Sustainable Development* 41, 1–14.

Adiele, J.G., et al., 2021b. A recalibrated and tested LINTUL-Cassava simulation model provides insight into the high yield potential of cassava under rainfed conditions. *European Journal of Agronomy* 124, 126242.

Adiele, J.G., et al., 2022. LINTUL-Cassava-NPK: a simulation model for nutrient-limited cassava growth. *Field Crops Research* 281, 108488.

Ågren, G.I., 1988. Ideal nutrient productivities and nutrient proportions in plant growth. *Plant, Cell & Environment* 11, 613–620.

Arqués Viver, I. (2022). Translation and evaluation of the LINTUL2-Banana crop growth model in R. MSc thesis, Wageningen University, Wageningen, the Netherlands.

Barrett, P.D., et al., 2005. GrazeGro: a European herbage growth model to predict pasture production in perennial ryegrass swards for decision support. *European Journal of Agronomy* 23, 37–56.

Berghuijs, H.N.C., et al., 2023. Catching-up with genetic progress: simulation of potential production for modern wheat cultivars in the Netherlands. *Field Crops Research* 296, 108891.

Berghuijs, H.N.C., et al., 2024. Expanding the WOFOST crop model to explore options for sustainable nitrogen management: a study for winter wheat in the Netherlands. *European Journal of Agronomy* 154, 127099.

Bouman, B.A.M., et al., 1996. The 'School of de Wit' crop growth simulation models: a pedigree and historical overview. *Agricultural Systems* 52, 171–198.

Bouman, B.A.M., et al., 2001. *ORYZA2000: modeling lowland rice.* International Rice Research Institute, Wageningen University and Research, Los Baños (Philippines) and Wageningen (the Netherlands).

Ceglar, A., et al., 2019. Improving WOFOST model to simulate winter wheat phenology in Europe: Evaluation and effects on yield. *Agricultural Systems* 168, 168–180.

Chevuru, S., et al., 2023. Copernicus global crop productivity indicators: an evaluation based on regionally reported yields. *Climate Services* 30, 100374.

Connor, D.J., et al., 2011. *Crop ecology: productivity and management in agricultural systems.* Cambridge University Press, Cambridge.

de Koeijer, K.J., & van der Werf, W. 1995. *Effect of beet yellowing viruses on light interception and light use efficiency of the sugarbeet crop. Crop Protection*, 14, 291–297. https://doi.org/10.1016/0261-2194(94)00003-Q.

Enders, A., et al., 2023. SIMPLACE–a versatile modelling and simulation framework for sustainable crops and agroecosystems. *In Silico Plants* 5, diad006.

Ezui, K.S., et al., 2018. Simulating drought impact and mitigation in cassava using the LINTUL model. *Field Crops Research* 219, 256–272.

Farré, I., et al., 2000. Analysis of maize growth for different irrigation strategies in northeastern Spain. *European Journal of Agronomy* 2000, 225–238.

Gastal, F., et al., 2015. Chapter 8 - Quantifying crop responses to nitrogen and avenues to improve nitrogen-use efficiency. In: Sadras, V.O. and Calderini, D.F. (Eds.), *Crop physiology* (2nd ed.). Academic Press, San Diego, pp. 161–206.

Gimplinger, D.M. and Kaul, H.P., 2009. Calibration and validation of the crop growth model LINTUL for Grain amaranth (Amaranthus sp.). *Journal of Applied Botany and Food Quality* 82, 183–192.

Greenwood, D.J., et al., 1990. Decline in percentage N of C3 and C4 crops with increasing plant mass. *Annals of Botany* 66, 425–436.

Groot, J.J.R., 1987. *Simulation of nitrogen balance in a system of winter wheat and soil.* Centre for Agrobiological Research (CABO): Department of Theoretical Production Ecology, Agricultural University, Wageningen.

Groot, J.J.R. and de Willigen, P., 1991. Simulation of the nitrogen balance in the soil and a winter wheat crop. *Fertilizer Research* 27, 261–271.

Habekotté, B., 1996. *Winter oilseed rape: analysis of yield formation and crop type design for higher yield potential.* Landbouwuniversiteit Wageningen, Wageningen.

Haverkort, A.J., et al., 2013. Climate change and potato production in contrasting South African agro-ecosystems 1. Effects on land and water use efficiencies. *Potato Research* 56, 31–50.

Heinen, M., et al., 2024. SWAP 50 years: advances in modelling soil-water-atmosphere-plant interactions. *Agricultural Water Management* 298, 108883.

Hoogenboom, G., et al., 2023. *Decision support system for agrotechnology transfer (DSSAT) version 4.8.2.* DSSAT Foundation, Gainesville, Florida, FL. www.DSSAT.net

Janssen, B.H., 2011. Simple models and concepts as tools for the study of sustained soil productivity in long-term experiments. II. Crop nutrient equivalents, balanced supplies of available nutrients, and NPK triangles. *Plant Soil* 339, 17–33.

Jones, J.W., et al., 2003. The DSSAT cropping system model. *European Journal of Agronomy* 18, 235–265.

Justes, E., et al., 1994. Determination of a critical nitrogen dilution curve for winter wheat crops. *Annals of Botany* 74, 397–407.

Kooman, P.L. and Haverkort, A.J., 1995. Modelling development and growth of the potato crop influenced by temperature and daylength: LINTUL-POTATO. In: Haverkort, A.J. and MacKerron, D.K.L. (Eds.), *Potato Ecology and modelling of crops under conditions limiting growth*. Kluwer Academic Publishers, Netherlands, pp. 41–60.

Lemaire, G., 1997. *Diagnosis of the nitrogen status in crops*. Springer, Berlin.

Makowski, D., et al., 2020. Analyzing uncertainty in critical nitrogen dilution curves. *European Journal of Agronomy* 118, 126770.

Monteith, J.L., 1977. Climate and efficiency of crop production in Britain. *Philosophical Transactions of the Royal Society of London Series B-Biological Sciences* 281, 277–294.

Monteith, J.L., 1996. The quest for balance in crop modeling. *Agronomy Journal* 88, 695–697.

Peng, S., et al., 1995. Relationship between leaf photosynthesis and nitrogen content of field-grown rice in tropics. *Crop Science* 35, 1627–1630.

Penning de Vries, F.W.T., et al., 1989. *Simulation of ecophysiological processes of growth in several annual crops*. Pudoc, Wageningen.

Penman, H.L., 1948. Natural evaporation from open water, bare soil and grass. *Proceedings of the Royal Society of London. Series A, Mathematical and Physical Sciences* 193, 120–145.

Rappoldt, C. and van Kraalingen, D.W.G., 1990. *Reference manual of the FORTRAN utility library TTUTIL with applications*. CABO-DLO, Wageningen.

Rodriguez, D., et al., 1999. LINGRA-CC: a sink-source model to simulate the impact of climate change and management on grassland productivity. *New Phytologist* 144, 359–368.

Schapendonk, A.H.C.M., et al., 1998. LINGRA, a sink/source model to simulate grassland productivity in Europe. *European Journal of Agronomy* 9, 87–100.

Shibu, M.E., et al., 2010. LINTUL3, a simulation model for nitrogen-limited situations: application to rice. *European Journal of Agronomy* 32, 255–271.

Sinclair, T.R. 1994. Limits to Crop Yield?, *Physiology and Determination of Crop Yield*, 509–532. https://doi.org/10.2134/1994.physiologyanddetermination.c33.

Sinclair, T.R. and Muchow, R.C., 1999. Radiation use efficiency. *Advances in Agronomy* 65, 215–265.

Soltani, A. and Sinclair, T.R., 2012. *Modeling physiology of crop development*, Growth and Yield. CABI, Wallingford.

Spitters, C.J.T. and Schapendonk, A.H.C.M., 1990. Evaluation of breeding strategies for drought tolerance in potato by means of crop growth simulation. *Plant Soil* 123, 193–203.

Taulya, G., 2015. *Ky'osimba onaanya: understanding productivity of East African highland banana. Plant production systems*. Wageningen University, Wageningen.

van Haren, R.J.F. and Haverkort, A.J., 1998. Description and application of the LINTUL-POTATO crop growth model. In: Stoorvogel, J.J., et al. (Eds.), *Information technology as a tool to assess land use options in space and time: proceedings of an international workshop, Lima, September 28–October 4, 1997*. AB-DLO, Wageningen.

van Ittersum, M.K. and Rabbinge, R., 1997. Concepts in production ecology for analysis and quantification of agricultural input-output combinations. *Field Crops Research* 52, 197–208.

van Ittersum, M.K., et al., 2003. On approaches and applications of the Wageningen crop models. *European Journal of Agronomy* 18, 201–234.

van Keulen, H. and Seligman, N.G., 1987. *Simulation of water use, nitrogen nutrition and growth of a spring wheat crop*. Pudoc, Wageningen.

van Kraalingen, D.W.G., 1995. *The FSE system for crop simulation, version 2.1*. AB-DLO, Wageningen.

van Kraalingen, D.W.G., et al., 2003. The Fortran simulation translator, a simulation language. *European Journal of Agronomy* 18, 359–361.

Van Laar, H.H., et al., 1997. *SUCROS97: simulation of crop growth for potential and water-limited production situations: as applied to spring wheat*. AB-DLO, Wageningen, p. 52.

Wang, E., et al., 2017. The uncertainty of crop yield projections is reduced by improved temperature response functions. *Nature Plants* 3, 17102.

Weiss, M., et al., 2020. S2ToolBox level 2 products: LAI, FAPAR, FCOVER. Version 2.1. https://step.esa.int/docs/extra/ATBD_S2ToolBox_V2.1.pdf

Wolf, J., 2002. Comparison of two potato simulation models under climate change. I. Model calibration and sensitivity analyses. *Climate Research* 21, 173–186.

Wolf, J., 2012. *User guide for LINTUL4 and LINTUL4V: simple generic model for simulation of crop growth under potential, water limited and nitrogen limited conditions*. Wageningen University, Wageningen.

Yetik, A.K., et al., 2023. Simulating the yield responses of sugar beet to different climate change scenarios by LINTUL-MULTICROP model. *Black Sea Journal of Engineering and Science* 6, 53–59.

Chapter 14

SWAPWOFOST: a combined ecohydrological and crop growth simulation model

I. Supit, P. Groenendijk, J. van Dam, M. Mulder and M. Heinen, Wageningen University and Research, The Netherlands

1 Introduction

Water and nitrogen represent the two principal constraining elements influencing crop growth that can potentially be managed. Given the increasing competition for water resources and growing concerns regarding nitrate pollution in the environment, agricultural practices necessitate the optimization of modifiable growth factors. The intricate cycling of nitrogen in the soil/plant system encompasses a multitude of pathways, states, and regulatory processes. All associated fluxes and transformations are intertwined

http://dx.doi.org/10.19103/AS.2025.0155.17

with water availability and moisture conditions. Due to the potential leaching of nitrate beyond the root zone through irrigation, a holistic optimization approach is imperative, precluding the consideration of individual variables in isolation.

The evaluation of the combinations involving temporal and volumetric aspects of both water and fertilizer applications demands the integration of a mathematical model. A secondary objective of such model development is to furnish a research tool for scrutinizing the dynamics of water and nitrogen behaviour within the soil–plant system. While empirical relationships enable the construction of a predictive model tailored to specific crops and regions, a universally applicable model necessitates the substitution of empirical relations with robust scientific principles.

Over the past four decades, a series of simulation models have been developed and utilized to simulate crop growth under varying conditions of water and nutrient availability. The initial model description of SWATR, which preceded SWAP, dates back to 1978 (Feddes et al., 1978), while the precursor to WOFOST was documented by Van Keulen and Wolf (1986) and subsequently expanded by Van Diepen et al. (1989). These models are typically designed for specific climatic zones and soil conditions, reflecting the data used for their parameterization and validation. The structural framework of these models is often tailored to the particular geographic region of interest to the developers. Developers with expertise in botany and agronomy tend to provide detailed representations of crop development while treating soil water flow in a more generalized manner. In contrast, modelers with a background in soil science often simplify the descriptions of crop development, focusing more on the organic matter cycle and mineralization processes. In the 1990s, the aboveground routines of WOFOST were integrated into SWAP (Van Dam et al., 1997), augmenting the existing simplified crop growth routines. This integration allowed crop growth simulations to benefit from a more detailed modelling of soil moisture, including capillary rise and nutrient availability, while the simulation of water and solute transport processes gained from a more accurate depiction of aboveground soil cover, incorporating factors such as leaf area index (LAI), transpiration, and soil evaporation.

Our perspective underscores the conviction that, for an effective portrayal of water and nitrogen constraints on crop growth, a comprehensive depiction of both soil water flow and crop development is imperative. Additionally, addressing nitrogen availability in the soil necessitates consideration of the organic matter cycle and the dynamic processes of nitrogen mineralization.

The soil nitrogen module exchanges information with the soil water model SWAP and the crop growth model WOFOST on a daily basis. The interactions and dependencies between the module soil nitrogen, crop nitrogen module, and the SWAP and WOFOST models are illustrated in Fig. 1.

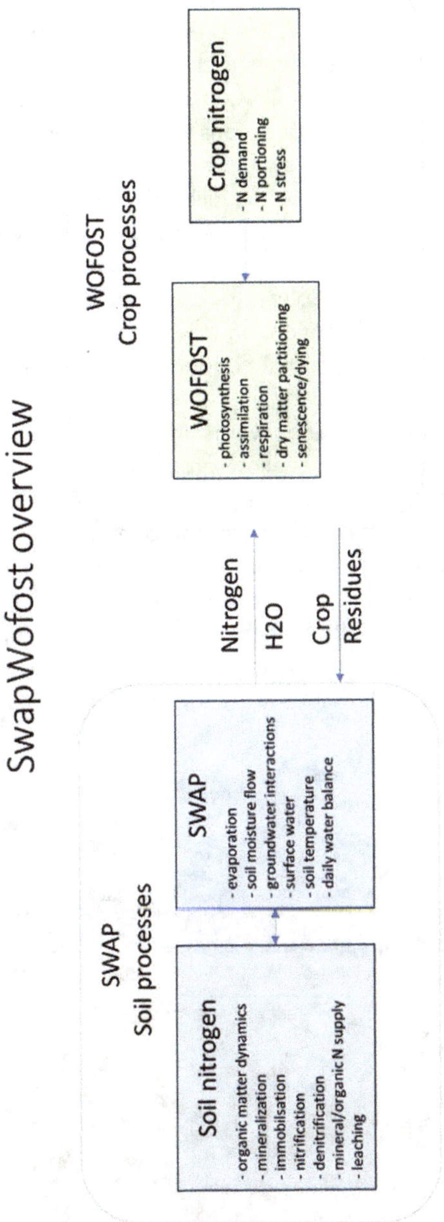

Figure 1 Overview of the SwapWofost model combination with its main components.

2 SWAP model overview

The ecohydrological model SWAP and its predecessors (Feddes et al., 1978; Belmans et al., 1983; Wesseling et al., 1991; Kabat et al., 1992; Van den Broek et al., 1994; Van Dam et al., 1997; Kroes et al., 2001, 2003, 2008) are used by researchers, engineers, and students in the field of environmental science to study the interactions between soil, water, plant, and atmosphere. SWAP can be adapted to specific research questions in an easy way. It simulates transport of water, solutes, and heat in the vadose zone in interaction with vegetation development. The model employs the Richards equation including root water extraction to simulate soil moisture movement in variably saturated soils. In addition to ordinary matrix flow, flow through macropores as in clay and peat soils can be accounted for. Solute transport includes the basic processes convection, dispersion, adsorption, and decomposition. Taking into account actual heat capacities and thermal conductivities the soil heat flow can be simulated as well.

The model domain extends vertically from a plane just above the canopy to a plane within the shallow groundwater, as illustrated in Fig. 2. Within this zone, transport processes primarily occur in the vertical direction. Consequently, SWAP is designed as a one-dimensional, vertically oriented model in this context. The flow below the groundwater level may include lateral drainage fluxes, which should be prescribed with analytical drainage formulas. The model exhibits high flexibility regarding input data for both the upper and lower extents of the soil column. Typically, daily weather conditions are adequate for input at the top. In the case of Nordic conditions,

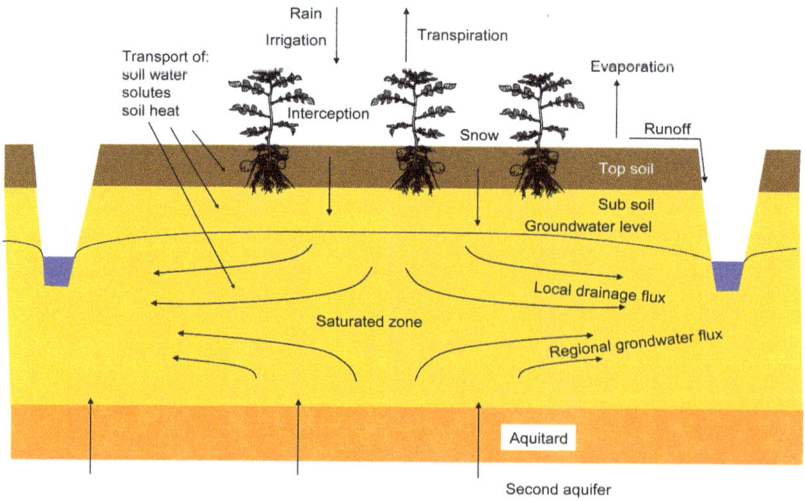

Figure 2 SWAP model domain and transport processes.

a straightforward snow-storage module has been incorporated to enhance accuracy. For specific studies that include, for example, runoff or diurnal transpiration fluxes, evapotranspiration and rainfall data can be specified in more detail. At the bottom, various forms of head and flux-based conditions are used.

In the horizontal direction, SWAP's main focus is the field scale. At this scale, most transport processes can be described in a deterministic way, as a field generally can be represented by one microclimate, one vegetation type, one soil type, and one drainage condition. Also cultivation practices occur at field scale, which means that many management options apply to this scale. To extend the analysis from the field to the regional scale for comprehensive management policy studies, geographical information systems can be employed for upscaling.

3 SWAP modelling of water movement through soil

3.1 Basic equations

Soil water potential gradients cause water movement through the soil and Darcy's equation is commonly used to quantify this movement. For one-dimensional vertical flow, Darcy's equation can be written as:

$$q = -K(h)\frac{\partial(h+z)}{\partial z} \tag{1}$$

where q is soil water flux density (positive upward) ($cm^3\ cm^{-2}\ d^{-1} = cm\ d^{-1}$), K(h) is hydraulic conductivity ($cm\ d^{-1}$), h is soil water pressure head (cm), and z is the vertical coordinate (cm), taken positively upward. Considering an amount of water in a tiny bit of soil, the continuity equation for soil water can be described by

$$\frac{\partial \theta}{\partial t} = -\frac{\partial q}{\partial z} - S_a(h) - S_d(h) - S_m(h) \tag{2}$$

where θ is volumetric water content ($cm^3\ cm^{-3}$), t is time (d), $S_a\,(h)$ is soil water extraction rate by plant roots ($cm^3\ cm^{-3}\ d^{-1}$), $S_d\,(h)$ is the extraction rate by drain discharge in the saturated zone ($cm^3\ cm^{-3}\ d^{-1}$), and $S_m\,(h)$ is the exchange rate with macro pores ($cm^3\ cm^{-3}\ d^{-1}$). Combination of Equations (1) and (2) provides the general water flow equation in variably saturated soils, known as the Richards equation:

$$\frac{\partial \theta}{\partial t} = \frac{\partial\left[K(h)\left(\frac{\partial h}{\partial z}+1\right)\right]}{\partial z} - S_a(h) - S_d(h) - S_m(h) \tag{3}$$

SWAP applies Richards equation integrally for the unsaturated-saturated zone, including possible transient and perched groundwater levels. SWAP solves Equation (3) numerically, using specified boundary conditions and relations between θ, *h*, and *K*.

3.2 Soil hydraulic functions

SWAP uses the Mualem-Van Genuchten functions (Mualem, 1976; Van Genuchten, 1980) which have been used in numerous studies. The analytical θ(*h*) function proposed by Van Genuchten (1980) reads:

$$\theta = \theta_{res} + \left(\theta_{sat} - \theta_{res}\right)\left(1 + |\alpha h|^{n}\right)^{-m} \tag{4}$$

where θ_{sat} is the saturated water content ($cm^3\ cm^{-3}$), θ_{res} is the residual water content in the very dry range ($cm^3\ cm^{-3}$) and α (cm^{-1}), n (−) and m (−) are empirical shape factors. Without losing much flexibility, *m* can be taken equal to

$$m = 1 - \frac{1}{n} \tag{5}$$

Using the above θ(*h*) relation, with *m* = 1 − 1/n, and applying the theory on unsaturated hydraulic conductivity by Mualem (1976), the following *K*(θ) function results:

$$K = K_{sat} S_e^{\lambda} \left[1 - \left(1 - S_e^{\frac{1}{m}}\right)^{m}\right]^2 \tag{6}$$

where K_{sat} is the saturated conductivity ($cm\ d^{-1}$), λ is a shape parameter (−) depending on flow path tortuosity, and S_e is the relative degree of saturation defined as

$$S_e = \frac{\theta - \theta_{res}}{\theta_{sat} - \theta_{res}} \tag{7}$$

The numerical solution to the Richards equation requires an expression of the differential water capacity *C* (cm^{-1}), which is defined as the derivative of θ to *h* (Equation (4)):

$$C = \frac{\partial \theta}{\partial h} = \alpha\, m\, n\, |\alpha h|^{n-1} \left(\theta_{sat} - \theta_{res}\right)\left(1 + |\alpha h|^{n}\right)^{-(m+1)} \tag{8}$$

4 SWAP modelling of wetting and drying cycles

Hysteresis refers to the non-uniqueness of the θ(*h*) relation and is caused by variations of the pore diameter (inkbottle effect), differences in radii of advancing and receding meniscus, entrapped air, thermal gradient,s and swelling/shrinking processes (Hillel, 1980; Feddes et al., 1988). Gradual desorption of an initially saturated soil sample gives the main drying curve, while slow absorption of an initially dry sample results in the main wetting curve. In the field partly wetting and drying occurs in numerous cycles, resulting in so-called drying and wetting scanning curves lying between the main drying and the main wetting curves (Fig. 3).

In simulation practices, the primary drying curve is frequently used to depict the relationship between water content (θ) and soil suction (*h*). This simplification is primarily driven by the significant time and cost investments required for measuring the complete θ(*h*) relationship. This comprehensive assessment involves capturing the main wetting, main drying, and scanning curves, particularly in the dry range. However, relying solely on the main drying curve for simulating infiltration events may lead to inaccuracies. To counter act this deficiency, the scaling method of Scott et al. (1983) has been implemented. This method rescaled the main wetting or the main drying curve to the actual water content to derive the scanning curves.

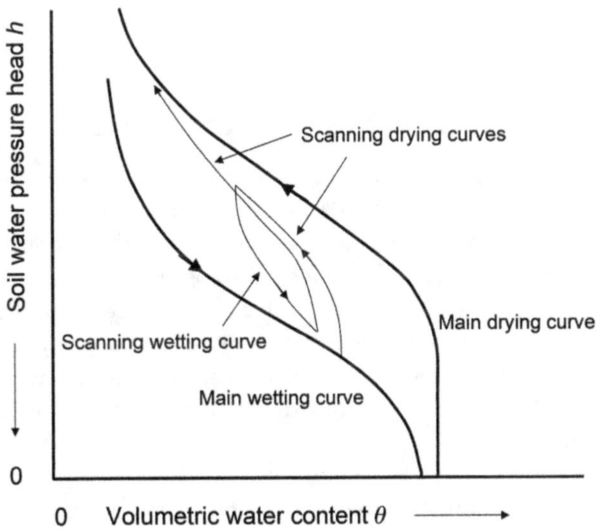

Figure 3 Water retention function with hysteresis, showing the main wetting, main drying, and scanning curves.

Defining θ_{md} as the water content of the main drying curve at h_{act} and θ^*_{sat} as the saturated water content of the drying scanning curve. Linear scaling of the main drying curve with respect to the vertical axis $\theta = \theta_{res}$ gives:

$$\frac{\theta^*_{sat} - \theta_{res}}{\theta_{sat} - \theta_{res}} = \frac{\theta_{act} - \theta_{res}}{\theta_{md} - \theta_{res}} \quad \Rightarrow \quad \theta^*_{sat} = \theta_{res} + \left(\theta_{sat} - \theta_{res}\right)\frac{\theta_{act} - \theta_{res}}{\theta_{md} - \theta_{res}} \qquad (9)$$

The only unknown in Equation (9) is θ^*_{sat}, which can be solved directly. The drying scanning curve is described accordingly with the parameters (α_{dry}, n, θ_{res}, θ^*_{sat}). As long as the soil keeps drying, this drying scanning curve is valid.

When the soil gets wetter, an arbitrary actual water content θ_{act} at the soil water pressure head h_{act} is defined. Subsequently, θ_{mw} is defined as the water content of the main wetting curve at h_{act}, and θ_{res*} as the residual water content of the wetting scanning curve. Linear scaling of the main wetting curve with respect to the vertical axis $\theta = \theta_{sat}$ leads to

$$\frac{\theta_{sat} - \theta^*_{res}}{\theta_{sat} - \theta_{res}} = \frac{\theta_{sat} - \theta_{act}}{\theta_{sat} - \theta_{mw}} \quad \Rightarrow \quad \theta^*_{res} = \theta_{sat} - \left(\theta_{sat} - \theta_{res}\right)\frac{\theta_{sat} - \theta_{act}}{\theta_{sat} - \theta_{mw}} \qquad (10)$$

From Equation (10), θ^*_{res} can be directly solved. The wetting scanning curve is accordingly described with the parameters (α_{wet}, n, θ^*_{res}, θ_{sat}) and is valid as long as the soil keeps wetting. As the wetting–drying history is different at each soil depth, each node may show a different scanning curve. The unique $K(\theta)$ relation of a soil layer always follows from the parameter set (n, θ_{res}, θ_{sat}, K_{sat}, λ) according to Eq. (6).

5 SWAP model: lower boundary conditions

Solving the Richards' equation requires starting values for the unknown pressure head (initial condition) and requires information on what applies at the boundaries of the isolated system that is being modelled. The bottom boundary of the one-dimensional column is either in the unsaturated zone or in the upper part of the saturated zone where the transition takes place to three-dimensional groundwater flow. The lower boundary conditions can be specified, depending on the application and the relevant spatial scale. Three general types and some special cases of lower boundary conditions are distinguished. The general types are as follows:

1 The Dirichlet condition refers to a known status (pressure head) at the boundary and involves the imposing of a pressure head h_{bot} at the lower boundary. A special case involves the use of a recorded groundwater elevation. The pressure head at the groundwater elevation ϕ_{avg} is defined

as $h = 0$. This yields a linear relation between the pressure heads at the grid points above and below ϕ_{avg}:

$$h_{i+1} = -h_i \; \frac{\phi_{avg} - z_{i+1}}{z_i - \phi_{avg}} \tag{11}$$

2 The Neumann condition refers to a known flux perpendicular to the boundary. It prescribes a flux q_{bot} at the bottom. Since the model uses an explicit linearization scheme, the flux–groundwater level relations are treated as a Neumann condition, where the actual flux is calculated from the groundwater level of the previous time step. The relation between flux and groundwater level can be obtained from regional groundwater flow models (e.g. Van Bakel, 1986).

3 The Cauchy condition is combination of the former two conditions, the flux q at the lower boundary is defined as a function of the prevailing pressure head. This condition can be used when unsaturated flow models are combined with models for regional groundwater flow and when an implicit handling of q_{bot} in the iterative computation scheme is required. The flux through the bottom boundary is defined by the difference of the total hydraulic head (h + z) and the hydraulic head ϕ (cm) of the regional groundwater outside the flow domain described by the model, divided by a flow resistance c (d).

$$q_{bot} = \frac{\left[h + z\right]_{z=bot} - \phi}{c} \tag{12}$$

6 The WOFOST model and its combination with SWAP

WOFOST (Van Keulen and Wolf, 1986; Spitters et al., 1989; Supit et al., 1994; Hijmans et al., 1994; Boogaard et al., 1998, de Wit et al., 2018) simulates potential and limited production, considering water, salinity, and nutrient stress (Fig. 4). Canopy-absorbed energy, influenced by incoming radiation, crop leaf area, and photosynthetic leaf characteristics, determines potential gross photosynthesis, which is subsequently reduced by water shortage, salinity, and nutrient stress, resulting in actual gross photosynthesis.

Produced carbohydrates are used for maintenance and converted into structural matter, incurring energy costs and weight loss (growth respiration). Partitioning factors, dependent on phenological development stage, determine the distribution of dry matter among roots, leaves, stems, and storage organs (Spitters et al., 1989). Leaf fraction affects light interception dynamics. Integrating growth rates yields organ dry weight. Senescence causes biomass reduction during the season. Light interception and CO_2 assimilation are key growth processes, influenced by temperature (e.g. maximum photosynthesis

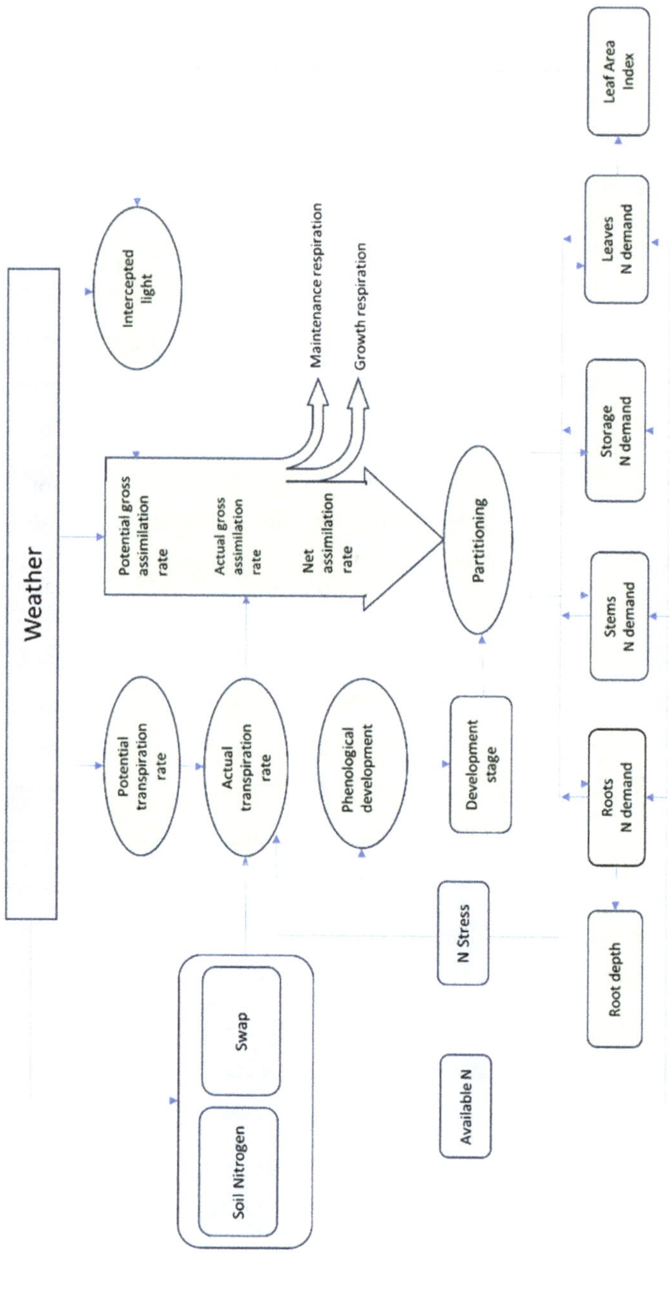

Figure 4 Overview of WOFOST and its interaction with SWAP.

and maintenance respiration). Processes like assimilate partitioning and tissue decay depend on phenological development stage, linked to thermal time.

7 SWAPWOFOST functions: phenological development and CO_2 assimilation

7.1 Phenological development stage

Phenological development quantification is crucial in crop growth simulation models, given the variation in physiological and morphological processes. For many annual crops, the development stage (D_s) is used, with values of 0 at seedling emergence, 1 at anthesis, and 2 at maturity (Van Heemst, 1986a; 1986b). The transition from vegetative ($0 < D_s < 1$) to reproductive stage ($1 < D_s < 2$) significantly alters dry matter allocation to organs.

Simulation begins at sowing or emergence, specified by the user. Phenological stages, influenced by temperature, progress from floral initiation or anthesis (controlled by temperature and sometimes day length) to maturity. After anthesis, the development rate depends solely on temperature, with higher temperatures accelerating it. WOFOST employs temperature sums based on effective temperature (T_{eff}), calculated from daily average temperature (T_{day}) using a tabular function. D_s is defined as

$$D_s^{j+1} = D_s^j + \frac{T_{eff}}{T_{sum,i}} \tag{13}$$

where superscript j is the day number and $T_{sum,i}$ is the temperature sum required to complete either the vegetative or the reproductive stage. For some species or cultivars day length during the vegetative stage has to be considered. Quantitative approaches for describing these effects are provided by Weir et al. (1984), Hadley et al. (1984), and Reinink et al. (1986).

7.2 CO_2 assimilation

The daily gross CO_2 assimilation rate is determined by calculating the absorbed radiation and utilizing the photosynthesis–light response curve, which is influenced by temperature and leaf age. Non-linear photosynthetic response to light intensity is considered, accounting for variations in radiation levels. Vertical canopy variation is addressed by separately calculating sunlight absorption for upper and lower leaves, assuming a spherical leaf angle distribution. The assimilation of leaves across the relative canopy depth is computed based on individual leaf response curves. Horizontal plane variation (e.g. plant rows) is not considered. Temporal variation, linked to the daily sun cycle, is simulated by

WOFOST using total daily radiation, distinguishing between direct and diffuse radiation.

The daily gross CO_2 assimilation rate is obtained by integrating assimilation rates through the crop canopy and over the day. A sinusoidal course of incoming radiation is assumed for daily integration. Both canopy and daily integration use a three-point Gaussian method with nine iterations (Goudriaan, 1986). Additionally, daily gross CO_2 assimilation rate may be reduced based on daily minimum temperature. Low temperatures at night hinder the transformation of assimilates into structural biomass, leading to accumulation and diminished assimilation rates over consecutive days. Figure 5 illustrates the CO_2 assimilation rates at different sunlight intensities, as observed across various crops. Notably, C4 crops exhibit higher assimilation rates compared to C3 plants which can be attributed to the higher efficiency CO_2 exchange rate of C4 plants.

The effects of CO_2 concentrations on the assimilation are implemented, using a CO_2-dependent correction factor that modifies leaf-level maximum assimilation rate and initial light-use efficiency (Wolf et al., 2010; Vanuytrecht and Thorburn, 2017).

8 SWAPWOFOST functions: respiration/partioning and growth/senescence

8.1 Respiration and partioning

WOFOST incorporates two forms of respiration losses. Firstly, a portion of the assimilates formed is allocated to maintenance respiration, which is

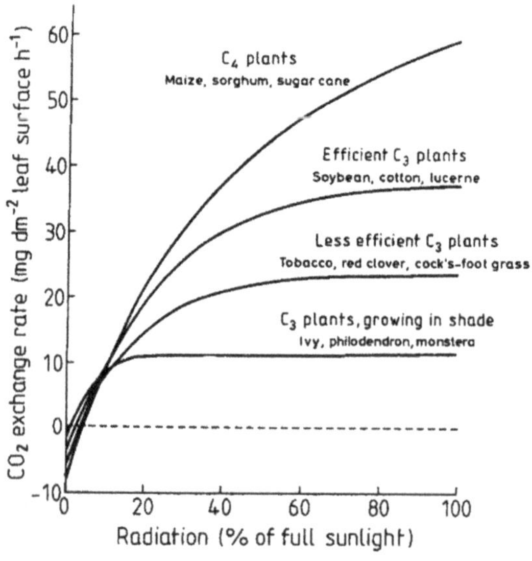

Figure 5 CO_2 exchange rate as function of radiation amount for C3 and C4 plants.

estimated based on the dry weight and chemical composition of various organs. Maintenance respiration is temperature-dependent, with its magnitude doubling for each 10-degree increase (Q_{10}). It is defined as follows (Kase and Catsky, 1984; Penning de Vries and Van Laar, 1982):

$$R_m = f_{senes} R_{mref} Q_{10}^{\frac{T_{avg}-25}{10}} \tag{14}$$

where R_m is the actual maintenance respiration rate (kg ha^{-1} d^{-1}), f_{senes} is the reduction factor for senescence, R_{mref} (kg ha^{-1} d^{-1}) reference maintenance requirements, and T_{avg} is the average temperature (°C). The model imposes a constraint, ensuring that the maintenance respiration rate does not surpass the actual gross CO_2 assimilation rate, preventing vegetation from 'self-consuming' carbohydrates. Secondly, growth respiration occurs during the conversion of carbohydrates into structural plant material, requiring energy. Instead of separately computing growth respiration, WOFOST utilizes a conversion efficiency factor. The conversion efficiency varies among plant organs.

Partitioning is not treated as an independent process; rather, it is directly linked to the crop's developmental stage. Static partitioning tables outline the assimilate fraction allocated to different plant organs based on the crop's developmental stage. Figure 6 depicts a typical example.

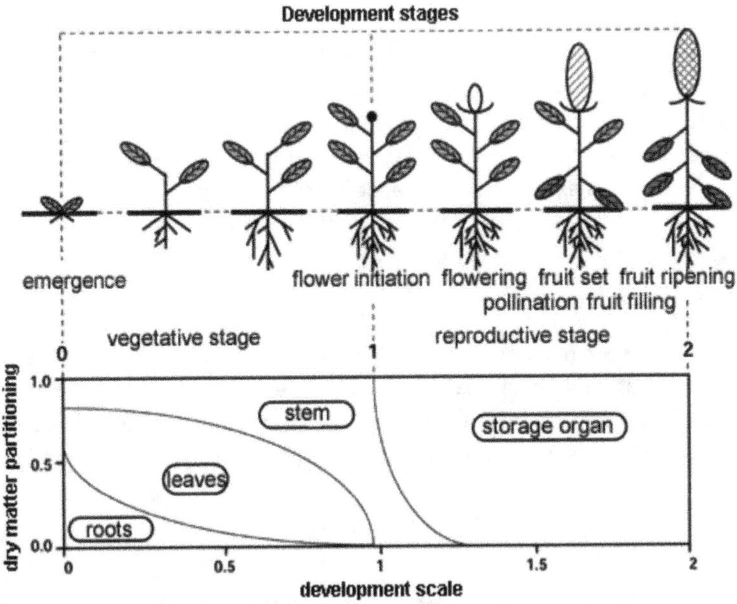

Figure 6 Typical partitioning of assimilated dry matter among leaves, stem, roots, and storage as function of development stage.

While this mechanism is simple and robust, its implementation in WOFOST depend on accurate crop phenological development descriptions. Incorrect phenological development can lead to excessively high leaf area indices or minimal leaf growth. Another limitation of static partitioning tables is their lack of consideration for sink limitations and the absence of environmental impact on partitioning. To address these limitations, a new partitioning mechanism could be introduced, accounting for sink limitations and maintaining root–shoot functional balance during organ allocation. The approach suggested by Xinyou and Van Laar (2005) and implemented in GECROS could be an option.

To account for losses during the conversion of carbohydrates into specific plant tissues, it is essential to specify conversion factors, denoted as $C_{e,i}$ (kg kg^{-1}) for leaf, storage organ, stem, and root biomass. WOFOST then computes a weighted average, denoted as C_e (kg kg^{-1}), by multiplying the organ-specific values with the corresponding partitioning factors.

$$C_e = \frac{1}{\left(\frac{\xi_{leaf}}{C_{e,leaf}} + \frac{\xi_{stor}}{C_{e,stor}} + \frac{\xi_{stem}}{C_{e,stem}}\right)(1-\xi_{root}) + \frac{\xi_{root}}{C_{e,root}}} \tag{15}$$

where ξ_i is the partitioning factor for the leaves, stems, root, and storage organs.

8.2 Growth and senescence

WOFOST places significant emphasis on precise leaf area simulation, crucial for light absorption and photosynthesis. Initial leaf area is derived from the initial dry weight, with early leaf development being sink-limited, transitioning to source-limited growth later in the growing season. During initial growth, temperature primarily influences leaf development, constrained by cell division and extension. An exponential curve, defined by a crop-specific relative growth rate, describes the early LAI development. The LAI growth rate, w_{LAI} (ha ha^{-1} d^{-1}), during this stage, is described by

$$w_{LAI} = LAI\, w_{LAI,max}\, T_{eff} \tag{16}$$

where $w_{LAI,max}$ is the maximum relative increase of LAI (°C^{-1} d^{-1}). T_{eff} is the effective temperature and LAI is the leaf area index. Later stages witness source-limited leaf area increase due to branching and tillering. The exponential growth rate is maintained until source-limited increase equals it. The source limited growth rate is described by

$$w_{LAI} = w_{net,leaf}\, S_{la} \tag{17}$$

where S_{la} is the specific leaf area (ha kg^{-1}). Leaf biomass, specific leaf area, and leaf age intricately are simulated throughout the growth cycle. A daily cohort system tracks leaf biomass, specific leaf area, and physiological age, contributing to total living leaf biomass and computed leaf area. Adjustments consider stems and storage organs absorbing radiation.

Leaf senescence involves multiple factors. Leaves die due to physiological ageing, self-shading at high LAI, and water and nitrogen stress. WOFOST computes a total leaf death rate, removing biomass from cohorts based on age until it matches the death rate, maintaining carbon balance. The potential death rate of leaves due to water stress $\zeta_{leaf,water}$ (kg ha^{-1} d^{-1}) is calculated as

$$\zeta_{\text{leaf,water}} = W_{\text{leaf}}\left(1-\frac{T_a}{T_p}\right)\zeta_{\text{leaf,p}} \tag{18}$$

where W_{leaf} is the leaf dry matter weight (kg ha^{-1}), T_a and T_p are the actual and potential transpiration rates (cm d^{-1}), respectively, and $\zeta_{leaf,p}$ is the maximum relative death rate of leaves due to water stress (kg kg^{-1} d^{-1}). The latter is crop specific and should be provided by the user. Death rate due to selfshading, $\zeta_{leaf,shade}$ (kg ha^{-1} d^{-1}) is defined which increases linearly from zero at a critical LAI LAI_c (-), to its maximum value at 2*LAI_c:

$$\zeta_{\text{leaf,shade}} = 0.03\,W_{\text{leaf}}\left(\frac{LAI-LAI_c}{LAI_c}\right) \quad \text{with} \quad 0<\left(\frac{LAI-LAI_c}{LAI_c}\right)<1 \tag{19}$$

with LAI_c = 3.2/$_{qdf}$. Typical values for $\zeta_{leaf,p}$ and LAI_c are 0.03 d^{-1} and 4 m^{2}m^{-2}, respectively (Spitters et al., 1989). The highest value of $\zeta_{leaf,water}$ and $\zeta_{leaf,shade}$ is selected for the combined effect of water/nitrogen stress and mutual shading. Leaves that have escaped from premature death due to water stress or mutual shading, inevitably die due to exceedance of the life span for leaves. Life span is defined as the maximum time a leaf can live at a constant temperature of 35°C. The net growth rates of the plant organs, $w_{net,i}$ (kg ha^{-1} d^{-1}) result from the gross growth rates and the senescence rates, ζ_i (kg kg^{-1} d^{-1}) is given by

$$w_{\text{net},i} = w_{\text{gross},i} - \zeta_i W_i \tag{20}$$

The dry matter weight of organ i, W_i (kg ha^{-1}), is calculated by integrating $w_{net,i}$ over time. Root extension is calculated, considering the crop's maximum rooting depth and the soil's influence, along with the maximum daily increase in rooting depth denoted as, $d_{root,max}$ (cm). The WOFOST implementation in SwapWofost provides two user-dependent methods for determining actual rooting depth. The first method assumes a consistent daily increase, constrained by either the maximum rooting depth or assimilate availability. The second method characterizes the daily increase in rooting depth as a function of relative dry matter increase.

9 SWAPWOFOST functions: carbon balance and crop nitrogen

9.1 Carbon balance check

The actual gross CO_2 assimilation rate has to be identical to the amount of structural plant material produced plus the amounts used for maintenance respiration and conversion. The carbon balance has to be zero and is implemented as

$$0 = \frac{R_d - R_{m,T} - R_g * (\xi_{root} + \left(\xi_{leaf} + \xi_{stem} + \xi_{stor}\right) * \left(1 - \xi_{root}\right) * \left(D_{mi} / C_e\right)}{R_d} \tag{21}$$

Where R_g is the growth respiration rate (kg ha^{-1} d^{-1}), R_d is the actual daily CH_2O assimilation rate (kg ha^{-1} d^{-1}), $R_{m,T}$ is the maintenance respiration rate (kg ha^{-1} d^{-1}), and ξ_i are partitioning factors (–) of various plant organs. D_{mi} is the dry matter increase. WOFOST assumes that maintenance respiration cannot exceed the actual gross assimilation rate. However, in case the daily CH_2O assimilation rate comes close to zero, the maintenance respiration rate and the growth respiration rate also becomes small and a situation may occur that the numerator of the above-mentioned equation becomes very small and negative. To address this issue, introducing a division by R_d in the carbon check will identify such an event.

9.2 Crop nitrogen

The crop nitrogen routines, implemented in SwapWofost, are based on Shibu et al. (2010). It is assumed that the overall crop nitrogen demand is equivalent to the cumulative nitrogen demands of its distinct organs. A schematical overview is depicted in Fig. 7.

The nitrogen demand for individual organs is computed as the disparity between their maximum and current nitrogen contents. The maximum nitrogen content is defined as a function of the canopy development stage, as established by Drenth et al. (1994). Total N demand (TN_{dem}: kg ha^{-1} d^{-1} N) of the crop is

$$TN_{dem} = \sum_{i=1}^{n} \frac{W_i N_{max,i} - AN_i}{\Delta t} \tag{22}$$

where $N_{max,i}$ is the maximum nitrogen concentration of organ i (kg N kg^{-1} biomass), with i referring to the plant organs, W_i is the weight of organ i (kg biomass ha^{-1}), AN_i is the actual nitrogen content of organ i (kg N ha^{-1}), and Δt is the timestep. Nitrogen uptake processes such as mass flow and diffusion are not explicitly simulated. Instead, crop nitrogen uptake is estimated using

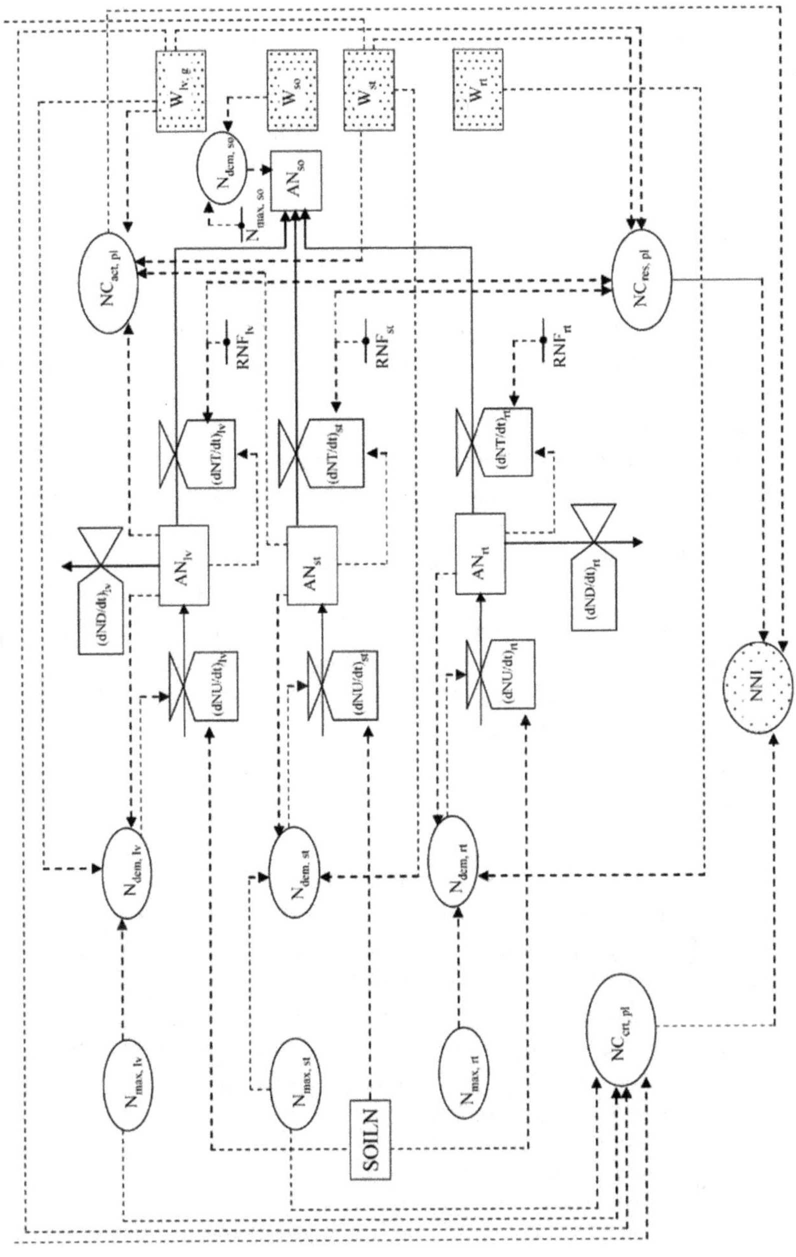

Figure 7 Overview of the main data flow in the Crop nitrogen routines. Source: Adapted from Shibu et al. (2010).

a straightforward bookkeeping approach. Nitrogen derived from indigenous sources is presumed to have a higher recovery rate (almost 100%) compared to applied fertilizers, representing the amount of nitrogen actually absorbed by a crop in the absence of nitrogen fertilizers. Consequently, the model assumes that the crop initially assimilates nitrogen mineralized from indigenous organic matter, followed by uptake from fertilizers. The total nitrogen absorbed by the crop (d*NU*/dt) is distributed among leaves, stems, roots, and storage organs in proportion to their respective nitrogen demands:

$$\left(\frac{dNU}{dt}\right)_i = \left(\frac{N_{dem,i}}{TN_{dem}}\right)\left(\frac{dNU}{dt}\right) \tag{23}$$

where $(dNU/dt)_i$, and $N_{dem,i}$ are the rate of nitrogen uptake (g m^{-2} d^{-1}) and nitrogen demand (g m^{-2} d^{-1}) of organ *i* (*i* refers to leaves, stems, roots, and storage organs), respectively.

SwapWofost assumes nitrogen uptake halts after anthesis, given the minimal increase in nitrogen content in vegetative parts (Groot, 1987; Sinclair and Amir, 1992). Storage organs exclusively rely on translocation from leaves, stems, and roots during grain formation. Nitrogen accumulation in storage organs is influenced by nitrogen demand, calculated from the difference between maximum and actual nitrogen content, and total translocatable nitrogen from other organs. Total translocatable nitrogen equals overall organ nitrogen content minus non-transferable nitrogen in structural components. The net rate of change of nitrogen $(dN/dt)_i$ in each of the organs AN_i, where *i* refers to leaves, stem and roots, is

$$\left(\frac{dN}{dt}\right)_i = \left(\frac{dNU}{dt}\right)_i - \left(\frac{dNT}{dt}\right)_i - \left(\frac{dND}{dt}\right)_i \tag{24}$$

where $(dNU/dt)_i$, $(dNT/dt)_i$, and $(dND/dt)_i$ are the contributions of nitrogen uptake to the organ, translocation from the organ and loss of nitrogen due to the death of the organ, respectively. It is assumed that the stem does not die and therefore $(dND/dt)_{stem}$ equals zero and the outflow rate is not included.

N stress occur at N concentrations below a critical value for unrestricted growth. To quantify crop response to nitrogen shortage, a Nitrogen Nutrition Index (NNI) is defined, ranging from 0 (maximum N shortage) to 1 (no shortage) (Lemaire et al., 1989; Van Delden, 2001):

$$NNI = \frac{actual\ crop[N] - residual[N]}{critical[N] - residual[N]} \tag{25}$$

The critical crop nitrogen concentration is defined as the lower limit of canopy nitrogen concentration in leaves and stems required for unrestricted growth.

It is assumed to be half of the maximum nitrogen concentration (Porter, 1993; Jamieson et al., 1998). Similar to the carbon balance, a check is carried out on the closure of the N balance and the simulation is interrupted when the N balance exceeds 0.001 kg ha^{-1} N.

Nitrogen fixation was based on a simple approach, assuming that N_2 fixation from the air is potentially unlimited. The crop defines the demand for Nitrogen and a simple user defined fraction indicates the amount of Nitrogen demand which met by N-fixation from air and soil. Crop growth and the corresponding N demand are already limited by drought, salinity, etc. which is simulated with SWAP.

10 SWAPWOFOST functions: actual transpiration and stress factors

10.1 Actual transpiration

Transpiration and evaporation fluxes are based on atmospheric conditions and vegetation properties, assuming favourable soil moisture conditions. The maximum possible root water extraction rate, integrated over the rooting depth, should equal T_p (cm d^{-1}). Taking into account the root length density distribution (Bouten, 1992), the potential root water extraction rate at a certain depth, $S_p(z)$ (d^{-1}), can be calculated by

$$S_p(z) = \frac{\ell_{root}(z)}{\int_{-D_{root}}^{0} \ell_{root}(z)dz} T_p \tag{26}$$

where D_{root} is the root layer thickness (cm). Under suboptimal soil conditions, such as excessive wetness, dryness, salinity, or low temperatures, the fluxes towards the roots decrease. SwapWofost computes stress factors for these conditions. In cases of simultaneous water and salt stress, the method proposed by Skaggs et al. (2006) is implemented, involving the multiplication of stress factors for water and salt stress. The actual water extraction rate $S_a(z)$ (d^{-1}) is calculated as

$$S_a(z) = \alpha_{rw}\alpha_{rd}\alpha_{rs}\alpha_{rf}S_p(z) \tag{27}$$

where $_{\alpha rd}$ (-), $_{\alpha rw}$ (-), $_{\alpha rs}$ (-), and $_{\alpha rf}$ (-) are the reduction factors for conditions which are too wet, too dry, too saline, or too cold, respectively. Integration of $S_a(Z)$ over the root layer yields the actual transpiration rate T_a (cm d^{-1}):

$$T_a = \int_{-D_{root}}^{0} S_a(z)\partial z \tag{28}$$

10.2 Water stress

Two options to assess water stress are provided, a simple macroscopic (Feddes et al., 1978), which is depicted in Fig. 8. In the range $h_3 < h < h_2$ root water uptake is optimal. Below h_3 root water uptake linearly declines due to drought until zero at h_4 (wilting point). The critical pressure head h_3 increases for higher potential transpiration rates T_p. Above h_2 root water uptake linearly declines due to insufficient aeration until zero at h_1, i.e. representing threshold values for minimum and maximum oxygen deficiency.

Sometimes only some parts of the root zone are stressed and show reduced water uptake, while other parts have favourable conditions for root water uptake. In these conditions, the reduction in the stressed parts can be compensated by extra root water uptake in the parts with favourable conditions. This compensation was not part of the original Feddes reduction function. The user can add this compensation to the Feddes function. The compensation concept implemented has been described by Jarvis (1989; 2011).

In addition to the macroscopic Feddes model, two microscopic models describing the radial flow of soil water towards individual roots (Fig. 9) have been implemented in SWAP. The first model proposed by De Willigen and Noordwijk (1987) and the second by De Jong van Lier et al. (2013) both assume a cylindrical soil column around each root (Fig. 10). However, they differ in their descriptions of the water flux from the soil into the root. For detailed information see Heinen et al. (2024).

The water balance of the radial flow pattern towards root can be described as

$$\frac{\partial \theta}{\partial t} = -\frac{q}{r} - \frac{\partial q}{\partial r} \tag{29}$$

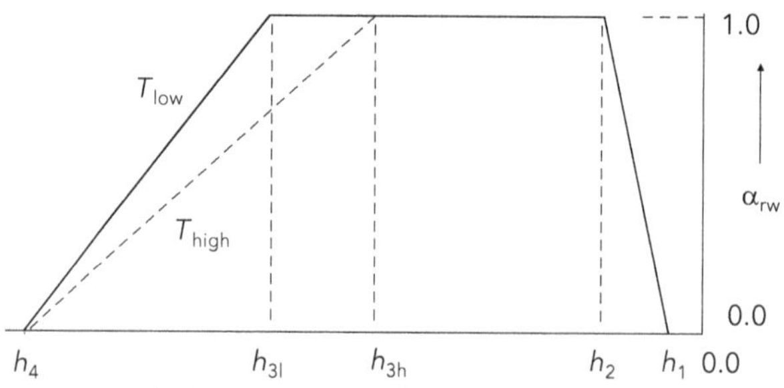

Figure 8 Reduction coefficient for root water uptake, α_{rw}, as function of soil water pressure head h and potential transpiration rate T_p. Source: After Feddes et al. (1978).

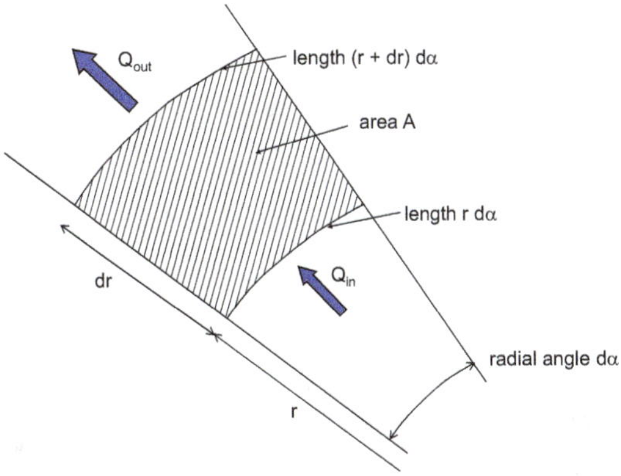

Figure 9 Water balance of a radial segment. Where Q_{in} is the inflow ($m^2\ d^{-1}$) and Q_{out} is the outflow ($m^2\ d^{-1}$).

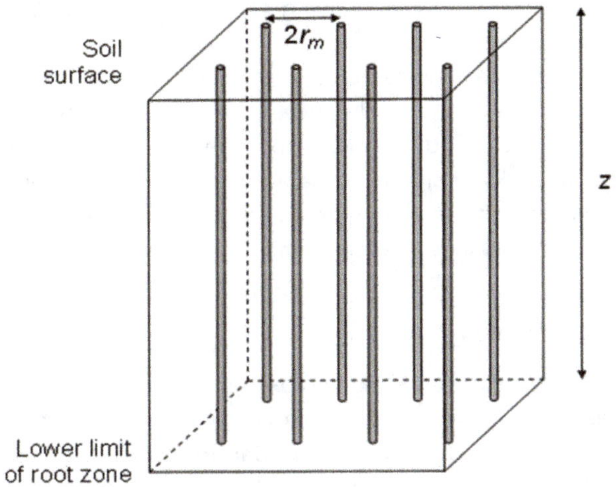

Figure 10 Schematization of root system by equally spaced tubes. Density in the vertical may differ.

The soil water flux itself can be described by the Darcy equation without the gravity component:

$$q = -k\frac{\partial h}{\partial r} \tag{30}$$

where k is hydraulic conductivity and h is soil water pressure head (cm). When realistic values for extraction rates and soil hydraulic properties are applied to solve Equation (29), pronounced gradients $\partial h/\partial r$ near the root surface are observed. This is attributed to the rapid decline in hydraulic conductivity at lower h values and increasing flux density due to converging flow lines near

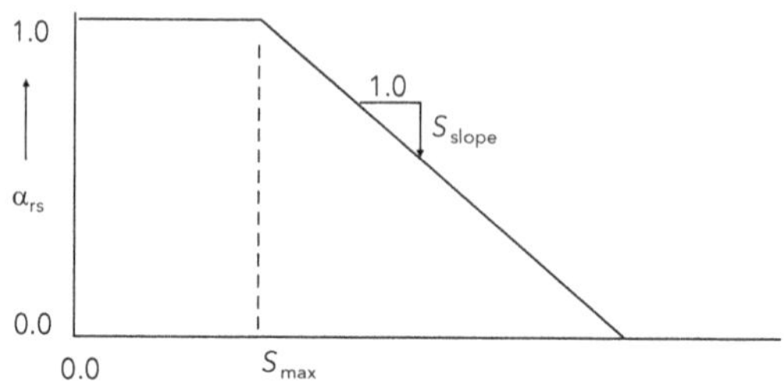

Figure 11 Reduction coefficient for root water uptake, $_{\alpha rs}$, as a function of soil water salinity concentration according to Maas and Hoffman (1977).

the roots. Historically, numerical solutions of Equation (30) involved significant simplifications. However, a more accurate resolution is achievable by utilizing the matric flux potential as the driving variable, rather than the soil water pressure head.

The root system is considered as a set of such individual tubes, regularly spaced in the soil at definable distances. The density of the tubes may vary with depth, similar to root density in a root zone. De Jong van Lier et al. (2013) describe the analytical solution in detail and did upscale the microscopic approach to entire root zones with variable root densities and soil moisture and radial and axial hydraulic resistances within the plant system.

10.3 Oxygen stress

Constant critical pressure heads for oxygen stress seem not realistic, as oxygen stress depends on various abiotic and biotic factors. Therefore, as alternative, the process-based method of Bartholomeus et al. (2008) has been implemented. Onthe dry side, the input values, which are not commonly available, for h_{3l} and h_{3h} depend on plant type, soil texture, root density, and atmospheric demand. Also the linear decline of the reduction factor in the range $h_4 < h < h_3$ may deviate from reality.

10.4 Salinity stress

Salinity stress reduces root water uptake due to the solute concentration in soil water. SWAP simulates solute transport and salinity levels in the soil by accounting for diffusion, convection, and dispersion processes. Two options are available for modelling this: the first is the linear reduction function (Maas and Hoffman, 1977) shown in Fig. 11, and the second is the osmotic head approach (De Jong van Lier et al., 2009).

The reduction of root water uptake (α_{rs}) when the salt concentration exceeds a threshold value is linearly related to the salinity concentration:

$$\alpha_{rs} = 1.0 - (c - S_{max}) S_{slope} \tag{31}$$

where c is the salinity concentration (mg cm^{-3}), S_{max} is the salinity threshold value (mg cm^{-3}) and S_{slope} is the root water uptake decline factor per unit increase of salinity concentration (cm^3 mg^{-1}). De Jong van Lier et al. (2009) demonstrated how the osmotic head due to salts in soil water can be included in microscopic root water uptake. To apply this concept, the root water uptake under dry conditions should be based on the microscopic approach. For detailed information, see the Swap manual (https://swap.wur.nl/) and the reference mentioned above. Note that this option should be applied with the utmost care, it requires tuning of the drought input parameters of the crop input file.

10.5 Crop growth and stress

Nitrogen stress as well as stress caused by water uptake reduction (Equation 27) reduces the crop's ability to assimilate. In SwapWofost this effect is described as

$$A_{net} = \alpha_{wn} A_{gross} \tag{32}$$

Where A_{net} is the net assimilation rate (kg DM ha^{-1} d^{-1}), α_{wn} the stress factor, either nitrogen or water uptake stress and A_{gross} is the gross Assimilation rate (kg DM ha^{-1} d^{-1}). In case water uptake stress occurs the nitrogen uptake is hampered. On the other hand, if the nitrogen stress occurs the water uptake is reduced. To account for this effect the most dominant of the two is selected. In Section 17.4, the water uptake stress factor is described. The nitrogen stress factor is given by

$$\alpha_n = 1 - c(1 - NN_i)^2 \tag{33}$$

Where c is the coefficient for the reduction of the radiation use efficiency, NN_i is the Nitrogen Nutrition index. α_n is constrained to values between 0 and 1. Similar to the assimilation reduction, the most dominant stress factor decides the impact on the partitioning of the newly formed dry matter production. If water uptake stress is the dominant of the two, then the dry matter fraction partitioned to the roots is reduced with a fraction depending on the magnitude of the stress factor. In case the nitrogen stress is the larger one, the newly produced leaf dry matter is reduced with a fraction that depends on the magnitude of the stress. This fraction is subsequently added to the newly formed stem dry matter.

11 SWAPWOFOST functions: soil nitrogen

Organic compounds undergo relatively rapid decomposition when a sufficient supply of molecular oxygen is available. In cases where oxygen is limited, degradation occurs through the utilization of oxygen from nitrates and sulfates. Under conditions of ample oxygen, these reactions conclude with the production of carbon dioxide (CO_2) and water (H_2O). The decomposition of organic matter primarily involves oxidation processes. Microorganisms participating in this decomposition are predominantly aerobic or facultative aerobic, indicating their ability to thrive in both aerobic and anaerobic environments. It is noteworthy that the anaerobic decomposition process is significantly less efficient, ranging from 100 to 1000 times less efficient compared to the aerobic conditions (Hämäläinen, 1991).

Deceased plant components and all other organic materials introduced into the soil are considered as additions of fresh organic matter. Living plant roots release soluble organic substances into the soil solution, including the products of dead root cells during growth. These substances become available for decomposition, contributing to both the carbon and nutrient cycles. Upon initiation of decomposition, this material undergoes partial oxidation to form CO_2 and H_2O, alongside a partial transformation into biomass. The ratio between the produced soil biomass and the total amount of transformed material is defined as assimilation efficiency. Some of these transformations occur through the intermediate stage of dissolved organic material. The initial phase of the decomposition process, involving molecules such as cellulose, hemicellulose, pectin, and lignin, entails the breakdown of these compounds into smaller constituents. Microorganisms employ exo-enzymes, operating externally to the biomass cells, to facilitate this process. Generally, smaller compounds exhibit higher solubility, and these diminutive molecules are subsequently assimilated by the microorganism cell for further transformation.

The structure of the RothC-26.3 model (Coleman and Jenkinson, 2014) has been taken as the starting point for the development of an organic matter module within SwapWofost. It is extensively used to investigate the evolution of organic matter stocks in relation to the sustainability of agricultural systems. It is relatively simple but has the flexibility to address various soil and crop types. The model describes the organic matter cycle by four biological transformable pools and an inert organic matter (IOM) pool as depicted in Fig. 12.

The DPM- and RPM-pools transform into the BIO-(microbial biomass) and HUM(humified organic material)-pools through assimilation, a process concomitant with the degradation of organic molecules into carbon dioxide. Subsequently, the BIO- and HUM-pools undergo a further transformation,

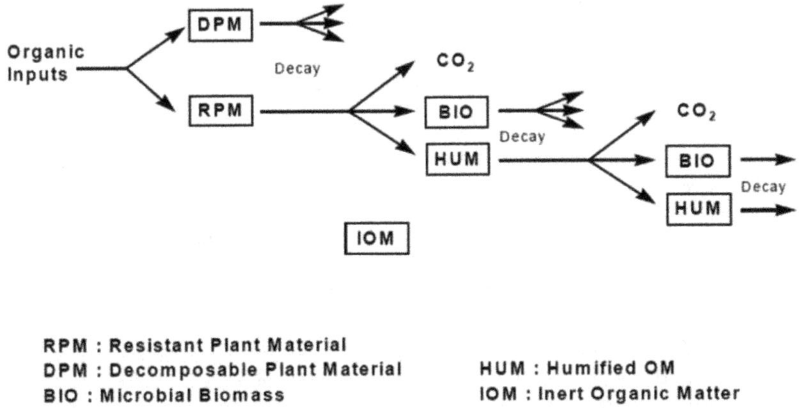

Figure 12 Organic carbon pool structure of the RothC26.3 model. Source: Adapted from Groenendijk et al. (2016).

approximated by a partial return to the parent pool and a partial transformation to the other pool. This conversion is also accompanied by the degradation of organic molecules to carbon dioxide. The introduction of rapid decomposable plant material (DPM) and slowly decomposing material (resistant plant material–RPM) pools into the soil occurs through various means, including:

- the application of farmyard manure and cattle slurry;
- the incorporation of crop residues, encompassing leaves, stems, and roots that have died off during the growing season, as well as residues left after harvest; and
- the application of diverse organic materials such as compost, straw for frost protection, tree shred, and others.

To enable the parametrization for a wide variety of different organic materials, the one-parameter model of Janssen (1984) was utilized to construct time-dependent decomposition rate relations of peat and other organic materials using the data published by Kolenbrander (1969, 1974).

The IOM pool is considered as a fraction that is found in measurements but does not participate in the transformations and is not affected by the amendments. This pool has no further effect on nitrogen mineralization. When model parameterization is based on measured organic matter contents, one should take this pool into account. The four biological active pools are schematically represented in Fig. 13.

The transformations are described as first-order rate processes where the rates depend on the organic matter amounts in the pools. The interdependence of the BIO-pool and the HUM-pool can be mathematically expressed by

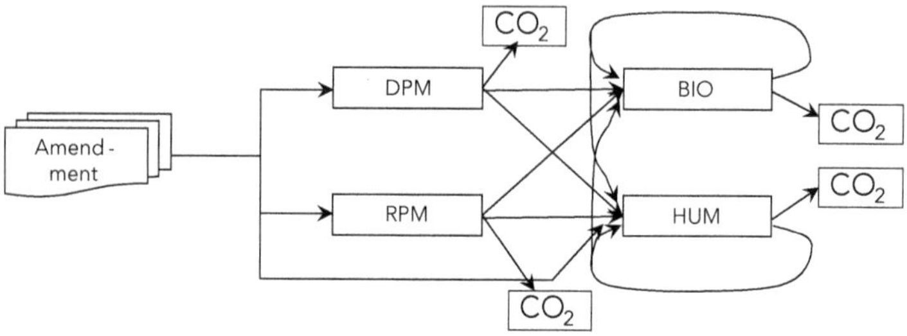

Figure 13 Schematic representation of organic matter stocks and transformation in the RothC-26.3 model. Where DPM is the decomposable plant material, RPM is the resistant plant material, BIO is the microbial biomass, HUM is the humified organic material, and CO_2 the amount of CO_2 that leaves the system. Source: Adapted from Groenendijk et al. (2016).

describing the processes as a set of linear differential equations (see also Groenendijk et al., 2005; Heinen and De Willigen, 2005; De Willigen et al., 2008):

$$\begin{pmatrix} \frac{d\,DPM}{dt} \\ \frac{d\,RPM}{dt} \\ \frac{d\,BIO}{dt} \\ \frac{d\,HUM}{dt} \end{pmatrix} = \begin{pmatrix} -k_1 & 0 & 0 & 0 \\ 0 & -k_2 & 0 & 0 \\ f\varepsilon k_1 & f\varepsilon k_2 & -(1-f\varepsilon)k_3 & f\varepsilon k_4 \\ (1-f)\varepsilon k_1 & (1-f)\varepsilon k_2 & (1-f)\varepsilon k_3 & -(1-(1-f)\varepsilon)k_4 \end{pmatrix} \begin{pmatrix} DPM \\ RPM \\ BIO \\ HUM \end{pmatrix} \quad (34)$$

Where k_1 ... k_4 are transformation rate constants, ε is the assimilation fraction (fraction of total decayed C that is newly incorporated in the BIO and HUM pool), and f is the fraction of assimilated carbon that is incorporated in the BIO pool. Inputs are imposed by an instantaneous increase of the DPM and RPM pools at the beginning of the time step. The state variables DPM and RPM are recalculated according to

$$\begin{aligned} DPM(t+\delta t) &= DPM(t) + pQ \\ RPM(t+\delta t) &= RPM(t) + (1-p)Q \end{aligned} \quad (35)$$

where Q is the quantity of added organic matter, and p and $(1 - p)$ are the partitioning factors of the amendment to either the DPM or the RPM pool.

The set of linear differential equations is solved analytically by

- deriving the solutions for the DPM(t) and RPM(t) values at the end of the time step;
- calculating the eigenvalues for the BIO and HUM subsystem;
- deriving the solutions for the BIO(t) and HUM(t) values at the end of the time step, taking into account the quantities assimilated from the DPM- and RPM-pool and the internal assimilation and dissimilation rates of this subsystem;
- deducing the dissociation, assimilation, and dissimilation rates from the quantities of DPM, RPM, BIO, and HUM at the start and the end of the time step;
- calculating the mineralized nitrogen quantities and CO_2 produced by multiplying the different process rates by a nitrogen fraction and carbon fraction, respectively, and setting up balances according to the transformation scheme; and
- nitrogen supplied is to soil through fertilizer applications and organic matter decay is stored in the soil (Fig. 13). The processes that control the nitrogen mineralization and immobilization in relation to the processes in the organic matter cycle can be described by the following:
  - The net total NH_4 mineralization rate follows from the formation/ decomposition balance of the different organic materials, taking into account their diverse N contents.
  - The decomposition NO_3 rate is determined by the part of the total decomposition of organic material that takes place under anaerobic conditions. The decomposition rates under strict anaerobic conditions are much lower and in the topsoil and complete anaerobic decomposition can generally be neglected.

11.1 Ammonium balance

The ammonium balance is defined by a number of inputs and outputs and the change of stocks. Most processes are time dependent; however, fertilizer applications and volatilization are formulated as event-based functions. Only the NH_3 gaseous emission (i.e. volatilization) during fertilizer application is accounted for. The fraction of ammonium dosage for each fertilization event is user specified. For the time-dependent functions, the mass conservation equation is taken as a starting point:

$$K_{sorp}\rho_d \frac{dc_{NH4}}{dt} + \frac{d\theta c_{NH4}}{dt} = q_{in}c_{NH4,in} - q_{out}c_{NH4} + R_{Nmin} - R_{NH4,upt} - R_{nitr} \qquad (36)$$

Where C_{NH4} is the ammonium concentration of soil water (kg m^{-3}), $C_{NH4,in}$ is the ammonium concentration of inflowing water (kg m^{-3}), q_{in} inflowing water flux (m d^{-1}), q_{out} is the outflowing water flux (m d^{-1}), Θ is the moisture volume fraction (m^3 m^{-3}), K_{sorp} is the linear sorption constant (m^3 kg^{-1}), ρ_d is the dry bulk density (kg m^{-3}), R_{Nmin} is the mineralization rate (kg m^{-3} d^{-1}), $R_{NH4,upt}$ is the ammonium-N uptake rate (kg m^{-3} d^{-1}), and R_{nitr} is the nitrification rate (kg m^{-3} d^{-1}).

The linear sorption coefficient (K_{sorp}) and the dry bulk density (ρ_d) are user specified. The water and the soil water contents are calculated by SWAP. The ammonium concentration at the lower boundary should be specified as ammonium inflow from deeper soil layers. This inflow may occur when the evaporative demand is met by capillary rise. The mineralization rate is treated as a zero-order production. Nitrification is formulated as a first-order rate process according to

$$R_{nitr} = k_{nitr}\left(\theta c_{NH4}\right) \tag{37}$$

The first-order rate constant k_{nitr} depends on soil temperature, soil pH, and the relative water saturation degree. The ammonium amount that can be absorbed by the plant root system is defined as

$$R_{NH4,upt} = \delta_{NH4}\frac{q_{evtr}}{\Delta z}c_{NH4} \tag{38}$$

Where q_{evtr} is the transpiration flux (m d^{-1}), Δz is the depth of the root zone (m), and δ_{NH4} is a so-called transpiration – concentration stream factor and C_{NH4} is the time-averaged ammonium concentration for the current time step.

From crop emergence to anthesis, the limitation in ammonium uptake rate emerges as a crucial factor influenced by the precision of plant transpiration simulations. Achieving complete alignment of parameter sets between the WOFOST and SWAP models poses practical challenges. Consequently, an alternative approach is implemented to estimate nitrogen uptake by recently emerged crops. Before crop anthesis and in case the crop LAI is less than a user-defined critical value (LAI_{CNU}), a factor F_{CNU} is computed according to

$$F_{CNU} = \frac{LAI_{CNU} - LAI}{LAI_{CNU}} \tag{39}$$

The available uptake rate is subsequently calculated as

$$R_{NH4,upt} = \left(1 - F_{CNU}\right)\delta_{NH4}\frac{q_{evtr}}{\Delta z}c_{NH4} \tag{40}$$

11.2 Nitrate balance

Similar to the ammonium balance, the nitrate balance is defined by a number of inputs and outputs and the change of stocks. The fertilizer application is defined as an event-based function, all other processes are computed as time-dependent functions. The mass conservation equation is taken as a starting point and can be described by

$$\frac{d\theta c_{NO3}}{dt} = q_{in}c_{NO3,in} - q_{out}c_{NO3} + R_{nitr} - R_{NO3,upt} - R_{denitr} \tag{41}$$

where C_{NO3} is the ammonium concentration of the soil water (kg m^{-3}), $C_{NO3,in}$ is the ammonium concentration of inflowing water (kg m^{-3}), $R_{NO3,upt}$ is the ammonium-N uptake rate (kg m^{-3} d^{-1}), and R_{denitr} is nitrification rate (kg m^{-3} d^{-1}).

The nitrification rate is treated as a zero-order production. From the resulting ammonium concentration values, the time-averaged nitrification rate is calculated and its value is passed to the nitrate balance. Denitrification is also formulated as a first-order rate process:

$$R_{denitr} = k_{denitr}\left(\theta\, c_{NO3}\right) \tag{42}$$

where the value of the first-order rate constant k_{denitr} depends on soil temperature, soil pH, and the relative water saturation degree. Denitrification is defined as a first-order rate process where the rate constant depends on the organic matter dissimilation which in its turn is described by a Michealis–Menten equation. The nitrate amount that can be absorbed by the plant root system is defined as

$$R_{NO3,upt} = \delta_{NO3}\frac{q_{evtr}}{\Delta z}c_{NO3} \tag{43}$$

where δ_{NO3} is a so-called transpiration – concentration stream factor for nitrate uptake C_{NO3} and is the time-averaged nitrate concentration for the current time step. If both ammonium and nitrate concentrations in soil are too low, the crop demand cannot be fulfilled and WOFOST adjust its growth rates according to the soil conditions. In the early crop growth stages, the LAI-based procedure calculates the uptake as the sum of the demand-driven uptake, multiplied by the factor F_{CNU}, and the availability driven uptake, multiplied by (1 – F_{CNU}), is maintained. The mineral nitrogen amount that can be supplied by the soil is implemented as

$$N_{supply} = \min\left(R_{NO3,upt}, N_{demand} - \min\left(R_{NH4,upt}, N_{demand}\right)\right) + \min\left(R_{NH4,upt}, N_{demand}\right) \tag{44}$$

Where N_{demand} is the crop nitrogen demand for the current time step as calculated by WOFOST. The presented calculation scheme implies a preference for ammonium uptake. If too little ammonium is available, the remaining demand is being met by the nitrate uptake.

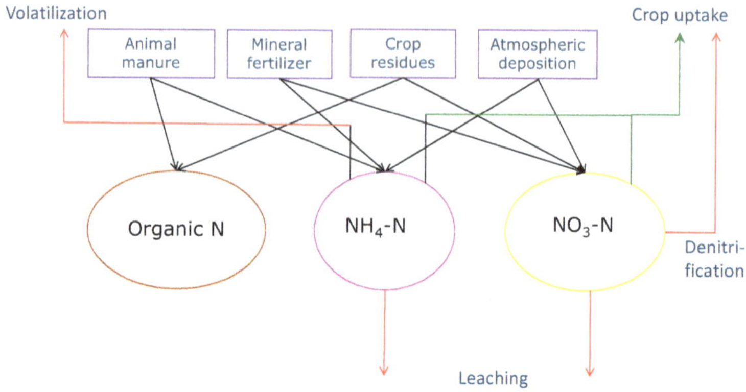

Figure 14 Schematic representation of the nitrogen pools, transformations and transport as described in the soil–N module. Source: Adapted from Groenendijk et al. (2016).

Please note that the soil N-module in SWAP provides a simplified representation of the nitrogen balance in the root zone. A more detailed description of the nitrogen balance is available in the ANIMO model (Groenendijk et al., 2005), which uses the water balance terms simulated by SWAP as input (Fig. 14).

12 Computational sequence and information exchange

The computational sequence and information exchange among modules are illustrated in Fig. 15 and detailed as follows (Groenendijk et al., 2016):

1 WOFOST calculates potential rates of gross assimilation, growth, maintenance respiration, and dry matter production in various plant parts. New values for state variables at the end of the day are computed without constraints from moisture deficit or nitrogen insufficiency. Growth rates of plant organs are then adjusted for the ratio between actual and potential plant evaporation, and nutritional demand is determined based on these adjusted growth rates. The portion supplied by biological nitrogen fixation is subtracted, and the resulting demand is forwarded to the Soil nitrogen module.

2 SWAP computes the water balance by considering the initial moisture profile at the beginning of the day, daily precipitation, and potential evaporation rates. Water extraction by plant roots is determined based on the pressure head distribution in the root zone, indicating the extent to which soil moisture conditions meet the evaporative demand. The ratio of actual to potential plant evaporation is calculated and transmitted to the WOFOST model.

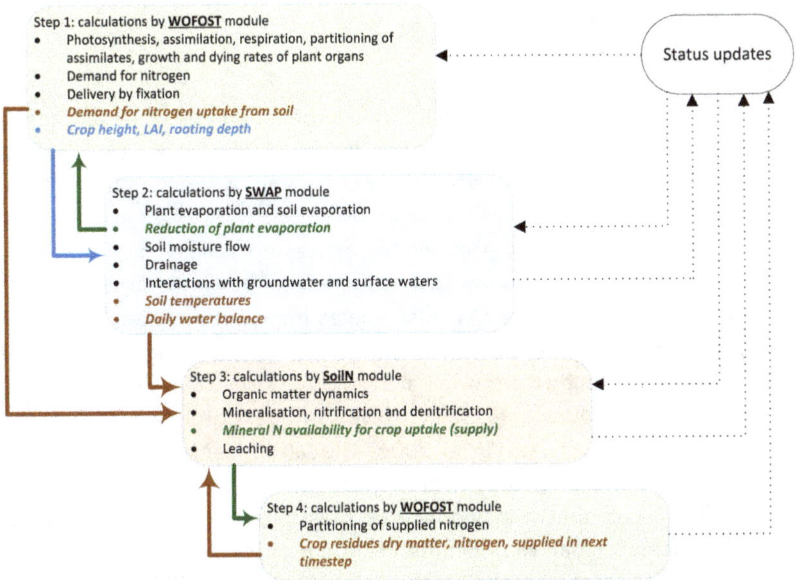

Figure 15 Computational sequence and information exchange between modules implemented in the innovated model.

3 The soil nitrogen module calculates daily changes in organic matter and nitrogen stocks in various soil pools. It factors in the previous day's end stocks, current-day fertilizer applications (organic + mineral), and crop residues. Ammonia volatilization is considered with a user-defined factor. Decomposition and transformation yield mineralized and immobilized nitrogen. The ammonium balance accounts for adsorption, nitrification, crop demand, and leaching. Nitrified ammonium is a production term in the nitrate balance, considering crop demand, denitrification, and leaching. The resulting plant uptake terms determine the soil's deliverable mineral nitrogen.

4 The N-delivery from the soil is utilized by the WOFOST model to adjust growth rates computed in step 2. WOFOST then calculates final values for state variables at the end of the day, establishing values for LAI, rooting depth, dry matter quantities of senescent leaves, stems, and roots, along with corresponding nitrogen quantities in these crop residues.

13 Case studies

13.1 The Netherlands

Based on: Kroes, J. G. and Supit, I. 2011. Impact analysis of drought, water excess and salinity on grass production in The Netherlands using historical and future climate data, *Agriculture, Ecosystems & Environment*, 144(1), 370–381.

This paper examines the effects of increasing salinity of groundwater, drought, and water excess on grass production in The Netherlands. WOFOST simulates crop growth, while SWAP simulates transport of water, solutes, and heat in the vadose zone. The model was tested using several datasets from field experiments. We applied the models at a regional scale where we quantified the impact of various groundwater salinity levels on grass growth and production using historical weather data (1980–2006). The salt concentrations in the subsoil were derived from the National Hydrological Instrument.

Results indicate that salinity effects on grass production are limited. In wet years, the excess rainfall will infiltrate the soil and reduce saltwater seepage. In the next step we used future weather data for the year 2050, derived from three Global Circulation Models. From each model we used data from two CO_2 emission scenarios. As expected higher temperatures increased drought stress; however, the production reduction as a result of salt water in the root zone is limited. Salt stress mainly occurred when irrigation was applied with saline water. The increased CO_2 concentration in combination with the limited drought stress resulted in increasing simulated actual and potential yields. Overall conclusion for grassland in The Netherlands is that drought stress is stronger than stress caused by water excess which on its turn is stronger than salinity stress. Future water demand for irrigation may increase by 11–19% and result in water scarcity if water supply is insufficient.

13.2 Argentina

Based on: Agrohydrological analysis of groundwater recharge and land use changes in the Pampas of Argentina. J. Kroes, J. van Dam, I. Supit, D. de Abelleyra, S. Verón, A. de Wit, H. Boogaard, M. Angelini, F. Damiano, P. Groenendijk, J. Wesseling, A. Veldhuizen, 2019. Agricultural Water Management Agricultural Water Management, 213, 843–857.

This paper examines the shifts in groundwater, climate, and land use in the Pampas region of Argentina, highlighting both the opportunities and challenges these changes present. A decline in groundwater levels, in the absence of irrigation, can lead to drought, causing successive crop failures and yield losses. Conversely, rising groundwater levels can mitigate drought by enhancing root water uptake through capillary action, thereby supporting crop

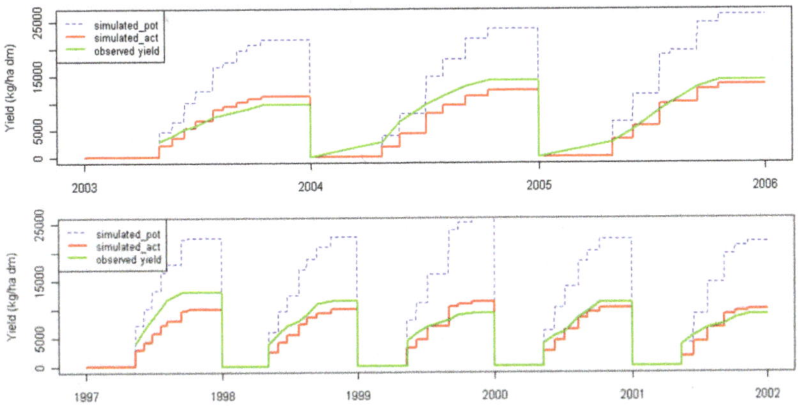

Figure 16 Results for soybeans at the site Zavalla: top left graph shows simulated and observed yields (kg ha^{-1}DM) from official statistics at county level for the period 1990–2015. Top right shows boxplot with median, quartile and extreme values. Lower graph shows the actual yields (kg ha^{-1}DM) during the period 1990–2015.

growth and reducing the gap between potential and actual yields. However, elevated groundwater levels may also reduce soil water storage capacity, increase the risk of flooding in urban areas, and negatively affect crop yields. Transitioning land use from continuous soybean cultivation to crop rotations or natural vegetation may reduce groundwater recharge, subsequently lowering groundwater levels. However, in the case of crop rotation, nutrient leaching, such as nitrates, may increase.

SwapWofost was validated at the field scale using local data from Argentina and subsequently applied at the regional scale to assess the effects on groundwater recharge and crop yields over long-term weather patterns. The findings suggest that continuous, uniform land use poses significant risks, whereas a more varied approach to land use that balances soil water supply and demand offers substantial opportunities. Expanding land use types with higher evapotranspiration rates, such as permanent grasslands and forests, can lead to a more stable hydrologic system with increased soil water storage and lower groundwater levels.

13.3 Bangladesh

Based on: Effect of soil and water salinity on dry season boro rice production in the south-central coastal area of Bangladesh. M.I. Bhuyan, I. Supit, S. Mia, M. Mulder, F. Ludwig, 2023. Heliyon 9 (8).

This study investigated the influence of spatio-temporal variations in soil and water salinity on boro rice production in the south-central coast of Bangladesh. The study area was categorized into three zones (low, moderate, and high salinity) determined by their proximity to the sea in kilometres. Soil

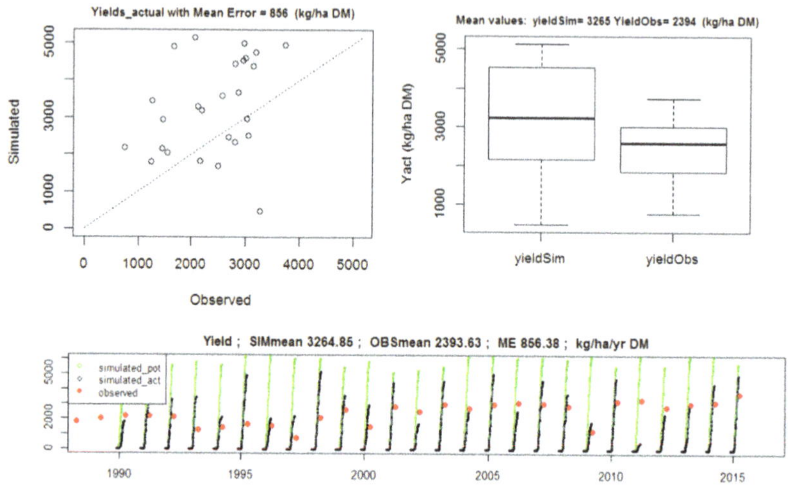

Figure 17 The left panel shows the comparison between observed and simulated groundwater level (m). The right presents the comparison between observed and simulated boro rice yields (t ha^{-1}).

and river water salinity were monitored from November 2020 to June 2021. In the period from March to April 2021, water samples were collected from various canals and crop fields. Groundwater levels, physical soil parameters, and yield data for salt-tolerant rice varieties were obtained from secondary sources. The collected data were validated using the SwapWofost model.

The simulations revealed that salinity levels varied across different soil textural classes and depths. Soil textural classes also exerted an influence on boro rice production, particularly in the high salinity zone, where coarse-textured soil yielded higher boro rice production compared to fine-textured soil. Additionally, boro rice yields were found to vary across different salinity zones. Near the seafront, the actual yield ranged from 668 kg ha^{-1} to 1239 kg ha^{-1}, significantly lower than the moderate (2098–4843 kg ha^{-1}) and low salinity zones (4213–4843 kg ha^{-1}).

To optimize boro rice yield in high salinity areas, it is recommended to sow the crop approximately 15 days earlier than the currently recommended date and cultivate it within polder boundaries. Although boro rice is less affected by salinity in moderate and low salinity zones, early sowing still results in higher yields due to extended radiation capture. The insights gained from this study can aid stakeholders in developing crop-level adaptation plans and formulating policies in coastal districts.

14 Conclusion and recent developments

SwapWofost documentation and model code (Fortran 90) can be found at https://swap.wur.nl/. Note that the WOFOST 7.2 version is implemented in SwapWofost. Contrary to the standalone python WOFOST version, SwapWofost crop files do not take cultivars, crop varieties, etc. into account, meaning that for every variety or cultivar a separate crop file has be to constructed. For those interested in the latest python WOFOST developments detailed information can be found at https://pcse.readthedocs.io/en/stable/. Past and current Swap developments discussed by Heinen et al. (2024).

To run SwapWofost, various input files are required:

1. SWAP file (*.swp): Contains the parameters necessary for the main SWAP routine, such as input files, output files, paths, simulation periods, etc.
2. Drainage file (*.dra): Includes the variables needed to simulate drainage processes. It is only needed if the drainage subroutine is activated in *.swp.
3. Crop file (*.crp): Contains the variables required to run the WOFOST routines.
4. Soil nitrogen routine input files:
   - Soil management events file (*.sme): Contains information on soil management events.
   - Soil management materials file (*.snm): Defines the applied soil management materials, such as urea, cattle slurry, etc.
   - Initial state variables file (*.snp): Includes initial state variables for soil organic matter pools, ammonium-N liquid concentration, and nitrate-N liquid concentration.

Each input file is subdivided in subsections that contain the parameter sets that drive specific model processes, activate the processes that will be included or provide names of input and output, etc. The following daily meteorological information is needed, solar radiation, air temperature (min and max), air humidity, and wind speed. As output daily as well as overall water and soil nutrient balances are produced. The crop simulations output are provided daily base and contain information regarding LAI, development stage, dry weight of the leaves, stems, storage organs, root, etc.

New developments to enhance its accuracy and applicability in addressing emerging environmental challenges focus on the following subjects (Heinen et al., 2024):

1. Improved root water uptake and oxygen stress modelling: The need for a more precise root growth simulations has been identified, leading to plans to incorporate adaptive root growth simulations.
2. Climate change adaptation: The cumulative effects of extreme weather conditions, such as drought, water excess, and heat waves, on crops are not yet fully understood and will be addressed.
3. Soil subsidence and salinization: With the increasing severity of soil subsidence and salinization, particularly in delta regions and areas with peat soils, there is a growing need to integrate soil subsidence processes and enhance the salinity simulation within the model. This will enable more accurate predictions of their effects on root development and crop response.
4. Temporal variation of soil conditions: Currently seasonal soil condition changes, such as ploughing and temperature fluctuations are not well addressed. There is a need for better integration of these processes to reflect their impact on root and crop development.
5. Userfriendliness and case study applications: Enhancements are planned to improve accessibility and adaptability for future agricultural transitions, extreme weather events, and land-use changes. This includes modelling mixed crops, agroforestry, and nutrient-related crop growth, beyond the current water-related focus.
6. Extended nutrient modelling: Plans to expand solute transport functions to include nitrogen source and sink processes which will allow more realistic modelling of agricultural yields and water-nitrogen interactions, improving its use in land evaluation, climate impact studies, and resilient system design.

15 References

Bartholomeus, R.P., et al. 2008. Critical soil conditions for oxygen stress to plant roots: substituting the Feddes-function by a process-based model. *J. Hydrol.*, 360, 147–165.

Belmans, C., et al. 1983. Simulation of the water balance of a cropped soil: SWATRE. *J. Hydrol.*, 63, 271–286.

Boogaard, H.L., et al. 1998. *WOFOST 7.1. User guide for the WOFOST 7.1 crop growth simulation model and WOFOST Control Center 5.1*. Techn. Doc. 52, Alterra, WUR, Wageningen, The Netherlands, p. 144.

Bouten, W. 1992. *Monitoring and modelling forest hydrological processes in support of acidification research*. Diss. Univ. A'dam. 218 pp.

Coleman, K. and Jenkinson, D.S. 2014. RothC – A model for the turnover of carbon in soil. Model description and windows user guide. (updated June 2014). Rothamsted Research, Harpenden.

De Jong van Lier, Q., et al. 2009. Root water extraction under combined water and osmotic stress. *Soil Sci. Soc. Am. J.*, 73, 862–875.

De Jong van Lier, Q., et al. 2013. Modeling water potentials and flows in the soil-plant system comparing hydraulic resistances and transpiration reduction functions. *Vadose Zone J.*, 12(3), 1–20. doi:10.2136/vzj2013.02.0039.

De Willigen, P., et al. 2008. *Decomposition and accumulation of organic matter in soil; comparison of some models*. Alterra Rapport, 1726, Wageningen.

De Willigen, P. and van Noordwijk, M. 1987. *Roots, plant production and nutrient use efficiency*. Ph.D. Thesis. Wageningen Agricultural University, Wageningen, the Netherlands. Available at: https://edepot.wur.nl/202228 (Accesses 12.03.2024).

Drenth, H., et al. 1994. *ORYZA simulation modules for potential and nitrogen limited rice production. In: ORYZA simulation modules for potential and nitrogen limited rice production*. SARP Research Proceedings, CABO-TT, Wageningen, The Netherlands, p. 223.

De Wit, A., et al. 2018. 25 years of the WOFOST cropping systems model. *Agri. Sys.*, 168, 154–167.

Feddes, R.A., et al. 1978. *Simulation of field water use and crop yield*. Simulation Monographs. Pudoc. Wageningen, 189 pp.

Feddes, R.A., et al. 1988. Modelling soil water dynamics in the unsaturated zone – state of the art. *J. Hydrol.*, 100, 69–111.

Groot, J.J.R. 1987. *Simulation of nitrogen balance in a system of winter wheat and soil*. Simulation Reports CABO-TT, No. 13. Wageningen, The Netherlands, 69 pp.

Goudriaan, J. 1986. A simple and fast numerical method for the computation of daily totals of crop photosynthesis. *Agric. For. Meteorol.*, 38, 249–254.

Groenendijk, P., et al. 2005. *Prediction of nitrogen and phosphorus leaching to groundwater and surface waters; process descriptions of the ANIMO4.0 model*. Wageningen, Alterra, Report 983.

Groenendijk, P., et al. 2016. Wageningen: Wageningen Environmental Research. 59 p. (Wageningen Environmental Research rapport; no. 2721).

Hadley, P., et al. 1984. Effects of temperature and photoperiod on flowering in soya bean: a quantitative model. *Ann. Bot.*, 53, 669–681.

Hämäläinen, M. 1991. *Principal variations in the chemical composition of peat*. Swedish University of Agricultural Sciences, Department of Chemistry, Uppsala, pp. 21–22.

Heinen, M. and De Willigen, P. 2005. *Vergelijking van de organische stofmodellen MOTOR en MINIP*. Alterra Rapport, 1260, Wageningen.

Heinen, M., et al. 2024. SWAP 50 years: Advances in modelling soil-water-atmosphere-plant interactions. *Agr. Water Manag.*, 298, 108883. https://doi.org/10.1016/j.agwat.2024.108883.

Hijmans, R.J., et al. 1994. *User's guide for the WOFOST 6.0 crop growth simulation model*. Technical Document 12, Winand Staring Centre, Wageningen, The Netherlands, 144 p.

Hillel, D. 1980. *Fundamentals of soil physics*. Academic Press, San Diego, CA, 412 p.

Jamieson, P.D., et al. 1998. Sirius: a mechanistic model of wheat response to environmental variation. *Eur. J. Agron.*, 8, 161–179.

Janssen, B.H. 1984. A simple method for calculating deomposition and accumulation of "young" soil organic matter. *Plant Soil*, 304, 297–304. http://doi.org/10.1007/bf02205588.

Jarvis, N.J. 1989. A simple empirical model of root water uptake. *J. Hydrol.*, 107, 57–72.

Kabat, P., et al. 1992. SWACROP: a water management and crop production simulation model. *ICID Bulletin 92*, 41(2), 61–84.

Kase, M. and Catský, J. 1984. Maintenance and growth components of dark respiration rate in leaves of C_3 and C_4 plants as affected by leaf temperature. *Biologia Plantarum*, 26, 461–470.

Jarvis, N.J. 2011. Simple physics-based models of compensatory plant water uptake: concepts and eco-hydrological consequences. *Hydrol. Earth Syst. Sci.*, 15, 3431–3446.

Kolenbrander, G.J. 1969. *Establishing the contribution of various ypes of applied organic materials to the soil organic matter content in arable soils* (in Dutch). Rapport C 6988, Institute for Soil Fertility, Haren.

Kolenbrander, G.J. 1974. *Efficiency of organic manure in increasing soil organic matter content*. No. 2 in Trans. 10th Int. Congr. Soil Sci. Moscow, USSR.

Kroes, J.G., et al. 2001. *Actualisatie van de hydrologie voor STONE 2.0*. Rapport 298, Alterra, Wageningen, 68 p.

Kroes, J.G. and Van Dam, J.C. (Eds.). 2003. *Reference manual SWAP version 3.03*. Alterra report 773. Wageningen, 211 p.

Kroes, J.G., et al. 2008. *SWAP version 3.2. Theory description and user manual*. Alterra report 1649, Wageningen University and Research Centre, Wageningen.

Lemaire, G., et al. 1989. *Analysis of the effect of nitrogen nutrition on dry matter yield of a sward by reference to potential yield and optimum nitrogen content*. Proceedings of 16th International Grassland Congress, Nice, France, October 4–11.

Maas, E.V. and Hoffman, G.J. 1977. Crop salt tolerance-current assessment. *J. Irrig. Drainage Div., ASCE*, 103, 115–134.

Mualem, Y. 1976. A new model for predicting the hydraulic conductivity of unsaturated porous media. *Water Resour. Res.*, 12, 513–522.

Penning de Vries, F.W.T. and Van Laar, H.H. 1982. Simulation of growth processes and the model BACROS. In F.W.T Penning de Vries and H.H. van Laar (Eds.), *Simulation of plant growth and crop production*. Simulation Monographs, Pudoc, Wageningen, The Netherlands, pp. 114–135.

Porter, J.R. 1993. AFRCWHEAT2: a model of the growth and development of wheat incorporating responses to water and nitrogen. *Eur. J. Agron.*, 2, 69–82.

Reinink, K., et al. 1986. Adaption of the AFRC wheat phenology model for Dutch conditions. *Neth. J. Agric. Sci.*, 34, 1–13.

Scott, P.S., et al. 1983. *Hysteretic effects on net infiltration*. 'Advances in infiltration', American Society of Agricultural Engineers, St. Joseph, MI, pp. 163–170.

Shibu, M.E., et al. 2010. LINTUL3, a simulation model for nitrogen-limited situations: application to rice. *European Journal of Agronomy*, 32(4), 255–271.

Sinclair, T.R. and Amir, J., 1992. A model to assess nitrogen limitations on the growth and yield of spring wheat. *Field Crops Res.*, 30, 63–78.

Skaggs, T.H., et al. 2006. Macroscopic approaches to root water uptake as a function of water and salinity stress. *Agric. Water Man.*, 86, 140–149.

Spitters, C.J.T., et al. 1989. A simple and universal crop growth simulator: SUCROS87. In R. Rabbinge, et al. (Eds.), *Simulation and systems management in crop protection*. Simulation Monographs, Pudoc, Wageningen, The Netherlands, pp. 147–181.

Supit, I., et al. (Eds.). 1994. *System description of the WOFOST 6.0 crop simulation model implemented in CGMS. Vol. 1: theory and algorithms*. EUR Publication 15956, Agricultural Series, Luxembourg, 146 p.

Van Bakel, P.J.T. 1986. *Planning, design and operation of surface water management systems; a case study*. PhD thesis, Wageningen Agricultural University.

Van Dam, J.C., et al. 1997. *Theory of SWAP version 2.0. Simulation of water flow, solute transport and plant growth in the Soil-Water-Atmosphere-Plant environment.* Wageningen University and Alterra. Technical Document 45.

Van Delden, A. 2001. *Yielding ability and weed suppression of potato and wheat under organic nitrogen management.* Ph.D. Thesis. Wageningen Agricultural University, Wageningen, The Netherlands, 197 pp.

Van den Broek, B.J., et al. 1994. *SWAP 1993, input instructions manual.* Report 45, Dep. Water Resources, Wageningen Agricultural University.

Van Diepen, C.A., et al. 1989. WOFOST: a simulation model of crop production. *Soil Use Manage.*, 5, 16–24.

Van Genuchten, M.T. 1980. A closed form equation for predicting the hydraulic conductivity of unsaturated soils. *Soil Sci. Soc. Am. J.*, 44, 892–898.

Van Heemst, H.D.J. 1986a. The distribution of dry matter during growth of a potato crop. *Potato Res.*, 29, 55–66.

Van Heemst, H.D.J. 1986b. Crop phenology and dry matter distribution. In H. van Keulen and J. Wolf (Eds.), *Modelling of agricultural production: soil, weather and crops.* Pudoc, Wageningen, The Netherlands, pp. 13–60.

Vanuytrecht, E. and Thorburn, P.J. 2017. Responses to atmospheric CO2 concentrations in crop simulation models: a review of current simple and semicomplex representations and options for model development. *Global Change Biol.*, 23, 1806–1820.

Van Keulen, H. and Wolf, J. 1986. *Modelling of agricultural production: weather, soil and crops.* Simulation Monographs. Pudoc, Wageningen, The Netherlands, pp. 479.

Weir, A.H., et al. 1984. A winter wheat crop simulation model without water and nutrient limitations. *J. Agric. Sci.*, 102, 371–382.

Wesseling, J.G., et al. 1991. *SWATRE; instructions for input.* Internal note. Winand Staring Centre, Wageningen.

Wolf, J., et al. 2010. *Methodologies for analyzing future farming systems in Flevoland as applied within the AgriAdapt project.* Wageningen UR.

Xinyou, Y. and Van Laar, H.H. 2005. *Crop systems dynamics: an ecophysiological simulation model for genotype-by-environment interactions.* Wageningen Academic Pub, Wageningen.

Chapter 15

The SIMPLACE crop and ecosystem management modelling platform

M. Vianna, A. Enders, T. Gaiser, G. Krauss and T. H. Nguyen, University of Bonn, Germany; S. J. Seidel, University of Natural Resources and Life Sciences, Austria; E. E. Rezaei and H. Webber, Leibniz Centre for Agricultural Landscape Research, Germany; A. Tewes, University of Bonn and xarvioTM BASF Digital Farming GmbH, Germany; A. K. Srivastava and F. Ewert, University of Bonn and Leibniz Centre for Agricultural Landscape Research, Germany

1 A brief introduction to the SIMPLACE framework

Process-based crop models have become a central tool for agricultural systems analysis as they enable the explicit formulation and testing of how physiological and soil processes interact and explain emerging system behaviour (Jones et al., 2017). Owing to the multifaceted and multidimensional nature of crops and agroecosystems, the development of a unified model is often implausible and practically non-existent. Thereby, models are predominantly employed through the "fit-for-purpose" concept, wherein the contextualisation of scientific questions assumes an important role in the delineation of model boundaries, methods and processes (Hamilton et al., 2022). Moreover, the modularisation of well-established biological processes (i.e. photosynthesis) in the form of algorithms or sub-modules has become increasingly prevalent with the advent of programming paradigms. This trend has allowed for the representation of the same process with varying levels of physiological detail (Adam et al., 2011),

http://dx.doi.org/10.19103/AS.2025.0155.18

a typical example can be given by the differing methodologies associated with radiation use efficiency and the Farquhar approach (Monteith et al., 1977; Farquhar et al., 1980).

Given the intricate and multifaceted nature of crops and agroecosystems, as well as the often-complex trade-offs between crop production and environmental impact, there has been a growing demand for interdisciplinary and integrated assessments (Ewert et al., 2015). Moreover, the burgeoning accessibility to vast amounts of new data and increased processing capacity have opened new avenues for the modelling community. Furthermore, the development of a "next generation" of modelling platforms, coupled with capacity building and close collaboration with the software development community, remains imperative (Antle et al., 2017; Janssen et al., 2017).

In this context, the Scientific Impact Assessment and Modelling Platform for Advanced Crop and Ecosystem Management (SIMPLACE) has been conceived (Enders et al., 2023). The framework has been under continuous development since the early 2010s, based on the concept of formulating an appropriate model structure for a specific research or analytical question. In SIMPLACE, a model structure is referred to as a "ModelSolution" and consists of discrete, replaceable and interchangeable software units, or sub-models, called "SimComponents". The complexity of ModelSolutions differs according to the user needs and input data availability. One ModelSolution may, for example, need a detailed description of soil water, nitrogen and phosphorus dynamics, while another may need to be more general with respect to the description of soil processes but must have a detailed description of carbon assimilation, respiration and biomass partitioning. As a flexible modelling framework, SIMPLACE can accommodate both cases.

The implementation of SIMPLACE infrastructure primarily attempts to balance three paradigms: flexibility, performance and transparency (Enders et al., 2023). Therefore, it contains several packages developed for advanced input/output (I/O) handling, relational databases for internal data resource structure and exchange across components and multi-threading simulations. This allows efficient integration of various data formats within a simulation workflow, maintaining the computational performance and scalability (e.g. in High-Performance Computing systems). A native graphical user interface (GUI) can be used to assist in the development, visualisation and interpretation of a ModelSolution structure and simulation outputs. In addition, a set of wrapper packages is maintained in Python, R, Matlab and Octave languages to provide further control over SIMPLACE simulations, enabling for example, the use of complex statistical methods for sensitivity and calibration analysis as well as coupling with external platforms (Midingoyi et al., 2021; Seidel et al., 2022).

By default, the framework operates at daily time steps, but sub-daily calculations are also possible within the algorithms of a SimComponent (Enders et al., 2023). Typical input information required to simulate agroecosystems consists of daily or sub-daily meteorological data (e.g. solar radiation, air temperature, precipitation, etc.), soil profile characteristics (e.g. water retention parameters, bulk density, texture, carbon and nutrients content, etc.), crop parameters (see Section 2) and management (dates, amounts and properties of fertilisation, manure and irrigation applications, tillage, rotation schemes, etc.). The level of detail and resolution of the input data must conform with the timescale and units' ontology required by all SimComponents that are combined in a ModelSolution. As such, daily or sub-daily outputs describing soil-crop-atmosphere status can be obtained depending on the ModelSolution structure. Modellers can check the required inputs, constants, state and rate variables and outputs for each SimComponent in the modules documentation page: https://simplace.net/doc/simplace_modules/.

The framework is implemented in the Java(R) programming language based on the object-oriented paradigm (Gamma et al., 1995). It's an open-source software project based on the GNU Lesser General Public License (Free Software Foundation, 2007). The source code, ancillary interfacing packages and installable application can be accessed at http://www.simplace.net, and an in-depth description of the framework is provided by Enders et al. (2023). In this chapter, we present an overview of how SIMPLACE (v5.1) core engine and modularisation scheme can be used to represent crops and their genetic differences in contrasting environments. We also show SIMPLACE applications with emphasis on genotype x environment (G x E) interactions and areas for further model enhancement in this topic.

2 Modelling different crops and genotypes with SIMPLACE

As briefly mentioned, SimComponents are interchangeable software units containing the mathematical formulation about the functioning of one or more processes and can be considered sub-models (Enders et al., 2023). A set of SimComponents for the major crop growth and agroecosystem processes are available in SIMPLACE, i.e. crop phenology and development, shoot growth, biomass accumulation, yield formation, crop nitrogen demand, crop water demand, root growth, soil and canopy temperature, root water uptake, soil water, carbon and nitrogen dynamics (Fig. 1). Many of these SimComponents were derived from widely tested models, which are also available in SIMPLACE as whole group components, including simple to quite detailed generic crop models such as LINTUL (Van Oijen & Lefelaar, 2008; Wolf, 2012a), GECROS (Xinyou & van Laar, 2005) and SUCROS (Goudriaan & Van Laar, 2012).

Therefore, users can define and test various combinations of SimComponents that better represent the target system, conform with input

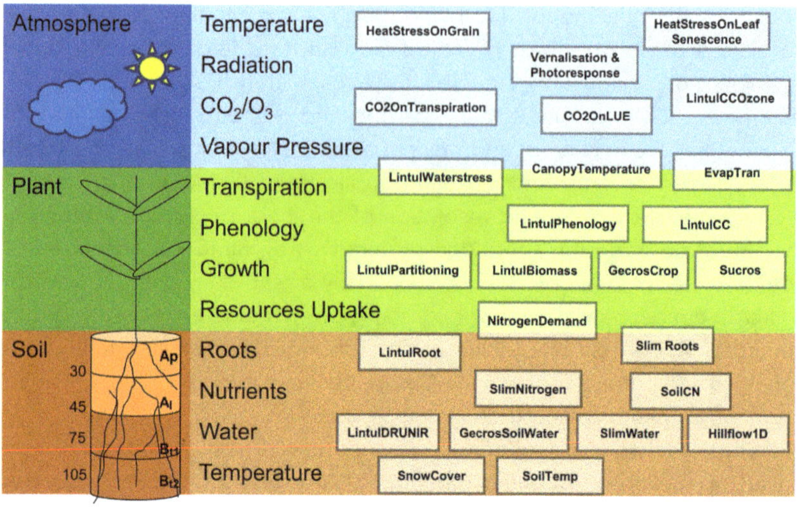

Figure 1 Conceptual overview of the main set of sub-models (SimComponents) that are currently available for crop and agroecosystem modelling in the SIMPLACE platform (v5.1). Most of the SimComponents listed here are related to crop growth and development processes but are positioned next to their main driver within the soil and atmosphere boundaries. Model algorithms coded into these components are previously tested in the literature (Addiscott & Whitmore, 1991; Bronstert & Plate, 1997; Corbeels et al., 2005; Xinyou & van Laar, 2005; Goudriaan & Van Laar, 2012; Wolf, 2012a; Webber et al., 2016b; Nguyen et al., 2022). The full list of SimComponents can be found at: https://simplace.net/doc/simplace_modules/. Adapted from Enders et al. (2023).

data availability, as well as quantify the underlying uncertainties of model structure and detail. As each ModelSolution can be understood as a stand-alone simulation model composed of a distinct set and order of SimComponents, a naming convention has been created which lists the names of the key model components as an array. For example, the ModelSolution named SIMPLACE <LINTUL5, Hillflow1D> is a model which combines the physically based soil water balance model component Hillflow1D (Bronstert and Plate 1997) with the generic crop growth and phenology-related SimComponents of LINTUL5 (Wolf, 2012a).

Special cases of SimComponents are needed to accommodate the users' need for testing diverse modelling structures and crop management practices. For these purposes, two special cases of SimComponents are available in SIMPLACE, namely DefaultManagement and FWSimpleSimComponent. They allow scientists to parse customised equations and small algorithms in between SimComponents without the need for source-code (Java) implementation or modification. A typical application for such SimComponents is triggering management interventions or events, such as sowing and harvesting, tillage, fertilisation, irrigation and control of crop rotations. Data harmonisation, including unit conversion, aggregation and disaggregation between SimComponents, are also examples of common usages for the FWSimpleSimComponents.

It should be mentioned here that the use of such structures should be minimalistic since the computational performance may drop considerably. For more complex implementations, modellers can add custom SimComponents. Adding new SimComponents requires a basic understanding of the modularisation scheme of SIMPLACE and Java programming language as described in Enders et al. (2023). Transpiler platforms such as Crop2ML can be potentially used in the future for bilateral translation of components' code across modelling platforms (Midingoyi et al., 2021). However, adding a new SimComponent may still require finer and some intricate adjustments to become seamlessly operable with the existing SimComponents, especially concerning the (re)initialisation and step-rate discretisation, as well as the state/rate and variables ontologies.

The unified and standard module documentation is an important tool for model development in SIMPLACE (https://simplace.net/doc/simplace_modules/). It contains an explicit description of each variable, underlying units and reference values (default, min, max), which promotes transparency and assists scientists in rapid model development. A help and manual webpage support model installation, setup and guidance tutorials (https://simplace.net/doc/). The control of model versions and an open channel to report model issues facilitates model developers' collaboration and results' reproducibility. A set of template simulation experiments and protocols are also provided to document and maintain the quality of SIMPLACE components (https://simplace.net/doc/index.html?quality_assurance.htm).

With this framework setup, scientists can represent a variety of crops by parameterizing the underlying process governing crop growth and development to the target genotype. Although having some SimComponents with distinct modelling approaches, the generic crop model LINTUL has served as the backbone for most of the modelling developments within SIMPLACE over the last years (Enders et al., 2023). Therefore, in this section, we focus on the main functioning of LINTUL models and how their parameter values can be adjusted to represent different crop species and cultivars.

The first version of the Light INTerception and UtiLisation (LINTUL) model was initially developed to simulate the potential growth of potatoes from daily intercepted photosynthetically active radiation and light use efficiency under optimal growth conditions (Spitters, 1990). It was thereafter parameterised for many annual crops, sharing many similarities with other generic crop models such as the WOrld FOod STudies (WOFOST) (van Ittersum et al., 2003; de Wit et al., 2019). The latest version of the model (LINTUL5) simulates crop phenology, light interception, carbon fixation and partitioning under potential conditions and when soil water, nitrogen, phosphorus and potassium are limiting factors (Wolf, 2012a).

The model assumes that crop phenology is mainly driven by the accumulation of thermal heat units which are normalised into a development stage index (DVS) according to important phenological stages, e.g. emergence (DVS=0), anthesis (DVS=1) and physiological maturity (DVS=2). Therefore, different rates of crop development can be controlled by the parameters defining the cardinal temperatures of crop development and the thermal sums necessary for crop emergence (TSUMEM), anthesis (TSUM1) and physiological maturity (TSUM2). By default, the representation of the photoperiod and vernalisation effects on crop development is switched off but can be parameterised as described in LINTUL4V (Wolf, 2012b).

Light interception is simulated by the classical Beer-Lambert law approach, considering a one-layered crop canopy represented by the leaf area index (LAI). Canopy formation and the proportion of intercepted radiation for carbon fixation can be controlled by parameters describing the specific leaf area (e.g. SLA) and canopy light extinction (e.g. KDIF). Intercepted light is then converted to biomass using the radiation use efficiency (RUE) of the crop, which is directly affected by the current state of atmospheric CO_2, air temperature, soil moisture, nitrogen, phosphorus and/or potassium availability (Wolf, 2012a).

The total amount of biomass converted in the step rate is then partitioned into four organ pools (roots, stems, leaves and storage organs), which are mainly driven by DVS but also can be affected by water and nitrogen shortage. The partitioning fractions over DVS are defined by parameter tables, which are linearly interpolated during the model execution. This approach is also employed in other crop parameters (e.g. SLA, KDIF, RUE, etc.), allowing for high flexibility to represent various shapes of crop response. Senescence is simulated as a function of crop ageing, water and nitrogen stress, and overshadowing. The model also simulates crop nutrient demand and balance on different organs, including translocation to storage organs.

In total, the model has nearly 100 crop parameters, but only the phenology (e.g. TSUMEM, TSUM1, TSUM2) and growth (e.g. RUE) parameters are typically modified to represent different genotypes in each species. In SIMPLACE, model parameter values and input data are normally documented on individual project folders, for which the data structure is maintained by each leader modeller(s) according to their needs. Thus, a unique database matching ModelSolution structure with crop and cultivar parameter values is still a work in progress for future versions of SIMPLACE. Table 1 illustrates a subset of crop parameters of LINTUL5 that controls most of the aforementioned physiological processes. The full definition and ontology for each of the parameters is given on the documentation webpage of SimComponents: https://simplace.net/doc/simplace_modules/

Table 1 Crop physiological processes and respective relevant parameters of the LINTUL5 model implemented in SIMPLACE. Their definitions, units and reference values for winter wheat and maize are also provided after (Van Heemst, 1988; Boons-Prins et al., 1993; Wolf, 2012a). Parameter tables (marked with superscript letters) are named according to the original LINTUL5 implementation (Wolf, 2012a); their reference values are omitted for simplicity but can be found in SIMPLACE's SVN server. The full list of parameters, their definitions and ontology, including the required inputs, state and rates and outputs can be found in the LINTUL5 SimComponent documentation page: https://simplace.net/doc/simplace_modules/net/simplace/sim/components/models/lintul5/Lintul5.html

Physiological processes		Parameter	Definition	Units	Wheat	Maize
Phenology	Development	TSUMEM, TSUM1, TSUM2	Temperature sum for crop emergence (TSUMEM), between emergence-anthesis (TSUM1), and between anthesis-maturity (TSUM2). See Fig. 3a	°C d	60, 1050, 1000	110, 695, 860
		DTSMTB[T]	Increase in temperature sum as dependent on mean daily temperature (cardinal temperatures)	°C	Par. Table	Par. Table
	Photoperiod-response	PHOTTB[D]	Table of development rate reduction until flowering by day length	0–1	Par. Table	Par. Table
	Vernalisation	VBASE	Vernalisation base thermal day	d	2.8	Not used
		VERNRT[T]	Vernalisation factor dependent on average temperature	0–1	Par. Table	Not used
		VERSAT	Vernalisation max thermal day	d	14	Not used
		VernalisationDevStage	Maximal Development stage as DVS* up to which vernalisation is applied	0–2	0.3	Not used
Growth	Light Interception	KDIFTB[D]	Table of light extinction factor as a function of DVS	–	Par. Table	Par. Table
	Radiation use efficiency	RUETB[D]	Radiation use efficiency as a function of DVS (see Fig. 3b)	g MJ^{-1}	Par. Table (Fig. 3b)	Par. Table (Fig. 3b)
		COTB[C]	Table with correction factor of RUE as a function of atmospheric CO_2 concentration	0–1	Par. Table	Par. Table
		TMNFTB[T], TMPFTB[T]	Reduction factor of RUE as a function of low minimum (TMNFTB) and mean (TMPFTB) daily air temperature	0–1	Par. Table	Par. Table
		NLUE	Coefficient for the reduction of RUE due to nutrient (N-P-K) stress	–	1.1	1.1

(*Continued*)

Table 1 (*Continued*)

Physiological processes		Parameter	Definition	Units	Wheat	Maize
Partitioning	Carbon allocation	FRTB[D], FLTB[D], FSTB[D] FOTB[D]	Fraction of biomass partition as a function of DVS to roots (FRTB), leaves (FLTB), stems (FSTB) and storage organs (FOTB) (see Fig. 2)	0–1	Par. Table (Fig. 2)	Par. Table (Fig. 2)
		NPART	Coefficient for N stress-effect on leaf biomass reduction	0–1	1.0	1.0
Canopy formation	Leaf expansion	SLATB[D]	Table of specific leaf area as dependent on DVS	g m^{-2}	Par. Table	Par. Table
		NLAI, NSLA	Coefficient for the reduction due to nutrient (N-P-K) stress of the LAI increase during juvenile phase (NLAI), and in SLA (NSLA)	0–1	1.0	1.0
		RGRLAI	Maximal relative increase in LAI	m^2 m^{-2} d^{-1}	0.00817	0.0294
Senescence	Overshadowing	LAICR	Critical LAI for overshadowing	m^2 m^{-2}	4.0	4.0
		RDRSHM	Relative death rate of leaves due to shading (when LAILAICR)	0–1	0.03	0.03
	W+N stress ageing	DVSDR, DVSDLT	Development stage (DVS scale) above which death of roots and stems starts (DVSDR); and leaves (DVSDLT)	0–2	1.0, 1.0	1.0, 1.0
		RDRL, RDRNS	Maximum relative death rate of leaves due to water (RDRL) and nutrient (RDRNS) stress	0–1	0.05, 0.05	0.05, 0.05
		RDRRTB[D], RDRSTB[D], RDRLTB[T]	Table with relative death rates as a function of DVS for roots (RDRRTB), stems (RDRSTB) and leaves (RDRLTB)	0–1	Par. Table	Par. Table
Root growth	Rooting depth	RDM	Maximal rooting depth	m	1.25	1.8
		RRI	Maximum daily increase in rooting depth	m d^{-1}	0.012	0.012
Nutrients	Demand and balance	FRNX, FRPX, FRKX	Optimal N-P-K concentration in biomass as a fraction of maximal	0–1	1.0,1.0,1.0	1.0,1.0,1.0
		NMAXSO, PMAXSO, KMAXSO, NMXLV[D], PMXLV[D], KMXLV[D]	Maximal N-P-K concentration in storage organs (SO), and in crop leaves (LV) as a function of DVS	g g^{-1}	0.0176, 0.0026, 0.0048, Par. Table	0.0152, 0.0027, 0.0032, Par. Table

Translocation	DVSNT	Development stage above which N-P-K translocations to storage organs occur	0–2	0.8	0.8
	FNTRT	N-P-K translocations from roots as a fraction of resp. total N-P-K amounts translocated from leaves and stems	0–1	0.15	0.15
	TCNT, TCPT, TCKT	Time constant for N-P-K translocation to storage organs	d	10, 10, 10	10, 10, 10

[D,C,T] Indicate parameter tables that are specified as a function of development stage (D), CO_2 concentration (C) or temperature (T); * DVS for grain crops: 0=emergence, 1=anthesis, 2=maturity

Nevertheless, the LINTUL5 model includes a pre-defined set of parameter values that can be used as a starting point for model calibration, including 22 annual crop species (Wolf, 2012a). The parameter values were derived from the WOFOST model, which was originally based on Boons-Prins et al. (1993) for European cropping systems and Van Heemst (1988) for tropical cropping systems. Figure 2 illustrates the differences in carbon partitioning fractions between each crop species as a function of DVS. A common pattern observed in most crops is that carbon allocation is high for roots at early development stages, shifting to storage organs (i.e. grain, tubers, sucrose) at later stages. The total temperature sum for most annual crops falls between 910 and 2280 °C days, except for cassava (550 °C days), cotton (2381 °C days), tobacco (3896 °C days) and sugarcane (6523 °C days) (Fig 3a). The RUE for most crops is higher between vegetative growth and anthesis (2.7 to 5.0 g[DM]/MJ) and progressively drops until physiological maturity (0.0 to 3.5 g[DM]/MJ) (Fig 3b).

It must be noted that the LINTUL model was initially developed and extensively tested for annual temperate crops, thus, the default parameterisation for crops such as cassava, sugarcane or tobacco should be carefully revised before usage (Srivastava et al., 2022; Vianna et al., 2022). In addition, the simulated storage organ pool for some tropical and tuber crops may not necessarily represent grains but rather harvestable tubers and or sucrose. Conversely, the development scale does not represent a typical cereal where anthesis is reached at DVS=1, but rather crop-specific stages, such as tuber initiation or sucrose accumulation development stages. Therefore, the default parameters of LINTUL may not represent the whole development cycle for non-cereal crops (e.g. sugarcane or biennial sugarbeet), but rather their production cycles from germination to the harvestable parts.

The representation of different cultivars within a crop species is typically done by modifying the parameter values controlling phenology (e.g. TSUM1, TSUM2) and/or light use efficiency (e.g. RUE). Other parameter values may also be modified but only provided that the corresponding crop data characteristics are available and show strong evidence for such modification. For example, biomass partitioning fractions, rooting biomass and depth information, specific leaf area, light interception coefficient by the canopy, etc. The definition and units for each crop parameter of the LINTUL5 version implemented in SIMPLACE can be found on their respective SimComponent page (Table 1).

The original LINTUL soil water balance considers only one soil compartment, which generally does not capture the vertical heterogeneity of soil properties and root dynamics (Jarvis et al., 2022). To better represent the vertical heterogeneity of soil properties and root dynamics the LINTUL modules in SIMPLACE can be coupled to more detailed soil water balance routines, such as SlimWater, a "tipping-bucket" type of soil water balance (Addiscott & Whitmore, 1991), or the Hillflow1D which is based on the Richard's equation

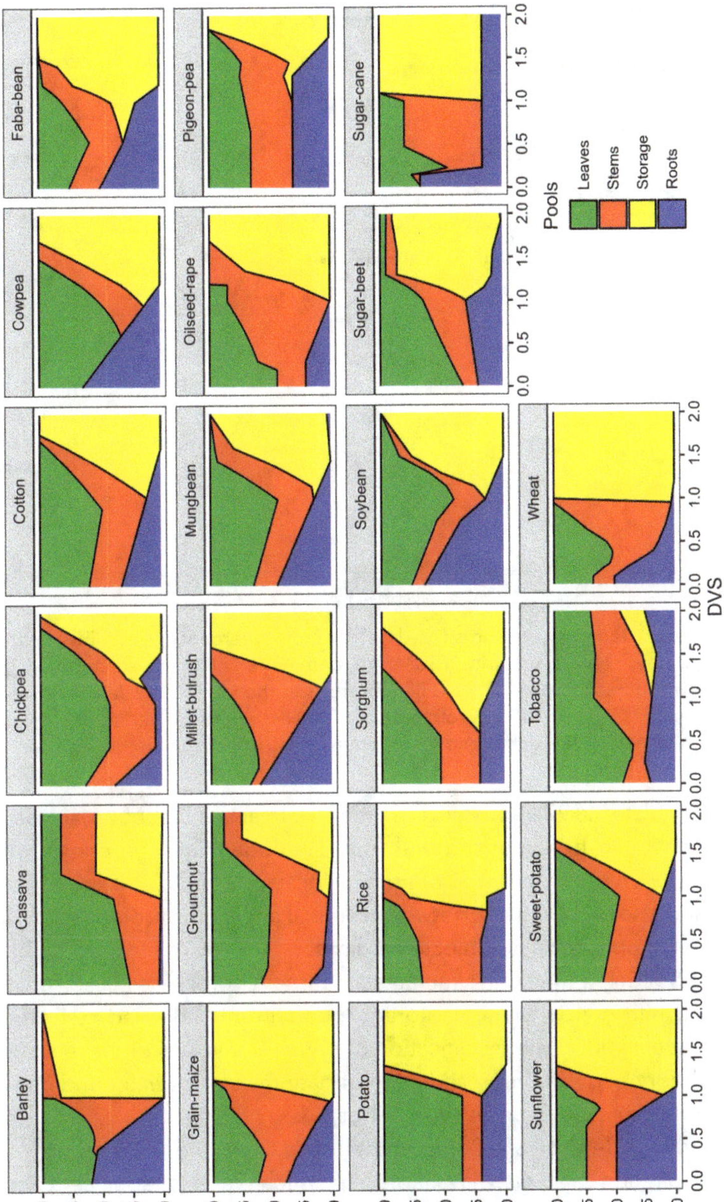

Figure 2 Default carbon partitioning fractions parameters tables (0–1) for different organ pools (roots, stems, leaves, storage) and different crops as a function of the crop development stage index (DVS) in the LINTUL model (Van Heemst, 1988; Boons-Prins et al., 1993; Wolf, 2012a).

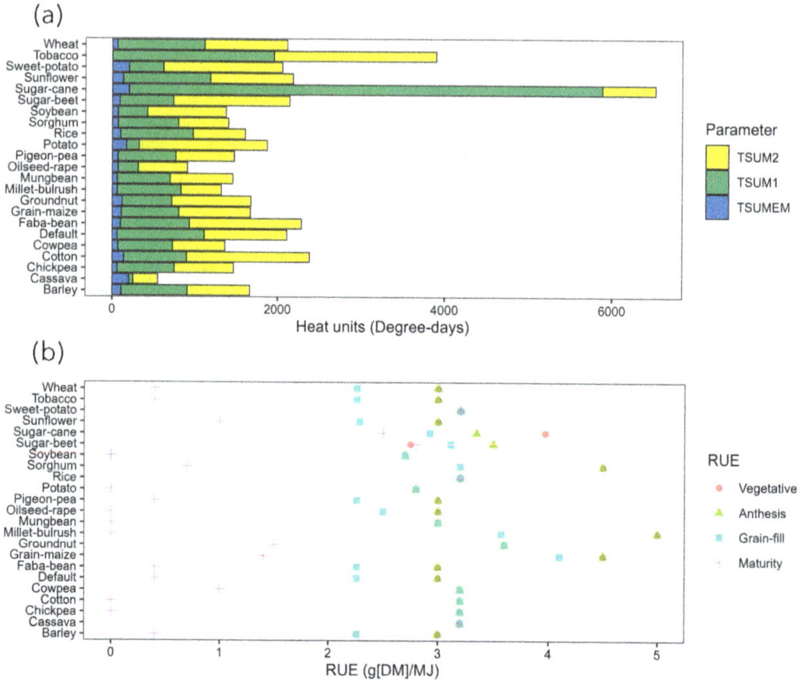

Figure 3 Default temperature sums (°C days) for crop emergence (TSUMEM), anthesis (TSUM1) and physiological maturity (TSUM2) (a) and the radiation use efficiency (RUE) at different development stages (b) for different crops in the LINTUL model (Van Heemst, 1988; Boons-Prins et al., 1993; Wolf, 2012a). These parameter values are typically calibrated to represent different genotypes within a crop species in SIMPLACE.

(Bronstert & Plate, 1997). Vertical growth of roots and distribution across soil compartments can be simulated with the SlimRoots SimComponent, which takes into account the amount of biomass allocated to roots, soil strength and root age and also differentiates between seminal and lateral roots (Gaiser et al., 2013). When crop growth is subjected to limited nutrients, it is also possible to couple the SoilCN routine that simulates the turnover processes of soil organic carbon and nitrogen in multiple storage pools in multi-layered soil profiles (Corbeels et al., 2005). The modification of parameter values of these routines would in theory, allow the representation of patterns of root water and nutrient uptake for different crop species and genotypes.

Similarly, different SimComponents dedicated to simulating crop evapotranspiration are also available in which various species can be parameterised following the principles of reference evapotranspiration and crop coefficients as defined in the FAO56 bulletin (Allen et al., 1998). Additionally, hourly-based canopy temperature and gas exchange may also be included to simulate the effect of heat stress, which was developed and tested for LINTUL5 using the Monin–Obukhov Similarity Theory (Webber et al., 2016b). An extension of LINTUL for water and ozone stress is also available in SIMPLACE. The model was initially developed by Van Oijen and Ewert (1999) (LINTULCC) and updated by Rodriguez et al. (2001) (LINTULCC2) to include crop response to CO_2 using the biochemical model of Farquhar et al. (1980), combined with an hourly-based stomatal model (Leuning, 1995). LINTULCC2 was further extended for ozone effects on leaf duration and senescence as in Ewert et al. (1999) and Ewert and Porter (2000) and to include plant hydraulic conductance and direct linkage of water potential gradient at the soil-root-plant system to simulate stomatal regulation and water stress (Nguyen et al., 2020; Feng et al., 2022). The parameterisation of hourly ozone flux and ozone effects (instantaneous and cumulative) on the maximum carboxylation rate of Rubisco and leaf senescence were introduced to the LINTULCC2 in SIMPLACE (Feng et al., 2022). Recently, the LINTULCC2 – Ozone SimComponent has been tested and applied at the European level (Feng et al., 2022; Nguyen et al., 2024).

Besides LINTULCC2, the current version of SIMPLACE (v5.1) includes SimComponents for LINTUL2 and LINTUL5 models. For both models, modellers can use either its monolithic (original structure) form or modular (e.g. Phenology, RadiationUseEfficiency, Biomass, WaterStress, NPKDemand, etc.), which gives more flexibility in combining with other SimComponents in a ModelSolution. A summary of all LINTUL models implemented into the SIMPLACE framework is given in Table 2. Different model structures may require substantial revision of model parameterisation to ensure numerical accuracy, and it's at the modellers' discretion to check the plausibility of the model solution for the given application (Enders et al., 2023). Furthermore, experimental SimComponents with LINTUL modifications for grassland and cassava are also available but require further validation and testing before usage.

3 Recent improvements and applications of SIMPLACE for GxE interactions

The SIMPLACE framework has been applied to a range of studies in agricultural systems in recent years, as summarised by Enders et al. (2023). A large proportion of these studies contributes to the understanding of climate change impacts and adaptation options for crop production under various spatiotemporal scales, geographies and crops, but also to a variety of applications quantifying the crop and agroecosystem responses to agronomic management to support policy-making, crop breeding strategies and resource use efficiency. It must be noted that in many studies, SIMPLACE was also employed to simulate optimal management (M) practices (e.g. irrigation, fertilisation, sowing, crop rotations) in conjunction with genotype and environment conditions, i.e.

Table 2 Different versions of the LINTUL model implemented into the SIMPLACE framework (v5.1).

Model name	Description	SimComponent page
LINTUL2	Implementation of the FST-Lintul2 on its original monolithic forms (Van Oijen and Lefelaar, 2008). The model only simulates crop growth under potential and water limited conditions.	https://simplace.net/doc/5.1/simplace_modules/net/simplace/sim/components/models/lintul2/Lintul2.html
LINTUL (2Plus)	• Same as LINTUL2, but modular and extended with: N-response subroutines from LINTUL3 (Shibu et al., 2010) • Drought tolerance factor • Atmospheric CO_2 effect on RUE and crop transpiration	https://simplace.net/doc/5.1/simplace_modules/net/simplace/sim/components/models/lintul/package-summary.html
LINTUL (2Plus) Grassland	Modular SimComponents for grassland simulations adapted from LINTUL (2Plus). Still in the experimental phase and not fully validated.	https://simplace.net/doc/5.1/simplace_modules/net/simplace/sim/components/experimental/grassland/lintul/package-summary.html
LINTUL5	• An improved version was developed by Wolf (2012a) for crop growth simulations under potential, water limited and N-P-K limited conditions. It was implemented in SIMPLACE in both monolithic and modular forms, including: Vernalisation routines from LINTUL4 (Wolf 2012b) • Heat stress parameterisation • N Fixation	https://simplace.net/doc/5.1/simplace_modules/net/simplace/sim/components/models/lintul5/Lintul5.html
LINTUL5 Cassava	Monolithic LINTUL5 implementation adapted for cassava simulations. Still on experimental phase and not fully validated.	https://simplace.net/doc/5.1/simplace_modules/net/simplace/sim/components/models/lintul5/extended/cassava/Lintul5Cassava.html
LINTULCC2	A detailed version of LINTUL developed in the early 2000s (Oijen and Ewert, 1999; Rodriguez et al., 2001) that simulates crop development using phyllochron interval and photosynthesis (Farquhar et al., 1980), including stomatal response to atmospheric CO_2 and O_3 concentrations. The inputs/outputs of this SimComponent are at daily step, but calculations are performed at hourly timesteps with internal aggregation and estimation of the diurnal time course of input variables.	https://simplace.net/doc/5.1/simplace_modules/net/simplace/sim/components/experimental/lintulcc/LintulCC.html
LINTULCC2 Diurnal	Same as LINTULCC2, but hourly input/outputs can be explicitly specified or extracted.	https://simplace.net/doc/5.1/simplace_modules/net/simplace/sim/components/experimental/lintulcc/LintulCCDiurnal.html
LINTULCC2 Diurnal Hillflow	Same as LINTULCC2Diurnal, but tightly coupled to the Hillflow SimComponent for sub-daily soil water and crop interactions solving the Richards' equation.	https://simplace.net/doc/5.1/simplace_modules/net/simplace/sim/components/experimental/lintulcc/HillFlow1DLintulCCDiurnal.html

GxExM interactions. In this section, we emphasise the framework application to represent different crops and genetic differences in contrasting environments.

Cereal crops such as wheat, maize and barley are among the most frequently simulated crops with SIMPLACE. Wheat is by far the most studied crop with SIMPLACE, with more than 77 genotypes represented using the LINTUL models across all the continents (Fig. 4 and Table 3). Most of these studies owe to a strong international collaboration within the Agricultural Model Intercomparison and Improvement Project (AgMIP) (Rosenzweig et al., 2013), which put together a number of modelling groups and datasets to accelerate model evaluation, testing and improvements over the last years.

Asseng et al. (2015) assessed the wheat response to temperature in 30 locations across the globe, then later doubled the number of sites to quantify the impact of climate change and adaptation strategies in wheat protein (Asseng et al., 2019). In their study, SIMPLACE was employed within an ensemble of crop models to simulate 29 wheat genotypes under contrasting pedoclimatic conditions. Guarin et al. (2022) expanded this dataset and used an ensemble of process-based crop models (SIMPLACE included) to investigate the limits of current high-yielding wheat genotypes across the world.

Using the same dataset, Stella et al. (2023) used SIMPLACE to explore trait characteristics of wheat (RUE, fruiting efficiency [FE] and light KDIF) conferring high yield potential and stability under climate change. For their analysis, they used an adapted version of LINTUL5 with canopy temperature and heat-stress function (Webber et al., 2016b), and tested over 1700 virtual genotypes per site with different combinations of trait levels for RUE, KDIF and FE. Rezaei et al. (2018) used SIMPLACE to quantify the effect of using the phenology traits of cultivars planted in the 1950s and 1960s as compared to modern cultivars on wheat growth simulations in Germany. Linking model parameters and genotypic traits, Tao et al. (2017) employed a crop mode ensemble (SIMPLACE, included) to design future crop ideotypes for barley in Finland and Spain. To design different barley ideotypes for Boreal and Mediterranean climatic zones, they modified not only the parameters controlling crop phenology (TSUM1 and TSUM2) and growth (RUE) but also initial canopy formation, specific leaf area and sensitivity to water stress.

Maize is the second most studied crop with SIMPLACE, with 17 different genotypes tested in the literature (Table 3). Some genotypes were tested in Germany and Spain (21%), Ecuador and Argentina (21%), but the majority (58%) originated from Africa (Table 3. and Fig. 4). It must be noted that in more than half of the studies (52%) the maize crop was calibrated using a "cultivar-mix" concept, where one set of parameters may represent a maturity group or was calibrated using pooled data of phenology, biomass and yield from a certain location or region. This approach is mainly used in regional or global simulation studies (Webber et al., 2016a, 2018; Siebert et al., 2017; Constantin

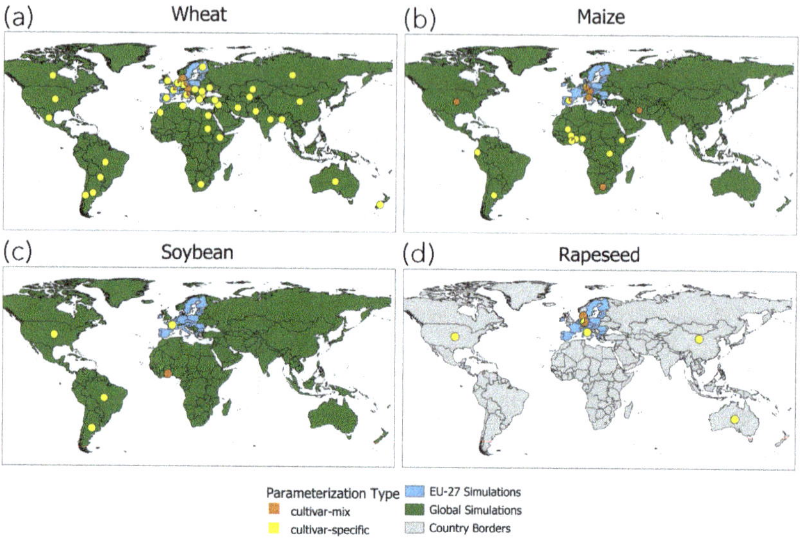

Figure 4 Spatial distribution of the SIMPLACE framework applications for wheat (a), maize (b), soybean (c) and rapeseed (d). Piecharts on the maps represent the frequency of times the respective crop was simulated using cultivar-specific (yellow) parameters or cultivar-mix (orange) parameters in each country. Blue countries represent when the crop was simulated at least once at EU-27 scale, whereas green represents crops simulated at the global scale (Jägermeyr et al., 2021). Results were generated based on the information extracted from a total of 87 studies using SIMPLACE from the years 2013 to 2023 (https://www.simplace.net/index.php/simplace-features/publications)

et al., 2019; Jägermeyr et al., 2021), but also to represent distinct maturity groups as adaptation options under climate change (Kamali et al., 2022). In some of these studies, aspects related to the spatiotemporal representativity of heat and drought events were tested in spatial scales ranging from 1 km x 1 km; 10 km x 10 km; 25 km x 25 km; 50 km x 50 km; and 100 km x 100 km (Eyshi Rezaei et al., 2015; Zhao et al., 2015; Hoffmann et al., 2016). Yet, the crop model parameter values for the maize genotypes "Obatanpa", "BH660" and "BH540" were frequently employed to represent crop production in many countries of Sub-Saharan Africa (Srivastava et al., 2017, 2019, 2020; Faye et al., 2018a; Falconnier et al., 2020).

Data from both wheat and maize crops were used to develop and test new modelling approaches, including canopy temperature, water and heat stress on cereals, and ozone stress. The genotype data of maize ("Nidera AX 842 MG" and "FAO 700") and wheat ("Bihilifa YDT") were also used to test new modules of canopy temperature and heat-stress on cereals in experimental sites located in Argentina, Spain and Burkina Faso (Gabaldón-Leal et al., 2016; Webber et al., 2016b). Two coupled root–shoot models were evaluated in terms of CO_2 and H_2O fluxes and the growth of a maize cultivar called "Zoey" (Nguyen et al., 2020,

Table 3 Crop types, respective list of cultivars and number (n) and predominant regions and scales on which the SIMPLACE framework was applied from 2013 to 2023 (total of 87 studies). Cultivar-mix represents parameterisations considering either maturity groups or large datasets composed of data from various genotypes (e.g. government statistics). The nCS represents the frequency (%) of cultivar-specific parameters employed over the total number of applications per crop.

Crop	Cultivars	n	nCS (%)	Predominant region and scale
Wheat	AC Radiant, Ambello, Anda, Anza, Apache, Arminda, Attila, Avalon, Avocet, Bacanora 88, Bandit, Basri Bey, Basri Rey, Benchmark, Bermude, Bolac, Bologna, Brigadier, Califa sur, Capo, Carimulti, Colt, Debeira, Dekan, Enorm, Faisalabad-2008, Fuller, Gamenya, Greif, Hanno, Hartog, HD 2687, HD2009, Heines VII, HI 1544, HS Burgweizen, Janz, JB Asano, Jenga, Jonny, Jubilar, Julius, Kanchan, Kruunu, LM26, Lyrik, Milaneco, Nesser, NM13, Oasis, Oassis, Okapi, Pishtaz, Poppelsdorfer Begrannter, Premio, RGT Reform, SB062, Scirocco, Seri M 82, SM15, Soissons, Solstice, Sperber, Spitfire, SsrT65, Steklov-24, Tacupeto C2001, Thasos, Tommi, Trebelir, Urho, Wakanui, Wenmai, Wyalkatchem, Yecora RojO, Yitpi, cultivar-mix	77	84.8	Global and EU-27 (gridded, site-based and homogeneous zones)
Maize	BH540, BH660, Bihilifa YDT, DK7088, Dodzie, EMBLEMA, EVDT-97 STR, Nidera AX 842 MG, Obatanpa, Pioneer 31N28, Pioneer 33Y72, Romario, Suwan 1-SR, TRUENO, ZM607, Zoey, cultivar-mix	17	54.1	Global, EU-27 and Sub-Saharan Africa (gridded, site-based and homogeneous zones)
Rapeseed	46Y78, Big Yellow, Hyola50, Taurus, Kabel, Oasis, Visby, cultivar-mix	8	37.5	Global (site-based), EU-27 (gridded, site-based and homogenous zones)
Soybean	42-M1, BRS5980, Don Mario, ISIDOR, Pioneer 92Y75R, cultivar-mix	6	71.4	Global (gridded and site-based)
Barley	Annabell, Eunova, Hispanic, Salome, Sidney, cultivar-mix	6	31.0	EU-27 (gridded, site-based and homogenous zones)
Cassava	Oko-iyawo, TME419, TMS91/02324, TMS96/1632, TMS98/0505	5	100.0	Nigeria (gridded)
Cowpea	Asontem, GH6060, Hans adua, Nketewade	4	100.0	Ghana and Burkina Faso (site-based)
Peanut	73-33, Fleur11, cultivar-mix	3	66.7	Senegal (site-based)

(*Continued*)

Table 3 (*Continued*)

Crop	Cultivars	n	nCS (%)	Predominant region and scale
Sorghum	CSM335, CSM63E, Kadaga	3	100.0	Mali, Burkina Faso, Ghana, Benin (gridded)
Millet	Bolga-local, CIVT	2	100.0	Mali, Burkina Faso, Ghana, Benin (gridded)
Potato, Rice, Rye, Sugar beet, Grass, Mustard, Oat, Pea, Radish, Clover	cultivar-mix	1	0.0	EU-27 (gridded and site-based)
Rangeland	species-mix	1	0.0	South Africa (site-based)

2022). Feng et al. (2022) used the data from an experimental site in Germany of four selected wheat varieties which were either tolerant ("Jenga" and "Apache") or sensitive ("Solstice" and "Enorm") to ozone exposure to parametrise and test the LINTULCC2 crop model in SIMPLACE. The combination of the dataset and model framework enabled the identification of key physiological traits that confer tolerance or sensitivity to ozone in winter wheat. More recently, Nguyen et al. (2024) calibrated and evaluated three process-based models (SIMPLACE, included) extended to simulate the impact of ozone on wheat production across Europe. In their study, open-top chamber data collected for a commercial cultivar named "Califa sur" in Spain was used to calibrate the models, while Sentinel-2 derived green leaf area and statistical grain yield of soft and durum wheat across Europe were used to evaluate the performance of the models.

Soybean and rapeseed crops were parameterised for LINTUL5 across contrasting environments and genotypes (Fig. 4). Kothari et al. (2022) assessed the uncertainty of an ensemble of process-based crop models in simulating crop growth under climate change conditions. The models, including SIMPLACE, were calibrated and tested using genotype data from five producing regions including Argentina (cv. "Don Mario"), France (cv. "ISIDOR"), USA (cv. "Pioneer 92Y75R" and "42-M1") and Brazil (cv. "BRS5980"). The LINTUL model was among the models that consistently showed the highest yield gain under higher temperatures and increased CO_2. In a similar approach, (Wang et al., 2022) examined the reliability of current crop models (SIMPLACE, included) for crop growth simulations of rapeseed at contrasting producing regions. In their study, SIMPLACE was calibrated and evaluated for seven genotypes located in Australia, China, Germany, Italy and the USA (Table 3 and Fig. 4). However, results showed high uncertainty, which was particularly in the crop response to temperature and CO_2, thereby, new comprehensive field experiments are still needed to further improve the representation of these physiological processes for rapeseed.

The parameterisation of various other annual crops, such as barley, rye, oat, clover, potato, sugar beet, rapeseed and peas was predominantly achieved in crop rotation studies in Europe (Fig. 5). Seidel et al. (2021) simulated six different crops (sugar beet, wheat, rye, clover, oat and potato) in a long-term experiment in Dikosphof, Germany, between 1906 and 2010. To simulate all of these crops and genotypic variations, the temperature sums and RUE parameter values of LINTUL5 were modified to represent older to modern variety groups: winter wheat – five groups, winter rye – four groups, sugar beet – four groups, clover – two groups and potato – three groups. Kollas et al. (2015) conducted the first model inter-comparison on crop rotations where more than 12 crops were simulated in five sites across Germany, France, Denmark and Austria. They found that the model ensemble mean minimised the simulation errors, but the representation of intermediate crops (e.g. radish and grassland) and carry-over effects in the models required further research efforts. Five crops (wheat, maize, barley, potato and sugar beet) were extensively used at the national (Webber et al., 2020) and European level (Zimmermann et al., 2017), including crop rotations (Faye et al., 2023). In these studies, either the models were calibrated to simulate distinct maturity groups or the genotypes were not specified; therefore, here we classify them as "cultivar-mix" studies (Table 3).

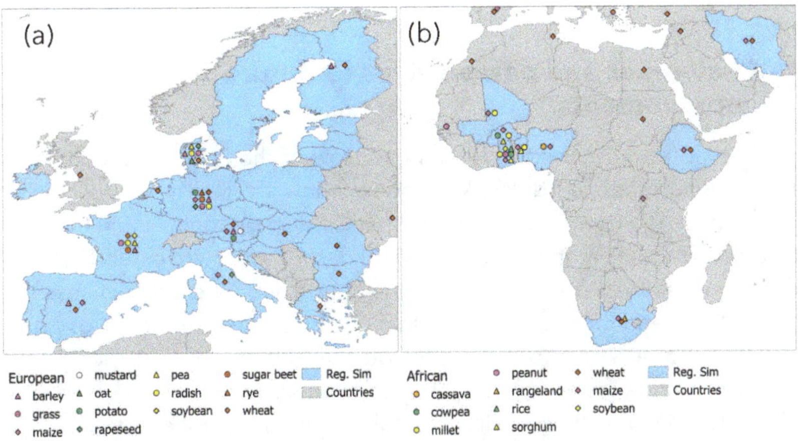

Figure 5 Spatial distribution of crops simulated with the SIMPLACE framework in Europe (a) and Africa + Iran (b). Blue-coloured countries represent studies where SIMPLACE was applied at the national or regional level (e.g. EU-27). Results were generated based on the information extracted from a total of 87 studies using SIMPLACE from the years 2013 to 2023 (https://www.simplace.net/index.php/simplace-features/publications)

SIMPLACE was also used to represent tropical crops such as cassava, cowpea, peanut, sorghum and millet in Sub-Saharan Africa (Fig. 5). Faye et al. (2018b) calibrated and used SIMPLACE for climate change impact on peanut production in Senegal, considering two genotypes under different levels of water regimes and fertilisation. A similar approach was used to estimate the impact of increasing temperature on cereal yields in West Africa (Faye et al., 2018a), where two genotypes of maize, two of millet and three of sorghum were represented with the SIMPLACE and DSSAT platforms. Four cowpea genotypes were represented by Adusei et al. (2023) in Ghana and Burkina Faso, with SIMPLACE as an option for climate change adaptation. Using LINTUL5, Srivastava et al. (2022) calibrated five different cassava varieties based on field experiments to determine the magnitude of the yield gap in Nigeria. To parameterise cassava, many crop parameters had to be modified controlling crop phenology, carbon fixation, canopy formation and root growth.

In an attempt to simulate rangeland ecosystems, Oomen et al. (2016) developed a SimComponent (Linrange) in SIMPLACE. The SimComponent was calibrated and tested for rangeland at different levels of degradation in South Africa and represents species-mix through one set of community parameters. Except for rangeland (Oomen et al., 2016), all other crops and respective genotypes have been simulated with the LINTUL models within SIMPLACE. This not only shows the ability of the LINTUL models to simulate a range of annual crops but also the appropriate level of complexity to link and test several methods of soil and crop processes. All of these examples demonstrate the wide range of applications and how SIMPLACE can be utilised to represent GxE interactions in different cropping systems.

4 Challenges and perspectives for future improvements

In this chapter, we provided a brief overview of the SIMPLACE framework's main structure and modules. We showed how LINTUL modules could be modified to represent various crop species and genotypes within a given species via model parameterisation. It also described how to extend the model for detailed soil water and nutrients simulations, and we cited some examples of how to extend the model to represent other abiotic stresses (e.g. extreme heat and ozone). A compilation of studies from the years 2013 to 2023 was conducted, showcasing the range of crops and genotypes represented with SIMPLACE in contrasting environmental conditions for different applications and scales.

Although SIMPLACE has shown high flexibility and potential for incorporating a variety of algorithms, most of the studies predominantly used LINTUL-based components to represent crop and genotypic variability. The parameterisation of tropical crops in generic crop models like LINTUL may also be possible but with greater alteration on the parameter values as exemplified for cassava (Srivastava et al., 2022). Yet, this also limits the representation of intrinsic processes of non-cereal crops, such as dormancy (grassland), sucrose accumulation (sugarcane), tuber initiation and formation (potato) or source-sink relations. The latter may also be important for cereals, such as wheat, for estimating crop-environmental interactions that affect assimilate supply, including breeding, industrialisation-induced or geo-engineered solar dimming, genetically and atmospheric CO_2-related increased RUE, and source manipulations, such as pest and disease impacts (Asseng et al., 2017).

Furthermore, while LINTUL5 provides reference parameterisation for rice, there is still no paddy and flooding system explicitly simulated in SIMPLACE, which limits its application to only upland rice (Adelesi et al., 2023). Future versions of SIMPLACE may incorporate the water balance scheme, nutrient limitation (NPK) and crop-specific modules of rice to overcome these limitations, such as those from the ORYZA2000 model (Bouman and van Laar, 2006). It may also have a good link with the LINTUL models since its predecessor version (LINTUL-3) was originally developed to simulate the growth of flooded rice in Asia (Shibu et al., 2010). Adding crop-specific models may be an option for other crops, which have dedicated crop models embedded in modelling platforms like potatoes (Raymundo et al., 2017), cassava (Moreno-Cadena et al., 2020) and sugarcane (Jones & Singels, 2018; Dias et al., 2019; Vianna et al., 2020). Due to its flexibility, LINTUL models may also be extended to represent crop-specific species, for example, LINTUL-Cassava (Ezui et al., 2018) or LINTUL-Banana (Taulya, 2015).

Some studies have used SIMPLACE to explore different crop traits for the identification of ideotypes that are more productive and resilient to climate change (Tao et al., 2017; Stella et al., 2023), but the crop parameters of most process-based models (including LINTUL) do not represent the true genetic expression in a given crop, but rather its phenotypic characteristics under one or several environment realisations (depending on the calibration dataset). Recent attempts to overcome this issue include the coupling of multiple linear models representing the interactions of quantitative trait loci (QTLs) and environment (E) with phenological parameters of crop models, such as CROPGRO-Drybean (Oliveira et al., 2021). This is still an area to be explored in SIMPLACE.

The need for an improved representation of extreme weather events in crop growth and their consequent effects on food provision remains substantial (Raymond et al., 2020; Ruane et al., 2022). While the impact of heat stress has been studied for cereals, the effects of excessive rainfall on crop production have still not been tested in SIMPLACE and have gained attention with the increasing frequency of extreme rainfall and flooding events (Nóia Júnior et al., 2023; Kim et al., 2024). Adding new physiological crop responses may also substantially increase the complexity of these dynamic models. While abiotic stressors are widely employed in many crop models, there's still limited knowledge on the mechanisms of simultaneous abiotic stressors in crops, as well as the representation of crop acclimation or priming or memory effects (Webber et al., 2022).

Modelling biotic stressors in process-based crop models is still at a preliminary stage. Integrating pest and disease models with dynamic crop models may necessitate not only the development of new modules to portray host–pathogen interactions but also the incorporation of spatially distributed and interactive fields to simulate pathogen population dynamics (Donatelli et al., 2017). The simulation of resource dynamics and the provision of ecosystem services have gained momentum in recent years and will demand an enhanced portrayal of belowground dynamics, spatial soil heterogeneity and the lateral movement of water and matter across the landscape (Vereecken et al., 2016). Additionally, ensuring product quality in agricultural production is imperative for human health and nutrition (Asseng et al., 2019). Yet, underutilised and less common crops, such as vegetables, tree fruits and nuts, are seldom incorporated into modelling platforms (Holzworth et al., 2015).

New diversified systems derived from intercropping, grasslands, agroforestry and integrated livestock practices have also become important pathways to promote ecosystem services and more sustainable and resilient agricultural systems (Ewert et al., 2023). This poses challenges to the current crop and agroecosystem models in particular regarding the representation of the multi-species interactions and competition for resources (above and belowground), the inclusion of more plant and animal species options, the incorporation of soil legacy effects and biodiversity estimations (Hernández-Ochoa et al., 2022). While some of these examples are still to be unravelled and documented by pure experimentation, modelling platforms like SIMPLACE will need to adapt to accommodate such diversified systems, spatiotemporal scales and multi-species interactions.

5 Conclusion

In this chapter we described the SIMPLACE modeling framework. The literature review listed in this chapter reflects the interdisciplinary foundation and scientific rigor underpinning the SIMPLACE modeling framework. By drawing from a wide array of studies on crop growth simulation, climate impact analysis, and agroecosystem modeling, SIMPLACE has established itself as a versatile and adaptable tool aimed for addressing complex agricultural challenges. The studies sourced in this chapter demonstrate not only the model's capacity for integrating various components–such as soil, weather, crop, and management interactions–but also its relevance in both regional and global contexts in representing GxE interactions. Continued engagement with such multidisciplinary research will be crucial for enhancing model accuracy, expanding its application domains, and supporting evidence-based decision-making in sustainable agriculture and climate resilience planning.

6 Where to look for further information

For a detailed description about the SIMPLACE framework see Enders et al. (2023). An entry-point for the SIMPLACE framework together with a "Getting started" documentation tutorial can be found in the URLs below:

- Official website: https://www.simplace.net/
- Getting started tutorial: https://simplace.net/doc/index.html?getting_started.htm

7 References

Adam, M., et al. (2011). Effects of modelling detail on simulated potential crop yields under a wide range of climatic conditions. *Ecological Modelling*, *222*(1), 131–143.

Addiscott, T. M. & Whitmore, A. P. (1991). Simulation of solute leaching in soils of differing permeabilities. *Soil Use and Management*, *7*(2), 94–102.

Adelesi, O. O., et al. (2023). Accounting for weather variability in farm management resource allocation in Northern Ghana: An integrated modeling approach. *Sustainability: Science Practice and Policy*, *15*(9), 7386.

Adusei, G., et al. (2023). Model-based climate change adaptational potential and productivity of some cowpea genotypes and its sensitivity to bias adjustment. *Frontiers in Agronomy*, *5*. https://doi.org/10.3389/fagro.2023.1144219.

Allen, R. G., et al. (1998). Crop evapotranspiration-Guidelines for computing crop water requirements-FAO Irrigation and drainage paper 56. *Fao, Rome*, *300*(9), D05109.

Antle, J. M., et al. (2017). Next generation agricultural system models and knowledge products: Synthesis and strategy. *Agricultural Systems*, *155*, 179–185.

Asseng, S., et al. (2017). Simulating the impact of source-sink manipulations in wheat. *Field Crops Research*, *202*, 47–56.

Asseng, S., et al. (2019). Climate change impact and adaptation for wheat protein. *Global Change Biology*, *25*(1), 155–173.

Asseng, S., et al. (2015). Rising temperatures reduce global wheat production. *Nature Climate Change*, *5*(2), 143–147.

Boons-Prins, E. R., et al. (1993). Crop-specific simulation parameters for yield forecasting across the European Community Simulation Reports CABO-TT, no 32.

Bouman, B. A. M. & van Laar, H. H. (2006). Description and evaluation of the rice growth model ORYZA2000 under nitrogen-limited conditions. Agricultural Systems, 87(3), 249–273. https://doi.org/10.1016/j.agsy.2004.09.011

Bronstert, A. & Plate, E. J. (1997). Modelling of runoff generation and soil moisture dynamics for hillslopes and micro-catchments. *Journal of Hydrology*, *198*(1), 177–195.

Constantin, J., et al. (2019). Management and spatial resolution effects on yield and water balance at regional scale in crop models. *Agricultural and Forest Meteorology*, *275*, 184–195.

Corbeels, M., et al. (2005). A process-based model of nitrogen cycling in forest plantations: part I. Structure, calibration and analysis of the decomposition model. *Ecological Modelling*, *187*(4), 426–448.

de Wit, A., et al. (2019). 25 years of the WOFOST cropping systems model. *Agricultural Systems*, *168*, 154–167.

Dias, H. B., et al. (2019). New APSIM-Sugar features and parameters required to account for high sugarcane yields in tropical environments. *Field Crops Research*, *235*, 38–53.

Donatelli, M., et al. (2017). Modelling the impacts of pests and diseases on agricultural systems. *Agricultural Systems*, *155*, 213–224.

Enders, A., et al. (2023). SIMPLACE - A versatile modelling and simulation framework for sustainable crops and agroecosystems. *Silico Plants*, diad006.

Ewert, F., et al. (2015). Crop modelling for integrated assessment of risk to food production from climate change. *Environmental Modelling & Software*, *72*, 287–303.

Ewert, F., et al. (2023). Agroecology for a sustainable agriculture and food system: from local solutions to large-scale adoption. *Annual Review of Resource Economics*, *15*(1), 351–381.

Ewert, F. & Porter, J. R. (2000). Ozone effects on wheat in relation to CO_2: modelling short-term and long-term responses of leaf photosynthesis and leaf duration. *Global Change Biology*, *6*(7), 735–750.

Ewert, F., et al. (1999). Simulation of growth and development processes of spring wheat in response to CO_2 and ozone for different sites and years in Europe using mechanistic crop simulation models. *European Journal of Agronomy: The Journal of the European Society for Agronomy*, *10*(3), 231–247.

Eyshi Rezaei, E., et al. (2015). Impact of data resolution on heat and drought stress simulated for winter wheat in Germany. *European Journal of Agronomy: The Journal of the European Society for Agronomy*, *65*, 69–82.

Ezui, K. S., et al. (2018). Simulating drought impact and mitigation in cassava using the LINTUL model. *Field Crops Research*, *219*, 256–272.

Falconnier, G. N., et al. (2020). Modelling climate change impacts on maize yields under low nitrogen input conditions in Sub-Saharan Africa. *Global change biology*, *26*(10), 5942–5964.

Farquhar, G. D., et al. (1980). A biochemical model of photosynthetic CO_2 assimilation in leaves of C 3 species. *Planta*, *149*(1), 78–90.

Faye, B., et al. (2023). Climate change impacts on European arable crop yields: sensitivity to assumptions about rotations and residue management. *European Journal of Agronomy: The Journal of the European Society for Agronomy*, *142*, 126670.

Faye, B., et al. (2018a). Impacts of 1.5 versus 2.0 °C on cereal yields in the West African Sudan Savanna. *Environmental Research Letters: ERL [Web Site]*, *13*(3), 034014.

Faye, B., et al. (2018b). Potential impact of climate change on peanut yield in Senegal, West Africa. *Field Crops Research*, *219*, 148–159.

Feng, Y., et al. (2022). Identifying and modelling key physiological traits that confer tolerance or sensitivity to ozone in winter wheat. *Environmental Pollution*, *304*, 119251.

Free Software Foundation. (2007). GNU Lesser General Public License, Version 3. https://www.gnu.org/licenses/lgpl-3.0.html

Gabaldón-Leal, C., et al. (2016). Modelling the impact of heat stress on maize yield formation. *Field Crops Research*, *198*, 226–237.

Gaiser, T., et al. (2013). Modeling biopore effects on root growth and biomass production on soils with pronounced sub-soil clay accumulation. *Ecological Modelling*, *256*, 6–15.

Gamma, E., et al. (1995). Design Patterns: Elemente wiederverwendbarer objektorientierter Software. Pearson Deutschland GmbH. https://books.google.com.br/books?hl=en&lr=&id=jUvf7wMUGcUC

Goudriaan, J. & Van Laar, H. H. (2012). *Modelling Potential Crop Growth Processes: Textbook with Exercises*. Springer Science & Business Media, 239 pp.

Guarin, J. R., et al. (2022). Evidence for increasing global wheat yield potential. *Environmental Research Letters: ERL [Web Site]*, *17*(12), 124045.

Hamilton, S. H., et al. (2022). Fit-for-purpose environmental modeling: Targeting the intersection of usability, reliability and feasibility. *Environmental Modelling & Software*, *148*, 105278.

Hernández-Ochoa, I. M., et al. (2022). Model-based design of crop diversification through new field arrangements in spatially heterogeneous landscapes: a review. *Agronomy for Sustainable Development*, *42*(4), 74.

Hoffmann, H., et al. (2016). Impact of spatial soil and climate input data aggregation on regional yield simulations. *PLoS One*, *11*(4), e0151782.

Holzworth, D. P., et al. (2015). Agricultural production systems modelling and software: current status and future prospects. *Environmental Modelling & Software*, *72*, 276–286.

van Ittersum, M. K., et al. (2003). On approaches and applications of the Wageningen crop models. *European Journal of Agronomy: The Journal of the European Society for Agronomy*, *18*(3), 201–234.

Jägermeyr, J., et al. (2021). Climate impacts on global agriculture emerge earlier in new generation of climate and crop models. *Nature Food*, *2*(11), 873–885.

Janssen, S. J. C., et al. (2017). Towards a new generation of agricultural system data, models and knowledge products: information and communication technology. *Agricultural Systems*, *155*, 200–212.

Jarvis, N., et al. (2022). Improved descriptions of soil hydrology in crop models: the elephant in the room? *Agricultural Systems*, *202*, 103477.

Jones, M. R. & Singels, A. (2018). Refining the Canegro model for improved simulation of climate change impacts on sugarcane. *European Journal of Agronomy: The Journal of the European Society for Agronomy*, *100*, 76–86. https://doi.org/10.1016/j.eja.2017.12.009.

Jones, J. W., et al. (2017). Brief history of agricultural systems modeling. *Agricultural Systems*, *155*, 240–254.

Kamali, B., et al. (2022). Uncertainty in climate change impact studies for irrigated maize cropping systems in southern Spain. *Scientific Reports*, *12*(1), 4049.

Kim, Y.-U., et al. (2024). Mechanisms and modelling approaches for excessive rainfall stress on cereals: waterlogging, submergence, lodging, pests and diseases. *Agricultural and Forest Meteorology*, *344*, 109819.

Kollas, C., et al. (2015). Crop rotation modelling–A European model intercomparison. *European Journal of Agronomy: The Journal of the European Society for Agronomy, 70*, 98–111.

Kothari, K., et al. (2022). Are soybean models ready for climate change food impact assessments? *European Journal of Agronomy: The Journal of the European Society for Agronomy, 135*, 126482.

Leuning, R. (1995). A critical appraisal of a combined stomatal-photosynthesis model for C3 plants. *Plant, Cell & Environment, 18*(4), 339–355.

Midingoyi, C. A., et al. (2021). Crop2ML: an open-source multi-language modeling framework for the exchange and reuse of crop model components. *Environmental Modelling & Software, 142*, 105055.

Monteith, J. L., et al. (1977). Climate and the efficiency of crop production in Britain. *Philosophical transactions of the Royal Society of London. Series B, Biological Sciences, 281*(980), 277–294.

Moreno-Cadena, L. P., et al. (2020). Importance of genetic parameters and uncertainty of MANIHOT, a new mechanistic cassava simulation model. *European Journal of Agronomy: The Journal of the European Society for Agronomy, 115*, 126031.

Nguyen, T. H., et al. (2020). Comparison of root water uptake models in simulating CO_2 and H2O fluxes and growth of wheat. *Hydrology and Earth System Sciences, 24*(10), 4943–4969.

Nguyen, T. H., et al. (2022). Expansion and evaluation of two coupled root–shoot models in simulating CO_2 and H2O fluxes and growth of maize. *Vadose Zone Journal, n/a*(n/a), e20181.

Nguyen, T. H., et al. (2024). Assessing the spatio-temporal tropospheric ozone and drought impacts on leaf growth and grain yield of wheat across Europe through crop modeling and remote sensing data. *European Journal of Agronomy: The Journal of the European Society for Agronomy, 153*, 127052.

Nóia Júnior, R. S., et al. (2023). A call to action for global research on the implications of waterlogging for wheat growth and yield. *Agricultural Water Management, 284*, 108334.

Oliveira, F. A. A., et al. (2021). Incorporating a dynamic gene-based process module into a crop simulation model. *In Silico Plants, 3*(1), diab011.

Oomen, R. J., et al. (2016). Modelling rangeland productivity in response to degradation in a semi-arid climate. *Ecological Modelling, 322*, 54–70.

Raymond, C., et al. (2020). Understanding and managing connected extreme events. *Nature Climate Change, 10*(7), 611–621.

Raymundo, R., et al. (2017). Performance of the SUBSTOR-potato model across contrasting growing conditions. *Field Crops Research, 202*, 57–76.

Rezaei, E. E., et al. (2018). Climate change effect on wheat phenology depends on cultivar change. *Scientific Reports, 8*(1). https://doi.org/10.1038/s41598-018-23101-2

Rodriguez, D., et al. (2001). Modelling the response of wheat canopy assimilation to atmospheric CO_2 concentrations. *The New Phytologist, 150*(2), 337–346.

Rosenzweig, C., et al. (2013). The Agricultural Model Intercomparison and Improvement Project (AgMIP): protocols and pilot studies. *Agricultural and Forest Meteorology, 170*, 166–182.

Ruane, A. C., et al. (2022). The climatic impact-driver framework for assessment of risk-relevant climate information. *Earth's Future*. https://doi.org/10.1029/2022ef002803.

Seidel, S. J., et al. (2021). Crop response to P fertilizer omission under a changing climate - Experimental and modeling results over 115 years of a long-term fertilizer experiment. *Field Crops Research*, *268*, 108174.

Seidel, S. J., et al. (2022). Simulating root growth as a function of soil strength and yield with a field-scale crop model coupled with a 3D architectural root model. *Frontiers in Plant Science*, *13*, 865188.

Shibu, M. E., et al. (2010). LINTUL3, a simulation model for nitrogen-limited situations: application to rice. *European Journal of Agronomy: The Journal of the European Society for Agronomy*, *32*(4), 255–271.

Siebert, S., et al. (2017). Heat stress is overestimated in climate impact studies for irrigated agriculture. *Environmental Research Letters: ERL [Web Site]*, *12*(5), 054023.

Spitters, C. J. T. (1990). Crop growth models: their usefulness and limitations. *Acta Horticulturae*, *267*, 349–368.

Srivastava, A. K., et al. (2017). Impact of climatic variables on the spatial and temporal variability of crop yield and biomass gap in Sub-Saharan Africa- a case study in Central Ghana. *Field Crops Research*, *203*, 33–46.

Srivastava, A. K., et al. (2019). Effect of mineral fertilizer on rain water and radiation use efficiencies for maize yield and stover biomass productivity in Ethiopia. *Agricultural Systems*, *168*, 88–100.

Srivastava, A. K., et al. (2020). The implication of different sets of climate variables on regional maize yield simulations. *Atmosphere*, *11*(2), 180.

Srivastava, A. K., et al. (2022). Cassava yield gap–A model-based assessment in Nigeria. *Frontiers in Sustainable Food Systems*, *6*. https://doi.org/10.3389/fsufs.2022.1058775.

Stella, T., et al. (2023). Wheat crop traits conferring high yield potential may also improve yield stability under climate change. *In Silico Plants*, *5*(2), diad013.

Tao, F., et al. (2017). Designing future barley ideotypes using a crop model ensemble. *European Journal of Agronomy: The Journal of the European Society for Agronomy*, *82*, 144–162.

Taulya, G. (2015). *Ky'osimba Onaanya: Understanding Productivity of East African Highland Banana*. Wageningen University, 167 pp.

Van Heemst, H. D. J. (1988). *Plant Data Values Required for Simple Crop Growth Simulation Models: Review and Bibliography*. CABO.

Van Oijen, M. & Ewert, F. (1999). The effects of climatic variation in Europe on the yield response of spring wheat cv. Minaret to elevated CO_2 and O3: an analysis of open-top chamber .... *European Journal of Agronomy: The Journal of the European Society for Agronomy*, *0*(3–4), 249–264.

Van Oijen, M. & Lefelaar, P. (2008). *Lintul-2: Water Limited Crop Growth: A Simple General Crop Growth Model for Water-Limited Growing Conditions*. Wageningen University, Plant Sciences.

Vereecken, H., et al. (2016). Modeling soil processes: review, key challenges, and new perspectives. *Vadose Zone Journal*, *15*(5). https://doi.org/10.2136/vzj2015.09.0131.

Vianna, M. S., et al. V. (2022). Improving the representation of sugarcane crop in the Joint UK Land Environment Simulator (JULES) model for climate impact assessment. *Global Change Biology. Bioenergy*, *14*(10), 1097–1116.

Vianna, M. S., et al. (2020). Modelling the trash blanket effect on sugarcane growth and water use. *Computers and Electronics in Agriculture*, *172*, 105361.

Wang, E., et al. (2022). How reliable are current crop models for simulating growth and seed yield of canola across global sites and under future climate change? *Climatic Change*, *172*(1), 20.

Webber, H., et al. (2016a). Uncertainty in future irrigation water demand and risk of crop failure for maize in Europe. *Environmental Research Letters: ERL [Web Site]*, *11*(7), 074007.

Webber, H., et al. (2020). No perfect storm for crop yield failure in Germany. *Environmental Research Letters: ERL [Web Site]*, *15*(10), 104012.

Webber, H., et al. (2022). Framework to guide modeling single and multiple abiotic stresses in arable crops. *Agriculture, Ecosystems & Environment*, *340*, 108179.

Webber, H., et al. (2016b). Simulating canopy temperature for modelling heat stress in cereals. *Environmental Modelling & Software*, *77*, 143–155.

Webber, H., et al. (2018). Diverging importance of drought stress for maize and winter wheat in Europe. *Nature Communications*, *9*(1), 4249.

Wolf, J. (2012a). *User Guide for LINTUL5, Simple Generic Model for Simulation of Crop Growth under Potential, Water Limited and Nitrogen, Phosphorus and Potassium Limited Conditions*. Wageningen University.

Wolf, J. (2012b). *User Guide for LINTUL4 and LINTUL4V, Simple Generic Model for Simulation of Crop Growth under Potential, Water Limited and Nitrogen, Phosphorus and Potassium Limited Conditions*. Wageningen University.

Xinyou, Y. I. N. & van Laar, H. H. (2005). *Crop Systems Dynamics: An Ecophysiological Simulation Model of Genotype-by-Environment Interactions*. Wageningen Academic Publishers, 156 pp.

Zhao, G., et al. (2015). Demand for multi-scale weather data for regional crop modeling. *Agricultural and Forest Meteorology*, *200*, 156–171.

Zimmermann, A., et al. (2017). Climate change impacts on crop yields, land use and environment in response to crop sowing dates and thermal time requirements. *Agricultural Systems*, *157*, 81–92.

Chapter 16

The MONICA agroecosystem model

Claas Nendel, Michael Berg-Mohnicke, Rachel Escueta, Gohar Ghazaryan, Leonardo Inforsato, Bahareh Kamali, Marlene Palka, Ahsan Raza, Ehsan Eyshi Rezaei, Xenia Specka and Jing Yu, Leibniz Centre for Agricultural Landscape Research (ZALF), Germany

1 Introduction

The MOdel for NItrogen and Carbon in Agroecosystems (MONICA) has been developed to simulate the interplay of biophysical processes in the soil–plant–atmosphere continuum with a focus on the interaction between soil and plant under the influence of weather and management. Emerging from HERMES, a simulation tool for advising nitrogen fertilisation in crop production, MONICA has developed into a software that offers a wide range of applications across scales, elucidating crop responses and related ecosystem services and disservices to weather, climate and a range of management strategies, including the most recent climate change mitigation and adaptation strategies in crop production. With a focus on central Europe, MONICA sees a rapidly growing user community, spanning from policymaking over environmental administration to research.

In this chapter, we discuss the historical evolution of HERMES and its descendant, MONICA, and introduce MONICA's capabilities to simulate cropping systems from the sub-field to the continental scale. We discuss not only unique features but also current limitations and illustrate example applications for a range of different purposes and areas for further improvement.

http://dx.doi.org/10.19103/AS.2025.0155.19

2 The MONICA legacy

2.1 The dawn of agroecosystem modelling

The development of mechanistic simulation models for processes in agroecosystems has its roots in hydrology, where the movement of water in porous media was the first applied problem to be described using mathematics (Luthin and Gaskell, 1950). Scholars achieved first milestones with the analytical (Brutsaert, 1968) and numerical (Hanks, 1962; Philip, 1957) solution of the Richards equation (Richardson, 1922; Richards, 1931), the transport of dissolved substances in the soil (Biggar and Nielsen, 1962), the cation exchange at substrate surfaces (Dutt and Doneen, 1963) and the implementation of chemical equilibria (Dutt and Tanji, 1962). As computer performance increased, more complex approaches were developed for modelling water quality on the one hand (Tanji, 1967) and the water balance in the soil–plant–atmosphere system on the other (Nimah, 1973) to better understand the feedback mechanisms in hydrological systems. For modelling the effect of timber harvesting on the hydrology of a forested catchment in the Rocky Mountains, for example, Leaf and Alexander (1975) already used a density model for the tree canopy, which also made it possible to map reflectivity, precipitation interception and the redistribution of snow.

As a result of an improved understanding of how water and dissolved ions move in hydrological systems, the role of agriculture as a source of substances that are transported with the water to places where they are undesirable also came into focus (Hagin and Amberger, 1974). In the US, in particular, the development of comprehensive models for nitrogen dynamics in the agroecosystem advanced (Dutt et al., 1972). In 1974, Mehran (1974) presented a model for processes of the nitrogen cycle based on first-order kinetics. It was the first model to include denitrification.

In Europe, the core cell of the model developers was located in Wageningen, The Netherlands. Building on the work of Kees de Wit (de Wit and van Keulen, 1972), Beek (1973) presented a model for the decomposition of organic matter in soil that did not, however, consider plants. The model was characterised by a high process resolution; however, a test of the model against measured data was only published much later with the successor models (Frissel and van Veen, 1978; van Veen and Frissel, 1981). At the same time, de Wit's (1965) work on leaf photosynthesis was advanced in Wageningen towards simulation models for whole plants (de Wit, 1970; de Wit and Brouwer, 1968). The SUCROS model (van Keulen, 1982) was the first model of this developmental series with which a plant responding to water stress could be simulated. A reaction to nitrogen undersupply was realised in the PAPRAN model (Seligman and van Keulen, 1981).

In the German Democratic Republic, a comprehensive concept for an agroecosystem model with the sub-models plant growth, water and N balance was presented as early as 1969 (Kaltofen, 1971), but this was no longer pursued after Herbert Kaltofen retired from the research station at Paulinenaue. The state of knowledge in Eastern Europe at this time is best reflected by Unger (1968).

The first European model for soil processes that also took into account the nitrogen removal of the plant was the 'nitrogen dynamics in soil' (NDS) model by Hagin and Amberger (1974). Bosatta (1981) presented the first nitrogen balance model for forest ecosystems, but without taking into account the water and nitrogen uptake of the trees. As the first co-production on both sides of the Atlantic, McGill (1981) developed the PHOENIX model for grassland sites, probably the most comprehensive process model for an ecosystem at the time.

This pioneering work gave rise to a large number of models in the 1980s, the first representatives of which were compared in an overview by Frissel and van Veen (1981). Further developments were documented by Addiscott and Wagenet (1985), Groot et al. (1991) and Engel et al. (1993). While model development in the US in this period focused mainly on crop physiological processes ('crop models'; Wilkerson et al. (1983), Ritchie et al. (1984), Jones and Kiniry (1986), Boote et al. (1986)), model development in Europe primarily addressed the soil–plant system ('agroecosystem models'), with an initial focus on winter wheat sites (Groot, 1987; Matthäus, 1986; Vereecken, 1989; Whitmore and Parry, 1988). The simulation of winter nitrogen leaching under spring wheat, winter barley and sugar beet for the purpose of fertiliser advice was pursued by Richter et al. (1978, 1985), initially with a simple removal algorithm for the plant. The same model was used by Duynisveld and Strebel (1986) for the first continuous simulation of nitrate translocation under a 3-year crop rotation. After integrating the SUCROS approach into their model, this working group produced the first regionalisation of simulated nitrate losses to groundwater (Anlauf et al., 1988). At the same time, models for modelling the nitrogen cycle in multi-annual crop rotations were developed elsewhere and published almost simultaneously (Grant, 1989; Hansen et al., 1991).

2.2 The HERMES agroecosystem model

In the group of Jörg Richter at Hanover University, Germany, Kurt-Christian Kersebaum developed the HERMES model for applications in on-farm agriculture, water protection and land use planning (Kersebaum, 1989). For this purpose, he selected a range of simple model approaches that could operate under restricted data availability (Kersebaum, 1995; Kersebaum and Beblik, 2001). The core of the HERMES model was the SUCROS crop model, which uses a generic structure for all plants, which then require only specific parameters to represent individual annual crops. For the soil, Kersebaum formulated a capacity approach (Burns, 1974) that employed capacity parameters from the German soil survey manual (AG Bodenkunde, 2005). The manual also provided empirical parameters for the effective rooting depths for each texture class. This look-up table approach made HERMES much more robust for hydraulic simulations at the field scale as compared to models that used pedo-transfer functions (PTFs). To use this information properly, HERMES required only the texture class, organic matter content and bulk density as the minimum soil information to run. As for all simulation models of that era, HERMES could only simulate vertical processes (one-dimensional).

For potential evapotranspiration (PET, ET_0), HERMES was equipped with two options: (1) the empirical HAUDE formula (Haude, 1954) using crop-specific monthly coefficients of Heger (1978) and the daily vapour pressure deficit at 2 pm, and (2) the Turc-Wendling formula (Wendling et al., 1991) for a grass reference, using monthly k_c factors to calculate crop-specific PET for crops.

Unique to HERMES is the description of the nitrogen dynamics in soil. While contemporary models for soil organic matter (SOM) were all based on carbon pools (e.g. Century (Parton et al., 1987), RothC (Jenkinson, 1990), DAISY (Hansen et al., 1991), CANDY (Franko et al., 1995)), HERMES was constructed on two organic nitrogen pools with different standard decay rates. Daily net mineralisation of nitrogen was then calculated as a function of temperature and soil moisture (Nordmeyer and Richter, 1985). As a standard pool initialisation, 13% of the total soil organic nitrogen was set as the initial value of the slowly decomposable nitrogen pool (Nuske, 1983). Denitrification is calculated for the top soil (30 cm) using Michaelis-Menten kinetics, modified by water-filled pore space and temperature (Kersebaum, 1995). Since HERMES does not represent carbon in soils, any simulation of SOM dynamics requires an inverse calculation of carbon from nitrogen using a fixed carbon-to-nitrogen ratio (Bohuslav et al., 2025).

HERMES calculates root water uptake from soil using the ET_0 as a driver. Water extraction from soil is distributed over the soil layers according to their plant-available water content, modulated by a proportion factor (van Keulen, 1975) for the evaporation part, and by the root distribution over depth and a root water uptake efficiency for the transpiration part of ET.

For calculating the nitrogen uptake of crops from soil, HERMES applies the critical nitrogen concentration as a function of the accumulated thermal sum (Ritchie et al., 1989; Vielemeyer et al., 1983) as a driver and as a control, a diffusion approach for the nitrate transport to the root. The transport of nitrate in the soil is calculated using the convection-dispersion equation for non-stationary conditions, which makes the nitrogen nutrition of the crop dependent on the soil water status.

2.3 Integrating DAISY

Towards the twenty-first century, the carbon cycle in agroecosystems gradually moved into focus, replacing the nitrate debate for a while. This triggered the need for adding the carbon cycle to HERMES. At this time, the most elaborated carbon and nitrogen soil model was found in DAISY (Abrahamsen and Hansen, 2000) and, based on experiences with constructing the EU-Rotate_N model for field vegetables (Rahn et al., 2010), it was decided to replace the original nitrogen pools with the DAISY SOM turn-over module. Since this was a major change in the concept of the model, the new offspring was renamed to MONICA, the MOdel for NItrogen and Carbon dynamics in Agroecosystems (Nendel et al., 2011). In the course of this reconstruction, MONICA was equipped with slightly adjusted algorithms for the nitrogen stress response (Salo et al., 2016) and for the critical nitrogen concentration (Greenwood et al., 1980).

2.4 Cherry-picking

There were a few more functionalities that HERMES was lacking, and that were gradually added to MONICA. One of them was the ability to simulate soil temperature, which was borrowed from the 4C forest simulation model (Bugmann et al., 1997). Based on a Cholesky decomposition of the soil temperature matrix, this algorithm computes the temperature of the distinct soil layers based on a lower boundary and a soil surface temperature, which in turn is influenced by incoming radiation and air temperature. Convective energy transport, as implemented in the original model, was, however, ignored for simplicity.

In addition, further knowledge on soil water transport was added to MONICA from the THESEUS modelling toolbox (Wegehenkel, 2000), enabling the control of the water flux via an empirical conductivity parameter to stop any water flux in the case the soil was frozen. This allowed the addition of a module for soil water freezing, thawing and for a snow layer. Groundwater dynamics were introduced, enabling the simulation of ascending water from surface-near groundwater (Khaledi et al., 2024b).

The crop module was enhanced by integrating a more sophisticated plant physiology module from ARCWHEAT1 (Mitchell et al., 1995), which employed a set of algorithms for calculating the maximum photosynthesis rate based on the ideas of Farquhar and von Caemmerer (1982) and Long (1991). It allows to simulate the effect of elevated atmospheric CO_2 concentrations on both the photosynthesis and the transpiration (Nendel et al., 2009). For this purpose, the Penman-Monteith equation for evapotranspiration (Allen et al., 1998) was introduced to MONICA, linking photosynthesis and transpiration via the stomata. Additionally, respiration algorithms were borrowed from the TRITSIM/AGROSIM model (Mirschel et al., 1987a, 1987b), which allowed to differentiate between daytime and dark respiration to simulate effects of an increased night-time temperature on the plant's assimilate resources (Mirschel and Wenkel, 2007). Temperature effects on various plant physiological processes in MONICA are represented using the Wang-Engel temperature function (Wang et al., 2017).

Root growth, as part of MONICA linking the soil and plant part, has been taken from the EU-Rotate_N model (Rahn et al., 2010), but remaining with the 1D approach. The representation of rooting systems with different morphology originates from Pedersen et al. (2009) for annual or from Nendel and Kersebaum (2004) for perennial crops. The rate of root penetration through the soil is modified by the soil's clay content, although evidence for this behaviour is scarce. The target depth of the root is defined through a crop-specific parameter and modified by texture. Under drought stress, the root may exceed this lower boundary and continue growing if the lower soil layers contain moisture.

An overview of MONICA's parameter and their importance in the calibration process is provided by Specka et al. (2015), also including an analysis of the temporal sensitivity of the most important parameters (Specka et al., 2019).

3 Important MONICA features

3.1 The generic SUCROS crop module

In contrast to other advanced modelling systems (e.g. DSSAT, APSIM, STICS), MONICA has remained with a generic crop module that represents all crops just by using different parameter sets and remains with a one-leaf approach as its predecessor SUCROS. As a consequence, some crop physiological processes are not represented as detailed as in other crop models, e.g. the yield formation of cereals based on grain number and single grain weight. The main advantage of this simple approach was seen in the technically straightforward computation of cropping system time series, where many different crops were grown in rotation and where water and nitrogen dynamics and their carry-over effect from crop to crop were followed over the course of several years. It also supported the design of a very fast C++ computer code, which enabled the model to provide results of a 1-year simulations within milliseconds on a regular personal computer. In the context of an application with a graphical user interface, the swift response time between execution and presentation of the results was an essential precondition to allow for an interactive use. Not having to wait until the next morning to see the results of the simulation but being able to test several alternative scenarios within minutes was one of the success factors of the LandCare2020 platform, in which the first MONICA versions were embedded (Wenkel et al., 2013).

3.2 Simulating crop development and growth

Accurate simulation of the phenological development of crops is critical to the success of simulating almost all other physiological processes in the plant. HERMES uses a simple heat summation approach and some key phenological stages. While most phenological stages are clearly visible in the field and often recorded in experiments, the double-ridge stage in cereals – an indicator of the transition from the vegetative to the generative stage and only visible under the microscope – was used in HERMES to initiate the storage organ. The heat sums required for the completion of each phenological phase are associated with specific base temperatures, below which no development takes place. The response to day length and vernalisation requirements are two processes that control plant development and are integrated into the heat sum generation algorithm. As a further development, MONICA was later equipped with emergence control functions that introduced soil moisture as an additional factor that prevents seedling emergence when the soil is too dry or saturated with water. In addition, seedling emergence was now controlled by soil temperature, which had previously been introduced in MONICA. Table 1 shows examples of the phenological stages and the parameters required for two contrasting crops, winter wheat and silage maize.

3.3 Considering genotypic differences

MONICA crop parameter sets are the result of calibrations against field- or plot-scale experimental data, and they represent the crop with its genetic traits as used in the experiment. For crops for which only a limited number of experiments has been available, the parameters represent a theoretical genotype that assembles the average development and growth behaviour as observed in the trials. This approach acknowledges that for large-area applications, the actual genotypes grown on individual fields are also not known. However, for scenarios that support crop breeding, it was necessary to test different genetic traits against each other. For this purpose, some MONICA parameters have been declared being variable among important genotypes. These parameters have been extracted into cultivar files, while the remaining parameters are collected in a species file (Table 2).

3.4 Simulating perennial crops

First experiences with simulating perennial crops date back to HERMES and the attempt to simulate the development and growth of grapevine (Nendel and Kersebaum, 2004). For this purpose, a perennial crop compartment was introduced, adding a fifth to the already existing leaf, stem, root and fruit carbon pools. This knowledge was later used to build a grassland version of MONICA

Table 1 Crop phenological stages and example parameters in MONICA for winter wheat and silage maize

	Stage	BBCH	Minimum temperature	Temperature sum	Vernalisation requirement	Day length requirement	Base day length
				Winter wheat			
1	Sowing	00	0.0	146	0	0	0
2	Emergence	09	1.0	276	47	20.0	7.0
3	Double ridge	na	1.0	336	0	20.0	7.0
4	Heading	51	1.0	112	0	20.0	7.0
5	Anthesis (full)	65	9.0	586	0	0	0
6	Maturity	89	9.0	15	0	0	0
				Silage maize			
1	Sowing	00	7.0	66	0	0	0
2	Emergence	09	6.0	326	0	0	0
3	Shooting	31	6.0	244	0	0	0
4	Tasseling	51	6.0	233	0	0	0
5	Begin female flowering	61	9.0	144	0	0	0
6	Begin grain filling	71	9.0	518	0	0	0
7	Black layer	87	9.0	21	0	0	0

Parameters are valid for the phase following the stage. Note that MONICA is flexible on the definition of phenological phases for crops.

Table 2 List of cultivar-level parameters in MONICA

Parameter name	Unit	Description
CultivarName	String	Name identifier of the cultivar
WinterCrop	Boolean	Defines if the cultivar is grown as a winter crop
AssimilatePartitioningCoeff	[ ]	Matrix of assimilate partitioning to the different plant organs per developmental stage
BaseDaylength	[h]	Defines the base daylength below which the plant will not develop
BeginSensitivePhaseHeatStress	[°C days]	Defines a temperature sum after which the plant is particularly sensitive to floret sterility through heat
CriticalTemperatureHeatStress	[°C]	Defines the temperature threshold above which the florets take damage
CropHeightP1	[ ]	Defines the growth pattern over time
CropHeightP2	[ ]	Defines the growth pattern over time
CropSpecificMaxRootingDepth	[m]	Defines the maximum rooting depth potential of the plant
DaylengthRequirement	[h]	Defines the daylength requirement per developmental stage for the plant's development. Negative values denote long-night plants.
DroughtStressThreshold	[ ]	Defines the relation between actual and potential transpiration below which the plant experiences drought stress
EndSensitivePhaseHeatStress	[°C days]	Defines a temperature sum after which the plant is not any more sensitive to floret sterility through heat
FrostDehardening	[$°C^{-1}$ day^{-1}]	Modulates the daily change in LT50 due to deacclimation
FrostHardening	[$°C^{-1}$ day^{-1}]	Modulates the daily change in LT50 due to acclimation
LT50cultivar	[°C]	Threshold temperature for the cultivar, below which 50% of the population is killed by frost
MaxAssimilationRate	[kg CO_2 ha $Leaf^{-1}$]	Maximum assimilation rate
MaxCropHeight	[m]	Maximum height of the crop
OrganIdsForCutting	[ ]	Defines the organs that are affected by cutting/mowing
OrganIdsForPrimaryYield	[ ]	Defines the organs that are primarily harvested

(*Continued*)

Table 2 (*Continued*)

Parameter name	Unit	Description
OrganIdsForSecondaryYield	[ ]	Defines the organs that are harvested as secondary products
OrganSenescenceRate	[day^{-1}]	Defines the rate of senescence per organ and developmental phase
ResidueNRatio	[ ]	Ratio between the N concentration in the residues and in the marketable yield at harvest
RespiratoryStress	[ ]	Modulates the daily change in LT50 due to respiratory stress under snow cover
SpecificLeafArea	[ha kg^{-1}]	Defines the leaf surface per leaf mass for each developmental stage
StageKcFactor	[ ]	Modulates the reference grass evapotranspiration to represent the crop, specific for each developmental stage
StageTemperatureSum	[°C days]	Defines the temperature sum per developmental stage that needs to accumulate to reach the next stage
VernalisationRequirement	[day]	Defines the number of low-temperature days the crop requires to induce flowering

(Kamali et al., 2022). Here, the perennial compartment takes the role of the grass rhizome, from which previously stored assimilates could be remobilised to cover respiration expenses over winter and fuel regrowth in spring. The daily litter fall and root die-back necessitated an enhanced management of the added organic matter (AOM) pools, which were newly initiated every time organic matter with different chemical properties was added to the soil. The advanced module now merges pools with similar properties, reducing the number of active AOM pools significantly. For the simulation of phenology for permanent crops, it was necessary to introduce 'budbreak' as a phenological event to initiate the crop's development, replacing the 'sowing' event. The heat summation for 'budbreak' starts in late February by default (Nendel, 2010) but can be adjusted (Table 3).

3.5 The representation of mixed cropping systems

The MONICA-Intercropping model incorporates the 'homogeneous horizontal canopy' (HHC) sub-model (Tsubo and Walker, 2002) to simulate intercropping systems, focusing primarily on light competition in relay-row intercropping with two partner species. For this purpose, the canopy of the two crops is divided into two layers based on plant height differences, allowing for dynamic modelling of light competition as crop heights change over time. The model calculates light

Table 3 Annual crops available in MONICA

Crop		References
	Well-tested parameter sets	
Barley	*Hordeum vulgare*	(Nendel et al., 2011; Rötter et al., 2012; Salo et al., 2016; Tao et al., 2020; Tao et al., 2017)
Maize	*Zea mays*	(Bassu et al., 2014; Durand et al., 2018; Falconnier et al., 2020; Kimball et al., 2019; Kumudini et al., 2014; Nendel et al., 2011)
Oil-seed rape	*Brassica napus*	(Nendel et al., 2011; Wang et al., 2022)
Potato	*Solanum tuberosum*	(Fleisher et al., 2017; 2021; Nendel et al., 2011)
Rye	*Secale cereale*	(Nendel et al., 2011)
Soybean	*Glycine max* (L.) MERR.	(Battisti et al., 2017; Battisti et al., 2018; da Silva et al., 2024; Kothari et al., 2022; Kothari et al., 2024; Nendel et al., 2023b; Yu et al., 2024)
Wheat	*Triticum aestivum*	(Asseng et al., 2015; Asseng et al., 2013; Asseng et al., 2019; Dueri et al., 2022; Nendel et al., 2011; Nendel et al., 2013; Pirttioja et al., 2015)
	Reasonably well-tested parameter sets	
Cotton	*Gossypium*	(Carauta et al., 2021)
Chickpea	*Cicer arietinum*	(Triacca et al., 2026)
Durum wheat	*Triticum durum* DESF.	
Emmer	*Triticum dicoccum* SCHRANK	(Baum et al., 2016)
Lentil	*Lens culinaris*	(Triacca et al., 2026)
Oat	*Avena*	
Sugar beet	*Beta vulgaris* subsp. *vulgaris*	(Nendel et al., 2011)
Sunflower	*Helianthus annuus*	(Carauta et al., 2021)
Tomato	*Solanum lycopersicum*	(Yang et al., 2025)
	Not well-tested parameter sets	
Carrot	*Daucus carota*	
Faba bean	*Vicia faba*	
Field pea	*Pisum sativum*	
Mustard, white	*Sinapis alba*	(Nendel et al., 2011)
Lucerne	*Medicago sativa*	
Lupin, narrow-leaved	*Lupinus angustifolius*	
Oil radish	*Raphanus sativus* var. *oleiformis* PERS.	(Nendel et al., 2011)
Sudan grass	*Sorghum × drummondii*	(Nendel et al., 2011)
Sorghum	*Sorghum bicolor*	
Triticale	×*Triticosecale*	
Grassland vegetation		(Kamali et al., 2022; Khaledi et al., 2024a)

interception for each layer using species-specific light extinction coefficients, determining photosynthetically active radiation interception separately for the upper and lower canopy layers, while accounting for light distribution through the canopy (Yu et al., 2024). Water competition, while important, is handled indirectly as a secondary consideration. The model assumes the dominant crop has primary access to available water, with remaining water allocated to the relayed crop. Changes in leaf area index resulting from light competition indirectly affect evapotranspiration and water uptake, allowing the model to respond to different water regimes without explicitly modelling direct water competition (Yu et al., 2025). This approach has been calibrated and validated for both rainfed and irrigated conditions, demonstrating the model's ability to capture the primary effects of intercropping on resource utilisation, with light competition as the main driver and water dynamics as an indirect consequence (Yu et al., 2024). Nitrogen dynamics are calculated independently for each crop without direct interactions. The current version of the model does not incorporate facilitation processes, focusing instead on competition for resources.

3.6 Soil-plant-atmosphere water dynamics

3.6.1 The capacity approach and its hydraulic parameters

MONICA remains with the simple but robust capacity (tipping bucket) approach that HERMES used before. The capacity of the distinct soil layers (default: 20 layers of 10 cm thickness) is defined using the empirical relationship between soil texture classes of the German Soil Survey Manual (AG Bodenkunde, 2005) and soil hydraulic properties as described in Wessolek et al. (2009). While PTFs sometimes result in implausible values for soil hydraulic properties at the threshold range of soil particle distributions, the look-up table provides a very robust solution. However, smaller discontinuities in soil hydraulic properties have to be accepted. Since the SOM classes in this framework are few, the effect of changes in SOM on soil hydraulic properties causes significant discontinuities, which is an unwelcomed behaviour in long-term simulations of SOM dynamics (e.g. Farina et al., 2021). For this reason, the PTF of Toth et al. (2015) has been implemented in MONICA to provide a smoother change in soil hydraulic properties as a consequence soil SOM change, as an alternative to the Wessolek look-up table.

Saturated soil water conductivity has not previously existed in HERMES. For MONICA, this lever to control water flux rates from one layer to the next below, an empirical factor λ has been introduced. It computes a conductivity coefficient from texture (Nendel, 2014). An overview of MONICA's soil moisture module in comparison to other models is given by Tenreiro et al. (2020).

3.6.2 Groundwater dynamics and the capillary fringe

Ascending water from groundwater into the root zone of a crop has a significant effect on crop growth. Especially grasslands are often found in lowland areas, where this additional water resource plays an important role. HERMES used empirical capillary rise rates from the German Soil Survey Manual (AG Bodenkunde, 2005) to add a daily amount of water to the soil layers above the groundwater table, bypassing the capacity approach that allows only downward movement of water. This look-up table contains rise rates according to the distance to the groundwater table and the soil texture. For MONICA, a dynamic groundwater table was introduced as a sine wave, defined by the maximum and the minimum groundwater table distance and the month of the lowest distance (Khaledi et al., 2024b). Percolating rainwater can modify the groundwater table. For this purpose, MONICA calculates backwater rise, allowing for modification of the generic groundwater table dynamics through significant amounts of rainfall, as far as to permanently flooded conditions (Khaledi et al., 2024a). Using a groundwater inflow rate parameter would also allow the groundwater table to drop in times of excessive drought.

3.6.3 Precipitation interception, ponding and infiltration

The standing crop acts as an interception storage for water from precipitation and overhead irrigation. The interception storage is a function of crop height and the degree of soil coverage, the latter being a function of the leaf area index. Water from precipitation that is not intercepted and water from melting snow ponds at the soil surface before it is allowed to infiltrate into the soil. From this reservoir, run-off may occur, depending on the slope, the fraction of vegetation cover and the roughness of the soil surface. Remaining water infiltrates as much as the topmost soil layer has pore space to be filled. Water that cannot infiltrate remains ponded at the surface. Evaporation from the soil does not occur in this case, but open-water evaporation rates are assumed. The issue of rapid infiltration through preferential flow paths, e.g. cracks in shrinking clay soils, has not been solved yet.

3.6.4 Snow

Snow layers are simulated according to Riley and Bonesmo (2005). MONICA assumes falling precipitation below a temperature of 1.8°C as containing an increasing share of snow, which is added to a virtual snow layer. The density and the resulting height of this snow layer is continuously recalculated, assuming snow to melt at above +0.3°C and to refreeze at below −1.7°C. The snow layer has water-holding capacity that allows also liquid precipitation to be stored in the snow layer. Melting snow adds water to the surface reservoir, subject to subsequent infiltration or run-off.

3.6.5 Freezing and thawing in soils

Soil frost algorithms originate from Olsen and Haugen (1997), whereby uniform thermal properties are assumed over the entire soil profile. The temperatures calculated in the soil temperature module are not used, as this does not take into account the heat transfer in the water-ice phase transition. Parameter values are taken from the SOIL model (Jansson, 1991). The module requires the surface temperature modified by the snow cover as input information. The approach used to describe thawing is that of the ECOMAG model (Motovilov et al., 1999). For the calculation of the thermal conductivity of the frozen soil, an average water content at field capacity is assumed. It is furthermore assumed that in the event of ground frost, no water movement takes place in the affected soil layer ($\lambda = 0$). Infiltrating water from precipitation or from the melting snow reservoir fills the soil up to the saturation limit. Excess water is retained in the surface reservoir or added to the surface runoff. When the ground has thawed again, i.e. when the thaw front has reached the frost depth, λ is set back to the original value. Consequently, gravitational water can again pass through the profile and water from the surface reservoir can follow (Fig. 1).

3.7 Soil organic matter and the carbon cycle

The DAISY SOM module consists of 2 × 3 conceptual soil organic carbon (SOC) pools (Fig. 2). The first couple of pools describe freshly AOM, from either organic fertilisers or crop residues, and distinguish two different qualities, differentiated by the C:N ratio. A wide C:N ratio (e.g. cereal straw) comes with a slow decomposition rate, a narrow ratio with a fast rate. The same concept applies also to the two SOM pools, where young SOM gets a fast, and matured SOM a slow decomposition rate. For the soil microbial biomass, this distinction compares to r-strategists and k-strategists. While r-strategists reproduce rapidly on easily available feed (sugars, amino acids) and appropriate conditions, the k-strategists feed on chemically more stable material.

MONICA is initialised using a standard distribution of the SOC pools, which assumes an equilibrium at the beginning of the simulation (Bruun and Jensen, 2002). Inert SOC is subtracted from the total SOC (Falloon et al., 1998). The SOM and soil microbial biomass (SMB) pools are initialised by multiplying the total SOC content by the given distribution parameters. The AOM pools

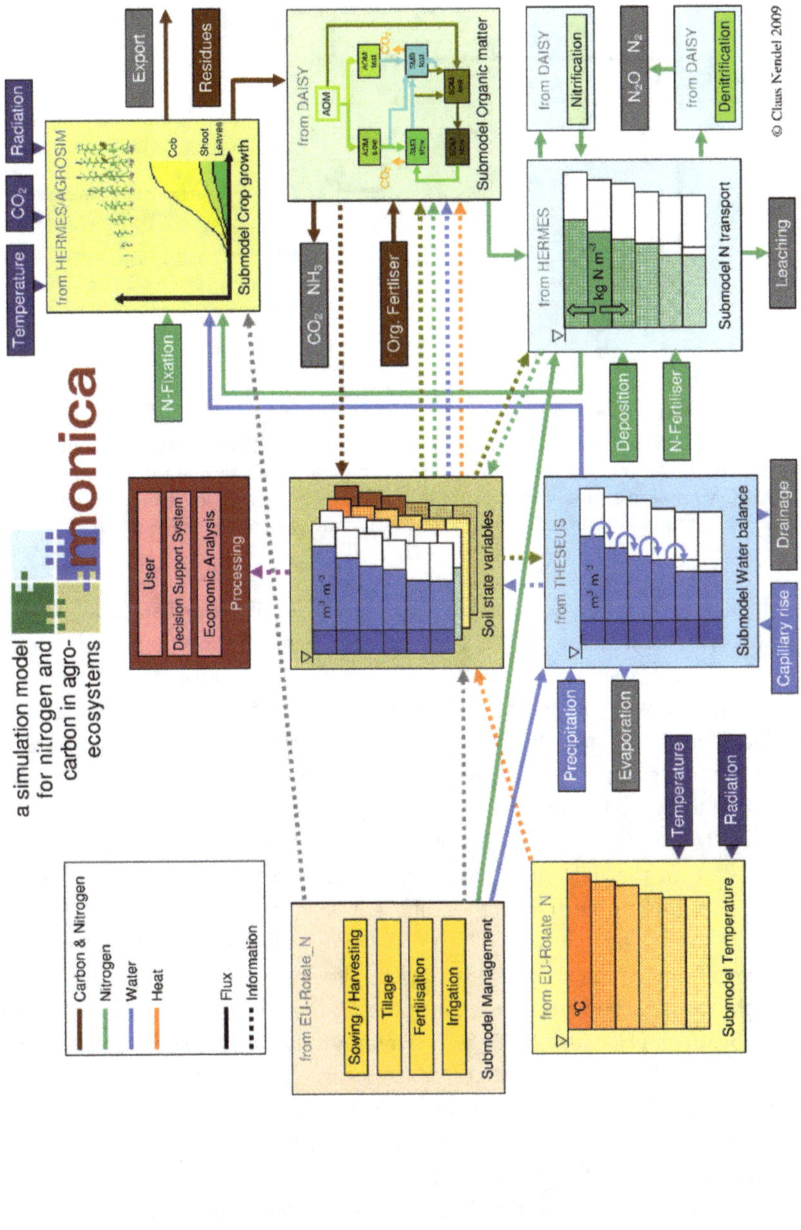

Figure 1 MONICA modules and their interconnection through matter and information fluxes.

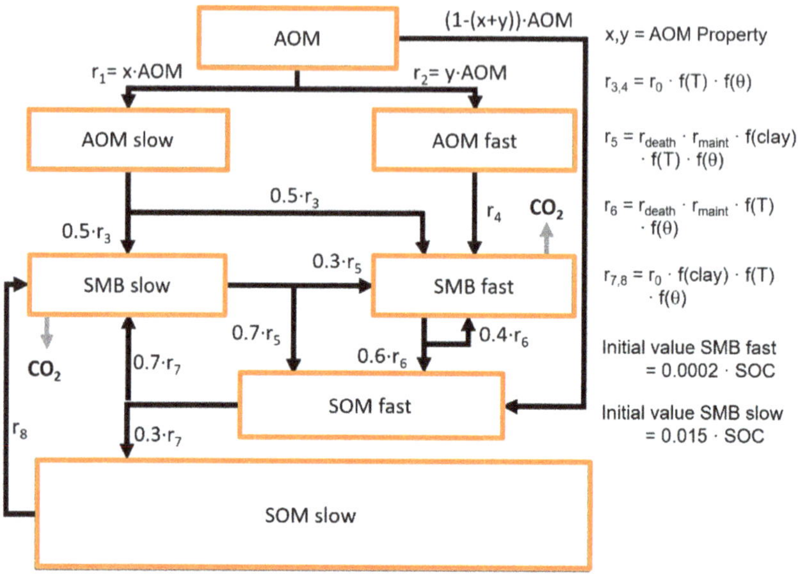

Figure 2 Soil organic matter module as implemented in MONICA showing conceptual pools and turnover rates for organic matter decomposition. Pools include added organic matter (AOM), soil microbial biomass (SMB) and particulate and mineral-associated soil organic matter (SOM). Source: Adapted from Abrahamsen and Hansen (2000).

are initialised as soon as crop residues or organic fertilisers are added. Organic fertiliser and residue parameters describe specific ammonium and nitrate contents, decomposition rate coefficients and pool distributions.

During growth, plants accumulate carbon in up to five compartments: root, leaf, shoot, fruit and sugar. For perennial plants, an additional permanent structure is considered. Each compartment comes with its own parameters, controlling the growth rate and the biomass that can potentially be harvested or added to the AOM pools. Following a harvest command, accrued crop residues left on the field are immediately assigned to the AOM pools. Root remains are fragmented into the different soil layers, depending on the root mass distribution in the soil profile. During tillage, the SOM content is distributed homogenously to the soil layers affected by tillage depth.

Each pool's carbon losses are controlled by first-order decomposition rates (Abrahamsen and Hansen, 2000). These rates are modified by soil temperature, soil water and – in the case of the SMB pools – clay contents (Bruun et al., 2003; Farina et al., 2021), which have recently been refurbished and tested by Aiteew et al. (2024). Decayed AOM is transferred to the SMB pools. A pool-specific carbon-use efficiency controls which fraction is then respired to the atmosphere as CO_2 or converted further to SOM. SMB also constantly feeds on the SOM pools, with additional CO_2 being respired and rapidly decomposing SOM slowly turning into slowly decomposing SOM.

3.8 The nitrogen cycle and related emissions

Reproducing the nitrogen cycle within crop fields has been the prime motivation to build the HERMES model. When MONICA was developed, it was therefore still an important ambition to remain competitive with HERMES in reproducing N dynamics in the soil–plant–atmosphere system. Different from HERMES, which uses two explicit N pools to simulate N turnover in soils, the release of nitrogen from SOM in MONICA is the consequence of the microbial C turnover. This requires defined C to N ratios to keep the balances of both elements right. The SMB ratios are fixed (Müller et al., 1997) and the SOM ratios are calculated from user-defined inputs, but the AOM ratios are assembled in a database, in which various organic matter inputs (crop residues and organic fertilisers) have their C to N ratio and other properties defined. For fresh crop residues, however, this database provides only C to N ratios for the slowly decomposing fraction of the material (and how large that fraction typically would be). The rapidly decomposing fraction remains dynamic, retrieving the nutritional status of the crop at harvest for the calculation. For this purpose, MONICA uses an empirical parameter (Rahn et al., 2010) to determine how much of the N in the total biomass is located in the marketable part of the crop, and how much in the remaining residues. While the N in the marketable yield can be used to calculate protein contents of harvested cereal grains using a fixed conversion factor, the N amount in the residues defines their C to N ratio and thus influences the release rate of N under the decomposition processes.

The N content in the total biomass, in turn, is a result of a daily N uptake from the soil. For this purpose, MONICA employs the dilution curve approach, computing the critical N concentration in relation to the fresh aboveground biomass (Greenwood et al., 1990). With this approach, implemented also in the EU-Rotate_N model (Rahn et al., 2010), MONICA deviates from the HERMES critical N calculation, which is based on the crop's development. Learning from EU-Rotate_N, MONICA additionally applies a target N curve to allow for luxury N uptake, which is typical for some vegetables. The root N concentration is a crop-specific parameter that remains constant. Per definition of the N dilution curve, the daily growth increment of the crop's biomass in terms of C gain thus defines the requirement for daily N uptake. The N demand is checked against the availability of nitrates in the different soil layers, in accordance with the root biomass distribution over depth, and the resulting root length density. A maximum daily flux of water into the root limits the daily uptake. Outside the root, in the rhizosphere, the Convection-Dispersion Equation (Baldwin et al., 1973) is numerically solved using the finite difference method to simulate the transport of water and nitrates from the surrounding soil volume to the root. As a consequence of this approach, no nitrates can be taken up when soil moisture levels are too low to allow ion movement.

The availability of nitrates in each soil layer finally is the consequence of the nitrogen release from SOM turn-over and the subsequent nitrification of ammonia (NH_4^+) into nitrite (NO_2^-) and then nitrate (NO_3^-); both processes depend on temperature and moisture in the same manner as the SOM turn-over. Nitrates are furthermore subject to denitrification, while N_2O production is calculated from the NO_2^- pool (Dueri et al., 2023). NH_3 volatilisation of manure and slurry is described using an empirical function of the ALFAM model (Søgaard et al., 2002). A function to describe the dependency of gaseous NH_3 emissions on soil pH was derived from data published by He et al. (1999). A Michaelis-Menten kinetic was fitted to their data and the relation between pH and N volatilisation half-life was normalised to pH 7.0. The hydrolysis of applied urea and subsequent volatilisation of NH_3 is simulated using the AMOVOL model (Sadeghi et al., 1988).

Dueri et al. (2023) tested MONICA and other models against a number of N_2O emission experiments. The simulation of nitrogen uptake and of protein contents in cereal grains at harvest has been evaluated in the studies of Salo et al. (2016), Yin et al. (2017) and Asseng et al. (2019). The reproduction of nitrate concentration dynamics in different soil depths under various crop rotations hast been evaluated in the Nendel et al. (2011) study, while Khaledi et al. (2024a) provided a test for various variables, including CO_2 gas exchange, N_2O emission, nitrate leaching and ET against grassland lysimeter data.

3.9 Input requirements and output

MONICA has retained the relatively low demand for input data from the original HERMES idea. Only the introduction of the Penman-Monteith evapotranspiration algorithms has increased the weather data requirements to include wind speed and relative humidity, two variables that are increasingly recorded at standard weather stations around the world. Daily minimum and maximum temperatures, the sum of precipitation and the sum of global radiation or hours of sunshine complete MONICA's need for daily weather data. Soil information must include organic matter content, texture and bulk density per horizon, with the top (tilled) horizon being the most important. The information on texture can be given as proportions of clay, silt, sand and skeleton or as texture class according to the German soil classification. Soil pH and the C-to-N ratio of the organic matter are facultative information. If the site is affected by groundwater, the minimum and maximum distances of the groundwater table to the surface can be supplied, together with the month in which the minimum distance occurs. Mandatory site information includes latitude, altitude and slope. The cropping system is described by the plant species and cultivar. Sowing and harvest dates as well as fertilisation and irrigation schedules can be provided, but MONICA would also be able to generate these events based on predefined rules, which is particularly useful for large-scale applications. Current atmospheric CO_2-concentrations are calculated internally but can be provided to reproduce historic experimental data.

MONICA output is flexible and can be adjusted to any decision-support platform where MONICA is to be embedded. The command-line version of MONICA provides a simple way to define the output, while the student version is restricted to a set of crop and soil state variables that are supported by the user interface and are in the interest of the student education programs.

4 Assessing model performance

4.1 Calibration to field trials

As for most of the existing agroecosystem and crop models, also MONICA has been calibrated and tested first of all against data from field experiments. Already when the model was first presented to the scientific community (Nendel et al., 2011), it came with a test against data that has not been previously used for calibration. This test for transferability, which soon after became a research focus in large community-driven model improvement activities such as COST 734 CLIVAGRI (2006–2011) AgMIP (2010–today), MACSUR (2012–2017) is a standard today for model developers to understand the uncertainty afflicted with the simulation results and the limitations of their model. Multivariate data sets (Kersebaum et al., 2015) have been preferably used to demonstrate MONICA's fitness to simulate the interrelation between soil and plant processes and to prove that important target variables were predicted in accordance with underlying processes (Fig. 3). This also includes test cases for greenhouse gas (GHG) emissions and nitrate leaching (Fig. 4). Over the years, MONICA has been calibrated and tested for a range of crops and cropping systems all over the world (Table 3) and also for its ability to reproduce temperature (Asseng et al., 2013; Wang et al., 2017), moisture (Groh et al., 2020; Khaledi et al., 2024a; Nendel et al., 2025), CO_2 (dos Santos et al., 2025; Durand et al., 2018; Fleisher et al., 2021; Nendel et al., 2009) and nitrogen (Falconnier et al., 2020; Salo et al., 2016; Yin et al., 2017) impact on crops.

4.2 Ubiquity

The most important feature of a mechanistic agroecosystem model is its ubiquity. Being based on physical principles, mechanistic simulation models should be able to describe the functioning of each and every cropping

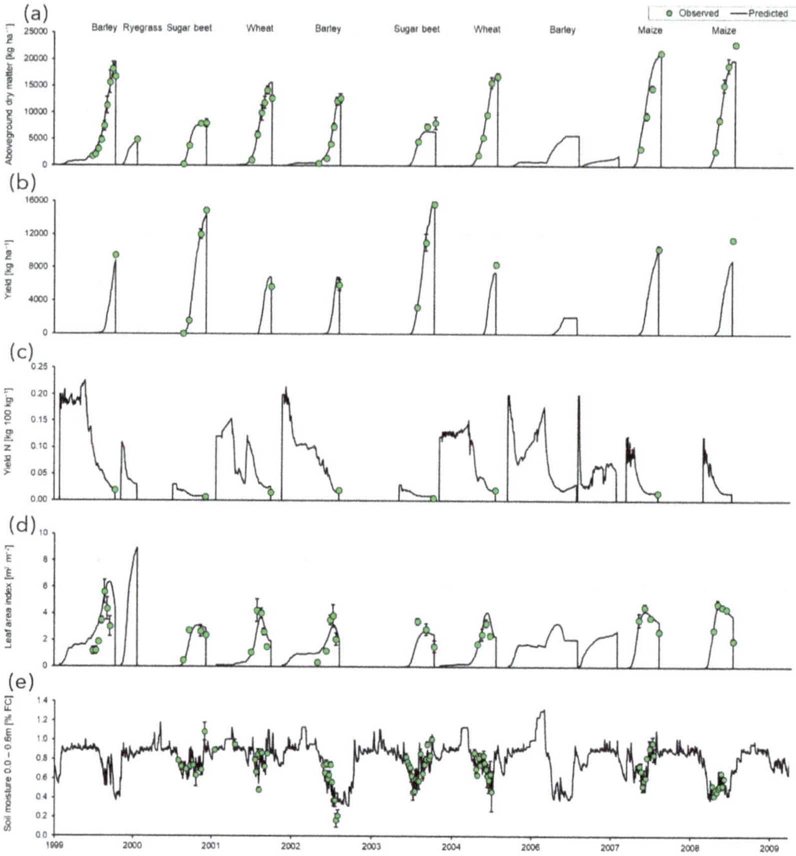

Figure 3 Example simulation (line) of MONICA calibrated to experimental data (dots) from a 9-year crop rotation grown in Braunschweig, Germany (Weigel and Dämmgen, 2000). All variables originate from the same simulation, representing aboveground biomass (a) marketable dry-matter yield, (b) nitrogen concentration in the marketable yield, (c) leaf area index (d) and soil moisture (e) across a range of crops typically grown in Central Europe (Nendel et al., 2011).

system in the world, according to theory. However, since many models include simplifications in the form of empirical functions, the representation of physical processes may not always be accurate. In some models, important processes are missing or have not been described in a sufficiently generic manner. The description of crops is merely a matter of identifying the local genotype being grown and parameterising it. Further, it would require considering growth-affecting factors like pest and diseases, the application of growth regulators and chemical crop protection agents, or management failures; processes that are not included in the models. To overcome this shortcoming, MONICA offers the possibility to dampen photosynthesis using a field condition modifier. This

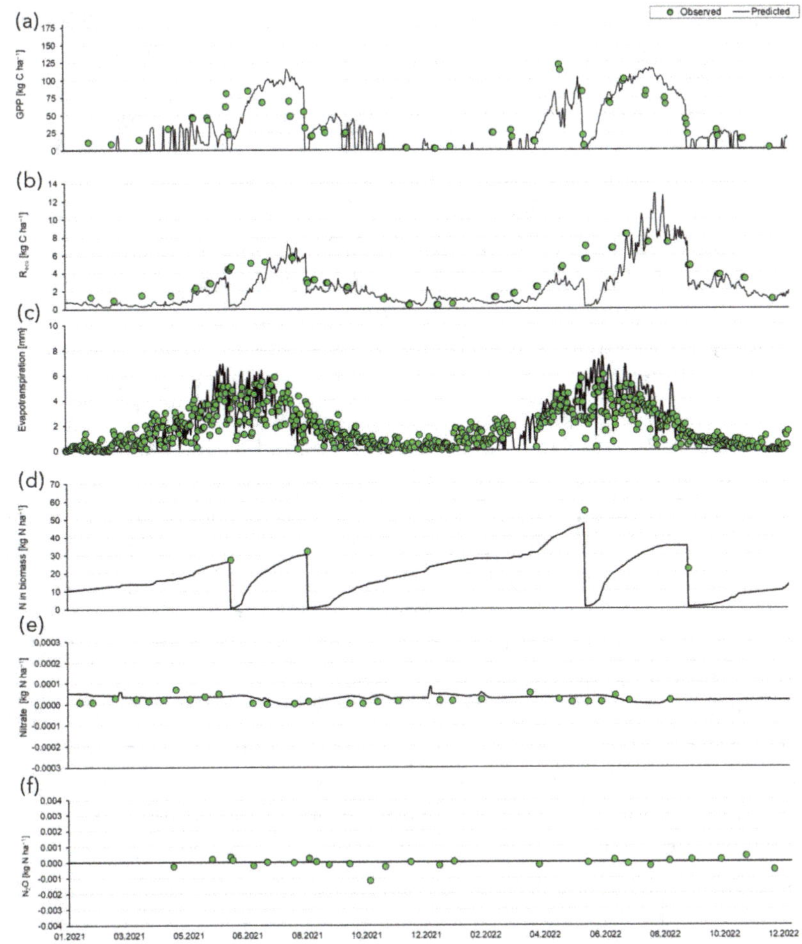

Figure 4 Example simulation of MONICA (line) applied to reproduce experimental data (dots) from a 2-year shallow-groundwater grassland lysimeter (Dietrich et al., 2016) equipped with gas chambers. Dominant species were *Carex acuta*, *Festuca arundinacea*, *Festuca rubra*, *Holcus lanatus*, *Plantago lanceolata* and *Poa pratensis*. All variables originate from the same simulation, representing gross primary production (a) ecosystem respiration, (b) evapotranspiration, (c) nitrogen concentration in the aboveground biomass, (d) nitrate leaching below the rooted zone (e) and emission of nitrous oxide (f) (Khaledi et al., 2024a).

modifier is used to reduce the yield level of on-farm simulations in contrast to reference experiments, which often deliver higher yields because of the greater care that is taken by the experimentalist for their crop. It assumes that most of the yield reducing factors in the end impact photosynthesis, either directly through interfering with the crop's physiology (chemical agents) or by limiting any of the incoming resources, e.g. radiation (covering or destroying leaf area), water or nitrogen (destroying the root). The field condition modifier can also be employed to mimic the phytosanitary effects of pre-crops on the following crop.

The description of soil processes is more critical. In most agroecosystem models, the concepts of SOM turnover have been formulated decades ago (Le Noë et al., 2023) and remained unaltered until today. In the meanwhile, soil science has further advanced the understanding of how SOM turnover is controlled in soils (Manzoni and Cotrufo, 2024; Rocci et al., 2024) and it was shown that, e.g., the natural nitrogen release of agricultural soils depends a lot on their mineralogy, which in turn is a consequence of the long-term climate prevailing (Nendel et al., 2019). This includes the consideration of aluminium and iron oxides at sites where SOM can associate with and where it is protected from decomposition, and of soil aggregates, which play a vital role in the accessibility of SOM to microbes. At the moment, MONICA and most other models are fit for simulating water and nitrogen dynamics in temperate soils but lack performance in tropical and subtropical soils. With respect to carbon, MONICA has demonstrated satisfying performances in reproducing CO_2 gas fluxes within the soil–plant–atmosphere nexus at the crop level in a daily resolution (Khaledi et al., 2024a; Specka et al., 2016) and at the same time in reproducing long-term dynamics of SOC for fallow soils (Farina et al., 2021) and long-term monitoring sites (Aiteew et al., 2024).

At the interface between crop and soil, MONICA has also been intensively tested against a large range of water supply experiments (irrigation, rain-out shelter, groundwater lysimeters) for its performance in reproducing the response of crops to variations in water supply, from extreme drought to occasional flooding (Groh et al., 2020, 2022; Khaledi et al., 2024a; Nendel et al., 2025).

4.3 Performance across scales

MONICA has been applied in various studies at different spatial scales, ranging from single fields to continents. A few of these studies provided the opportunity for performance testing. Figure 3 shows three cases: a field of approximately 50 ha, for which high-resolution yield records and soil samples were available, and two regions of 34 000 km² and 357 000 km², respectively, for which yield statistics were available at county level, spatial units ranging from 222 km² to 5500 km². The smaller of the two regions, which was subject to a MACSUR study on scaling effects (Zhao et al., 2016), illustrates the modelling practice at that time, with the assumption that for large-area simulations every single pixel represents a space where the respective crop could grow, and is grown. The larger region illustrates the current state of art, where remote sensing (RS) informs about where the respective crops are grown, and where not

(Blickensdörfer et al., 2022). Disregarding poor production sites (dark red areas in Fig. 3c), where models would potentially simulate very low yields and where in many cases there is even no farmland, allows for a better representation of the environmental production conditions and facilitates a model calibration much closer to realistic productivity. What remains is the necessity to flag that after aggregation to larger spatial units, some of the values represent aggregates from a large number of pixels, while the value of others is the result of only a few.

Key to the satisfactory simulation results of MONICA across field to nation scales is the quality of the data that feeds into the model. The single-field case in Fig. 5 demonstrates that if high-resolution soil data is available, the uncalibrated simulation model is able to reproduce the observed yield pattern within a field across the whole range, from 2.5 t ha^{-1} to 9.5 t ha^{-1}. The mean error of prediction is 1.2 t ha^{-1}, which is approximately the same error that the model produces when predicting yields at county level as aggregated from 1 km^2 pixels, masked by crop and not far away from errors reported by studies where experimental data was available for calibration, which usually range between 10% and 15% for calibrated models (Asseng et al., 2013; Hao et al., 2021; Wallor et al., 2018; Yang et al., 2024). If, in addition, terrain effects on the redistribution of water in the soil (flowing from hilltops into depressions, on the surface or subcutaneously) are accounted for, the mean error of prediction is reduced to 0.8 t ha^{-1}. This demonstrates that the model performance is almost always limited by the quality and resolution of the data that the model is informed with and that any provisioning of better data leads to an improvement of the simulation performance. When crop growth and related processes in the soil are to be simulated, many agroecosystem models have already demonstrated a very convincing performance at the point scale. This is the precondition for any application at larger scales. For large-scale model applications, the consequence of this is that a successful simulation requires a huge investment into data provisioning, while the most of the established agroecosystem models do not require substantial improvements.

4.4 Limitations

As for all other mechanistic point-scale models, also MONICA suffers from a range of limitations that any user of the model must be aware of. Some of the limitations are general and are rooted in the method, some others are more specific to the approach that has been chosen for the development of SUCROS/HERMES/MONICA. In general terms, point-scale models such as MONICA perform well at the single-plant scale, as most of the processes integrated in the model are designed to represent plant physiological responses to environmental factors. The representation of the soil processes is adjusted to this conceptual approach, limiting transport processes in the soil to the vertical dimension. However, since such models are not meant to be used

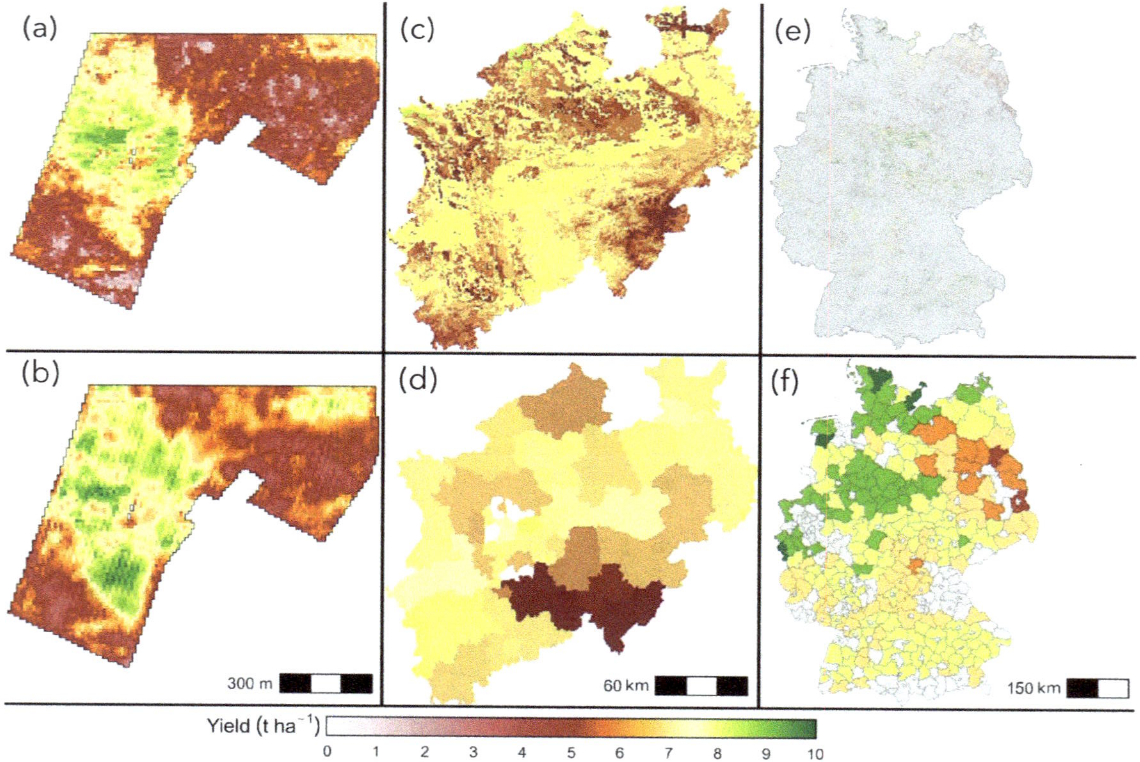

Figure 5 MONICA simulations for winter wheat yields at different spatial scales (top row) against observed winter wheat yield patterns from different sources (bottom row): (a) MONICA simulation for a field near Booßen, Brandenburg, Germany, in 10 m resolution for the 2021 season. Precipitation input has been inversely calibrated to account for water redistribution along the terrain. (b) 2021 winter wheat yield across the Booßen field as detected using a through-flow grain sensor mounted on a combine harvester, corrected for outliers and vehicle speed (Donat et al., 2022) and interpolated to 10 m resolution. (c) MONICA simulation for the federal state of North Rhine-Westphalia, Germany, 1 km resolution, average yields for 1982–2011, assuming winter wheat being grown in the entire area (Zhao et al., 2016). (d) Observed winter wheat yields as reported at district levels in official statistics, average for 1999–2011 (Zhao et al., 2016). (e) MONICA simulations for Germany, 1 km resolution, average yields for 1999–2019. A winter wheat mask derived from remote sensing (Blickensdörfer et al., 2022) informs in which pixel winter wheat

to investigate the plant in detail and elucidate plant physiology but rather to use it at the field scale and beyond, many processes are described in a semi-empirical manner which would not withstand testing against laboratory-scale observations. Nevertheless, modellers feel obliged to keep their models within the basic rules of physics, taking care of balanced energy and matter budgets and their conservation during turn-over processes.

When using the model at larger scales, e.g. at the scale of a single field, the inability to reproduce spatial patterns of plant growth and soil variables is grounded in the fact that lateral transport processes are not included in the model. Various modelling solutions are at hand to do this for soil transport processes, but only very few approach the field representation in a fully 3D manner (e.g. Šimůnek et al., 2024). Main reason for this is that three-dimensional simulations come at a high computational cost and are therefore mainly used for hypothesis testing in research. At coarser resolutions, within-field patterns are overshadowed by the aggregation process and inadequate data availability gradually takes over as the main reason behind model inabilities to reproduce landscape-scale patterns (Fig. 6).

More specifically to MONICA, the generic crop module inherited from SUCROS proofs very useful in simulations that include multiple crop seasons. However, it comes at the expense of a number of drawbacks. One of them is related to the big-leaf approach, representing the crop as a combination of four major organs: root, stem leaf and fruit. For the investigation of how cereal yields develop during plant growth, a single 'fruit' is insufficient, as cereals yields form as a result of the number of ear-bearing tillers × grain number per ear × individual grain

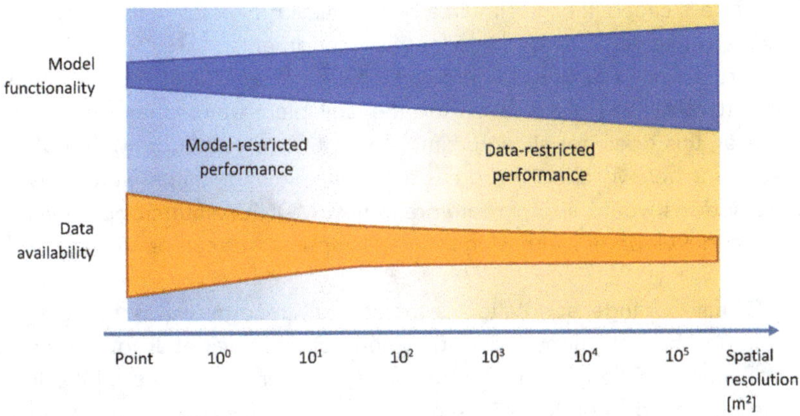

Figure 6 Relation of model functionality and data availability as restrictions for a mechanistic agro-ecosystem model's ability to reproduce spatial patterns of environmental variables to the spatial resolution of model application. This relation may vary across models, depending on their level of process detail.

size, and environmental factors can affect any of these three variables during the season (Nuttall et al., 2017; Reynolds et al., 2022). At this process level, MONICA is not suited to reproduce environmental effects on cereal yields, especially when it comes to comparing breeding lines of the same species. Secondly, the one-leaf approach of SUCROS comes with a circular reference that emerges from leaf area defining the absorbed radiation for photosynthesis, photosynthesis defining the amount of assimilates available for leaf biomass growth and leaf biomass defining the leaf area. This circular reference sometimes causes the leaf area simulation to run out of physically plausible ranges, especially when the crop is limited in growth due to environmental stresses. While a maximum effective leaf area index parameter prevents feedback to radiation absorption, the leaf area simulation in this case becomes implausible.

4.5 Plans for model improvements

Even though the current focus of the MONICA community lies on improving the applications by understanding better the role of the data to initialise and drive the model, there are still some activities scheduled to further improve the model itself. These include the representation of GHG emissions, intercropping systems, genetic traits of crop varieties and the role of soils in climate adaptation strategies.

MONICA has been tested against a few data set for its ability to simulate N_2O emissions in crop fields around the world (Dueri et al., 2023). Now that MONICA is seriously considered as a tool to inform selected national GHG reporting schemes in Europe, the testing will be intensified and accompanied by model development towards microbial CH_4 production and consumption. The model's capabilities to simulate biochemical processes in water-saturated soils is an important precondition for this amendment.

The recently presented intercropping version of MONICA conserves the one-dimensional representation of the soil–plant–atmosphere continuum. However, this does not allow to simulate root competition between different species for belowground resources and dynamic exploitation strategies that these species would apply. Therefore, a new MONICA intercropping version is under development with which the above processes can be investigated *in silico*.

Climate change adaptation is increasingly recognised as being tightly bound to the selection and deployment of crop varieties optimised for specific local challenges that farmers face. To address this need, MONICA is being enhanced to better represent genetic variability and support breeding applications (Gogna et al., 2025). Current investigations on phenological plasticity in cereals are building the foundation for a new MONICA version that incorporates detailed developmental and yield-forming processes at a level that allows meaningful differentiation between genotypes. This enhancement will enable the model to simulate genotype × environment × management interactions, providing crucial insights for breeding programs targeting climate resilience. The improved model will incorporate physiological trait parameters that can be linked to quantitative trait loci (QTLs) and genetic markers, facilitating integration with genomic selection approaches. Additionally, the model will be capable of simulating stress memory and acclimation processes that are essential for identifying varieties with both high yield potential and enhanced multiple stress tolerance. This development will position MONICA as a valuable tool for *in-silico* evaluation of breeding strategies, phenotyping target prioritisation and assessment of varietal performance across diverse current and future climate scenarios.

Finally, current research looks at progress that has been made in the recent years to understand the turn-over and protection mechanisms of SOM. Following the example of Tougma et al. (2025), a make-over of the DAISY pool concept in MONICA becomes likely. However, at the moment there is no appropriate data set available that would allow testing the benefit of a more sophisticated approach using measurable pools at the field scale and beyond.

5 Large-scale applications

A one-dimensional model, MONICA can only be applied across larger areas if being executed in a gridded mode, i.e. one MONICA simulation per grid cell. These simulations do no exchange information across grid cells in a way three-dimensional catchment models would do. Departing from the single plot-scale experiment, where the data obtained for model parameterisation, calibration and driving is aggregated to one single point, information for large-area applications is required for each grid cell. Reproducing spatial patterns now adds to the list of challenges for the model. Since one-dimensional models do not come with a fixed definition for the space that they may represent, the size of the grid cell can be freely selected, if the available information in the grid cells allows for a plausible interpretation of the model's results. Here, much care has to be taken, as this interpretation often becomes overstretched. Typical grids contain cells with a side length of 3–10 m (precision agriculture) and represent a single field with all its heterogeneities in soil and terrain, or 1 km, representing nations or continents and their climatic and geologic gradients. Applications with a coarser resolution are applied at the global scale, however, with hugely uncertain input. With a resolution between 3 m and 1000 m, RS can supply valuable data to inform the model.

5.1 Remote sensing products

For large-area model applications, RS needs to supply or supplement three categories of data: soil, crop and management. Soil information is often available in forms of maps, which originate from decades of soil survey work and digitalisation of formerly analogue data recordings. For MONICA applications across Germany, the digital Soil Overview Map (Bodenübersichtskarte) at a scale of 1:250.000 is used, which provides soil profile information for each grid cell within 3 m and 1000 m side length, adjusted for the respective land use. At the field scale, vehicle-mounted electromagnetic sensors can provide much higher resolved texture information even for the subsoil. Here, electrical resistivity as amplitude and phase is interpreted as soil moisture and texture signals (Lück and Rühlmann, 2013). SOM contents can be derived from optical sensors, also at much larger scales (Zepp et al., 2023). Spatially distributed data on soil bulk density and other information is often derived from soil sampling points and spatialised using geostatistical methods or machine learning (Hengl et al., 2017).

Primary crop information to be used with agroecosystem models is the information about which is grown where in an area. This has led to a rapid development of wall-to-all crop type maps, benefiting from the technique of building best-pixel composites from multiple optical sensors and the resulting dense time series of cloud-free images (Griffiths et al., 2019). Recently, sensor fusion included also radar imagery, and other data sources, such as a digital elevation map and climate data, have been introduced to the classification algorithm, enabling the identification of 22 different crop types across Germany at 10 m resolution (Blickensdörfer et al., 2022). Further data on crops, such as leaf area, leaf angle distribution, and chlorophyll content allow the derivation of biomass and tissue nitrogen content. While not useful in the initialisation or driving of a model, these variables could potentially be used to update running simulations on crop growth in a data assimilation approach (Hayman et al., 2024; Nendel et al., 2023a), but could also help to calibrate a model for a better spatial representativeness.

The most important contribution of RS to large-area model applications is in the field of agricultural management. For MONICA, a suite of RS products is under preparation to support the latest generation of simulations across Germany and other regions in Europe, also preparing for data assimilation efforts (Kamali et al., 2024). From the crop type maps of sequential years, crop rotation patterns can be derived (Jänicke et al., 2022). Using artificial intelligence methods, these patterns can then be extrapolated into the past and future, providing cropping information for every single field from 1990 to 2100 based on observed patterns and climatic, pedologic and economic variables (Palka et al., 2026). Potentially, information on the positioning of intermediate crops in the rotation can be added from RS (Jensch et al., 2025). Sowing date products for wheat and maize are available (Flores et al., 2025; Main-Knorn et al., 2025; Rezaei et al., 2021) and will be expanded to other crops in due time. The fact whether a crop has been irrigated or has access to ascending groundwater, makes a huge difference in simulating its productivity. Ghazaryan et al. (2025) provide the first map of Germany with remotely sensed information of irrigation being applied on fields. In combination with the irrigation trigger (a fixed irrigation rate being applied as soon as the simulated soil moisture falls beneath a threshold), MONICA is able to better simulate the spatial pattern of cereal yields across Germany (Fig. 7) and also quantify the amount of water being used for irrigation in a region (Nendel et al., 2025). Likewise, the provisioning of high-resolution groundwater distance data (Raza et al., 2025) allows to consider ascending groundwater as additional water resource to crops.

5.2 Climate change outlooks

Many functionalities of MONICA have been built into the model for the purpose of making it fit for simulating the multiple effects of climate change on crop growth, yield formation and related processes in the soil. Increasingly, MONICA is called for climate change outlooks, especially in relation to future crop productivity and related GHG sinks and sources. Earlier studies used to evaluate climate change impacts and potential adaptation measures at the point scale (Nendel et al., 2014)

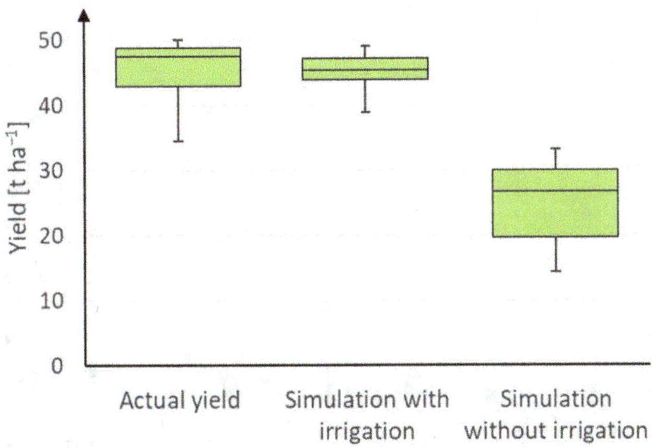

Figure 7 Summary of spatial simulations in 1 km² resolution across a district in Germany in which irrigation of maize is common. The boxplot shows the actual yield as provided by district-level statistics and compares to MONICA simulations with and without the irrigation information. Irrigation is implemented using MONICA's automatic irrigation feature with a defined soil moisture threshold.

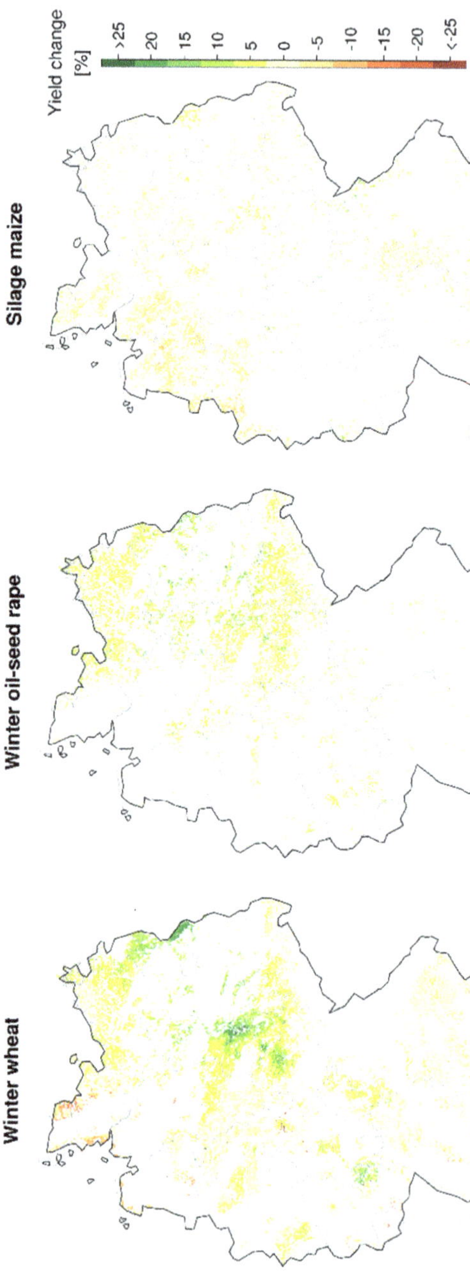

Figure 8 MONICA simulations for winter wheat, winter oil-seed rape and silage maize yields across Germany based on CMIP5 future climate simulations from the core ensemble of the German Weather Service. The maps display the simulated yield change from 2021 to 2050 under a radiative forcing scenario of 8.5 W m^{-2} (RCP 8.5). The respective crop's distribution is extracted from a crop classification using satellite imagery (Blickensdörfer et al., 2022). Note: upcoming downscaled CMIP6 runs from EURO-CORDEX are expected to come out drier during the spring months, which would result in lower yield expectation.

or provided user-defined simulations at field level for different climate scenarios via an online decision-support platform (Wenkel et al., 2013). Here, the user was able to define crop rotations for his own fields and retrieve simulation results for a range of environmental variables, including a translation into economic terms (Münch et al., 2014). The most recent generation of studies was already presented at the national scale for Germany (Söder et al., 2022). They were still based on single crops but applied already the true (observed) spatial distribution of crops (Fig. 8). The next generation of climate change outlooks for agriculture, which are currently under preparation for Germany (impact and adaptation/mitigation options) and Czechia (GHG emissions monitoring and accounting) will build on observed crop rotations (Palka et al., 2026) and observed supplementary irrigation (Ghazaryan et al., 2025) and will also include grasslands.

5.3 *Applications in policy support and decision-making*

Climate change impact information is currently the most demanded service that MONICA offers to support. Being embedded in a first online decision-support platform (Wenkel et al., 2013), several other tools are under development, e.g. irrigation planning and control software. Through this pathway, individual users have access to MONICA and can use the model independently to address their own questions in a guided way or benefit from the specific services offered through the platforms. Water use quantification across agricultural lands and estimates for water losses through evapotranspiration are emerging topics that regional authorities are eager to solve and for which MONICA offers a solution.

At the national scale, ministries are currently occupied with quantifying GHG emissions and developing measures that have the potential to mitigate these emissions without impairing other environmental, economic or social targets. One prominent example is the rewetting of bogs and fens for the purpose of re-establishing the mire ecosystems and reactivate their carbon-sequestering function. Here, MONICA is used in the framework of advisory services to the governments in different countries. These applications illustrate how MONICA has evolved from a research tool into an operational decision support tool, helping farmers, advisors, authorities and policymakers to address pressing environmental and agricultural challenges from field to national levels.

6 Conclusion

By design, the MONICA agroecosystem model provides a balanced view of crop and soil processes. It has proven its ability to simulate cropping systems in a variety of environments and management regimes, including precision agriculture applications at the sub-field scale and large-scale applications using remote sensing to provide essential information. Focusing on Central

Europe, MONICA is increasingly being used to answer research and decision-making questions, providing input for future scenarios, trade-off analyses and strategic development in crop production and related ecosystem services and disservices in the context of climate change

7 Where to look for further information

The MONICA model is documented on GitHub, where also all relevant resources can be found, including the latest release for download, a wiki and example files (https://github.com/zalf-rpm/monica). MONICA is distributed under the Mozilla Public License, v. 2.0 (https://www.mozilla.org/en-US/MPL/2.0/). A student version that comes with a web-based user interface for playing with basic crop responses to environmental changes is available at https://monica.zalf.de.

8 References

Abrahamsen, P. and Hansen, S., 2000. Daisy: an open soil-crop-atmosphere system model. *Environmental Modelling & Software*, 15(3): 313–330.

Addiscott, T.M. and Wagenet, R.J., 1985. Concepts of solute leaching in soils: a review of modeling approaches. *Journal of Soil Science*, 36: 94–102.

AG Bodenkunde, 2005. *Bodenkundliche Kartieranleitung* (5. Aufl.). Hannover: Bundesanstalt für Geowissenschaften und Rohstoffe.

Aiteew, K., et al., 2024. Evaluation and optimisation of the soil carbon turnover routine in the MONICA model (version 3.3.1). *Geoscientific Model Development*, 17(3): 1349–1385.

Allen, R.G., et al., 1998. *Crop evapotranspiration - Guidelines for computing crop water requirements*. Rome: Food and Agriculture Organization of the United Nations.

Anlauf, R., et al., 1988. Regional modelling of nitrate leaching in the winter half-year. *Mitteilungen der Deutschen Bodenkundlichen Gesellschaft*, 57: 123–130.

Asseng, S., et al., 2013. Uncertainty in simulating wheat yields under climate change. *Nature Climate Change*, 3(9): 827–832.

Asseng, S., et al., 2015. Rising temperatures reduce global wheat production. *Nature Climate Change*, 5(2): 143–147.

Asseng, S., et al., 2019. Climate change impact and adaptation for wheat protein. *Global Change Biology*, 25(1): 155–173.

Baldwin, J.P., et al., 1973. Uptake of solutes by multiple root systems from soil. 3. Model for calculating solute uptake by a randomly dispersed root system developing in a finite volume of soil. *Plant and Soil*, 38(3): 621–635.

Bassu, S., et al., 2014. How do various maize crop models vary in their responses to climate change factors? *Global Change Biology*, 20(7): 2301–2320.

Battisti, R., et al., 2017. Gauging the sources of uncertainty in soybean yield simulations using the MONICA model. *Agricultural Systems*, 155: 9–18.

Battisti, R., et al., 2018. Sensitivity and requirement of improvements of four soybean crop simulation models for climate change studies in Southern Brazil. *International Journal of Biometeorology*, 62(5): 823–832.

Baum, T., et al., 2016. "Slash and burn" or "weed and manure"? A modelling approach to explore hypotheses of late Neolithic crop cultivation in pre-alpine wetland sites. *Vegetation History and Archaeobotany*, 25(6): 611–627.

Beek, J.F., 1973. *Simulation of nitrogen behaviour in soils*. Wageningen: Centre for Agricultural Publishing and Documentation (PUDOC).

Biggar, J.W. and Nielsen, D.R., 1962. Miscible displacement. II. Behavior of tracers. *Soil Science Society of America Proceedings*, 26: 125–128.

Blickensdörfer, L., et al., 2022. Mapping of crop types and crop sequences with combined time series of Sentinel-1, Sentinel-2 and Landsat 8 data for Germany. Remote Sensing of Environment, 269: 112830.

Bohuslav, J., et al., 2025. Observed changes in the physical parameters of soils under no-till and conventional management in temperate climates and application in selected agro-ecosystem models. *Agricultural Systems*, 163 (3), 320–333.

Boote, K.J., et al., 1986. Modeling growth and yield of groundnut, Agrometeorology of groundnut. Proc. Int. Symp., ICRISAT Sahelian Center, 21–26 August 1985. ICRISAT, Patancheru, Andhra Pradesh, India, Niamey, Niger, pp. 243–254.

Bosatta, E., 1981. *Plant-soil system of an old Scots Pine forest in Central Sweden. Simulation of nitrogen behaviour of soil-plant systems*. Wageningen: Pudoc, pp. 155–170.

Brutsaert, W., 1968. A solution for vertical infiltration into a dry porous medium. *Water Resources Research*, 4: 1031–1038.

Bruun, S. and Jensen, L.S., 2002. Initialisation of the soil organic matter pools of the Daisy model. *Ecological Modelling*, 153(3): 291–295.

Bruun, S., et al., 2003. Calibration and validation of the soil organic matter dynamics of the DAISY model with data from the Askov long-term experiments. *Soil Biology and Biochemistry*, 35(1): 67–76.

Bugmann, H., et al., 1997. A new forest gap model to study the effects of environmental change on forest structure and functioning. In: G.M.J. Mohren, et al. (Eds.), *Impacts of global change on tree physiology and forest ecosystem*. Dordrecht, The Netherlands: Kluwer Academic Publisher, pp. 255–261.

Burns, I.G., 1974. Model for predicting redistribution of salts applied to fallow soils after excess rainfall or evaporation. *Journal of Soil Science*, 25(2): 165–178.

Carauta, M., et al., 2021. No more double cropping in Mato Grosso, Brazil? Evaluating the potential impact of climate change on the profitability of farm systems. *Agricultural Systems*, 190: 103104.

da Silva, E.H.F.M., et al., 2024. Inter-comparison among soybean models for simulation of evapotranspiration for a humid continental climate. *Agricultural and Forest Meteorology*, in press.

de Wit, C.T., 1965. Photosynthesis of leaf canopies. *Agricultural Research Report, Pudoc - Wageningen*, 663: 57.

de Wit, C.T., 1970. *Dynamic concepts in biology. Prediction and management of photosynthetic productivity*. Wageningen: Pudoc, pp. 17–23.

de Wit, C.T. and Brouwer, R., 1968. Über ein dynamisches Modell des vegetativen Wachstums von Pflanzenbeständen. *Angewandte Botanik*, 42: 1–12.

de Wit, C.T. and van Keulen, H., 1972. *Simulation of transport processes in soil*. Wageningen: Centre for Agricultural Publishing and Documentation.

Dietrich, O., et al., 2016. Behavior of water balance components at sites with shallow groundwater tables: possibilities and limitations of their simulation using different ways to control weighable groundwater lysimeters. *Agricultural Water Management*, 163: 75–89.

Donat, M., et al., 2022. Patch cropping- a new methodological approach to determine new field arrangements that increase the multifunctionality of agricultural landscapes. *Computers and Electronics in Agriculture*, 197: 106894.

dos Santos, M.L., et al., 2025. Evaluation of soybean multi-model responses to elevated CO2 with experimental data. *Field Crops Research*, in preparation.

Dueri, S., et al., 2022. Simulation of winter wheat response to variable sowing dates and densities in a high-yielding environment.Journal of Experimental Botany | Oxford Academic *Journal of Experimental Botany*, 73(16): 5715–5729.

Dueri, S., et al., 2023. Sources of uncertainty in simulating crop N2O emissions under contrasting environmental conditions. *Agricultural and Forest Meteorology*, 340: 109619.

Durand, J.L., et al., 2018. How accurately do maize crop models simulate the interactions of atmospheric CO2 concentration levels with limited water supply on water use and yield? *European Journal of Agronomy*, 100: 67–75.

Dutt, G.R. and Doneen, L.D., 1963. Predicting the solute composition of the saturation extract from soil undergoing salinization. *Soil Science Society of America Proceedings*, 27: 627–630.

Dutt, G.R. and Tanji, K.K., 1962. Predicting concentrations of solutes in water percolated through a column of soil. *Journal of Geophysical Research*, 67(9): 3437–3439.

Dutt, G.R., et al., 1972. *Computer simulation model of dynamic bio-physicochemical processes in soils*. Arizona: Agricultural Experimental Station, University of Arizona.

Duynisveld, W.H.M. and Strebel, O., 1986. *Ermittlung der Nitrat-Verlagerung aus Ackerböden ins Grundwasser mit Hilfe von Simulationsmodellen*. Berlin: Umweltbundesamt, p. 129.

Engel, T., et al., 1993. Simulationsmodelle zur Stickstoffdynamik. *Analyse und Vergleich. Agrarinformatik*, 25: 484.

Falconnier, G.N., et al., 2020. Modelling climate change impacts on maize yields under low nitrogen input conditions in sub-Saharan Africa. *Global Change Biology*, 26(10): 5942–5964.

Falloon, P., et al., 1998. Estimating the size of the inert organic matter pool from total soil organic carbon content for use in the Rothamsted carbon model. *Soil Biology and Biochemistry*, 30(8–9): 1207–1211.

Farina, R., et al., 2021. Ensemble modelling, uncertainty and robust predictions of organic carbon in long-term bare-fallow soils. *Global Change Biology*, 27(4): 904–928.

Farquhar, G.D. and von Caemmerer, S., 1982. Modelling of photosynthetic response to environmental conditions. In: O.L. Lange, et al. (Eds.), *Physiological plant ecology, encyclopedia of plant physiology*. Berlin: Springer Verlag, pp. 549–587.

Fleisher, D.H., et al., 2017. A potato model intercomparison across varying climates and productivity levels. *Global Change Biology*, 23(3): 1258–1281.

Fleisher, D.H., et al., 2021. Yield response of an ensemble of potato crop models to elevated CO2 in continental Europe. *European Journal of Agronomy*, 126: 126265.

Flores, L., et al., 2025. Potential of Sentinel-1 time series for large-scale assessment of maize and wheat phenology across Germany. *GIScience and Remote Sensing*, 62 (1), 2531593.

Franko, U., et al., 1995. Simulation of temperature, water and nitrogen dynamics using the model CANDY. *Ecological Modelling*, 81(1–3): 213–222.

Frissel, M.J. and van Veen, J.A., 1978. Critique of computer simulation modeling for nitrogen in irrigated croplands. Nitrogen in the environment - Nitrogen behavior in field soil. *Academic Press*, 1: 145–162.

Frissel, M.J. and van Veen, J.A., 1981. *Simulation of nitrogen behaviour of soil-plant systems*. Wageningen: Centre for Agricultural Publishing and Documentation (PUDOC).

Ghazaryan, G., et al., 2025. Large-scale irrigation mapping at field level in Northern Germany with integrated use of Sentinel-2, Landsat 8 and Sentinel-1 time series. *Remote Sensing Applications: Society and Environment*, 38: Article 101593.

Gogna, A., et al., 2025. Predicting enviromically adapted varieties with big data. *Genome Biology*, under review.

Grant, R.F., 1989. Test of a simple biochemical model for photosynthesis of maize and soybean leaves. *Agricultural and Forest Meteorology*, 48: 59–74.

Greenwood, D.J., et al., 1980. Relationships between the critical concentrations of nitrogen, phosphorus and potassium in 17 different vegetable crops and duration of growth. *Science of Food and Agriculture*, 31: 1343–1353.

Greenwood, D.J., et al., 1990. Decline in percentage N of C3 and C4 crops with increasing plant mass. *Annals of Botany*, 66(4): 425–436.

Griffiths, P., et al., 2019. Intra-annual reflectance composites from Sentinel-2 and Landsat for national-scale crop and land cover mapping. *Remote Sensing of Environment*, 220: 135–151.

Groh, J., et al., 2020. Crop growth and soil water fluxes at erosion-affected arable sites: Using weighing lysimeter data for model intercomparison. *Vadose Zone Journal*, 19(1): e20058.

Groh, J., et al., 2022. Same soil, different climate: crop model intercomparison on translocated lysimeters. *Vadose Zone Journal*, 21(4): e20202.

Groot, J.J.R., 1987. *Simulation of nitrogen balance in a system of winter wheat and soil*. Wageningen: Centre for Agrobiological Research and Department of Theoretical Production Ecology, Agricultural University.

Groot, J.J.R., et al., 1991. *Nitrogen turnover in the soil/crop system*. Dordrecht: Kluwer Academic Publishers.

Hagin, J. and Amberger, A., 1974. *Contribution of fertilizers and manures to the N and P load of waters. A computer simulation*. Munich: Technische Universität München, 123 pp.

Hanks, R.J.B., 1962. Numerical solution of the moisture flow equation for infiltration into layered soils. *Soil Science Society of America Proceedings*, 26: 530–534.

Hansen, S., et al., 1991. Simulation of nitrogen dynamics and biomass production in winter-wheat using the Danish simulation-model DAISY. *Fertilizer Research*, 27: 245–259.

Hao, S.R., et al., 2021. Performance of a wheat yield prediction model and factors influencing the performance: a review and meta-analysis. *Agricultural Systems*, 194: 103278.

Haude, W., 1954. Zur praktischen Bestimmung der aktuellen und potentiellen Evaporation und Evapotranspiration. *Mitteilungen des Deutschen Wetterdienstes*, 8: 3–32.

Hayman, G., et al., 2024. A framework for improved predictions of the climate impacts on potential yields of UK winter wheat and its applicability to other UK crops. *Climate Services*, 34: 100479.

He, Z.L., et al., 1999. Ammonia volatilization from different fertilizer sources and effects of temperature and soil pH. *Soil Science*, 164(10): 750–758.

Heger, K., 1978. Bestimmung der potentiellen Evapotranspiration u¨ber unterschiedlichen landwirtschaftlichen Kulturen. *Mitteilungen der Deutschen Bodenkundlichen Gesellschaft*, 26: 21–40.

Hengl, T., et al., 2017. SoilGrids250m: global gridded soil information based on machine learning. *PLoS One*, 12(2): e0169748.

Jänicke, C., et al., 2022. Field-level land-use data reveal heterogeneous crop sequences with distinct regional differences in Germany. *European Journal of Agronomy*, 141: 126632.

Jansson, P.E., 1991. *Simulation model for soil water and heat conditions*. Uppsala: Description of the SOIL model, Swedish University of Agricultural Sciences.

Jenkinson, D.S., 1990. The turnover of organic-carbon and nitrogen in soil. *Philosophical Transactions of the Royal Society of London Series B-Biological Sciences*, 329(1255): 361–368.

Jensch, K., et al., 2025. Integrating Landsat Sentinel-1 and Sentinel-2 time series for mapping of intermediate crops. *European Journal of Remote Sensing*, 8 (1): 2507738.

Jones, C.A. and Kiniry, J.R., 1986. *CERES-Maize: a simulation model of maize growth and development*. College Station: Texas A&M University Press.

Kaltofen, H., 1971. Probleme der Systemanalyse und Modellierung auf dem Gebiet der Ertragsbildung. In: H. Kaltofen, et al. (Eds.), *Problematik der Modellierung der Ertragsbildung*. Berlin: Akademie-Verlag, pp. 5–40.

Kamali, B., et al., 2022. Improving simulations of permanent grasslands across Germany by multi-objective uncertainty-based calibration of plant-water dynamics. *European Journal of Agronomy*, 134: 126464.

Kamali, B., et al., 2024. Quest to find compromised spatial and temporal resolutions for integrating remote sensing data with an agro-ecosystem model for grasslands. *International Journal of Applied Earth Observation and Geoinformation*, 128: 103705.

Kersebaum, K.C., 1989. *Die Simulation der Stickstoff-Dynamik von Ackerböden*. Germany: Hanover University, p. 141.

Kersebaum, K.C., 1995. Application of a simple management model to simulate water and nitrogen dynamics. *Ecological Modelling*, 85: 145–156.

Kersebaum, K.C. and Beblik, A.J., 2001. Performance of a nitrogen dynamics model applied to evaluate agricultural management practices. In: M. Shaffer, et al. (Eds.), *Modeling carbon and nitrogen dynamics for soil management*. Boca Raton, FL: Lewis Publishers, pp. 549–569.

Kersebaum, K.C., et al., 2015. Analysis and classification of data sets for calibration and validation of agro-ecosystem models. *Environmental Modelling & Software*, 72: 402–417.

Khaledi, V., et al., 2024a. Evaluating MONICA's capability to simulate water, carbon and nitrogen fluxes in a wet grassland at contrasting water tables. *Science of the Total Environment*, 949: 174995.

Khaledi, V., et al., 2024b. Challenges of including wet grasslands with variable groundwater tables in large-area crop production simulations. *Agriculture-Basel*, 14(5): 679.

Kimball, B.A., et al., 2019. Simulation of maize evapotranspiration: an inter-comparison among 29 maize models. *Agricultural and Forest Meteorology*, 271: 264–284.

Kothari, K., et al., 2022. Are soybean models ready for climate change food impact assessments? *European Journal of Agronomy*, 135: 126482.

Kothari, K., et al., 2024. Evaluating differences among crop models in simulating soybean in-season growth. *Field Crops Research*, 309: 109306.

Kumudini, S., et al., 2014. Predicting maize phenology: intercomparison of functions for developmental response to temperature. *Agronomy Journal*, 106(6): 2087–2097.

Le Noë, J., et al., 2023. Soil organic carbon models need independent time-series validation for reliable prediction. *Communications Earth & EnvironmentCommunications Earth & Environment*, 4(1): 158.

Leaf, C.F. and Alexander, R.R., 1975. Simulating timber yields and hydrologic impact resulting from timber harvest on subalpine watersheds. USDA Forest Service Research Paper.

Long, S.P., 1991. Modification of the response of photosynthetic productivity to rising temperature by atmospheric CO2 concentrations - Has its importance been underestimated? *Plant, Cell & Environment*, 14(8): 729–739.

Lück, E. and Rühlmann, J., 2013. Resistivity mapping with geophilus electricus - information about lateral and vertical soil heterogeneity. *Geoderma*, 199: 2–11.

Luthin, J.N. and Gaskell, R.A., 1950. Numerical solution for tile drainage of layered soils. *Transactions of the American Geophysical Union*, 31: 595–602.

Main-Knorn, M., et al., 2025. Crop phenology assessment with multiscale and multisource time series. *European Journal of Remote Sensing*, under review.

Manzoni, S. and Cotrufo, M.F., 2024. Mechanisms of soil organic carbon and nitrogen stabilization in mineral-associated organic matter - insights from modeling in phase space. *Biogeosciences*, 21(18): 4077–4098.

Matthäus, E.M.W.K.H., et al., 1986. *The winter wheat crop model TRITSIM of the agroecosystem AGROSIM-W. Computer aided modelling and simulation of the winter wheat agroecosystem AGROSIM-W for integrated pest management*. Müncheberg: Akademie-Verlag, pp. 43–74.

McGill, W.B.H., et al., 1981. *PHOENIX: a model of carbon and nitrogen dynamics in grassland soils*. Stockholm: Swedish Natural Science Research Council, pp. 49–115.

Mehran, M.T., 1974. Computer modeling of nitrogen transformations in soils. *Journal of Environmental Quality*, 3: 391–396.

Mirschel, W., et al., 1987a. Dynamic yield formation and development model TRITSIM for winter wheat. 2. Model validation and verification. *Archiv für Acker- und Pflanzenbau und Bodenkunde*, 31(4): 259–267.

Mirschel, W., et al., 1987b. Dynamic yield formation and development model TRITSIM for winter wheat. 1. Model description. *Archiv für Acker- und Pflanzenbau und Bodenkunde*, 31(4): 249–257.

Mirschel, W., and Wenkel, K.O. 2007. Modelling soil–crop interactions with AGROSIM model family. In: Kersebaum, K.C., Hecker, JM., Mirschel, W., Wegehenkel, M. (Eds.), *Modelling water and nutrient dynamics in soil–crop systems*. Springer, Dordrecht. https://doi.org/10.1007/978-1-4020-4479-3_6

Mitchell, R.A.C., et al., 1995. Effects of elevated CO2 concentration and increased temperature on winter wheat - Test of Arcwheat1 simulation model. *Plant, Cell & Environment*, 18(7): 736–748.

Motovilov, Y.G., et al., 1999. Validation of a distributed hydrological model against spatial observations. *Agricultural and Forest Meteorology*, 98–9: 257–277.

Müller, T., et al., 1997. Temporal variation of C and N turnover in soil after oilseed rape straw incorporation in the field: Simulations with the soil-plant-atmosphere model DAISY. *Ecological Modelling*, 99(2–3): 247–262.

Münch, T., et al., 2014. Considering cost accountancy items in crop production simulations under climate change. *European Journal of Agronomy*, 52: 57–68.

Nendel, C., 2010. Grapevine bud break prediction for cool winter climates. *International Journal of Biometeorology*, 54(3): 231–241.

Nendel, C., 2014. MONICA – a simulation model for nitrogen and carbon dynamics in agro-ecosystems. In: L. Müller, et al. (Eds.), *Novel measurement and assessment tools for monitoring and management of water and soil resources in agricultural landscapes of Central Asia*. Environmental Science and Engineering. Switzerland: Springer International Publishing, pp. 389–405.

Nendel, C. and Kersebaum, K.C., 2004. A simple model approach to simulate nitrogen dynamics in vineyard soils. *Ecological Modelling*, 177(1–2): 1–15.

Nendel, C., et al., 2009. Testing different CO2 response algorithms against a FACE crop rotation experiment. *NJAS: Wageningen Journal of Life Sciences*, 57(1): 17–25.

Nendel, C., et al., 2011. The MONICA model: testing predictability for crop growth, soil moisture and nitrogen dynamics. *Ecological Modelling*, 222(9): 1614–1625.

Nendel, C., et al., 2013. Simulating regional winter wheat yields using input data of different spatial resolution. *Field Crops Research*, 145: 67–77.

Nendel, C., et al., 2014. Testing farm management options as climate change adaptation strategies using the MONICA model. *European Journal of Agronomy*, 52: 47–56.

Nendel, C., et al., 2019. The nitrogen nutrition potential of arable soils. *Scientific Reports*, 9: 5851.

Nendel, C., et al., 2023a. Improving data flow and integration in models assessing the impact of climate change on agriculture. In: C. Nendel (Ed.), *Modelling climate change impacts on agricultural systems*. Burleigh Dodds Series in Agricultural Science. Cambridge, UK: Burleigh Dodds Science Publishing Ltd, pp. 39–74.

Nendel, C., et al., 2023b. Future area expansion outweighs increasing drought risk for soybean in Europe. *Global Change Biology*, 29(5): 1340–1358.

Nendel, C., et al., 2025. Projections of irrigation water use for crop production in Brandenburg. *Agricultural Water Management*, under review.

Nimah, M.N.H., 1973. Model for estimating soil-water, plant, and atmospheric interrelations. *Soil Science Society of America Journal*, 37: 522–527.

Nordmeyer, H. and Richter, J., 1985. Incubation experiments on nitrogen mineralization in loess and sandy soils. *Plant and Soil*, 83(3): 433–445.

Nuske, A., 1983. i*Modell für die Stickstoff-Dynamik von Acker-Lößböden im Winterhalbjahr - Messungen und Simulationen*. Dissertation Universität Hannover.

Nuttall, J.G., et al., 2017. Models of grain quality in wheat: a review. *Field Crops Research*, 202: 136–145.

Olsen, P.A. and Haugen, L.E., 1997. *Jordas termiske egenskaper*. Ås, Norway: Agricultural University of Norway.

Palka, M., et al., 2026. Analysing drivers for cropping decisions and predicting crop rotation patterns for Central Europe using machine learning. *Agricultural Systems*, 231, 104522.

Parton, W.J., et al., 1987. Analysis of factors controlling soil organic-matter levels in great-plains grasslands. *Soil Science Society of America Journal*, 51(5): 1173–1179.

Pedersen, A., et al., 2009. Simulating nitrate retention in soils and the effect of catch crop use and rooting pattern under the climatic conditions of Northern Europe. *Soil Use Manage*, 25(3): 243–254.

Philip, J.R., 1957. The theory of infiltration: 1. The infiltration equation and its solution. *Soil Science*, 83: 345–357.

Pirttioja, N., et al., 2015. Temperature and precipitation effects on wheat yield across a European transect: a crop model ensemble analysis using impact response surfaces. *Climate Research*, 65: 87–105.

Rahn, C.R., et al., 2010. EU-Rotate_N - a decision support system - to predict environmental and economic consequences of the management of nitrogen fertiliser in crop rotations. *European Journal of Horticultural Science*, 75(1): 20–32.

Raza, A., et al., 2025. Predicting groundwater levels at high spatial resolution for Brandenburg using machine learning models. *Agricultural Water Management*, under review.

Reynolds, M.P., et al., 2022. A wiring diagram to integrate physiological traits of wheat yield potential. *Nature Food*, 3: 318–324.

Rezaei, E.E., et al., 2021. The use of remote sensing to derive maize sowing dates for large-scale crop yield simulations. *International Journal of Biometeorology*, 65(4): 565–576.

Richards, L.A. 1931. Capillary conduction of liquids through porous medium. *Physics* 1: 318–333. https://doi.org/10.1063/1.1745010

Richardson, L.F. 1922. *Weather prediction by numerical process*. University Press, Cambridge. p. 262.

Richter, J.N., 1985. Simulation of nitrogen regime in loess soils in the winter half-year - Comparison between field measurements and simulations. *Plant and Soil*, 83: 419–431.

Richter, J.S., 1978. Simulation der winterlichen Nitratverlagerung in Böden. *Plant and Soil*, 49: 381–393.

Riley, H. and Bonesmo, H., 2005. *Modelling of snow and freeze-thaw cycles in the EU-Rotate_N decision support system*. Planteforsk Grønn Kunnskap.

Ritchie, J.T., et al., 1984. *CERES-Wheat: a user oriented wheat yield model*. Michigan: Preliminary documentation, AGRISTARS Publication, East Lansing, pp. 252.

Ritchie, J.T., et al., 1989. *CERES-wheat: a simulation model for wheat growth and development*. College Station, TX: Texas A&M University Press.

Rocci, K.S., et al., 2024. Bridging 20 years of soil organic matter frameworks: empirical support, model representation, and next steps. *Journal of Geophysical Research: Biogeosciences*, 129(6): e2023JG007964.

Rötter, R.P., et al., 2012. Simulation of spring barley yield in different climatic zones of Northern and Central Europe: a comparison of nine crop models. *Field Crops Research*, 133: 23–36.

Sadeghi, A.M., et al., 1988. Mechanistic model for predicting ammonia volatilization from urea. Ammonia volatilization from urea fertilizers, National Fertilizer Development Center, Muscle Shoals.

Salo, T.J., et al., 2016. Comparing the performance of 11 crop simulation models in predicting yield response to nitrogen fertilization. *Journal of Agricultural Science*, 154(7): 1218–1240.

Seligman, N.C. and van Keulen, H., 1981. PAPRAN: a simulation model of annual pasture production limited by rainfall and nitrogen. In: M.J. Frissel and J.A. van Veen (Eds.), *Simulation of nitrogen behaviour of soil-plant systems*. Wageningen: PUDOC, pp. 192–221.

Šimůnek, J., et al., 2024. Developments and applications of the HYDRUS computer software packages since 2016. Vadose Zone Journal, 23(4): e20310.

Söder, M., et al., 2022. *Klimawandelbedingte Ertragsveränderungen und Flächennutzung (KlimErtrag), Braunschweig*. Braunschweig: Johann Heinrich von Thünen-Institut.

Søgaard, H.T., et al., 2002. Ammonia volatilization from field-applied animal slurry - the ALFAM model. *Atmospheric Environment*, 36: 3309–3319.

Specka, X., et al., 2015. Analysing the parameter sensitivity of the agro-ecosystem model MONICA for different crops. *European Journal of Agronomy*, 71: 73–87.

Specka, X., et al., 2016. Reproducing CO2 exchange rates of a crop rotation at contrasting terrain positions using two different modelling approaches. *Soil & Tillage Research*, 156: 219–229.

Specka, X., et al., 2019. Temporal sensitivity analysis of the MONICA model: application of two global approaches to analyze the dynamics of parameter sensitivity. *Agriculture-Basel*, 9(2), e20310.

Tanji, K.K., et al., 1967. I. Quality of waters percolating through stratified substrata as predicted by computer analyses. *Hilgardia*, 38: 319–347.

Tao, F.L., et al., 2017. Designing future barley ideotypes using a crop model ensemble. *European Journal of Agronomy*, 82: 144–162.

Tao, F., et al., 2020. Why do crop models diverge substantially in climate impact projections? A comprehensive analysis based on eight barley crop models. *Agricultural and Forest Meteorology*, 281, 107851.

Tenreiro, T.R., et al., 2020. Water modelling approaches and opportunities to simulate spatial water variations at crop field level. *Agricultural Water Management*, 240: 106254.

Toth, B., et al., 2015. New generation of hydraulic pedotransfer functions for Europe. *European Journal of Soil Science - Wiley Online LibraryEuropean Journal of Soil Science*, 66(1): 226–238.

Tougma, I.A., et al., 2025. AMPSOM: a measureable pool soil organic carbon and nitrogen model for arable cropping systems. *Environmental Modelling & Software*, 185, 106291.

Triacca, A., et al. 2026. Lentil and chickpea yield prediction across Europe. *European Journal of Agronomy*, in press.

Tsubo, M. and Walker, S., 2002. A model of radiation interception and use by a maize-bean intercrop canopy. *Agricultural and Forest Meteorology*, 110(3): 203–215.

Unger, K., 1968. Biophysik pflanzlicher Systeme. *Studia Biophysica*, 11: 1–10.

van Keulen, H.P.D.V., 1975. *Simulation of water use and herbage growth in arid regions*. Wageningen: PUDOC.

van Keulen, H.P.D.V., et al., 1982. *A summary model for crop growth. Simulation of plant growth and crop production*. Wageningen: PUDOC, pp. 87–97.

van Veen, J.A. and Frissel, M.J., 1981. *Simulation model of the behaviour of N in soil. Simulation of nitrogen behaviour of soil-plant systems*. Wageningen: PUDOC, pp. 126–144.

Vereecken, H.M., et al., 1989. Estimating the soil moisture retention characteristic from texture, bulk density, and carbon content. *Soil Science*, 148: 389–403.

Vielemeyer, H.P., et al., 1983. A new method of determining limiting values for assessing the level of nutrient supply in agricultural crops on the basis of plant analysis. *Archiv fur Acker und Pflanzenbau und Bodenkunde-Archives of Agronomy and Soil Science*, 27: 445–453.

Wallor, E., et al., 2018. The response of process-based agro-ecosystem models to within-field variability in site conditions. *Field Crops Research*, 228: 1–19.
Wang, E.L., et al., 2017. The uncertainty of crop yield projections is reduced by improved temperature response functions. Nature Plants, 3, 17102.
Wang, E.L., et al., 2022. How reliable are current crop models for simulating growth and seed yield of canola across global sites and under future climate change? *Climatic Change*, 172(1–2), Article 33.
Wegehenkel, M., 2000. Test of a modelling system for simulating water balances and plant growth using various different complex approaches. *Ecological Modelling*, 129: 39–64.
Weigel, H.J. and Dämmgen, U., 2000. The Braunschweig Carbon project: atmospheric flux monitoring and free air carbon dioxide enrichment (FACE). *Journal of Applied Botany and Food Quality*, 74(1–2): 55–60.
Wendling, U., et al., 1991. Bereitstellung von täglichen Informationen zum Wasserhaushalt des Bodens für die Zwecke der agrarmeteorologischen Beratung. *Zeitschrift für Meteorologie*, 41: 468–475.
Wenkel, K.O., et al., 2013. LandCaRe DSS - An interactive decision support system for climate change impact assessment and the analysis of potential agricultural land use adaptation strategies. *Journal of Environmental Management*, 127: S168–S183.
Wessolek, G., et al., 2009. *Bodenphysikalische Kennwerte und Berechnungsverfahren für die Praxis*. Bodenökologie und Bodengenese, 40. Berlin: Technische Universität Berlin.
Whitmore, A.P. and Parry, L.C., 1988. *Computer simulation of the behaviour of nitrogen in soil and crop in the Broadbalk continuous wheat experiment. Nitrogen efficiency in agricultural soils*. London: Elsevier, pp. 141–157.
Wilkerson, G.G., et al., 1983. Modeling soybean growth for crop management. *Transactions of the ASAE*, 26(1): 63–73.
Yang, C.P., et al., 2024. Prediction of drought trigger thresholds for future winter wheat yield losses in China based on the DSSAT-CERES-Wheat model and Copula conditional probabilities. *Agricultural Water Management*, 299: 108881.
Yang, X., et al., 2025. Investigating the appropriateness of two different modelling approaches for the estimation of nitrogen leaching under field-grown tomato. *European Journal of Agronomy*, under review.
Yin, X.G., et al., 2017. Performance of process-based models for simulation of grain N in crop rotations across Europe. *Agricultural Systems*, 154: 63–77.
Yu, J., et al., 2024. Modelling crop yield in a wheat-soybean relay intercropping system: a simple routine in capturing competition for light. *European Journal of Agronomy*, 153.
Yu, J., et al., 2025. Exploring wheat-soybean intercropping in conjunction with shifting sowing dates as a strategy for climate change adaptation of crop production in Germany. *Field Crops Research*, 322: 109695.
Zepp, S., et al., 2023. Optimized bare soil compositing for soil organic carbon prediction of topsoil croplands in Bavaria using Landsat. *ISPRS Journal of Photogrammetry and Remote Sensing*, 202: 287–302.
Zhao, G., et al., 2016. Evaluating the precision of eight spatial sampling schemes in estimating regional means of simulated yield for two crops. *Environmental Modelling & Software*, 80: 100–112.

Chapter 17

The AquaCrop model

Dirk Raes, KU Leuven University, Belgium; Elias Fereres, IAS-CSIC and University of Cordoba, Spain; Theodore C. Hsiao and Pasquale Steduto, University of California, USA; Gabriëlle De Lannoy, KU Leuven University, Belgium; Eline Vanuytrecht, European Environment Agency, Denmark; Lee Heng, International Atomic Energy Agency, Austria; Margarita Garcia Vila, IAS-CSIC, Spain; Joost Wellens, Université de Liège, Belgium; Patricia Mejias Moreno and Maher Salman, Food and Agriculture Organization of the United Nations, Italy

1 Origin and initial development

The response of crop yield to variations in the water supply has been a topic of great interest in agricultural sciences for a long time. A major contribution to the quantification of such response was the development of empirical water production functions in the 1960s and 1970s, which related yield to crop consumptive use as linear functions. Such information was fitted by Stewart (Stewart and Hagan, 1973) to a linear model which was then used by Doorenbos and Kassam to compile all the information available on various crops into a FAO (the Food and Agriculture Organisation of the United Nations) publication, the Irrigation & Drainage no. 33 paper (ID33; Doorenbos and Kassam, 1979).

In 2002, Martin Smith, senior officer of the Land and Water division of FAO, called an expert consultation to discuss the possible revision of ID33

http://dx.doi.org/10.19103/AS.2025.0155.20

and whether there could be approaches that were different from the empirical production functions. Two options were discussed: the use of existing crop simulation models (e.g. DSSAT) and the development of a new simulation model focused on crop yield response to water. Following several international meetings in FAO, Rome, Pasquale Steduto, then recently appointed head of the water division, decided to develop a new water-driven crop simulation model with a minimum of variables but with solid foundations on the relations between crop production and water use. For the development of such a model, called AquaCrop, a core group of four was formed, namely, P. Steduto, T. C. Hsiao, E. Fereres, and D. Raes, who designed and tested different versions of the model from its inception. Finally, a working version of the model was presented officially at the American Society of Agronomy meetings of 2008 and was published in 2009 (Steduto et al., 2009; Raes et al., 2009; Hsiao et al., 2009).

With the contribution of several scientists worldwide, the model has been calibrated and partially validated to simulate a broad range of field crops, and it has been regularly updated and expanded with additional crops and new features (e.g. Vanuytrecht et al., 2014a; Van Gaelen et al., 2016; Raes et al., 2023a). AquaCrop is freely available, and key documents, training material, reference manuals, and software can be downloaded from the FAO AquaCrop website (https://www.fao.org/aquacrop/en/).

2 Calculation scheme

AquaCrop differs from other crop models, which are driven by the intercepted solar radiation in that it has a 'water-driven' growth engine that simulates crop production based on the conservative relation between crop transpiration and biomass production (Steduto et al., 2007). To take into consideration the differential sensitivity of key physiological processes affected by water stress, AquaCrop uses three different levels of soil water content for the onset of the reduction of canopy expansion and stomatal conductance, and for the triggering of early senescence (Hsiao, 1973; Bradford and Hsiao, 1982).

AquaCrop successively simulates on each day of the growing cycle, the (1) green canopy cover, (2) crop transpiration, (3) above-ground biomass, and (4) crop yield, as schematically depicted in the infographic (Fig. 1) and described in detail in the AquaCrop Reference Manual (Raes et al., 2023b).

Briefly, in each daily time step the model starts with the simulation of the degree of the green canopy cover (CC_i), which is affected by the prevailing water, fertility, and salinity stress (Fig. 1). This leads to the calculation of crop transpiration (Tr_i), which transpiration coefficient (Kc_i) is proportional to the simulated CC_i. Next, a normalized water productivity (WP*) is introduced to convert the amount of water transpired that day into biomass production

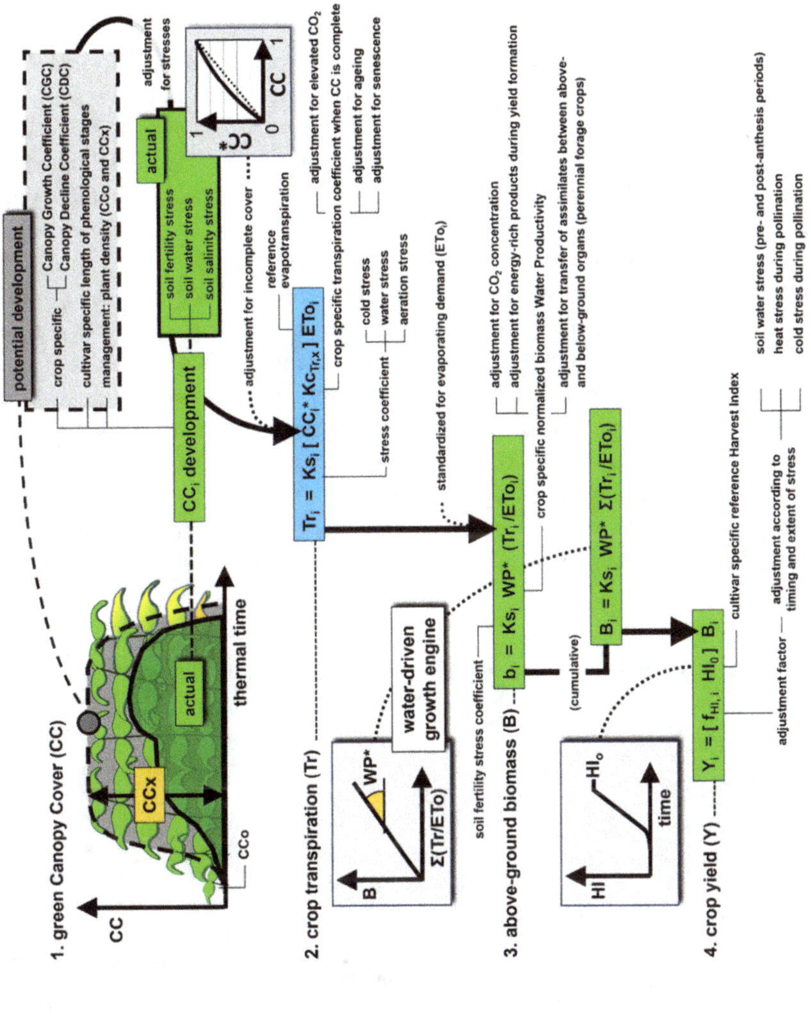

Figure 1 Infographic of the calculation scheme of AquaCrop. For details, see text or consult the Reference Manual (Raes et al., 2023b).

(b_i). Finally, a daily estimate of crop yield (Y_i) is obtained by considering the cumulative amount of biomass produced to date ($B = \Sigma b_i$), and the effect of water and temperature stresses on the harvest index (HI).

The effect of stress is described in AquaCrop by a stress coefficient (Ks) which varies in value from one (no stress) above an upper threshold to zero (full stress) below a lower threshold of the stress indicator (Fig. 2). Between the two thresholds, Ks follows a curve whose shape determines the impact on the simulated process for the various degrees of stress. AquaCrop considers air temperature, soil water, soil salinity, and soil fertility stress coefficients. The stress indicator (abscissa) for soil water stress is depletion (shortage versus field capacity), which is expressed as a fraction (p) of the total available soil water (TAW). The indicator for air temperature stress is growing degrees (affecting crop transpiration) or a minimum or maximum air temperature (affecting pollination). For soil salinity, the indicator is the electrical conductivity of the saturated soil-paste extract (EC_e).

The two presented Ks coefficients in Fig. 2 illustrate the differential sensitivity of physiological processes to water stress. Leaf growth by area expansion (expansive growth) and, therefore, canopy development are the highest in sensitivity to water stress among all the plant processes described by the model (Hsiao, 1973; Bradford and Hsiao, 1982). The decrease in canopy expansion is simulated by multiplying the Canopy Growth Coefficient (CGC) by Ks_{exp}. Stomata have been shown to be much less sensitive to water stress in comparison to leaf expansive growth. The multiplication of the crop transpiration coefficient ($Kc_{Tr,x}$) by Ks_{sto} simulates stomatal closure.

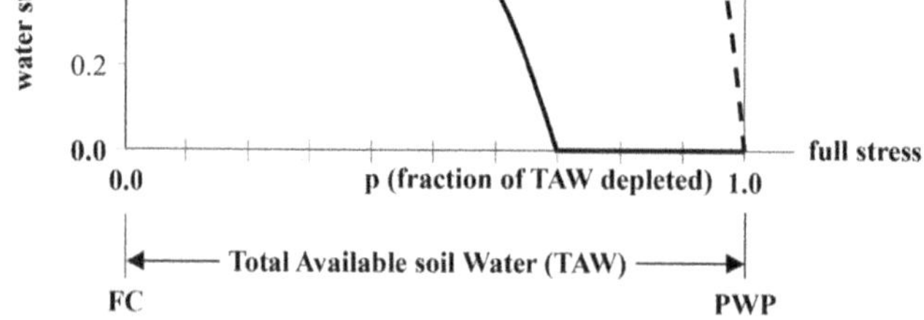

Figure 2 Variation of the water stress coefficient for leaf expansion (Ks_{exp}, full line) and stomatal conductance (Ks_{sto}, dotted line) for various soil water depletions (p) in the root zone.

2.1 Green Canopy Cover (CC)

The green Canopy Cover (CC) is the fraction of the soil surface covered by the green part of the canopy. In the absence of stress (potential development), the canopy development increases from a minimum value at emergence (CCo) to a maximum canopy cover (CCx), both of which are determined by the plant density. The potential canopy development, which is driven by thermal time (expressed in Growing Degree Days), is simulated by two equations, one describing the exponential growth from emergence till half of CCx, and the other the decelerating growth till CCx is reached. The Canopy Growth Coefficient (CGC) determines the development rate in both equations, while the Canopy Decline Coefficient (CDC) determines the rate of the green canopy decline in the late season stage. Both coefficients are conservative crop parameters and are crop specific. Contrary to other models, AquaCrop does not make use of the leaf area index (LAI) as a measure of canopy development, but CC and LAI have been shown to be closely related (e.g. Hsiao et al., 2009).

Soil water, soil fertility, and soil salinity stress affect canopy expansion. As a result, the expected CCx might not be achieved or might be achieved much later in the season. Severe soil water stress can even result in early canopy senescence as simulated in the model (Fig. 1).

2.2 Crop transpiration (Tr)

In the absence of stress, crop transpiration (Tr) is calculated by multiplying the reference evapotranspiration (ETo) with a crop transpiration coefficient (Kc_{Tr}, Eq.1), which considers (i) the fraction by which the green canopy covers the ground ($0 \leq CC \leq 1$) and (ii) specific characteristics that distinguish the crop from the reference grass ($Kc_{Tr,x}$):

$$Kc_{Tr} = CC^{*}\, Kc_{Tr,x} \tag{Eq.1}$$

where $Kc_{Tr,x}$ is the crop coefficient for maximum crop transpiration (when CC = 1), and CC* the green canopy cover adjusted for incomplete cover. To estimate crop transpiration, the green canopy cover when incomplete (CC < 1) is increased to a CC* (see insert in Fig. 1), accounting for interrow micro-advection and sheltering effect by partial canopy cover (Adams et al., 1976; Villalobos and Fereres, 1990).

Transpiration is reduced when $Kc_{Tr,x}$ declines (i) under elevated CO_2 concentrations, (ii) with the ageing of the canopy, and (iii) once senescence is triggered. Further on, crop transpiration is restricted (by several Ks stress coefficients targeting $Kc_{Tr,x}$) when there are insufficient growing degrees for full crop transpiration (cold stress) and when there is too much (aeration stress) or too little water in the root zone (soil water stress).

2.3 Above-ground biomass (B)

The above-ground biomass (B) is derived from crop transpiration by means of the normalized biomass water productivity WP* (see insert in Fig. 1), which has been shown to be conservative under water and salinity stress, along with a low sensitivity to nutrient deficiency (Steduto et al., 2000; Steduto and Albrizio, 2005). Since AquaCrop does not simulate biomass partitioning among various organs, variations in partitioning (in the case of alfalfa) between the shoot and the root (in Spring and Fall) are simulated in the alfalfa model by adjusting the WP* (Raes et al., 2023a). Furthermore, AquaCrop adjusts WP* when the mean annual atmospheric CO_2 concentration differs from its reference value (369.41 ppm in the year 2000). The effect of soil fertility stress is simulated by a set of Ks coefficients, which gradually decreases CC and the WP* as discussed below.

2.4 Crop yield (Y)

The partition of above-ground biomass into yield (Y) is simulated by means of a Harvest Index (HI), which increases with time (see insert in Fig. 1) after flowering (fruit/grain producing crops) or tuber formation (root/tuber crops). There is an adjustment of HI to soil water and air temperature stresses, which depends on the timing and the intensity of the stresses.

3 Input requirements

AquaCrop uses a relatively small number of explicit and largely intuitive parameters and input variables, either widely used or requiring simple methods for their determination. Input consists of weather data, crop and soil characteristics, and field and irrigation management practices that characterize the environment in which the crop will develop (Fig. 3).

The weather data can be daily, 10-daily, or monthly. In the absence of daily data AquaCrop invokes built-in procedures to approximate the required daily data (time step of the simulation) from the 10-day or monthly means (Raes et al., 2023b). However, the use of 10-day or monthly total rainfall should be avoided since the dynamic nature of crop responses to water will be lost. A calculator, which computes the reference evapotranspiration (ETo) from imported weather data by means of the FAO Penman-Monteith method (Allen et al., 1998), is incorporated into AquaCrop. For the adjustment of WP* to the average annual CO_2 concentration, AquaCrop uses historically recorded or one of the projected future series of annual atmospheric concentrations available in the database.

The database of AquaCrop contains default values for the required soil profile characteristics for the 12 textural classes. The required input can also be

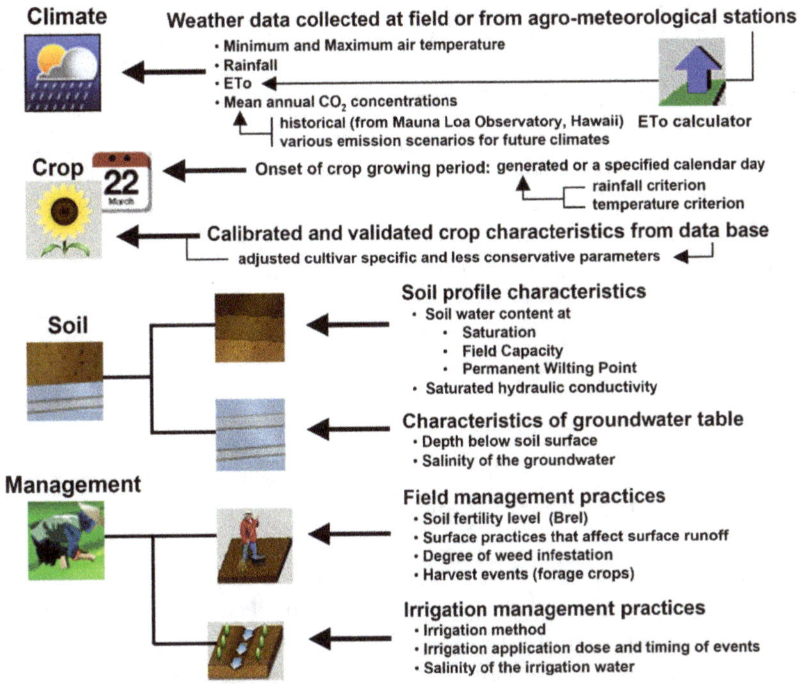

Figure 3 Schematic overview of the required input for AquaCrop.

collected in the field or derived from soil texture with the help of pedotransfer functions (Saxton et al., 1986). Since the groundwater table is not part of the simulated system, its characteristics (depth and water quality) are required as inputs.

4 Parameterization of AquaCrop for crops

4.1 Crop parameters

The database of AquaCrop contains calibrated and validated crop parameters for a number of herbaceous crops. A distinction is made between:Conservative crop parameters. These are parameters which do not change materially with time, management practices, geographic location, or climate, and are mostly genetically dependent. They are also assumed not to change among cultivars unless shown otherwise. In fact, as the model has been used more with time, our experience is that some of these parameters may vary among genotypes of the same species, more so than what was originally anticipated. Table 1 lists specific conservative crop parameters (genotype dependent) for some selected crops that have been already calibrated and validated in AquaCrop.

Table 1 Some specific conservative crop parameters of AquaCrop for selected crops: Crop transpiration coefficient for a full canopy ($Kc_{Tr,x}$); normalized biomass water productivity (WP*); reference Harvest Index (HI_o); upper and lower water stress thresholds for canopy expansion (p_{exp}) and for stomatal closure (p_{sto}), expressed as fractions of total available water (f_{TAW}).

	Cotton	Maize	Potato	Rice	Soybean	Sunflower	Wheat
$Kc_{Tr,x}$ [–]	**1.10**	1.05	1.10	1.10	1.10	1.10	1.10
WP* [gr.m⁻²]	15 (*1)	33.7	18–20	19	15	18 (*2)	15
HI_o [%]	25–40	48–52	70–85	35–50	40	35	45–50
p_{exp}[f_{TAW}]	0.2–0.7	0.15–0.7	0.2–0.6	0–0.4	0.15–0.65	0.15–0.65	0.2–0.65
p_{sto} [f_{TAW}]	0.75-1	0.69-1	0.6-1	0.50-1	0.6-1	0.6-1	0.65

WP* of cotton drops to 70% of its value during yield formation.
WP* of sunflower drops to 60% of its value during yield formation.

The parameters chosen in Table 1 are among the most relevant in those that characterize AquaCrop simulations. The variations indicated in some of the parameters (HI, for example) account for the genotypic differences within a specie.

Cultivar-specific and less-conservative parameters. These are parameters which might require an adjustment when selecting a cultivar which is different from the one considered for crop calibration or when the planting method, field management, or conditions in the soil profile differ from the conditions assumed at calibration.

The reader is referred to Annex 1 of the Reference Manual of AquaCrop for the entire list of conservative and less conservative parameters (management and genotype dependent) of each of the 17 crops that have already been calibrated and validated. The Reference Manual can be downloaded from the FAO AquaCrop website (https://www.fao.org/aquacrop/en/).

From the start of reporting about model performance, most AquaCrop publications listed the specific parameters for the different species and cultivars (e.g. Hsiao et al., 2009; Heng et al., 2009; Farahani et al., 2009; Todorovic et al., 2009; Geerts et al., 2009; García-Vila et al., 2009, Wellens et al., 2022b).

4.2 Required adjustment of WP* for a correct simulation of above-ground biomass

As shown in Fig. 1, Aquacrop calculates the above-ground biomass (B) from crop transpiration (Tr) normalized for climate (ETo) by using a normalized biomass water productivity factor (WP*). Since intermediary steps in the accumulation of biomass are not simulated, a single value for WP* can be used for the entire crop cycle (Steduto et al., 2009).

A single value of WP* is, however not justified when the harvestable yield has a high proportion of lipids and protein, since more energy is required per unit of dry weight produced (Penning de Vries et al., 1983; Azam-Ali and Squire, 2002). To consider the formation of energy-rich products, WP* declines during yield formation. This is simulated by adjusting the daily produced mass of the above-ground biomass (b_i):

$$B = \sum f_i b_i = f_i \; WP^* \sum \left(\frac{Tr_i}{ETo_i} \right) \quad \text{(Eq. 2)}$$

where B is the above-ground biomass produced during the growing cycle, and f_i a correction factor which has the value of one until flowering, and exponentially decreases during yield formation (Fig. 4a). The f_i values were obtained through calibration, giving f_i of 0.70 for cotton, and 0.60 for soybean and sunflower at the end of the reproductive period (Tab. 1). With such f_i values, the attainable yield could be well simulated (Garcia-Vila et al., 2009; Farahani et al., 2009; Todorovic et al., 2009; Mbangiwa et al., 2019).

The AquaCrop model, originally designed for annual crops, was expanded in 2022 to simulate perennial herbaceous forage crops (Raes et al., 2023a). Perennial herbaceous forage crops allocate the carbon assimilated through photosynthesis to above- (leaves and stems) and below-ground organs (crowns and roots). Since AquaCrop does not simulate biomass partitioning among various organs, variations in partitioning along the season are simulated by increasing (remobilisation) or reducing (storage) the daily produced mass of the above-ground biomass (b_i). After mid-season, perennial herbaceous forage crops such as alfalfa transfer a considerable fraction of the assimilates below ground (Teixeira et al., 2008; Moot et al., 2012). This is simulated by an exponential decrease of b_i with the help of the correction factor f_i during the storage stage (Eq. 2). The biomass partitioning below ground in late summer/autumn is cultivar specific. For example, it is greater for higher latitude cultivars (in response to lower temperatures and shorter day length) than for Mediterranean alfalfa cultivars (Moot et al., 2012). For the remobilization of some of the stored assimilates, b_i is increased by f_i at the start of the next season

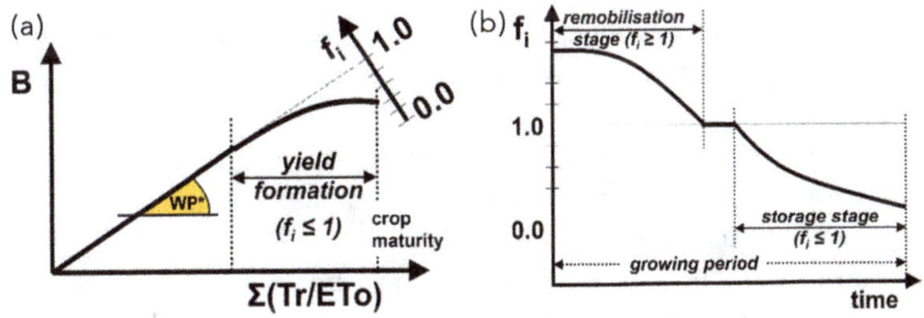

Figure 4 Daily correction (f_i) for WP* (Eq. 2) for (a) crops where the harvestable yield contains energy-rich products, and (b) perennial herbaceous forage which transfer assimilates between above- and below-ground organs.

(remobilisation stage). The rise of f_i gradually decreases to 1, as more and more assimilates are remobilized (Fig. 4b).

To evaluate the simulation of alfalfa, yield data was collected in several different climates, for different alfalfa cultivars, over various years of field and irrigation management strategies (Raes et al., 2023a). There was a good agreement between the observed and simulated cumulative dry above-ground biomass during the growing cycle, and a systematic over- or underestimation by the model of the measured yield was not observed.

4.3 Dealing with genetic differences among species and cultivars

AquaCrop was originally designed to simulate a generic herbaceous crop. All the parameters that are needed for the simulation were then adjusted by calibration to simulate the performance of about a dozen different herbaceous species. However, within any species, different cultivars exist which have vastly different genetic makeup. In AquaCrop, there is no systematic way of including such genetic differences in the form of 'genetic coefficients' that are used in other models. The cultivar differences must be taken into account in the simulations by varying any of the crop parameters to reflect a cultivar-specific feature. Among commercially available cultivars of the major crops, the genotypic variations are relatively small and mostly tied to a few features. Without a doubt, season length (determining biomass accumulation and water use) is the most diverse feature within a single crop that is genetically controlled. In AquaCrop, the duration of the developmental phases is an input to the model and must be known for each specific cultivar. The harvest index also varies within a species (often tied to season length; long season varieties tend to have lower HI). To take into account such differences, the reference HI may be modified in AquaCrop accordingly. Table 1 shows that other parameters, such as WP* vary little outside the obvious difference between C_3 and C_4 crops. There must be sufficient experimental evidence before modifying certain parameters in the simulations of specific genotypes. It is possible, however, to introduce genetic diversity in AquaCrop simulations by varying parameters to reflect features such as a deeper root system, a more drought tolerant genotype, or an early season cultivar.

As an example of the simulation of different cultivars, different conservative crop parameters were considered for three potato cultivars by the Belgian Collaborative Agriculture Monitoring (BELCAM) initiative (Table 2). The data were used in an interactive web-based geo-platform to help farmers monitor their crops and estimate yields. To calibrate and validate the AquaCrop model for potato under Belgian conditions, data were collected in 2015 (a particular dry year) for 80 plots and for 67 fields in 2016 (a rather wet year) in the potato

Table 2 The default values in AquaCrop of conservative crop parameters for potato and the adjusted values for the cultivars Bintje, Fontane and Nicola: normalized biomass water productivity (WP*); reference Harvest Index (HI_o); upper and lower water stress thresholds for canopy expansion (p_{exp}) and for stomatal closure (p_{sto}), expressed as fraction of Total Available Water (f_{TAW}); (adapted from Wellens et al., 2022a).

Conservative crop parameter	Default	Bintje	Fontane	Nicola
WP* [$gr.m^{-2}$]	18.0	17.5	18.5	18.5
Hlo [%]	75	90	80	90
$p_{exp}[f_{TAW}]$	0.20–0.60	0.30–0.65	0.30–0.65	0.30–0.60
p_{sto} [f_{TAW}]	0.60–1	0.45–1	0.65–1	0.55–1

growing areas in Belgium. Calibration and validation results for the simulation of tuber dry weights are plotted in Fig. 5 (Wellens et al., 2022a).

While all the crop parameters may be adjusted to a specific genotype, it is important to start adjusting those which are known to vary among cultivars,

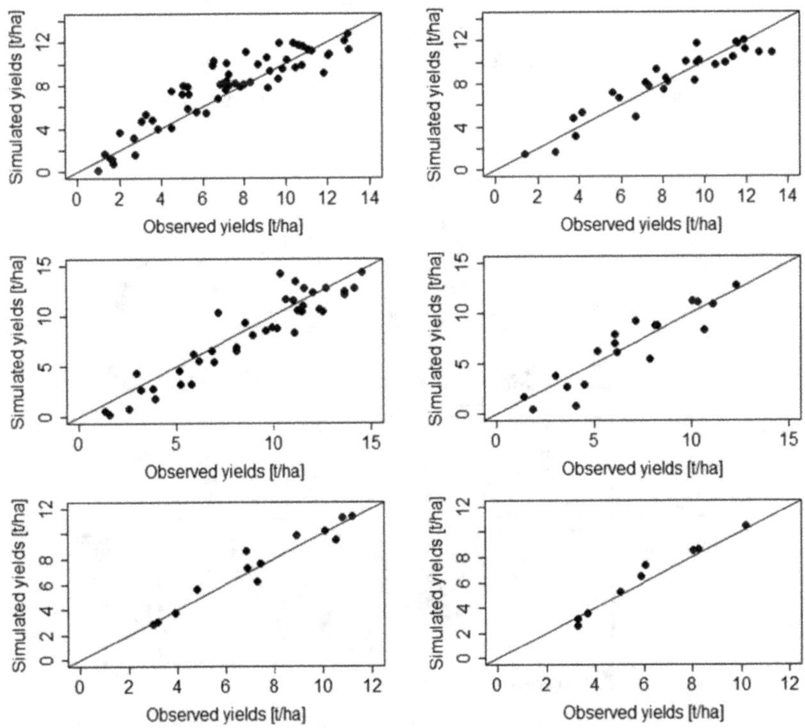

Figure 5 Observed versus simulated dry tuber weight throughout the growing season for Bintje (Row1), Nicola (Row 2), and Fontane (Row 3). Data used for calibration (column 1) and validation (column 2) are presented (from Wellens et al., 2022a).

such as those related to phenological development or harvest index (Fig. 6). These and less conservative crop parameters might require an adjustment when running a simulation for a specific cultivar or in a climate or management different from the environment in which the crop was calibrated and validated.

The plant density, which varies largely with management, determines the initial canopy cover (CCo, the canopy cover at 90% germination) and the maximum canopy cover (CCx) that can be reached at mid-season in the absence of stress. The time to reach 90% germination might require an adjustment since it is affected by field preparation, soil temperature, and soil water content.

As stated above, most differences among commercial crop cultivars are related to the length of development stages, while the maximum effective rooting depth (Z_x) and its deepening rate (or the time to reach Z_x) are also affected by soil physical (temperature, mechanical impedance, aeration) and soil chemical (pH, salinity, high levels of aluminium, or manganese) characteristics. The adjustment of the time to reach a particular stage and/or its duration for the local cultivar can be specified in AquaCrop in calendar days (Fig. 6). After loading the climatic data and specifying the planting date for the year of calibration, AquaCrop converts the specified calendar days into thermal time by considering the corresponding growing degree days. This automatically adjusts the length (number of days) of the various crop development stages and crop cycle to the temperature regime in other years or regions (Fig. 7).

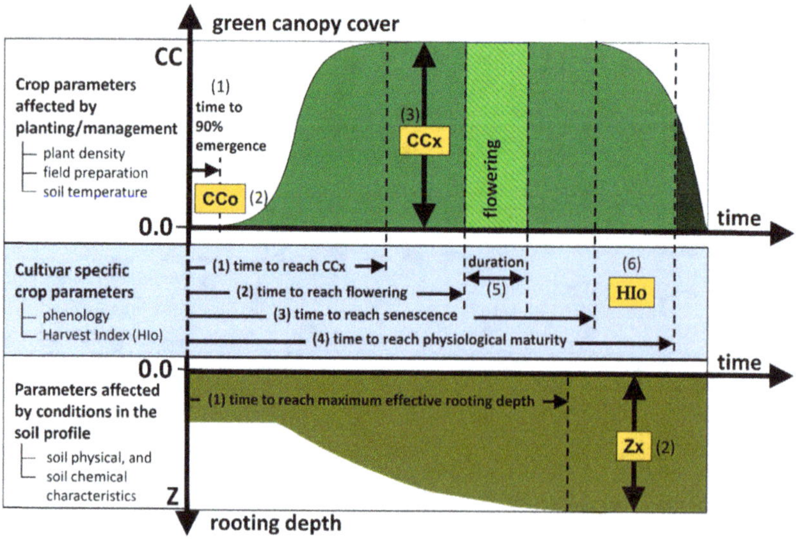

Figure 6 Non-conservative crop parameters affecting the development of the green canopy cover (CC), the extension of the rooting depth (Z), and crop yield (HIo), which might require an adjustment when selecting a cultivar or environment different from the one considered for crop calibration.

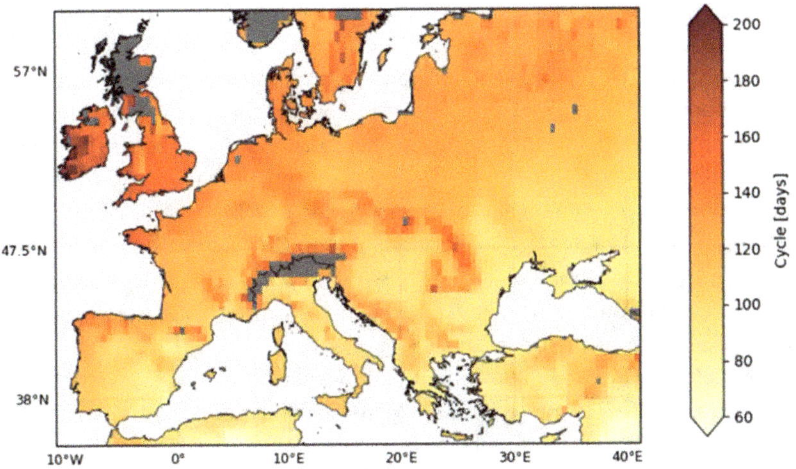

Figure 7 Variation of the average length (days) of the growing cycle of maize (requiring 1,112 growing degree days to reach maturity) across the European continent for the period 1985–2014. Planting dates were generated when the minimum air temperature exceeded 8 °C for four consecutive days from March 1 (from Deketelaere, 2024).

The reference Harvest Index (HI_0) is a representative HI reported in the literature for the crop species under non-stress conditions. Although HI_0 is conservative to a large extent, plant breeding has improved HI over the years as a major target, and its improvement has altered the reported values in the database of AquaCrop, which was put together in 2006, using published values which were obtained around 2000. For example, Ruiz et al. (2023) reported an average increase of 0.26% per year for the HIo of maize since 1964, while Hay and Gilbert (2001) provided values for maize landraces that differ widely from the HIo of the high-yielding commercial cultivars in AquaCrop's database. As stated earlier, some of the parameters considered conservative at the launching of the model have shown to vary among genotypes and management conditions. We recommend seeking several sources of knowledge before agreeing on the value of HI_0 as an input to the simulations, emphasizing the use of results obtained in local experimentation.

4.4 Calibration and simulation of the crop response to soil fertility stress

In AquaCrop, a soil water and salt balance keep track of the daily incoming and outgoing water and salt fluxes at the boundaries of the root zone. The shape of the various stress response curves determines daily the magnitude of the stresses (Ks) between its upper and lower threshold, and the crop's response to those stresses as described in detail in the AquaCrop Reference Manual (Raes et al., 2023b).

It is common in developing countries that the application of fertilizers is limited, and this adds another constraint on production relative to the limited water supply. Other crop models often use a mass balance for modeling the crop response to soil fertility. However, to counter the vast input requirement for simulating quantitatively the major nutrients and the effects of soil fertility management, AquaCrop adopts a simpler, semi-quantitative approach based on the crop response to nutrient stress. Instead of all the details of nutrient concentrations and the chemical forms, AquaCrop requires as input the relative crop production (with reference to the potential production) that can be achieved with the current soil fertility management in farmers' field in the absence of water or any other stress (Fig. 8).

AquaCrop simulates the effect of soil fertility stress by (a) restricting the canopy growth (CC) and (b) a gradual decrease of the normalized biomass water productivity (WP*) as more and more biomass is produced (Fig. 9). Because AquaCrop does not simulate nutrient cycles and balances but only mimics the effect of soil fertility stress, the crop response to soil fertility management needs to be calibrated.

The crop is calibrated for the relative crop production ($Brel_{cal}$) that can be reached with the soil fertility management in farmers' fields. When specifying $Brel_{cal}$ for calibration, AquaCrop will plot a limited canopy development and reduced WP* for which the integrated effect yields the specified $Brel_{cal}$ (Fig. 9). With observations in the well-watered fertility-stressed field of (1) the

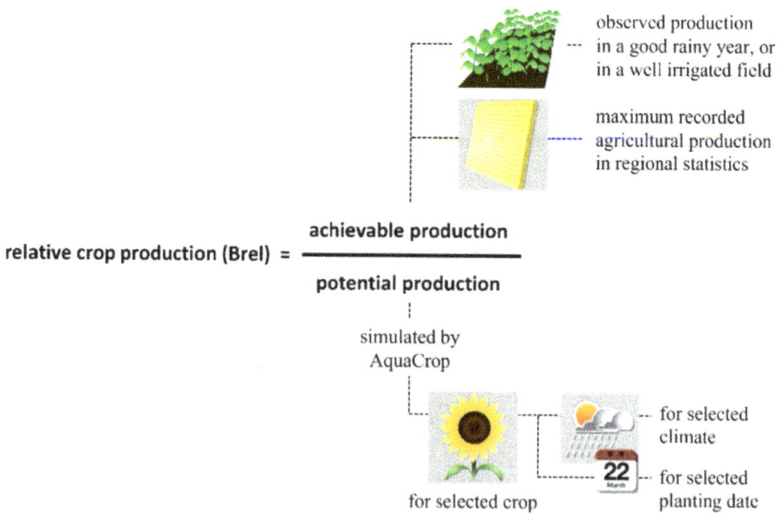

Figure 8 The required input for soil fertility stress, expressed as the relative crop production (Brel) that can be achieved with the actual soil fertility management in famers' field.

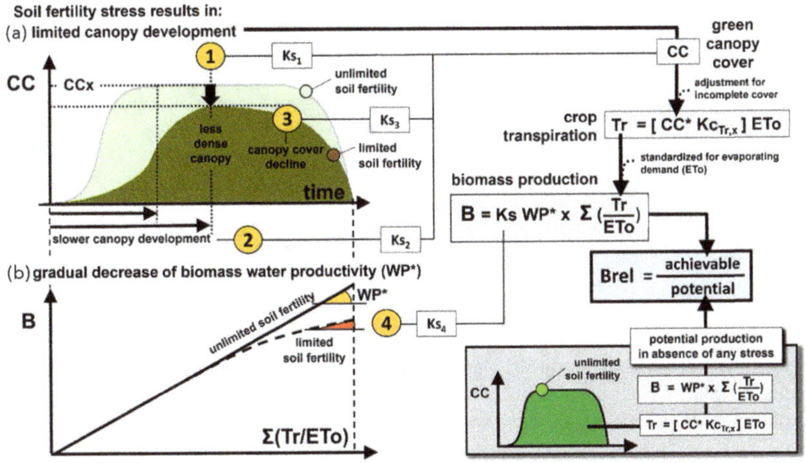

Figure 9 Simulation of the achievable biomass production when limited by soil fertility stress (Brel), resulting from three impacts (Ks 1 to 3) on the green canopy development (CC), and (Ks 4) the gradual decrease of the normalized biomass water productivity (WP*).

maximum canopy cover that can be reached, and/or (2) the time when the canopy closes, and/or (3) the selection of a class (small/medium/high) of the canopy decline in the season, the user can fine-tune the displayed canopy development during the calibration process, while (4) AquaCrop will make the required adjustment to WP* to ensure that the integrated effect still matches with the specified $Brel_{cal}$. The calibration process for soil fertility stress ultimately fixes the shape of four fertility stress coefficients (Ks), which are the modifiers targeting respectively (1) CCx, (2) CGC, (3) in-seasonal canopy-decline, and (4) WP* (Fig. 9).

Once calibrated, AquaCrop simulates the effects on crop production of water and fertility stress, independently or combined. When running a simulation, the level of soil fertility, expressed as a relative crop production ($Brel_{input}$), is available as input in Field Management (Fig. 3). If $Brel_{input}$ differs from $Brel_{cal}$, AquaCrop will jointly increase or decrease the stress indicator of the four Ks coefficients until the integrated effect yields $Brel_{input}$. However, the specified $Brel_{input}$ might not be reached if other stresses also occur during the season. When soil water or soil salinity stress also starts to limit the biomass production, AquaCrop automatically decreases the stress for the four Ks coefficients in the next daily time step(s), since more nutrients would remain in the soil reservoir (the biomass production is less than expected due to water or salinity stress). The higher the other stresses, the greater the Ks adjustments. If the other stresses are relieved by rainfall or irrigation, soil fertility stress increases, and the four Ks coefficients adjust once more to another level. This dynamic adjustment of the Ks coefficients makes the effect

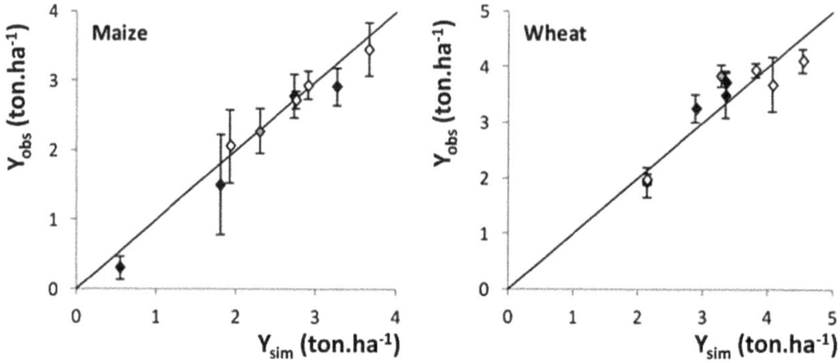

Figure 10 Observed versus simulated yield for maize and wheat in Nepal for various levels of soil fertility and water treatments: white symbols represent irrigation; grey, deficit irrigation, and black symbols represent rainfed agriculture (from Van Gaelen et al. (2015)).

of soil fertility stress automatically adjust to the result of other stresses (Raes et al., 2023b).

Following case-specific calibration for tef in Ethiopia, maize and wheat in Nepal (Fig. 10), and quinoa in Bolivia, it has been shown that AquaCrop was able to accurately simulate the soil water balance, crop development, biomass production and grain yield for several soil fertility levels in different environments (Van Gaelen et al., 2015).

4.5 Parameterization of crops not (yet) available in the database of AquaCrop

In the database of AquaCrop, calibrated and validated values of the conservative crop parameters are provided for 14 major annual herbaceous crops, two less common crops (teff and quinoa), and one perennial herbaceous forage crop (alfalfa). The widespread use of AquaCrop has resulted in its applications for crops not yet available in the database. A literature review in 2019 revealed that the model has already been implemented for 46 different crops (Salman et al., 2021). Although not clearly presented in all publications, detailed data are needed for the validation and parameterization of the model for a particular crop. A guide for the parameterization, calibration, and validation of AquaCrop is available in Steduto et al. (2012). An example for maize is worked out by Hsiao et al. (2009) and Heng et al. (2009).

For varying environmental conditions, guidance on which crop and soil parameters should be given priority during calibration is provided by Vanuytrecht et al. (2014b). Several root and soil parameters have a higher impact on yield simulation under diverse conditions, particularly when water availability is limited. Hence, even though root and soil water characteristics are not the easiest to measure in the field, they require meticulous calibration to ensure good simulation results.

Although applications of AquaCrop for fruit trees are reported in the literature, the model has only been designed for herbaceous crops. The complex behavior of fruit trees makes it difficult to simulate the yield response with AquaCrop's simple calculation procedure. Mixed cropping (intercropping) cannot be simulated either, since uniform large fields are assumed.

5 Simulation of the effects of climate change

When studying the impact of climate change on crop production, both the changing weather conditions and the effects of CO_2 fertilization need to be considered. The structure of AquaCrop allows to assess the combined effect of increased temperature and ETo, altered rainfall patterns, and elevated atmospheric CO_2 concentrations.

In accordance with physiological plant responses, the elevated atmospheric CO_2 concentration induces in AquaCrop (i) a small reduction in transpiration at field level and (ii) a strong increase of the normalized biomass water productivity (WP*) for C3 crops:

- The decrease of the crop transpiration coefficient ($Kc_{Tr,x}$) with increased CO_2 concentration follows the interplay between the decrease in stomatal conductance (due to increased CO_2) and the increased transpiration due to the increased vapor pressure gradient affected by higher temperatures. The simulated decline in transpiration mimics the field-scale effects of elevated CO_2 concentration on agricultural crops in free air CO_2 enrichment (FACE) conditions for C3 and C4 crops (Vanuytrecht et al. 2011, 2012).
- Elevated CO_2 concentration enhances photosynthesis (particularly for C3 crops since C4 crops are considered to be CO_2 saturated at ambient $[CO_2]$ when other resources are adequately present) and therefore crop productivity. This effect of CO_2 fertilization is simulated in AquaCrop by adjusting WP* when the average annual atmospheric CO_2 concentration deviates from its reference value of 369.41 ppm in the year 2000 for which WP* has been normalized (Fig. 11). By means of a sink term, the user can select between a relatively high theoretical adjustment (Hsiao and Xu, 2000; Steduto et al., 2007) corresponding to an assumed strong crop sink capacity, versus a lower adjustment based on the analysis of crop responses in FACE environments (Vanuytrecht et al., 2011) corresponding to an assumed weak crop sink capacity. Lower response levels in FACE

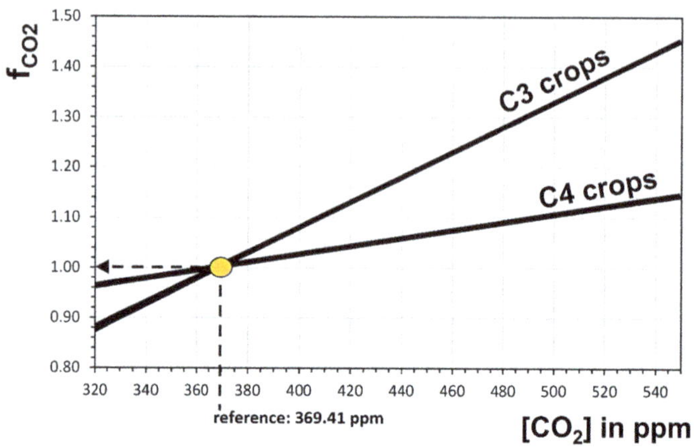

Figure 11 Multiplier for WP* (f_{CO2}) for the theoretical adjustment (assuming strong crop sink capacity) as specified by Steduto et al. (2007) for various atmospheric CO_2 concentrations for C3 crops and for C4 crops (e.g. maize and sorghum).

experiments have been attributed to various causes, including a lack of sink strength to optimally capture the additional atmospheric carbon, partly due to limited crop nitrogen (Vanuytrecht et al., 2012 and 2014a). More recently, it has been further suggested that the experimental setups of the FACE experiments could cause rapid and wide fluctuations in CO_2 concentration and, hence, an unrealistically low crop response to high CO_2 concentration (Allen et al., 2020). This hypothesis suggests that the crop responses observed in FACE experiments may be underestimating the real responses and that any modeling exercises should be performed with a high-value sink term.

To avoid unrealistically high modelled effects of CO_2 fertilization at high atmospheric CO_2 concentrations, the adjustment of WP* starts to decline at concentrations above 550 ppm, and it no longer increases beyond 900 ppm. This is in line with the decreasing trend in crop response with increasing elevated CO_2 levels (e.g. Akita and Moss, 1973; Chapin et al., 2011; Kirschbaum, 2011; Kromdijk and Long, 2016; Yin and Struik, 2009). AquaCrop uses the well-documented and general trend described by Yin and Struik (2009) to model the decreasing trend above 550 ppm. This is, however a purely theoretical adjustment, without any validation by experimental data as FACE experiments generally do not reach these high levels of atmospheric CO_2 concentration.

Compared to the possible positive effects of CO_2 fertilization, the changed weather conditions in a future climate are likely to have a more negative effect on crop production. As a result of the increase in air temperature and therefore

ETo, water stress may become more severe. Also, the plant cycle will be shorter, resulting from its specification in thermal time. The effect of cold stress on crop transpiration and on pollination will be less, but heat stress during flowering can become more severe. Additionally, the changing rainfall patterns and air temperature increase will likely result in different planting dates, which can be generated in AquaCrop by built-in air temperature and rainfall criteria.

The considering of the altered weather conditions and the incorporation of a flexible and realistic crop response to atmospheric CO_2 concentration has allowed the study of future impacts of climate change on crop production in a wide range of studies (e.g. Vanuytrecht et al., 2015; Bouras et al., 2019; Busschaert et al., 2022).

The effect of climate change on crop yield in smallholder farms in West Africa was simulated with AquaCrop (Raes et al., 2021). For the calibration, crop yields, cultivar characteristics, and information on current soil fertility and weed management were collected. For the simulations under future climatic conditions, the three best performing RCM-GCM combinations for 8 sites in the study area were retrieved from the CORDEX-Africa dataset and two emission scenarios were considered. The projected climate change corresponded well with the findings of Sylla et al. (2016), who reported significant warming and a wide range of precipitation uncertainty in the Sahel.

The relative yield change under future climatic conditions, with the current field management, is negative and about –5 to –10% for the irrigated crops and –10 to –20% for the rainfed crops. The simulated negative yield impact corresponds with the median simulated yield loss around 11% found in West Africa based on a review of 16 existing studies (Roudier et al., 2011). The simulations by Raes et al. (2021) revealed that unless soil fertility is addressed, benefits from CO_2 fertilization resulting from elevated CO_2 in the future are insufficient to compensate for the negative impact of the changes in precipitation patterns and the increase in air temperature and the associated increase in evapotranspiration. By increasing soil fertility by 15% to 20% in the simulations, the yield impact of C3 crops became positive (+8 to +10%). Since C4 crops profit much less from the CO_2 fertilization effect (e.g. Leakey, 2009), the response of maize yield remained mostly negative (–2 to 0%).

6 Output

The standard AquaCrop program with a graphical user interface (GUI) has a well-developed user interface. Multiple graphs and schematic displays in the menus help the user to follow the effect of water, salinity, soil fertility, and temperature stresses on crop development and production during the crop cycle. Results are plotted in several graphs, which are updated at the end of each daily time step. Figure 12 presents the result of the simulation of

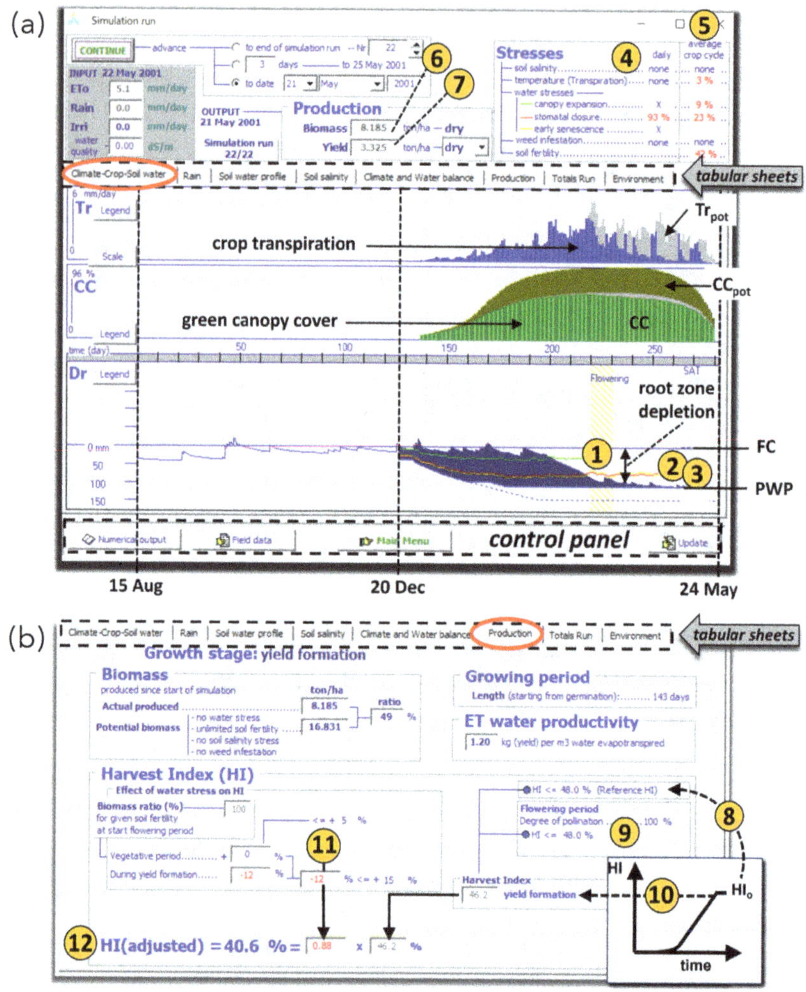

Figure 12 The 'Simulation run' menu of AquaCrop, with (a) the 'Climate-Crop-Soil water' tabular sheet with its three panels showing daily crop transpiration (Tr), green canopy cover (CC), and root zone depletion (Dr) with indication of (1) the upper threshold for canopy expansion (green line), (2) stomatal closure (red line), and (3) early senescence (yellow line). Daily updates are displayed of (4) the degree of several stresses and (5) their averages, (6) the cumulative biomass production, and (7) the crop yield. Information on the Harvest Index adjustment is available in (b) the 'Production' tabular sheet, with indication of (8) the reference Harvest Index (HI_o), (9) the effect of water, cold, and heat stress on pollination, (10) the value of the Harvest Index for that day, (11) the cumulative effect of water stress on HI, and (12) the simulated Harvest Index for that day.

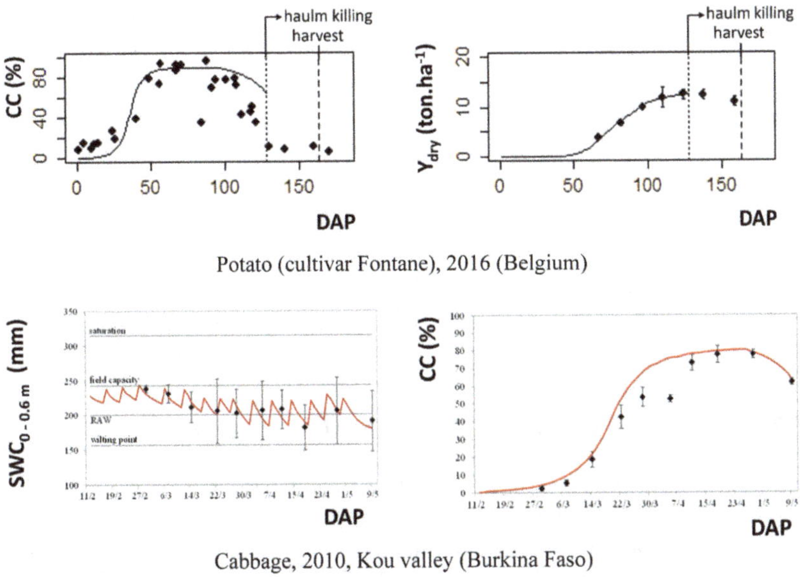

Figure 13 Observed (dots) and simulated (line) green canopy cover (left) and dry tuber yield (right) for potato (cultivar Fontane in Belgium (data from Wellens et al., 2022a); and observed (dots) and simulated (line) soil water content in the top 0.6 m (left) and green canopy cover (right) of Cabbage in Burkina Faso (from Wellens et al., 2013). Potato (cultivar Fontane), 2016 (Belgium)

winter wheat in Tunis (Tunisia), three days before crop maturity (24 May). The simulation started with dry bare soil on 15 August. Germination was generated by AquaCrop on 20 December when the topsoil was sufficiently wet. Using command keys in the control panel at the bottom of the menu (Fig. 12a), the user can overlay the simulations with observed field data and access numerical output.

The simulation results are recorded in output files and can be retrieved in spreadsheet programs for further processing and analysis. Figure 13 provides some examples comparing simulated data with observed data.

7 Recent model improvements

7.1 Release of open-source and stand-alone programs for various operating systems

The AquaCrop model is freely available and fully transparent to the scientific community and society. Furthermore, users can contribute to the model improvement by submitting suggestions. While the original (up to version 6) AquaCrop source code was written in Delphi/Pascal, version 7 was converted line by line to Fortran90, optimized, and released as open source on GitHub to stimulate a wide use and long-term community-based maintenance.

Along with the open-source code distribution, 'stand-alone' (or plugin) AquaCrop programs are provided for the three most used operating systems, i.e. Windows, MacOS, and Linux. The stand-alone program can be readily plugged into any application, including geographical information systems, data assimilation tools, calibration tools, visualization tools, web applications, etc. The stand-alone programs, the open-source code, and the standard program with a graphical user interface (GUI) are the three official AquaCrop assets currently distributed by FAO (https://www.fao.org/aquacrop/en/, Fig. 14).

7.2 Upscaling of AquaCrop for regional applications

For regional spatially distributed simulations at any spatial resolution, the AquaCrop stand-alone program can be run within a Python wrapper (de Roos et al., 2021, Busschaert et al., 2022, https://github.com/KUL-RSDA/RegionalAC_Py). AquaCrop can also be run within NASA's Land Information System (LIS, Kumar et al., 2008, https://github.com/KUL-RSDA/LISF). Both options allow to run AquaCrop in a highly efficient and parallelized way for a large number of pixels and for long-term (e.g. historical or future climate) scenarios.

Large-scale AquaCrop simulations over Europe were performed by de Roos et al. (2021) and Busschaert et al. (2022). In the first study, AquaCrop

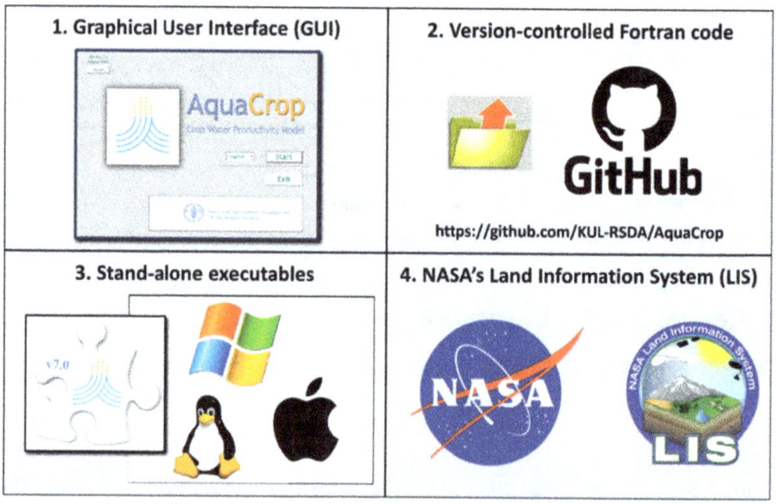

Figure 14 Three official FAO AquaCrop assets available from v7 onwards: (i) the standard GUI, (ii) open-source code on GitHub, and (iii) stand-alone programs for Windows, MacOS, and Linux. A fourth asset is AquaCrop within NASA's LIS.

was run at 1-km resolution using meteorological input from the Modern-Era Retrospective analysis for Research and Applications version 2 (MERRA-2), a global re-analysis product, and spatially distributed soil parameters linked to soil texture information from the Harmonized World Soil Database. The simulations assumed a generic C3 crop for all agricultural pixels. Despite the latter limitation, the biomass and soil moisture results were similar to satellite-based vegetation and surface soil moisture retrievals in terms of short-term and interannual variability. In follow-up research by de Roos et al. (2023), the soil moisture and biomass simulations from AquaCrop were converted to microwave backscatter estimates, and these were also favorably evaluated against the radar signals from the Sentinel-1 satellite. This confirms that AquaCrop can be reliably upscaled for regional applications and that the output is comparable to a range of satellite observations.

Busschaert et al. (2022) ran AquaCrop at the 50-km resolution over Europe with meteorological data from the Inter-Sectoral Impact Model Intercomparison Project phase 3 (ISIMIP3), both for historical and future climate scenarios. Also for this coarse resolution, the historical AquaCrop simulations compared well against coarse-scale satellite retrievals of soil moisture and vegetation. The climate scenarios included various Shared Socioeconomic Pathways (SSP) and were produced with various global climate models (GCM). In this study, the net irrigation requirement was estimated by AquaCrop to increase by 18% to 35% by the end of the 21st century (relative to 1985–2014), for the mitigated (SSP1-2.6) and unmitigated (SSP5-8.5) scenario respectively, but the results strongly varied with the underlying GCM.

7.3 Upscaling AquaCrop towards satellite-based data assimilation

The NASA LIS framework (Fig. 14) supports AquaCrop simulations with perturbed meteorological or parameter input to estimate the ensemble uncertainty of the simulations. LIS also allows intermittent updates simulated AquaCrop variables when observations become available, i.e. via data assimilation (De Lannoy et al., 2022).

In the last two decades, a wealth of large-scale spatially distributed meteorological, soil, and vegetation data from satellite and re-analysis products has become available. This information allows to force and evaluate large-scale spatial AquaCrop simulations, as illustrated further below. A key to further advance regional crop simulations is the creation of spatially distributed high-resolution crop and management parameter sets that evolve in time with cropping practices. Another pathway to optimize crop estimates is to use frequent satellite observations to adjust the model trajectory of AquaCrop, either via input, state, or parameter updating. Figure 15 shows that

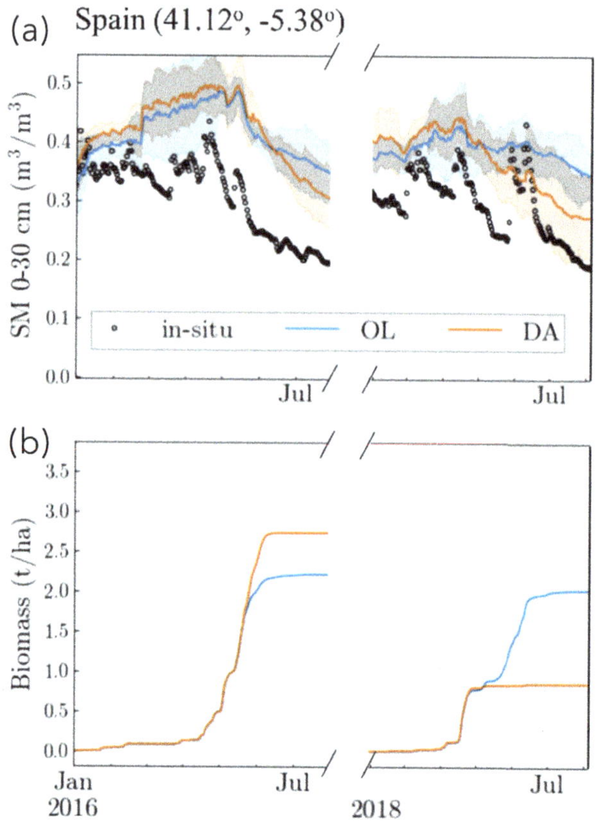

Figure 15 Example time series of (a) surface soil moisture (SM) and (b) biomass. Both figures show open loop AquaCrop simulations (OL, model only, blue) and the results with assimilation of satellite-based soil moisture retrievals (DA, orange). Also shown are *in situ* data for SM (black dots).

satellite-based updates in time series variability of soil moisture can have a significant impact on the simulation of vegetation development and biomass production in water-limited regions.

8 Further model enhancement

Currently, AquaCrop simulates responses to extremely high temperatures through an adjustment of the reference Harvest Index when heat stress affects pollination (Fig. 1). A further adjustment of model routines for crop responses to extreme high temperatures could possibly improve the simulations, as also indicated by Maiorano et al (2017). Particularly, responses to heat under compound impacts, e.g. a combination of drought and heat stress (Webber et al, 2022), which are becoming more common under climate change conditions, could contribute to further model development. These model developments could for example, introduce a heat stress effect on canopy senescence. While heat stress is an obvious and relevant stressor where model development can contribute to more realistic simulations, modeling other responses to climate extremes (Toreti et al, 2020), e.g. late frosts or extreme rainfall, could also be the subject of further research to improve model routines and simulations.

Differences in simulated crop responses across different crop models are common, particularly for responses related to transpiration and water use, and just as other models, AquaCrop would benefit from both further validation and optimization of CO_2 responses in the range of concentrations anticipated in the future, to improve our predictive understanding of these responses (Vanuytrecht and Thorburn, 2017).

9 Conclusion

Aquacrop simulates the yield of annual herbaceous and perennial forage crops by considering the effect of air temperature, soil water, salinity and fertility stress on crop development, biomass production, and harvest index. AquaCrop uses a relatively small number of explicit and largely intuitive parameters and input variables to characterize crop and soil properties and field and irrigation management practices. By varying cultivar-specific crop parameters, differences among varieties can be taken into account in the simulations. AquaCrop is a robust model that has been tested widely worldwide and is very easy to use. The simulation results are recorded in output files and can be retrieved in spreadsheet programs for further processing and analysis. Since its publication in 2009, it has been widely adopted by many stakeholders who have calibrated it for numerous additional crops in different agro-ecological environments, and it has been applied to simulate crop production under current and future climatic conditions.

10 Where to look for further information

AquaCrop is freely available. The software, documentation, training material, and a reference manual containing a user's guide and a detailed description of all calculation procedures can be downloaded from the FAO AquaCrop website (https://www.fao.org/aquacrop). Forty-three training modules (MP4 videos) are posted in an 'AquaCrop Training' channel of YouTube (https://www.youtube.com/playlist?list=PLzp5NgJ2-dK7H85cyEmGc8KSodqm8gCf2). Further information on crop responses, including a guide for the parameterization, calibration, and validation of AquaCrop is available in FAO Irrigation and Drainage Paper No. 66, 501 p. (http://www.fao.org/nr/water/aquacrop.html).

11 References

Adams, J.E., et al. (1976). Influence of row spacing and straw mulch on first stage drying. *Soil Science Society of America Journal* 40: 436–442.

Allen, L.H., et al. (2020). Fluctuations of CO_2 in Free-Air CO_2 Enrichments (FACE) depress plant photosynthesis, growth and yield. *Agricultural and Forest Meteorology* 284. https://doi.org/10.1016/j.agrformet.2020.107899

Allen, R., et al. (1998). *Crop Evapotranspiration – Guidelines for Computing Crop Water Requirements*. FAO Irrigation and Drainage Paper N° 56. Rome, Italy.

Akita, S. and Moss, D.N. (1973). The effect of an oxygen-free atmosphere on net photosynthesis and transpiration of Barley (Hordeum vulgare L.) and Wheat (Triticum aestivum L.) leaves. *Plant Physiology* 52: 601–603.

Azam-Ali, S.N. and Squire, G.R. (2002). *Principles of Tropical Agronomy*. CABI Publishing, Wallingford, UK, 236 pp.

Bouras, E., et al. (2019). Assessing the impact of global climate changes on irrigated wheat yields and water requirements in a semi-arid environment of Morocco. *Scientific Reports* 9(1): 19142. https://doi.org/10.1038/s41598-019-55251-2.

Bradford, K.J. and Hsiao, T.C. (1982). Physiological responses to moderate water stress. In: Lange, O.L., et al. Editors, *Encyclopedia of Plant Physiology New Series*, Volume 12A.

Busschaert, L., et al. (2022). Net irrigation requirement under different climate scenarios using AquaCrop over Europe. *Hydrology and Earth System Sciences* 26: 3731–3752. https://doi.org/10.5194/hess-26-3731-2022.

Chapin, F.S., et al. (2011). *Principles of Terrestrial Ecosystem Ecology*, 2nd edn. Springer-Verlag, New York.

Deketelaere, V. (2024). *Potential Yield Gap of Maize in Europe under Current and Future Climates*. Master of Science in Water Resources Engineering dissertation, KU Leuven.

De Lannoy, G.J.M., et al. (2022). Perspective on Satellite-Based Land Data Assimilation to Estimate Water Cycle Components in an Era of Advanced Data Availability and Model Sophistication. *Frontiers of Water*. https://doi.org/10.3389/frwa.2022.981745.

de Roos, S., et al. (2021). Performance analysis of regional AquaCrop (v6.1) biomass and surface soil moisture simulations using satellite and in situ observations. *Geoscientific Model Development* 14(12): 7309–7328. https://doi.org/10.5194/gmd-14-7309-2021.

de Roos, S., et al. (2023). Optimisation and uncertainty assessment of AquaCrop backscatter simulations using Sentinel-1 observations. *Remote Sensing of Environment* 294: 113621.

Doorenbos, J. and Kassam, A.H. (1979). *Yield Response to Water*. FAO Irrigation and Drainage Paper N° 33. Rome, Italy.

Farahani, H.J., et al. (2009). Parametrization and evaluation of Aquacrop for full and deficit irrigated cotton. *Agronomy Journal* 101(3): 469–476.

Garcia-Vila, M., et al. (2009). Deficit irrigation optimization of cotton with AquaCrop. *Agronomy Journal* 101(3): 477–487.

Geerts, S., et al. (2009). Simulating yield response of quinoa to water availability with AquaCrop. *Agronomy Journal* 101(3): 499–508.

Hay, R.K.M. and Gilbert, R.A. (2001). Variation in the harvest index of tropical maize: evaluation of recent evidence from Mexico and Malawi. *Annals of Applied Biology* 138: 103–109.

Hsiao, T.C. (1973). Plant responses to water stress. *Annual Review of Plant Physiology* 24: 519–570. https://doi.org/10.1146/annurev.pp.24.060173.002511.

Hsiao, T.C. and Xu, L.K. (2000). Predicting water use efficiency of crops. *Acta Horticulturae* 537: 199–206.

Hsiao, T.C., et al. (2009). III. Parameterization and testing for maize. *Agronomy Journal* 101: 448–459.

Heng, L.K., et al. (2009). Validating the FAO AquaCrop model for irrigated and water deficient field maize. *Agronomy Journal* 101: 488–498.

Kirschbaum, M.U.F. (2011). Does enhanced photosynthesis enhance growth? Lessons learned from CO2 enrichment studies. *Plant Physiology* 155: 117–124.

Kromdijk, J. and Long, S.P. (2016). One crop breeding cycle from starvation? How engineering crop photosynthesis for rising CO2 and temperature could be one important route to alleviation. *Proceedings of the Royal Society B* 283: 20152578.

Kumar, S.V., et al. (2008). An integrated hydrologic modeling and data assimilation framework enabled by the Land Information System (LIS). *IEEE Computer* 41(12): 52–59.

Leakey, A.D.B. (2009). Rising atmospheric carbon dioxide concentration and the future of C4 crops for food and fuel. *Proceedings of the Royal Society B: Biological Sciences* 276: 2333–2343.

Maiorano, A., et al. (2017). Crop model improvement reduces the uncertainty of the response to temperature of multi-model ensembles. *Field Crops Research* 202: 5–20. https://doi.org/10.1016/j.fcr.2016.05.001.

Mbangiwa, N.C., et al. (2019). Modelling and measurement of water productivity and total evaporation in a dryland soybean crop. *Agricultural and Forest Meteorology* 266–267: 65–72.

Moot, D., et al. (2012). Alfalfa. In: Steduto, P., et al. (eds), *Crop Yield Response to Water*. FAO Irrigation and Drainage Paper Nr. 66. Rome, Italy, 212–219. https://researcharchive.lincoln.ac.nz/bitstream/handle/10182/6262/MootAlfalfachapter.pdf;sequence=1.

Penning de Vries F.W.T., et al. (1983). Bioenergetics of growth of seeds, fruits and storage organs. In: Smith, W.H. and Banta, S.J. (eds), *Productivity of Field Crops Under Different Environments*. IRRI, Los Baños, The Philippines, 37–59.

Raes, D., et al. (2023a). Simulation of alfalfa yield with AquaCrop. *Agricultural Water Management* 284: 108341. https://doi.org/10.1016/j.agwat.2023.108341.

Raes, D., et al. (2023b). *AquaCrop Reference manual, Version 7.1*. FAO. Rome, Italy. http://www.fao.org/nr/water/aquacrop.html.

Raes, D., et al. (2009). AquaCrop-The FAO crop model to simulate yield response to water: II. Main algorithms and software description. *Agronomy Journal* 101: 438–447.

Raes, D., et al. (2021). Improved management may alleviate some but not all the adverse effects of climate change on crop yields in smallholder farms in West Africa. *Agricultural and Forest Meteorology* 308–309: 108563. https://doi.org/10.1016/j.agrformet.2021.108563, 2021.

Roudier, P., et al. (2011). The impact of future climate change on West African crop yields: what does the recent literature say? *Global Environmental Change, Symposium on*

Social Theory and the Environment in the New World (Dis)Order 21: 1073–1083. https://doi.org/10.1016/j.gloenvcha.2011.04.007.

Ruiz, A., et al. (2023). Harvest index has increased over the last 50 years of maize breeding. *Field Crops Research* 300: 108991. https://doi.org/10.1016/j.fcr.2023.108991.

Salman, M., et al. (2021). *The AquaCrop Model: Enhancing Crop Water Productivity*. FAO Water reports 47. Rome, Italy.

Saxton, K.E., et al. (1986). Estimating generalized soil-water characteristics from texture. *Soil Science Society of America Journal* 50(4): 1031–1036.

Stewart, J.I. and Hagan, R.M. (1973). Functions to predict effects of crop water deficits. *Journal of the Irrigation and Drainage Division, ASCE* 99(4): 421–439.

Sylla, M.B., et al. (2016). Climate change over west africa: recent trends and future projections. In: Yaro, J. and Hesselberg, J. (eds), *Adaptation to Climate Change and Variability in Rural West Africa*. Springer International Publishing, Switzerland.

Steduto, P. and Albrizio, R. (2005). Resource use efficiency of field-grown sunflower, sorghum, wheat and chickpea. II. Water use efficiency and comparison with radiation use efficiency. *Agricultural and Forest Meteorology* 130: 269–281.

Steduto, P., et al. (2000). Gas-exchange and stomatal and non-stomatal limitations to carbon assimilation of sunflower under salinity. *Environmental and Experimental Botany* 44(3): 243–255. https://doi.org/10.1016/s0098-8472(00)00071-x.

Steduto, P., et al. (2007). On the conservative behaviour of biomass water productivity. *Irrigation Science* 25: 189–207.

Steduto, P., et al. (2012). *Crop Yield Response to Water*. FAO Irrigation and Drainage Paper Nr. 66. Rome, Italy. http://www.fao.org/nr/water/aquacrop.html.

Steduto, P., et al. (2009). AquaCrop-The FAO crop model to simulate yield response to water: I. Concepts and underlying principles. *Agronomy Journal* 101: 426–437.

Teixeira, E.I., et al. (2008). Defoliation frequency and season affected radiation use efficiency and dry matter partitioning to roots of lucerne (Medicago sativa, L.) crops. *European Journal of Agronomy* 28(2): 103–111.

Todorovic, M., et al. (2009). Assessment of AquaCrop, CropSyst, and WOFOST models in the simulation of sunflower growth under different water regimes. *Agronomy Journal* 101(3): 509–521.

Toreti, A., et al. (2020). Narrowing uncertainties in the effects of elevated CO2 on crops. *Nature Food* 1: 775–782. https://doi.org/10.1038/s43016-020-00195-4.

Van Gaelen, H., et al. (2015). A semi-quantitative approach for modelling crop response to soil fertility: evaluation of the AquaCrop procedure. *Journal of Agricultural Science* 153: 1218–1233. https://doi.org/10.1017/S0021859614000872.

Van Gaelen, H., et al. (2016). Simulation of crop water productivity in weed-infested fields for data-scarce regions. *Journal of Agricultural Science* 154(6): 1026–1039.

Vanuytrecht, E. and Thorburn, P.J. (2017). Responses to atmospheric CO2 concentrations in crop simulation models: a review of current simple and semicomplex representations and options for model development. *Global Change Biology* 23(5): 1806–1820. https://doi.org/10.1111/gcb.13600.

Vanuytrecht, E., et al. (2011). Considering sink strength to model crop production under elevated atmospheric CO2. *Agricultural and Forest Meteorology* 151: 1753–1762.

Vanuytrecht, E., et al. (2012). Quantifying field-scale effects of elevated carbon dioxide concentration on crops. *Climate Research* 54: 35–47. https://doi.org/10.3354/cr01096.

Vanuytrecht, E., et al. (2014a). AquaCrop: FAO's crop water productivity and yield response model. *Environmental Modelling & Software* 62: 351–360.

Vanuytrecht, E., et al. (2014b). Global sensitivity analysis of yield output from the water productivity model. *Environmental Modelling & Software* 51: 323–332. https://doi.org/10.1016/j.envsoft.2013.10.017.

Vanuytrecht, E., et al. (2015) Regional and global climate projections increase mid-century yield variability and crop productivity in Belgium. *Regional Environmental Change* 16(3): 659–672. https://doi.org/10.1007/s10113-015-0773-6.

Villalobos, F.J. and Fereres, E. (1990). Evaporation measurements beneath corn, cotton, and sunflower canopies. *Agronomy Journal* 82: 1153–1159.

Webber, H., et al. (2022). Framework to guide modeling single and multiple abiotic stresses in arable crops. *Agriculture, Ecosystems & Environment* 340: 108179. https://doi.org/10.1016/j.agee.2022.108179.

Wellens, J., et al. (2022a). *AquaCrop and the Belgian collaborative agriculture monitoring (BELCAM) platform*. International Symposium on Managing Land and Water for Climate-Smart Agriculture, Joint FAO/IAEA Centre, 25–29 July 2022, Vienna, Austria. Proceedings in press.

Wellens, J., et al. (2022b). Calibration and validation of the FAO AquaCrop water productivity model for cassava (Manihot esculenta Crantz). *Agricultural Water Management* 263: 107491.

Wellens, J., et al. (2013). Performance assessment of the FAO AquaCrop model for irrigated cabbage on farmer plots in a semi-arid environment. *Agricultural Water Management* 127: 40–47.

Yin, X. and Struik, P.C. (2009). C3 and C4 photosynthesis models: an overview from the perspective of crop modelling. *NJAS - Wageningen Journal of Life Sciences* 57: 27–38.

Chapter 18

The Daisy crop model

Kiril Manevski[1], Aarhus University, Denmark and Chinese Academy of Sciences and University of the Chinese Academy of Sciences, China; Shaohui Zhang, Aarhus University, Denmark and University Centre for Circular Bioeconomy, Denmark/Northwest A&F University, China; Yanmin Yang, Chinese Academy of Sciences, China; Yonghui Yang, University of the Chinese Academy of Sciences and Chinese Academy of Sciences, China; and Mohamed Jabloun, The James Hutton Institute, UK

1 Introduction: brief history and description of the Daisy model

The mathematical model Daisy was originally developed in 1986 by the Royal Veterinarian and Agricultural University, now University of Copenhagen, in Denmark, as a response to the need to study nitrogen (N) transport from agricultural land to groundwater amid increasing crisis in the Baltic region of soil, ground- and surface-water pollution with N, especially nitrate (Hansen et al. 1993). As such and with knowledge on linked biogeochemical cycles, the model is able to mechanistically simulate water, heat, N, and carbon (C) flows, including leaching and gaseous fluxes (CO_2, N_2/N_2O), crop production and fate of agro-chemicals (e.g. pesticides) in agroecosystems at daily time scale subjected to various field management strategies and driven by daily weather

1 Corresponding author. K.M. (kiril.manevski@agro.au.dk).

http://dx.doi.org/10.19103/AS.2025.0155.21

data (Fig. 1). Actually, since the development of the model, it has been a priority for operational reasons to keep the number of necessary driving variables at a minimum and to limit the requirement of temporal resolution of the driving variables to one day. In Europe, N and water are the two most common limiting factors in crop production to obtain high yields. Therefore, the Daisy model simulates plant growth as a function of only water and N availability, although interspecies plant competition as a limiting factor can also be simulated. This however means, according to the law of the minimum (the Liebig's Law), the model cannot simulate plant growth limited by other nutrients on, for instance, the phosphorus-deprived soils in much of tropical Africa or the potassium-limited soils in Southeast Asia. In the early 1990s, the theory was disseminated to a wider international audience for the description of the model simulation of crop production, water dynamics, and N dynamics in crop production at various agricultural management practices and strategies, integrating at that time a substantial knowledge from various disciplines including soil science and crop science (Hansen et al. 1993). In the late 1990s, Daisy has been re-written from the original Fortran to C++ object-oriented structure code in about 170.000 code lines. A major rewrite in year 2000 allowed a number of new models

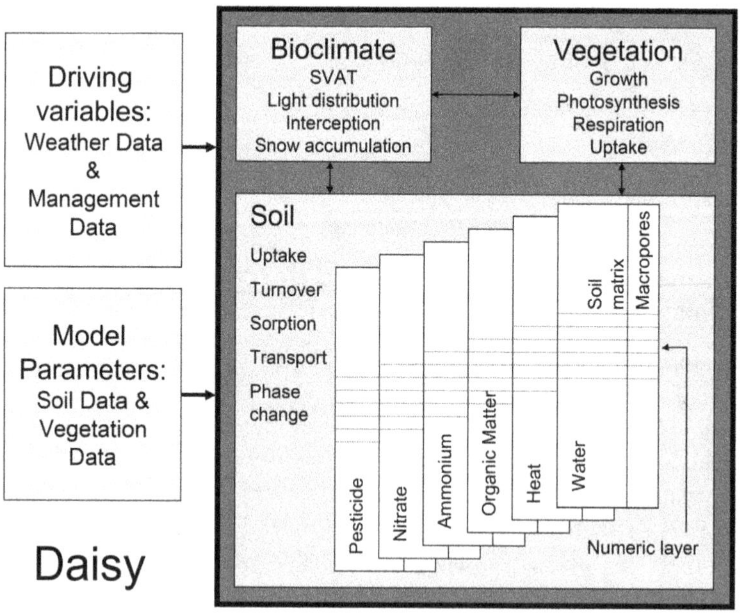

Figure 1 Main compartments of the Daisy model of bioclimate, vegetation, and soil. The bioclimate compartment simulates surface processes, exchange with the atmosphere, and the aerial environment of the plant. The vegetation compartment simulates plant-related processes. The soil module simulates soil-related processes. Source: Adapted from Hansen et al. (2012; with permission).

within Daisy to be implemented, including the simulation of intercropping systems and multiple soil columns (Abrahamsen and Hansen 2000). The theory behind the model is documented with a comprehensive set of user's manuals available at the Daisy homepage (www.daisy.ku.dk).

Even though Daisy contains process descriptions mimicking numerous physical, chemical or biological processes, as any mathematical "crop" model, it represents a significant simplification of the complex soil–plant–atmosphere continuum governed in reality by a vast number of complex and often uknown physical, chemical, and biological processes. The model is nevertheless deemed among the best to consider the most important processes with sound process descriptions and a relatively balanced level of complexity (Gavasso-Rita et al. 2024; Manevski et al. 2016a). Also, originally parameterized to simulate the growth and development of major grain cereals in Europe, such as wheat and barley, Daisy over time included an ever-increasing number of parameterized crops, including legumes and vegetables and their varieties, as well as a model for tree growth. Questions are often asked and are answered in the Daisy scholarly discussion forum (https://groups.google.com/forum/#!forum/daisy-model). Daisy is integrated as a teaching tool in the MSc and PhD curricula of several universities and in various research and education capacity-building programs. The model program is open source, compiled and sourced from the Daisy homepage (http://www.daisy.ku.dk/) and Github (https://github.com/perabrahamsen/daisy-model), respectively. The model's exact equations are available in the literature (Abrahamsen and Hansen 2000; Hansen 2002; Hansen et al. 2012), including the model homepage (https://daisy.ku.dk/publications/).

The remaining of this chapter provides an overview of the main processes simulated with the Daisy model. Following brief description of the main crop and soil processes described in Daisy, Section 2 describes the main crops and varieties simulations in the model, including genotype × environment × management interactions, followed by Section 3 showing recent model improvements and examples of current more advanced applications, while Section 4 revealing areas for further model enhancement and how these are being addressed, especially incorporation of crop genetics into the model. Section 5 finally concludes with a perspective on the direction of the model development and usage.

1.1 Crop development and growth

Vegetation is one of the main modules in the Daisy structure, together with soil and bioclimate (Fig. 1). The module was developed to describe crop growth and crop production on well-drained agricultural soils in Denmark with relatively high yield. The potential production in a given year is thus controlled by plant genetics, local environment (global radiation and temperature), and limiting factors considered by Daisy to be water and N. The Daisy model does not include any other limitations on crop growth such as lack of phosphorus, potassium or micronutrients, pests and diseases attacks, or serious weed competition; however, weed or other interspecies competition can be handled as a separate crop in the field as the model has an intercropping module (Manevski et al. 2015). Crop development from sowing to emergence is by default simulated by soil temperature sum calculated from the time of sowing, whereas development stages (DSs) in the vegetative (from emergence, DS = 0, to flowering, DS = 1) and the reproductive period (from flowering to maturity, DS = 2) are simulated with respective development rates, i.e. development per day. Users can opt to simulate the DSs in the vegetative and the reproductive periods with thermal time models. Either way, the development is also affected by temperature, photoperiod, and vernalization effects. Crop growth is simulated by temporal variation in leaf area index (LAI), dry matter production, crop N demand/content, and rooting depth/density, all of which are used by other parts of the model related to water and N transport and transformation. The LAI (m^2 m^{-2}) is a development-dependent product of leaf weight (g m^{-2}) and specific leaf area (m^{-2} g, fixed value). Dry matter production is calculated from photosynthesis and respiration (growth and maintenance), the former based on the light distribution within the canopy and leaf light response curves. The light distribution (and adsorption) within the canopy layers is derived from Beer's law as the model calculates leaf area distribution (has a starting value) in the function of development. When intercropping, a composite canopy is simulated by adding the leaf area distributions of the crops (Fig. 2). The competition for light is then simulated by allocating light in a layer to the canopy of each crop proportionally to the considered crop's contribution to the total LAI within the layer of the composite canopy.

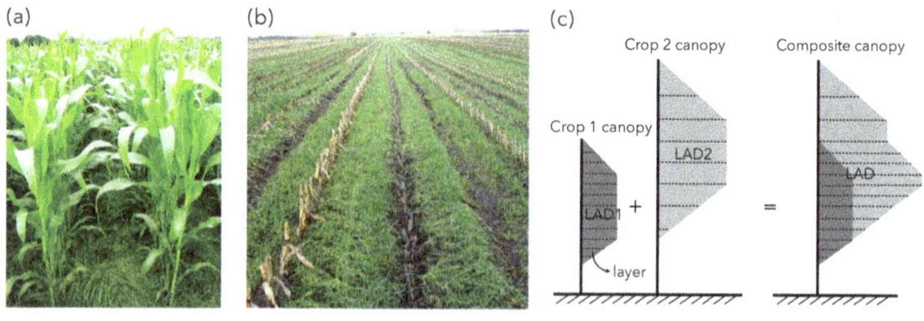

Figure 2 Intercropping in the Daisy model shown by (a) maize and red fescue under field conditions, (b) red fescue growth after maize harvest, and (c) composite canopy schemed under intercropping conditions. LAD is leaf area distribution. Source: Adapted from: Manevski et al. (2015; with permission).

Under both sole and intercrop conditions, leaf photosynthesis (g CO_2 m^{-2}) is by default calculated according to the leaf light response curve of Goudriaan and van Laar (1978) for each canopy layer (LAI/n, n is the number of layers), involving maximum photosynthetic rate (F_m paramerer, g CO_2 m^{-2}/W m^{-2}) and light-use efficiency at low intensity. However, users can select among several 'Farquhar' formulations, including specifically for C3 or C4 crops, a coupled photosynthesis–stomata conductance formulation. The photosynthesis is constrained by environmental factors of temperature, DS and senescence (days after planting), coded, respectively, TempEff (T), DSEff (DS), and DAPEff (DAP).

1.2 Evapotranspiration and water balance

Evapotranspiration is a derived variable of major importance in the Daisy model, particularly the potential evapotranspiration needed to estimate actual evapotranspiration. Potential evapotranspiration is the maximum water loss by evaporation and transpiration from a surface (crop+soil) that is amply supplied with water. This definition implies the basic assumption that potential evapotranspiration constitutes the upper limit of the soil water balance. Also, potential evapotranspiration differs for different crops due to the fact that the quantity of energy available for evapotranspiration is influenced by the particular surface properties. To bypass this energetic variation, the concept of a reference evapotranspiration from short- and well-irrigated grass has been introduced and is also adopted in the Daisy model. Reference evapotranspiration can be simulated by different mathematical expressions, including the FAO-modified Penman–Monteith equation (Allen et al. 1998) but only by the Makkink equation (Hansen 1984) when sole data of temperature, global radiation, and precipitation are at hand. Afterwards, evapotranspiration (evaporation and transpiration) can occur either at a potential rate or at a lower rate than the potential rate, depending on the soil water availability. Soil water (redistribution) is simulated by the Richard's equation (Richards 1931) based on measured precipitation, applied irrigation and reference evapotranspiration constituting the system upper boundary, including water ponding on the soil surface due to, e.g., rain, irrigation, or melting snow at a rate exceeding the soil infiltrability and freezing or thawing. The lower boundary of water transport is the groundwater being either with a known table or free drainage, assuming gravity flow. Noteworthy is that Daisy can estimate the water flow in the fast-medium of the soil, i.e. macropores, which are vertically oriented features in the model characterized by physical properties, such as length and diameter, which need to be specified (Mollerup 2010).

1.3 Organic matter turnover and inorganic nitrogen transport

Carbon input to the agroecosystem simulated with Daisy is through photosynthesis and through organic fertilizers, whereas C is lost to the atmosphere through plant and microbial respiration (growth and maintenance). The major C pools in the model are thus soil organic matter (SOM; native humus), soil microbial biomass (SMB), and added organic matter (AOM), each subdivided with a slow and a fast turnover rate, in addition to one 'inert' (recalcitrant) SOM. The C flows between these individual pools based on first-order degradation rates and the maintenance respiration rates are modified according to soil temperature and soil water, the latter expressed in terms of the soil water pressure potential. Also, the turnover rates are affected by N (in) availability.

Vertical transport of inorganic N is numerically solved (convection–dispersion equation) where upper boundary condition is always a flux condition, either zero flux or a positive flux for instance when N dissolved in precipitation or irrigation water enters the soil through the surface. The lower boundary is a concentration gradient equal to zero as it is impossible to know *a priori* the conditions at the bottom of the soil column under consideration. The turnover of organic N is described by 'added organic matter' pools from litter, residues roots and organic fertilizer, in addition to two SMB pools and two SOM pools, each with fast and slow turnover and parameterized in long-term experiments (Bruun et al. 2003). Based on the actual amount of N (and C) in these pools using their C/N ratio, net N mineralization (or immobilization) is derived from the overall N balance. The N balance also considers nitrification and denitrification as well as the effects of soil temperature and soil water content. Biological/atmospheric N fixation is also mimicked in the model as a fraction (80%) of the potential crop N uptake after it has been adjusted for the soil N uptake. The robust performance of Daisy in terms of short- and long-term simulation of nutrient and SOM dynamics when compared to field data has been documented in a number of publications (e.g. Bruun et al. 2006; Gyldengren et al. 2020b), including simulation of the turnover of organic matter from leguminous residues (Manevski et al., 2015), soil water and nitrate leaching out of the root zone in actual and scenario systems (Børgesen et al. 2001; Manevski et al. 2015; Rashid et al. 2022; Salazar et al. 2013), and organic matter turnover and soil C sequestration (Bruun et al. 2003; Mueller et al. 1998; Murindangabo et al. 2023). Studies with conservation and precision agriculture and soil compaction on the SOM turnover have also been conducted (Gyldengren et al. 2020a; Pedersen et al. 2021).

2 Overview of crop and variety simulations in the Daisy model

Upon model installation, a library is created containing all crops currently parameterized and readily available for simulation (Table 1). From the early days of the model, grain cereals such as winter wheat and spring barley have received the largest attention in terms of field experiments and data for model parameterization, which is also reflected in the quality of the simulations in terms of lower uncertainty of the parameters and higher accuracy of the results for these crops (Gavasso-Rita et al. 2024; Kollas et al. 2015). Many crop parameter values are conservative and do not vary significantly between different regions (e.g., root water uptake rate; Wasson et al. 2012) and different time periods (e.g. time of flowering and maturity observed as remarkably stable over a wide range of conditions; McMaster et al. 2008). For parameters that are region-specific,

Table 1 Overview of the crop library in the Daisy model as of January 2024 involving generic crop parameterizations and parameterizations of varieties suited to the pedo-climatic conditions in Denmark. The parameterizations are listed as input files appearing in a text editor linked with the Daisy model

Crop	Variety/Comment
Maize	Silage, Pioneer, Ikuwala
Grass	Grass to grain
Spring barley	
Winter barley	
Spring rape	
Winter rape	
Spring wheat	
Winter wheat	
Red fescue	
Rye	
Ryegrass	
Broccoli	Including transplanted
Brussels sprouts	
Celeriac	Including transplanted
Onion	
White cabbage	Including transplanted
Pea	Including green pea
White clover	
Potato	Many
Sugar beet	Fodder beet, beetroot
Sunflower	Not yet in the model library; Pouryazdankhah et al. (2022)

their application in considerably different pedo-climatic and management conditions requires calibration (Jabloun et al. 2018; Manevski et al. 2016a). Daisy requires rigorous calibration in order to consider genotype × environment × management interactions affecting crop productivity and the dynamics of hydrology and N. Jabloun et al. (2018) emphasized that the sensitivity of the parameters in the model depends also on management and designed a comprehensive framework for determining parameters for calibration under different genotype × environment × management interactions. From model structure perspective, the temperature response functions in the Daisy model has been pointed as important when simulating a range of genotype × environment interactions (Manevski et al. 2016a; Wang et al. 2017). Maize and wheat are also some of the main crops which have been re-parameterized (Fig. 3) to reflect changes in varieties with larger biomass production and management practices compared to the original parameterization in the 1980s (Manevski et al. 2015) and has undergone re-calibration under genotype × environment interactions (Jabloun et al. 2018; Manevski et al. 2016a; Manevski et al. 2016b). Users can always parameterize different cultivars for different regions through calibration.

Under temperate humid climatic conditions and sandy agricultural soils, leaching of nitrate is a problem, and growing 'catch crop' outside the main crop growth season (in northern Europe from May to September) is a major agro-technical measure to control this reactive N flow. As a result, the Daisy model has several crops parameterized in a role of catch crops, such as red clover (*Trifolium pratense L.*) and white clover (*Trifolium repens L.*) in mixture with ryegrass (*Lolium perenne L.*). However, the accuracy of their simulation is questionable and requires good data and dedicated calibration. Clovers

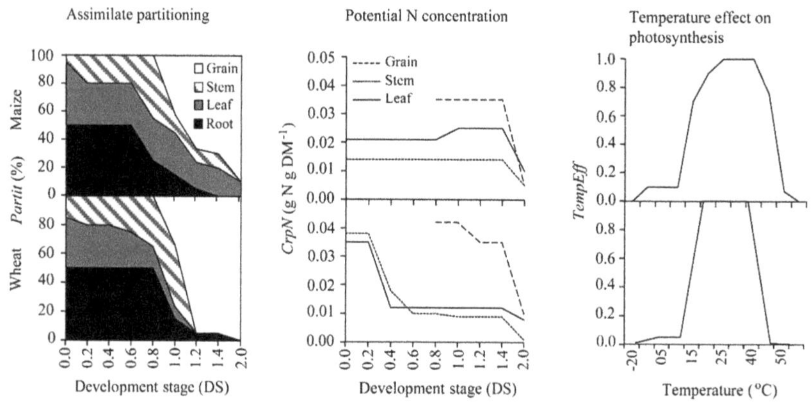

Figure 3 Crop parameters used in Daisy originally calibrated for pedo-climatic and management conditions in Northern Europe and re-parameterized for the North China Plain. Development stage (DS) equals 0, 1, and 2 for, respectively, emergence, flowering, and physiological maturity. Source: Adapted from: Manevski et al. (2016b; with permission).

are leguminous plants capable of biological N fixation. Accurate modelling of organic matter decomposition rates for leguminous residues has been in focus of Daisy model development and application for decades (de Neergaard et al. 2002; Magid et al. 2004; Müller et al. 2003; Müller et al. 2006) and organic matter pools of legumes and non-legume catch crops have been parameterized based on field experiments (Bruun et al. 2003; Müller et al. 2003).

Intercropping is one of the most powerful features of the model. Manevski et al. (2015) applied Daisy for analyses of crop growth processes and soil N dynamics in complex systems of maize intercropped with grass catch crop and found the model coukd correctly capture the effect of interspecies competition depicted by suppressed grass growth during the maize main season and improved grass growth toward the maize harvest. Another study simulated cereal–legume system and showed that LAI of the involved crops throughout the growing season is among the most important plant trait for correct intercropping simulation, as it affects the plant light-capturing capability (and the ability of one crop to compete with the other) (Karley et al. 2019). More modelling studies are needed to identify the key traits and mechanisms driving the interaction of different crop species under intercropping in field conditions across different pedoclimatic regions.

Daisy also contains routine for 'permanent' vegetation such as forest cover, which gained research momentum in the last decade amid the need to investigate systems with agroforestry, short rotation coppice, and alley cropping for effects on soil C and N pools and C sequestration and nitrate leaching. In Daisy the user defines the extent of the permanent canopy development (e.g. permanent forest *versus* short rotation coppice) through an elegant LAI-day of year (DOY) array input and this is of great flexibility among different trees as many trees of same leaf category (broadleaf or evergreen) and taxonomy (family) share botanical and morpho-physiological similarities when coppiced, including water fluxes and LAI dynamics (Hall et al. 1998; Proe et al. 2002). This avoids the (often costly and time-consumping) need for actual LAI measurements but uses literature values and approximations. Manevski et al. (2022; 2019b) and Manevski et al. (2024) parameterized the *permanent* vegetation module described by Boegh et al. (2009) to simulate the water balance and estimate nitrate leaching from willow and poplar trees (Fig. 4), following a global sensitivity analysis to select parameters for calibration according to Jabloun et al. (2018). The description of this permanent vegetation (forest, tree) requires values for leaf interception capacity, minimum transpiration resistance, and root depth, and same values of, respectively, 0.5 mm, 30 s m^{-1}, and 1.2 m were used for poplar and willow based on measurements or literature as these are both broadleaf *Salicaceae* trees with botanical and morpho-physiological similarities when coppiced, including water fluxes and LAI dynamics (Hall et al. 1998; Proe et al. 2002).

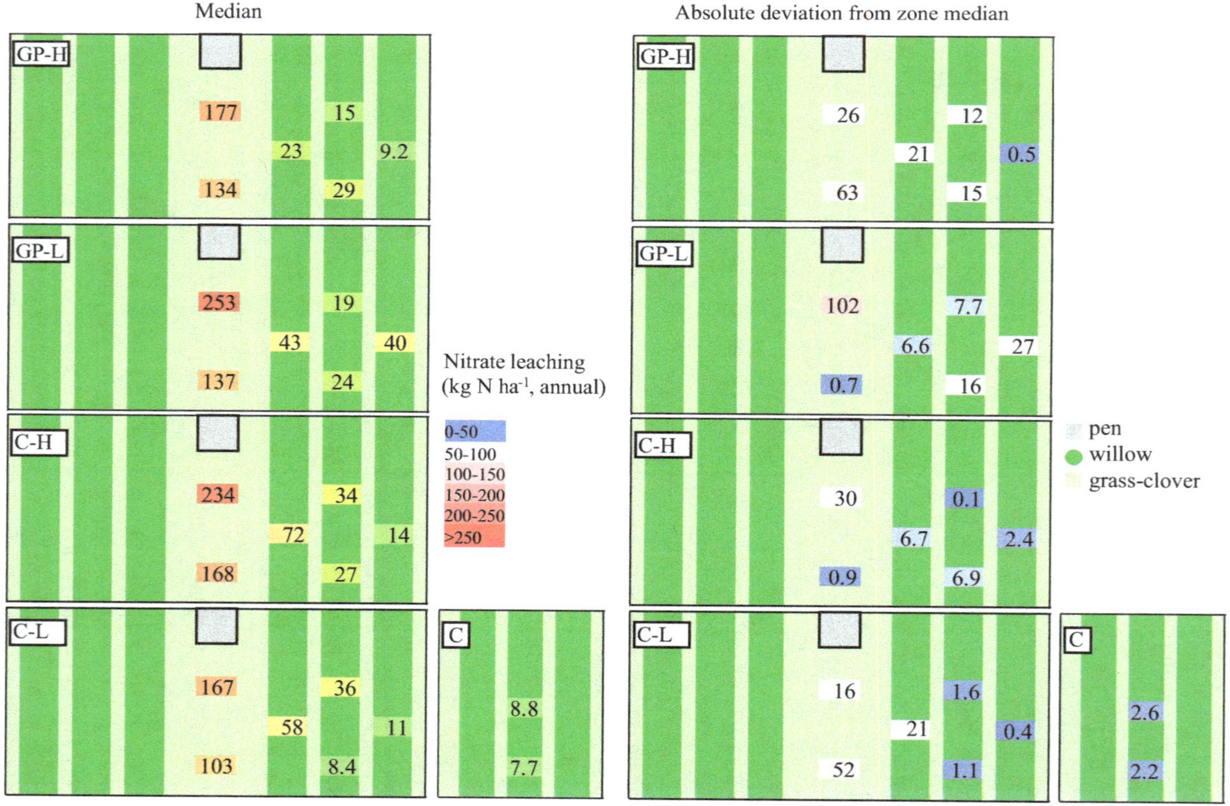

Figure 4 Illustration of spatial variation of median nitrate leaching from 1 m depth in two zones of willow and grass-clover in experimental paddocks of a study involving hens having feed with either green protein (GP) and control (C) diet, stocked at high (H, 4 m^2 hen^{-1}) or low density (L, 6 m^2 hen^{-1}) and compared to a background control without hens (C). Nitrate leaching was estimated by the percolation-weighted concentration method similar to Lord and Shepherd (1993), where nitrate concentrations were measured and water percolation was simulated with the Daisy model. Source: Adapted from Manevski et al. (2024; with permission).

Even under potential growth conditions, crop growth can show considerable differences between varieties (uncrossed types as found in its natural environment) and cultivars (hybrids made by crossing two or more plants) in relation to traits performance (e.g. Larsen et al. 2018; Porter and Semenov 2005; Tang et al. 2018). In the Daisy model, genotypic specifities of the crop can be defined by the modeller by utilizing an already parameterized crop and modifying or calibrating selected parameters to eventually save the parameterization as a 'new' crop variety or cultivar (Manevski et al. 2016b; Palosuo et al. 2011). Heidmann et al. (2008) describes a procedure for handling different varieties in the Daisy model through case studies of potatoes tested across Europe, and how to build a generic parameterization that can be used as a base for parameterizations of different crop genetic profiles. Technical document 'Guide on crop calibration' also exists on the model homepage to ease users in managing the crop compartment, including new crops and characterization of genetic variations. It should be noted that different combinations of parameters can result in equally good simulations, and the study of Heidmann et al. (2008) selected important parameters for genetic-level calibration based on expert/ modeller judgment. Newer and unbiased computational approaches are nowadays available based on sensitivity analysis under different genotype × environment × management interactions (Jabloun et al. 2018; 2014).

3 Model applications for genotype × environment interactions

The Daisy model had improved over the years and can accurately predict crop growth and yield for most of the locations, management conditions, and genotypes tested (Gavasso-Rita et al. 2024). The model is constantly being updated to include the latest relevant knowledge and also to remove programing bugs and update the quality of the code, with official releases for all platforms available on the model website. Many scientific projects in the field of agriculture, environment and climate change include parameterization of a large number of Daisy model setups, either to cover geographical areas, local variation in parameters (heterogeneity), or uncertainty of parameters. New functions have been introduced in the model for running many simulations on available cores in single computers ('spawn') and summarizing many simulations ('nwaps').

The model has already been involved in large-scale model intercomparison exercises on simulating continuous multi-year crop rotations compared to simulation of single years with good performance, including the European Modelling European Agriculture with Climate Change for Food Security (MACSUR; Kollas et al. 2015; Yin et al. 2020; Yin et al. 2017) and the global

Agricultural Model Intercomparison and Improvement Project (AgMIP; Asseng et al., 2015). Daisy is also often used for parameterization of new crops driven by the research and the innovation need for botanically diversified agricultural systems in space (mixed-/inter-/alley cropping) and time (double-cropping, crop rotations, vegetables, trees) with a combination of major and less-known crops for multi-purpose biomass (food, feed, bioenergy, biorefining, etc.). Many vegetables are parameterized in the model, including broccoli, brussels sprouts, celery, and onion, based on earlier dedicated study (Olesen and Grevsen 1997), but up to authors knowledge, these have not be tested or published in any simulation exercises. Manevski et al. (2018) used maize, ryegrass, and grass parameterizations as respective proxies to simulate the growth of hemp, festulolium/reed canary/cocksfoot, and miscanthus, which are not parameterized in the model but highly relevant in the context of agroecosystems targeting local sourcing of protein (Manevski et al. 2017). This approach is convenient and easy to apply with satisfactory accuracy in long-term simulations (e.g. a decade; Fig. 5), however, dedicated field experiments with these less-known crops are needed. From Fig. 5, the highest root mean square error (RMSE) for aboveground biomass is seen for perennial crops (RMSE = 2.1–2.5 Mg ha^{-1} against measured variation of 2–12 Mg ha^{-1}), which can be explained by the lack of accurate simulation of not only Daisy but also other models (in this case DSSAT) to reproduce the regrowth of perennial grasses after in-season defoliation. The mean of the two models reduces the simulation errors compared to both single models, and this concept of ensembled modelling has indeed been pointed as an approach to improve simulations of multiple growth and yield variables under genotype × environment × management interactions (Martre et al. 2015).

Pouryazdankhah et al. (2022) parameterized sunflower grown under partial root-zone drying conditions in Iran, which is one of the few applications of the 2-D versions of the model simulating ridge-furrow water transport in vertical and lateral direction (Plauborg et al. 2010). García-Jorgensen et al. (2020) also used Daisy to show a concept for modelling the fate of ptaquiloside (PTA), a carcinogenic phytotoxin produced by bracken (*Pteridium aquilinum*), in the soil–plant–atmosphere continuum. This study parameterized bracken in Daisy using mostly literature data and the model was adapted to realistically reproduce the observed dynamics of PTA in the bracken plants as a function of biomass, including processes such as PTA generation in the canopy, wash off by precipitation, and toxin recovery in the canopy after depletion events (Fig. 6); transport of the PTA in the soil was simulated by the advection–dispersion equation assuming weak sorption and degradation for two Danish soils. The study pinpointed Daisy as a useful tool to customize new processes and understand their dynamics, with further need for model calibration and validation.

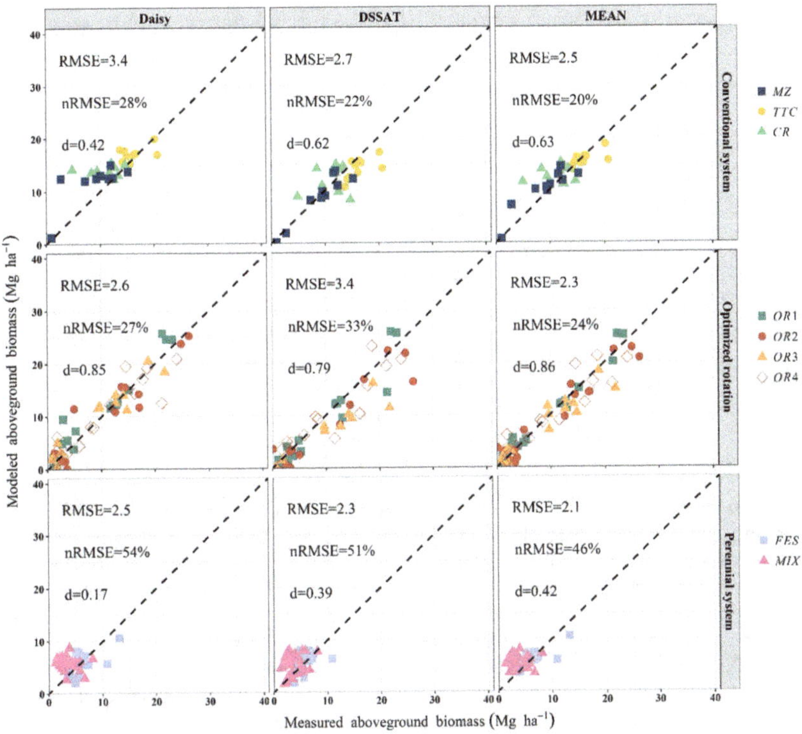

Figure 5 Measured and modeled aboveground biomass for 10 years (2013–2023) of different crops simulated by Daisy and DSSAT. Crops involve maize (MZ), triticale (TTC), cereal rotation (CR) consisting of spring barley, winter barley, winter rape, and winter wheat, optimized rotations (OR, numbers 1–4 denote rotation replicate) consisting of maize with catch crop, beet, hemp, and triticale with catch crop, and perennial crops festulolium (FES), and grass legume mixture (MIX). Further details on the systems can be found in Manevski et al. (2018, 2019). Simulations and data shown on the plots are from Zhang et al. (unpublished, in preparation).

Artificial drainage as an effective water management practice to enable and improve crop production on otherwise water-logged soils, but also increases the risk for leaching of nutrients such as N and phosphorous to the groundwater. Motarjemi et al. (2021) parameterized Daisy to simulate the essential processes for nitrate leaching risk assessment through artificially drained soil (such as the N and the C cycles), in addition to activated tile drain system to simulate groundwater dynamics. Their study simulated actual crop rotation of winter wheat, sugar beet, spring barley, winter rape, and maize readily parameterized in the model and emphasized the role of the frozen soil conditions during the winter when soil hydraulic conductivity decreases and the water flow is slower. Other studies with Daisy quantified the degree of preferential flow to drainage in an agricultural clay till field and the effects on denitrification and

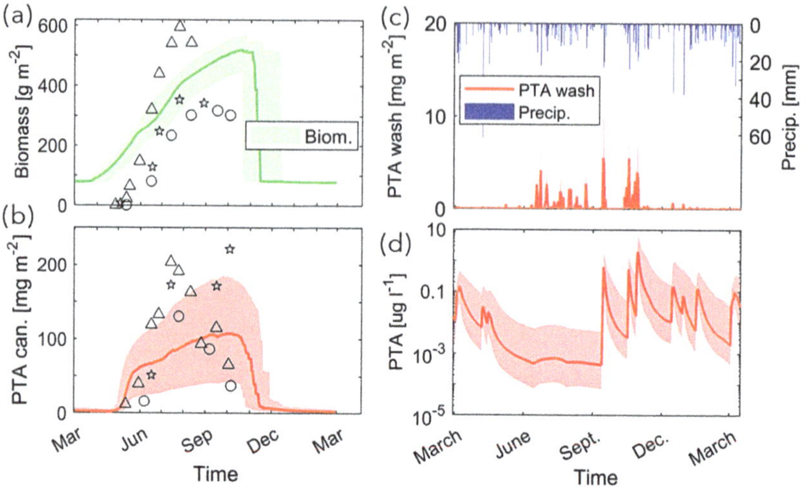

Figure 6 Simulated results for biomass, PTA content in the canopy, and wash off and leaching concentrations are shown in the figure. On the left side is shown the surface processes implemented in the model, the median biomass (a) and PTA in the canopy (b) for all the years and simulations (n = 2100). Symbols in the (a) and (b) subplots are the real data for bracken biomass and PTA content in the canopy, respectively. Each symbol represents a different location in Denmark (data derived from the reports of Rasmussen et al., 2003 and Rasmussen and Hansen, 2004). On the right side of the figure is shown the mass of PTA washed off by precipitation (c) and the PTA concentration at a 100 cm depth (d) for a specific year in Humleore (n = 100). The lines represent the median value, while the shaded area represents the upper (90th) and lower (10th) percentiles. Source: Adapted from: García-Jorgensen et al. (2020; with permission).

nitrate leaching to the drainage (Nagy et al. 2020a; Nagy et al. 2020b). These activities revealed valuable insights into the role of preferential transport and coherent denitrification for nitrate leaching from fine-textured soils and the risk of underestimating leaching when models such as Daisy exclude these soil physical conditions.

In response to an emerging N crisis in the 1980s, research institutions started to investigate the potential of the Daisy model for regional applications, despite its field-scale (1D) nature (Christian et al. 2001) whereas risk maps of nitrate leaching to groundwater and reduction potentials are prerequisites to guide legislation and spatial planning. For instance, Børgesen et al. (2001) used Daisy to upscale field-scale simulations of nitrate leaching for different combinations of weather clusters, soil types, and crop systems (including fertilization and irrigation regimes) to municipality level in Denmark, using soil and farm data as input to a statistical model. This was one of the first studies to quantify the effect of policy implementation involving changes in N fertilization schedules on the nitrate leaching from common crop cultivars in Europe and thus on water quality and safety. Manevski et al. (2019a) used similar approach using calibrated winter wheat and maize cultivars for semi-arid and sub-humid environments, respectively, and remote sensing derived cropping area in the upscaling to 5 × 5 km grid over an intensive agricultural region in China. One of the main advantages of upscaling procedure involving 'typical' or dominant combinations of weather – soil – management combinations is the simplicity and the availability of data. Improvements in computational power and statistical algorithms also drove Daisy to be linked with a variety of software (R, Python, Java) to perform simulations on hardware, ranging from workstations to High Performance Computing clusters, and analyse the output with various statistical methods, including deep learning (Vuaille et al. 2022); examples of assimilation of satellite data also exist (Boegh et al. 2004; van der Keur et al. 2001). While most users of the model use ad-hoc solutions, Jabloun et al. (2014) created reusable frameworks (RDaisy toolbox), paving the way for convenient and fully automated sensitivity and uncertainty analysis and calibration of the Daisy model (Fig. 7). This is very useful when crops cultivars need to be assesed and calibrated under genotype × environment interactions.

Coupling different process models is difficult due to the requirement to obtain values for the numerous parameters required, and a link to a remote sensing data derived variable might provide more convenient avenue to verify and update such a model system. Earlier studies have investigated more

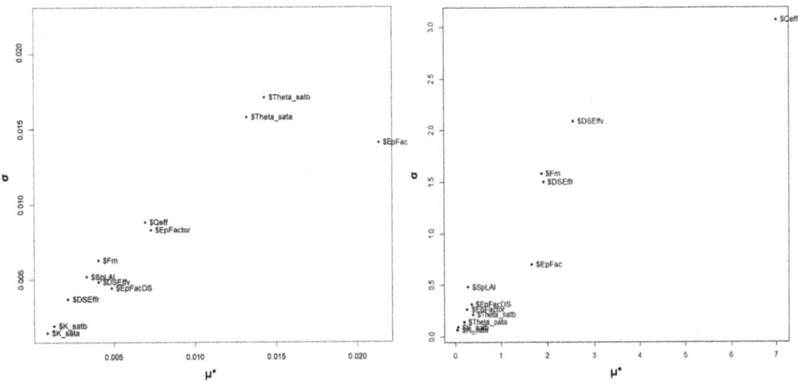

Figure 7 Outcome of global sensitivity analysis conducted by the screening method of Morris (1991) in the RDaisy infrastructure of Jabloun et al. (2014). The method of Morris (1991) calculates a set of incremental ratios ($\Delta_{output}/\Delta_{parameter}$) at various points of the parameters space and obtains means (μ^*; calculated on absolute values) and standard deviations (σ) of these ratios, where large μ^* denotes a parameter with an important overall influence (total effect), and large σ indicates nonlinearities in model response or interactions with other parameters. A subset of parameters is chosen and parameters with large μ^* and σ are typically selected for calibration. The plots show no interaction between soil (left panel) and crop parameters (right panel), supporting separate calibration of crop biomass and soil moisture. The data used in the sensitivity analysis are from Shang et al. (2024).

explicitly the potential for more direct integration of remote sensing data with the Daisy model (Boegh et al. 2004; van der Keur et al. 2001). LAI or comparable canopy variable of interest is one of the few canopy variables possible to derive from sensed canopy spectral reflectance regardless of the sensor altitude (field, air- or spaceborne). Indeed Veihe et al. (2006) found that Daisy is sensitive to changes in LAI with increases of more than 200% on nitrate leaching rates when reducing LAI by 15%. However, major challenges have been outlined in relation to adjusting the simulated variable, typically LAI, to the remotely sensed derived value, which in the case of models such as Daisy includes synchronization of several vegetation parameters to maintain congruity of the model canopy representation. Comparable studies also pointed on uncertainties in the remotely sensed variable due to noise from atmospheric vapor, e.g. in the summer (Mo et al. 2005). Also, the crop encounters subtle stress dynamics from lack of water, hours of high vapor pressure deficit (low moisture-high temperature, also known as heat stress), micro-biological stresses, etc., which all integrate over time as a multi-factorial response of canopy growth. On the other hand, utilizing LAI for calibrating the N balance thus assumes that the crop is affected only by N.

The Daisy model is often used to estimate daily soil water percolation out of the root zone from different crops (e.g. Frederiksen et al. 2023; Manevski et al. 2016b; Manevski et al. 2019b), which is needed to estimate daily nitrate leaching according to the percolation-weighted concentration method of Shepherd and Lord (1996). Daisy is also integrated with a hydrologically distributed MIKE-SHE model in a system to describe nitrate transport in a catchment (Hoang et al. 2014; Seidenfaden et al. 2022). Daisy simulates water transport in the root zone and nitrate leaching out of it, while MIKE-SHE is used to simulate the transport and the reduction of nitrate in the saturated zone. Both models are forced by the same daily inputs of precipitation, temperature, and potential evapotranspiration. Jiali et al. (2025) coupled hydrological-N internal loading modeling framework involving Daisy through MIKE-SHE to simulate N fluxes across the water-sediment interface in rivers under different water diversion conditions.

4 Further model enhancements and incorporation of crop genetics

From an end-user perspective, Daisy is relatively easy to operate given a basic understanding of the general form of the language syntax (name *value* [unit]), where name is the name of a specific parameter, *value* is the value of the parameter with specific unit, for instance (T 24 [dg C]), and the principle of calling names, functions, operations, etc. Students at MSc and PhD level and employees in relevant institutions such as environmental agencies can find the model challenging to manipulate in relation to the lack of more intuitive 'user-friendly' interface (Manevski and Andersen 2021). From 1991 onwards, the development and maintenance of Daisy has been funded on a project basis mostly focusing on new developments. Such funding structure does not typically provide resources for maintenance and adequate user support, which is problematic for new users and when results are used for education and policy support.

The field scale 1D nature of the Daisy requires site-specific calibration, demands input data and multi-disciplinary operator expertise to face the current challenges of coupled climate-environmental problems, not least with respect to upscaling, as users must calibrate the model based on their own data and build their own data management systems for integrated analyses. Quantification of impact of unstudied or new agronomic measures (e.g. diversified cropping system) on agricultural production and environment (i.e. soil C sequestration and N_2O emission, leaching of nitrate and chemicals such as pesticides and phytotoxins), under present and future climatic conditions is a challenge, due to the requirements for detailed process understanding and data (Hernández-Ochoa et al. 2022). Soil-Vegetation-Atmosphere Transfer (SVAT) models such as Daisy quantifying such impacts require multidisciplinary expertise, considerable time-consuming data acquisition and organization, and model calibration and have limitations in prediction and upscaling. Going beyond state of the art through establishing an advanced computational and georeferenced platform could increase the quality of the simulations given access to geo-referenced and temporal data sets involving weather, topography, soil, land use, and remote-sensing-derived canopy state data (Hernández-Ochoa et al. 2022; Manevski et al. 2019a). Daisy has recently been embedded in a modelling framework 'OurSmartFarm' (https://oursmartfarm.com/) testing potential for both advice to farmers and teaching activities. Moreover, larger modelling infrastructure should ideally include a range of advanced statistical and data science tools that can process the original or the simulated data, or a mixture of the two in fast and user-friendly manner through automatic data flows from databases, coupled with High-Performance Computing (HPC) to allow region- or country-wide high-resolution simulations and comprehensive scenario analyses. Such comprehensive infrastructure can also be a valuable learning platform to ease dissemination and provide support to scholars and researchers to work faster and more consistently, share data, and combine data, modelling, and artificial intelligence methods in novel ways. This is the purpose of the year 2023-started project 'A high-performance data-driven agroecosystem modelling platform for developing agricultural systems with minimum environmental impact' (AgroEco-HPM; https://daisy.ku.dk/about-daisy/agroeco-hpc/; Fig. 8) to facilitate upscaling, strengthen cooperation between institutions and thus contribute to reliable, countrywide assessments and predictions required for the

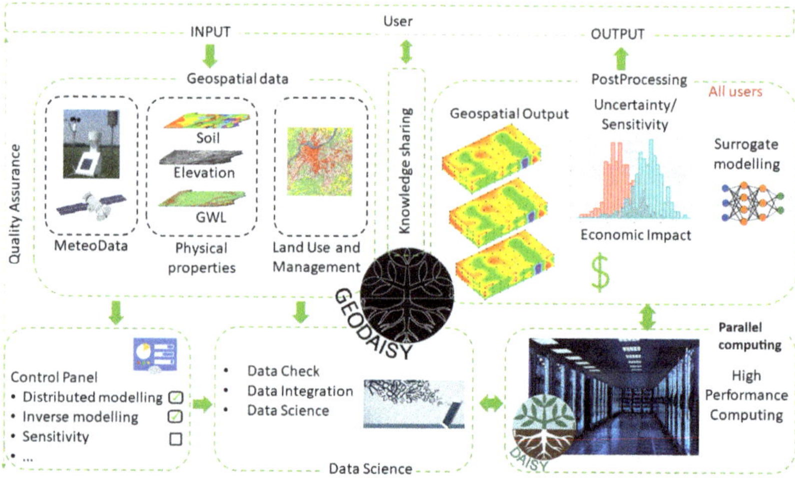

Figure 8 Overview of suggested Daisy model integration into larger infrastructure consisting of parallel computing, access to geospatial input data, and advanced methods in model simulations, post-processing, knowledge sharing and gaps, quality assurance, dissemination, outreach, and teaching. Figure adopted from project documentation Data Science Research Infrastructure 2022 – 'A high-performance data-driven agroecosystem modelling platform for developing agricultural systems with minimum environmental impact (AgroEco-HPM),' Application number 0077631, available online at https://daisy.ku.dk/about-daisy/agroeco-hpc/Proposal_short.pdf.

green transition in Denmark and generally in Europe. The project is envisaged to support research benefitting authorities (quantification, regulation), agricultural advisors, farmers, and related industries, but also other topics of public, environmental, and agricultural interest (Fig. 9). Actually, integration of relevant stakeholders into science-driven model-based framework has been developed and used by AgMIP to generate actionable information for adaptation planning processes. For a world likely to be hotter and more variable, stakeholders are invaluable for both data inputs and developing plausible socio-economic scenarios (Representative Agricultural Pathways, RAPs) to portray a sustainable development scenario and a rapid economic growth pathway.

In the context of a 'green transition' in Europe, Daisy is linked with funding for a project developing a new modelling approach which should be able to calculate more accurately estimates of N_2O emissions at the field level from soil biogeophysical conditions and farm activity data, especially N fertilization. The project integrates the latest knowledge on key controlling factors on N_2O emissions, such as weather, soil, crop and crop residues, fertilizer type and their management in the existing N turnover routines which currently include N_2O production from a simple nitrification process, and a denitrification process producing only N_2 but not N_2O. Moreover, Grosz et al. (2023) argue that studies of reactive N flows such as N_2O fluxes and nitrate leaching should

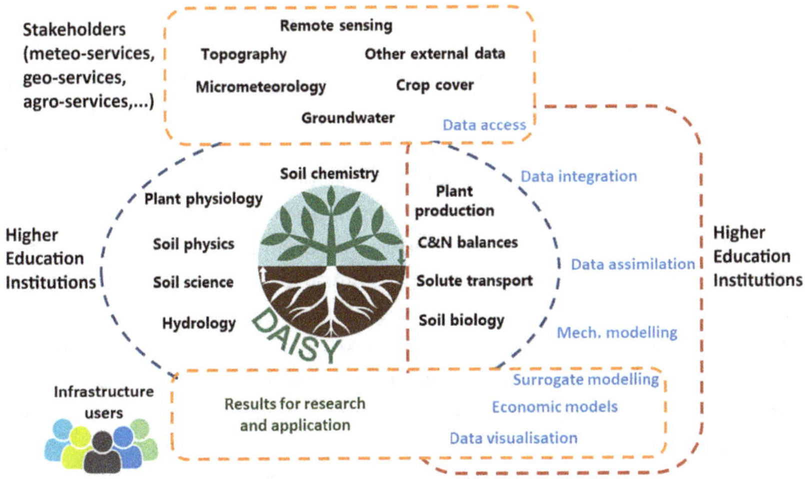

Figure 9 Overview of the different scientific areas and the various stakeholders within advanced geo-referenced modelling platform that links multivariate, spatio-temporal data access from stakeholders while providing services to these stakeholders and higher education institutions for advanced data science and mechanistic model simulations of water, carbon, and nitrogen pools and fluxes with automatic data flows coupled with other types of models and data presentation. Figure adopted from project documentation Data Science Research Infrastructure 2022 – 'A high-performance data-driven agroecosystem modelling platform for developing agricultural systems with minimum environmental impact (AgroEco-HPM),' Application number 0077631, available online at https://daisy.ku.dk/about-daisy/agroeco-hpc/Proposal_short.pdf.

communicate the complete N cycle as calculated by the models, despite lack of measurements, in order to detect anomalies, advance model development, facilitate model intercomparison, and ensure quality control.

5 Conclusion

This chapter presents an overview of the capabilities of the SVAT agroecosystem Daisy model for a wide range of applications, including genotype × environment × management interactions. For more than three decades now the model has been extensively used to study crop growth transport phenomena for soil water and nitrate leaching out of the root zone in actual and scenario systems and organic matter turnover and soil C sequestration. The last decade has reflected diversified model application, especially parameterization of different crops and their cultivars under vastly different genetic x environmental x management conditions, including formation, losses, and fate of phytotoxins, effects of conservation and precision agriculture and soil compaction, and climate change and future yield stability. Few efforts combined the model with data science for parameter optimization and identifying (via machine learning) optimum conditions for pesticide spraying. Also, there are few examples of assimilation of satellite data feeding the model.

Stakeholders other than researchers, such as agricultural advisory services, may employ the model for economic assessments and insights on precision agriculture given proper capacity building on model usage. The model is also used as a teaching tool on MSc and PhD level for soil hydraulic processes, crop science, and computational methods in agronomy. Future uses of Daisy comprehend spatially distributed mode of running with the aim to more directly support the needs of the agricultural extension organizations and wider with provisions of data and integration of the model into their 'crop manager' system solutions used by farmers, as well as those dealing with meteorology. To overcome the limitations of mechanistic modelling regarding predictability and scaling, 'artificial intelligence' methods and modelling must be combined. The mixture of scientific fields and scales makes it necessary to ensure that new process parameterizations, datasets, and documentation are shared and available to the multidisciplinary user community, enhancing research efficiency and quality. Finally, despite the suitable structure and fair complexity of Daisy, many explorations remain to be done for predicting the impact of climate change, innovative crop management practices, and new trait- or gene-based breeding technologies on crop production and the N and the C cycles in agroecosystems. New hypotheses and concepts from (eco) physiology, fueled by phenomics data and used to bridge the gap between genes/genotypes and phenotypes. As for any model, quality assurance on long term for the code system is required to support model credibility, development, and management.

6 Where to look for further information

The most complete single description of the Daisy model and its core processes can be found in the Report A10 from the Danish National Agency of Environmental Protection (Hansen et al., 1990).

Given the crop library, Rashid et al. (2022; 2025) used many different crops and varieties parameterized in Daisy in comprehensive scenario analyses to assess the performance of biobased-recovered N fertilizer in European cropping systems on crop production and mitigation potential nitrate leaching. Delhez et al. (2025) provides insight into the most influential parameters driving crop yield, CO_2, N_2O and energy fluxes as seen through global sensitivity analysis of the models' SVAT Sun-Shade Open Canopy energy transfer mode.

The Department of Plant and Environmental Sciences at University of Copenhagen hosts the official website as the single source of information on the Daisy model, including the model versions, documentation, tools and guides, projects and collaboration, publications and academic courses (https://daisy.ku.dk/). From 2025 the model can be downloaded from github (https://github.com/daisy-model/daisy/releases/tag/v7.0.7), where it is possible to access the code, follow the development of the new versions and add comments and wishes to future versions

The Daisy model is also included in the International Soil Modeling Consortium (https://soil-modeling.org/resources-links/model-portal/daisy) towards integrated and advanced soil systems modeling, data gathering, and observational capabilities.

7 Acknowledgments

Authors acknowledge Merete Elisabeth Styczen and Per Abrahamsen at University of Copenhagen for continuous and precious work with the model ensuring funds and updates targeting wide and diverse modelling community. First and last authors also acknowledge *post mortem* Søren Hansen, the mastermind behind the Daisy model for the introduction to the process-based modelling during their PhD studies and providing profound overview on the soil science and the concept of 'effective' parameter values.

8 References

Abrahamsen P and Hansen S (2000) Daisy: an open soil-crop-atmosphere system model. *Environ Model Softw* 15: 313–330. https://doi.org/10.1016/s1364-8152(00)00003-7.

Allen RG, et al. (1998) Crop evapotranspiration-Guidelines for computing crop water requirements-FAO Irrigation and drainage paper 56. FAO, Rome.

Boegh E, et al. (2004) Incorporating remote sensing data in physically based distributed agro-hydrological modelling. *J Hydrol* 287: 279–299. https://doi.org/10.1016/j.jhydrol.2003.10.018.

Boegh E, et al. (2009) Remote sensing based evapotranspiration and runoff modeling of agricultural, forest and urban flux sites in Denmark: from field to macro-scale. *J Hydrol* 377: 300–316. https://doi.org/10.1016/j.jhydrol.2009.08.029.

Børgesen CD, et al. (2001) Estimating the effect of legislation on nitrogen leaching by upscaling field simulations. *Ecol Modell* 136: 31–48.

Bruun S, et al. (2003) Calibration and validation of the soil organic matter dynamics of the Daisy model with data from the Askov long-term experiments. *Soil Biol Biochem* 35: 67–76. https://doi.org/10.1016/s0038-0717(02)00237-7.

Bruun S, et al. (2006) Application of processed organic municipal solid waste on agricultural land – a scenario analysis. *Environ Model Assess* 11: 251–265. https://doi.org/10.1007/s10666-005-9028-0.

Christian T, et al. (2001) *Modeling Nitrate Leaching at Different Scales? Application of the DAISY Model*. Modeling Carbon and Nitrogen Dynamics for Soil Management. CRC Press, Boca Raton, FL.

Delhez L, et al. (2025) Leveraging temporal variability in global sensitivity analysis of the Daisy soil-plant-atmosphere model. *Eur J Agron* 165: 127533. https://doi.org/10.1016/j.eja.2025.127533.

de Neergaard A, et al. (2002) Decomposition of white clover (Trifolium repens) and ryegrass (Lolium perenne) components: C and N dynamics simulated with the DAISY soil organic matter submodel. *Eur J Agron* 16: 43–55. https://doi.org/10.1016/s1161-0301(01)00118-6.

Frederiksen RR, et al. (2023) Development and application of a parsimonious statistical model to predict tile flow in minerogenic soils. *Agric Water Manag* 281: 108244. https://doi.org/10.1016/j.agwat.2023.108244.

García-Jorgensen DB, et al. (2020) A novel model concept for modelling the leaching of natural toxins: results for the case of ptaquiloside. *Environ Sci Process Impacts*. https://doi.org/10.1039/D0EM00182A.

Gavasso-Rita YL, et al. (2024) Crop models and their use in assessing crop production and food security: a review. *Food Energy Secur* 13: e503. https://doi.org/10.1002/fes3.503.

Goudriaan J and van Laar HH (1978) Calculation of daily totals of the gross C02 assimilation of leaf canopies. *Neth J Agr Sci* 26: 10.

Grosz B, Matson A., Butterbach-Bahl K, Clough T, Davidson EA, Dechow R, et al. (2023). Modeling denitrification: Can we report what we don't know? *AGU Advances*, 4, e2023AV000990. https://doi.org/10.1029/2023AV000990.

Gyldengren JG, et al. (2020a) Field scale agronomic and environmental consequences of overlapping N fertilizer application by disc spreaders. *Field Crops Res* 255: 107901. https://doi.org/10.1016/j.fcr.2020.107901.

Gyldengren JG, et al. (2020b) Effects of winter wheat N status on assimilate and N partitioning in the mechanistic agroecosystem model DAISY. *J Agron Crop Sci* 206: 784–805. https://doi.org/10.1111/jac.12412.

Hall RL, et al. (1998) Transpiration from coppiced poplar and willow measured using sap-flow methods. *Agr Forest Meteorol* 90: 275–290. https://doi.org/10.1016/S0168-1923(98)00059-8.

Hansen S (1984) Estimation of potential and actual evapotranspiration. *Nord Hydrol* 15: 205–212.

Hansen S., Jensen H.E., Nielsen N.E., Svendsen H., 1990. Daisy - Soil Plant Atmosphere System Model. Report A10, NPO research program of the Danish National Agency of Environmental Protection. https://daisy.ku.dk/pdfs/A10.pdf

Hansen S (2002) *Daisy Description - Equation Section One*. Copenhagen University, Copenhagen, Denmark. https://daisy-model.googlecode.com/files/DaisyDescription.pdf.

Hansen S, et al. (1993) *The Soil Plant System Model DAISY: Basic Principles and Modelling Approach. Simulation Model for Transformation and Transport of Energy and Matter in the Soil Plant Atmosphere System*. The Royal Veterinary and Agricultural University of Copenhagen, Jordbrugsforlaget.

Hansen S, et al. (2012) Daisy: model use, calibration and validation. *T ASABE* 55: 1315–1333. https://doi.org/10.13031/2013.42244.

Heidmann T, et al. (2008) Calibration procedure for a potato crop growth model using information from across Europe. *Ecol Modell* 211: 209–223. https://doi.org/10.1016/j.ecolmodel.2007.09.008.

Hernández-Ochoa IM, et al. (2022) Model-based design of crop diversification through new field arrangements in spatially heterogeneous landscapes: a review. *Agron Sustain Dev* 42: 74. https://doi.org/10.1007/s13593-022-00805-4.

Hoang L, et al. (2014) Comparison and evaluation of model structures for the simulation of pollution fluxes in a tile-drained river basin. *J Environ Qual* 43: 86–99. https://doi.org/10.2134/jeq2011.0398.

Jabloun M, et al. (2014) *RDAISY: A Comprehensive Modelling Framework for Automated Calibration, Sensitivity and Uncertainty Analysis of the DAISY Model*. Modelling Climate Change Impacts on Crop Production for Food Security, Oslo, Norwa.

Jabloun M, et al. (2018) Sensitivity of simulated crop yield and nitrate leaching of the wheat-maize cropping system in the North China Plain to model parameters. *Agr Forest Meteorol* 263: 25–40. https://doi.org/10.1016/j.agrformet.2018.08.002.

Jiali L, et al. (2025) Regulating nitrogen fluxes in nitrogen-rich rivers through optimized water diversion timing and duration (submitted).

Kollas C, et al. (2015) Crop rotation modelling–A European model intercomparison. *Eur J Agron* 70: 98–111. https://doi.org/http://dx.doi.org/10.1016/j.eja.2015.06.007.

Larsen SU, et al. (2018) Biomass yield, nutrient concentration and nutrient uptake by SRC willow cultivars grown on different sites in Denmark. *Biomass Bioenergy* 116: 161–170. https://doi.org/10.1016/j.biombioe.2018.06.011.

Lord EI and Shepherd MA (1993) Developments in the use of porous ceramic cups for measuring nitrate leaching. *J Soil Sci* 44: 435–449. https://doi.org/10.1111/j.1365-2389.1993.tb00466.x.

Magid J, et al. (2004) Decomposition of plant residues at low temperatures separates turnover of nitrogen and energy rich tissue components in time. *Plant Soil* 258: 351–365. https://doi.org/10.1023/B:Plso.0000016565.14718.4b.

Manevski K and Andersen MN (2021) Performing simulation modelling in the classroom: students enact the Daisy model. In: ME Styczen (ed) *Daisy 2021 Workshop*. University of Copenhagen, Copenhagen.

Manevski K, et al. (2015) Reduced nitrogen leaching by intercropping maize with red fescue on sandy soils in North Europe: a combined field and modeling study. *Plant Soil* 388: 67–85. https://doi.org/10.1007/s11104-014-2311-6.

Manevski K, et al. (2016a) Integrated modelling of crop production and nitrate leaching with the Daisy model. *MethodsX* 3: 350–363. http://dx.doi.org/10.1016/j.mex.2016.04.008.

Manevski K, et al. (2016b) Optimising crop production and nitrate leaching in China: measured and simulated effects of straw incorporation and nitrogen fertilisation. *Eur J Agron* 80: 32–44. http://dx.doi.org/10.1016/j.eja.2016.06.009.

Manevski K, et al. (2017) Biomass productivity and radiation utilisation of innovative cropping systems for biorefinery. *Agr Forest Meteorol* 233: 250–264. http://dx.doi.org/10.1016/j.agrformet.2016.11.245.

Manevski K, et al. (2018) Nitrogen balances of innovative cropping systems for feedstock production to future biorefineries. *Sci Total Environ* 633: 372–390. https://doi.org/10.1016/j.scitotenv.2018.03.155.

Manevski K, et al. (2019a) Modelling agro-environmental variables under data availability limitations and scenario managements in an alluvial region of the North China Plain. *Environmental Modelling & Software* 111: 94–107. https://doi.org/10.1016/j.envsoft.2018.10.001.

Manevski K, et al. (2019b) Effect of poplar trees on nitrogen and water balance in outdoor pig production – A case study in Denmark. *Sci Total Environ* 646: 1448–1458. https://doi.org/10.1016/j.scitotenv.2018.07.376.

Manevski K, et al. (2022) *Agroforestry reduces the nitrogen footprint of organic livestock production*. Paper (accepted) submitted to the 9 European Grassland Federation General Meeting: Grassland at the heart of circular and sustainable food systems – 26 to 30 June 2022, Caen, France.

Manevski K, et al. (2024) *Nitrate Leaching and Nitrogen Balances for Integrated Willow-Poultry Organic Systems in Denmark*. Agricultural Systems 221, 104149. https://doi.org/10.1016/j.agsy.2024.104149.

Martre P, et al. (2015) Multimodel ensembles of wheat growth: many models are better than one. *Glob Change Biol* 21: 911–925. https://doi.org/10.1111/gcb.12768.

McMaster GS, et al. (2008) Simulating Crop Phenological Responses to Water Deficits. In: LR Ahuja, et al. (eds) *Response of Crops to Limited Water: Understanding and Modeling Water Stress Effects on Plant Growth Processes*. American Society of Agronomy, ASA, CSSA, SSSA, 677 S. Segoe Rd., Madison, WI, USA.

Mo X, et al. (2005) Prediction of crop yield, water consumption and water use efficiency with a SVAT-crop growth model using remotely sensed data on the North China Plain. *Ecol Modell* 183: 301–322. http://dx.doi.org/10.1016/j.ecolmodel.2004.07.032.

Mollerup M (2010) Daisy 2d numerics. http://daisy.ku.dk/publications/Num2D-v2.pdf.

Morris MD (1991) Factorial sampling plans for preliminary computational experiments. *Technometrics* 33: 161–174. https://doi.org/10.2307/1269043.

Motarjemi SK, et al. (2021) Important factors when simulating the water and nitrogen balance in a tile-drained agricultural field under long-term monitoring. *Sci Total Environ* 787: 147610. https://doi.org/https://doi.org/10.1016/j.scitotenv.2021.147610.

Mueller T, et al. (1998) Soil C and N turnover after incorporation of chopped maize, barley straw and blue grass in the field: evaluation of the DAISY soil–organic-matter submodel. *Ecol Modell* 111: 1–15. https://doi.org/10.1016/s0304-3800(98)00094-5.

Müller T, et al. (2003) Decomposition of plant residues of different quality in soil–DAISY model calibration and simulation based on experimental data. *Ecol Modell* 166: 3–18. https://doi.org/10.1016/s0304-3800(03)00114-5.

Müller T, et al. (2006) Catch crops affect nitrogen dynamics in organic farming systems without livestock husbandry–Simulations with the DAISY model. *Ecol Modell* 191: 538–544. https://doi.org/10.1016/j.ecolmodel.2005.05.026.

Murindangabo YT, et al. (2023) Quantitative approaches in assessing soil organic matter dynamics for sustainable management. *Agronomy* 13: 1776.

Nagy D, et al. (2020a) Estimating the degree of preferential flow to drainage in an agricultural clay till field for a 10-year period. *Hydrol Earth Syst Sci Discuss* 2020: 1–42. https://doi.org/10.5194/hess-2019-665.

Nagy D, et al. (2020b) Effect of preferential transport and coherent denitrification on leaching of nitrate to drainage. *Hydrol Earth Syst Sci Discuss* 2020: 1–40. https://doi.org/10.5194/hess-2019-666.

Olesen JE and Grevsen K (1997) Effects of temperature and irradiance on vegetative growth of cauliflower (Brassica oleracea L. botrytis) and broccoli (Brassica oleracea L. italica). *J Experimen Bot* 48: 1591–1598. https://doi.org/10.1093/jxb/48.8.1591.

Palosuo T, et al. (2011) Simulation of winter wheat yield and its variability in different climates of Europe: a comparison of eight crop growth models. *Eur J Agron* 35: 103–114. https://doi.org/10.1016/j.eja.2011.05.001.

Pedersen MF, et al. (2021) A simulation of variable rate nitrogen application in winter wheat with soil and sensor information: an economic feasibility study. *Agr Syst* 192: 103147. https://doi.org/10.1016/j.agsy.2021.103147.

Plauborg F, et al. (2010) Modelling of root ABA synthesis, stomatal conductance, transpiration and potato production under water saving irrigation regimes. *Agric Water Manag* 98: 425–439. http://dx.doi.org/10.1016/j.agwat.2010.10.006.

Porter JR and Semenov MA (2005) Crop responses to climatic variation. *Philos T Roy Soc B* 360: 2021–2035. https://doi.org/10.1098/rstb.2005.1752.

Pouryazdankhah H, et al. (2022) Calibration and validation of daisy model for sunflower under partial root-zone drying. *Irrigat Sci Engi* 45: 15–30. https://doi.org/10.22055/jise.2021.37452.1973.

Proe MF, et al. (2002) Effects of spacing, species and coppicing on leaf area, light interception and photosynthesis in short rotation forestry. *Biomass Bioenergy* 23: 315–326. https://doi.org/10.1016/S0961-9534(02)00060-0.

Rashid MA, et al. (2022) Scenario analysis using the Daisy model to assess and mitigate nitrate leaching from complex agro-environmental settings in Denmark. *Sci Total Environ* 816: 151518. https://doi.org/10.1016/j.scitotenv.2021.151518.

Rashid MA, et al. (2025) Evaluating the performance of biobased, recovered nitrogen fertilizers in European cropping systems using modelling. *Farming System* 3 (2): 100141. https://doi.org/10.1016/j.farsys.2025.100141.

Rasmussen LH, et al. (2003) Occurrence of the carcinogenic bracken constituent ptaquiloside in fronds, topsoils and organic soil layers in Denmark. *Chemosphere* 51(2): 117–127. https://doi.org/10.1016/s0045-6535(02)00694-x.

Rasmussen L, Hansen H (2004). Growth of bracken in Denmark and the content of Ptaquiloside in fronds, rhizomes and roots. In *Poisonous plants and related toxinspp*. 354–360. CABI. https://doi.org/10.1079/9780851996141.0354.

Richards LA (1931) Capillary conduction of liquids through porous mediums. *Physics-J Gen Appl P* 1: 318–333. https://doi.org/10.1063/1.1745010.

Salazar O, et al. (2013) Changes in soil water balance following afforestation of former arable soils in Denmark as evaluated using the DAISY model. *J Hydrol* 484: 128–139. https://doi.org/http://dx.doi.org/10.1016/j.jhydrol.2013.01.036.

Seidenfaden IK, et al. (2022) Are maps of nitrate reduction in groundwater altered by climate and land use changes? *Hydrol Earth Syst Sci* 26: 955–973. https://doi.org/10.5194/hess-26-955-2022.

Shang Y, et al. (2024) Perennial cropping systems increased topsoil carbon and nitrogen stocks over annual systems–a nine-year field study. *Agric Ecosyst Environ* 365: 108925. https://doi.org/https://doi.org/10.1016/j.agee.2024.108925.

Shepherd MA and Lord EI (1996) Nitrate leaching from a sandy soil: the effect of previous crop and post-harvest soil management in an arable rotation. *J Agric Sci* 127: 215–229. https://doi.org/10.1017/S002185960007800X.

Tang RM, et al. (2018) Physiological and growth responses of potato cultivars to heat stress. *Botany* 96: 897–912. https://doi.org/10.1139/cjb-2018-0125.

van der Keur P, et al. (2001) Modification of DAISY SVAT model for potential use of remotely sensed data. *Agr Forest Meteorol* 106: 215–231.

Veihe A, et al. (2006) Sensitivity of DaisyGIS to LAI: Implications for land use based modelling of nitrate leaching. *Geografisk Tidsskrift* 106: 12.

Vuaille J, et al. (2022) Can information on past and near-future weather and field conditions predict the safest pesticide application day? *Comput Elect Agric* 203: 107454. https://doi.org/10.1016/j.compag.2022.107454.

Wang E, et al. (2017) The uncertainty of crop yield projections is reduced by improved temperature response functions. *Nat Plants* 3. https://doi.org/10.1038/nplants.2017.102.

Wasson AP, et al. (2012) Traits and selection strategies to improve root systems and water uptake in water-limited wheat crops. *J Exper Bot* 63: 3485–3498. https://doi.org/10.1093/jxb/ers111.

Weih M, et al. (2019) Summary report on mechanisms underpinning beneficial plant associations based on APSIM and DAISY. https://researchprofiles.ku.dk/files/382496619/DIVERSifyWP3Deliverable.pdf.

Yin X, et al. (2020) Uncertainties in simulating N uptake, net N mineralization, soil mineral N and N leaching in European crop rotations using process-based models. *Field Crops Res* 255: 107863. https://doi.org/10.1016/j.fcr.2020.107863.

Yin X, et al. (2017) Performance of process-based models for simulation of grain N in crop rotations across Europe. *Agr Syst* 154: 63–77. http://dx.doi.org/10.1016/j.agsy.2017.03.005.

Part 3

Asia and the Pacific

Chapter 19

InfoCrop: a process-based dynamic crop simulation model for analysing genotype x environment interactions in tropical crops

S. Naresh Kumar, Bidisha Chakrabarti, Arti Bhatia, Shweta Panjwani, D. N. Swaroopa Rani and Vikas Kumar, ICAR – Indian Agricultural Research Institute, India; P. K. Aggarwal, ICAR - Indian Agricultural Research Institute/currently South Asia Climate Adaptation of Agriculture Programme - CIMMYT, India; H. Pathak, ICAR - Indian Agricultural Research Institute/currently ICRISAT, India; Subhash Chander, ICAR - Indian Agricultural Research Institute/currently ICAR - National Centre for Integrated Pest Management, India; and N. Kalra, ICAR - Indian Agricultural Research Institute, India

http://dx.doi.org/10.19103/AS.2025.0155.23

1 Introduction

Interest in crop simulation modelling in India started between 1985 and 1995 with the involvement of Indian scientists in the international project "Simulation and Systems Analysis in Rice Production (SARP)" steered by the International Rice Research Institute (IRRI – The Philippines) and the Research Institute for Agrobiology at Wageningen University (The Netherlands). Indian scientists also collaborated with the International Benchmark Sites for Agrotechnology Transfer (IBSNAT) Project based at the University of Hawaii, USA.

Prior to InfoCrop, a simple wheat crop model was developed in 1982 to forecast yield at the regional scale, based on radiation use efficiency and satellite remote sensing-based thermal reflectance by the Indian Agricultural Research Institute (IARI) in collaboration with the Space Applications Centre, Ahmedabad. This model was further developed as WTGROWS by incorporating the major interactions between weather factors, soil factors, variety, water, and nitrogen (N) management (Aggarwal et al., 1994). A dynamic model for mustard (BRASSICA) was also developed at the IARI to pre-estimate phenology, growth, and yield in different zones. However, most models developed globally in 1980s and 1990s had a range of drawbacks, including the following: they were not able to perform satisfactorily in tropical conditions, since they were developed using data from temperate and high input conditions; they were complex with large input data requirements; and they did not account for biotic stresses such as insect pests and weeds.

InfoCrop was developed by IARI in collaboration with a range of crop-specific Indian research groups as a process-based dynamic model to simulate the growth, development and yield of major tropical crops (Aggarwal et al., 2006). The basic motivation behind the development of InfoCrop was to have a model that

- required simple and easy-to-measure input data;
- could simulate the crop performance in tropical agricultural environments;
- is user-friendly;
- could be easily integrated with decision support systems (DSS) for use in the field; and
- could provide easily accessible training on crop simulation modelling.

InfoCrop simulates the effects of weather, soils, agronomic management (including planting, nitrogen, organic matter, and irrigation) and major pests on crop growth and yield. It is based in part on the earlier crop-specific MACROS (Penning de Vries et al., 1989), WTGROWS (Aggarwal et al., 1994), ORYZA1 (Kropff et al., 1994), and SUCROS (Laar et al., 1997) models.

The basic crop model has been written in Fortran Simulation Translator programming language (FST/FSE; Graduate School of Production Ecology,

Wageningen, The Netherlands; Van Kraalingen, 1995). The latest version (InforCrop V2.1) is compatible with various programming languages such as C++, .Net and is also compatible with Windows 7 and above. The model requires inputs on varietal coefficients, which fall mainly into three categories: phenology, growth, and source–sink balance (a minimum of nine coefficients). Apart from these, input parameters include details of crop management, soil characteristics (at different layers), and daily weather data for the relevant location. The model also has provision for changing atmospheric CO_2 concentrations.

The InfoCrop model has been calibrated and validated for simulating the growth and yield of a range of crops, including

- wheat and rice (Mall and Aggarwal, 2002; Aggarwal et al., 2006; Naresh Kumar et al., 2013, 2014a);
- maize (Byjesh et al., 2010);
- sorghum (Srivastava et al., 2010);
- mustard (Boomiraj et al., 2010; Naresh Kumar et al., 2015);
- potato (Singh et al., 2005; Naresh Kumar et al., 2014b);
- cotton (Hebbar et al., 2008, 2013);
- perennial crops such as coconut (Naresh Kumar et al., 2008, 2013, 2015);
- soybean, green gram, and groundnut (Naresh Kumar et al., under preparation);
- sugarcane (Naresh Kumar et al., 2025 under preparation);
- Spatial InfoCrop v1. (Naresh Kumar et al., 2025 under preparation); and
- Simulation Model for Assessing Cropping Systems (SMACS), (Naresh Kumar et al., 2025 under preparation).

Calibration and validation results have, e.g., been published for wheat and rice (Aggarwal et al., 2006), maize (Byjesh et al., 2010), sorghum (Srivastava et al., 2010), mustard (Boomiraj et al., 2010), potato (Singh et al., 2005), and cotton (Hebbar et al., 2008).

The model can be used at a point (Mall and Aggarwal, 2002; Srivastava et al., 2010, Boomiraj et al., 2010, Byjesh et al., 2010) or at a spatial scale (Naresh Kumar et al., 2011, 2013, 2014a, 2015, 2019, 2022, 2023; Naresh Kumar and Aggarwal, 2013). Applications of the model include

- climate change impact assessments and assessment of adaptation options for key crops such as rice (Mall and Aggarwal, 2002; Naresh Kumar et al., 2013, 2015), wheat (Naresh Kumar et al., 2013), maize (Byjesh et al., 2010, Naresh Kumar 2023), sorghum (Srivastava et al., 2010; Naresh Kumar et al., 2011), Pearl Millet (Naresh Kumar, 2023), mustard (Naresh Kumar et al., 2015), soybean (Naresh Kumar, 2023), Chick pea (Naresh Kumar 2023), Pigeon pea (Naresh Kumar 2023), potato (Naresh Kumar et al., 2014b),

cotton (Hebber et al., 2013; Naresh Kumar, 2023), and coconut (Naresh Kumar and Aggarwal, 2013);
- modelling cropping combinations (Naresh Kumar et al., 2011);
- simulating nitrogen dynamics in soil (Ebrayi et al., 2007);
- calculating greenhouse gas (GHG) emissions (N_2O, CH_4) from agricultural fields (Bhatia et al., 2010, 2012); and
- assessing crop pest dynamics (Chander et al., 2006).

There is ongoing work to model the effects of

- ozone (O_3) on crops such as rice;
- phosphorus effects;
- refining the management options/ addition of tillage and mulching effects (Naresh Kumar et al., unpublished); and
- refining soil water balance (Paresh et al., unpublished), etc.

Use of the model beyond India (as well as enhancement of the model itself) has been facilitated by participation in the Agricultural Model Intercomparison and Improvement Project (AgMIP), e.g., in comparing different wheat and maize models. The model is being used in some parts of Southeast Asia, Africa, and South America. For example, the InfoCrop rice model and the Oryza-2000 models have been utilized to develop the SimulArroz model for use in Brazil. InfoCrop v2.1 has been downloaded in as many as 46 countries, covering all major regions, with over 2500 primary users. Fig. 1 shows results from a Web of Science literature search on the use of InfoCrop in peer-reviewed research on different crops.

2 Model inputs and outputs

InfoCrop requires relatively a fewer number of input parameters. *Minimum input requirements* are shown in **bold**. Inputs are

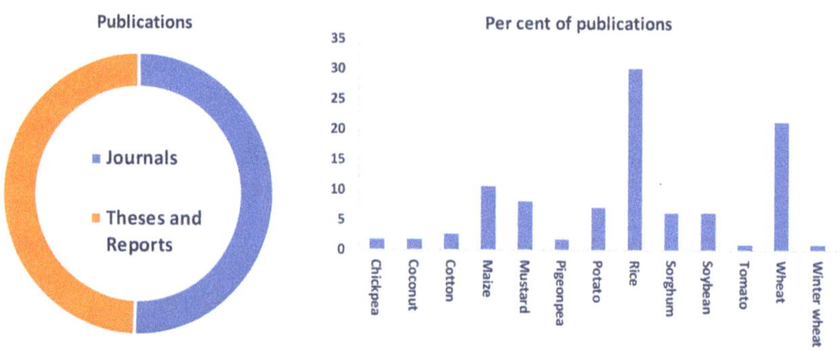

Figure 1 Peer-reviewed publications using InfoCrop.

- Daily weather: **Tmin (°C), Tmax (°C), Rainfall (mm)**, Solar Radiation (kJ. m^2), Wind speed (m/s), Vapor pressure;
- Location geo-coordinates (**latitude, longitude**, **and altitude**) and atmospheric **CO_2** (ppm, single value);
- Varietal coefficients:
    - **thermal time to flowering**;
    - **thermal time to physiological maturity**;
    - **potential grain weight**;
    - potential grain number per m^2;
    - **specific leaf area at the seedling stage**;
    - maximum radiation use efficiency;
    - light extinction coefficient of the canopy;
    - base temperatures for phenological stages (vegetative and reproductive phases); and
    - optimum and maximum temperatures for growth and development sensitivity of variety to day length, high and low temperatures, and flooding.
- Soil characteristics:
    - **Soil depth (3 layers)**;
    - **soil texture** (sand, silt, clay);
    - hydraulic characteristics (**permanent wilting point, field capacity, saturation fraction**) (NB. Where a user does not know soil hydraulic characteristics, InfoCrop can generate them using pedo-transfer functions);
    - pH;
    - electric conductivity;
    - soil organic carbon;
    - bulk density;
    - initial soil moisture; and
    - initial soil available nitrogen (ammonia and nitrate).
- Crop management inputs:
    - **date of sowing or transplantation** with seedling age;
    - seed rate and amount;
    - timing and type of **nitrogen and organic matter application**;
    - timing and amount of **irrigation**; and
    - insect pest, disease, and weed management.

InfoCrop can provide daily and summary outputs on several crop growth, development, and yield parameters, such as

- total and partitioning of dry matter, yield, gain number, grain weight, phenology, crop nitrogen (N) uptake, and partitioning, transpiration, etc;
- soil N balance in terms of mineralization, nitrification, denitrification, volatilization, immobilization, leaching, etc.;

- soil water balance in terms of runoff, drainage, evapotranspiration (ET, PET), and soil available water;
- GHG emissions (N_2O, CH_4, CO_2); and
- insect pest and disease damage.

These can be taken as daily outputs and/or as a summary output at the physiological maturity of a crop. InfoCrop can be used to carry out the following:

- Sensitivity Analysis, e.g. by varying soil, weather, management, or varietal characteristics.
- Seasonal Analysis – multiple year/seasonal simulations.
- Spatial Analysis – variable soil, weather, management characteristics, and different cultivars across a region.
- Spatial and seasonal analysis – spatial analysis can be carried out for multiple years.

3 Simulation of soil processes

InfoCrop simulates a number of soil processes.

3.1 Soil-water balance

The soil water dynamics (dW) at any given time is estimated using the following equation:

$$dW = I + Rf + C - dW\ E - T - Ic - P - D - R$$

where, dW is the rate of change in soil water, I is irrigation, Rf is rainfall, C is upward flux of water, E is evaporation from soil surface, T is transpiration of crops and weeds, Ic is the rainfall intercepted by the crops, P is percolation, R is runoff, and D is drainage.

Based on applied water as irrigation (surface as well as sub-surface), effective rainfall (typically >2.5 mm), downward and upward moisture fluxes, actual evaporation and transpiration and initial water content, and the net water content in three soil layers is updated daily (Fig. 2). It is recommended to use the maximum root depth for each crop/variety rather than recoverable root length. When the maximum root depth is shorter, the crop cannot take up water from lower soil layers, and it will face water stress.

The model simulates the potential ETnbased on Penman-Monteith or Priestley-Taylor methods, depending on the availability of input data.

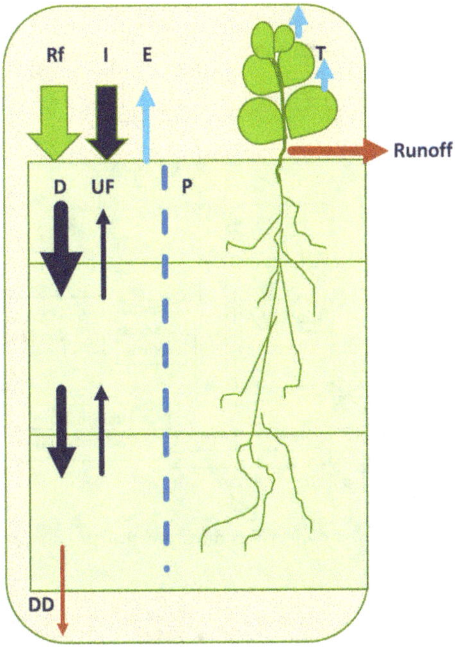

Figure 2 Soil–water balance in InfoCrop.

3.2 Soil nitrogen

InfoCrop calculates nitrogen transformation in soil daily. Processes considered in the model are mineralization, immobilization, hydrolysis of urea, nitrification, denitrification, volatilization, N2O emission, N movement in soil, and nitrogen uptake by crops. Fig. 3 presents a relational diagram illustrating various soil nitrogen processes in the InfoCrop model. The total soil nitrogen pool was divided into organic N, ammonium-N, and nitrate-N in the model. N sources from indigenous soil, inorganic fertilizer, organic manure, irrigation, rainfall, and biological N fixation are considered. Sinks of nitrogen include N uptake by plants and N losses through nitrate leaching, ammonia volatilization, and N_2O emissions.

3.3 Soil carbon

The InfoCrop model has a carbon module that considers major processes of soil carbon dynamics, i.e. mineralization, immobilization, and methane and CO_2 emissions. The relational diagram showing different soil carbon processes in InfoCrop is shown in Fig. 4.

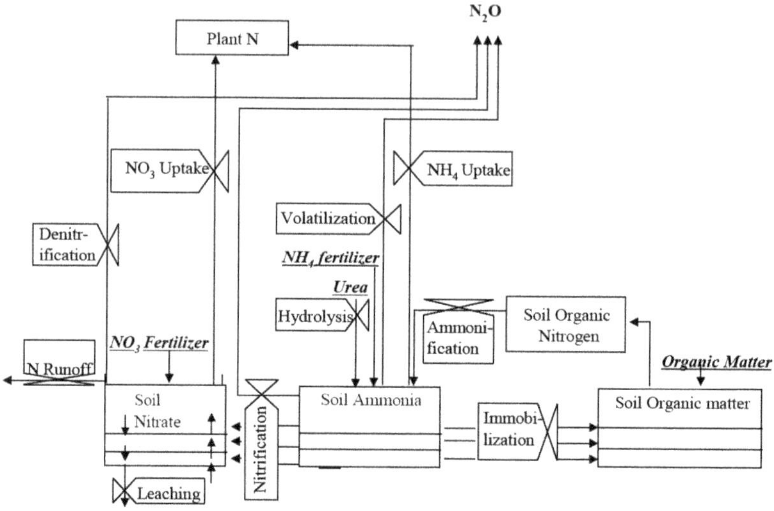

Figure 3 Relational diagram showing the nitrogen transformation processes in soil in InfoCrop.

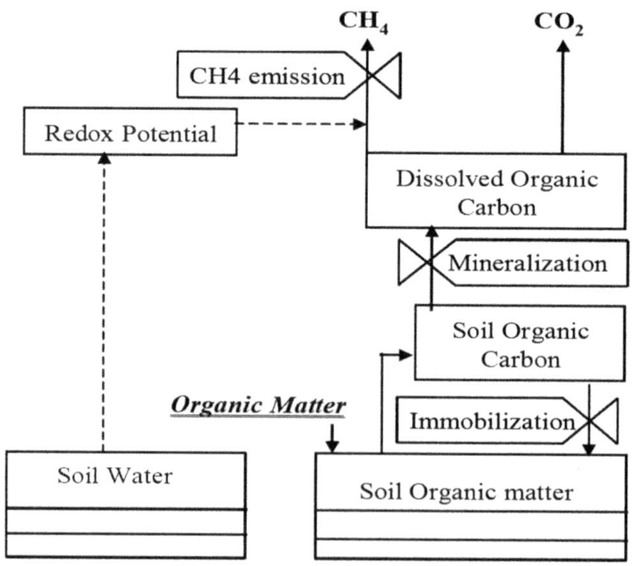

Figure 4 Relational diagram showing the processes relating to soil carbon dynamics in InfoCrop.

Soil organic matter is divided into two major pools, i.e. fresh organic matter (FOM) and humified organic matter (HOM). HOM is native soil organic matter, while application of organic matter as green manure, farmyard manure, or crop residue adds to the fresh organic matter pool. The FOM pool was divided into three sub-pools of carbohydrates, cellulose plus hemicellulose, and lignin. Root exudates and dying roots of the standing crop and roots of the previous crop are also inputs to the FOM pool. InfoCrop allows a maximum of three types of external organic matter additions in a crop cycle. Composition of the major organic sources is included in the model (Fig. 4).

4 Simulation of greenhouse gas emissions

Based on the soil carbon and nitrogen balance as influenced by the biochemical composition of organic matter applied to soil, dissolved organic carbon substrate, modified by the soil redox potential. CO_2 emission is dependent on the remaining dissolved organic carbon. Slimily, the N_2O emission is dependent on nitrification and denitrification, which in turn are influenced by temperature, soil pH, moisture status of soil, etc. The total GHG emissions per hectare are given as the sum of CO_2, CH_4, and N_2O expressed as the CO_2 equivalent based on the respective gas global warming potential.

5 Simulation of plant processes

InfoCrop is a generic model that can simulate the processes of most annual crops. Simulations of major processes are presented below, followed by a brief account of how different crops are simulated.

The major plant processes used for simulating growth, development and yield of a crop as follows (summarized from Aggarwal et al., 2004):

- Phenology;
- Dry matter production and partitioning;
- Leaf area growth;
- Source–sink balance;
- Nitrogen uptake and distribution;
- Transpiration;
- Abiotic stress effects; and
- Biotic stress effects.

The relationships between these plant, soil and other processes are shown in Fig. 5.

5.1 Phenology

The total development of a crop is quantified as development stages (DS), a dimensionless variable having a value of 0 at sowing, 0.1 at seedling

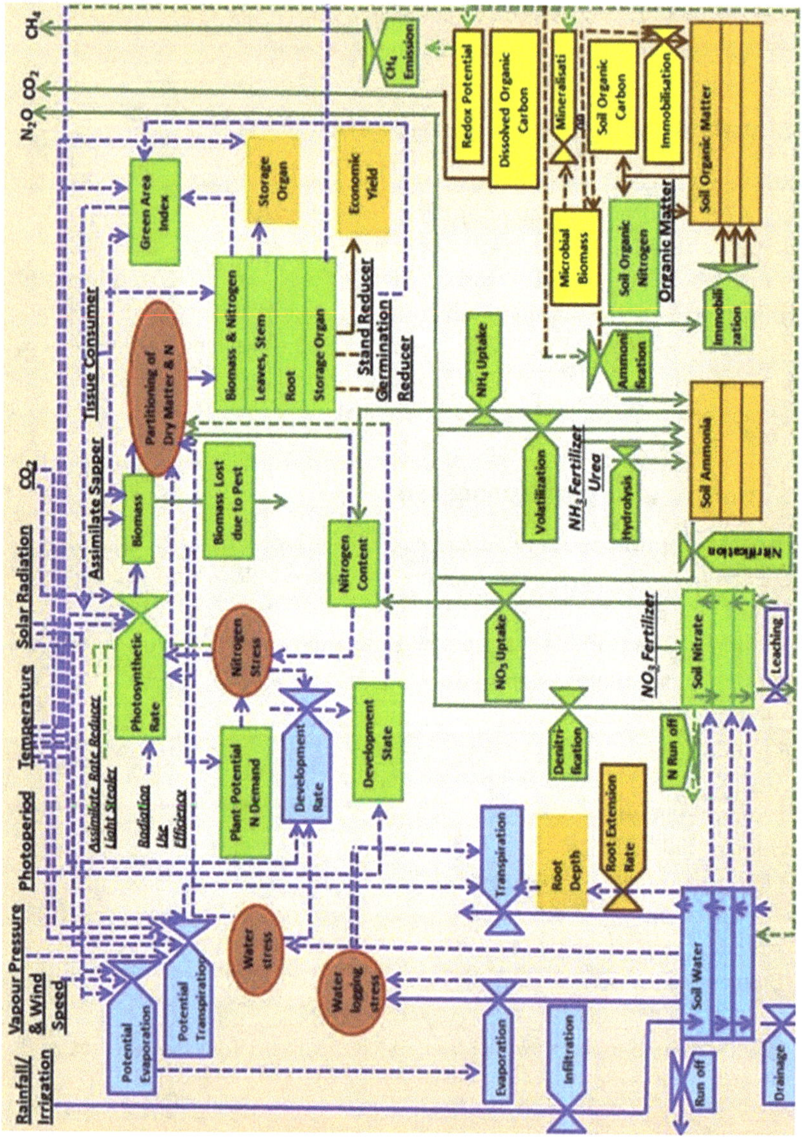

Figure 5 Relational diagram of plant and soil processes and the effects of external factors on various processes in InfoCrop. Source: Aggarwal et al. (2004).

emergence, 1.0 at flowering and 2.0 at maturity. The DS is calculated by integrating the temperature-driven development rates of phases from sowing to seedling emergence, seedling emergence to anthesis, and storage organ filling phases. Each of these phases are driven by the thermal time as specified by the user while considering the base, optimum and maximum temperatures. The DS is also modulated by water and nitrogen stress and photoperiod sensitivity. The effects of stratification and vernalization are simplified by integration into the thermal time for germination and vegetative phase respectively. In transplanted crops, seedling transplantation shock is also simulated.

The grain filling phase is simulated as a function of crop/variety-specific thermal time which is modified by water and nitrogen stress. The grain filling period is also modified by the source–sink relationship in terms of translocatable dry matter, current photosynthates, grain filling rate, number of grains, and potential grain weight.

A combination of nine phenology parameters can be used to characterize varieties:

- Base temperature for sowing to germination;
- Thermal time for sowing to germination;
- Base temperature for germination to 50% flowering stage (tuber initiation in case of potato);
- Thermal time for germination to 50% flowering stage;
- Base temperature for 50% flowering stage to physiological maturity;
- Thermal time for germination to 50% flowering stage;
- Optimal temperature for development;
- Maximum temperature for development; and
- Sensitivity to photoperiod.

5.2 Dry matter production and partitioning

Shoot and root weights at seedling emergence are initialized based on the user-specified seeding rate. Leaf dry matter is converted to leaf area index (LAI) using the specific leaf area of the variety. Dry matter per day is determined by the maximum radiation use efficiency (RUE), light interception by canopy and effective leaf area. This is modulated by the crop/variety-specific DS, leaf nitrogen content, ambient temperatures, photosynthetically active radiation, CO_2, and water stress as well as biotic stresses (weeds and pests). Contribution of green stalks and awns to photosynthesis is also taken into conservation for dry matter production. The dry matter is partitioned into root, leaf, stem, and storage organs (generic name given for grain, seed/tuber, etc.) as determined by the developmental stage of crop as well as the N and water stress and remobilization of reserves.

5.3 Leaf area growth

The leaf lamina area (LAI) changes proportionally with leaf area growth rate, a function of increment in leaf weight and the specific leaf area. The net leaf area is determined using the initial LAI, leaf growth rate and leaf area death due to senescence and or due to pest, nitrogen and water stresses. Leaf senescence is accelerated by adverse temperatures, developmental stage, and other biotic and abiotic stresses. The effective leaf area is a sum of leaf areas and non-lamina green area after subtracting all losses due to senescence and losses from insect pests.

5.4 Source-sink balance

The source-sink balance is determined by the

1 number of storage organs (grains or seeds or tubers), which are formed during a crop-specific period, limited by a maximum number of storage organs (SO), affected by floral sterility due to adverse temperature and pest attack, and lack of dry matter for SO growth;
2 The weight of storage organs; and
3 temperature-dependant potential filling rate and the level of dry matter available for growth.

The growth of storage organs is terminated when their weights reach potential weights (sink limitation) or if no more dry matter is available (source limitation) or when the thermal time-dependent DS is reached. Crop/cultivar-specific energy content and the moisture content of the storage organs is also taken into consideration.

5.5 Nitrogen uptake and distribution

The rate of N-uptake is dependent upon crop N demand, phenological age, soil N availability, transpiration, rooting depth, and soil water status. The total soil N-supply to the crop is the sum of N-uptake from each soil layer and nitrogen fixed in case of legumes. In estimating crop nitrogen demand, the potential nitrogen that different plant parts require is determined depending on developmental stage and plant nitrogen balance. Distribution of nitrogen amongst root, stem, and leaves is done in proportion to their relative demand. Actual rate of N-accumulation in storage organs is limited by potential N demand of storage organ and the availability of mobilizable nitrogen pool.

5.6 Evapotranspiration

Potential ET is estimated from radiation balance and wind speed and air-drying power. Temperature and elevated CO_2 also affect ET rates; the latter is accounted for by modifying potential ET by a factor dependent on CO_2 concentration.

6 Simulation of abiotic stresses

InfoCrop is capable of simulating various abiotic stresses and their effects on crop growth and development.

6.1 Temperature

The InfoCrop model simulates the effect of temperature on phenology by accelerating the developmental rate, net photosynthetic rates, ET rates, leaf growth and senescence rates, pollen viability, and grain-filling rates.

The effects of temperature on crop growth and development are simulated at various stages of crop development. Temperature influences the developmental rates through thermal time. It also influences leaf area development and senescence, with a greater effect of temperature on leaf area formation in the initial stages of development when LAI is <0.75. Temperatures higher or lower than the optimum accelerate rate of senescence, depending upon crop sensitivity to temperature. Temperature also influences photosynthetic rates depending on the developmental stage, thus affecting radiation use efficiency. Crop-specific responses of photosynthesis to temperature and CO_2 are updated as new information becomes available. A percentage of grains formed could be lost due to adverse temperatures. Adverse temperatures during the meiosis stage could also significantly increase sterility. A part of the storage organ becomes sterile if either the maximum or minimum temperatures of the day deviate from their respective threshold values during anthesis and a few days afterwards.

6.2 Frost/low-temperature stress

Similar to flooding (see below), days of frost/low temperature were calculated when the minimum temperature on any given day was below a crop-specific threshold. Frost/low temperatures, depending upon their duration, are assumed to affect leaf and non-leaf green area index in proportion to their sensitivity. Irrigation is known to alleviate frost damage partially.

6.3 Water

Water stress is determined as the ratio of actual water uptake and potential transpiration (Fig. 6). Stress is assigned a value of zero (maximum water stress) when soil water content is depleted, and linearly approaches a value of 1.0 when actual water uptake approaches potential ET (no water stress). Water stress affects germination, accelerating or delaying pre- and post-anthesis development rates. It also accelerates senescence, reduces radiation-use efficiency and nitrogen fixation in legumes, and increases the root-to-shoot ratio.

6.4 Flooding

Soil is considered waterlogged on a day if the available water fraction in the surface layer exceeds a threshold level of the field capacity, or if there is

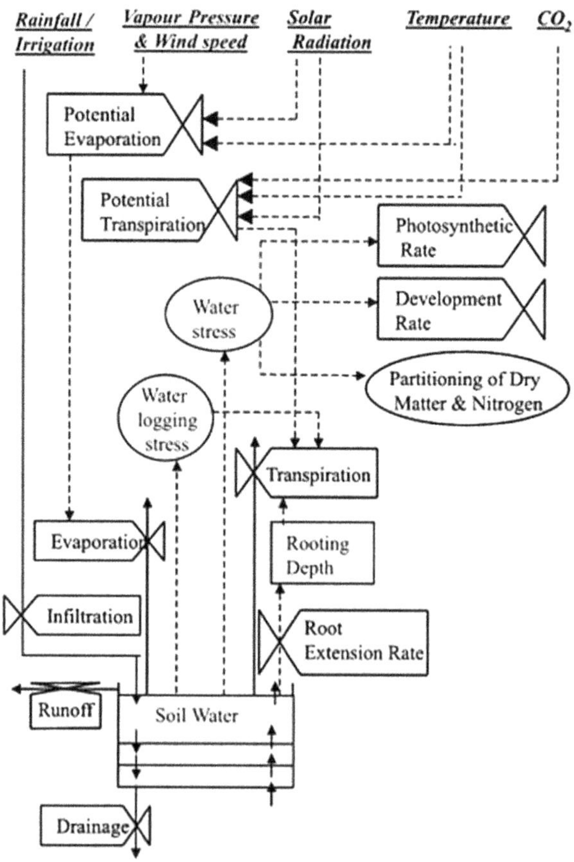

Figure 6 Parameters determining water stress.

free-standing water (ponding) on the soil surface. This starts to affect crop growth processes, depending on the DS and the crop/variety's sensitivity to flooding of different durations.

6.5 Nitrogen

Nitrogen stress is determined based on the potential and current levels of N in different plant parts, analogous to the actual/potential transpiration ratio used to determine the water stress factor (Fig. 7). The N stress effect was assigned a value of zero (maximum N stress) when actual mobilizable nitrogen is zero, and linearly approached a value of 1.0 when actual nitrogen approached the maximum value (no N-stress). The N-stress effect is the same as that of water stress. Amongst water and nitrogen stresses, the more severe of the two is assumed to affect the rate of crop development.

7 Simulation of biotic stresses: pests

InfoCrop has the capability to simulate the effects of weeds on crop dry matter production. Weed effects are simulated using a simple approach of a reduction in the radiation interception of crops due to weed infestation. The InfoCrop model can simulate crop damage due to insect pests and diseases (Fig. 8).

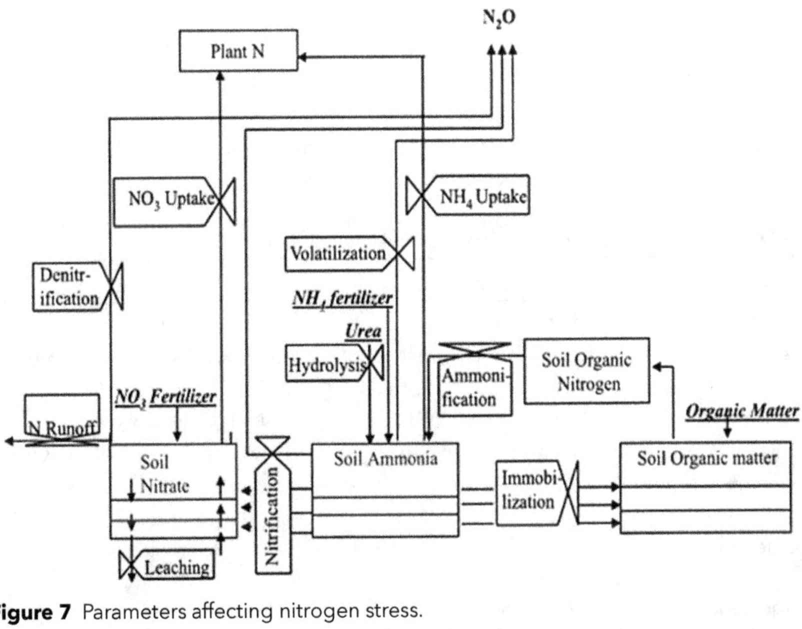

Figure 7 Parameters affecting nitrogen stress.

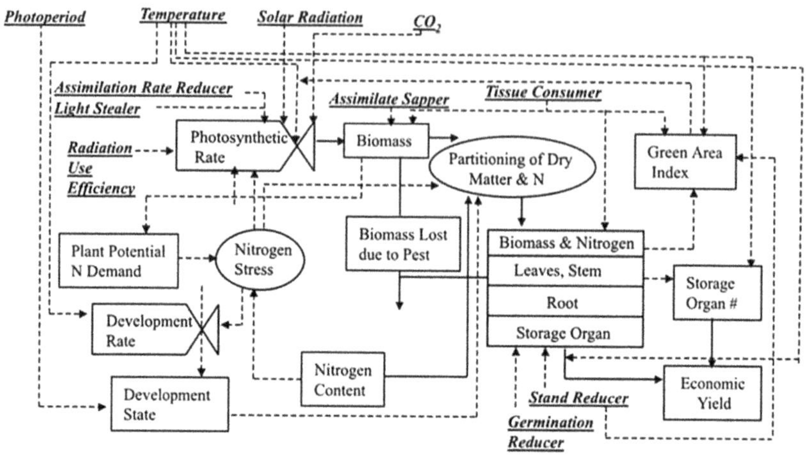

Figure 8 Relational diagram showing pest damage mechanism linked to plant physiological processes.

Pest (insect pest and disease) damage to crops is categorized broadly into the following categories (Aggerwal et al., 2004, 2006a, 2006b) (Table 1):

- Germination reducers;
- Stand reducers;
- Light stealers;
- Assimilation rate reducers;
- Assimilate sappers;
- Tissue consumers; and
- Turgor reducers.

The InfoCrop model uses these damage mechanisms to simulate the effects of pest damage. The damage is dependent on crop stage, magnitude and

Table 1 Damage mechanisms coupled to generic models of InfoCrop

Damage mechanism	Processes affected	Pests
Germination reducer	Germination	Mole cricket, diseases
Stand reducer	Leaf area, plant organ weights	Stem borer, cutworms, damping off
Light stealer	Interception of PAR	Sucking insects, weeds
Assimilation rate reducer	Radiation use efficiency	Powdery mildew, sucking insects
Assimilate sapper	Increase the maintenance cost of plants	Sucking insects
Tissue consumer	Leaf area, stem weight, panicle weight	Defoliators, blights
Turgor reducer	Nutrient and water uptake	Nematodes

duration of pest infestation which can be provided by user input data. It is essential to give the start date (in terms Days After Sowing -DAS) and end date (DAS) of pest incidence and the magnitude of incidence. The pest damage on leaf area (if any), dry matter, and yield can be obtained as an output.

InfoCrop was calibrated and validated for different types of pest damage mechanisms involving yield loss assessment resulting from rice pests such as leaf folder, stem borer, brown planthopper (BPH), and Russian wheat aphids. Field experiments for damage mechanism validation were conducted at variable pest stress levels ranging from un-infested crop to highly damaged crops. Variable damage levels were generated through pesticide application at differential frequencies and intervals. Data on phenology, growth, development and agronomic management of the crop, and pest damage/ population were collected. The model was first calibrated and validated for growth, yield, and dry matter of uninfested crop and then for pest effects. The leaf folder, e.g., was seen to act as a tissue consumer and assimilation rate reducer as it reduced leaf weight, leaf area, and radiation use efficiency of the crop (Satish et al., 2007). Rice planthoppers were categorized as assimilate sappers as they sucked sap from plant stems and leaf sheaths, which then reduced plant growth (Yadav and Chander, 2010). The rice stem borer was characterized as a stand reducer, causing tiller loss, thereby affecting leaf area, leaf weight, stem weight, panicle weight, and crop nutrients (Reji et al., 2008). Simulation of Russian wheat aphid, *Diuraphis noxia* (Mordvilko) damage on winter wheat allowed it to be categorized as an assimilate sapper and light stealer (Chander et al., 2006).

In addition to the pest-load-based damage simulations, a pest population dynamics model is also included. The population dynamics model simulates the growth and development as well as stage-specific feeding rates of pests (Fig. 9). The model consists of bio-ecological parameters such as fecundity, sex ratio, abiotic and biotic mortality factors, and thresholds of development and thermal constants for different insect species (Rabbinge et al., 1994; Pinnschmidt et al., 1995; Sutherst et al., 1996; Rappoldt and van Kraalingen, 1996). Thermal constants (K) and development thresholds (T_0) for insect DSs were determined by assessing their development rate against temperature (Kipyatkov and Lopatina, 2010), where:

that $K = 1/b$ and $T_0 = -a/b$

where a and b are the intercept and regression coefficients.

Insect populations convert from one stage to another (e.g. from egg to nymph) based on fulfilment of the relevant thermal constant, with the magnitude of the population entering the next stage depending on an estimate of stage-specific mortality.

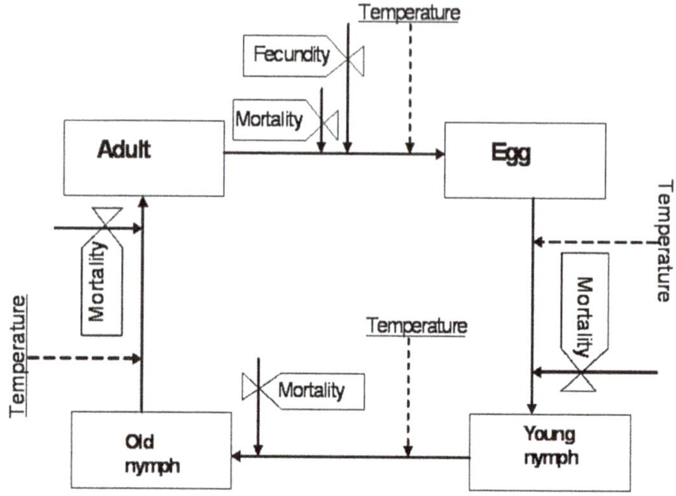

Figure 9 Relational diagram of the pest population dynamics model.

This model is crucial for establishing economic injury levels (EILs) and iso-loss curves, which are vital decision-making tools in pest management, enabling more targeted pesticide applications. Although EILs can be determined empirically, this is time-consuming to achieve in practice for each location and set of local conditions. Crop simulation models can be easily adapted for local crop management, weather, and pest situations and used to devise location-specific decision support tools (Fig. 10).

The InfoCrop model has been used to devise EILs and iso-loss curves for rice leaf folder (Satish et al., 2007), the rice gundhi bug (or stink bug) (Reji and

Figure 10 InfoCrop v2.1 graphical user interface home page.

Chander, 2008), stem borer (Reji et al., 2008), brown planthopper (Yadav and Chander, 2010; Sujithra and Chander, 2013), and Russian wheat aphid (Chander et al., 2006). EILs were determined by comparing the monetary yield loss due to pests with the expenditure on pest control. Simulated EILs were found to be comparable with empirically derived EILs, suggesting that the simulation model could help to determine location-specific EILs.

Iso-loss curves have also been formulated by plotting combinations of pest incidence and crop age that caused the same yield loss. InfoCrop has been used, e.g., to devise iso-loss curves for joint incidence of stem borer and bacterial leaf blight (BLB) in rice (Daryaei et al., 2003). These curves can be used for pest monitoring and, based on assessing the cost of expected yield loss due to pest damage at a given crop stage, can inform planning of management interventions.

Crop-pest models have also been used to analyze the potential impact of climate change on insect dynamics as well as crop-pest interactions. Studies have been undertaken, e.g., on brown planthopper (BPH) populations in rice and on pink borer and maize stem borer populations in maize (Sujithra and Chander, 2013). These showed potential population declines due to the effects of CO_2 and temperature rise, though this did not account for potential adaptation by these pest species to new conditions.

InfoCrop can also be used to simulate the effect of winter annual grass weed species, such as downy brome, jointed goat-grass, kochia, and volunteer rye, on the yield of winter wheat (Chander et al., 2006) and results compared with those of the "Great Plains Framework for Agricultural Resource Management Decision Support System" (GPFARM) (Canner et al., 2002). In InfoCrop, the light-stealing effect of weeds was simulated by reducing the interception of photosynthetically active radiation in proportion to the relative weed cover. Relative weed cover is the proportion of the leaf area of weeds and that of the crop. Many crop models require biomass of weeds as an input for simulating their effect on crop growth and yield. However, we preferred relative weed cover to weed biomass because it was easier to estimate.

8 Adapting the generic model to simulate growth and yield of different crops

InfoCrop is a generic process-based dynamic simulation model. It is adopted to simulate the major crops (Table 2). In this section, we briefly describe how variations among crop species are captured in the model.

Crop-specific and variety data: InfoCrop is a generic model that simulates the basic crop phenological and physiological processes. However, the crop-specific information is given as the crop-specific data, which is used to simulate these processes. In addition, varietal difference in a crop species is determined

Table 2 Crop-specific data used in the model to simulate differential responses of crop species to weather, water, and nitrogen conditions

Process	Major parameter	wheat	Rice	Maize	Sorghum	Pearl millet	Mustard	Soybean	Groundnut	Chickpea	Pigeon pea	Potato	Cotton
Crop-specific data													
Sowing	Fixed day sowing	*	*	*	*	*	*	*	*	*	*	*	*
	Soil moisture-dependent sowing	*	*	*	*	*	*	*	*	*	*	*	*
	Transplanted	x	*	x	x	x	x	x	x	x	x	x	x
Phenology	Sprouting-related modification of thermal time for germination	x	x	x	x	x	x	x	x	x	x	*	x
	Photoperiod-sensitive stage	*	*	*	*	*	*	*	*	*	*	*	*
	Daylength acceleration of flowering	*	*	*	*	*	*	*	*	*	*	*	*
	Water stress modification of the development rate	*	*	*	*	*	*	*	*	*	*	*	*
	Nitrogen stress modification of the development rate	*	*	*	*	*	*	*	*	*	*	*	*
Leaf area and other green area development	Fraction of seed (or potato seed tuber) reserves in leaves at emergence	*	*	*	*	*	*	*	*	*	*	*	*
	Leaf area reaches a maximum at the development stage	*	*	*	*	*	*	*	*	*	*	*	*
	Leaf senescence rate -temperature effect	*	*	*	*	*	*	*	*	*	*	*	*

	Leaf senescence rate -development stage effect	*	*	*	*	*	*	*	*	*	*	*	*
	Development stage-based specific leaf area	*	*	*	*	*	*	*	*	*	*	*	*
	Non-laminar photosynthetic areas	*	*	*	*	*	*	*	*	*	*	*	*
	Development stage-dependent senescence rate	*	*	*	*	*	*	*	*	*	*	*	*
Dry matter accumulation	Radiation Use Efficiency Maximum	*	*	*	*	*	*	*	*	*	*	*	*
	Canopy light interception -development stage dependent	*	*	*	*	*	*	*	*	*	*	*	*
	Radiation use efficiency modification -crop stage dependent	*	*	*	*	*	*	*	*	*	*	*	*
	Radiation use efficiency modification -Leaf N content dependent	*	*	*	*	*	*	*	*	*	*	*	*
	Radiation use efficiency modification -temperature dependent	*	*	*	*	*	*	*	*	*	*	*	*
	Radiation use efficiency modification -CO2 dependent	*	*	*	*	*	*	*	*	*	*	*	*
	Radiation use efficiency modification -water stress dependent	*	*	*	*	*	*	*	*	*	*	*	*

(Continued)

Table 2 (*Continued*)

Process	Major parameter	wheat	Rice	Maize	Sorghum	Pearl millet	Mustard	Soybean	Groundnut	Chickpea	Pigeon pea	Potato	Cotton
Dry matter partitioning	Dry matter partitioning to the leaf- crop-stage dependent	*	*	*	*	*	*	*	*	*	*	*	*
	Dry matter partitioning to the stem- crop-stage dependent	*	*	*	*	*	*	*	*	*	*	*	*
	Dry matter partitioning to root- crop-stage dependent	*	*	*	*	*	*	*	*	*	*	*	*
	Dry matter remobilization from plant parts to the storage organ	*	*	*	*	*	*	*	*	*	*	*	*
Rooting depth	Rooting depth -maximum	*	*	*	*	*	*	*	*	*	*	*	*
Frost sensitivity	Frost sensitivity-magnitude	*	*	*	*	*	*	*	*	*	*	*	*
Flooding sensitivity	Flooding sensitivity-magnitude	*	* (except puddled rice)	*	*	*	*	*	*	*	*	*	*
	Puddling-related effects	x	* (for puddled rice)	x	x	x	x	x	x	x	x	x	x

Water relations	Crop water stress effect	*	*	*	*	*	*	*	*	*	*	*	*
	Sensitivity to water stress	*	*	*	*	*	*	*	*	*	*	*	*
Nitrogen uptake and fixation	Maximum rate of N uptake	*	*	*	*	*	*	*	*	*	*	*	*
	Fraction of N remobilization	*	*	*	*	*	*	*	*	*	*	*	*
	Carbon cost of N fixation	x	x	x	x	x	x	*	*	*	*	x	x
	Development stage-based maximum N content in plant tissue	*	*	*	*	*	*	*	*	*	*	*	*
	Maximum rate of N accumulation in leaves, stem and root	*	*	*	*	*	*	*	*	*	*	*	*
	Nitrogen fixation rate with the development stage	*	*	*	*	*	*	*	*	*	*	*	*
Nitrogen partitioning	Nitrogen partitioning to leaf-crop-stage dependent	*	*	*	*	*	*	*	*	*	*	*	*
	Nitrogen partitioning to stem- crop-stage dependent	*	*	*	*	*	*	*	*	*	*	*	*
	Nitrogen partitioning to root- crop-stage dependent	*	*	*	*	*	*	*	*	*	*	*	*
	Nitrogen remobilization from plant parts to the storage organ	*	*	*	*	*	*	*	*	*	*	*	*

(Continued)

Table 2 (*Continued*)

Process	Major parameter	wheat	Rice	Maize	Sorghum	Pearl millet	Mustard	Soybean	Groundnut	Chickpea	Pigeon pea	Potato	Cotton
Source-sink balance	DS of the beginning of grain/seed/ storage organ fill	*	*	*	*	*	*	*	*	*	*	*	*
	Potential grain/seed/ storage organ determination stages (beginning and end)	*	*	*	*	*	*	*	*	*	*	*	*
	Potential grain/seed/ storage organ number /m2	*	*	*	*	*	*	*	*	*	*	*	*
	Potential grain/seed/ storage organ filling rate	*	*	*	*	*	*	*	*	*	*	*	*
	Sterility of reproductive organs -high and low temperature thresholds	*	*	*	*	*	*	*	*	*	*	*	*
	Shelling percentage	x	*	x	x	x	*	*	*	*	*	x	*
	Grain/seed/ reproductive organ moisture content	*	*	*	*	*	*	*	*	*	*	*	*
	Energy content in the storage organ	*	*	*	*	*	*	*	*	*	*	*	*
	The reproductive organ sterility sensitive stage of the crop	*	*	*	*	*	*	*	*	*	*	*	*

using variety-specific data. Additionally, for rice, the puddling effect and seedling age at transplantation are taken into account. The older the seedling at transplant, the greater the transplantation shock.

Each crop model has a specific model.exe and fetches data from the crop-specific data file. These are updated based on the varietal coefficient data. In addition, data on crop management, soil characteristics, and daily weather are input to simulate the growth, development, and yield of the chosen variety(ies) of a particular crop.

9 The InfoCrop graphical user interface

All crop-specific model.exe and crop-specific data files, along with data on varietal characteristics, soil, weather, and management, are bundled as a graphical user interface (GUI) and released as InfoCrop v1. In 2004. Since the updating of Windows OS did not support the execution of version 1, version 2.0 was developed and released in 2014, and version 2.1 was released in 2015.

The InfoCrop model GUI is user-friendly, allowing users to select a crop model, variety, management, soil type, and weather conditions. The GUI provides soil characteristics of the major soils of each district of India. However, there is also an option for the user to give user-defined data. In the absence of soil hydraulic conductivity details, the user can depend on InfoCrop-derived parameters, based on pedo-transfer functions. Weather files are provided for some locations. The user can upload the weather files in InfoCrop format. The weather conversion tool offers options to convert weather data in Excel or in formats such as .txt, .csv, or DSSAT format to the InfoCrop format. The user can also convert the Excel or .txt, .csv, or InfoCrop format weather files to DSSAT format. The user can choose any existing variety of a given crop, or can add a variety for a specific crop.

Details on how to add variety, soil, weather, organic matter, pests, and diseases are given in the InfoCrop 2.1 user manual. Step-wise details on project creation, weather file conversion, simulation setup, data visualization, and saving options are given in the manual.

10 Case study: simulating growth and development of the perennial crop coconut using InfoCrop

The InfoCrop model was initially developed to simulate annual crops, with around 700 maximum days of simulation. Model simulation capacity was subsequently extended to 60 years, provided input weather data are available, to simulate growth, development and yield of perennial crops such as coconut (Naresh Kumar et al., 2008). Models of plantation crops are scarce. Modeling of perennial tree growth and tree architecture has developed subsequently from early work such as Reffye et al. (1997) and Godin et al. (1999). A completed model was released as InfoCrop-COCONUT in 2008 (Naresh Kumar et al., 2008). A GUI for the model has subsequently been developed and released as the CocoSim v1 to allow use by growers (Fig. 11).

InfoCrop-COCONUT is a dynamic simulation model which runs at daily steps. It simulates the complete life cycle of coconut in three developmental phases:

1 seed nut planting to emergence;
2 juvenile period (emergence to first flowering); and
3 nut yield phase.

The last growth phase can be up to 60 years, the typical economic life of a coconut tree. Nuts mature about 12 months after pollination. The period between first flowering and the stable nut yield stage is defined as the unstable yield phase, which varies between 10 years and 14 years after planting, depending on type of variety.

Details about the model specifics are given in Naresh Kumar et al. (2008). Since the model simulates at a daily step, daily outputs on stem dry matter stem (including inflorescence parts other than nuts), leaves and nuts are summed and can be viewed at monthly or yearly intervals at any age of the tree Table 3.

The optimum minimum and maximum temperatures for growth are 20°C and 35°C. Low temperatures and short days delay the emergence of leaves and inflorescences. Prolonged temperatures of less than 10°C for 1 month

Figure 11 The CocoSim homepage.

Table 3 Variety-specific data used in the model to simulate differential responses of varieties within a crop species to weather, water, and nitrogen conditions

Parameter	Wheat	Rice	Maize	Sorghum	Pearl millet	Mustard	Soybean	Groundnut	Chickpea	Pigeon pea	Potato	Cotton
Sensitivity to photoperiod	*	*	*	*	*	*	*	*	*	*	*	*
Sensitivity to high temperature for reproductive part sterility	*	*	*	*	*	*	*	*	*	*	*	*
Sensitivity to low temperature for reproductive part sterility	*	*	*	*	*	*	*	*	*	*	*	*
Nitrogen fixation efficiency	x	x	x	x	x	x	*	*	*	*	x	x
Thermal time from sowing to seedling emergence	*	*	*	*	*	*	*	*	*	*	*	*
Thermal time from seedling emergence to 50% flowering	*	*	*	*	*	*	*	*	*	*	*	*
Thermal time from seedling emergence to 50% first flowering	*	*	*	*	*	*	*	*	*	*	*	*
Thermal time for grain/ seed/ stirage organ filling	*	*	*	*	*	*	*	*	*	*	*	*
Potential relative growth rate of leaf area	*	*	*	*	*	*	*	*	*	*	*	*
Specific leaf area of the variety	*	*	*	*	*	*	*	*	*	*	*	*

(*Continued*)

Table 3 (*Continued*)

Parameter	Wheat	Rice	Maize	Sorghum	Pearl millet	Mustard	Soybean	Groundnut	Chickpea	Pigeon pea	Potato	Cotton
Canopy extinction coefficient (maximum)	*	*	*	*	*	*	*	*	*	*	*	*
Potential grain/ seed/ storage organ number /m2	*	*	*	*	*	*	*	*	*	*	*	*
Potential grain/seed/ storage organ weight	*	*	*	*	*	*	*	*	*	*	*	*
Potential root growth rate	*	*	*	*	*	*	*	*	*	*	*	*
Maximum rooting depth	*	*	*	*	*	*	*	*	*	*	*	*
N content in the storage organ	*	*	*	*	*	*	*	*	*	*	*	*
Sensitivity to frost	*	*	*	*	*	*	*	*	*	*	*	*
Sensitivity to flooding	*	*	*	*	*	*	*	*	*	*	*	*
Age of the seedling at transplantation	x	* (for transplanted rice)	x	x	x	x	x	x	x	x	x	x
Sprouting	x	x	x	x	x	x	x	x	x	x	*	x
grain/seed/ storage organ filling rate	*	*	*	*	*	*	*	*	*	*	*	*

trigger nut fall. Temperatures above 40°C, which prevail during April–July in the tropics, result in low vapor pressure deficits leading to increased leaf mortality, thus reducing the functional LAI. These temperature effects are accounted for in the model by decreasing the rate of development if night-time temperatures fall below 10°C or if daytime temperatures exceed 40 °C (Naresh Kumar et al., 2008). Increasing LAI above 4 through the continued growth of leaves hastens the death of shaded leaves. The source–sink relationship is influenced by soil water availability (Naresh Kumar et al., 2008). Low temperatures, water and nitrogen limitations hasten leaf senescence. Dry mass partitioning to root, stem, leaf and nut as a function of DS is simulated in the model through empirical interpolation functions developed from field experiments. The model can generate the cumulative DM of a tree of any age. Inflorescence DM, excluding nut DM, is summed on a cumulative basis with stem DM. Total nut DM on tree, formed every day, is partly harvested every month.

Factors influencing pistillate flower production and button shedding include varietal variations, irrigation, nutrition, leaf area, and DM production (Naresh Kumar et al., 2006). Nut dry matter is converted to partially dry nut by multiplying by a variety-specific nut fresh:dry mass ratio. Nut dry matter is further partitioned into copra by a specific correction factor. Copra is further converted to oil by multiplying a variety-specific oil percentage value. Nut DM is converted to partially dry nut by multiplying by a variety-specific nut fresh: dry mass ratio. Nut DM is partitioned into copra, which is the solid endosperm in coconut fruit, shell, and husk (Siju Thomas, 2003). The number of nuts is obtained by providing the variety-specific mean nut dry weight. In the model, the number of nuts is simulated, but not the mass of individual nuts, which is user-defined.

Model inputs include variety, soil, daily weather, and crop management data. Coconut model-specific input data on parameters for the crop and variety include thermal time for germination, germination to juvenile phase and economic yielding phase, potential nut number per ha, potential nut weight, and fraction of copra and oil in nut. Field-level parameters include plants per hectare.

Standard outputs include development time and stage, daily crop growth, DM of roots, stem, and leaves, total DM, and cumulative values of leaf mass, stem mass, yield of nuts, copra, and oil, and nut number. Other outputs include LAI, crop nitrogen uptake, soil water and nitrogen availability, ET, nitrogen and water stress, amounts of irrigation and fertilizer. All the outputs can be obtained as monthly or yearly data. However, it is recommended to use yearly output data for better accuracy. Using these inputs, the model has been used for assessing the

- yield potential in different agro-climatic zones (Naresh Kumar et al., 2008);
- impact of climate change on coconut plantations in India (Naresh Kumar and Aggarwal, 2013);
- adaptation gains in current and future climates (Naresh Kumar et al., 2013); and
- carbon sequestration potential of coconut plantations in different agro-climatic zones of India.

In addition to India, requests for use have come from other major coconut-growing regions, such as Sri Lanka, as well as countries in Southeast Asia.

11 Conclusion and future trends

The InfoCrop Model has proved to be a versatile model for both researchers and growers. Strengths include

- Development specifically for tropical crops;
- User-friendliness, including a simple, Windows-based GUI for growers and a range of readily available input data requirements;
- Ease in uploading weather and soil data as well as other information such as new variety data;
- Simulation of a wide range of variables, including effects of pests; and
- Versatile applications, including yield gap analysis, estimating GHG emissions, and impacts of climate change on crop productivity across different spatial and temporal scales, as well as informing improvements in crop management practices.

Areas for further model enhancement include the following:

- InfoCrop is a trait-based model. There are efforts to develop gene/quantitative trait loci (QTL) linkage to traits using phenomics data, particularly for rice and wheat.
- Developing an integrated model with climate, hydrological, crop management, and economic components brought under a single GUI framework, with seamless data flow between the models to help carry out integrated assessments on impacts of climate change and technology adoption.
- Linking of InfoCrop models with remote sensing data for use in a web-based weather forecast module. A module has been developed for wheat.
- Calibration of ozone impacts on the growth and development of different crops.

- InfoCrop can address the impact of elevated temperature on insect population, and the impact of elevated CO_2 and temperature on crop yield. However, likely changes in insect feeding rates owing to increased C:N ratio in plants under elevated CO_2 also need to be incorporated to ensure a realistic simulation of crop-pest interactions under a changed climate. Furthermore, continuous exposure of insect species to higher temperatures over time may help them adapt to new situations and minimize the adverse effects of climate change. This also needs to be taken into account during climate change impact assessment.
- Inclusion of the effect of phosphorus on plant growth and development. A prototype module has been developed and tested. However, more details are needed before the modules can be fully incorporated.
- Improving crop–weed interaction dynamics.

Looking forward, the impact of climate change also suggests that a crop may now face multiple abiotic stresses within a single season. Based on the experimental data on phenology, LAI, dry matter partitioning, yield, and yield components, the ability of InfoCrop to simulate the effects of sequential stresses in a crop season is being tested for maize, soybean, and groundnut. The maize model, e.g., was found to be robust in simulating the effects of sequential multiple stresses except in a few situations such as when maize was exposed to excess water/flooding in the vegetative period and then to terminal droughts, or where the crop was exposed to droughts during the vegetative period and then to excess water/flooding coinciding with the flowering and initial grain filling period. The groundnut model was found to be good in simulating crop phenology while requiring improvement for simulating dry matter and yield under sequential stresses. Overall results on phenology, LAI, biomass accumulation and distribution, yield and yield components indicate that plants respond differentially to sequential stresses depending on the combination and sequential pattern of stresses. More experimentation is required to test and improve the performance of InfoCrop in these situations.

12 Source code

The source code of the models can be obtained from Prof PK Aggarwal and Prof Soora Naresh Kumar after due documentation of the intended use and IPR-related issues.

13 Model description and documentation

1 Aggarwal, P.K., et al. 2004. *InfoCrop: a generic simulation model for annual crops in tropical environments*. Indian Agricultural Research Institute, New Delhi, p. 132.

2 Aggarwal, P.K., et al. 2006. InfoCrop: a dynamic simulation model for the assessment of crop yields, losses due to pests, and environmental impact of agro-ecosystems in tropical environments. 1. Model description. *Agric. Sys.* 89(1): 1–25.
3 Naresh Kumar, S., et al. 2008. Simulating coconut growth, development and yield using InfoCrop-coconut model. *Tree Physiol.* 28: 1049–1058.

14 References

Aggarwal, P.K., et al. 2004. *InfoCrop: a generic simulation model for annual crops in tropical environments*. Indian Agricultural Research Institute, New Delhi, pp. 132.

Aggarwal, P.K., et al. 2006. InfoCrop: A dynamic simulation model for the assessment of crop yields, losses due to pests, and environmental impact of agro-ecosystems in tropical environments. 1. Model description. *Agric. Sys.* 89(1): 1–25.

Attri, S.D. and Naresh Kumar, S. 2024. Implications of climate change on agriculture and adaptation. In N. Khare (ed.) *Climate Change in India: Impacts and Assessments.* CRC Press, Boca Raton, FL. p. 266. https://doi.org/10.1201/9781003485995

Aggarwal, P.K. and Mall, R.K. 2002. Climate change and rice yields in diverse agro-environments of India. II. Effect of uncertainties in scenarios and crop models on impact assessment. *Clim. Chang.* 52: 331–343.

Aggarwal, P.K. and Rani, S. 2009. Assessment of climate change: impacts on wheat production in India. In: P.K. Aggarwal (ed.) *Global climate change and Indian agriculture*. Chapter 2. Indian Agricultural Research Institute, New Delhi, pp. 5–12.

Aggarwal, P.K., et al. 1994. Analyzing the limitations set by climatic factors, genotype, and water and nitrogen availability on productivity of wheat. I. The model documentation, parameterization and validation. *Field Crops Res.* 38: 73–91.

Aggarwal, P.K., et al. 2006a. InfoCrop: A dynamic simulation model for the assessment of crop yields, losses due to pests, and environmental impact of agro-ecosystems in tropical environments. 1. Model description. *Agric. Sys.* 89(1): 1–25.

Aggarwal, P.K., et al. 2006b. InfoCrop: A dynamic simulation model for the assessment of crop yields, losses due to pests, and environmental impact of agro-ecosystems in tropical environments. II. Performance of the model. *Agric. Sys.* 89: 47–67

Bhatia, A., et al. 2010. Trade-off between productivity enhancement and global warming potential of rice and wheat in India. *Nutr. Cycl. Agroecosyst.* 86: 413–424.

Bhatia, A., et al. 2012. Greenhouse gas emission from rice and wheat growing areas in India: spatial analysis and upscaling. Greenhouse Gases Sci. echnol. 2(2): 115–125.

Boomiraj, K., et al. 2010. Assessing the vulnerability of Indian mustard to climate change. *Agric. Ecosyst. Environ.* 138: 265–273.

Byjesh, K., et al. 2010. Simulating impacts, potential adaptation and vulnerability of maize to climate change in India. *Mitn. Adaptn. Strat. Global Chang.* 15: 413–431.

Canner, R., et al. 2002. Weed reproduction model parameters may be estimated from crop yield loss data. *Weed Sci.* 50: 763–772.

Chander, S., et al. 2006. Modelling the effect of Russian wheat aphid, Diuraphis noxia (Mordvilko) and weeds in winter wheat as guide to management. *Agric. Syst.* 88(2–3): 494–513.

Daryaei, M.G., et al. 2003. Use of simulation mode to formulate iso-loss curves for stem borer and bacterial leaf blight (BLB) incidence in rice. *Shashpa* 10(2): 105–114.

Ebrayi, K.N., et al. 2007. Simulation of nitrogen dynamics in soil using InfoCrop model. *Environ Monit Assess*. 131(1–3): 451–465.

Godin, C., et al. 1999. A Method for Describing Plant Architecture which Integrates Topology and Geometry. *Annals of Botany* 84(3): 343–357.

Hebbar, K.B., et al. 2008. Predicting cotton production using InfoCrop-cotton simulation model, remote sensing and spatial agro-climatic data. *Curr. Sci*. 95(11), 1570–1579.

Hebbar, K.B., et al. 2013. Simulating the impacts of climate change on cotton production in India. *Clim Chang*. 118: 701–713. https://doi.org/10.1007/s10584-012-0673-4

Kipyatkov, V.E. and Lopatina, E.B. 2010. Intraspecific variation of thermal reaction norms for development in insects: new approaches and prospects. *Entomol. Rev*. 89: 33–61.

Kropff, M. J., et al. 1994. *ORYZA1: An ecophysiological model for irrigated rice production*, SARP Research Proceedings (DLO/WUR/IRRI) – Wageningen University and Research, Wageningen, The Netherlands.

van Laar, H. H. et al. 1997. *SUCROS 97: Simulation of crop growth for potential and water-limited production situations* (Quantitative Approaches in Systems Analysis No. 14), Wageningen University and Research, Wageningen, The Netherlands.

Mall, R.K. and Aggarwal, P.K. 2002. Climate change and rice yields in diverse agro-environments of India. I. Evaluation of impact assessment models. *Clim Chang*. 52: 315–330.

Naresh Kumar, S. 2022. *Integrated modelling for assessing impacts of climate change and adaptation strategies*. IARI Publishing, New Delhi, 8 p.

Naresh Kumar, S. 2023. Modelling the impact of climate change on agriculture in South Asia. In C. Nendel (ed.) *Modelling climate change impacts on agricultural systems*. Burleigh Dodds Science Publishing, Cambridge, UK, pp. 541–588. http://dx.doi.org/10.19103/AS.2022.0115.18

Naresh Kumar, S., and Aggarwal, P.K. 2013. Climate change and coconut plantations in India: Impacts and potential adaptation gains. *Agric. Sys*. http://dx.doi.org/10.1016/j.agsy.2013.01.001

Naresh Kumar, S., et al. 2008. Simulating coconut growth, development and yield using InfoCrop-coconut model. *Tree Physiol*. 28: 1049–1058.

Naresh Kumar, S., et al. 2011. Impact of climate change on crop productivity in Western Ghats, coastal and northeastern regions of India. *Curr. Sci*. 101(3): 33–42.

Naresh Kumar, S., et al. 2013. An assessment of regional vulnerability of rice to climate change in India. *Clim. Chang*. 18: 683–699. http://dx.doi.org/10.1007/s10584-013-0698-3

Naresh Kumar, S., et al. 2014a. Vulnerability of Indian mustard (*Brassica juncea* (L.) Czernj. Cosson) to climate variability and future adaptation strategies. *Mitig. Adapt. Strateg. Global Chang*. 21(3): 403–420.

Naresh Kumar, S., et al. 2014b. Vulnerability of wheat production to climate change in India. *Clim. Res*. 59: 173–187.

Naresh Kumar S., et al. 2015. Assessment of impact of climate change on potato and potential adaptation gains in the Indo-Gangetic Plains of India. *Int. J. Plant Prod*. 9(1): 151–170.

Naresh Kumar, S., et al. 2019. *Seasonal climate change scenarios for India: impacts and adaptation strategies for wheat and rice*. ICAR- IARI Pub, p 105.

Penning de Vries, F. W. T. 1989. *Simulation of ecophysiological processes of growth in several annual crops*: Simulation Monograph 29, Wageningen University and Research, Wageningen, The Netherlands.

Pinnschmidt, H.O., et al., 1995. Simulation of multiple species pest damage in rice using CERES-rice. *Agric. Syst*. 48: 193–222.

Rabbinge, R., et al. 1994. Systems approaches in pest management: the role of production ecology. In A. Rajan and Y. Ibrahim (eds.) *Proc. 4th Internat. Conf. on Plant Protection in the Tropics*. Malaysian Plant Protection Society, Kuala Lumpur, pp. 25–46.

Rappoldt, C. and van Kraalingen, D.W.G. 1996. The fortran simulation translator: FST version 2.0. *Introduction and Reference Manual. Quantitative Approaches in Systems Analysis* No. 5, C.T. de Wit Graduate School for Production Ecology, Wageningen University, The Netherlands, 178.

de Reffye, P. et al. 1997. A functional model of tree growth and architecture. *Silva Fennica* 31(3): 297–312.

Reji, G. and Chander, S. 2008. A degree-day simulation model for the population dynamics of the rice bug, *Leptocorisa acuta*. *J. App. Entomol*. 132(8): 646–653.

Reji, G., et al. 2008. Simulating rice stem borer, *Scirpophaga incertulas*, Damage for developing decision support tools. *Crop Protect*. 27: 1194–1199.

Satish, D., et al. 2007. Simulation of economic injury levels for leaf folder on rice. *J. Sci. Ind. Res*. 66: 905–911.

Siju Thomas, T. 2003. *Photosynthetic characteristics, dry matter production and nut yield in coconut*. Thesis submitted to Mangalore University, Karnataka State, India.

Singh, J.P., et al., 2005. Increasing the efficiency of agronomy experiments in potato using InfoCrop-POTATO model. *Potato Res*. 48: 131–152.

Srivastava, A., et al. 2010. Assessment on vulnerability of sorghum to climate change in India. *Agric. Ecosyst. Environ*. 138: 160–169.

Sujithra, M. and Chander, S. 2013. Simulation of rice brown planthopper, *Nilaparvata lugens* (Stal.) population and crop-pest interactions to assess climate change impact. *Clim. Chang*. (in Press) https://doi.org/10.1007/s10584-013-0878-1

Sutherst, R.W., et al. 1996. A generic approach to defining impacts of climate change on pests, weeds and diseases in Australasia. In W.J. Bouma, et al. (eds.) *Greenhouse, coping with climate change*. CSIRO, Melbourne, pp. 281–307.

Yadav, D.S. and Chander, S. 2010. Simulation of rice plant hopper damage for developing decision support tools. *Crop Protect*. 29: 267–276.

Chapter 20

The CropGrow Model

Yan Zhu, Leilei Liu, Bing Liu, Liang Tang and Liujun Xiao, Nanjing Agricultural University, China

1 Introduction

The CropGrow series model, comprising WheatGrow and RiceGrow, was developed by the Engineering and Technology Center for Information Agriculture at Nanjing Agricultural University. This model follows a systematic workflow, incorporating physiological mechanism analysis, model algorithm development, dynamic productivity prediction, and quantitative effect assessment via simulation platforms. It integrates cutting-edge methodologies, such as system analysis, dynamic modeling, scenario simulation, and a decision support system.

The series model consists of multiple sub-models that comprehensively address wheat or rice growth and development processes, including

http://dx.doi.org/10.19103/AS.2025.0155.24

phasic development and phenology, organ development and population establishment, photosynthetic production and biomass accumulation, assimilate partitioning, yield and quality formation, nutrient dynamics, and water balance. This holistic approach provides a detailed, dynamic understanding of wheat and rice response to different factors, supporting precision agriculture and optimized crop management.

2 The evolution and development of the CropGrow Model

The CropGrow series model (http://www.cropgrow.net/) was initially developed in the 1990s, drawing from the frameworks of the CERES and SUCROS models. Extensive field experiments were conducted across China, involving various wheat and rice varieties, sowing dates, planting densities, water regimes, and nitrogen levels. These experiments aimed to uncover how crop physiological processes respond to environmental factors and management practices. This research led to the creation of simulation algorithms that accurately represent the interactions and relationships among different variables in crop growth dynamics and yield formation.

Since 2010, CropGrow series model has been capable of simulating a wide range of processes, including phenology, morphological and organ development, photosynthesis, biomass production, partitioning, yield, grain protein and starch content, and soil water and nitrogen balance in wheat and rice (Fig. 1). However, evaluations of CropGrow under climate change scenarios revealed significant uncertainties (Feng et al., 2019). To address this,

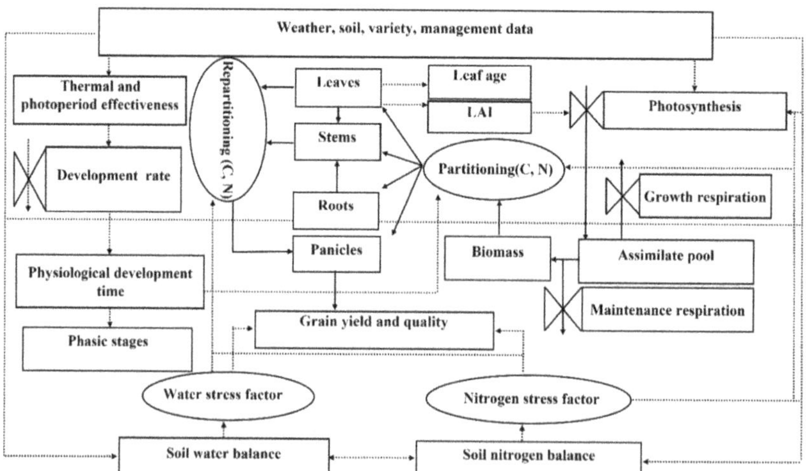

Figure 1 Structural flowchart of crop growth model CropGrow.

new algorithms were developed to account for the effects of high and low-temperature stress on crop growth, development, yield, and quality (Kang et al., Liu et al., 2020; Sun et al., 2023a; Xiao et al., 2022). These algorithms were refined through experiments conducted in controlled temperature environments known as phytotrons.

Ongoing research is focused on simulating the impacts of interactions among temperature, water, and light on crop growth (Liu et al., 2018, 2020; Lv et al., 2013). In the 2020s, recognizing that many crop models struggle to accurately simulate greenhouse gas emissions, efforts are underway to incorporate a carbon and nitrogen module related to greenhouse gas formation into CropGrow. Additionally, current work aims to bridge the gap between genomics and crop modeling by integrating single nucleotide polymorphisms (SNPs) information into the model's genetic parameters (Zhang et al., 2025).

3 Modeling of crop growth dynamics in CropGrow

The CropGrow series model consists of multiple sub-models that comprehensively address crop growth and development processes, including phasic development and phenology, organ development and population establishment, photosynthetic production and biomass accumulation, assimilate partitioning and yield and quality formation, nutrient dynamics, and water balance. The model is driven by daily weather data, including maximum and minimum temperatures, rainfall, and sunshine hours. In addition, model execution requires crop genetic parameters (Table 1, taking WheatGrow as an example), soil physicochemical properties (bulk density, water content, pH, organic matter, total nitrogen, available phosphorus, and available potassium,

Table 1 Genetic parameters in WheatGrow

Names of parameters	Explanations of parameters	Ranges of parameters	Unit
PVT	Physiological vernalization time	0–60	Day
PS	Photoperiod sensitivity	0–0.001	–
BFF	Basic filling factor	0.6–1.0	–
TS	Temperature sensitivity	1.0–3.0	–
IE	Intrinsic earliness	0.6–1.0	c
AMAX	Leaf maximum photosynthetic rate under optimum conditions	10.0–60.0	kg CH_2O ha^{-1} h^{-1}
Rp	Leaf area index Relative Speed	0.0030–0.0080	°Cd^{-1}
SLA	Specific leaf area	100–500	10^{-4} m^2 g^{-1}
HI	Harvest index	0.35–0.55	–

etc.), and management practice data (sowing/transplanting date and density, fertilizer type, amount and timing, irrigation amount and timing, straw return, etc.).

3.1 The phasic development and phenology sub-model

In the CropGrow series model, phenology is scaled using physiological development time (PDT), which is quantified by the effects of temperature and photoperiod. The PDT represents the development time accumulated under optimal environmental conditions. Genetic parameters such as temperature sensitivity (TS), photoperiod sensitivity, optimum temperature, intrinsic earliness, and basic filling factor are used to quantify the response of phenological development to temperature, photoperiod, and intrinsic characteristics. Similar to other models, such as ORYZA v3 and CERES, PDT serves as a state variable for different growth stages. For instance, in rice, PDT values are 0 at emergence, 8 at the start of the photoperiod-sensitive phase, 13 at panicle initiation, 18 at the end of the photoperiod-sensitive phase, 32 at heading, and 57 at maturity, as determined by literature observations.

3.2 The organ formation sub-model

The organ formation sub-model simulates the growth processes of roots, stems, leaves, and ears, influenced by thermal time and affected by nitrogen and water factors quantified by the soil nitrogen and water module. The simulation of the leaf area index (LAI) occurs in two phases: an exponential growth phase during early growth, where leaf growth is driven by temperature without water or nutrient limitations, and a non-exponential growth phase, where LAI increase is calculated as a product of the increment in leaf dry weight and specific leaf area (SLA) of green leaves. The SLA is calculated using a quadratic function based on thermal time, with a minimum SLA as a genetic parameter.

3.3 The photosynthetic production sub-model

Photosynthesis is modeled using Gaussian integration to calculate daily canopy photosynthesis, integrating the instantaneous photosynthetic rate over the day along with the LAI. Key genetic parameters for photosynthesis simulation include the extinction coefficients for visible light in the canopy, the initial light-use efficiency (EFF) of a single leaf, and the maximum rate of CO_2 assimilation of a single leaf. Growth respiration is accounted as a daily loss of total dry matter. Maintenance respiration is calculated from biomass and a maintenance coefficient, which declines with physiological age and is expressed as a function of PDT. Growth respiration is determined from daily total gross canopy assimilation.

3.4 The assimilate partitioning sub-model

A partition index is introduced to simulate the time-course dynamics of dry matter distribution among organs during development. The shoot and root partitioning indices are defined as a fraction of dry weight in plant biomass, and for green leaves, stems, and ears as the proportion of their dry weights in shoot mass. The potential partitioning index for shoot, root, and organs is quantified by PDT and influenced by water and nitrogen factors. The average moisture content of grains is about 14%, used as a default value to calculate final yields. The potential harvest index is regarded as a genetic parameter for varietal differences in panicle partitioning.

3.5 The grain quality formation sub-model

Grain quality, including starch and protein formation, depends on carbon and nitrogen assimilation processes pre-anthesis and translocation post-anthesis. Algorithms for grain starch and protein accumulation are based on the relationship between plant carbon-nitrogen supply and grain demand, governed by the source-sink balance. Plant nitrogen uptake varies during pre-anthesis and post-anthesis, driven by vegetative organs pre-anthesis and by grains post-anthesis. Plant carbon flow is measured based on biomass growth from photosynthesis and biomass partition among organs. The carbon and nitrogen flow ultimately impacts the amount available to grains, interacting with sink capacity to regulate grain protein and starch accumulation.

3.6 The water balance sub-model

The water balance sub-model is based on soil water balance principles, dividing the soil profile into layers based on the crop root system's depth. It calculates the water balance for each soil layer, simulating dynamic processes such as water interception by crops, irrigation, rainfall, surface drainage or runoff, field evaporation and transpiration, and root water uptake. The evapotranspiration is estimated by the Priestley–Taylor method, which takes account of radiation, temperature, LAI, extinction coefficient, and soil parameters. The drought stress factor is based on critical soil water content, indicating limited soil water availability. The waterlogging stress factor is quantified by integrating soil water content, waterlogging duration, and plant sensitivity at different growth stages.

3.7 The soil nutrients dynamics sub-model

The soil nitrogen nutrition simulation includes processes such as soil organic matter mineralization, nitrification, denitrification, ammonium adsorption and desorption, ammonia volatilization, nitrogen leaching, and nitrogen uptake and distribution within crops. Nitrogen absorption is determined by calculating critical nitrogen concentrations for the whole plant and individual organs, allowing for the estimation of nitrogen demand and transport across plant organs. Critical plant-nitrogen content, nitrogen uptake, and allocation are quantified to describe the relationships between soil-N supply, plant uptake, dry matter production, and N accumulation in grains.

4 Recent improvements: modeling wheat response to low-temperature stress

To reduce the uncertainty of CropGrow under climate change, the model has been improved with predictive algorithms for crop responses to low-temperature stress and high-temperature stress based on multiple years of experimental observation data.

4.1 Low-temperature stress indicators

To quantify the effects of low-temperature stress on wheat growth, development, and yield formation, two key indicators are utilized: the semi-lethal temperature (LT50) and the Low-Temperature Stress Damage Index (LDI). LT50, defined as the daily minimum temperature at which 50% of the leaves are killed, is a vital parameter for assessing a crop's cold resistance under low-temperature stress (Xiao et al., 2022). It provides a direct measure of the physical damage sustained by plants due to freezing injury. To further quantify the effects of low-temperature stress, a quantification algorithm for the low-temperature damage index (LDI) was developed, building on the LT50 simulation methodology.

The semi-lethal temperature (LT50) is a critical indicator for quantifying crop cold resistance under low-temperature stress, as it represents the temperature at which 50% mortality occurs in wheat leaves. In modeling studies, the logistic curve equation is often employed to characterize the relationship between the survival rate of winter wheat leaves and the intensity of low-temperature exposure (Xiao et al., 2022):

$$f_{surv} = \frac{1}{1 + \exp\left[-r \times \left(T - LT50\right)\right]} \tag{1}$$

where f_{surv} represents the survival rate of plant leaves under low-temperature stress, and r is a model constant that indicates the rate of variation in cold resistance among crop populations.

At the point where the temperature T equals LT50, the survival rate of plant leaves is most sensitive to temperature fluctuations per unit change. By fitting

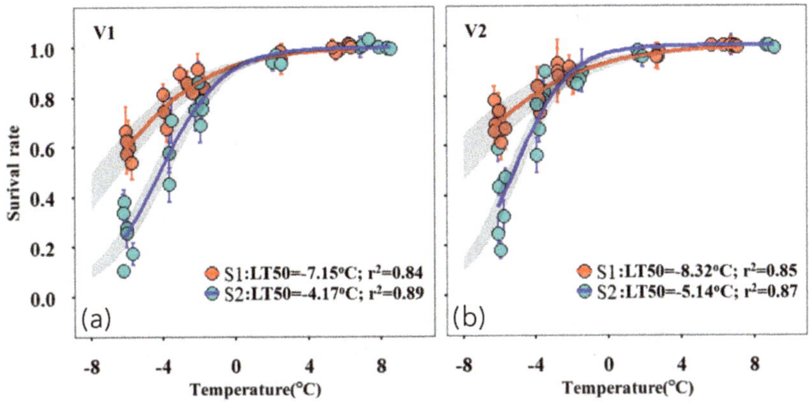

Figure 2 The relationship between minimum temperature and observed leaf survival rate (symbols) for Yangmai16 (V1) and Xumai30 (V2) in elongation (S1) and booting stage (S2) in phytotron experiments.

experimental data to the relationship between the daily minimum temperature and leaf survival rate, the LT50 values for 'Yangmai 16' were determined to be −7.15°C during the elongation stage and −4.17°C during the booting stages (Fig. 2). Similarly, for 'Xumai 30', the LT50 values were found to be −8.32°C and −5.14°C at the respective developmental stages.

The LT50 value increases linearly from the elongation to flowering stages as the crop develops, indicating a gradual decline in cold resistance with advancing developmental stages. To simulate LT50 at various PDTs, Equation (2) is employed:

$$LT50 = LT50_{\text{slope}} \times \left(PDT - PDT_{\text{elongation}}\right) + LT50_{\text{elongation}}$$

$$LT50_{\text{slope}} = \frac{LT50_{\text{booting}} - LT50_{\text{elongation}}}{PDT_{\text{booting}} - PDT_{\text{elongation}}} \quad (2)$$

where $PDT_{\text{elongation}} = 16.1$ and $PDT_{\text{booting}} = 21.4$ represent the PDT at the elongation and booting stages, respectively.

The parameter $LT50_{\text{slope}} = 0.58$ indicates the rate of change in LT50 with respect to PDT. Given the minimal variation in slopes between different cultivars, an average value of 0.58 is utilized (Xiao et al., 2021). $LT50_{\text{elongation}}$ and $LT50_{\text{booting}}$ refer to the semi-lethal temperatures at elongation and booting stages, and can be used as cultivar-specific and stage-specific parameters, respectively.

The extent of plant damage under low-temperature stress is determined by both the intensity and duration of the stress. When plants are subjected to prolonged low-temperature conditions, the daily rate of damage decreases over time, and this reduction rate is linked to the specific cultivar's adaptability or cold resistance. In winter wheat, the relationship between mortality due to low-temperature stress and its duration at the elongation stage follows a negative exponential trend across different stress intensities (Xiao et al., 2022). To effectively quantify frost damage (mortality) in winter wheat, the LDI was developed. This index utilizes a negative exponential formula to describe the relationship between LDI and duration under varying stress intensities, with LDI being represented as leaf mortality (Equation (3)):

$$LDI = M_{max} \times \left(1 - \frac{\exp(-kx \times Duration)}{M_{max}}\right) \tag{3}$$

Parameters M_{max} and *kx* under different low-temperature treatments can be determined through experimental fitting (Fig. 3). Subsequently, exponential formulas are employed to establish relationships between M_{max}/*kx* and the daily minimum temperature (T_{min}) or the cold resistance parameter LT50, as shown in Equations (4) and (5):

$$M_{max} = k1 \times \exp\left[-a \times (T_{min} - 2)\right] \quad (k1 = 0.05, a = 0.026 \times LT50 + 0.5) \tag{4}$$

$$kx = k2 \times \exp\left[-b \times (T_{min} - 2)\right] \quad (k2 = 0.017, b = 0.026 \times LT50 + 0.49) \tag{5}$$

where M_{max} represents the potential maximum frost damage rate of plants under daily minimum temperature (T_{min}); *kx* is the frost damage rate coefficient of plants; M_{max} and *kx* are related to the daily minimum temperature (T_{min}) and cold resistance parameter (LT50); Duration is the low-temperature stress duration (days); *k*1, *k*2, *a* and *b* are parameters in the M_{max} and *kx* algorithms, determined through experimental data fitting.

4.2 Modeling low-temperature stress effects on developmental stages

Observations of leaf age dynamics following low-temperature treatment indicate a significant deceleration in wheat's developmental progression, suggesting a diminished daily thermal effect (DTS) under stress conditions (Xiao et al., 2021). This slowdown is likely due to the suppressive impact of low-temperature stress experienced after the elongation stage, which affects wheat's TS. Once the stress subsides, the DTS does not immediately return to normal but gradually recovers as ambient temperatures rise and time progresses. To model this delayed recovery, a post-elongation relative low-temperature limitation factor (LF) is introduced, which serves to constrain DTS. This factor is defined in Equations (6) and (7), providing a framework to

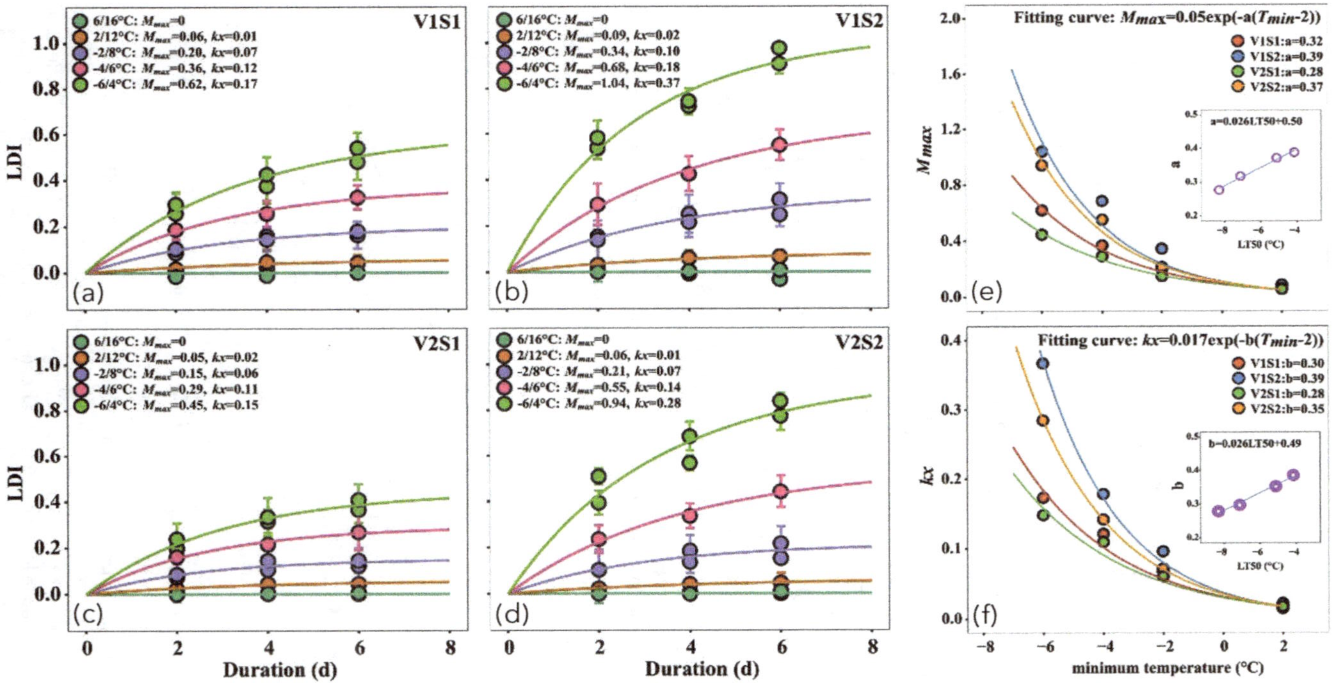

Figure 3 Leaf mortality (LDI) of winter wheat under low-temperature stress at elongation (S1) and booting (S2) stages for Yangmai16 (V1) and Xumai30 (V2) (a–d), and relationships between M_{max} (e), *kx* (f) and minimum temperature of low-temperature stress.

better understand and predict the impact of low-temperature stress on wheat development.

$$DTS = RPE \times \min(VP, LF) \tag{6}$$

$$LF=\min\left(\frac{\sum DTE_f}{T_{\text{mutiplier}} \times \sum DLE_f}, 1\right) \tag{7}$$

where LF represents the low-temperature stress LF on DTS, with values ranging from 0 to 1.

Due to the absence of experimental validation regarding the effects of low-temperature stress on DTS during the vernalization period, LF is set to 1 for the pre-elongation stage. The vernalization effect factor VP is assigned a value of 1 post-elongation. The variable *f* denotes the number of days elapsed since the onset of low-temperature stress. The DLE_f represents the daily relative low-temperature effect on day *f* after stress initiation, while $\sum DLE_f$ and $\sum DTE_f$ denote the cumulative low-temperature effect and cumulative normal DTS by day *f*, respectively. The $T_{multiplier}$ is the daily relative low-temperature effect coefficient, which quantifies cultivar-specific sensitivity to low-temperature stress and serves as an ecotype-specific parameter within the model (Fig. 4b).

The relationship between T_{min} and DLE_f is quantified using a Weibull cumulative probability function, which is expressed as follows:

$$DLE_f = 1 - \exp\left(-\ln(2) \times \left(\frac{T_{min}}{LT50}\right)^k\right) \tag{8}$$

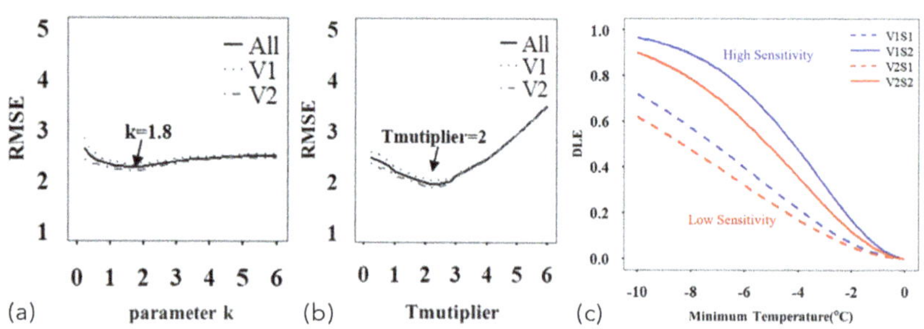

Figure 4 Relationships between root mean square error (*RMSE*) and parameter *k* (a), *Tmultiplier* (b) with the $DTS_{improved}$ routine for simulating the elongation-anthesis durations under low temperature stress (T_{min}<0) for Yangmai16 (V1) and Xumai30 (V2). Relationships between daily relative low temperature effectiveness (DLE) and minimum temperature (c) for V1 and V2 under low temperature treatments during elongation (S1) and booting (S2) stages.

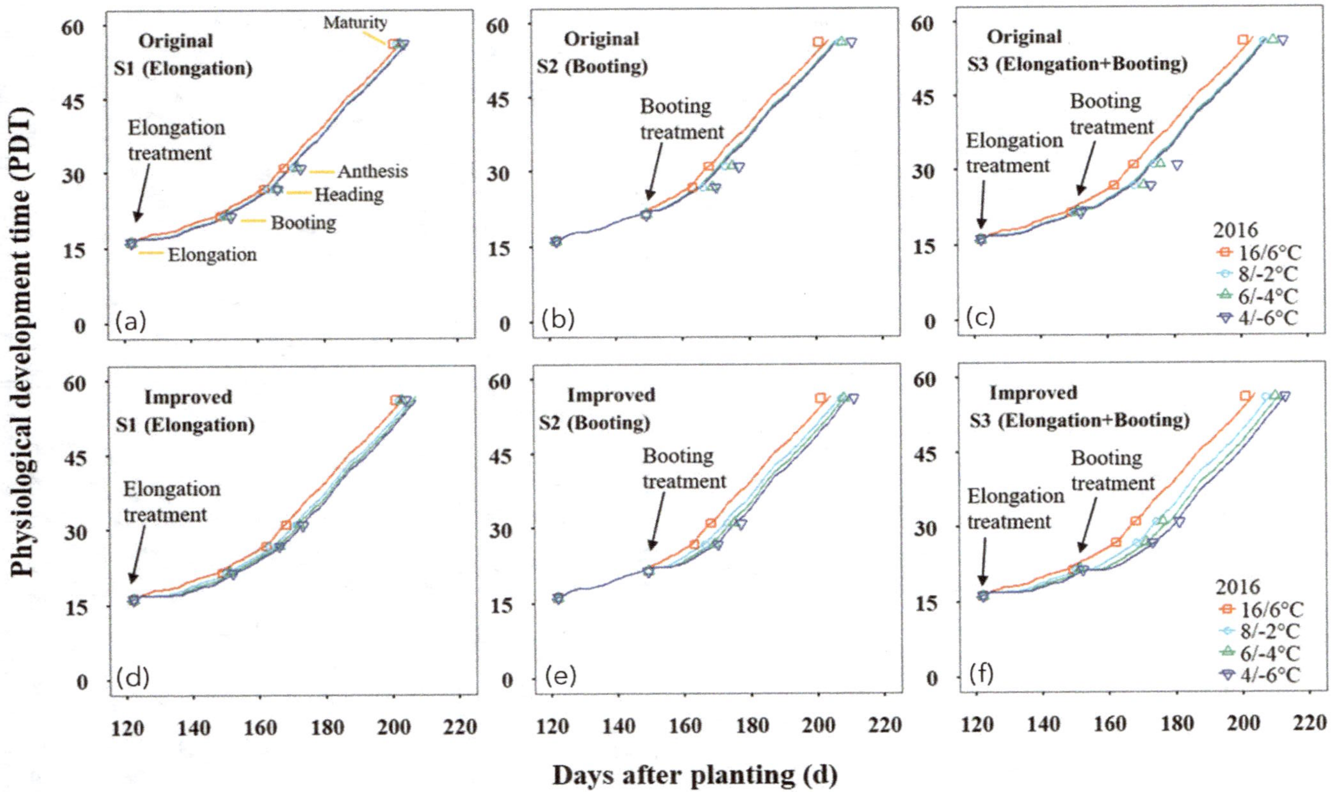

Figure 5 Comparison of observed and simulated physiological development time (PDT) with original (a–c) and improved models (d–f) for Yangmai 16 from environment-controlled phytotron experiments in model validation (a,d: Elongation treatment S1; b,e: Booting treatment S2; c,f: Elongation and booting treatment, S3).

where k determines the shape of this curve and quantifies the rate of change of DLE_f with temperature. k is a model constant and is cultivar-insensitive (Fig. 4a). When T_{min} = LT50, DLE_f = 0.5.

The original WheatGrow model tended to underestimate the PDT when subjected to low-temperature stress and showed a lack of sensitivity to different temperature levels and treatment timings (Fig. 5). However, the improved WheatGrow phenology model, incorporating the revised DTS algorithm, has shown improved accuracy in simulating phenological responses to low-temperature stress.

4.3 Modeling the low-temperature stress effects on dry matter production

Numerous studies have shown that low-temperature stress significantly inhibits the maximum photosynthetic rate (A_{max}) and initial EFF in leaves (Kang et al., 2024). Once the non-lethal low-temperature stress ends, the photosynthetic activity of crop leaves partially recovers. To simulate the effects of low-temperature stress on crop dry matter production, four key aspects are considered: the impact of low-temperature treatment on EFF and A_{max} during the stress period, and the post-stress recovery effects on A_{max} and the LAI.

4.3.1 Simulating the impact of low-temperature stress effects on EFF

Initial EFF is a critical parameter in models of photosynthetic light-response curve and is generally stable across variations in leaf age, nutritional status, or CO_2 concentration. However, extreme temperature stress can significantly reduce EFF due to the decreased activity of photosynthesis-related enzymes. To quantify the relationship between EFF and temperature in wheat leaves

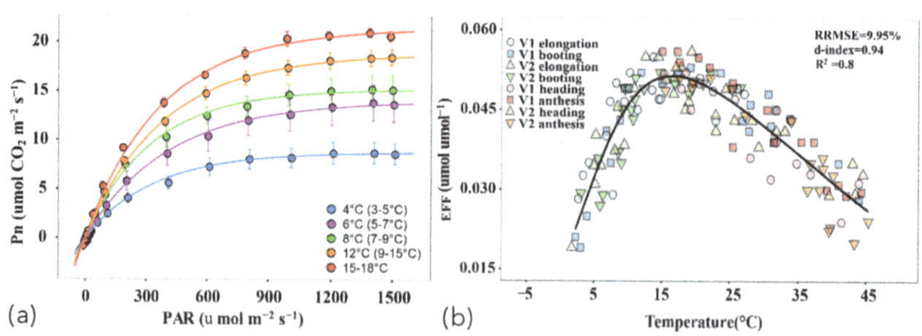

Figure 6 Measured light-response curves of flag leaves with different temperature levels (a). Relationships between measured leaf initial light-use efficiency (ε) and temperature in the growing season of 2013–2014 and 2014–2015 (b).

under different low-temperature stress conditions, EFF values were determined by measuring leaf light-response curves at various controlled temperatures (Fig. 6a). A dual-logistic equation was then employed to model the relationship between EFF and temperature across cultivars (Fig. 6b), resulting in the following findings.

$$EFF = \frac{0.078}{1 + \exp\left[-0.214 \times (T - 3.3)\right]} - \frac{0.078}{1 + \exp\left[-0.055 \times (T - 32.9)\right]} \tag{9}$$

4.3.2 Simulating the impact of low-temperature stress effects on A_{max}

Under low-temperature stress, leaves undergo dehydration and shrinkage, gradual yellowing, accelerated degradation of photosynthetic pigments, and disruption of normal cell membrane structures, all of which led to significant declines in photosynthetic rate (Kang et al., 2024). Empirical data show a negative correlation between leaf chlorophyll content and the severity of low-temperature stress (Xiao et al., 2022). To quantify the impact of low-temperature stress on photosynthetic efficiency, a low-temperature stress impact factor (FLT) was developed. This factor reflects the combined effects of frost-induced leaf structural damage, chlorophyll degradation, and reduced photosynthetic rates. The revised formula for calculating Amax in the improved WheatGrow model is as follows:

$$A_{max} = AMX \times FCO_2 \times FA \times FT \times \min(FN, WDF, F_{HT}, F_{LT}) \tag{10}$$

$$ALDI_{max} = LTS_p \times LDI \tag{11}$$

$$F_{LT} = \max(0, 1 - ALDI_{max}) \tag{12}$$

where AMX represents the maximum photosynthetic rate under optimal conditions. ALDI is the cumulative photosynthetic damage factor, while LDI indicates the leaf lethality rate across cultivars with varying cold resistance, influenced by different temperature conditions and exposure durations. LTSp represents the proportion of photosynthetic rate reduction per unit increase in the LDI. LTS_p is observed to depend solely on LDI, remaining unaffected by cultivar differences, and is considered an ecotype-specific parameter with a fixed value of 1.1 (Fig. 7). $ALDI_{max}$ represents the peak cumulative photosynthetic damage factor, quantifying the extent of photosynthetic capacity impairment under low-temperature stress. FLT ranges from 0 to 1, dynamically adjusts photosynthetic efficiency in response to the severity of stress.

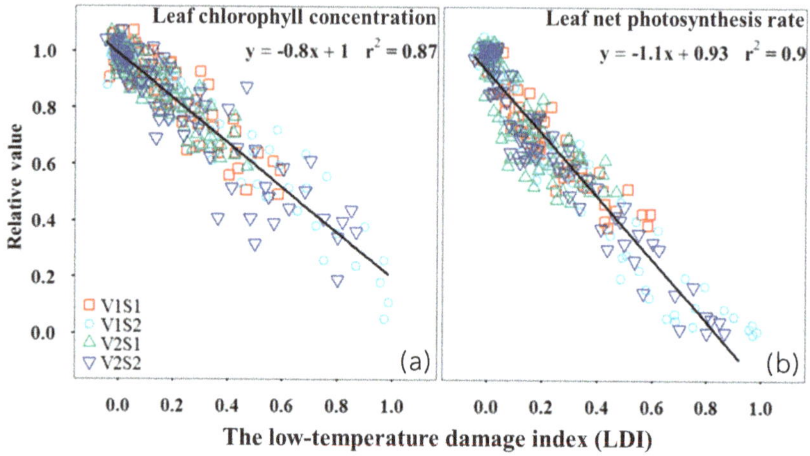

Figure 7 Relationships between relative leaf chlorophyll concentration (Chl a + Chl b) (a) or relative leaf net photosynthesis rate (b) and the low-temperature damage index (LDI) after cold stress treatments in growing season 2013–2015.

4.3.3 Simulating the recovery effects on A_{max} following low-temperature stress

After exposure to non-lethal low-temperature stress, wheat leaves experience a gradual recovery in physiological traits such as stomatal conductance and photosynthetic capacity, which can take from several days to weeks to return to normal levels (Xiao et al., 2022). Notably, shorter durations of stress are associated with quicker recovery times. To quantify the gradual restoration of photosynthetic capacity from accumulated damage due to low-temperature stress, a photosynthetic recovery factor ($FLT_{recover}$) algorithm was developed, as detailed in Equation (13). By analyzing photosynthetic data collected during post-stress recovery periods across different cultivars and treatment timings, the relationship between the relative recovery ratio of photosynthetic capacity and the growing degree days (GDD) with temperatures above 0°C during recovery was modeled, along with determining its parameter values (Fig. 8), as shown in Equations (13)–(16).

$$FLT_{recover} = \frac{ALDI_{max}}{1 + 0.05 \times \exp\left(\dfrac{r \times \left(GDD_{T>0} - GDD_{95}\right)}{ALDI_{max}}\right)} \tag{13}$$

$$GDD_{95} = 102.2 \times ALDI_{max} + 40 \tag{14}$$

$$r = 0.06 \times ALDI_{max} + 0.055 \tag{15}$$

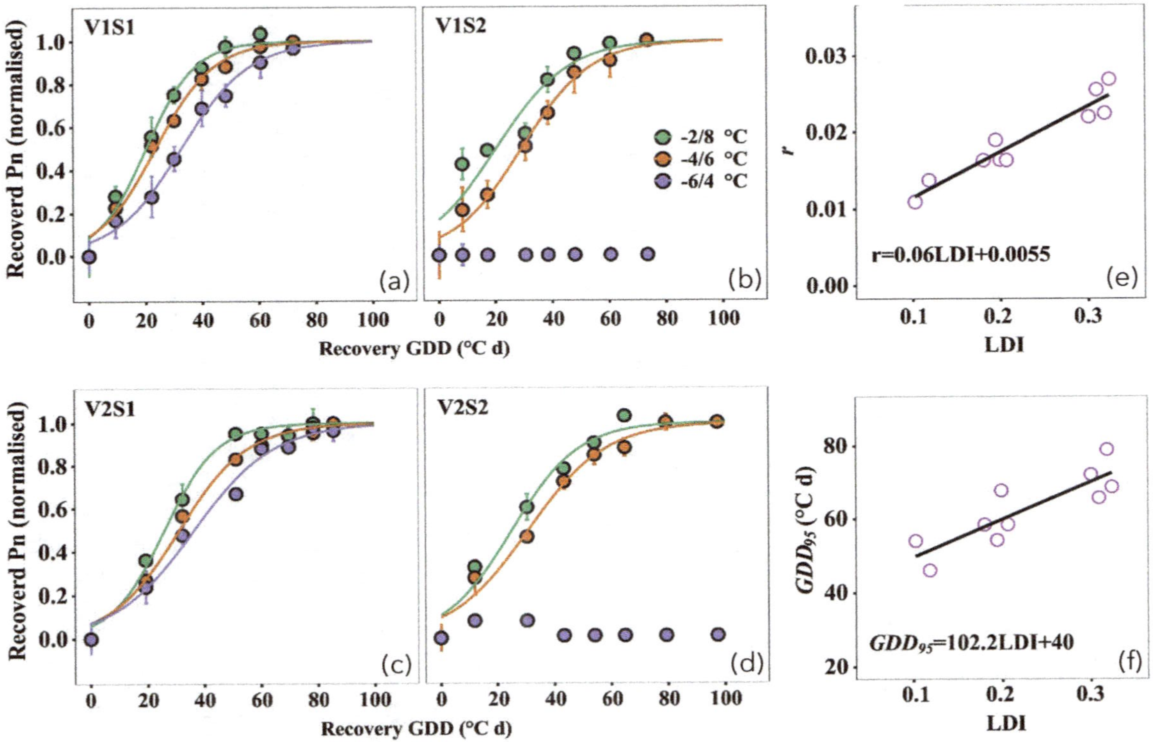

Figure 8 Relationship between normalized photosynthesis rate and GDD (a–d) during recovery period after low-temperature stress treatments, and relationship between parameters in Eq. (13) (GDD95 or r) and LDI (e-f) in growing season of 2013–2015. V1: Yangmai16, V2: Xumai30, S1: elongation, S2: booting.

$$F_{LT} = \min\left[\max\left(0, 1 - ALDI_{max} + FLT_{recover}\right), 1\right] \quad (16)$$

where $FLT_{recover}$ represents the proportional recovery of the accumulated photosynthetic damage factor after low-temperature stress treatment; $GDD_{T>0}$ denotes the GDD with temperatures above 0°C accumulated post-stress; *r* is the rate of recovery; GDD_{95} is the GDD required for photosynthetic rate recovery to 95% of its pre-stress level. Both *r* and GDD_{95} are model parameters dependent on the LDI. FLT is the low-temperature stress impact factor incorporating the recovery effect.

4.3.4 Simulating the impact of low-temperature stress on leaf area index

Low-temperature stress accelerates the degradation of chlorophyll in wheat leaves, resulting in frost damage, yellowing, and rapid dehydration-induced mortality. This degradation leads to a reduction in leaf area, which in turn decreases the effective photosynthetic surface area and reduces canopy-level photosynthetic productivity. By integrating the LDI, the model effectively quantifies the impact of low-temperature stress on green leaf biomass and its subsequent effect on LAI, as described in Equation (17):

$$WLVG_S = WLVG \times LDI \quad (17)$$

where WLVG represents leaf biomass without low-temperature stress, and WLVGS denotes leaf biomass after low-temperature stress treatment.

4.3.5 Modeling the effects of low-temperature stress on dry matter partitioning

Under low-temperature stress, different wheat organs display varying levels of cold tolerance. Leaves are the most sensitive, showing lower frost resistance compared to stems. If frost damage to stems persists without recovery, the translocation of dry matter to the stems remains limited. To address this issue, an algorithm was integrated into the WheatGrow model to quantify stem damage caused by low-temperature stress, calculated as follows:

$$WST_i = WST_i \times LTS_{stem} \times LDI \quad (18)$$

$$LTS_{stem} = 0.11 \times LT50 + 1.45 \quad (19)$$

where WST_i represents the stem biomass on day *i*; LDI is the LDI; LTS_{stem} denotes the stem-specific sensitivity parameter to LDI. Based on experimental fitting, LTS_{stem} exhibits a linear relationship with the semi-lethal temperature (LT50).

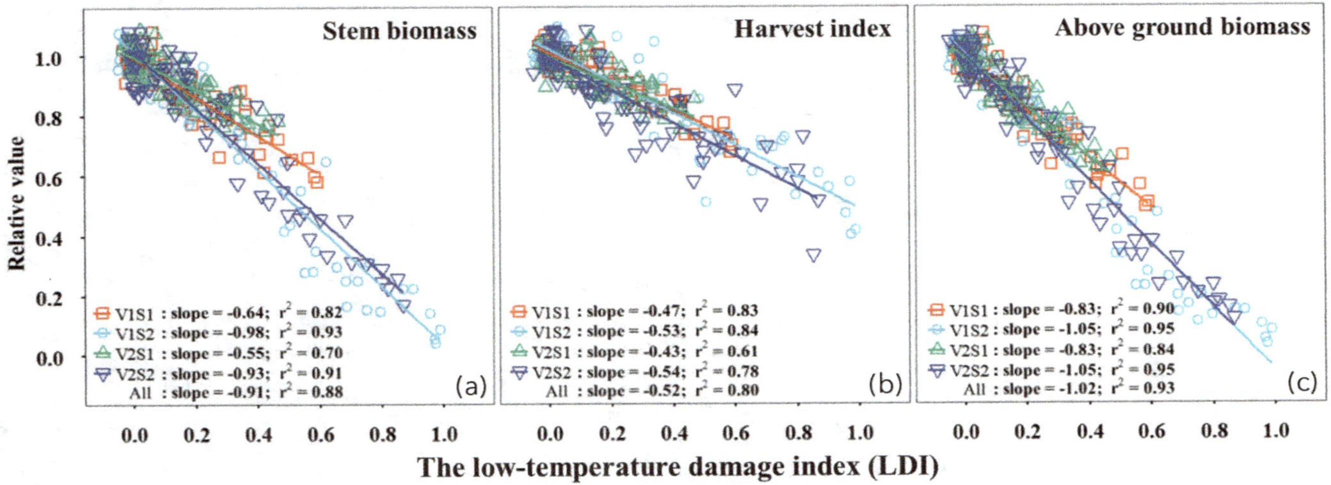

Figure 9 Relationships between relative stem biomass (a), relative harvest index (HI) (b), or relative above ground biomass (c) and LDI after cold stress treatments in growing season 2013–2015. V1: Yangmai16, V2: Xumai 30, S1: elongation, S2: booting, all: all data together.

After the jointing stage, wheat enters the phase of young panicle differentiation. The jointing and booting stages are crucial for stamen-pistil differentiation and the meiotic phase development, respectively. Low-temperature stress during these stages can cause significant damage to the young panicle, resulting in reduced floret differentiation, increased floret abortion, a lower harvest index, and restricted dry matter translocation to sink organs, ultimately leading to decreased grain yield (Fig. 9). To quantify the effects of low-temperature stress on partitioning indices, an algorithm was developed to model its impact on the harvest index, utilizing the LDI as described in Equation (20).

$$HI = HI_{potential} \times LTS_{HI} \times LDI \tag{20}$$

$$LTS_{HI} = 0.026 \times LT50 + 0.66 \tag{21}$$

where LDI is the LDI; HI represents the harvest index; $HI_{potential}$ denotes the cultivar-specific potential harvest index (a genotype-dependent parameter); LTS_{HI} is the sensitivity parameter of HI to LDI. Based on experimental fitting, LTS_{HI} exhibits a positive linear correlation with LT50, as defined in Equation (21).

By integrating the newly developed low-temperature stress algorithms into the WheatGrow model, the predictive capabilities of the original and enhanced models regarding dry matter production, partitioning, and yield formation under low-temperature stress conditions were compared. The improved model was validated using low-temperature experimental data in Rugao (120.33°E, 32.23°N) during 2014–2015 in Jiangsu Province of China. In WheatGrow, management practices were set to match actual operations: sowing 3 November at a depth of 2 cm; basal applications of 183.3 kg ha^{-1}

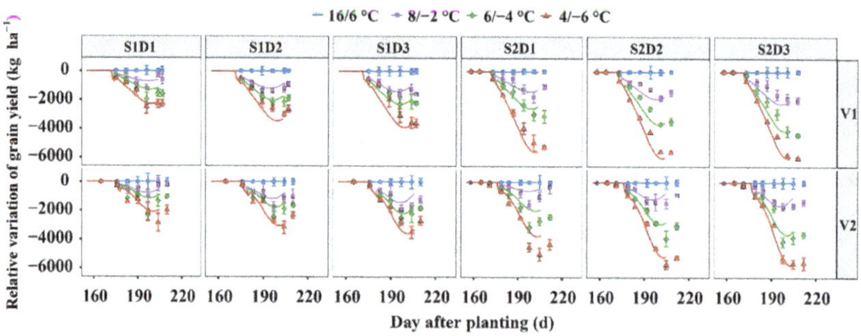

Figure 10 Comparison of simulated (line) and observed (point) grain yield dynamics of winter wheat with original WheatGrow models under low-temperature stress at elongation and booting in the growing season 2014–2015. V1: Yangmai16, V2: Xumai30. S1D1, S1D2, and S1D3 represent low-temperature treatments of 2, 4, and 6 days during the elongation stage, respectively, while S2D1, S2D2, and S2D3 represent the same treatments during the booting stage.

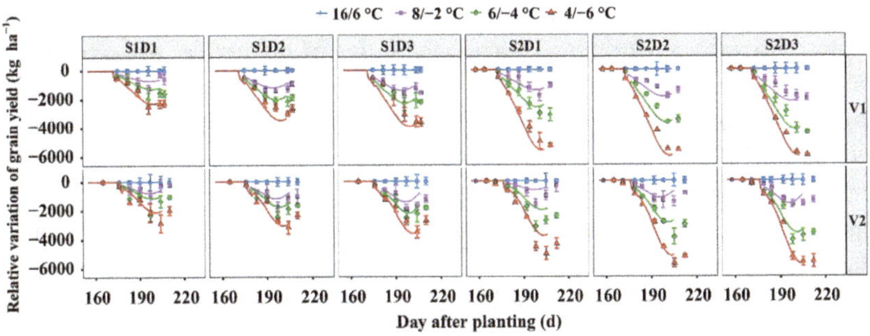

Figure 11 Comparison of simulated (lines) and observed (symbols) relative variation of grain yield with the improved WheatGrow model under low-temperature stress at elongation and booting stages in growing season 2014–2015. V1: Yangmai16, V2: Xumai30. S1D1, S1D2, and S1D3 represent low-temperature treatments of 2, 4, and 6 days during the elongation stage, respectively, while S2D1, S2D2, and S2D3 represent the same treatments during the booting stage.

N, 101.9 kg ha^{-1} P_2O_5, and 183.3 kg ha^{-1} K_2O; irrigation with a 30 mm water layer; and an additional application of the same amount of nitrogen before the elongation stage. The results showed that the improved model significantly reduced prediction errors for critical parameters under low-temperature stress (Figs. 10 and 11). These parameters included the single-leaf photosynthetic rate, LAI, total aboveground biomass, organ-specific biomass, and grain yield across both cultivars.

5 Recent improvements: modeling crop response to high-temperature stress

Global climate change has resulted in rising average temperatures and increased climatic variability, leading to more frequent extreme high-temperature events (Sun et al., 2021). Crop growth, development, and grain yield formation are particularly sensitive to high-temperature stress, especially during the reproductive growth phase, where such stress can significantly threaten yield improvement. Therefore, it is crucial to enhance the ability of crop growth models to simulate responses to high-temperature stress. This enhancement will provide digital tools essential for ensuring food security and developing adaptive strategies to cope with future climatic conditions.

5.1 Modeling the effects of high-temperature stress on phenological development

Experimental studies have shown that prolonged exposure to high temperatures and increased stress levels significantly reduce the grain days after maturity in rice and wheat (Shi et al., 2015a). When subjected to the same high-temperature conditions, the advancement of maturity is less pronounced during the grain-filling stage compared to the flowering stage. This highlights the varying sensitivity to high-temperature stress across different post-flowering phenological phases in these crops. Additionally, there are varietal differences in how post-flowering phenological development responds to high-temperature stress. In rice and wheat, post-flowering phenological development mainly involves senescence processes, which are highly temperature-dependent.

Based on previous research (Shi et al., 2015b), the thermal effects on post-flowering phenological development under high-temperature stress can be divided into two components: one related to the acceleration of senescence and the other to the alteration of physiological processes. Taking rice as an example, the algorithm for phenological development under high-temperature stress is as follows:

$$TE = DTE + HTE \tag{22}$$

where DTE is calculated using the original algorithm in the RiceGrow model. To reflect the combined effects of high-temperature stress intensity and duration, the accumulated high-temperature degree days (AHDD) are adopted to quantify HTE, as detailed in Equations (23) and (24):

$$HTE_i = \frac{HTS \times AHDD_i}{GDD_{RGP}} \tag{23}$$

$$HDD_i = \sum_{j=1}^{i} HD_j \tag{24}$$

where GDD_{RGP} represents the GDD (°C·d) required for rice or wheat to complete the reproductive growth phase (RGP) under optimal conditions; $AHDD_i$ is the accumulated high-temperature degree days (°C·d), calculated as the sum of daily high-temperature degree days (HDD) post-heading; HTS is a genetic parameter for heat tolerance, ranging from 0 to 1, with higher HTS values indicating greater cultivar-specific heat tolerance; HDD (°C·d) is defined as the mean hourly temperature exceeding a critical high-temperature threshold over a 24-hour period.

5.2 Modeling the effects of high-temperature stress on dry matter production

In the CropGrow model, the photosynthetic rate at the single-leaf level is simulated using an exponential model that captures the response of leaf photosynthesis to absorbed photosynthetically active radiation intensity (Xiao et al., 2022; Kang et al., 2024). This process is governed by two critical parameters: initial light energy utilization efficiency (EFF) and maximum single-leaf photosynthetic rate (A_{max}). To effectively quantify the impacts of high-temperature stress on photosynthesis at the single-leaf scale, it is crucial to understand how high-temperature stress influences both EFF and A_{max}. The relationship between EFF and temperature can be calculated using Equation 9.

Extensive studies have demonstrated that high-temperature stress accelerates crop chlorophyll degradation, leading to post-stress reductions in photosynthetic rates (Shi et al., 2015a; Sun et al., 2023b). Notably, the prolonged duration of photosynthetic suppression after stress removal may exert a greater impact on dry matter assimilation than the acute decline during the stress period. In the original RiceGrow model, A_{max} is primarily influenced by CO_2 concentration, temperature, nitrogen status, and water availability (Liu et al., 2014). The temperature impact factor (FT) decreases significantly when daily mean temperatures exceed the optimum range. While the combined effects of FT and high-temperature-induced reductions in EFF can effectively quantify A_{max} declines during stress exposure, they fail to account for persistent post-stress photosynthetic inhibition.

To address this limitation, an improved temperature impact factor (F_{HT}) is introduced into the CropGrow model. F_{HT} quantifies the residual effects of high-temperature stress on net photosynthetic rate post-stress, primarily mediated by chlorophyll degradation, and its calculation is given in Equation (25). The calculation is defined as:

$$F_{HT} = \max\left(0, 1 - HTS \times \frac{HDD}{GDD_{RGP}}\right) \tag{25}$$

where HDD represents the accumulated high-temperature degree days (°C·d) exceeding a critical threshold since heading; GDD_{RGP} denotes the effective accumulated thermal time (°C·d) required for the wheat reproductive growth phase (RGP), with a fixed value of 520°C·d for winter wheat (Liu et al., 2017); HTS is the cultivar-specific sensitivity coefficient to high-temperature stress, reflecting varietal differences in leaf senescence responses to such stress.

5.3 Modeling the effects of high-temperature stress on leaf senescence

During the reproductive growth phase, leaf senescence in crops is primarily driven by shading and the remobilization of nutrients (e.g. nitrogen) to reproductive organs (Liu et al., 2017). Most studies demonstrate that high-temperature stress significantly accelerates leaf senescence in crops. To quantify the impact of high-temperature stress on LAI, the post-flowering decline in LAI is partitioned into two components, that is, Baseline LAI Decline and High-Temperature-Accelerated. In WheatGrow, this accelerated senescence is simulated by introducing a high-temperature stress acceleration factor (S_{HT}) into the model:

$$LAI_s = WLVG_s \times SLA \times S_{HT} \tag{26}$$

where LAI_s is the daily senesced LAI accelerated by high-temperature stress; $WLVG_s$ represents the daily senesced leaf biomass (g·m^{-2}) under high-temperature stress; and SLA denotes the SLA (m^2·g^{-1}). The high-temperature senescence acceleration factor (S_{HT}) is quantified based on accumulated high-temperature degree days (AHDD), as shown in Equation (27):

$$S_{HT} = 1 + HTS \times \frac{AHDD}{GDD_{RGP}} \tag{27}$$

By coupling the high-temperature stress simulation algorithms for EFF, A_{max}, and LAI with the original WheatGrow model, an enhanced WheatGrow model

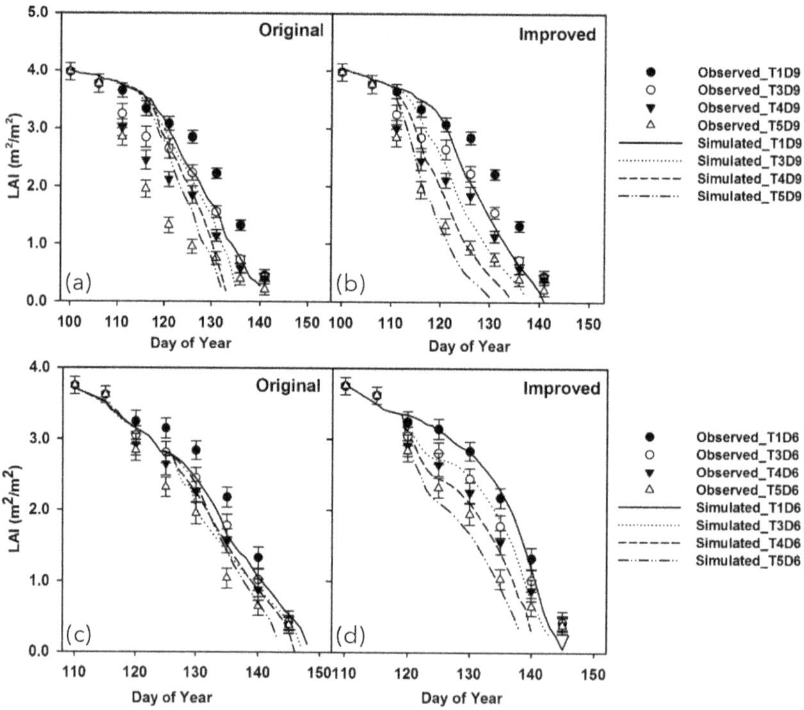

Figure 12 Comparison of simulated and observed leaf area index dynamics of Yangmai16 with the original and the improved WheatGrow model under heat stress at anthesis (a and b) and grain filling (c and d) in growing season 2013–2014.

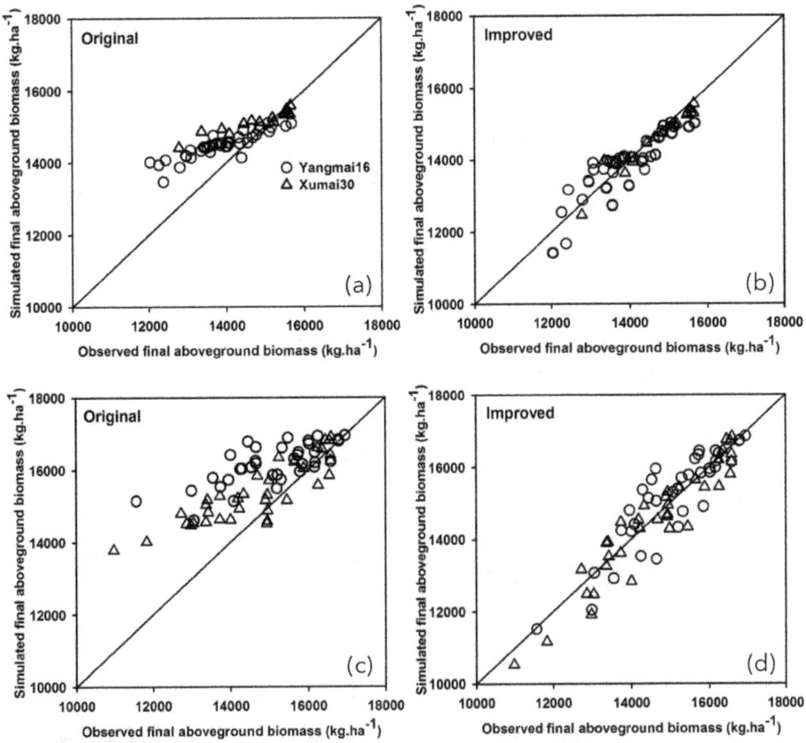

Figure 13 Comparison of simulated and observed final aboveground biomass with the original and the improved WheatGrow model in model calibration (a and b) and validation (c and d).

was developed. Validation using experimental data revealed that, compared to the original model, the enhanced model significantly reduced prediction errors for LAI and total aboveground dry matter under high-temperature stress conditions (Figs. 12 and 13).

However, experimental observations in rice demonstrate that leaf senescence is delayed under high-temperature stress (Fig. 14). To address this unique response, the post-flowering leaf growth sub-model was refined by incorporating improved dynamics of nitrogen accumulation and remobilization across rice organs under stress. In RiceGrow, this sub-model simulates leaf senescence patterns under varying environments, particularly post-flowering high-temperature stress, based on observed nitrogen accumulation and translocation patterns in leaves across treatments.

$$N_j = N_{max} e^{-K_n L},\ j = 1,2,3,\cdots,x \tag{28}$$

$$L = \frac{WLeaf}{x} \times \frac{j}{SLA} \tag{29}$$

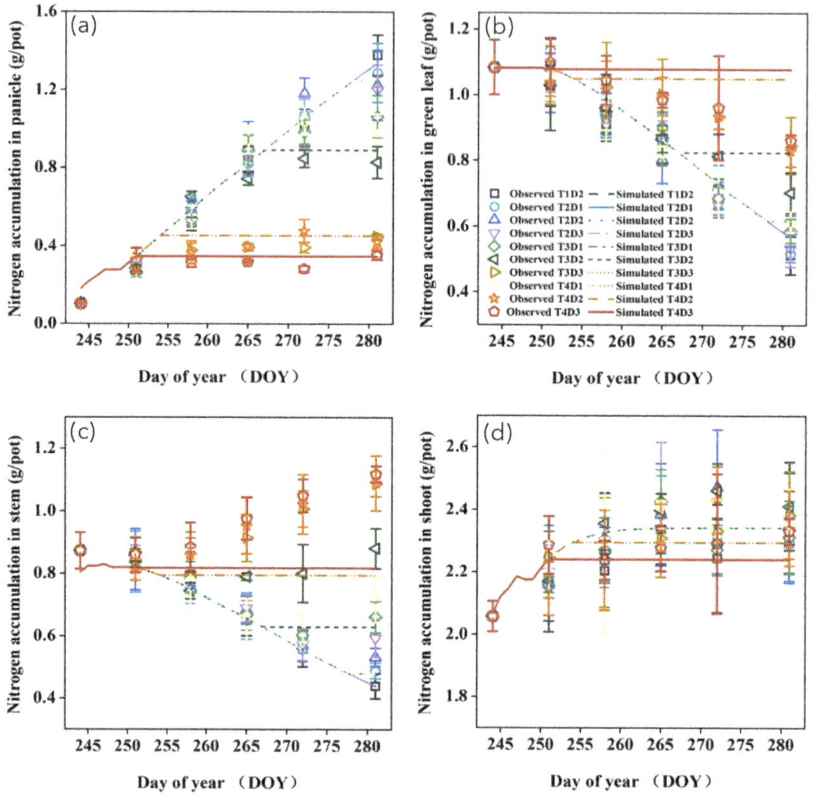

Figure 14 Simulated and observed N accumulation in panicle (a), green leaf (b), stem (c), and aboveground (d) of Nanjing41 in 2014 during the grain filling period. Simulations are conducted by the original RiceGrow. T1, T2, T3, and T4 indicate the temperature intensities (Tmax/Tmin) of 32/22°C, 36/26°C, 40/30°C, and 44/34°C, respectively. D1, D2, and D3 indicate the durations of 2 days (d), 4 d and 6 d under different temperature intensities, respectively.

$$N_{max} = NLeaf \times \frac{K_n}{\left(1 - e^{-K_n L}\right)} \tag{30}$$

where N_j represents the nitrogen content (g N·g^{-1} leaf) of the j*th* leaf segment; N_{max} denotes the nitrogen content of the first minimum leaf unit (segment 1); K_n is the nitrogen extinction coefficient; L is the cumulative leaf area from the first to the j*th* minimum leaf unit (m^2·m^{-2}); WLeaf/x represents the biomass of a single minimum leaf unit (g·m^{-2}); SLA is assumed constant post-flowering; and NLeaf is the total nitrogen content of the entire leaf.

During post-flowering leaf senescence, it is assumed that the daily nitrogen remobilization rate from each minimum leaf unit remains proportional to the overall nitrogen remobilization rate of the entire leaf. Specifically:

$$Tr_i = \frac{TrNLeaf_i}{AccNLeaf_{i-1}} \tag{31}$$

$$N_{i,j} = N_{i-1,j} \times (1 - Tr_i) \tag{32}$$

where $TrNLeaf_i$ is the leaf nitrogen translocation amount on day i; $AccNLeaf_{i-1}$ represents the accumulated leaf nitrogen content on day i−1; Tr_i denotes the daily nitrogen translocation rate on day i; and $N_{i,j}$ is the nitrogen content of the *jth* leaf segment on day i.

When leaf nitrogen content falls below a specific threshold (i.e. baseline nitrogen content, N_{base}), the photosynthetic rate of the leaf is reduced to zero. If the nitrogen content of the *jth* leaf segment (N_j) drops below N_{base}, that segment is defined as senesced. Based on these rules, post-flowering leaf biomass can be calculated using Equations (33)–(35).

$$WLeaf_i = WLeaf_{i-1} - \Delta WDLeaf_i, \quad PDT > 34 \tag{33}$$

$$\Delta WDLeaf_i = \sum_{j=1}^{x} \Delta WDLeaf_{i,j} \tag{34}$$

$$\Delta WDLeaf_{i,j} = \begin{cases} \frac{WLeaf_f}{x}, & N_{i,j} < N_b \\ 0, & N_{i,j} \geq N_b \end{cases} \tag{35}$$

where $WLeaf_i$ and $WLeaf_{i-1}$ represent the green leaf biomass on day i and day i−1, respectively; $\Delta WDLeaf_i$ denotes the senesced leaf biomass on day i; $\Delta WDLeaf_{i,j}$ is the senesced biomass of the *jth* leaf segment on day i; $WLeaf_f/x$ corresponds to the biomass of a single minimum leaf unit; $N_{i,j}$ indicates the nitrogen content of the j_{th} leaf segment on day i; and N_b is the baseline nitrogen content required for effective photosynthesis in leaves.

5.4 Modeling the effects of high-temperature stress on seed-setting rate

After heading and flowering in rice and wheat, high-temperature stress during the flowering stage primarily impacts grain yield formation by inducing spikelet sterility (Sun et al., 2018; Liu et al., 2014), significantly reducing the seed-setting rate (SR). This reduction alters assimilate partitioning and decreases final grain yield. Thus, accurately quantifying the effects of high-temperature stress on grain yield and dry matter allocation necessitates precise modeling of its impact on SR.

Extensive studies have shown that the rice SR can be described by a logistic function of high-temperature stress intensity; accordingly, in RiceGrow it is represented by Equation (36). In WheatGrow, a logistic function was likewise used to construct the algorithm for simulating the effect of high-temperature stress on floret fertility within spikelets, but its functional form differs slightly from Equation (36); the detailed algorithm can be found in Liu et al. (2020).

$$SR(HSI) = \frac{SR_{max}}{1 + e^{b(HSI - c)}} \tag{36}$$

where SR is the seed-setting rate (%); HSI denotes the high-temperature stress intensity index; SR_{max} represents the potential SR under optimal conditions; parameter *b* is the slope at the inflection point of the logistic curve; and parameter *c* corresponds to the value of HSI when the SR reaches the semi-lethal level (50% of SR_{max}).

When using four high-temperature quantifiers – maximum temperature (T_{max}), minimum temperature (T_{min}), mean temperature (T_{mean}), and high-temperature degree days (HDD) – as independent variables for logistic regression, all four metrics significantly fit the SR across different high-temperature treatment periods (Fig. 15). However, HDD exhibited the best overall fit. When quantifying high-temperature stress effects on SR using HDD, parameters *b* and *c* in Equation (36) were found to be highly correlated, with their product approximating 3 (Fig. 16). Consequently, Equation (37) can be simplified as follows when HDD is used.

$$SR(HDD) = \frac{SR_{max}}{1 + e^{b \times HDD - 3}} \tag{37}$$

where SR is the seed-setting rate (%); SR_{max} represents the potential SR; HDD denotes high-temperature degree days; and parameter b determines the slope of the logistic curve. Fitting experimental data from various high-temperature treatments revealed that parameter b increases after flowering, peaks at 2 days post-flowering, then declines rapidly before stabilizing after 10 days. Additionally, *b* exhibits significant variation depending on stress timing and spikelet positions.

To address this variability, empirical functions were developed for each spikelet position, with stress timing as the independent variable, to dynamically determine b. These functions are derived from field-measured data.

5.5 Modeling the effects of high-temperature stress on dry matter partitioning and yield formation

The source-sink relationship fundamentally governs post-flowering dry matter partitioning and translocation in crops. From the source perspective, high-temperature stress significantly reduces dry matter production via impaired

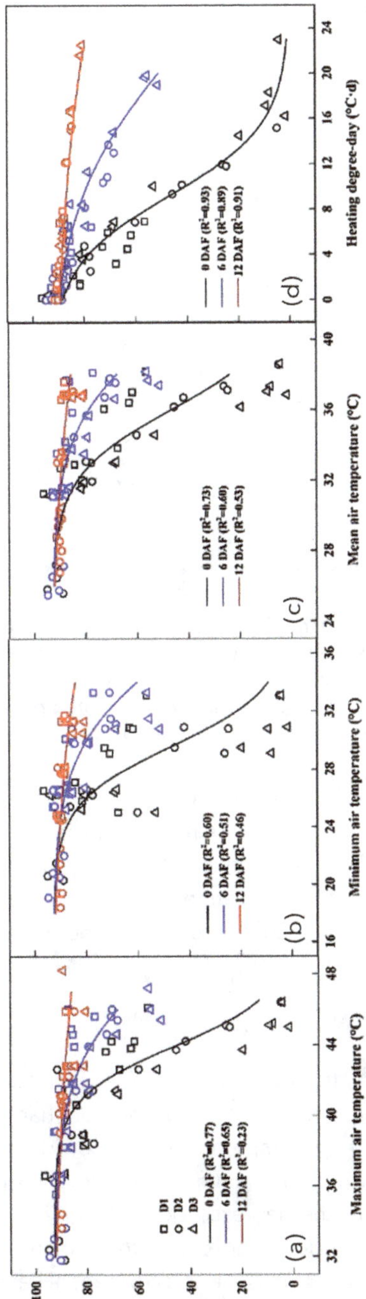

Figure 15 Observed and fitting seed-setting rates (%) in relation to maximum air temperature (a), minimum air temperature (b), mean air temperature (c) and heating degree-day (d). D1, D2, and D3 indicate heat treatments for 2, 4, and 6 days, respectively. 0 DAF, 6 DAF, and 12 DAF indicate the timing of treatments, which are 0, 6, and 12 days after flowering, respectively.

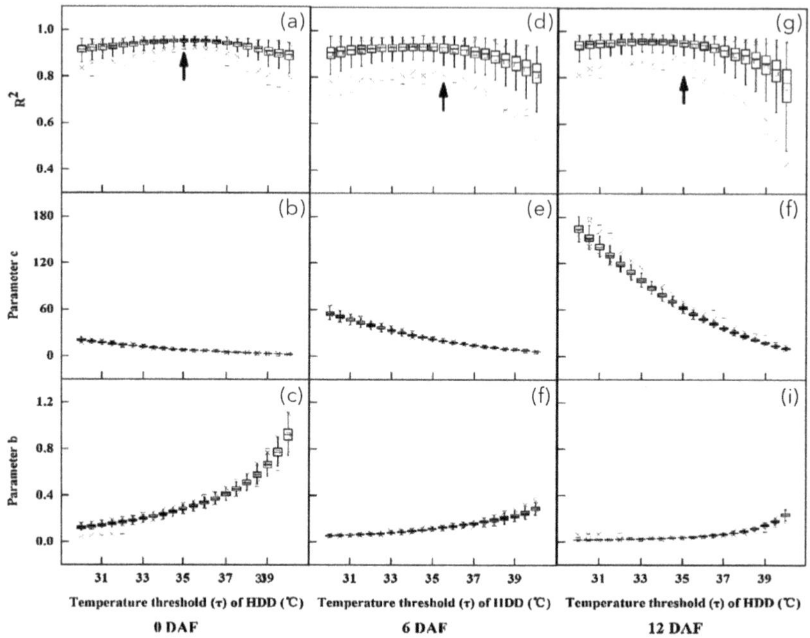

Figure 16 Parameter b (c, f, i), parameter c (b,e, h) and determination coefficient (a, d, g) of logistic curves to fit seed-setting rates (%) with heating degree-day (HDD) above diverse thresholds of heat stress occurred separately at 0 (a, b, c), 6 (d, e, f) and 12 days (g, h, i) after flowering. Variation of the parameters was estimated based on the 1000-group training datasets with the bootstrap resampling methods.

photosynthesis (Liu et al., 2020). From the sink perspective, heat stress diminishes two key determinants of sink strength, that is, grain number (GN) and maximum grain weight (GW_{max}) (Equation 38). Thus, high-temperature stress ultimately restricts dry matter translocation to sink organs while increasing the partitioning index to stems – a compensatory response to sustain structural integrity under stress.

$$Sink\_capacity_i = GN_i \times GW_{max,i} \tag{38}$$

In the Wheat Grow model, the partitioning index to spikes is determined by physiological development stage (PDS) and potential harvest index ($HI_{potential}$). To simulate the effects of high-temperature stress on dry matter partitioning, the improved WheatGrow model quantifies stress impacts on sink strength to derive daily adjustments to the harvest index (HI), which ultimately regulates the dry matter partitioning index to spikes. In the enhanced model, the daily maximum harvest index (HI_i) replaces the static potential harvest index ($HI_{potential}$) for calculating spike partitioning. HI_i is computed as:

$$HI_i = \frac{Sink\_capacity_i}{Sink\_capacity_{potential}} \times HI_{potential} \tag{39}$$

The potential sink capacity ($Sink\_capacity_{potential}$) is calculated as follows:

$$Sink\_capacity_{potential} = GN_{potential} \times GW_{potential} \tag{40}$$

where $GN_{potential}$ represents the potential number of fertile grains, and $GW_{potential}$ denotes the potential maximum grain weight.

The enhanced WheatGrow model was calibrated and validated using experimental data from both controlled-environment chambers and field warming trials (Liu et al., 2020). Results demonstrated that the improved model significantly enhanced the simulation accuracy of organ-specific dry matter partitioning, grain yield, and yield components under high-temperature stress (Fig. 17).

6 Applications of CropGrow: upscaling to predict regional crop productivity

Crop growth models are developed based on the assumption of consistent crop growth and environmental variables within specific regions, known as a single-point simulation system. However, there are various spatial differences in environmental and management variables (climate, soil, variety, technology, etc.) in the actual cropping region; hence, it is necessary to combine GIS with models to extend crop productivity prediction models from single-point simulation to regional applications. Therefore, there are two strategies for achieving regional productivity prediction through the coupling of CropGrow models and GIS, namely, upscaling strategies based on spatial interpolation and spatial zoning.

Spatial interpolation process involves generating rasterized datasets of input parameters essential for crop growth model computations, enabling simulations across each raster to derive regional simulation outcomes (Zhang et al., 2018). Spatial zoning approach divides the region into multiple assumed homogeneous simulation units based on spatial differences in environmental and management variables in the study area and runs the models for typical ecological points within each simulation unit to obtain simulated crop productivity for the entire region.

Due to the diverse spatial heterogeneity of specific regions, both spatial interpolation and spatial zoning approaches generally have uncertainties in regional productivity predictions. Therefore, by comparing the different methods on regional simulation results and selecting appropriate upscaling strategies, the accuracy of regional simulation can be improved. For example, by

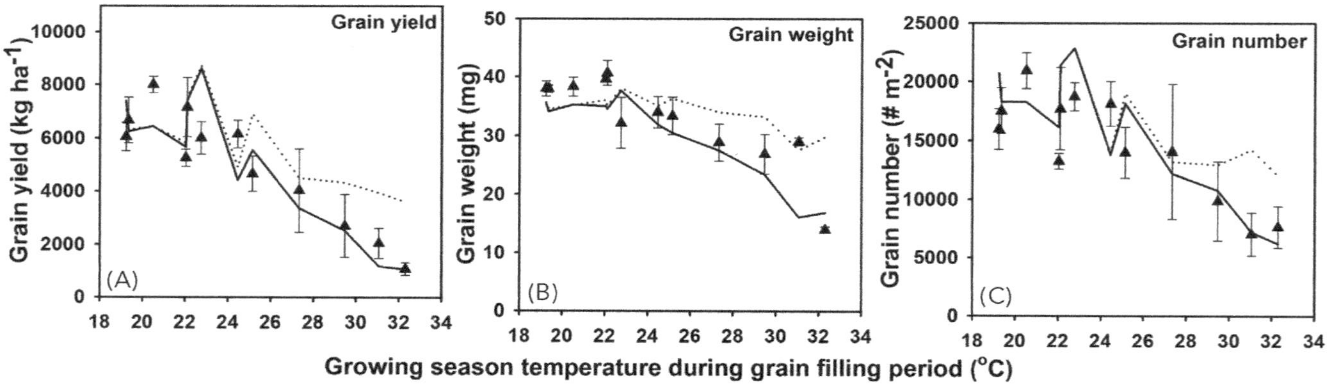

Figure 17 Simulated and measured grain yield (A), grain weight (B), and grain number (C) versus the mean temperature during the grain-filling period using data from the 'Hot Serial Cereal' experiments. The symbols indicate the observed values, dotted lines are values simulated using the original version of WheatGrow, and solid lines are values simulated using the improved version of the model. Data are means (±SD) for n = 3–6 replicates.

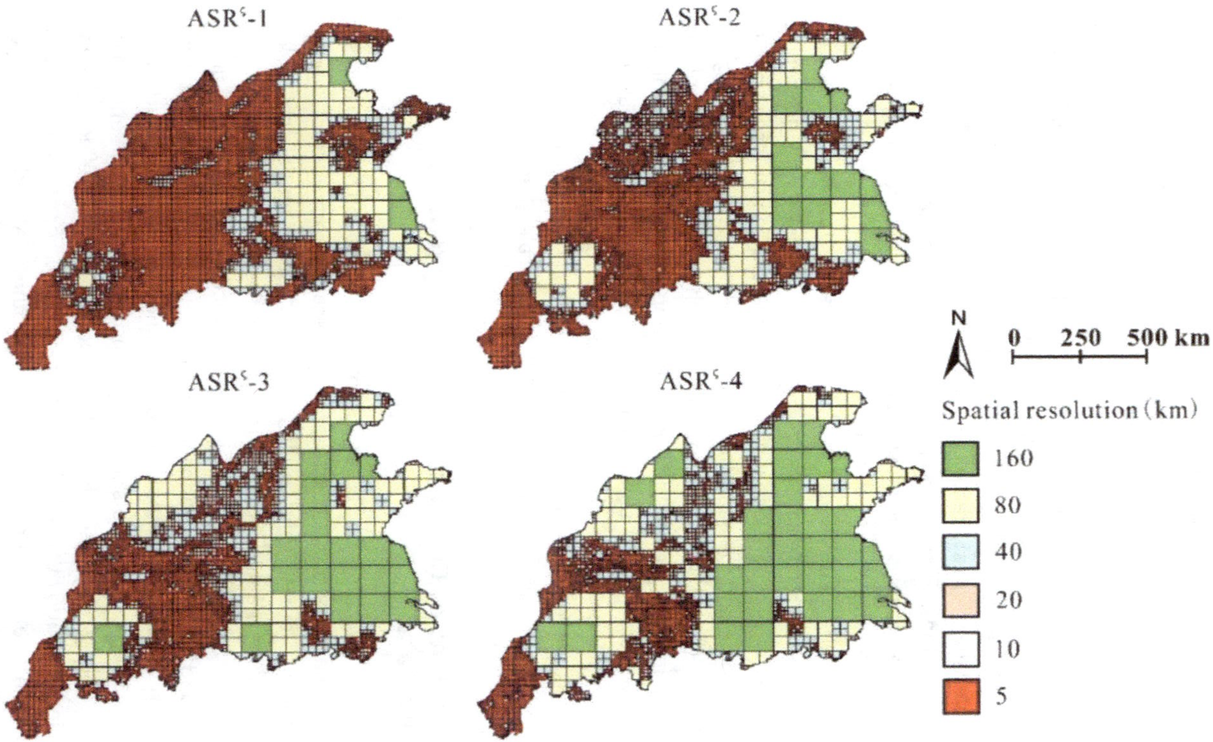

Figure 18 Appropriate spatial resolution of meteorological data for the regional potential productivity simulation determined based on the sill value of the DEM semi-variogram. (a) Appropriate spatial resolution ASRS-1 for the threshold of the scale effect index S of 100 kg ha^{-1} year^{-1}; (b) Appropriate spatial resolution ASRS-2 for the threshold of the scale effect index S of 200 kg ha^{-1} year^{-1}; (c) Appropriate spatial resolution ASRS-3 for the threshold of the scale effect index S of 300 kg ha^{-1} year^{-1}; and (d) Appropriate spatial resolution ASRS-4 for the threshold of the scale effect index S of 400 kg ha^{-1} year^{-1}.

investigating the influence of different spatial resolutions on simulation accuracy during the spatial interpolation process, the suitable interpolation resolution for winter wheat areas in China was determined, thereby simultaneously improving the accuracy and efficiency of regional simulation (Fig. 18).

In this study, we analyzed the influence of meteorological data with six spatial resolutions (from a 5 km resolution to a 160 km resolution) on the regional potential wheat productivity simulation with the WheatGrow model in the main winter wheat producing region of China over 10 years (2000–2009) (Zhang et al., 2018). We used the ANUSPLIN local thin plate smooth spline method to interpolate the daily meteorological data at 982 meteorological stations in the winter wheat area of China into the meteorological data with six spatial resolutions. Then, we simulated wheat potential wheat yield for the whole study region with the six spatial resolutions of meteorological data during 2000–2009, and a uniform cultivar (cv. Ningmai13) and management practices were used in the simulations. With simulated potential wheat yield, we constructed the scale effect index S, which was quantified by the differences between simulated potential wheat yield with different spatial resolutions of meteorological data and the simulated potential wheat yield with the highest spatial resolution (5 km) of meteorological data. When setting a threshold for the scale effect index S, we can determine the appropriate spatial resolution (ASRS) of the meteorological data required for the regional potential wheat productivity simulation in the whole study area. Fig. 18 showed the ASRS of meteorological data required for the regional potential productivity simulation of WheatGrow when using different thresholds of the scale effect index S.

7 Applications of CropGrow: integrating remote sensing data

Remote sensing can rapidly and accurately obtain information on crop and soil status using various sensors at different spatial scales, from field to regional levels. Combining remote sensing information with crop models can utilize the advantages of real-time monitoring by remote sensing and dynamic predictions by crop models, enhancing the prediction accuracy of regional crop growth and yield quality formation. When coupling the remote sensing data and crop models, commonly used coupling parameters include crop growth or physiological indicators such as LAI and leaf nitrogen content, as well as directly coupling remote sensing indices such as spectral reflectance and vegetation indices (Guo et al., 2019; Zhang et al., 2016). For example, by using a novel particle swarm optimization (PSO) algorithm, LAI and leaf nitrogen accumulation (LNA) obtained from spectral remote sensing and satellite images were used as coupling parameters with the RiceGrow model to optimize initial input parameters of crop models, thereby achieving reliable prediction of regional rice yield (Wang et al., 2014). The adjusted model input parameters were sowing date, sowing amount, and nitrogen application rate. Here we only coupled the RiceGrow with remote sensing data, not data assimilation. The optimization was accomplished by minimizing the differences between the remotely sensed and simulated LAI and LNA with the PSO algorithm.

In addition, we coupled remote sensing with the WheatGrow model and the PROSAIL radiation transfer model. By assimilating the temporal vegetation indices inverted by remote sensing data with the temporal vegetation indices simulated by the WheatGrow-PROSAIL model, the optimal LAI dynamic sequences were obtained, which were used to drive the WheatGrow model to accurately predict winter wheat growth indicators and grain yield (Fig. 19) (Guo et al., 2019). Meanwhile, we also determined the impacts of different vegetation index and growth stage for coupling remote sensing with the WheatGrow and found that the soil-adjusted vegetation index during the jointing to heading stage and the enhanced vegetation index after heading as coupling parameters can achieve the best predictive performance. When coupling remote sensing data for a single growth stage with the WheatGrow model, the heading-booting stage is the optimal window, and if there were

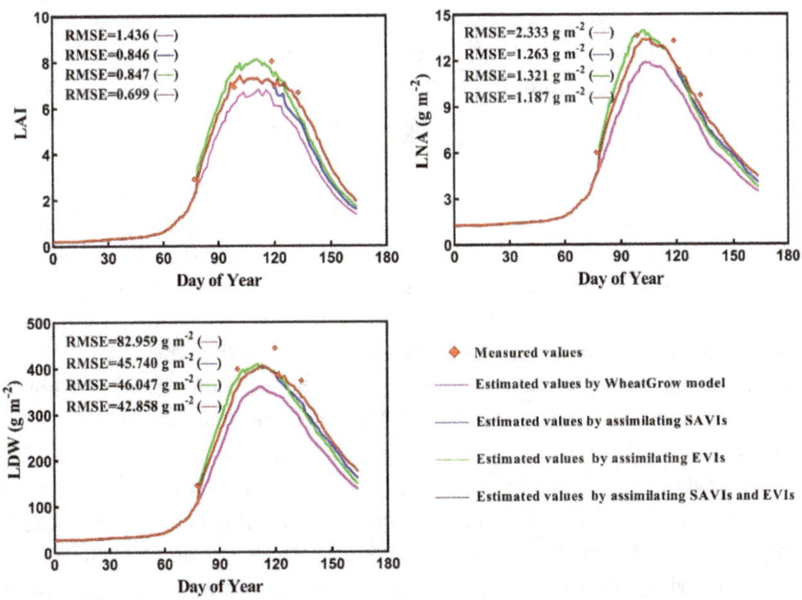

Figure 19 Comparison between estimated parameters and measured wheat LAI, leaf nitrogen accumulation (LNA) and leaf dry matter weight (LDW) by assimilating WheatGrow model with different vegetation indices as assimilating parameters in Yanhu farm in Jiangsu Province under local management. SAVI: soil-adjusted vegetation index. EVI: enhanced vegetation index.

multiple growth stage remote sensing data available, the highest prediction accuracy is achieved based on temporal remote sensing data from the jointing to grain filling stage (Zhang et al., 2016).

8 Applications of CropGrow: predicting crop yield potential and limitation factors

With limited land resources, increasing crop yield per unit area has become a feasible way to increase total crop production and ensure food security in China and across the globe. Therefore, it is urgently needed to determine the crop yield potential and yield gaps in different regions, further analyze the factors limiting yield, and propose improvement strategies to enhance crop productivity. For example, on rice crops, the potential yields at 294 sites in China's main rice-producing areas from 1981 to 2010 were simulated by a model ensemble of the RiceGrow, CERES-Rice, and ORYZA v3, and the yield gaps between potential yields and regional yield targets and actual yields were quantified (Fig. 20) (Wang et al., 2025). These results showed that the spatial distribution of yield gaps at different levels exhibited significant regional differences, but there was still considerable room for improvement of China's rice productivity. Among them, single rice has the largest yield gap, followed by late rice and early rice. In the single rice region, the southwestern regions exhibited higher values of both yield gaps, whereas the northeastern regions showed lower values. Combined with grid data on rice planting areas, the total yield gaps at different levels and their spatial distribution characteristics were also estimated.

Except for estimating crop yield potential and yield gaps, we have quantitatively evaluated the contributions of different factors such as climate change, soil improvement, variety renewal, and crop management optimization to crop productivity improvement, based on scenario simulation experiments with crop models. For example, on wheat crops, we used a multi-model ensemble (WheatGrow, DSSAT-CERES-Wheat, DSSAT-Nwheat, and APSIM-Wheat) to reveal the main limiting factors associated with yield enhancement during the last 40 years (from 1981 to 2020), including climate change, cultivar renewal, management optimization, and soil improvement) (Ye et al., 2023). In this study, we collected climate, cultivar, management and soil data from two periods (T1: from 1981 to 2000, and T2: from 2001 to 2020) across the main winter wheat producing region of China. Then we ran the wheat model ensemble with different combinations of climate, cultivar, management and soil data in the two periods. The yield differences between the two periods can be attributed to a single factor when keeping other factors constant while changing the corresponding factor from T1 to T2. The results show that the contribution of cultivar renewal to the wheat yield enhancement in China was

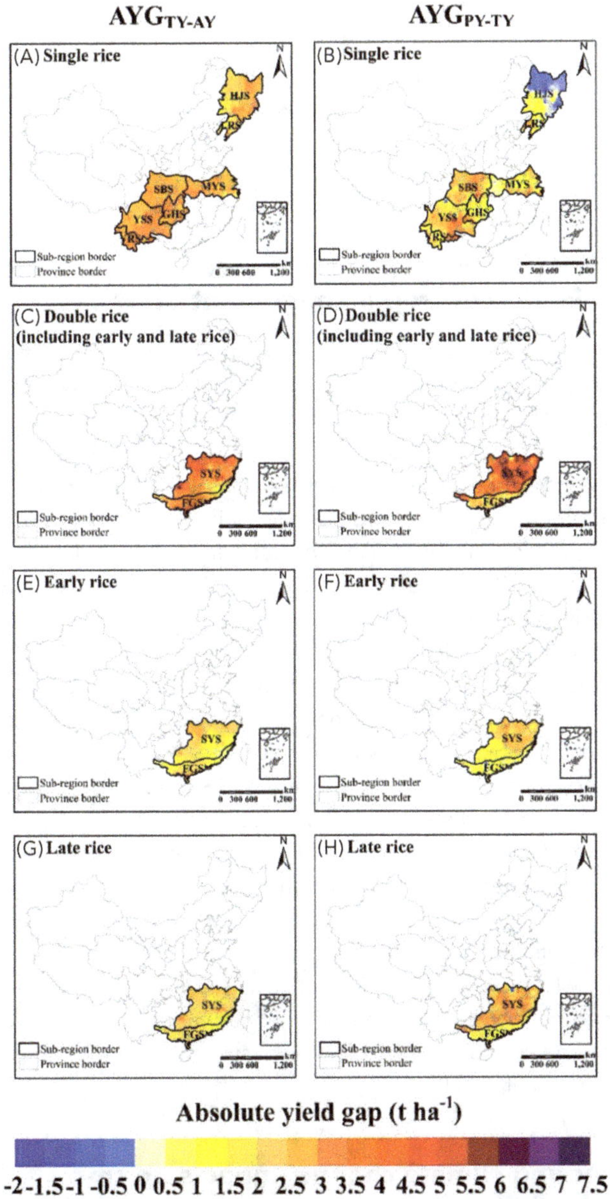

Figure 20 Spatial distribution of the absolute yield gap between actual and target yields (A, C, E, G) and between target and potential yields (B, D, F, H). Abbreviations for nine subregions are as follows: Heilongjiang-Jilin plain subregion (HJS), Liaohe River coastal plain subregion (LRS), middle-lower reaches of Yangtze River plain subregion (MYS), Sichuan basin subregion (SBS), Guizhou-Hunan mountain valley subregion (GHS), Yunnan-Sichuan plateau subregion (YSS), Yunnan river valley and basin subregion (YRS), south of Yangtze River plain and hills subregion (SYS), and Fujian-Guangdong-Guangxi plain and hills subregion (FGS). The spatial resolution is 0.025° × 0.025°.

the largest among the four factors in the whole study area (ranging from 5.1% to 46.2%) from T1 to T2, and the contribution of soil improvement to yield enhancement was the second largest factor (ranging from 0.4% to 33.0%). The contribution of management optimization to winter wheat yield enhancement ranged from 0.3% to 29.9%, while climate change had the smallest contribution to yield enhancement, ranging from −11.6% to 12.3%.

9 Applications of CropGrow: decision support and optimization of crop management practices

Through scenario-based simulation experiments involving different sowing dates, densities, nitrogen fertilization, irrigation, genotypes, and other management practices, the most suitable management schemes can be determined with crop models. For example, based on the RiceGrow model, the most suitable sowing date windows under current and future climate conditions were determined for different regions in China's main rice-producing areas (Wang et al., 2017). The GCM used for generating future climate scenario for 2030s, 2050s, and 2070s was the HadGEM2-ES model in the Coupled Model Intercomparison Project Phase 5 (CMIP5) under RCP8.5. The results showed that due to differences in temperature and radiation variations, the optimal sowing windows exhibited a gradually delayed pattern from south to north and from west to east across China's main rice-producing areas. Compared with the baseline period, optimal sowing windows in the future warming scenarios should be advanced by 1–18 days in single rice and by 1–10 days in double rice, with large variations across different ecological subregions (Fig. 21). For the selection of ideotypes, we could evaluate the performances of different crop genetic parameter combinations on crop phenology, photosynthesis, and yield, and obtain ideal genetic parameter combinations for optimizing crop breeding targets. For example, by changing the light extinction coefficients of different wheat varieties in WheatGrow, the light utilization efficiency of structure-compact and structure-losing wheat varieties was evaluated, laying the foundation for the design of ideal plant structures with high light utilization efficiency (Zhang et al., 2015).

10 Assessing climate change impacts on crop productivity

Using historical and future climate data, crop models can be used to assess climate change impact assessments at different spatial scales. For example, applying the WheatGrow model with a grid-based simulation approach, changes in wheat productivity and water use efficiency in China's main wheat producing

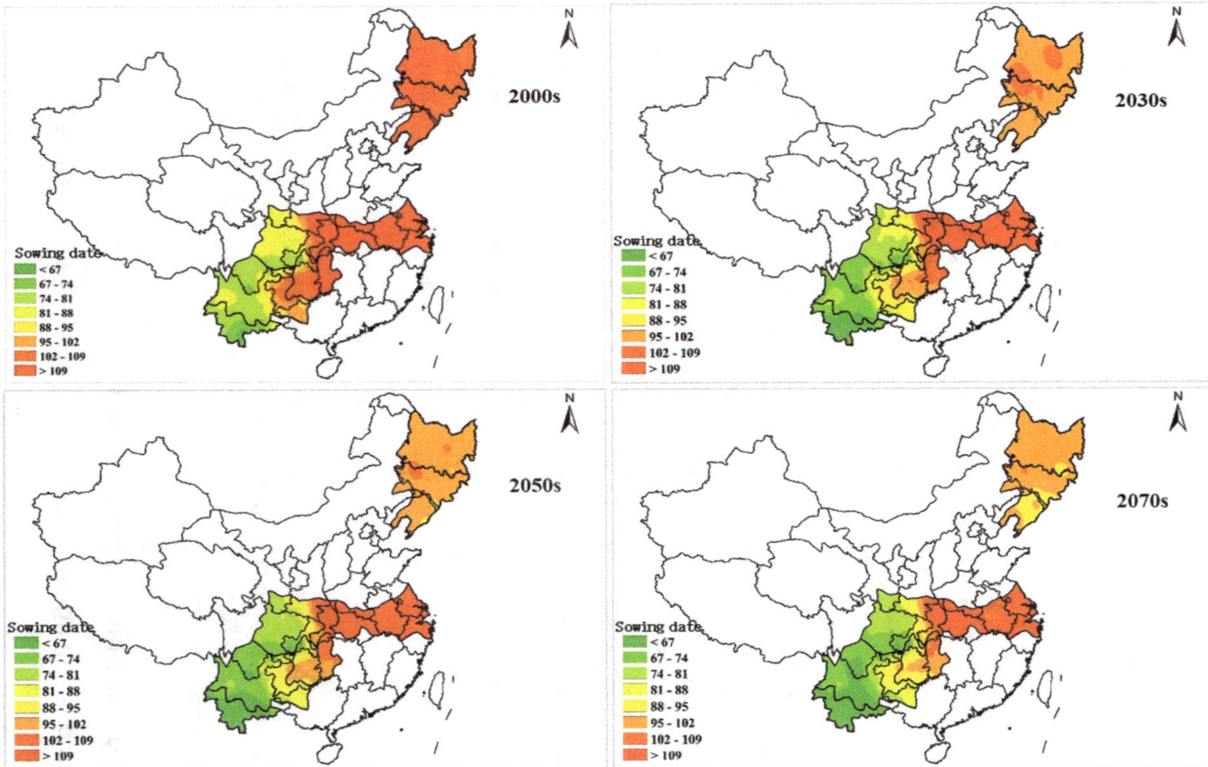

Figure 21 Spatial distribution of suitable sowing dates determined by the RiceGrow model under different climate scenarios in main rice production regions of China. The GCM used for generating future climate scenario for 2030s, 2050s, and 2070s was the HadGEM2-ES model in the Coupled Model Intercomparison Project Phase 5 (CMIP5) under RCP8.5. Source: Wang et al. (2017).

areas in different future decades (2030s, 2050s, 2070s) were evaluated under IPCC AR4 scenarios. Recently, combining the latest climate change projections from the IPCC, comprehensive assessments of rice productivity across China's main rice-producing areas under global warming scenarios of 1.5°C and 2.0°C were made using an ensemble of three rice models: RiceGrow, ORYZA2000, and DSSAT-Rice (Liu et al., 2020). We found that warming could alleviate the impacts of low-temperature stress on rice growth and development in high-latitude areas (Northeast China's single-season rice area), leading to increased production. However, climate warming was projected to result in a decreasing rice yield in other regions (Fig. 22). The fertilization effect of CO_2 could not fully offset the negative impact of climate warming on rice production.

Additionally, we used the WheatGrow model, which has incorporated the effects of extreme high-temperature stress, as described in Section 3.2, along with detailed information from 19 location-specific cultivars and local agronomic management practices at 129 research stations across the main wheat producing region of China, to quantify the regional impacts of temperature increase and heat stress separately on wheat in China (Liu et al., 2021). The cultivar parameters for the 19 cultivars used here were calibrated based on observed wheat phenology and yield data from multi-year field experiments conducted at agro-meteorological stations. The results showed that heat stress and its yield impacts were more severe in the cooler northern subregions than the warmer southern subregions with historical and future warming scenarios. Heat stress was estimated to reduce wheat yield in most of the northern subregions by 2.0–4.0% (up to 29% in extreme years) under the historical climate. Climate warming is projected to increase heat stress events in frequency and extent, especially in northern subregions (Fig. 23). Surprisingly, higher warming did not result in more yield-impacting heat stress compared to low warming, due to advanced phenology with mean warming and finally avoiding heat stress events during grain filling in summer. Most negative impacts of climate warming are attributed to increasing mean growing-season temperature, while changes in heat stress are projected to reduce wheat yields by an additional 1.0–1.5% in northern subregions. Adapting wheat production to climate change in China must consider the different regional and temperature impacts to be effective.

11 Conclusion and future trends

11.1 Improving agricultural big data acquisition for model applications

With the development of model parameter estimation approaches, model input data (e.g. soil data, management data, and crop data) have been increasingly

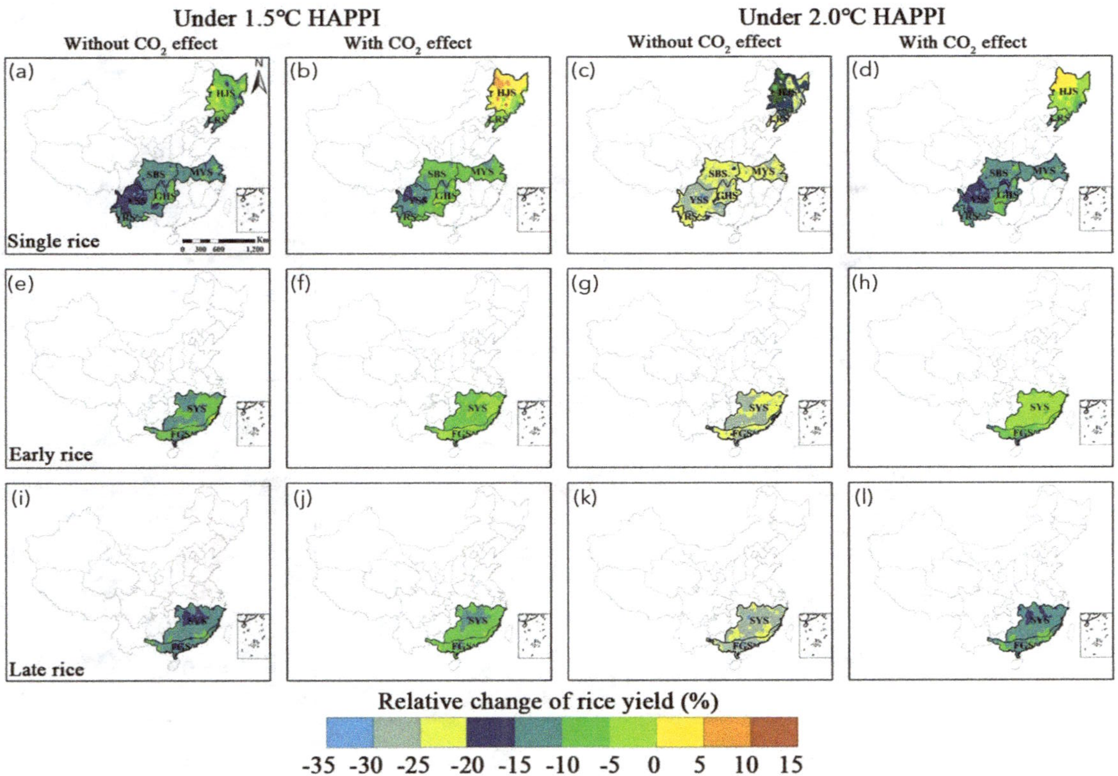

Figure 22 Spatial distribution of the ensemble means of simulated rice yield changes for single rice (a, b, c, d), early rice (e, f, g, h) and late rice (i, j, k, l) under 1.5°C and 2.0°C HAPPI. Values were measured in comparison to the 1981–2010 baseline, and measured with and without the effects of CO_2 fertilization. The rice yield changes under 1.5°C and 2.0°C HAPPI were calculated as the means of the simulated rice yield changes by three crop models (CERES-Rice, ORYZA v3, and RiceGrow), and from four global climate models (GCMs), including CanAM4, CAM4, MIROC5, and NorESM1. The set without CO_2 concentration effect was conducted with 390 ppm for 1.5°C and 2.0°C HAPPI, and the set with CO2 concentration effect was conducted with 423 ppm and 487 ppm for 1.5°C and 2.0°C HAPPI, respectively.

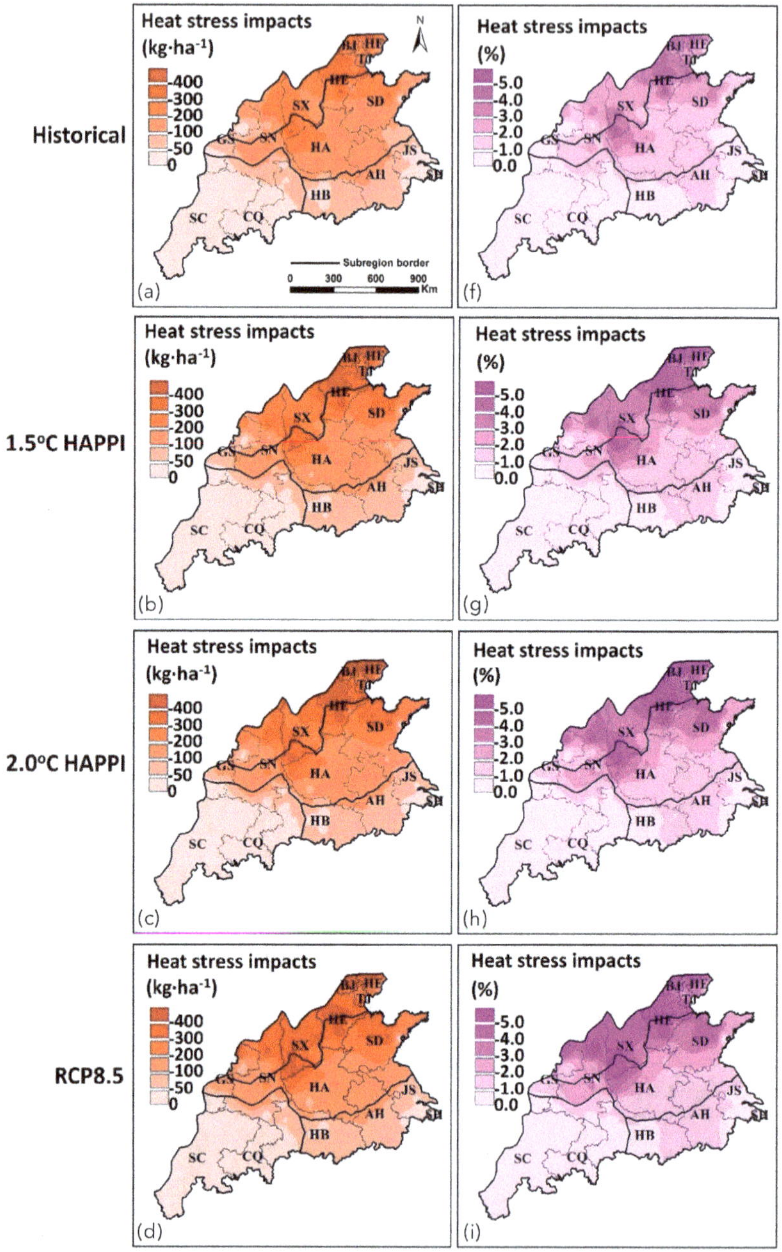

Figure 23 Estimated absolute (a–e, kg ha^{-1}) and relative (f–j, %) yield impacts of heat stress on wheat yields averaged over the 30-year period under historical (a, f), 1.5°C HAPPI (b, g), 2.0°C HAPPI (c, h), and RCP8.5 (d, i) scenarios across the main wheat producing region of China. NS: northern winter wheat subregion, HHS: Huang-Huai winter wheat subregion, MYS: middle and low Yangtze Valleys winter wheat subregion, SWS: Southwest winter wheat subregion.

a limitation during crop model application, especially at a regional scale. In the context of digital agriculture and smart agriculture, agricultural big data can play more important roles (Grassini et al., 2015). More available agricultural big data can significantly facilitate future crop model applications. Therefore, improving agricultural big data acquisition, including national agricultural census, soil surveys, multi-types of sensing data (satellite, unmanned aerial vehicles (USVs), Ground-based sensors, IoT (internet of things) sensors), and experimental data on climate conditions, soil characteristics, crop parameters, and management measures, will be future major targets. Applying technologies such as big data management and spatial data management to promote the integration of agricultural data resources will be needed. Meanwhile, more and more agricultural information-sharing cloud service platforms developed by government and public research institutions provide other opportunities for agricultural big data acquisition. These data acquisitions will provide reliable basic data support for improving crop model algorithms, simulating regional productivity, and applying multiscale applications.

11.2 Improving model algorithms under various conditions

The current CropGrow models predict crop growth and development under different production levels, such as photothermal potential, water-limited, and nutrient-limited conditions. However, other major factors affecting crop growth and development were not considered in current models. As for the nutrient-limited growth simulations, existing CropGrow models mostly considered the dynamics of soil nitrogen (N); the effects of other soil nutrients, e.g. potassium (K) and phosphorus (P), on crop growth and deployment are being investigated currently, especially for the N × P and N × K interactions. With global warming, the intensity and frequency of extreme climate events (such as high and low temperatures, droughts, and sparse sunlight) are increasing (Liu et al., 2014; Shi et al., 2015b). Future focuses of crop models include exploring the physiological mechanisms of extreme climatic conditions on crop growth, development, and yield, and developing algorithms for simulating these extreme climate events, especially the compound extreme events (e.g. compound heat and drought events), to improve simulation accuracy under future climatic conditions. Furthermore, with the improvement of agricultural mechanization in China, modern crop cultivation management practices (e.g. sowing methods) have shifted from traditional cultivation to simplified and mechanized cultivation. Therefore, it is necessary to accurately simulate the process of crop productivity formation under different management measures to better adapt to the development of modern crop production conditions. In addition, simulating the effects of pests, diseases, and weeds on crop yield and quality can also enable models to represent actual crop production conditions in the fields.

11.3 Incorporating genetic effects into the models

The rapid development of genetic sequencing technology has made it possible to rapidly obtain genetic information, enabling the quantification of the relationship between genetic parameters of crop models and genetic effects (Wang et al., 2019; Yin et al., 2018). This is expected to enhance the genetic mechanism of genetic parameters in traditional crop growth models. On the one hand, the quantification of these relationships will aim to develop new methods for estimating crop genetic parameters based on the physiological effects of functional genes. On the other hand, by quantifying the regulatory effects of important functional genes on major phenotypic traits, this would help to explore the interaction mechanisms between genetic effects and environmental effects and management practices. One of the widely used approaches for incorporating genetic effects into crop models is developing a gene-based model for predicting the genotype-specific parameters (GSPs) of crop models. Our recent study has developed SNP-based models for predicting phenology-related GSPs, and the rice models (ORYZA, CERES-Rice, and RiceGrow) driven by SNP-based GSPs showed good performance in simulating the rice flowering date, compared to the traditional model calibration approach (Zhang et al., 2025). By incorporating genetic effects into the models, integrated gene-based simulation models can facilitate efficient prediction of crop phenotypes and lay the digital technology foundation for rapid selection of breeding traits. For example, current research on molecular breeding in rice has made significant progress. Coupling gene effect models with crop three-dimensional visualization models holds promise for providing technical support for the molecular breeding design of rice ideotypes.

11.4 Expanding model functions for intelligent decision-making

The continuous improvement of crop models can further expand their decision support capabilities. In recent years, artificial intelligence (AI) technology has rapidly been applied in agriculture (Jin et al., 2019). By combining the dynamic prediction function of crop models with AI decision-making algorithms, an intelligent agricultural production management decision system can be constructed, thereby improving the intelligence of management decisions. For example, using crop models for simulating growth stages and yields under different conditions can provide training data for yield prediction based on satellite remote sensing data and AI algorithms, thereby enhancing the predictive and decision-making capabilities of regional crop productivity (Feng et al., 2019).

Methods for estimating model parameters will also shift from manual trial-and-error methods toward automatic estimation algorithms based on AI algorithms including genetic algorithms, PSO, and Monte Carlo methods, thereby achieving rapid and automatic estimation of crop growth model parameters for different application scenarios. Additionally, by combining crop growth models with crop phenotype monitoring technologies, they can be applied to dynamically simulate and predict high-throughput crop phenotype indicators under different climate and management scenarios. Furthermore, through coupling crop models with crop canopy structure and virtualization models, the three-dimensional simulation platforms with remote sensing data could provide digital twins of crops. Users can directly perceive intuitive and immersive scenes of crop growth with digital twins of crops.

12 References

Feng, P. Y., et al. 2019. Incorporating machine learning with biophysical model can improve the evaluation of climate extremes impacts on wheat yield in south-eastern Australia. *Agricultural and Forest Meteorology*, 275: 100–113.

Grassini, P., et al. 2015. How good is good enough? Data requirements for reliable crop yield simulations and yield-gap analysis. *Field Crops Research*, 177: 49–63.

Guo, C. L., et al. 2019. Predicting wheat productivity: integrating time series of vegetation indices into crop modeling via sequential assimilation. *Agricultural and Forest Meteorology*, 272: 69–80.

Jin, Z., et al. 2019. Smallholder maize area and yield mapping at national scales with Google Earth Engine. *Remote Sensing of Environment*, 228: 115–128.

Kang, M., et al. 2024. Simulating the effects of low-temperature stress during flowering stage on leaf-level photosynthesis with current rice models. *Agricultural and Forest Meteorology*, 354: 110087.

Liu, B., et al., 2014. Post-heading heat stress and yield impact in winter wheat of China. *Global Change Biology*, 20(2): 372–381.

Liu, B., et al., 2016a. Testing the responses of four wheat crop models to heat stress at anthesis and grain filling. *Global Change Biology*, 22(5): 1890–1903.

Liu, B., et al., 2016b. Modelling the effects of heat stress on post-heading durations in wheat: a comparison of temperature response routines. *Agricultural and Forest Meteorology*, 222: 45–58.

Liu, B., et al., 2017. Modelling the effects of post-heading heat stress on biomass growth of winter wheat. *Agricultural and Forest Meteorology*, 247: 476–490.

Liu, B., et al., 2020. Modelling the effects of post-heading heat stress on biomass partitioning, and grain number and weight of wheat. *Journal of Experimental Botany*, 71(19): 6015–6031.

Liu, B., et al. 2021. Separating the impacts of heat stress events from rising mean temperatures on winter wheat yield of China. *Environmental Research Letters*, 16(12), 124035.

Liu, L. L., et al. 2018. Uncertainty in wheat phenology simulation induced by cultivar parameterization under climate warming. *European Journal of Agronomy*, 94: 46–53.

Liu, Y. J., et al. 2020. Impacts of 1.5 and 2.0°C global warming on rice production across China. *Agricultural and Forest Meteorology*, 284: 107900.

Lv, Z. F., et al. 2023. Climate change impacts on regional winter wheat production in main wheat production regions of China. *Agricultural and Forest Meteorology*, 171/172: 234–248.

Shi, P., et al., 2015a. Modeling the effects of post-anthesis heat stress on rice phenology. *Field Crops Research*, 177: 26–36.

Shi, P. H., et al. 2015b. Post-heading heat stress in rice of south China during 1981–2010. *PLoS One*, 10(6): e0130642.

Sun, T., et al., 2018. Stage-dependent temperature sensitivity function predicts seed-setting rates under short-term extreme heat stress in rice. *Agricultural and Forest Meteorology*, 256–257: 196–206.

Sun, T., et al., 2021. Current rice models underestimate yield losses from short-term heat stresses. *Global Change Biology*, 27(2): 402–416.

Sun, T., et al., 2023a. Sink-source unbalance leads to abnormal partitioning of biomass and nitrogen in rice under extreme heat stress: an experimental and modeling study. *European Journal of Agronomy*, 142: 126678.

Sun, T., et al., 2023b. Improving the predictions of leaf photosynthesis during and after short-term heat stress with current rice models. *Plant, Cell & Environment*, 46(11): 3353–3370.

Wang, E. L., et al. 2019. Improving process-based crop models to better capture genotype×environment×management interactions. *Journal of Experimental Botany*, 70(9): 2389–2401.

Wang, L. H. 2017. *Determining Optimum Sowing Date of Rice Using Multiple Crop Models*. Nanjing: Nanjing Agricultural University.

Wang, L. H., et al. 2014. Integrating remotely sensed leaf area index and leaf nitrogen accumulation with RiceGrow model based on particle swarm optimization algorithm for rice grain yield assessment. *Journal of Applied Remote Sensing*, 8(1): 83674.

Wang, S. Y., et al. 2025. Rice yield stability and its determinants across different rice-cropping systems in China. *Agricultural and Forest Meteorology*, 364: 110452.

Xiao, L., et al., 2021. Modeling the response of winter wheat phenology to low temperature stress at elongation and booting stages. *Agricultural and Forest Meteorology*, 303: 108376.

Xiao, L., et al., 2022. Simulating the effects of low-temperature stress on wheat biomass growth and yield. *Agricultural and Forest Meteorology*, 326: 109191.

Ye, Z. 2023. *Study on the Yield Prediction and Yield Enhancement Strategy Based on a Multi-model Ensemble in Main Winter Wheat Production Region of China*. Nanjing: Nanjing Agricultural University.

Yin, X. Y., et al. 2018. Bringing genetics and biochemistry to crop modelling, and vice versa. *European Journal of Agronomy*, 100: 132–140.

Zhang, L., et al. 2016. Estimating wheat yield by integrating the WheatGrow and PROSAIL models. *Field Crops Research*, 192: 55–66.

Zhang, J., et al. 2025. Integrating crop models, single nucleotide polymorphism, and climatic indices to develop genotype-environment interaction model: a case study on rice flowering time. *Plant Phenomics*, 7(1): 100007.

Zhang, W. Y., et al. 2015. A simulation model for predicting canopy structure and light distribution in wheat. *European Journal of Agronomy*, 67: 1–11.

Zhang, X. H., et al. 2018. Selection of appropriate spatial resolution for the meteorological data for regional winter wheat potential productivity simulation in China based on WheatGrow model. *Agronomy*, 8(10): 198.

Zhuang, J. X., et al. 2013. Parameters optimization of rice development stages model based on individual advantages genetic algorithm. *Scientia Agricultura Sinica*, 46(11): 2220–2231.

Index